Linear Algebra and Optimization
with Applications to Machine Learning

Volume I

Linear Algebra for Computer Vision,
Robotics, and Machine Learning

Linear Algebra and Optimization
with Applications to Machine Learning

Volume I

Linear Algebra for Computer Vision, Robotics, and Machine Learning

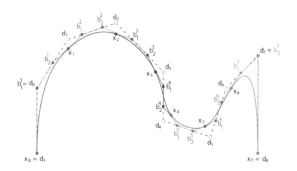

Jean Gallier
Jocelyn Quaintance

University of Pennsylvania, USA

World Scientific

NEW JERSEY · LONDON · SINGAPORE · BEIJING · SHANGHAI · HONG KONG · TAIPEI · CHENNAI · TOKYO

Published by

World Scientific Publishing Co. Pte. Ltd.

5 Toh Tuck Link, Singapore 596224

USA office: 27 Warren Street, Suite 401-402, Hackensack, NJ 07601

UK office: 57 Shelton Street, Covent Garden, London WC2H 9HE

British Library Cataloguing-in-Publication Data
A catalogue record for this book is available from the British Library.

LINEAR ALGEBRA AND OPTIMIZATION WITH APPLICATIONS TO
MACHINE LEARNING
Volume I: Linear Algebra for Computer Vision, Robotics, and Machine Learning

ISBN 978-981-120-639-9
ISBN 978-981-120-771-6 (pbk)

For any available supplementary material, please visit
https://www.worldscientific.com/worldscibooks/10.1142/11446#t=suppl

Desk Editor: Liu Yumeng

Preface

In recent years, computer vision, robotics, machine learning, and data science have been some of the key areas that have contributed to major advances in technology. Anyone who looks at papers or books in the above areas will be baffled by a strange jargon involving exotic terms such as kernel PCA, ridge regression, lasso regression, support vector machines (SVM), Lagrange multipliers, KKT conditions, *etc.* Do support vector machines chase cattle to catch them with some kind of super lasso? No! But one will quickly discover that behind the jargon which always comes with a new field (perhaps to keep the outsiders out of the club), lies a lot of "classical" linear algebra and techniques from optimization theory. And there comes the main challenge: in order to understand and use tools from machine learning, computer vision, and so on, one needs to have a firm background in linear algebra and optimization theory. To be honest, some probability theory and statistics should also be included, but we already have enough to contend with.

Many books on machine learning struggle with the above problem. How can one understand what are the dual variables of a ridge regression problem if one doesn't know about the Lagrangian duality framework? Similarly, how is it possible to discuss the dual formulation of SVM without a firm understanding of the Lagrangian framework?

The easy way out is to sweep these difficulties under the rug. If one is just a consumer of the techniques we mentioned above, the cookbook recipe approach is probably adequate. But this approach doesn't work for someone who really wants to do serious research and make significant contributions. To do so, we believe that one must have a solid background in linear algebra and optimization theory.

This is a problem because it means investing a great deal of time and energy studying these fields, but we believe that perseverance will be amply rewarded.

Our main goal is to present fundamentals of linear algebra and optimization theory, keeping in mind applications to machine learning, robotics, and computer vision. This work consists of two volumes, the first one being linear algebra, the second one optimization theory and applications, especially to machine learning.

This first volume covers "classical" linear algebra, up to and including the primary decomposition and the Jordan form. Besides covering the standard topics, we discuss a few topics that are important for applications. These include:

(1) Haar bases and the corresponding Haar wavelets.
(2) Hadamard matrices.
(3) Affine maps (see Section 5.4).
(4) Norms and matrix norms (Chapter 8).
(5) Convergence of sequences and series in a normed vector space. The matrix exponential e^A and its basic properties (see Section 8.8).
(6) The group of unit quaternions, $\mathbf{SU}(2)$, and the representation of rotations in $\mathbf{SO}(3)$ by unit quaternions (Chapter 15).
(7) An introduction to algebraic and spectral graph theory.
(8) Applications of SVD and pseudo-inverses, in particular, principal component analysis, for short PCA (Chapter 21).
(9) Methods for computing eigenvalues and eigenvectors, with a main focus on the QR algorithm (Chapter 17).

Four topics are covered in more detail than usual. These are

(1) Duality (Chapter 10).
(2) Dual norms (Section 13.7).
(3) The geometry of the orthogonal groups $\mathbf{O}(n)$ and $\mathbf{SO}(n)$, and of the unitary groups $\mathbf{U}(n)$ and $\mathbf{SU}(n)$.
(4) The spectral theorems (Chapter 16).

Except for a few exceptions, we provide complete proofs. We did so to make this book self-contained, but also because we believe that no deep knowledge of this material can be acquired without working out some proofs. However, our advice is to skip some of the proofs upon first reading, especially if they are long and intricate.

The chapters or sections marked with the symbol ⊛ contain material that is typically more specialized or more advanced, and they can be omitted upon first (or second) reading.

Acknowledgement: We would like to thank Christine Allen-Blanchette, Kostas Daniilidis, Carlos Esteves, Spyridon Leonardos, Stephen Phillips, João Sedoc, Stephen Shatz, Jianbo Shi, Marcelo Siqueira, and C.J. Taylor for reporting typos and for helpful comments. Special thanks to Gilbert Strang. We learned much from his books which have been a major source of inspiration. Thanks to Steven Boyd and James Demmel whose books have been an invaluable source of information. The first author also wishes to express his deepest gratitude to Philippe G. Ciarlet who was his teacher and mentor in 1970–1972 while he was a student at ENPC in Paris. Professor Ciarlet was by far his best teacher. He also knew how to instill in his students the importance of intellectual rigor, honesty, and modesty. He still has his typewritten notes on measure theory and integration, and on numerical linear algebra. The latter became his wonderful book Ciarlet [Ciarlet (1989)], from which we have borrowed heavily.

Contents

Chapter 1

Introduction

As we explained in the preface, this first volume covers "classical" linear algebra, up to and including the primary decomposition and the Jordan form. Besides covering the standard topics, we discuss a few topics that are important for applications. These include:

(1) Haar bases and the corresponding Haar wavelets, a fundamental tool in signal processing and computer graphics.

(2) Hadamard matrices which have applications in error correcting codes, signal processing, and low rank approximation.

(3) Affine maps (see Section 5.4). These are usually ignored or treated in a somewhat obscure fashion. Yet they play an important role in computer vision and robotics. There is a clean and elegant way to define affine maps. One simply has to define *affine combinations*. Linear maps preserve linear combinations, and similarly affine maps preserve affine combinations.

(4) Norms and matrix norms (Chapter 8). These are used extensively in optimization theory.

(5) Convergence of sequences and series in a normed vector space. Banach spaces (see Section 8.7). The matrix exponential e^A and its basic properties (see Section 8.8). In particular, we prove the Rodrigues formula for rotations in $\mathbf{SO}(3)$ and discuss the surjectivity of the exponential map $\exp\colon \mathfrak{so}(3) \to \mathbf{SO}(3)$, where $\mathfrak{so}(3)$ is the real vector space of 3×3 skew symmetric matrices (see Section 11.7). We also show that $\det(e^A) = e^{\operatorname{tr}(A)}$ (see Section 14.5).

(6) The group of unit quaternions, $\mathbf{SU}(2)$, and the representation of rotations in $\mathbf{SO}(3)$ by unit quaternions (Chapter 15). We define a homomorphism $r\colon \mathbf{SU}(2) \to \mathbf{SO}(3)$ and prove that it is surjective and that its kernel is $\{-I, I\}$. We compute the rotation matrix R_q associated

1

with a unit quaternion q, and give an algorithm to construct a quaternion from a rotation matrix. We also show that the exponential map $\exp\colon \mathfrak{su}(2) \to \mathbf{SU}(2)$ is surjective, where $\mathfrak{su}(2)$ is the real vector space of skew-Hermitian 2×2 matrices with zero trace. We discuss quaternion interpolation and prove the famous *slerp interpolation formula* due to Ken Shoemake.

(7) An introduction to algebraic and spectral graph theory. We define the graph Laplacian and prove some of its basic properties (see Chapter 18). In Chapter 19, we explain how the eigenvectors of the graph Laplacian can be used for graph drawing.

(8) Applications of SVD and pseudo-inverses, in particular, principal component analysis, for short PCA (Chapter 21).

(9) Methods for computing eigenvalues and eigenvectors are discussed in Chapter 17. We first focus on the QR algorithm due to Rutishauser, Francis, and Kublanovskaya. See Sections 17.1 and 17.3. We then discuss how to use an *Arnoldi iteration*, in combination with the QR algorithm, to approximate eigenvalues for a matrix A of large dimension. See Section 17.4. The special case where A is a symmetric (or Hermitian) tridiagonal matrix, involves a *Lanczos iteration*, and is discussed in Section 17.6. In Section 17.7, we present power iterations and inverse (power) iterations.

Five topics are covered in more detail than usual. These are

(1) Matrix factorizations such as LU, $PA = LU$, Cholesky, and reduced row echelon form (rref). Deciding the solvablity of a linear system $Ax = b$, and describing the space of solutions when a solution exists. See Chapter 7.

(2) Duality (Chapter 10).

(3) Dual norms (Section 13.7).

(4) The geometry of the orthogonal groups $\mathbf{O}(n)$ and $\mathbf{SO}(n)$, and of the unitary groups $\mathbf{U}(n)$ and $\mathbf{SU}(n)$.

(5) The spectral theorems (Chapter 16).

Most texts omit the proof that the $PA = LU$ factorization can be obtained by a simple modification of Gaussian elimination. We give a complete proof of Theorem 7.2 in Section 7.6. We also prove the uniqueness of the rref of a matrix; see Proposition 7.13.

At the most basic level, duality corresponds to transposition. But duality is really the bijection between subspaces of a vector space E (say finite-dimensional) and subspaces of linear forms (subspaces of the dual space E^*) established by two maps: the first map assigns to a subspace V of E the subspace V^0 of linear forms that vanish on V; the second map assigns to a subspace U of linear forms the subspace U^0 consisting of the vectors in E on which all linear forms in U vanish. The above maps define a bijection such that $\dim(V) + \dim(V^0) = \dim(E)$, $\dim(U) + \dim(U^0) = \dim(E)$, $V^{00} = V$, and $U^{00} = U$.

Another important fact is that if E is a finite-dimensional space with an inner product $u, v \mapsto \langle u, v \rangle$ (or a Hermitian inner product if E is a complex vector space), then there is a canonical isomorphism between E and its dual E^*. This means that every linear form $f \in E^*$ is uniquely represented by some vector $u \in E$, in the sense that $f(v) = \langle v, u \rangle$ for all $v \in E$. As a consequence, every linear map f has an adjoint f^* such that $\langle f(u), v \rangle = \langle u, f^*(v) \rangle$ for all $u, v \in E$.

Dual norms show up in convex optimization; see Boyd and Vandenberghe [Boyd and Vandenberghe (2004)].

Because of their importance in robotics and computer vision, we discuss in some detail the groups of isometries $\mathbf{O}(E)$ and $\mathbf{SO}(E)$ of a vector space with an inner product. The isometries in $\mathbf{O}(E)$ are the linear maps such that $f \circ f^* = f^* \circ f = \mathrm{id}$, and the direct isometries in $\mathbf{SO}(E)$, also called rotations, are the isometries in $\mathbf{O}(E)$ whose determinant is equal to $+1$. We also discuss the hermitian counterparts $\mathbf{U}(E)$ and $\mathbf{SU}(E)$.

We prove the spectral theorems not only for real symmetric matrices, but also for real and complex normal matrices.

We stress the importance of linear maps. Matrices are of course invaluable for computing and one needs to develop skills for manipulating them. But matrices are used to represent a linear map over a basis (or two bases), and the same linear map has different matrix representations. In fact, we can view the various normal forms of a matrix (Schur, SVD, Jordan) as a suitably convenient choice of bases.

We have listed most of the `Matlab` functions relevant to numerical linear algebra and have included `Matlab` programs implementing most of the algorithms discussed in this book.

Chapter 2

Vector Spaces, Bases, Linear Maps

2.1 Motivations: Linear Combinations, Linear Independence and Rank

In linear optimization problems, we often encounter systems of linear equations. For example, consider the problem of solving the following system of three linear equations in the three variables $x_1, x_2, x_3 \in \mathbb{R}$:

$$x_1 + 2x_2 - x_3 = 1$$
$$2x_1 + x_2 + x_3 = 2$$
$$x_1 - 2x_2 - 2x_3 = 3.$$

One way to approach this problem is introduce the "vectors" u, v, w, and b, given by

$$u = \begin{pmatrix} 1 \\ 2 \\ 1 \end{pmatrix} \qquad v = \begin{pmatrix} 2 \\ 1 \\ -2 \end{pmatrix} \qquad w = \begin{pmatrix} -1 \\ 1 \\ -2 \end{pmatrix} \qquad b = \begin{pmatrix} 1 \\ 2 \\ 3 \end{pmatrix}$$

and to write our linear system as

$$x_1 u + x_2 v + x_3 w = b.$$

In the above equation, we used implicitly the fact that a vector z can be multiplied by a scalar $\lambda \in \mathbb{R}$, where

$$\lambda z = \lambda \begin{pmatrix} z_1 \\ z_2 \\ z_3 \end{pmatrix} = \begin{pmatrix} \lambda z_1 \\ \lambda z_2 \\ \lambda z_3 \end{pmatrix},$$

and two vectors y and and z can be added, where

$$y + z = \begin{pmatrix} y_1 \\ y_2 \\ y_3 \end{pmatrix} + \begin{pmatrix} z_1 \\ z_2 \\ z_3 \end{pmatrix} = \begin{pmatrix} y_1 + z_1 \\ y_2 + z_2 \\ y_3 + z_3 \end{pmatrix}.$$

5

Also, given a vector

$$x = \begin{pmatrix} x_1 \\ x_2 \\ x_3 \end{pmatrix},$$

we define the *additive inverse* $-x$ of x (pronounced minus x) as

$$-x = \begin{pmatrix} -x_1 \\ -x_2 \\ -x_3 \end{pmatrix}.$$

Observe that $-x = (-1)x$, the scalar multiplication of x by -1.

The set of all vectors with three components is denoted by $\mathbb{R}^{3\times 1}$. The reason for using the notation $\mathbb{R}^{3\times 1}$ rather than the more conventional notation $\mathbb{R}^3$ is that the elements of $\mathbb{R}^{3\times 1}$ are *column vectors*; they consist of three rows and a single column, which explains the superscript 3×1. On the other hand, $\mathbb{R}^3 = \mathbb{R} \times \mathbb{R} \times \mathbb{R}$ consists of all triples of the form (x_1, x_2, x_3), with $x_1, x_2, x_3 \in \mathbb{R}$, and these are *row vectors*. However, there is an obvious bijection between $\mathbb{R}^{3\times 1}$ and $\mathbb{R}^3$ and they are usually identified. For the sake of clarity, in this introduction, we will denote the set of column vectors with n components by $\mathbb{R}^{n\times 1}$.

An expression such as

$$x_1 u + x_2 v + x_3 w$$

where u, v, w are vectors and the x_is are scalars (in $\mathbb{R}$) is called a *linear combination*. Using this notion, the problem of solving our linear system

$$x_1 u + x_2 v + x_3 w = b$$

is equivalent to *determining whether b can be expressed as a linear combination of u, v, w*.

Now if the vectors u, v, w are *linearly independent*, which means that there is *no* triple $(x_1, x_2, x_3) \neq (0, 0, 0)$ such that

$$x_1 u + x_2 v + x_3 w = 0_3,$$

it can be shown that *every* vector in $\mathbb{R}^{3\times 1}$ can be written as a linear combination of u, v, w. Here, 0_3 is the *zero vector*

$$0_3 = \begin{pmatrix} 0 \\ 0 \\ 0 \end{pmatrix}.$$

It is customary to abuse notation and to write 0 instead of 0_3. This rarely causes a problem because in most cases, whether 0 denotes the scalar zero or the zero vector can be inferred from the context.

In fact, every vector $z \in \mathbb{R}^{3 \times 1}$ can be written *in a unique way* as a linear combination

$$z = x_1 u + x_2 v + x_3 w.$$

This is because if

$$z = x_1 u + x_2 v + x_3 w = y_1 u + y_2 v + y_3 w,$$

then by using our (linear!) operations on vectors, we get

$$(y_1 - x_1)u + (y_2 - x_2)v + (y_3 - x_3)w = 0,$$

which implies that

$$y_1 - x_1 = y_2 - x_2 = y_3 - x_3 = 0,$$

by linear independence. Thus,

$$y_1 = x_1, \quad y_2 = x_2, \quad y_3 = x_3,$$

which shows that z has a unique expression as a linear combination, as claimed. Then our equation

$$x_1 u + x_2 v + x_3 w = b$$

has a *unique solution*, and indeed, we can check that

$$x_1 = 1.4$$
$$x_2 = -0.4$$
$$x_3 = -0.4$$

is the solution.

But then, *how do we determine that some vectors are linearly indepen-dent?*

One answer is to compute a numerical quantity $\det(u, v, w)$, called the *determinant* of (u, v, w), and to check that it is nonzero. In our case, it turns out that

$$\det(u, v, w) = \begin{vmatrix} 1 & 2 & -1 \\ 2 & 1 & 1 \\ 1 & -2 & -2 \end{vmatrix} = 15,$$

which confirms that u, v, w are linearly independent.

Other methods, which are much better for systems with a large num-ber of variables, consist of computing an LU-decomposition or a QR-decomposition, or an SVD of the *matrix* consisting of the three columns u, v, w,

$$A = \begin{pmatrix} u & v & w \end{pmatrix} = \begin{pmatrix} 1 & 2 & -1 \\ 2 & 1 & 1 \\ 1 & -2 & -2 \end{pmatrix}.$$

If we form the vector of unknowns

$$x = \begin{pmatrix} x_1 \\ x_2 \\ x_3 \end{pmatrix},$$

then our linear combination $x_1 u + x_2 v + x_3 w$ can be written in matrix form as

$$x_1 u + x_2 v + x_3 w = \begin{pmatrix} 1 & 2 & -1 \\ 2 & 1 & 1 \\ 1 & -2 & -2 \end{pmatrix} \begin{pmatrix} x_1 \\ x_2 \\ x_3 \end{pmatrix},$$

so our linear system is expressed by

$$\begin{pmatrix} 1 & 2 & -1 \\ 2 & 1 & 1 \\ 1 & -2 & -2 \end{pmatrix} \begin{pmatrix} x_1 \\ x_2 \\ x_3 \end{pmatrix} = \begin{pmatrix} 1 \\ 2 \\ 3 \end{pmatrix},$$

or more concisely as

$$Ax = b.$$

Now what if the vectors u, v, w are *linearly dependent*? For example, if we consider the vectors

$$u = \begin{pmatrix} 1 \\ 2 \\ 1 \end{pmatrix} \qquad v = \begin{pmatrix} 2 \\ 1 \\ -1 \end{pmatrix} \qquad w = \begin{pmatrix} -1 \\ 1 \\ 2 \end{pmatrix},$$

we see that

$$u - v = w,$$

a nontrivial *linear dependence*. It can be verified that u and v are still linearly independent. Now for our problem

$$x_1 u + x_2 v + x_3 w = b$$

it must be the case that b can be expressed as linear combination of u and v. However, it turns out that u, v, b are linearly independent (one way to see this is to compute the determinant $\det(u, v, b) = -6$), so b cannot be expressed as a linear combination of u and v and thus, our system has *no* solution.

If we change the vector b to

$$b = \begin{pmatrix} 3 \\ 3 \\ 0 \end{pmatrix},$$

then

$$b = u + v,$$

and so the system

$$x_1 u + x_2 v + x_3 w = b$$

has the solution

$$x_1 = 1, \quad x_2 = 1, \quad x_3 = 0.$$

Actually, since $w = u - v$, the above system is equivalent to

$$(x_1 + x_3)u + (x_2 - x_3)v = b,$$

and because u and v are linearly independent, the unique solution in $x_1 + x_3$ and $x_2 - x_3$ is

$$x_1 + x_3 = 1$$
$$x_2 - x_3 = 1,$$

which yields an infinite number of solutions parameterized by x_3, namely

$$x_1 = 1 - x_3$$
$$x_2 = 1 + x_3.$$

In summary, a 3×3 linear system may have a unique solution, no solution, or an infinite number of solutions, depending on the linear independence (and dependence) or the vectors u, v, w, b. This situation can be generalized to any $n \times n$ system, and even to any $n \times m$ system (n equations in m variables), as we will see later.

The point of view where our linear system is expressed in matrix form as $Ax = b$ stresses the fact that the map $x \mapsto Ax$ is a *linear transformation*. This means that

$$A(\lambda x) = \lambda(Ax)$$

for all $x \in \mathbb{R}^{3 \times 1}$ and all $\lambda \in \mathbb{R}$ and that

$$A(u + v) = Au + Av,$$

for all $u, v \in \mathbb{R}^{3 \times 1}$. We can view the matrix A as a way of expressing a linear map from $\mathbb{R}^{3 \times 1}$ to $\mathbb{R}^{3 \times 1}$ and solving the system $Ax = b$ amounts to determining whether b belongs to the image of this linear map.

Given a 3×3 matrix

$$A = \begin{pmatrix} a_{11} & a_{12} & a_{13} \\ a_{21} & a_{22} & a_{23} \\ a_{31} & a_{32} & a_{33} \end{pmatrix},$$

whose columns are three vectors denoted A^1, A^2, A^3, and given any vector $x = (x_1, x_2, x_3)$, we defined the product Ax as the linear combination

$$Ax = x_1 A^1 + x_2 A^2 + x_3 A^3 = \begin{pmatrix} a_{11}x_1 + a_{12}x_2 + a_{13}x_3 \\ a_{21}x_1 + a_{22}x_2 + a_{23}x_3 \\ a_{31}x_1 + a_{32}x_2 + a_{33}x_3 \end{pmatrix}.$$

The common pattern is that the ith coordinate of Ax is given by a certain kind of product called an *inner product*, of a *row vector*, the ith row of A, times the *column vector* x:

$$\begin{pmatrix} a_{i1} & a_{i2} & a_{i3} \end{pmatrix} \cdot \begin{pmatrix} x_1 \\ x_2 \\ x_3 \end{pmatrix} = a_{i1}x_1 + a_{i2}x_2 + a_{i3}x_3.$$

More generally, given any two vectors $x = (x_1, \ldots, x_n)$ and $y = (y_1, \ldots, y_n) \in \mathbb{R}^n$, their *inner product* denoted $x \cdot y$, or $\langle x, y \rangle$, is the number

$$x \cdot y = \begin{pmatrix} x_1 & x_2 & \cdots & x_n \end{pmatrix} \cdot \begin{pmatrix} y_1 \\ y_2 \\ \vdots \\ y_n \end{pmatrix} = \sum_{i=1}^{n} x_i y_i.$$

Inner products play a very important role. First, we quantity

$$\|x\|_2 = \sqrt{x \cdot x} = (x_1^2 + \cdots + x_n^2)^{1/2}$$

is a generalization of the length of a vector, called the *Euclidean norm*, or ℓ^2-*norm*. Second, it can be shown that we have the inequality

$$|x \cdot y| \leq \|x\| \, \|y\|,$$

so if $x, y \neq 0$, the ratio $(x \cdot y)/(\|x\| \, \|y\|)$ can be viewed as the cosine of an angle, the angle between x and y. In particular, if $x \cdot y = 0$ then the vectors x and y make the angle $\pi/2$, that is, they are *orthogonal*. The (square) matrices Q that preserve the inner product, in the sense that $\langle Qx, Qy \rangle = \langle x, y \rangle$ for all $x, y \in \mathbb{R}^n$, also play a very important role. They can be thought of as generalized rotations.

Returning to matrices, if A is an $m \times n$ matrix consisting of n columns $A^1, \ldots, A^n$ (in $\mathbb{R}^m$), and B is a $n \times p$ matrix consisting of p columns $B^1, \ldots, B^p$ (in $\mathbb{R}^n$) we can form the p vectors (in $\mathbb{R}^m$)

$$AB^1, \ldots, AB^p.$$

These p vectors constitute the $m \times p$ matrix denoted AB, whose jth column is AB^j. But we know that the ith coordinate of AB^j is the inner product of the ith row of A by the jth column of B,

$$\begin{pmatrix} a_{i1} & a_{i2} & \cdots & a_{in} \end{pmatrix} \cdot \begin{pmatrix} b_{1j} \\ b_{2j} \\ \vdots \\ b_{nj} \end{pmatrix} = \sum_{k=1}^{n} a_{ik} b_{kj}.$$

Thus we have defined a multiplication operation on matrices, namely if $A = (a_{ik})$ is a $m \times n$ matrix and if $B = (b_{jk})$ if $n \times p$ matrix, then their product AB is the $m \times n$ matrix whose entry on the ith row and the jth column is given by the inner product of the ith row of A by the jth column of B,

$$(AB)_{ij} = \sum_{k=1}^{n} a_{ik} b_{kj}.$$

Beware that unlike the multiplication of real (or complex) numbers, if A and B are two $n \times n$ matrices, in general, $AB \neq BA$.

Suppose that A is an $n \times n$ matrix and that we are trying to solve the linear system

$$Ax = b,$$

with $b \in \mathbb{R}^n$. Suppose we can find an $n \times n$ matrix B such that

$$BA^i = e_i, \quad i = 1, \ldots, n,$$

with $e_i = (0, \ldots, 0, 1, 0 \ldots, 0)$, where the only nonzero entry is 1 in the ith slot. If we form the $n \times n$ matrix

$$I_n = \begin{pmatrix} 1 & 0 & 0 & \cdots & 0 & 0 \\ 0 & 1 & 0 & \cdots & 0 & 0 \\ 0 & 0 & 1 & \cdots & 0 & 0 \\ \vdots & \vdots & \vdots & \ddots & \vdots & \vdots \\ 0 & 0 & 0 & \cdots & 1 & 0 \\ 0 & 0 & 0 & \cdots & 0 & 1 \end{pmatrix},$$

called the *identity matrix*, whose ith column is e_i, then the above is equivalent to

$$BA = I_n.$$

If $Ax = b$, then multiplying both sides on the left by B, we get

$$B(Ax) = Bb.$$

But is is easy to see that $B(Ax) = (BA)x = I_n x = x$, so we must have

$$x = Bb.$$

We can verify that $x = Bb$ is indeed a solution, because it can be shown that

$$A(Bb) = (AB)b = I_n b = b.$$

What is not obvious is that $BA = I_n$ implies $AB = I_n$, but this is indeed provable. The matrix B is usually denoted A^{-1} and called the *inverse* of A. It can be shown that it is the unique matrix such that

$$AA^{-1} = A^{-1}A = I_n.$$

If a square matrix A has an inverse, then we say that it is *invertible* or *nonsingular*, otherwise we say that it is *singular*. We will show later that a square matrix is invertible iff its columns are linearly independent iff its determinant is nonzero.

In summary, if A is a square invertible matrix, then the linear system $Ax = b$ has the unique solution $x = A^{-1}b$. In practice, this is not a good way to solve a linear system because computing A^{-1} is too expensive. A practical method for solving a linear system is Gaussian elimination, discussed in Chapter 7. Other practical methods for solving a linear system $Ax = b$ make use of a factorization of A (QR decomposition, SVD decomposition), using orthogonal matrices defined next.

Given an $m \times n$ matrix $A = (a_{kl})$, the $n \times m$ matrix $A^\top = (a_{ij}^\top)$ whose ith row is the ith column of A, which means that $a_{ij}^\top = a_{ji}$ for $i = 1, \ldots, n$ and $j = 1, \ldots, m$, is called the *transpose* of A. An $n \times n$ matrix Q such that

$$QQ^\top = Q^\top Q = I_n$$

is called an *orthogonal matrix*. Equivalently, the inverse Q^{-1} of an orthogonal matrix Q is equal to its transpose $Q^\top$. Orthogonal matrices play an important role. Geometrically, they correspond to linear transformation that preserve length. A major result of linear algebra states that every $m \times n$ matrix A can be written as

$$A = V\Sigma U^\top,$$

where V is an $m \times m$ orthogonal matrix, U is an $n \times n$ orthogonal matrix, and Σ is an $m \times n$ matrix whose only nonzero entries are nonnegative diagonal entries $\sigma_1 \geq \sigma_2 \geq \cdots \geq \sigma_p$, where $p = \min(m, n)$, called the *singular values* of A. The factorization $A = V \Sigma U^\top$ is called a *singular decomposition* of A, or *SVD*.

The SVD can be used to "solve" a linear system $Ax = b$ where A is an $m \times n$ matrix, even when this system has no solution. This may happen when there are more equations that variables $(m > n)$, in which case the system is overdetermined.

Of course, there is no miracle, an unsolvable system has no solution. But we can look for a *good approximate solution*, namely a vector x that minimizes some measure of the error $Ax - b$. Legendre and Gauss used $\|Ax - b\|_2^2$, which is the squared Euclidean norm of the error. This quantity is differentiable, and it turns out that there is a unique vector x^+ of minimum Euclidean norm that minimizes $\|Ax - b\|_2^2$. Furthermore, x^+ is given by the expression $x^+ = A^+ b$, where A^+ is the *pseudo-inverse* of A, and A^+ can be computed from an SVD $A = V \Sigma U^\top$ of A. Indeed, $A^+ = U \Sigma^+ V^\top$, where Σ^+ is the matrix obtained from Σ by replacing every positive singular value σ_i by its inverse σ_i^{-1}, leaving all zero entries intact, and transposing.

Instead of searching for the vector of least Euclidean norm minimizing $\|Ax - b\|_2^2$, we can add the penalty term $K \|x\|_2^2$ (for some positive $K > 0$) to $\|Ax - b\|_2^2$ and minimize the quantity $\|Ax - b\|_2^2 + K \|x\|_2^2$. This approach is called *ridge regression*. It turns out that there is a unique minimizer x^+ given by $x^+ = (A^\top A + K I_n)^{-1} A^\top b$, as shown in the second volume.

Another approach is to replace the penalty term $K \|x\|_2^2$ by $K \|x\|_1$, where $\|x\|_1 = |x_1| + \cdots + |x_n|$ (the ℓ^1-norm of x). The remarkable fact is that the minimizers x of $\|Ax - b\|_2^2 + K \|x\|_1$ tend to be *sparse*, which means that many components of x are equal to zero. This approach known as *lasso* is popular in machine learning and will be discussed in the second volume.

Another important application of the SVD is *principal component analysis* (or *PCA*), an important tool in data analysis.

Yet another fruitful way of interpreting the resolution of the system $Ax = b$ is to view this problem as an intersection problem. Indeed, each of the equations

$$x_1 + 2x_2 - x_3 = 1$$
$$2x_1 + x_2 + x_3 = 2$$
$$x_1 - 2x_2 - 2x_3 = 3$$

defines a subset of $\mathbb{R}^3$ which is actually a *plane*. The first equation

$$x_1 + 2x_2 - x_3 = 1$$

defines the plane H_1 passing through the three points $(1,0,0)$, $(0,1/2,0)$, $(0,0,-1)$, on the coordinate axes, the second equation

$$2x_1 + x_2 + x_3 = 2$$

defines the plane H_2 passing through the three points $(1,0,0)$, $(0,2,0)$, $(0,0,2)$, on the coordinate axes, and the third equation

$$x_1 - 2x_2 - 2x_3 = 3$$

defines the plane H_3 passing through the three points $(3,0,0)$, $(0,-3/2,0)$, $(0,0,-3/2)$, on the coordinate axes. See Figure 2.1.

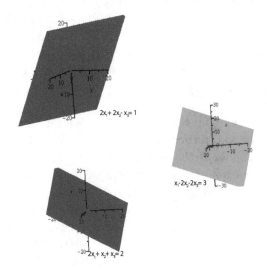

Fig. 2.1   The planes defined by the preceding linear equations.

The intersection $H_i \cap H_j$ of any two distinct planes H_i and H_j is a line, and the intersection $H_1 \cap H_2 \cap H_3$ of the three planes consists of the single point $(1.4, -0.4, -0.4)$, as illustrated in Figure 2.2.

The planes corresponding to the system

$$x_1 + 2x_2 - x_3 = 1$$
$$2x_1 + x_2 + x_3 = 2$$
$$x_1 - x_2 + 2x_3 = 3,$$

are illustrated in Figure 2.3. This system has no solution since there is no point simultaneously contained in all three planes; see Figure 2.4.

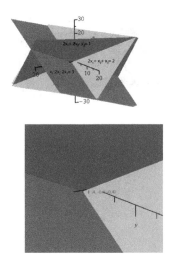

Fig. 2.2   The solution of the system is the point in common with each of the three planes.

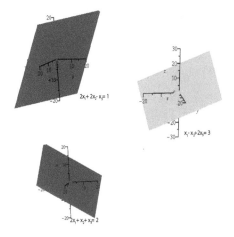

Fig. 2.3   The planes defined by the equations $x_1 + 2x_2 - x_3 = 1$, $2x_1 + x_2 + x_3 = 2$, and $x_1 - x_2 + 2x_3 = 3$.

Finally, the planes corresponding to the system

$$x_1 + 2x_2 - x_3 = 3$$
$$2x_1 + x_2 + x_3 = 3$$
$$x_1 - x_2 + 2x_3 = 0,$$

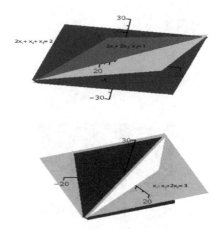

Fig. 2.4   The linear system $x_1 + 2x_2 - x_3 = 1$, $2x_1 + x_2 + x_3 = 2$, $x_1 - x_2 + 2x_3 = 3$ has no solution.

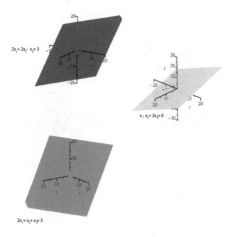

Fig. 2.5   The planes defined by the equations $x_1 + 2x_2 - x_3 = 3$, $2x_1 + x_2 + x_3 = 3$, and $x_1 - x_2 + 2x_3 = 0$.

are illustrated in Figure 2.5.

This system has infinitely many solutions, given parametrically by $(1 - x_3, 1 + x_3, x_3)$. Geometrically, this is a line common to all three planes; see Figure 2.6.

Under the above interpretation, observe that we are focusing on the *rows* of the matrix A, rather than on its *columns*, as in the previous interpretations.

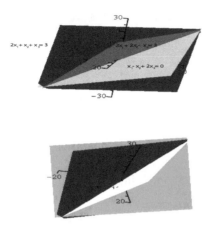

Fig. 2.6   The linear system $x_1 + 2x_2 - x_3 = 3$, $2x_1 + x_2 + x_3 = 3$, $x_1 - x_2 + 2x_3 = 0$ has the red line common to all three planes.

Another great example of a real-world problem where linear algebra proves to be very effective is the problem of *data compression*, that is, of representing a very large data set using a much smaller amount of storage.

Typically the data set is represented as an $m \times n$ matrix A where each row corresponds to an n-dimensional data point and typically, $m \geq n$. In most applications, the data are not independent so the rank of A is a lot smaller than $\min\{m, n\}$, and the the goal of *low-rank decomposition* is to factor A as the product of two matrices B and C, where B is a $m \times k$ matrix and C is a $k \times n$ matrix, with $k \ll \min\{m, n\}$ (here, $\ll$ means "much smaller than"):

$$\begin{pmatrix} A \\ m \times n \end{pmatrix} = \begin{pmatrix} B \\ m \times k \end{pmatrix} \begin{pmatrix} C \\ k \times n \end{pmatrix}.$$

Now it is generally too costly to find an exact factorization as above, so we look for a low-rank matrix A' which is a "good" *approximation* of A. In order to make this statement precise, we need to define a mechanism to determine how close two matrices are. This can be done using *matrix norms*, a notion discussed in Chapter 8. The norm of a matrix A is a nonnegative real number $\|A\|$ which behaves a lot like the absolute value

$|x|$ of a real number x. Then our goal is to find some low-rank matrix A' that minimizes the norm

$$\|A - A'\|^2,$$

over all matrices A' of rank at most k, for some given $k \ll \min\{m, n\}$.

Some advantages of a low-rank approximation are:

(1) Fewer elements are required to represent A; namely, $k(m+n)$ instead of mn. Thus less storage and fewer operations are needed to reconstruct A.

(2) Often, the process for obtaining the decomposition exposes the underlying structure of the data. Thus, it may turn out that "most" of the significant data are concentrated along some directions called *principal directions*.

Low-rank decompositions of a set of data have a multitude of applications in engineering, including computer science (especially computer vision), statistics, and machine learning. As we will see later in Chapter 21, the *singular value decomposition* (SVD) provides a very satisfactory solution to the low-rank approximation problem. Still, in many cases, the data sets are so large that another ingredient is needed: *randomization*. However, as a first step, linear algebra often yields a good initial solution.

We will now be more precise as to what kinds of operations are allowed on vectors. In the early 1900, the notion of a *vector space* emerged as a convenient and unifying framework for working with "linear" objects and we will discuss this notion in the next few sections.

2.2 Vector Spaces

A (real) vector space is a set E together with two operations, $+\colon E \times E \to E$ and $\cdot\colon \mathbb{R} \times E \to E$, called *addition* and *scalar multiplication*, that satisfy some simple properties. First of all, E under addition has to be a commutative (or abelian) group, a notion that we review next.

However, keep in mind that vector spaces are not just algebraic objects; they are also geometric objects.

Definition 2.1. A *group* is a set G equipped with a binary operation $\cdot\colon G \times G \to G$ that associates an element $a \cdot b \in G$ to every pair of elements $a, b \in G$, and having the following properties: $\cdot$ is associative, has an identity element $e \in G$, and every element in G is invertible (w.r.t. $\cdot$). More explicitly, this means that the following equations hold for all $a, b, c \in G$:

(G1) $a \cdot (b \cdot c) = (a \cdot b) \cdot c.$ (associativity);
(G2) $a \cdot e = e \cdot a = a.$ (identity);
(G3) For every $a \in G$, there is some $a^{-1} \in G$ such that
$$a \cdot a^{-1} = a^{-1} \cdot a = e.$$ (inverse).

A group G is *abelian* (or *commutative*) if
$$a \cdot b = b \cdot a \quad \text{for all } a, b \in G.$$

A set M together with an operation $\cdot : M \times M \to M$ and an element e satisfying only Conditions (G1) and (G2) is called a *monoid*. For example, the set $\mathbb{N} = \{0, 1, \ldots, n, \ldots\}$ of natural numbers is a (commutative) monoid under addition with identity element 0. However, it is not a group.

Some examples of groups are given below.

Example 2.1.

(1) The set $\mathbb{Z} = \{\ldots, -n, \ldots, -1, 0, 1, \ldots, n, \ldots\}$ of integers is an abelian group under addition, with identity element 0. However, $\mathbb{Z}^* = \mathbb{Z} - \{0\}$ is not a group under multiplication; it is a commutative monoid with identity element 1.

(2) The set $\mathbb{Q}$ of rational numbers (fractions p/q with $p, q \in \mathbb{Z}$ and $q \neq 0$) is an abelian group under addition, with identity element 0. The set $\mathbb{Q}^* = \mathbb{Q} - \{0\}$ is also an abelian group under multiplication, with identity element 1.

(3) Similarly, the sets $\mathbb{R}$ of real numbers and $\mathbb{C}$ of complex numbers are abelian groups under addition (with identity element 0), and $\mathbb{R}^* = \mathbb{R} - \{0\}$ and $\mathbb{C}^* = \mathbb{C} - \{0\}$ are abelian groups under multiplication (with identity element 1).

(4) The sets $\mathbb{R}^n$ and $\mathbb{C}^n$ of n-tuples of real or complex numbers are abelian groups under componentwise addition:
$$(x_1, \ldots, x_n) + (y_1, \ldots, y_n) = (x_1 + y_1, \ldots, x_n + y_n),$$
with identity element $(0, \ldots, 0)$.

(5) Given any nonempty set S, the set of bijections $f \colon S \to S$, also called *permutations of S*, is a group under function composition (i.e., the multiplication of f and g is the composition $g \circ f$), with identity element the identity function id_S. This group is not abelian as soon as S has more than two elements.

(6) The set of $n \times n$ matrices with real (or complex) coefficients is an abelian group under addition of matrices, with identity element the null matrix. It is denoted by $M_n(\mathbb{R})$ (or $M_n(\mathbb{C})$).

(7) The set $\mathbb{R}[X]$ of all polynomials in one variable X with real coefficients,
$$P(X) = a_n X^n + a_{n-1} X^{n-1} + \cdots + a_1 X + a_0,$$
(with $a_i \in \mathbb{R}$), is an abelian group under addition of polynomials. The identity element is the zero polynomial.

(8) The set of $n \times n$ invertible matrices with real (or complex) coefficients is a group under matrix multiplication, with identity element the identity matrix I_n. This group is called the *general linear group* and is usually denoted by $\mathbf{GL}(n, \mathbb{R})$ (or $\mathbf{GL}(n, \mathbb{C})$).

(9) The set of $n \times n$ invertible matrices with real (or complex) coefficients and determinant $+1$ is a group under matrix multiplication, with identity element the identity matrix I_n. This group is called the *special linear group* and is usually denoted by $\mathbf{SL}(n, \mathbb{R})$ (or $\mathbf{SL}(n, \mathbb{C})$).

(10) The set of $n \times n$ invertible matrices with real coefficients such that $RR^\top = R^\top R = I_n$ and of determinant $+1$ is a group (under matrix multiplication) called the *special orthogonal group* and is usually denoted by $\mathbf{SO}(n)$ (where $R^\top$ is the *transpose* of the matrix R, i.e., the rows of $R^\top$ are the columns of R). It corresponds to the rotations in $\mathbb{R}^n$.

(11) Given an open interval (a, b), the set $\mathcal{C}(a, b)$ of continuous functions $f \colon (a, b) \to \mathbb{R}$ is an abelian group under the operation $f + g$ defined such that
$$(f + g)(x) = f(x) + g(x)$$
for all $x \in (a, b)$.

It is customary to denote the operation of an abelian group G by $+$, in which case the inverse a^{-1} of an element $a \in G$ is denoted by $-a$.

The identity element of a group is *unique*. In fact, we can prove a more general fact:

Proposition 2.1. *If a binary operation* $\cdot \colon M \times M \to M$ *is associative and if* $e' \in M$ *is a left identity and* $e'' \in M$ *is a right identity, which means that*
$$e' \cdot a = a \quad \text{for all} \quad a \in M \tag{2.1}$$
and
$$a \cdot e'' = a \quad \text{for all} \quad a \in M, \tag{2.2}$$
then $e' = e''$.

Proof. If we let $a = e''$ in equation (2.1), we get

$$e' \cdot e'' = e'',$$

and if we let $a = e'$ in equation (2.2), we get

$$e' \cdot e'' = e',$$

and thus

$$e' = e' \cdot e'' = e'',$$

as claimed. $\square$

Proposition 2.1 implies that the identity element of a monoid is unique, and since every group is a monoid, the identity element of a group is unique. Furthermore, every element in a group has a *unique inverse*. This is a consequence of a slightly more general fact:

Proposition 2.2. *In a monoid M with identity element e, if some element $a \in M$ has some left inverse $a' \in M$ and some right inverse $a'' \in M$, which means that*

$$a' \cdot a = e \tag{2.3}$$

and

$$a \cdot a'' = e, \tag{2.4}$$

then $a' = a''$.

Proof. Using (2.3) and the fact that e is an identity element, we have

$$(a' \cdot a) \cdot a'' = e \cdot a'' = a''.$$

Similarly, using (2.4) and the fact that e is an identity element, we have

$$a' \cdot (a \cdot a'') = a' \cdot e = a'.$$

However, since M is monoid, the operation $\cdot$ is associative, so

$$a' = a' \cdot (a \cdot a'') = (a' \cdot a) \cdot a'' = a'',$$

as claimed. $\square$

Remark: Axioms (G2) and (G3) can be weakened a bit by requiring only (2.2) (the existence of a right identity) and (2.4) (the existence of a right inverse for every element) (or (2.1) and (2.3)). It is a good exercise to prove that the group axioms (G2) and (G3) follow from (2.2) and (2.4).

A vector space is an abelian group E with an additional operation $\cdot: K \times E \to E$ called scalar multiplication that allows rescaling a vector in E by an element in K. The set K itself is an algebraic structure called a *field*. A field is a special kind of structure called a *ring*. These notions are defined below. We begin with rings.

Definition 2.2. A *ring* is a set A equipped with two operations $+: A \times A \to A$ (called *addition*) and $*: A \times A \to A$ (called *multiplication*) having the following properties:

(R1) A is an abelian group w.r.t. $+$;
(R2) $*$ is associative and has an identity element $1 \in A$;
(R3) $*$ is distributive w.r.t. $+$.

The identity element for addition is denoted 0, and the additive inverse of $a \in A$ is denoted by $-a$. More explicitly, the axioms of a ring are the following equations which hold for all $a, b, c \in A$:

$$a + (b + c) = (a + b) + c \qquad \text{(associativity of +)} \qquad (2.5)$$
$$a + b = b + a \qquad \text{(commutativity of +)} \qquad (2.6)$$
$$a + 0 = 0 + a = a \qquad \text{(zero)} \qquad (2.7)$$
$$a + (-a) = (-a) + a = 0 \qquad \text{(additive inverse)} \qquad (2.8)$$
$$a * (b * c) = (a * b) * c \qquad \text{(associativity of *)} \qquad (2.9)$$
$$a * 1 = 1 * a = a \qquad \text{(identity for *)} \qquad (2.10)$$
$$(a + b) * c = (a * c) + (b * c) \qquad \text{(distributivity)} \qquad (2.11)$$
$$a * (b + c) = (a * b) + (a * c) \qquad \text{(distributivity)} \qquad (2.12)$$

The ring A is *commutative* if

$$a * b = b * a \quad \text{for all } a, b \in A.$$

From (2.11) and (2.12), we easily obtain

$$a * 0 = 0 * a = 0 \qquad (2.13)$$
$$a * (-b) = (-a) * b = -(a * b). \qquad (2.14)$$

Note that (2.13) implies that if $1 = 0$, then $a = 0$ for all $a \in A$, and thus, $A = \{0\}$. The ring $A = \{0\}$ is called the *trivial ring*. A ring for which $1 \neq 0$ is called *nontrivial*. The multiplication $a * b$ of two elements $a, b \in A$ is often denoted by ab.

The abelian group $\mathbb{Z}$ is a commutative ring (with unit 1), and for any field K, the abelian group $K[X]$ of polynomials is also a commutative ring (also with unit 1). The set $\mathbb{Z}/m\mathbb{Z}$ of residues modulo m where m is a positive integer is a commutative ring.

A field is a commutative ring K for which $K - \{0\}$ is a group under multiplication.

Definition 2.3. A set K is a *field* if it is a ring and the following properties hold:

(F1) $0 \neq 1$;

(F2) $K^* = K - \{0\}$ is a group w.r.t. $*$ (i.e., every $a \neq 0$ has an inverse w.r.t. $*$);

(F3) $*$ is commutative.

If $*$ is not commutative but (F1) and (F2) hold, we say that we have a *skew field* (or *noncommutative field*).

Note that we are assuming that the operation $*$ of a field is commutative. This convention is not universally adopted, but since $*$ will be commutative for most fields we will encounter, we may as well include this condition in the definition.

Example 2.2.

(1) The rings $\mathbb{Q}$, $\mathbb{R}$, and $\mathbb{C}$ are fields.

(2) The set $\mathbb{Z}/p\mathbb{Z}$ of residues modulo p where p is a prime number is field.

(3) The set of (formal) fractions $f(X)/g(X)$ of polynomials $f(X), g(X) \in \mathbb{R}[X]$, where $g(X)$ is not the zero polynomial, is a field.

Vector spaces are defined as follows.

Definition 2.4. A *real vector space* is a set E (of vectors) together with two operations $+\colon E \times E \to E$ (called *vector addition*)[1] and $\cdot\colon \mathbb{R} \times E \to E$ (called *scalar multiplication*) satisfying the following conditions for all $\alpha, \beta \in \mathbb{R}$ and all $u, v \in E$;

(V0) E is an abelian group w.r.t. $+$, with identity element 0;[2]

(V1) $\alpha \cdot (u + v) = (\alpha \cdot u) + (\alpha \cdot v)$;

(V2) $(\alpha + \beta) \cdot u = (\alpha \cdot u) + (\beta \cdot u)$;

---

[1] The symbol $+$ is overloaded, since it denotes both addition in the field $\mathbb{R}$ and addition of vectors in E. It is usually clear from the context which $+$ is intended.

[2] The symbol 0 is also overloaded, since it represents both the zero in $\mathbb{R}$ (a scalar) and the identity element of E (the zero vector). Confusion rarely arises, but one may prefer using $\mathbf{0}$ for the zero vector.

(V3) $(\alpha * \beta) \cdot u = \alpha \cdot (\beta \cdot u)$;
(V4) $1 \cdot u = u$.

In (V3), $*$ denotes multiplication in $\mathbb{R}$.

Given $\alpha \in \mathbb{R}$ and $v \in E$, the element $\alpha \cdot v$ is also denoted by αv. The field $\mathbb{R}$ is often called the *field of scalars*.

In Definition 2.4, the field $\mathbb{R}$ may be replaced by the field of complex numbers $\mathbb{C}$, in which case we have a *complex* vector space. It is even possible to replace $\mathbb{R}$ by the field of rational numbers $\mathbb{Q}$ or by any arbitrary field K (for example $\mathbb{Z}/p\mathbb{Z}$, where p is a prime number), in which case we have a *K-vector space* (in (V3), $*$ denotes multiplication in the field K). In most cases, the field K will be the field $\mathbb{R}$ of reals, but *all results in this chapter hold for vector spaces over an arbitrary field*.

From (V0), a vector space always contains the null vector 0, and thus is nonempty. From (V1), we get $\alpha \cdot 0 = 0$, and $\alpha \cdot (-v) = -(\alpha \cdot v)$. From (V2), we get $0 \cdot v = 0$, and $(-\alpha) \cdot v = -(\alpha \cdot v)$.

Another important consequence of the axioms is the following fact:

Proposition 2.3. *For any $u \in E$ and any $\lambda \in \mathbb{R}$, if $\lambda \neq 0$ and $\lambda \cdot u = 0$, then $u = 0$.*

Proof. Indeed, since $\lambda \neq 0$, it has a multiplicative inverse λ^{-1}, so from $\lambda \cdot u = 0$, we get

$$\lambda^{-1} \cdot (\lambda \cdot u) = \lambda^{-1} \cdot 0.$$

However, we just observed that $\lambda^{-1} \cdot 0 = 0$, and from (V3) and (V4), we have

$$\lambda^{-1} \cdot (\lambda \cdot u) = (\lambda^{-1}\lambda) \cdot u = 1 \cdot u = u,$$

and we deduce that $u = 0$. $\square$

Remark: One may wonder whether axiom (V4) is really needed. Could it be derived from the other axioms? The answer is **no**. For example, one can take $E = \mathbb{R}^n$ and define $\cdot : \mathbb{R} \times \mathbb{R}^n \to \mathbb{R}^n$ by

$$\lambda \cdot (x_1, \ldots, x_n) = (0, \ldots, 0)$$

for all $(x_1, \ldots, x_n) \in \mathbb{R}^n$ and all $\lambda \in \mathbb{R}$. Axioms (V0)–(V3) are all satisfied, but (V4) fails. Less trivial examples can be given using the notion of a basis, which has not been defined yet.

The field $\mathbb{R}$ itself can be viewed as a vector space over itself, addition of vectors being addition in the field, and multiplication by a scalar being multiplication in the field.

Example 2.3.

(1) The fields $\mathbb{R}$ and $\mathbb{C}$ are vector spaces over $\mathbb{R}$.
(2) The groups $\mathbb{R}^n$ and $\mathbb{C}^n$ are vector spaces over $\mathbb{R}$, with scalar multiplication given by

$$\lambda(x_1, \ldots, x_n) = (\lambda x_1, \ldots, \lambda x_n),$$

for any $\lambda \in \mathbb{R}$ and with $(x_1, \ldots, x_n) \in \mathbb{R}^n$ or $(x_1, \ldots, x_n) \in \mathbb{C}^n$, and $\mathbb{C}^n$ is a vector space over $\mathbb{C}$ with scalar multiplication as above, but with $\lambda \in \mathbb{C}$.
(3) The ring $\mathbb{R}[X]_n$ of polynomials of degree at most n with real coefficients is a vector space over $\mathbb{R}$, and the ring $\mathbb{C}[X]_n$ of polynomials of degree at most n with complex coefficients is a vector space over $\mathbb{C}$, with scalar multiplication $\lambda \cdot P(X)$ of a polynomial

$$P(X) = a_m X^m + a_{m-1} X^{m-1} + \cdots + a_1 X + a_0$$

(with $a_i \in \mathbb{R}$ or $a_i \in \mathbb{C}$) by the scalar λ (in $\mathbb{R}$ or $\mathbb{C}$), with $m \leq n$, given by

$$\lambda \cdot P(X) = \lambda a_m X^m + \lambda a_{m-1} X^{m-1} + \cdots + \lambda a_1 X + \lambda a_0.$$

(4) The ring $\mathbb{R}[X]$ of all polynomials with real coefficients is a vector space over $\mathbb{R}$, and the ring $\mathbb{C}[X]$ of all polynomials with complex coefficients is a vector space over $\mathbb{C}$, with the same scalar multiplication as above.
(5) The ring of $n \times n$ matrices $\mathrm{M}_n(\mathbb{R})$ is a vector space over $\mathbb{R}$.
(6) The ring of $m \times n$ matrices $\mathrm{M}_{m,n}(\mathbb{R})$ is a vector space over $\mathbb{R}$.
(7) The ring $\mathcal{C}(a, b)$ of continuous functions $f \colon (a, b) \to \mathbb{R}$ is a vector space over $\mathbb{R}$, with the scalar multiplication λf of a function $f \colon (a, b) \to \mathbb{R}$ by a scalar $\lambda \in \mathbb{R}$ given by

$$(\lambda f)(x) = \lambda f(x), \qquad \text{for all } x \in (a, b).$$

(8) A very important example of vector space is the set of linear maps between two vector spaces to be defined in Section 2.7. Here is an example that will prepare us for the vector space of linear maps. Let X be any nonempty set and let E be a vector space. The set of all functions $f \colon X \to E$ can be made into a vector space as follows: Given

any two functions $f\colon X \to E$ and $g\colon X \to E$, let $(f+g)\colon X \to E$ be defined such that
$$(f+g)(x) = f(x) + g(x)$$
for all $x \in X$, and for every $\lambda \in \mathbb{R}$, let $\lambda f\colon X \to E$ be defined such that
$$(\lambda f)(x) = \lambda f(x)$$
for all $x \in X$. The axioms of a vector space are easily verified.

Let E be a vector space. We would like to define the important notions of linear combination and linear independence.

Before defining these notions, we need to discuss a strategic choice which, depending how it is settled, may reduce or increase headaches in dealing with notions such as linear combinations and linear dependence (or independence). The issue has to do with using sets of vectors versus sequences of vectors.

2.3 Indexed Families; the Sum Notation $\sum_{i \in I} a_i$

Our experience tells us that *it is preferable to use sequences of vectors*; even better, indexed families of vectors. (We are not alone in having opted for sequences over sets, and we are in good company; for example, Artin [Artin (1991)], Axler [Axler (2004)], and Lang [Lang (1993)] use sequences. Nevertheless, some prominent authors such as Lax [Lax (2007)] use sets. We leave it to the reader to conduct a survey on this issue.)

Given a set A, recall that a *sequence* is an ordered n-tuple $(a_1, \ldots, a_n) \in A^n$ of elements from A, for some natural number n. The elements of a sequence need not be distinct and the order is important. For example, (a_1, a_2, a_1) and (a_2, a_1, a_1) are two distinct sequences in A^3. Their underlying set is $\{a_1, a_2\}$.

What we just defined are *finite* sequences, which can also be viewed as functions from $\{1, 2, \ldots, n\}$ to the set A; the ith element of the sequence $(a_1, \ldots, a_n)$ is the image of i under the function. This viewpoint is fruitful, because it allows us to define (countably) infinite sequences as functions $s\colon \mathbb{N} \to A$. But then, why limit ourselves to ordered sets such as $\{1, \ldots, n\}$ or $\mathbb{N}$ as index sets?

The main role of the index set is to tag each element uniquely, and the order of the tags is not crucial, although convenient. Thus, it is natural to define the notion of indexed family.

Definition 2.5. Given a set A, an *I-indexed family* of elements of A, for short a *family*, is a function $a\colon I \to A$ where I is any set viewed as an index set. Since the function a is determined by its graph

$$\{(i, a(i)) \mid i \in I\},$$

the family a can be viewed as the set of pairs $a = \{(i, a(i)) \mid i \in I\}$. For notational simplicity, we write a_i instead of $a(i)$, and denote the family $a = \{(i, a(i)) \mid i \in I\}$ by $(a_i)_{i \in I}$.

For example, if $I = \{r, g, b, y\}$ and $A = \mathbb{N}$, the set of pairs

$$a = \{(r, 2), (g, 3), (b, 2), (y, 11)\}$$

is an indexed family. The element 2 appears twice in the family with the two distinct tags r and b.

When the indexed set I is totally ordered, a family $(a_i)_{i \in I}$ is often called an I-*sequence*. Interestingly, sets can be viewed as special cases of families. Indeed, a set A can be viewed as the A-indexed family $\{(a, a) \mid a \in I\}$ corresponding to the identity function.

Remark: An indexed family should not be confused with a multiset. Given any set A, a *multiset* is a similar to a set, except that elements of A may occur more than once. For example, if $A = \{a, b, c, d\}$, then $\{a, a, a, b, c, c, d, d\}$ is a multiset. Each element appears with a certain multiplicity, but the order of the elements does not matter. For example, a has multiplicity 3. Formally, a multiset is a function $s \colon A \to \mathbb{N}$, or equivalently a set of pairs $\{(a, i) \mid a \in A\}$. Thus, a multiset is an A-indexed family of elements from $\mathbb{N}$, but not a $\mathbb{N}$-indexed family, since distinct elements may have the same multiplicity (such as c an d in the example above). *An indexed family is a generalization of a sequence, but a multiset is a generalization of a set.*

We also need to take care of an annoying technicality, which is to define sums of the form $\sum_{i \in I} a_i$, where I is any *finite* index set and $(a_i)_{i \in I}$ is a family of elements in some set A equiped with a binary operation $+ \colon A \times A \to A$ which is associative (Axiom (G1)) and commutative. This will come up when we define linear combinations.

The issue is that the binary operation $+$ only tells us how to compute $a_1 + a_2$ for two elements of A, but it does not tell us what is the sum of three of more elements. For example, how should $a_1 + a_2 + a_3$ be defined?

What we have to do is to define $a_1 + a_2 + a_3$ by using a sequence of steps each involving two elements, and there are two possible ways to do this: $a_1 + (a_2 + a_3)$ and $(a_1 + a_2) + a_3$. If our operation $+$ is not associative, these are different values. If it associative, then $a_1 + (a_2 + a_3) = (a_1 + a_2) + a_3$, but then there are still six possible permutations of the indices $1, 2, 3$, and if

+ is not commutative, these values are generally different. If our operation is commutative, then all six permutations have the same value. *Thus, if + is associative and commutative, it seems intuitively clear that a sum of the form $\sum_{i \in I} a_i$ does not depend on the order of the operations used to compute it.*

This is indeed the case, but a rigorous proof requires induction, and such a proof is surprisingly involved. Readers may accept without proof the fact that sums of the form $\sum_{i \in I} a_i$ are indeed well defined, and jump directly to Definition 2.6. For those who want to see the gory details, here we go.

First, we define sums $\sum_{i \in I} a_i$, where I is a finite sequence of distinct natural numbers, say $I = (i_1, \ldots, i_m)$. If $I = (i_1, \ldots, i_m)$ with $m \geq 2$, we denote the sequence $(i_2, \ldots, i_m)$ by $I - \{i_1\}$. We proceed by induction on the size m of I. Let

$$\sum_{i \in I} a_i = a_{i_1}, \quad \text{if } m = 1,$$

$$\sum_{i \in I} a_i = a_{i_1} + \left(\sum_{i \in I - \{i_1\}} a_i \right), \quad \text{if } m > 1.$$

For example, if $I = (1, 2, 3, 4)$, we have

$$\sum_{i \in I} a_i = a_1 + (a_2 + (a_3 + a_4)).$$

If the operation + is not associative, the grouping of the terms matters. For instance, in general

$$a_1 + (a_2 + (a_3 + a_4)) \neq (a_1 + a_2) + (a_3 + a_4).$$

However, if the operation + is associative, the sum $\sum_{i \in I} a_i$ should not depend on the grouping of the elements in I, as long as their order is preserved. For example, if $I = (1, 2, 3, 4, 5)$, $J_1 = (1, 2)$, and $J_2 = (3, 4, 5)$, we expect that

$$\sum_{i \in I} a_i = \left(\sum_{j \in J_1} a_j \right) + \left(\sum_{j \in J_2} a_j \right).$$

This indeed the case, as we have the following proposition.

Proposition 2.4. *Given any nonempty set A equipped with an associative binary operation $+ \colon A \times A \to A$, for any nonempty finite sequence I of distinct natural numbers and for any partition of I into p nonempty sequences*

$I_{k_1}, \ldots, I_{k_p}$, *for some nonempty sequence* $K = (k_1, \ldots, k_p)$ *of distinct natural numbers such that* $k_i < k_j$ *implies that* $\alpha < \beta$ *for all* $\alpha \in I_{k_i}$ *and all* $\beta \in I_{k_j}$, *for every sequence* $(a_i)_{i \in I}$ *of elements in* A, *we have*

$$\sum_{\alpha \in I} a_\alpha = \sum_{k \in K} \left(\sum_{\alpha \in I_k} a_\alpha \right).$$

Proof. We proceed by induction on the size n of I.

If $n = 1$, then we must have $p = 1$ and $I_{k_1} = I$, so the proposition holds trivially.

Next, assume $n > 1$. If $p = 1$, then $I_{k_1} = I$ and the formula is trivial, so assume that $p \geq 2$ and write $J = (k_2, \ldots, k_p)$. There are two cases.

Case 1. The sequence I_{k_1} has a single element, say β, which is the first element of I. In this case, write C for the sequence obtained from I by deleting its first element β. By definition,

$$\sum_{\alpha \in I} a_\alpha = a_\beta + \left(\sum_{\alpha \in C} a_\alpha \right),$$

and

$$\sum_{k \in K} \left(\sum_{\alpha \in I_k} a_\alpha \right) = a_\beta + \left(\sum_{j \in J} \left(\sum_{\alpha \in I_j} a_\alpha \right) \right).$$

Since $|C| = n - 1$, by the induction hypothesis, we have

$$\left(\sum_{\alpha \in C} a_\alpha \right) = \sum_{j \in J} \left(\sum_{\alpha \in I_j} a_\alpha \right),$$

which yields our identity.

Case 2. The sequence I_{k_1} has at least two elements. In this case, let β be the first element of I (and thus of I_{k_1}), let I' be the sequence obtained from I by deleting its first element β, let I'_{k_1} be the sequence obtained from I_{k_1} by deleting its first element β, and let $I'_{k_i} = I_{k_i}$ for $i = 2, \ldots, p$. Recall that $J = (k_2, \ldots, k_p)$ and $K = (k_1, \ldots, k_p)$. The sequence I' has $n - 1$ elements, so by the induction hypothesis applied to I' and the I'_{k_i}, we get

$$\sum_{\alpha \in I'} a_\alpha = \sum_{k \in K} \left(\sum_{\alpha \in I'_k} a_\alpha \right) = \left(\sum_{\alpha \in I'_{k_1}} a_\alpha \right) + \left(\sum_{j \in J} \left(\sum_{\alpha \in I_j} a_\alpha \right) \right).$$

If we add the left-hand side to a_β, by definition we get

$$\sum_{\alpha \in I} a_\alpha.$$

If we add the right-hand side to a_β, using associativity and the definition of an indexed sum, we get

$$a_\beta + \left(\left(\sum_{\alpha \in I'_{k_1}} a_\alpha\right) + \left(\sum_{j \in J}\left(\sum_{\alpha \in I_j} a_\alpha\right)\right)\right)$$

$$= \left(a_\beta + \left(\sum_{\alpha \in I'_{k_1}} a_\alpha\right)\right) + \left(\sum_{j \in J}\left(\sum_{\alpha \in I_j} a_\alpha\right)\right)$$

$$= \left(\sum_{\alpha \in I_{k_1}} a_\alpha\right) + \left(\sum_{j \in J}\left(\sum_{\alpha \in I_j} a_\alpha\right)\right) = \sum_{k \in K}\left(\sum_{\alpha \in I_k} a_\alpha\right),$$

as claimed.   □

If $I = (1, \ldots, n)$, we also write $\sum_{i=1}^n a_i$ instead of $\sum_{i \in I} a_i$. Since $+$ is associative, Proposition 2.4 shows that the sum $\sum_{i=1}^n a_i$ is independent of the grouping of its elements, which justifies the use the notation $a_1 + \cdots + a_n$ (without any parentheses).

If we also assume that our associative binary operation on A is commutative, then we can show that the sum $\sum_{i \in I} a_i$ does not depend on the ordering of the index set I.

Proposition 2.5. *Given any nonempty set A equipped with an associative and commutative binary operation $+: A \times A \to A$, for any two nonempty finite sequences I and J of distinct natural numbers such that J is a permutation of I (in other words, the underlying sets of I and J are identical), for every sequence $(a_i)_{i \in I}$ of elements in A, we have*

$$\sum_{\alpha \in I} a_\alpha = \sum_{\alpha \in J} a_\alpha.$$

Proof. We proceed by induction on the number p of elements in I. If $p = 1$, we have $I = J$ and the proposition holds trivially.

If $p > 1$, to simplify notation, assume that $I = (1, \ldots, p)$ and that J is a permutation $(i_1, \ldots, i_p)$ of I. First, assume that $2 \leq i_1 \leq p - 1$, let J' be the sequence obtained from J by deleting i_1, I' be the sequence obtained from I by deleting i_1, and let $P = (1, 2, \ldots, i_1 - 1)$ and $Q = (i_1 + 1, \ldots, p - 1, p)$. Observe that the sequence I' is the concatenation of the sequences P and Q. By the induction hypothesis applied to J' and I', and then by Proposition 2.4 applied to I' and its partition (P, Q), we have

$$\sum_{\alpha \in J'} a_\alpha = \sum_{\alpha \in I'} a_\alpha = \left(\sum_{i=1}^{i_1-1} a_i\right) + \left(\sum_{i=i_1+1}^{p} a_i\right).$$

If we add the left-hand side to a_{i_1}, by definition we get

$$\sum_{\alpha \in J} a_\alpha.$$

If we add the right-hand side to a_{i_1}, we get

$$a_{i_1} + \left(\left(\sum_{i=1}^{i_1-1} a_i \right) + \left(\sum_{i=i_1+1}^{p} a_i \right) \right).$$

Using associativity, we get

$$a_{i_1} + \left(\left(\sum_{i=1}^{i_1-1} a_i \right) + \left(\sum_{i=i_1+1}^{p} a_i \right) \right) = \left(a_{i_1} + \left(\sum_{i=1}^{i_1-1} a_i \right) \right) + \left(\sum_{i=i_1+1}^{p} a_i \right),$$

then using associativity and commutativity several times (more rigorously, using induction on $i_1 - 1$), we get

$$\left(a_{i_1} + \left(\sum_{i=1}^{i_1-1} a_i \right) \right) + \left(\sum_{i=i_1+1}^{p} a_i \right) = \left(\sum_{i=1}^{i_1-1} a_i \right) + a_{i_1} + \left(\sum_{i=i_1+1}^{p} a_i \right)$$

$$= \sum_{i=1}^{p} a_i,$$

as claimed.

The cases where $i_1 = 1$ or $i_1 = p$ are treated similarly, but in a simpler manner since either $P = ()$ or $Q = ()$ (where $()$ denotes the empty sequence). □

Having done all this, we can now make sense of sums of the form $\sum_{i \in I} a_i$, for any finite indexed set I and any family $a = (a_i)_{i \in I}$ of elements in A, where A is a set equipped with a binary operation $+$ which is associative and commutative.

Indeed, since I is finite, it is in bijection with the set $\{1, \ldots, n\}$ for some $n \in \mathbb{N}$, and any total ordering $\preceq$ on I corresponds to a permutation $I_\preceq$ of $\{1, \ldots, n\}$ (where we identify a permutation with its image). For any total ordering $\preceq$ on I, we define $\sum_{i \in I, \preceq} a_i$ as

$$\sum_{i \in I, \preceq} a_i = \sum_{j \in I_\preceq} a_j.$$

Then for any other total ordering $\preceq'$ on I, we have

$$\sum_{i \in I, \preceq'} a_i = \sum_{j \in I_{\preceq'}} a_j,$$

and since $I_{\preceq}$ and $I_{\preceq'}$ are different permutations of $\{1, \ldots, n\}$, by Proposition 2.5, we have

$$\sum_{j \in I_{\preceq}} a_j = \sum_{j \in I_{\preceq'}} a_j.$$

Therefore, the sum $\sum_{i \in I, \preceq} a_i$ does not depend on the total ordering on I. We define *the* sum $\sum_{i \in I} a_i$ as the common value $\sum_{i \in I, \preceq} a_i$ for all total orderings $\preceq$ of I.

Here are some examples with $A = \mathbb{R}$:

(1) If $I = \{1, 2, 3\}$, $a = \{(1, 2), (2, -3), (3, \sqrt{2})\}$, then $\sum_{i \in I} a_i = 2 - 3 + \sqrt{2} = -1 + \sqrt{2}$.
(2) If $I = \{2, 5, 7\}$, $a = \{(2, 2), (5, -3), (7, \sqrt{2})\}$, then $\sum_{i \in I} a_i = 2 - 3 + \sqrt{2} = -1 + \sqrt{2}$.
(3) If $I = \{r, g, b\}$, $a = \{(r, 2), (g, -3), (b, 1)\}$, then $\sum_{i \in I} a_i = 2 - 3 + 1 = 0$.

2.4   Linear Independence, Subspaces

One of the most useful properties of vector spaces is that they possess bases. What this means is that in every vector space E, there is some set of vectors, $\{e_1, \ldots, e_n\}$, such that *every* vector $v \in E$ can be written as a linear combination,

$$v = \lambda_1 e_1 + \cdots + \lambda_n e_n,$$

of the e_i, for some scalars, $\lambda_1, \ldots, \lambda_n \in \mathbb{R}$. Furthermore, the n-tuple, $(\lambda_1, \ldots, \lambda_n)$, as above is unique.

This description is fine when E has a finite basis, $\{e_1, \ldots, e_n\}$, but this is not always the case! For example, the vector space of real polynomials, $\mathbb{R}[X]$, does not have a finite basis but instead it has an infinite basis, namely

$$1, \ X, \ X^2, \ \ldots, X^n, \ \ldots$$

Given a set A, recall that an *I-indexed family* $(a_i)_{i \in I}$ of elements of A (for short, a *family*) is a function $a \colon I \to A$, or equivalently a set of pairs $\{(i, a_i) \mid i \in I\}$. We agree that when $I = \emptyset$, $(a_i)_{i \in I} = \emptyset$. A family $(a_i)_{i \in I}$ is finite if I is finite.

Remark: When considering a family $(a_i)_{i \in I}$, there is no reason to assume that I is ordered. The crucial point is that every element of the family is uniquely indexed by an element of I. Thus, unless specified otherwise, we do not assume that the elements of an index set are ordered.

Given two disjoint sets I and J, the union of two families $(u_i)_{i \in I}$ and $(v_j)_{j \in J}$, denoted as $(u_i)_{i \in I} \cup (v_j)_{j \in J}$, is the family $(w_k)_{k \in (I \cup J)}$ defined such that $w_k = u_k$ if $k \in I$, and $w_k = v_k$ if $k \in J$. Given a family $(u_i)_{i \in I}$ and any element v, we denote by $(u_i)_{i \in I} \cup_k (v)$ the family $(w_i)_{i \in I \cup \{k\}}$ defined such that, $w_i = u_i$ if $i \in I$, and $w_k = v$, where k is any index such that $k \notin I$. Given a family $(u_i)_{i \in I}$, a *subfamily* of $(u_i)_{i \in I}$ is a family $(u_j)_{j \in J}$ where J is any subset of I.

In this chapter, unless specified otherwise, *it is assumed that all families of scalars are finite (i.e., their index set is finite)*.

Definition 2.6. Let E be a vector space. A vector $v \in E$ is a *linear combination of a family $(u_i)_{i \in I}$ of elements of E* iff there is a family $(\lambda_i)_{i \in I}$ of scalars in $\mathbb{R}$ such that

$$v = \sum_{i \in I} \lambda_i u_i.$$

When $I = \emptyset$, we stipulate that $v = 0$. (By Proposition 2.5, sums of the form $\sum_{i \in I} \lambda_i u_i$ are well defined.) We say that a family $(u_i)_{i \in I}$ is *linearly independent* iff for every family $(\lambda_i)_{i \in I}$ of scalars in $\mathbb{R}$,

$$\sum_{i \in I} \lambda_i u_i = 0 \quad \text{implies that} \quad \lambda_i = 0 \text{ for all } i \in I.$$

Equivalently, a family $(u_i)_{i \in I}$ is *linearly dependent* iff there is some family $(\lambda_i)_{i \in I}$ of scalars in $\mathbb{R}$ such that

$$\sum_{i \in I} \lambda_i u_i = 0 \quad \text{and} \quad \lambda_j \neq 0 \text{ for some } j \in I.$$

We agree that when $I = \emptyset$, the family $\emptyset$ is linearly independent.

Observe that defining linear combinations for families of vectors rather than for sets of vectors has the advantage that *the vectors being combined need not be distinct*. For example, for $I = \{1, 2, 3\}$ and the families (u, v, u) and $(\lambda_1, \lambda_2, \lambda_1)$, the linear combination

$$\sum_{i \in I} \lambda_i u_i = \lambda_1 u + \lambda_2 v + \lambda_1 u$$

makes sense. Using sets of vectors in the definition of a linear combination does not allow such linear combinations; this is too restrictive.

Unravelling Definition 2.6, a family $(u_i)_{i \in I}$ is linearly dependent iff either I consists of a single element, say i, and $u_i = 0$, or $|I| \geq 2$ and some u_j in the family can be expressed as a linear combination of the other vectors in the family. Indeed, in the second case, there is some family $(\lambda_i)_{i \in I}$ of scalars in $\mathbb{R}$ such that

$$\sum_{i \in I} \lambda_i u_i = 0 \quad \text{and} \quad \lambda_j \neq 0 \text{ for some } j \in I,$$

and since $|I| \geq 2$, the set $I - \{j\}$ is nonempty and we get

$$u_j = \sum_{i \in (I - \{j\})} -\lambda_j^{-1} \lambda_i u_i.$$

Observe that one of the reasons for defining linear dependence for families of vectors rather than for sets of vectors is that our definition allows multiple occurrences of a vector. This is important because a matrix may contain identical columns, and we would like to say that these columns are linearly dependent. The definition of linear dependence for sets does not allow us to do that.

The above also shows that a family $(u_i)_{i \in I}$ is linearly independent iff either $I = \emptyset$, or I consists of a single element i and $u_i \neq 0$, or $|I| \geq 2$ and no vector u_j in the family can be expressed as a linear combination of the other vectors in the family.

When I is nonempty, if the family $(u_i)_{i \in I}$ is linearly independent, note that $u_i \neq 0$ for all $i \in I$. Otherwise, if $u_i = 0$ for some $i \in I$, then we get a nontrivial linear dependence $\sum_{i \in I} \lambda_i u_i = 0$ by picking any nonzero λ_i and letting $\lambda_k = 0$ for all $k \in I$ with $k \neq i$, since $\lambda_i 0 = 0$. If $|I| \geq 2$, we must also have $u_i \neq u_j$ for all $i, j \in I$ with $i \neq j$, since otherwise we get a nontrivial linear dependence by picking $\lambda_i = \lambda$ and $\lambda_j = -\lambda$ for any nonzero λ, and letting $\lambda_k = 0$ for all $k \in I$ with $k \neq i, j$.

Thus, the definition of linear independence implies that *a nontrivial linearly independent family is actually a set.* This explains why certain authors choose to define linear independence for sets of vectors. The problem with this approach is that linear dependence, which is the logical negation of linear independence, is then only defined for sets of vectors. However, as we pointed out earlier, it is really desirable to define linear dependence for families allowing multiple occurrences of the same vector.

Example 2.4.

(1) Any two distinct scalars $\lambda, \mu \neq 0$ in $\mathbb{R}$ are linearly dependent.
(2) In $\mathbb{R}^3$, the vectors $(1, 0, 0)$, $(0, 1, 0)$, and $(0, 0, 1)$ are linearly independent. See Figure 2.7.
(3) In $\mathbb{R}^4$, the vectors $(1, 1, 1, 1)$, $(0, 1, 1, 1)$, $(0, 0, 1, 1)$, and $(0, 0, 0, 1)$ are linearly independent.

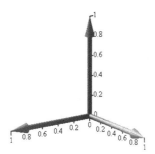

Fig. 2.7 A visual (arrow) depiction of the red vector $(1, 0, 0)$, the green vector $(0, 1, 0)$, and the blue vector $(0, 0, 1)$ in $\mathbb{R}^3$.

(4) In $\mathbb{R}^2$, the vectors $u = (1, 1)$, $v = (0, 1)$ and $w = (2, 3)$ are linearly dependent, since

$$w = 2u + v.$$

See Figure 2.8.

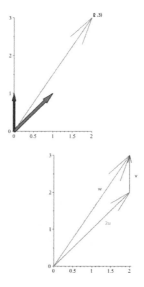

Fig. 2.8 A visual (arrow) depiction of the pink vector $u = (1, 1)$, the dark purple vector $v = (0, 1)$, and the vector sum $w = 2u + v$.

When I is finite, we often assume that it is the set $I = \{1, 2, \ldots, n\}$. In this case, we denote the family $(u_i)_{i \in I}$ as $(u_1, \ldots, u_n)$.

The notion of a subspace of a vector space is defined as follows.

Definition 2.7. Given a vector space E, a subset F of E is a *linear subspace* (or *subspace*) of E iff F is nonempty and $\lambda u + \mu v \in F$ for all $u, v \in F$, and all $\lambda, \mu \in \mathbb{R}$.

It is easy to see that a subspace F of E is indeed a vector space, since the restriction of $+ \colon E \times E \to E$ to $F \times F$ is indeed a function $+ \colon F \times F \to F$, and the restriction of $\cdot \colon \mathbb{R} \times E \to E$ to $\mathbb{R} \times F$ is indeed a function $\cdot \colon \mathbb{R} \times F \to F$.

Since a subspace F is nonempty, if we pick any vector $u \in F$ and if we let $\lambda = \mu = 0$, then $\lambda u + \mu u = 0u + 0u = 0$, so *every subspace contains the vector* 0.

The following facts also hold. The proof is left as an exercise.

Proposition 2.6.

(1) The intersection of any family (even infinite) of subspaces of a vector space E is a subspace.

(2) Let F be any subspace of a vector space E. For any nonempty finite index set I, if $(u_i)_{i \in I}$ is any family of vectors $u_i \in F$ and $(\lambda_i)_{i \in I}$ is any family of scalars, then $\sum_{i \in I} \lambda_i u_i \in F$.

The subspace $\{0\}$ will be denoted by (0), or even 0 (with a mild abuse of notation).

Example 2.5.

(1) In $\mathbb{R}^2$, the set of vectors $u = (x, y)$ such that
$$x + y = 0$$
is the subspace illustrated by Figure 2.9.

(2) In $\mathbb{R}^3$, the set of vectors $u = (x, y, z)$ such that
$$x + y + z = 0$$
is the subspace illustrated by Figure 2.10.

(3) For any $n \geq 0$, the set of polynomials $f(X) \in \mathbb{R}[X]$ of degree at most n is a subspace of $\mathbb{R}[X]$.

(4) The set of upper triangular $n \times n$ matrices is a subspace of the space of $n \times n$ matrices.

Proposition 2.7. *Given any vector space E, if S is any nonempty subset of E, then the smallest subspace $\langle S \rangle$ (or $\mathrm{Span}(S)$) of E containing S is the set of all (finite) linear combinations of elements from S.*

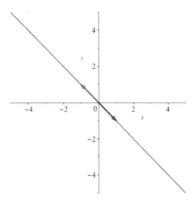

Fig. 2.9   The subspace $x+y = 0$ is the line through the origin with slope -1. It consists of all vectors of the form $\lambda(-1, 1)$.

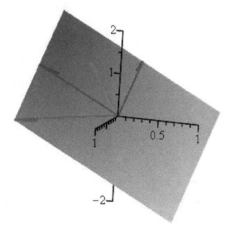

Fig. 2.10   The subspace $x + y + z = 0$ is the plane through the origin with normal $(1, 1, 1)$.

Proof. We prove that the set $\mathrm{Span}(S)$ of all linear combinations of elements of S is a subspace of E, leaving as an exercise the verification that every subspace containing S also contains $\mathrm{Span}(S)$.

First, $\mathrm{Span}(S)$ is nonempty since it contains S (which is nonempty). If $u = \sum_{i \in I} \lambda_i u_i$ and $v = \sum_{j \in J} \mu_j v_j$ are any two linear combinations in

Span(S), for any two scalars $\lambda, \mu \in \mathbb{R}$,

$$
\begin{aligned}
\lambda u + \mu v &= \lambda \sum_{i \in I} \lambda_i u_i + \mu \sum_{j \in J} \mu_j v_j \\
&= \sum_{i \in I} \lambda \lambda_i u_i + \sum_{j \in J} \mu \mu_j v_j \\
&= \sum_{i \in I-J} \lambda \lambda_i u_i + \sum_{i \in I \cap J} (\lambda \lambda_i + \mu \mu_i) u_i + \sum_{j \in J-I} \mu \mu_j v_j,
\end{aligned}
$$

which is a linear combination with index set $I \cup J$, and thus $\lambda u + \mu v \in$ Span(S), which proves that Span(S) is a subspace. $\square$

One might wonder what happens if we add extra conditions to the coefficients involved in forming linear combinations. Here are three natural restrictions which turn out to be important (as usual, we assume that our index sets are finite):

(1) Consider combinations $\sum_{i \in I} \lambda_i u_i$ for which

$$
\sum_{i \in I} \lambda_i = 1.
$$

These are called *affine combinations*. One should realize that every linear combination $\sum_{i \in I} \lambda_i u_i$ can be viewed as an affine combination. For example, if k is an index not in I, if we let $J = I \cup \{k\}$, $u_k = 0$, and $\lambda_k = 1 - \sum_{i \in I} \lambda_i$, then $\sum_{j \in J} \lambda_j u_j$ is an affine combination and

$$
\sum_{i \in I} \lambda_i u_i = \sum_{j \in J} \lambda_j u_j.
$$

However, we get new spaces. For example, in $\mathbb{R}^3$, the set of all affine combinations of the three vectors $e_1 = (1, 0, 0), e_2 = (0, 1, 0)$, and $e_3 = (0, 0, 1)$, is the plane passing through these three points. Since it does not contain $0 = (0, 0, 0)$, it is not a linear subspace.

(2) Consider combinations $\sum_{i \in I} \lambda_i u_i$ for which

$$
\lambda_i \geq 0, \quad \text{for all } i \in I.
$$

These are called *positive* (or *conic*) *combinations*. It turns out that positive combinations of families of vectors are *cones*. They show up naturally in convex optimization.

(3) Consider combinations $\sum_{i \in I} \lambda_i u_i$ for which we require (1) *and* (2), that is

$$
\sum_{i \in I} \lambda_i = 1, \quad \text{and} \quad \lambda_i \geq 0 \quad \text{for all } i \in I.
$$

These are called *convex combinations*. Given any finite family of vectors, the set of all convex combinations of these vectors is a *convex polyhedron*. Convex polyhedra play a very important role in convex optimization.

Remark: The notion of linear combination can also be defined for infinite index sets I. To ensure that a sum $\sum_{i \in I} \lambda_i u_i$ makes sense, we restrict our attention to families of finite support.

Definition 2.8. Given any field K, a family of scalars $(\lambda_i)_{i \in I}$ has *finite support* if $\lambda_i = 0$ for all $i \in I - J$, for some finite subset J of I.

If $(\lambda_i)_{i \in I}$ is a family of scalars of finite support, for any vector space E over K, for any (possibly infinite) family $(u_i)_{i \in I}$ of vectors $u_i \in E$, we define the linear combination $\sum_{i \in I} \lambda_i u_i$ as the finite linear combination $\sum_{j \in J} \lambda_j u_j$, where J is any finite subset of I such that $\lambda_i = 0$ for all $i \in I - J$. In general, results stated for finite families also hold for families of finite support.

2.5   Bases of a Vector Space

Given a vector space E, given a family $(v_i)_{i \in I}$, the subset V of E consisting of the null vector 0 and of all linear combinations of $(v_i)_{i \in I}$ is easily seen to be a subspace of E. The family $(v_i)_{i \in I}$ is an economical way of representing the entire subspace V, but such a family would be even nicer if it was not redundant. Subspaces having such an "efficient" generating family (called a basis) play an important role and motivate the following definition.

Definition 2.9. Given a vector space E and a subspace V of E, a family $(v_i)_{i \in I}$ of vectors $v_i \in V$ *spans* V or *generates* V iff for every $v \in V$, there is some family $(\lambda_i)_{i \in I}$ of scalars in $\mathbb{R}$ such that

$$v = \sum_{i \in I} \lambda_i v_i.$$

We also say that the elements of $(v_i)_{i \in I}$ are *generators* of V and that V is *spanned* by $(v_i)_{i \in I}$, or *generated* by $(v_i)_{i \in I}$. If a subspace V of E is generated by a finite family $(v_i)_{i \in I}$, we say that V is *finitely generated*. A family $(u_i)_{i \in I}$ that spans V and is linearly independent is called a *basis* of V.

Example 2.6.

(1) In $\mathbb{R}^3$, the vectors $(1,0,0)$, $(0,1,0)$, and $(0,0,1)$, illustrated in Figure 2.9, form a basis.

(2) The vectors $(1,1,1,1), (1,1,-1,-1), (1,-1,0,0), (0,0,1,-1)$ form a basis of $\mathbb{R}^4$ known as the *Haar basis*. This basis and its generalization to dimension 2^n are crucial in wavelet theory.

(3) In the subspace of polynomials in $\mathbb{R}[X]$ of degree at most n, the polynomials $1, X, X^2, \ldots, X^n$ form a basis.

(4) The *Bernstein polynomials* $\binom{n}{k}(1-X)^{n-k}X^k$ for $k = 0, \ldots, n$, also form a basis of that space. These polynomials play a major role in the theory of *spline curves*.

The first key result of linear algebra is that *every vector space E has a basis*. We begin with a crucial lemma which formalizes the mechanism for building a basis incrementally.

Lemma 2.1. *Given a linearly independent family $(u_i)_{i \in I}$ of elements of a vector space E, if $v \in E$ is not a linear combination of $(u_i)_{i \in I}$, then the family $(u_i)_{i \in I} \cup_k (v)$ obtained by adding v to the family $(u_i)_{i \in I}$ is linearly independent (where $k \notin I$).*

Proof. Assume that $\mu v + \sum_{i \in I} \lambda_i u_i = 0$, for any family $(\lambda_i)_{i \in I}$ of scalars in $\mathbb{R}$. If $\mu \neq 0$, then μ has an inverse (because $\mathbb{R}$ is a field), and thus we have $v = -\sum_{i \in I}(\mu^{-1}\lambda_i)u_i$, showing that v is a linear combination of $(u_i)_{i \in I}$ and contradicting the hypothesis. Thus, $\mu = 0$. But then, we have $\sum_{i \in I} \lambda_i u_i = 0$, and since the family $(u_i)_{i \in I}$ is linearly independent, we have $\lambda_i = 0$ for all $i \in I$. $\qquad\square$

The next theorem holds in general, but the proof is more sophisticated for vector spaces that do not have a finite set of generators. *Thus, in this chapter, we only prove the theorem for finitely generated vector spaces.*

Theorem 2.1. *Given any finite family $S = (u_i)_{i \in I}$ generating a vector space E and any linearly independent subfamily $L = (u_j)_{j \in J}$ of S (where $J \subseteq I$), there is a basis B of E such that $L \subseteq B \subseteq S$.*

Proof. Consider the set of linearly independent families B such that $L \subseteq B \subseteq S$. Since this set is nonempty and finite, it has some maximal element (that is, a subfamily $B = (u_h)_{h \in H}$ of S with $H \subseteq I$ of maximum cardinality), say $B = (u_h)_{h \in H}$. We claim that B generates E. Indeed, if B does not generate E, then there is some $u_p \in S$ that is not a linear combination of vectors in B (since S generates E), with $p \notin H$. Then by Lemma 2.1, the family $B' = (u_h)_{h \in H \cup \{p\}}$ is linearly independent, and since $L \subseteq B \subset B' \subseteq S$, this contradicts the maximality of B. Thus, B is a basis of E such that $L \subseteq B \subseteq S$. $\qquad\square$

Remark: Theorem 2.1 also holds for vector spaces that are not finitely generated. In this case, the problem is to guarantee the existence of a maximal linearly independent family B such that $L \subseteq B \subseteq S$. The existence of such a maximal family can be shown using Zorn's lemma; see Lang [Lang (1993)] (Theorem 5.1).

A situation where the full generality of Theorem 2.1 is needed is the case of the vector space $\mathbb{R}$ over the field of coefficients $\mathbb{Q}$. The numbers 1 and $\sqrt{2}$ are linearly independent over $\mathbb{Q}$, so according to Theorem 2.1, the linearly independent family $L = (1, \sqrt{2})$ can be extended to a basis B of $\mathbb{R}$. Since $\mathbb{R}$ is uncountable and $\mathbb{Q}$ is countable, such a basis must be uncountable!

The notion of a basis can also be defined in terms of the notion of maximal linearly independent family and minimal generating family.

Definition 2.10. Let $(v_i)_{i \in I}$ be a family of vectors in a vector space E. We say that $(v_i)_{i \in I}$ a *maximal linearly independent family* of E if it is linearly independent, and if for any vector $w \in E$, the family $(v_i)_{i \in I \cup_k \{w\}}$ obtained by adding w to the family $(v_i)_{i \in I}$ is linearly dependent. We say that $(v_i)_{i \in I}$ a *minimal generating family* of E if it spans E, and if for any index $p \in I$, the family $(v_i)_{i \in I - \{p\}}$ obtained by removing v_p from the family $(v_i)_{i \in I}$ does not span E.

The following proposition giving useful properties characterizing a basis is an immediate consequence of Lemma 2.1.

Proposition 2.8. *Given a vector space E, for any family $B = (v_i)_{i \in I}$ of vectors of E, the following properties are equivalent:*

(1) B is a basis of E.
(2) B is a maximal linearly independent family of E.
(3) B is a minimal generating family of E.

Proof. We will first prove the equivalence of (1) and (2). Assume (1). Since B is a basis, it is a linearly independent family. We claim that B is a maximal linearly independent family. If B is not a maximal linearly independent family, then there is some vector $w \in E$ such that the family B' obtained by adding w to B is linearly independent. However, since B is a basis of E, the vector w can be expressed as a linear combination of vectors in B, contradicting the fact that B' is linearly independent.

Conversely, assume (2). We claim that B spans E. If B does not span E, then there is some vector $w \in E$ which is not a linear combination of vectors in B. By Lemma 2.1, the family B' obtained by adding w to B is linearly independent. Since B is a proper subfamily of B', this contradicts the assumption that B is a maximal linearly independent family. Therefore, B must span E, and since B is also linearly independent, it is a basis of E.

Now we will prove the equivalence of (1) and (3). Again, assume (1). Since B is a basis, it is a generating family of E. We claim that B is a minimal generating family. If B is not a minimal generating family, then there is a proper subfamily B' of B that spans E. Then, every $w \in B - B'$ can be expressed as a linear combination of vectors from B', contradicting the fact that B is linearly independent.

Conversely, assume (3). We claim that B is linearly independent. If B is not linearly independent, then some vector $w \in B$ can be expressed as a linear combination of vectors in $B' = B - \{w\}$. Since B generates E, the family B' also generates E, but B' is a proper subfamily of B, contradicting the minimality of B. Since B spans E and is linearly independent, it is a basis of E. $\square$

The second key result of linear algebra is that *for any two bases* $(u_i)_{i \in I}$ *and* $(v_j)_{j \in J}$ *of a vector space* E, *the index sets* I *and* J *have the same cardinality.* In particular, if E has a finite basis of n elements, every basis of E has n elements, and the integer n is called the *dimension* of the vector space E.

To prove the second key result, we can use the following *replacement lemma* due to Steinitz. This result shows the relationship between finite linearly independent families and finite families of generators of a vector space. We begin with a version of the lemma which is a bit informal, but easier to understand than the precise and more formal formulation given in Proposition 2.10. The technical difficulty has to do with the fact that some of the indices need to be renamed.

Proposition 2.9. *(Replacement lemma, version 1) Given a vector space* E, *let* $(u_1, \ldots, u_m)$ *be any finite linearly independent family in* E, *and let* $(v_1, \ldots, v_n)$ *be any finite family such that every* u_i *is a linear combination of* $(v_1, \ldots, v_n)$. *Then we must have* $m \leq n$, *and there is a replacement of* *m of the vectors v_j by $(u_1, \ldots, u_m)$, such that after renaming some of the indices of the $v_j s$, the families $(u_1, \ldots, u_m, v_{m+1}, \ldots, v_n)$ and $(v_1, \ldots, v_n)$ generate the same subspace of E.*

Proof. We proceed by induction on m. When $m = 0$, the family $(u_1, \ldots, u_m)$ is empty, and the proposition holds trivially. For the induction step, we have a linearly independent family $(u_1, \ldots, u_m, u_{m+1})$. Consider the linearly independent family $(u_1, \ldots, u_m)$. By the induction hypothesis, $m \leq n$, and there is a replacement of m of the vectors v_j by $(u_1, \ldots, u_m)$, such that after renaming some of the indices of the vs, the families $(u_1, \ldots, u_m, v_{m+1}, \ldots, v_n)$ and $(v_1, \ldots, v_n)$ generate the same subspace of E. The vector u_{m+1} can also be expressed as a linear combination of $(v_1, \ldots, v_n)$, and since $(u_1, \ldots, u_m, v_{m+1}, \ldots, v_n)$ and $(v_1, \ldots, v_n)$ generate the same subspace, u_{m+1} can be expressed as a linear combination of $(u_1, \ldots, u_m, v_{m+1}, \ldots, v_n)$, say

$$u_{m+1} = \sum_{i=1}^{m} \lambda_i u_i + \sum_{j=m+1}^{n} \lambda_j v_j.$$

We claim that $\lambda_j \neq 0$ for some j with $m + 1 \leq j \leq n$, which implies that $m + 1 \leq n$.

Otherwise, we would have

$$u_{m+1} = \sum_{i=1}^{m} \lambda_i u_i,$$

a nontrivial linear dependence of the u_i, which is impossible since $(u_1, \ldots, u_{m+1})$ are linearly independent.

Therefore, $m + 1 \leq n$, and after renaming indices if necessary, we may assume that $\lambda_{m+1} \neq 0$, so we get

$$v_{m+1} = -\sum_{i=1}^{m} (\lambda_{m+1}^{-1} \lambda_i) u_i - \lambda_{m+1}^{-1} u_{m+1} - \sum_{j=m+2}^{n} (\lambda_{m+1}^{-1} \lambda_j) v_j.$$

Observe that the families $(u_1, \ldots, u_m, v_{m+1}, \ldots, v_n)$ and $(u_1, \ldots, u_{m+1}, v_{m+2}, \ldots, v_n)$ generate the same subspace, since u_{m+1} is a linear combination of $(u_1, \ldots, u_m, v_{m+1}, \ldots, v_n)$ and v_{m+1} is a linear combination of $(u_1, \ldots, u_{m+1}, v_{m+2}, \ldots, v_n)$. Since $(u_1, \ldots, u_m, v_{m+1}, \ldots, v_n)$ and $(v_1, \ldots, v_n)$ generate the same subspace, we conclude that $(u_1, \ldots, u_{m+1}, v_{m+2}, \ldots, v_n)$ and and $(v_1, \ldots, v_n)$ generate the same subspace, which concludes the induction hypothesis. $\square$

Here is an example illustrating the replacement lemma. Consider sequences (u_1, u_2, u_3) and $(v_1, v_2, v_3, v_4, v_5)$, where (u_1, u_2, u_3) is a linearly independent family and with the u_is expressed in terms of the v_js as follows:

$$u_1 = v_4 + v_5$$
$$u_2 = v_3 + v_4 - v_5$$
$$u_3 = v_1 + v_2 + v_3.$$

From the first equation we get

$$v_4 = u_1 - v_5,$$

and by substituting in the second equation we have

$$u_2 = v_3 + v_4 - v_5 = v_3 + u_1 - v_5 - v_5 = u_1 + v_3 - 2v_5.$$

From the above equation we get

$$v_3 = -u_1 + u_2 + 2v_5,$$

and so

$$u_3 = v_1 + v_2 + v_3 = v_1 + v_2 - u_1 + u_2 + 2v_5.$$

Finally, we get

$$v_1 = u_1 - u_2 + u_3 - v_2 - 2v_5.$$

Therefore we have

$$v_1 = u_1 - u_2 + u_3 - v_2 - 2v_5$$
$$v_3 = -u_1 + u_2 + 2v_5$$
$$v_4 = u_1 - v_5,$$

which shows that $(u_1, u_2, u_3, v_2, v_5)$ spans the same subspace as $(v_1, v_2, v_3, v_4, v_5)$. The vectors (v_1, v_3, v_4) have been replaced by (u_1, u_2, u_3), and the vectors left over are (v_2, v_5). We can rename them (v_4, v_5).

For the sake of completeness, here is a more formal statement of the replacement lemma (and its proof).

Proposition 2.10. *(Replacement lemma, version 2) Given a vector space E, let $(u_i)_{i \in I}$ be any finite linearly independent family in E, where $|I| = m$, and let $(v_j)_{j \in J}$ be any finite family such that every u_i is a linear combination of $(v_j)_{j \in J}$, where $|J| = n$. Then there exists a set L and an injection $\rho \colon L \to J$ (a relabeling function) such that $L \cap I = \emptyset$, $|L| = n - m$, and the families $(u_i)_{i \in I} \cup (v_{\rho(l)})_{l \in L}$ and $(v_j)_{j \in J}$ generate the same subspace of E. In particular, $m \leq n$.*

Proof. We proceed by induction on $|I| = m$. When $m = 0$, the family $(u_i)_{i \in I}$ is empty, and the proposition holds trivially with $L = J$ (ρ is the identity). Assume $|I| = m + 1$. Consider the linearly independent family $(u_i)_{i \in (I - \{p\})}$, where p is any member of I. By the induction hypothesis, there exists a set L and an injection $\rho \colon L \to J$ such that $L \cap (I - \{p\}) = \emptyset$, $|L| = n - m$, and the families $(u_i)_{i \in (I - \{p\})} \cup (v_{\rho(l)})_{l \in L}$ and $(v_j)_{j \in J}$ generate the same subspace of E. If $p \in L$, we can replace L by $(L - \{p\}) \cup \{p'\}$ where p' does not belong to $I \cup L$, and replace ρ by the injection ρ' which agrees with ρ on $L - \{p\}$ and such that $\rho'(p') = \rho(p)$. Thus, we can always assume that $L \cap I = \emptyset$. Since u_p is a linear combination of $(v_j)_{j \in J}$ and the families $(u_i)_{i \in (I - \{p\})} \cup (v_{\rho(l)})_{l \in L}$ and $(v_j)_{j \in J}$ generate the same subspace of E, u_p is a linear combination of $(u_i)_{i \in (I - \{p\})} \cup (v_{\rho(l)})_{l \in L}$. Let

$$u_p = \sum_{i \in (I - \{p\})} \lambda_i u_i + \sum_{l \in L} \lambda_l v_{\rho(l)}. \tag{2.15}$$

If $\lambda_l = 0$ for all $l \in L$, we have

$$\sum_{i \in (I - \{p\})} \lambda_i u_i - u_p = 0,$$

contradicting the fact that $(u_i)_{i \in I}$ is linearly independent. Thus, $\lambda_l \neq 0$ for some $l \in L$, say $l = q$. Since $\lambda_q \neq 0$, we have

$$v_{\rho(q)} = \sum_{i \in (I - \{p\})} (-\lambda_q^{-1} \lambda_i) u_i + \lambda_q^{-1} u_p + \sum_{l \in (L - \{q\})} (-\lambda_q^{-1} \lambda_l) v_{\rho(l)}. \tag{2.16}$$

We claim that the families $(u_i)_{i \in (I - \{p\})} \cup (v_{\rho(l)})_{l \in L}$ and $(u_i)_{i \in I} \cup (v_{\rho(l)})_{l \in (L - \{q\})}$ generate the same subset of E. Indeed, the second family is obtained from the first by replacing $v_{\rho(q)}$ by u_p, and vice-versa, and u_p is a linear combination of $(u_i)_{i \in (I - \{p\})} \cup (v_{\rho(l)})_{l \in L}$, by (2.15), and $v_{\rho(q)}$ is a linear combination of $(u_i)_{i \in I} \cup (v_{\rho(l)})_{l \in (L - \{q\})}$, by (2.16). Thus, the families $(u_i)_{i \in I} \cup (v_{\rho(l)})_{l \in (L - \{q\})}$ and $(v_j)_{j \in J}$ generate the same subspace of E, and the proposition holds for $L - \{q\}$ and the restriction of the injection $\rho \colon L \to J$ to $L - \{q\}$, since $L \cap I = \emptyset$ and $|L| = n - m$ imply that $(L - \{q\}) \cap I = \emptyset$ and $|L - \{q\}| = n - (m + 1)$. $\qquad\square$

The idea is that m of the vectors v_j can be *replaced* by the linearly independent u_is in such a way that the same subspace is still generated. The purpose of the function $\rho \colon L \to J$ is to pick $n - m$ elements $j_1, \ldots, j_{n-m}$ of J and to relabel them $l_1, \ldots, l_{n-m}$ in such a way that these new indices do not clash with the indices in I; this way, the vectors $v_{j_1}, \ldots, v_{j_{n-m}}$ who "survive" (i.e. are not replaced) are relabeled $v_{l_1}, \ldots, v_{l_{n-m}}$, and the other

m vectors v_j with $j \in J - \{j_1, \ldots, j_{n-m}\}$ are replaced by the u_i. The index set of this new family is $I \cup L$.

Actually, one can prove that Proposition 2.10 implies Theorem 2.1 when the vector space is finitely generated. Putting Theorem 2.1 and Proposition 2.10 together, we obtain the following fundamental theorem.

Theorem 2.2. *Let E be a finitely generated vector space. Any family $(u_i)_{i \in I}$ generating E contains a subfamily $(u_j)_{j \in J}$ which is a basis of E. Any linearly independent family $(u_i)_{i \in I}$ can be extended to a family $(u_j)_{j \in J}$ which is a basis of E (with $I \subseteq J$). Furthermore, for every two bases $(u_i)_{i \in I}$ and $(v_j)_{j \in J}$ of E, we have $|I| = |J| = n$ for some fixed integer $n \geq 0$.*

Proof. The first part follows immediately by applying Theorem 2.1 with $L = \emptyset$ and $S = (u_i)_{i \in I}$. For the second part, consider the family $S' = (u_i)_{i \in I} \cup (v_h)_{h \in H}$, where $(v_h)_{h \in H}$ is any finitely generated family generating E, and with $I \cap H = \emptyset$. Then apply Theorem 2.1 to $L = (u_i)_{i \in I}$ and to S'. For the last statement, assume that $(u_i)_{i \in I}$ and $(v_j)_{j \in J}$ are bases of E. Since $(u_i)_{i \in I}$ is linearly independent and $(v_j)_{j \in J}$ spans E, Proposition 2.10 implies that $|I| \leq |J|$. A symmetric argument yields $|J| \leq |I|$. $\square$

Remark: Theorem 2.2 also holds for vector spaces that are not finitely generated.

Definition 2.11. When a vector space E is not finitely generated, we say that E is of infinite dimension. The *dimension* of a finitely generated vector space E is the common dimension n of all of its bases and is denoted by $\dim(E)$.

Clearly, if the field $\mathbb{R}$ itself is viewed as a vector space, then every family (a) where $a \in \mathbb{R}$ and $a \neq 0$ is a basis. Thus $\dim(\mathbb{R}) = 1$. Note that $\dim(\{0\}) = 0$.

Definition 2.12. If E is a vector space of dimension $n \geq 1$, for any subspace U of E, if $\dim(U) = 1$, then U is called a *line*; if $\dim(U) = 2$, then U is called a *plane*; if $\dim(U) = n - 1$, then U is called a *hyperplane*. If $\dim(U) = k$, then U is sometimes called a *k-plane*.

Let $(u_i)_{i \in I}$ be a basis of a vector space E. For any vector $v \in E$, since the family $(u_i)_{i \in I}$ generates E, there is a family $(\lambda_i)_{i \in I}$ of scalars in $\mathbb{R}$, such that

$$v = \sum_{i \in I} \lambda_i u_i.$$

A very important fact is that the family $(\lambda_i)_{i \in I}$ is **unique**.

Proposition 2.11. *Given a vector space E, let $(u_i)_{i \in I}$ be a family of vectors in E. Let $v \in E$, and assume that $v = \sum_{i \in I} \lambda_i u_i$. Then the family $(\lambda_i)_{i \in I}$ of scalars such that $v = \sum_{i \in I} \lambda_i u_i$ is unique iff $(u_i)_{i \in I}$ is linearly independent.*

Proof. First, assume that $(u_i)_{i \in I}$ is linearly independent. If $(\mu_i)_{i \in I}$ is another family of scalars in $\mathbb{R}$ such that $v = \sum_{i \in I} \mu_i u_i$, then we have

$$\sum_{i \in I} (\lambda_i - \mu_i) u_i = 0,$$

and since $(u_i)_{i \in I}$ is linearly independent, we must have $\lambda_i - \mu_i = 0$ for all $i \in I$, that is, $\lambda_i = \mu_i$ for all $i \in I$. The converse is shown by contradiction. If $(u_i)_{i \in I}$ was linearly dependent, there would be a family $(\mu_i)_{i \in I}$ of scalars not all null such that

$$\sum_{i \in I} \mu_i u_i = 0$$

and $\mu_j \neq 0$ for some $j \in I$. But then,

$$v = \sum_{i \in I} \lambda_i u_i + 0 = \sum_{i \in I} \lambda_i u_i + \sum_{i \in I} \mu_i u_i = \sum_{i \in I} (\lambda_i + \mu_i) u_i,$$

with $\lambda_j \neq \lambda_j + \mu_j$ since $\mu_j \neq 0$, contradicting the assumption that $(\lambda_i)_{i \in I}$ is the unique family such that $v = \sum_{i \in I} \lambda_i u_i$. $\square$

Definition 2.13. If $(u_i)_{i \in I}$ is a basis of a vector space E, for any vector $v \in E$, if $(x_i)_{i \in I}$ is the unique family of scalars in $\mathbb{R}$ such that

$$v = \sum_{i \in I} x_i u_i,$$

each x_i is called the *component (or coordinate) of index i of v with respect to the basis* $(u_i)_{i \in I}$.

2.6   Matrices

In Section 2.1 we introduced informally the notion of a matrix. In this section we define matrices precisely, and also introduce some operations on matrices. It turns out that matrices form a vector space equipped with a multiplication operation which is associative, but noncommutative. We will explain in Section 3.1 how matrices can be used to represent linear maps, defined in the next section.

Definition 2.14. If $K = \mathbb{R}$ or $K = \mathbb{C}$, an $m \times n$-*matrix over* K is a family $(a_{ij})_{1 \leq i \leq m,\ 1 \leq j \leq n}$ of scalars in K, represented by an array

$$\begin{pmatrix} a_{11} & a_{12} & \cdots & a_{1n} \\ a_{21} & a_{22} & \cdots & a_{2n} \\ \vdots & \vdots & \ddots & \vdots \\ a_{m1} & a_{m2} & \cdots & a_{mn} \end{pmatrix}.$$

In the special case where $m = 1$, we have a *row vector*, represented by

$$(a_{11} \cdots a_{1n})$$

and in the special case where $n = 1$, we have a *column vector*, represented by

$$\begin{pmatrix} a_{11} \\ \vdots \\ a_{m1} \end{pmatrix}.$$

In these last two cases, we usually omit the constant index 1 (first index in case of a row, second index in case of a column). The set of all $m \times n$-matrices is denoted by $\mathrm{M}_{m,n}(K)$ or $\mathrm{M}_{m,n}$. An $n \times n$-matrix is called a *square matrix of dimension* n. The set of all square matrices of dimension n is denoted by $\mathrm{M}_n(K)$, or M_n.

Remark: As defined, a matrix $A = (a_{ij})_{1 \leq i \leq m,\ 1 \leq j \leq n}$ is a *family*, that is, a function from $\{1, 2, \ldots, m\} \times \{1, 2, \ldots, n\}$ to K. As such, there is no reason to assume an ordering on the indices. Thus, the matrix A can be represented in many different ways as an array, by adopting different orders for the rows or the columns. However, it is customary (and usually convenient) to assume the natural ordering on the sets $\{1, 2, \ldots, m\}$ and $\{1, 2, \ldots, n\}$, and to represent A as an array according to this ordering of the rows and columns.

We define some operations on matrices as follows.

Definition 2.15. Given two $m \times n$ matrices $A = (a_{ij})$ and $B = (b_{ij})$, we define their *sum* $A + B$ as the matrix $C = (c_{ij})$ such that $c_{ij} = a_{ij} + b_{ij}$;

that is,

$$
\begin{pmatrix}
a_{11} & a_{12} & \cdots & a_{1n} \\
a_{21} & a_{22} & \cdots & a_{2n} \\
\vdots & \vdots & \ddots & \vdots \\
a_{m1} & a_{m2} & \cdots & a_{mn}
\end{pmatrix}
+
\begin{pmatrix}
b_{11} & b_{12} & \cdots & b_{1n} \\
b_{21} & b_{22} & \cdots & b_{2n} \\
\vdots & \vdots & \ddots & \vdots \\
b_{m1} & b_{m2} & \cdots & b_{mn}
\end{pmatrix}
$$

$$
=
\begin{pmatrix}
a_{11}+b_{11} & a_{12}+b_{12} & \cdots & a_{1n}+b_{1n} \\
a_{21}+b_{21} & a_{22}+b_{22} & \cdots & a_{2n}+b_{2n} \\
\vdots & \vdots & \ddots & \vdots \\
a_{m1}+b_{m1} & a_{m2}+b_{m2} & \cdots & a_{mn}+b_{mn}
\end{pmatrix}.
$$

For any matrix $A = (a_{ij})$, we let $-A$ be the matrix $(-a_{ij})$. Given a scalar $\lambda \in K$, we define the matrix λA as the matrix $C = (c_{ij})$ such that $c_{ij} = \lambda a_{ij}$; that is

$$
\lambda
\begin{pmatrix}
a_{11} & a_{12} & \cdots & a_{1n} \\
a_{21} & a_{22} & \cdots & a_{2n} \\
\vdots & \vdots & \ddots & \vdots \\
a_{m1} & a_{m2} & \cdots & a_{mn}
\end{pmatrix}
=
\begin{pmatrix}
\lambda a_{11} & \lambda a_{12} & \cdots & \lambda a_{1n} \\
\lambda a_{21} & \lambda a_{22} & \cdots & \lambda a_{2n} \\
\vdots & \vdots & \ddots & \vdots \\
\lambda a_{m1} & \lambda a_{m2} & \cdots & \lambda a_{mn}
\end{pmatrix}.
$$

Given an $m \times n$ matrices $A = (a_{ik})$ and an $n \times p$ matrices $B = (b_{kj})$, we define their *product* AB as the $m \times p$ matrix $C = (c_{ij})$ such that

$$
c_{ij} = \sum_{k=1}^{n} a_{ik} b_{kj},
$$

for $1 \le i \le m$, and $1 \le j \le p$. In the product $AB = C$ shown below

$$
\begin{pmatrix}
a_{11} & a_{12} & \cdots & a_{1n} \\
a_{21} & a_{22} & \cdots & a_{2n} \\
\vdots & \vdots & \ddots & \vdots \\
a_{m1} & a_{m2} & \cdots & a_{mn}
\end{pmatrix}
\begin{pmatrix}
b_{11} & b_{12} & \cdots & b_{1p} \\
b_{21} & b_{22} & \cdots & b_{2p} \\
\vdots & \vdots & \ddots & \vdots \\
b_{n1} & b_{n2} & \cdots & b_{np}
\end{pmatrix}
=
\begin{pmatrix}
c_{11} & c_{12} & \cdots & c_{1p} \\
c_{21} & c_{22} & \cdots & c_{2p} \\
\vdots & \vdots & \ddots & \vdots \\
c_{m1} & c_{m2} & \cdots & c_{mp}
\end{pmatrix},
$$

note that the entry of index i and j of the matrix AB obtained by multiplying the matrices A and B can be identified with the product of the row matrix corresponding to the i-th row of A with the column matrix corresponding to the j-column of B:

$$
(a_{i1} \cdots a_{in})
\begin{pmatrix}
b_{1j} \\
\vdots \\
b_{nj}
\end{pmatrix}
= \sum_{k=1}^{n} a_{ik} b_{kj}.
$$

Definition 2.16. The square matrix I_n of dimension n containing 1 on the diagonal and 0 everywhere else is called the *identity matrix*. It is denoted by

$$I_n = \begin{pmatrix} 1 & 0 & \dots & 0 \\ 0 & 1 & \dots & 0 \\ \vdots & \vdots & \ddots & \vdots \\ 0 & 0 & \dots & 1 \end{pmatrix}.$$

Definition 2.17. Given an $m \times n$ matrix $A = (a_{ij})$, its *transpose* $A^\top = (a_{ji}^\top)$, is the $n \times m$-matrix such that $a_{ji}^\top = a_{ij}$, for all i, $1 \le i \le m$, and all j, $1 \le j \le n$.

The transpose of a matrix A is sometimes denoted by A^t, or even by $^t A$. Note that the transpose $A^\top$ of a matrix A has the property that the j-th row of $A^\top$ is the j-th column of A. In other words, transposition exchanges the rows and the columns of a matrix. Here is an example. If A is the 5×6 matrix

$$A = \begin{pmatrix} 1 & 2 & 3 & 4 & 5 & 6 \\ 7 & 1 & 2 & 3 & 4 & 5 \\ 8 & 7 & 1 & 2 & 3 & 4 \\ 9 & 8 & 7 & 1 & 2 & 3 \\ 10 & 9 & 8 & 7 & 1 & 2 \end{pmatrix},$$

then $A^\top$ is the 6×5 matrix

$$A^\top = \begin{pmatrix} 1 & 7 & 8 & 9 & 10 \\ 2 & 1 & 7 & 8 & 9 \\ 3 & 2 & 1 & 7 & 8 \\ 4 & 3 & 2 & 1 & 7 \\ 5 & 4 & 3 & 2 & 1 \\ 6 & 5 & 4 & 3 & 2 \end{pmatrix}.$$

The following observation will be useful later on when we discuss the SVD. Given any $m \times n$ matrix A and any $n \times p$ matrix B, if we denote the columns of A by $A^1, \dots, A^n$ and the rows of B by $B_1, \dots, B_n$, then we have

$$AB = A^1 B_1 + \cdots + A^n B_n.$$

For every square matrix A of dimension n, it is immediately verified that $AI_n = I_n A = A$.

Definition 2.18. For any square matrix A of dimension n, if a matrix B such that $AB = BA = I_n$ exists, then it is unique, and it is called the *inverse* of A. The matrix B is also denoted by A^{-1}. An invertible matrix is also called a *nonsingular* matrix, and a matrix that is not invertible is called a *singular* matrix.

Using Proposition 2.16 and the fact that matrices represent linear maps, it can be shown that if a square matrix A has a left inverse, that is a matrix B such that $BA = I$, or a right inverse, that is a matrix C such that $AC = I$, then A is actually invertible; so $B = A^{-1}$ and $C = A^{-1}$. These facts also follow from Proposition 5.10.

It is immediately verified that the set $\mathrm{M}_{m,n}(K)$ of $m \times n$ matrices is a *vector space* under addition of matrices and multiplication of a matrix by a scalar.

Definition 2.19. The $m \times n$-matrices $E_{ij} = (e_{h\,k})$, are defined such that $e_{i\,j} = 1$, and $e_{h\,k} = 0$, if $h \neq i$ or $k \neq j$; in other words, the (i, j)-entry is equal to 1 and all other entries are 0.

Here are the E_{ij} matrices for $m = 2$ and $n = 3$:

$$E_{11} = \begin{pmatrix} 1\ 0\ 0 \\ 0\ 0\ 0 \end{pmatrix}, \qquad E_{12} = \begin{pmatrix} 0\ 1\ 0 \\ 0\ 0\ 0 \end{pmatrix}, \qquad E_{13} = \begin{pmatrix} 0\ 0\ 1 \\ 0\ 0\ 0 \end{pmatrix}$$

$$E_{21} = \begin{pmatrix} 0\ 0\ 0 \\ 1\ 0\ 0 \end{pmatrix}, \qquad E_{22} = \begin{pmatrix} 0\ 0\ 0 \\ 0\ 1\ 0 \end{pmatrix}, \qquad E_{23} = \begin{pmatrix} 0\ 0\ 0 \\ 0\ 0\ 1 \end{pmatrix}.$$

It is clear that every matrix $A = (a_{i\,j}) \in \mathrm{M}_{m,n}(K)$ can be written in a unique way as

$$A = \sum_{i=1}^{m} \sum_{j=1}^{n} a_{i\,j} E_{ij}.$$

Thus, the family $(E_{ij})_{1 \leq i \leq m, 1 \leq j \leq n}$ is a basis of the vector space $\mathrm{M}_{m,n}(K)$, which has dimension mn.

Remark: Definition 2.14 and Definition 2.15 also make perfect sense when K is a (commutative) ring rather than a field. In this more general setting, the framework of vector spaces is too narrow, but we can consider structures over a commutative ring A satisfying all the axioms of Definition 2.4. Such structures are called *modules*. The theory of modules is (much) more complicated than that of vector spaces. For example, modules do not always have a basis, and other properties holding for vector spaces usually fail for modules. When a module has a basis, it is called a *free module*. For example, when A is a commutative ring, the structure A^n is a module such that the vectors e_i, with $(e_i)_i = 1$ and $(e_i)_j = 0$ for $j \neq i$, form a basis of A^n. Many properties of vector spaces still hold for A^n. Thus, A^n is a free module. As another example, when A is a commutative ring, $\mathrm{M}_{m,n}(A)$ is a free module with basis $(E_{i,j})_{1 \leq i \leq m, 1 \leq j \leq n}$. Polynomials over a commutative ring also form a free module of infinite dimension.

The properties listed in Proposition 2.12 are easily verified, although some of the computations are a bit tedious. A more conceptual proof is given in Proposition 3.1.

Proposition 2.12. *(1) Given any matrices $A \in \mathrm{M}_{m,n}(K)$, $B \in \mathrm{M}_{n,p}(K)$, and $C \in \mathrm{M}_{p,q}(K)$, we have*

$$(AB)C = A(BC);$$

that is, matrix multiplication is associative.

(2) Given any matrices $A, B \in \mathrm{M}_{m,n}(K)$, and $C, D \in \mathrm{M}_{n,p}(K)$, for all $\lambda \in K$, we have

$$(A + B)C = AC + BC$$
$$A(C + D) = AC + AD$$
$$(\lambda A)C = \lambda(AC)$$
$$A(\lambda C) = \lambda(AC),$$

so that matrix multiplication $\cdot : \mathrm{M}_{m,n}(K) \times \mathrm{M}_{n,p}(K) \to \mathrm{M}_{m,p}(K)$ is bilinear.

The properties of Proposition 2.12 together with the fact that $AI_n = I_n A = A$ for all square $n \times n$ matrices show that $\mathrm{M}_n(K)$ is a ring with unit I_n (in fact, an associative algebra). This is a noncommutative ring with zero divisors, as shown by the following example.

Example 2.7. For example, letting A, B be the 2×2-matrices

$$A = \begin{pmatrix} 1 & 0 \\ 0 & 0 \end{pmatrix}, \quad B = \begin{pmatrix} 0 & 0 \\ 1 & 0 \end{pmatrix},$$

then

$$AB = \begin{pmatrix} 1 & 0 \\ 0 & 0 \end{pmatrix} \begin{pmatrix} 0 & 0 \\ 1 & 0 \end{pmatrix} = \begin{pmatrix} 0 & 0 \\ 0 & 0 \end{pmatrix},$$

and

$$BA = \begin{pmatrix} 0 & 0 \\ 1 & 0 \end{pmatrix} \begin{pmatrix} 1 & 0 \\ 0 & 0 \end{pmatrix} = \begin{pmatrix} 0 & 0 \\ 1 & 0 \end{pmatrix}.$$

Thus $AB \neq BA$, and $AB = 0$, even though both $A, B \neq 0$.

2.7 Linear Maps

Now that we understand vector spaces and how to generate them, we would like to be able to transform one vector space E into another vector space F. A function between two vector spaces that preserves the vector space structure is called a homomorphism of vector spaces, or *linear map*. Linear maps formalize the concept of linearity of a function.

> *Keep in mind that linear maps, which are transformations of space, are usually far more important than the spaces themselves.*

In the rest of this section, we assume that all vector spaces are real vector spaces, but all results hold for vector spaces over an arbitrary field.

Definition 2.20. Given two vector spaces E and F, a *linear map* between E and F is a function $f \colon E \to F$ satisfying the following two conditions:

$$f(x + y) = f(x) + f(y) \qquad \text{for all } x, y \in E;$$
$$f(\lambda x) = \lambda f(x) \qquad \text{for all } \lambda \in \mathbb{R},\ x \in E.$$

Setting $x = y = 0$ in the first identity, we get $f(0) = 0$. *The basic property of linear maps is that they transform linear combinations into linear combinations.* Given any finite family $(u_i)_{i \in I}$ of vectors in E, given any family $(\lambda_i)_{i \in I}$ of scalars in $\mathbb{R}$, we have

$$f\left(\sum_{i \in I} \lambda_i u_i \right) = \sum_{i \in I} \lambda_i f(u_i).$$

The above identity is shown by induction on $|I|$ using the properties of Definition 2.20.

Example 2.8.

(1) The map $f \colon \mathbb{R}^2 \to \mathbb{R}^2$ defined such that
$$x' = x - y$$
$$y' = x + y$$
is a linear map. The reader should check that it is the composition of a rotation by $\pi/4$ with a magnification of ratio $\sqrt{2}$.

(2) For any vector space E, the *identity map* id$\colon E \to E$ given by
$$\text{id}(u) = u \quad \text{for all } u \in E$$
is a linear map. When we want to be more precise, we write id$_E$ instead of id.

(3) The map $D \colon \mathbb{R}[X] \to \mathbb{R}[X]$ defined such that

$$D(f(X)) = f'(X),$$

where $f'(X)$ is the derivative of the polynomial $f(X)$, is a linear map.

(4) The map $\Phi \colon \mathcal{C}([a, b]) \to \mathbb{R}$ given by

$$\Phi(f) = \int_a^b f(t)dt,$$

where $\mathcal{C}([a, b])$ is the set of continuous functions defined on the interval $[a, b]$, is a linear map.

(5) The function $\langle -, - \rangle \colon \mathcal{C}([a, b]) \times \mathcal{C}([a, b]) \to \mathbb{R}$ given by

$$\langle f, g \rangle = \int_a^b f(t)g(t)dt,$$

is linear in each of the variable f, g. It also satisfies the properties $\langle f, g \rangle = \langle g, f \rangle$ and $\langle f, f \rangle = 0$ iff $f = 0$. It is an example of an *inner product*.

Definition 2.21. Given a linear map $f \colon E \to F$, we define its *image (or range)* $\operatorname{Im} f = f(E)$, as the set

$$\operatorname{Im} f = \{ y \in F \mid (\exists x \in E)(y = f(x)) \},$$

and its *Kernel (or nullspace)* $\operatorname{Ker} f = f^{-1}(0)$, as the set

$$\operatorname{Ker} f = \{ x \in E \mid f(x) = 0 \}.$$

The derivative map $D \colon \mathbb{R}[X] \to \mathbb{R}[X]$ from Example 2.8(3) has kernel the constant polynomials, so $\operatorname{Ker} D = \mathbb{R}$. If we consider the second derivative $D \circ D \colon \mathbb{R}[X] \to \mathbb{R}[X]$, then the kernel of $D \circ D$ consists of all polynomials of degree ≤ 1. The image of $D \colon \mathbb{R}[X] \to \mathbb{R}[X]$ is actually $\mathbb{R}[X]$ itself, because every polynomial $P(X) = a_0 X^n + \cdots + a_{n-1} X + a_n$ of degree n is the derivative of the polynomial $Q(X)$ of degree $n + 1$ given by

$$Q(X) = a_0 \frac{X^{n+1}}{n+1} + \cdots + a_{n-1} \frac{X^2}{2} + a_n X.$$

On the other hand, if we consider the restriction of D to the vector space $\mathbb{R}[X]_n$ of polynomials of degree $\leq n$, then the kernel of D is still $\mathbb{R}$, but the image of D is the $\mathbb{R}[X]_{n-1}$, the vector space of polynomials of degree $\leq n - 1$.

Proposition 2.13. *Given a linear map $f \colon E \to F$, the set $\operatorname{Im} f$ is a subspace of F and the set $\operatorname{Ker} f$ is a subspace of E. The linear map $f \colon E \to F$ is injective iff $\operatorname{Ker} f = (0)$ (where (0) is the trivial subspace $\{0\}$).*

Proof. Given any $x, y \in \operatorname{Im} f$, there are some $u, v \in E$ such that $x = f(u)$ and $y = f(v)$, and for all $\lambda, \mu \in \mathbb{R}$, we have

$$f(\lambda u + \mu v) = \lambda f(u) + \mu f(v) = \lambda x + \mu y,$$

and thus, $\lambda x + \mu y \in \operatorname{Im} f$, showing that $\operatorname{Im} f$ is a subspace of F.

Given any $x, y \in \operatorname{Ker} f$, we have $f(x) = 0$ and $f(y) = 0$, and thus,

$$f(\lambda x + \mu y) = \lambda f(x) + \mu f(y) = 0,$$

that is, $\lambda x + \mu y \in \operatorname{Ker} f$, showing that $\operatorname{Ker} f$ is a subspace of E.

First, assume that $\operatorname{Ker} f = (0)$. We need to prove that $f(x) = f(y)$ implies that $x = y$. However, if $f(x) = f(y)$, then $f(x) - f(y) = 0$, and by linearity of f we get $f(x - y) = 0$. Because $\operatorname{Ker} f = (0)$, we must have $x - y = 0$, that is $x = y$, so f is injective. Conversely, assume that f is injective. If $x \in \operatorname{Ker} f$, that is $f(x) = 0$, since $f(0) = 0$ we have $f(x) = f(0)$, and by injectivity, $x = 0$, which proves that $\operatorname{Ker} f = (0)$. Therefore, f is injective iff $\operatorname{Ker} f = (0)$. $\qquad\square$

Since by Proposition 2.13, the image $\operatorname{Im} f$ of a linear map f is a subspace of F, we can define the *rank* $\operatorname{rk}(f)$ of f as the dimension of $\operatorname{Im} f$.

Definition 2.22. Given a linear map $f \colon E \to F$, the *rank* $\operatorname{rk}(f)$ of f is the dimension of the image $\operatorname{Im} f$ of f.

A fundamental property of bases in a vector space is that they allow the definition of linear maps as unique homomorphic extensions, as shown in the following proposition.

Proposition 2.14. *Given any two vector spaces E and F, given any basis $(u_i)_{i \in I}$ of E, given any other family of vectors $(v_i)_{i \in I}$ in F, there is a unique linear map $f \colon E \to F$ such that $f(u_i) = v_i$ for all $i \in I$. Furthermore, f is injective iff $(v_i)_{i \in I}$ is linearly independent, and f is surjective iff $(v_i)_{i \in I}$ generates F.*

Proof. If such a linear map $f \colon E \to F$ exists, since $(u_i)_{i \in I}$ is a basis of E, every vector $x \in E$ can written uniquely as a linear combination

$$x = \sum_{i \in I} x_i u_i,$$

and by linearity, we must have

$$f(x) = \sum_{i \in I} x_i f(u_i) = \sum_{i \in I} x_i v_i.$$

Define the function $f \colon E \to F$, by letting
$$f(x) = \sum_{i \in I} x_i v_i$$
for every $x = \sum_{i \in I} x_i u_i$. It is easy to verify that f is indeed linear, it is unique by the previous reasoning, and obviously, $f(u_i) = v_i$.

Now assume that f is injective. Let $(\lambda_i)_{i \in I}$ be any family of scalars, and assume that
$$\sum_{i \in I} \lambda_i v_i = 0.$$
Since $v_i = f(u_i)$ for every $i \in I$, we have
$$f\left(\sum_{i \in I} \lambda_i u_i\right) = \sum_{i \in I} \lambda_i f(u_i) = \sum_{i \in I} \lambda_i v_i = 0.$$
Since f is injective iff $\operatorname{Ker} f = (0)$, we have
$$\sum_{i \in I} \lambda_i u_i = 0,$$
and since $(u_i)_{i \in I}$ is a basis, we have $\lambda_i = 0$ for all $i \in I$, which shows that $(v_i)_{i \in I}$ is linearly independent. Conversely, assume that $(v_i)_{i \in I}$ is linearly independent. Since $(u_i)_{i \in I}$ is a basis of E, every vector $x \in E$ is a linear combination $x = \sum_{i \in I} \lambda_i u_i$ of $(u_i)_{i \in I}$. If
$$f(x) = f\left(\sum_{i \in I} \lambda_i u_i\right) = 0,$$
then
$$\sum_{i \in I} \lambda_i v_i = \sum_{i \in I} \lambda_i f(u_i) = f\left(\sum_{i \in I} \lambda_i u_i\right) = 0,$$
and $\lambda_i = 0$ for all $i \in I$ because $(v_i)_{i \in I}$ is linearly independent, which means that $x = 0$. Therefore, $\operatorname{Ker} f = (0)$, which implies that f is injective. The part where f is surjective is left as a simple exercise. $\qquad\square$

Figure 2.11 provides an illustration of Proposition 2.14 when $E = \mathbb{R}^3$ and $V = \mathbb{R}^2$.

By the second part of Proposition 2.14, an injective linear map $f \colon E \to F$ sends a basis $(u_i)_{i \in I}$ to a linearly independent family $(f(u_i))_{i \in I}$ of F, which is also a basis when f is bijective. Also, when E and F have the same finite dimension n, $(u_i)_{i \in I}$ is a basis of E, and $f \colon E \to F$ is injective, then $(f(u_i))_{i \in I}$ is a basis of F (by Proposition 2.8).

The following simple proposition is also useful.

Proposition 2.15. *Given any two vector spaces E and F, with F nontrivial, given any family $(u_i)_{i \in I}$ of vectors in E, the following properties hold:*

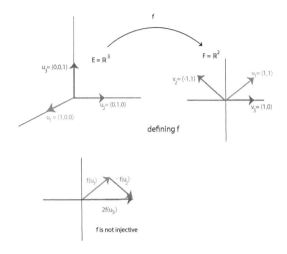

Fig. 2.11   Given $u_1 = (1,0,0)$, $u_2 = (0,1,0)$, $u_3 = (0,0,1)$ and $v_1 = (1,1)$, $v_2 = (-1,1)$, $v_3 = (1,0)$, define the unique linear map $f \colon \mathbb{R}^3 \to \mathbb{R}^2$ by $f(u_1) = v_1$, $f(u_2) = v_2$, and $f(u_3) = v_3$. This map is surjective but not injective since $f(u_1 - u_2) = f(u_1) - f(u_2) = (1,1) - (-1,1) = (2,0) = 2f(u_3) = f(2u_3)$.

(1) The family $(u_i)_{i \in I}$ generates E iff for every family of vectors $(v_i)_{i \in I}$ in F, there is at most one linear map $f \colon E \to F$ such that $f(u_i) = v_i$ for all $i \in I$.

(2) The family $(u_i)_{i \in I}$ is linearly independent iff for every family of vectors $(v_i)_{i \in I}$ in F, there is some linear map $f \colon E \to F$ such that $f(u_i) = v_i$ for all $i \in I$.

Proof. (1) If there is any linear map $f \colon E \to F$ such that $f(u_i) = v_i$ for all $i \in I$, since $(u_i)_{i \in I}$ generates E, every vector $x \in E$ can be written as some linear combination

$$x = \sum_{i \in I} x_i u_i,$$

and by linearity, we must have

$$f(x) = \sum_{i \in I} x_i f(u_i) = \sum_{i \in I} x_i v_i.$$

This shows that f is unique if it exists. Conversely, assume that $(u_i)_{i \in I}$ does not generate E. Since F is nontrivial, there is some some vector $y \in F$ such that $y \neq 0$. Since $(u_i)_{i \in I}$ does not generate E, there is some vector $w \in E$ that is not in the subspace generated by $(u_i)_{i \in I}$. By Theorem 2.2, there is a linearly independent subfamily $(u_i)_{i \in I_0}$ of $(u_i)_{i \in I}$ generating the same subspace. Since by hypothesis, $w \in E$ is not in the subspace generated by $(u_i)_{i \in I_0}$, by Lemma 2.1 and by Theorem 2.2 again, there is a basis $(e_j)_{j \in I_0 \cup J}$ of E, such that $e_i = u_i$ for all $i \in I_0$, and $w = e_{j_0}$ for some $j_0 \in J$. Letting $(v_i)_{i \in I}$ be the family in F such that $v_i = 0$ for all $i \in I$, defining $f \colon E \to F$ to be the constant linear map with value 0, we have a linear map such that $f(u_i) = 0$ for all $i \in I$. By Proposition 2.14, there is a unique linear map $g \colon E \to F$ such that $g(w) = y$, and $g(e_j) = 0$ for all $j \in (I_0 \cup J) - \{j_0\}$. By definition of the basis $(e_j)_{j \in I_0 \cup J}$ of E, we have $g(u_i) = 0$ for all $i \in I$, and since $f \neq g$, this contradicts the fact that there is at most one such map. See Figure 2.12.

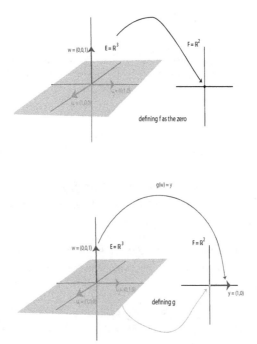

Fig. 2.12 Let $E = \mathbb{R}^3$ and $F = \mathbb{R}^2$. The vectors $u_1 = (1, 0, 0)$, $u_2 = (0, 1, 0)$ do not generate $\mathbb{R}^3$ since both the zero map and the map g, where $g(0, 0, 1) = (1, 0)$, send the peach xy-plane to the origin.

(2) If the family $(u_i)_{i \in I}$ is linearly independent, then by Theorem 2.2, $(u_i)_{i \in I}$ can be extended to a basis of E, and the conclusion follows by Proposition 2.14. Conversely, assume that $(u_i)_{i \in I}$ is linearly dependent.

Then there is some family $(\lambda_i)_{i \in I}$ of scalars (not all zero) such that

$$\sum_{i \in I} \lambda_i u_i = 0.$$

By the assumption, for any nonzero vector $y \in F$, for every $i \in I$, there is some linear map $f_i \colon E \to F$, such that $f_i(u_i) = y$, and $f_i(u_j) = 0$, for $j \in I - \{i\}$. Then we would get

$$0 = f_i\Big(\sum_{i \in I} \lambda_i u_i\Big) = \sum_{i \in I} \lambda_i f_i(u_i) = \lambda_i y,$$

and since $y \neq 0$, this implies $\lambda_i = 0$ for every $i \in I$. Thus, $(u_i)_{i \in I}$ is linearly independent. $\qquad \square$

Given vector spaces E, F, and G, and linear maps $f \colon E \to F$ and $g \colon F \to G$, it is easily verified that the composition $g \circ f \colon E \to G$ of f and g is a linear map.

Definition 2.23. A linear map $f \colon E \to F$ is an *isomorphism* iff there is a linear map $g \colon F \to E$, such that

$$g \circ f = \mathrm{id}_E \quad \text{and} \quad f \circ g = \mathrm{id}_F. \tag{2.17}$$

The map g in Definition 2.23 is unique. This is because if g and h both satisfy $g \circ f = \mathrm{id}_E$, $f \circ g = \mathrm{id}_F$, $h \circ f = \mathrm{id}_E$, and $f \circ h = \mathrm{id}_F$, then

$$g = g \circ \mathrm{id}_F = g \circ (f \circ h) = (g \circ f) \circ h = \mathrm{id}_E \circ h = h.$$

The map g satisfying (2.17) above is called the *inverse* of f and it is also denoted by f^{-1}.

Observe that Proposition 2.14 shows that if $F = \mathbb{R}^n$, then we get an isomorphism between any vector space E of dimension $|J| = n$ and $\mathbb{R}^n$. Proposition 2.14 also implies that if E and F are two vector spaces, $(u_i)_{i \in I}$ is a basis of E, and $f \colon E \to F$ is a linear map which is an isomorphism, then the family $(f(u_i))_{i \in I}$ is a basis of F.

One can verify that if $f \colon E \to F$ is a bijective linear map, then its inverse $f^{-1} \colon F \to E$, as a function, is also a linear map, and thus f is an isomorphism.

Another useful corollary of Proposition 2.14 is this:

Proposition 2.16. *Let E be a vector space of finite dimension $n \geq 1$ and let $f \colon E \to E$ be any linear map. The following properties hold:*

(1) If f has a left inverse g, that is, if g is a linear map such that $g \circ f = \mathrm{id}$, then f is an isomorphism and $f^{-1} = g$.

(2) If f has a right inverse h, that is, if h is a linear map such that $f \circ h =$ id, then f is an isomorphism and $f^{-1} = h$.

Proof. (1) The equation $g \circ f = $ id implies that f is injective; this is a standard result about functions (if $f(x) = f(y)$, then $g(f(x)) = g(f(y))$, which implies that $x = y$ since $g \circ f = $ id). Let $(u_1, \dots, u_n)$ be any basis of E. By Proposition 2.14, since f is injective, $(f(u_1), \dots, f(u_n))$ is linearly independent, and since E has dimension n, it is a basis of E (if $(f(u_1), \dots, f(u_n))$ doesn't span E, then it can be extended to a basis of dimension strictly greater than n, contradicting Theorem 2.2). Then f is bijective, and by a previous observation its inverse is a linear map. We also have

$$g = g \circ \text{id} = g \circ (f \circ f^{-1}) = (g \circ f) \circ f^{-1} = \text{id} \circ f^{-1} = f^{-1}.$$

(2) The equation $f \circ h = $ id implies that f is surjective; this is a standard result about functions (for any $y \in E$, we have $f(h(y)) = y$). Let $(u_1, \dots, u_n)$ be any basis of E. By Proposition 2.14, since f is surjective, $(f(u_1), \dots, f(u_n))$ spans E, and since E has dimension n, it is a basis of E (if $(f(u_1), \dots, f(u_n))$ is not linearly independent, then because it spans E, it contains a basis of dimension strictly smaller than n, contradicting Theorem 2.2). Then f is bijective, and by a previous observation its inverse is a linear map. We also have

$$h = \text{id} \circ h = (f^{-1} \circ f) \circ h = f^{-1} \circ (f \circ h) = f^{-1} \circ \text{id} = f^{-1}.$$

This completes the proof. $\square$

Definition 2.24. The set of all linear maps between two vector spaces E and F is denoted by $\text{Hom}(E, F)$ or by $\mathcal{L}(E; F)$ (the notation $\mathcal{L}(E; F)$ is usually reserved to the set of continuous linear maps, where E and F are normed vector spaces). When we wish to be more precise and specify the field K over which the vector spaces E and F are defined we write $\text{Hom}_K(E, F)$.

The set $\text{Hom}(E, F)$ is a vector space under the operations defined in Example 2.3, namely

$$(f + g)(x) = f(x) + g(x)$$

for all $x \in E$, and

$$(\lambda f)(x) = \lambda f(x)$$

for all $x \in E$. The point worth checking carefully is that λf is indeed a linear map, which uses the commutativity of $*$ in the field K (typically, $K = \mathbb{R}$ or $K = \mathbb{C}$). Indeed, we have

$$(\lambda f)(\mu x) = \lambda f(\mu x) = \lambda \mu f(x) = \mu \lambda f(x) = \mu (\lambda f)(x).$$

When E and F have finite dimensions, the vector space $\text{Hom}(E, F)$ also has finite dimension, as we shall see shortly.

Definition 2.25. When $E = F$, a linear map $f \colon E \to E$ is also called an *endomorphism*. The space $\text{Hom}(E, E)$ is also denoted by $\text{End}(E)$.

It is also important to note that composition confers to $\text{Hom}(E, E)$ a ring structure. Indeed, composition is an operation $\circ \colon \text{Hom}(E, E) \times \text{Hom}(E, E) \to \text{Hom}(E, E)$, which is associative and has an identity id_E, and the distributivity properties hold:

$$(g_1 + g_2) \circ f = g_1 \circ f + g_2 \circ f;$$
$$g \circ (f_1 + f_2) = g \circ f_1 + g \circ f_2.$$

The ring $\text{Hom}(E, E)$ is an example of a noncommutative ring.

It is easily seen that the set of bijective linear maps $f \colon E \to E$ is a group under composition.

Definition 2.26. Bijective linear maps $f \colon E \to E$ are also called *automorphisms*. The group of automorphisms of E is called the *general linear group (of E)*, and it is denoted by $\mathbf{GL}(E)$, or by $\text{Aut}(E)$, or when $E = \mathbb{R}^n$, by $\mathbf{GL}(n, \mathbb{R})$, or even by $\mathbf{GL}(n)$.

2.8 Linear Forms and the Dual Space

We already observed that the field K itself ($K = \mathbb{R}$ or $K = \mathbb{C}$) is a vector space (over itself). The vector space $\text{Hom}(E, K)$ of linear maps from E to the field K, the linear forms, plays a particular role. In this section, we only define linear forms and show that every finite-dimensional vector space has a dual basis. A more advanced presentation of dual spaces and duality is given in Chapter 10.

Definition 2.27. Given a vector space E, the vector space $\text{Hom}(E, K)$ of linear maps from E to the field K is called the *dual space (or dual)* of E. The space $\text{Hom}(E, K)$ is also denoted by E^*, and the linear maps in E^* are called *the linear forms*, or *covectors*. The dual space E^{**} of the space E^* is called the *bidual* of E.

As a matter of notation, linear forms $f\colon E \to K$ will also be denoted by starred symbol, such as u^*, x^*, etc.

If E is a vector space of finite dimension n and $(u_1, \ldots, u_n)$ is a basis of E, for any linear form $f^* \in E^*$, for every $x = x_1 u_1 + \cdots + x_n u_n \in E$, by linearity we have

$$f^*(x) = f^*(u_1)x_1 + \cdots + f^*(u_n)x_n$$
$$= \lambda_1 x_1 + \cdots + \lambda_n x_n,$$

with $\lambda_i = f^*(u_i) \in K$ for every i, $1 \le i \le n$. Thus, with respect to the basis $(u_1, \ldots, u_n)$, the linear form f^* is represented by the row vector

$$(\lambda_1 \ \cdots \ \lambda_n),$$

we have

$$f^*(x) = \begin{pmatrix} \lambda_1 & \cdots & \lambda_n \end{pmatrix} \begin{pmatrix} x_1 \\ \vdots \\ x_n \end{pmatrix},$$

a linear combination of the coordinates of x, and we can view the linear form f^* as a *linear equation*. If we decide to use a column vector of coefficients

$$c = \begin{pmatrix} c_1 \\ \vdots \\ c_n \end{pmatrix}$$

instead of a row vector, then the linear form f^* is defined by

$$f^*(x) = c^\top x.$$

The above notation is often used in machine learning.

Example 2.9. Given any differentiable function $f\colon \mathbb{R}^n \to \mathbb{R}$, by definition, for any $x \in \mathbb{R}^n$, the *total derivative* df_x of f at x is the linear form $df_x\colon \mathbb{R}^n \to \mathbb{R}$ defined so that for all $u = (u_1, \ldots, u_n) \in \mathbb{R}^n$,

$$df_x(u) = \begin{pmatrix} \dfrac{\partial f}{\partial x_1}(x) & \cdots & \dfrac{\partial f}{\partial x_n}(x) \end{pmatrix} \begin{pmatrix} u_1 \\ \vdots \\ u_n \end{pmatrix} = \sum_{i=1}^{n} \frac{\partial f}{\partial x_i}(x)\, u_i.$$

Example 2.10. Let $\mathcal{C}([0,1])$ be the vector space of continuous functions $f\colon [0,1] \to \mathbb{R}$. The map $\mathcal{I}\colon \mathcal{C}([0,1]) \to \mathbb{R}$ given by

$$\mathcal{I}(f) = \int_0^1 f(x)dx \quad \text{for any } f \in \mathcal{C}([0,1])$$

is a linear form (integration).

Example 2.11. Consider the vector space $M_n(\mathbb{R})$ of real $n \times n$ matrices. Let $\mathrm{tr}\colon M_n(\mathbb{R}) \to \mathbb{R}$ be the function given by

$$\mathrm{tr}(A) = a_{11} + a_{22} + \cdots + a_{nn},$$

called the *trace* of A. It is a linear form. Let $s\colon M_n(\mathbb{R}) \to \mathbb{R}$ be the function given by

$$s(A) = \sum_{i,j=1}^{n} a_{ij},$$

where $A = (a_{ij})$. It is immediately verified that s is a linear form.

Given a vector space E and any basis $(u_i)_{i \in I}$ for E, we can associate to each u_i a linear form $u_i^* \in E^*$, and the u_i^* have some remarkable properties.

Definition 2.28. Given a vector space E and any basis $(u_i)_{i \in I}$ for E, by Proposition 2.14, for every $i \in I$, there is a unique linear form u_i^* such that

$$u_i^*(u_j) = \begin{cases} 1 & \text{if } i = j \\ 0 & \text{if } i \neq j, \end{cases}$$

for every $j \in I$. The linear form u_i^* is called the *coordinate form* of index i w.r.t. the basis $(u_i)_{i \in I}$.

Remark: Given an index set I, authors often define the so called "Kronecker symbol" δ_{ij} such that

$$\delta_{ij} = \begin{cases} 1 & \text{if } i = j \\ 0 & \text{if } i \neq j, \end{cases}$$

for all $i, j \in I$. Then, $u_i^*(u_j) = \delta_{ij}$.

The reason for the terminology *coordinate form* is as follows: If E has finite dimension and if $(u_1, \ldots, u_n)$ is a basis of E, for any vector

$$v = \lambda_1 u_1 + \cdots + \lambda_n u_n,$$

we have

$$\begin{aligned} u_i^*(v) &= u_i^*(\lambda_1 u_1 + \cdots + \lambda_n u_n) \\ &= \lambda_1 u_i^*(u_1) + \cdots + \lambda_i u_i^*(u_i) + \cdots + \lambda_n u_i^*(u_n) \\ &= \lambda_i, \end{aligned}$$

since $u_i^*(u_j) = \delta_{ij}$. Therefore, u_i^* is the linear function that returns the ith coordinate of a vector expressed over the basis $(u_1, \ldots, u_n)$.

The following theorem shows that in finite-dimension, every basis $(u_1, \ldots, u_n)$ of a vector space E yields a basis $(u_1^*, \ldots, u_n^*)$ of the dual space E^*, called a *dual basis*.

Theorem 2.3. *(Existence of dual bases) Let E be a vector space of dimension n. The following properties hold: For every basis $(u_1, \ldots, u_n)$ of E, the family of coordinate forms $(u_1^*, \ldots, u_n^*)$ is a basis of E^* (called the dual basis of $(u_1, \ldots, u_n)$).*

Proof. (a) If $v^* \in E^*$ is any linear form, consider the linear form

$$f^* = v^*(u_1)u_1^* + \cdots + v^*(u_n)u_n^*.$$

Observe that because $u_i^*(u_j) = \delta_{ij}$,

$$
\begin{aligned}
f^*(u_i) &= (v^*(u_1)u_1^* + \cdots + v^*(u_n)u_n^*)(u_i) \\
&= v^*(u_1)u_1^*(u_i) + \cdots + v^*(u_i)u_i^*(u_i) + \cdots + v^*(u_n)u_n^*(u_i) \\
&= v^*(u_i),
\end{aligned}
$$

and so f^* and v^* agree on the basis $(u_1, \ldots, u_n)$, which implies that

$$v^* = f^* = v^*(u_1)u_1^* + \cdots + v^*(u_n)u_n^*.$$

Therefore, $(u_1^*, \ldots, u_n^*)$ spans E^*. We claim that the covectors $u_1^*, \ldots, u_n^*$ are linearly independent. If not, we have a nontrivial linear dependence

$$\lambda_1 u_1^* + \cdots + \lambda_n u_n^* = 0,$$

and if we apply the above linear form to each u_i, using a familiar computation, we get

$$0 = \lambda_i u_i^*(u_i) = \lambda_i,$$

proving that $u_1^*, \ldots, u_n^*$ are indeed linearly independent. Therefore, $(u_1^*, \ldots, u_n^*)$ is a basis of E^*. $\qquad\square$

In particular, Theorem 2.3 shows a finite-dimensional vector space and its dual E^* have the same dimension.

2.9   Summary

The main concepts and results of this chapter are listed below:

- The notion of a *vector space*.
- *Families* of vectors.

- *Linear combinations* of vectors; *linear dependence* and *linear independence* of a family of vectors.
- Linear *subspaces*.
- *Spanning* (or *generating*) family; *generators, finitely generated subspace*; *basis of a subspace*.
- *Every linearly independent family can be extended to a basis* (Theorem 2.1).
- A family B of vectors is a basis iff it is a maximal linearly independent family iff it is a minimal generating family (Proposition 2.8).
- The replacement lemma (Proposition 2.10).
- Any two bases in a finitely generated vector space E have the *same number of elements*; this is the *dimension* of E (Theorem 2.2).
- *Hyperplanes*.
- Every vector has a *unique representation* over a basis (in terms of its coordinates).
- *Matrices*
- *Column vectors, row vectors*.
- *Matrix operations*: addition, scalar multiplication, multiplication.
- The vector space $\mathrm{M}_{m,n}(K)$ of $m \times n$ matrices over the field K; The ring $\mathrm{M}_n(K)$ of $n \times n$ matrices over the field K.
- The notion of a *linear map*.
- The *image* Im f (or *range*) of a linear map f.
- The *kernel* Ker f (or *nullspace*) of a linear map f.
- The *rank* rk(f) of a linear map f.
- The image and the kernel of a linear map are subspaces. A linear map is injective iff its kernel is the trivial space (0) (Proposition 2.13).
- The *unique homomorphic extension property* of linear maps with respect to bases (Proposition 2.14).
- The vector space of linear maps $\mathrm{Hom}_K(E, F)$.
- Linear forms (covectors) and the *dual space* E^*.
- Coordinate forms.
- The existence of *dual bases* (in finite dimension).

2.10   Problems

Problem 2.1. Let H be the set of 3×3 upper triangular matrices given by

$$H = \left\{ \begin{pmatrix} 1 & a & b \\ 0 & 1 & c \\ 0 & 0 & 1 \end{pmatrix} \mid a, b, c \in \mathbb{R} \right\}.$$

(1) Prove that H with the binary operation of matrix multiplication is a group; find explicitly the inverse of every matrix in H. Is H abelian (commutative)?

(2) Given two groups G_1 and G_2, recall that a *homomorphism* if a function $\varphi \colon G_1 \to G_2$ such that

$$\varphi(ab) = \varphi(a)\varphi(b), \quad a, b \in G_1.$$

Prove that $\varphi(e_1) = e_2$ (where e_i is the identity element of G_i) and that

$$\varphi(a^{-1}) = (\varphi(a))^{-1}, \quad a \in G_1.$$

(3) Let S^1 be the unit circle, that is

$$S^1 = \{ e^{i\theta} = \cos\theta + i\sin\theta \mid 0 \leq \theta < 2\pi \},$$

and let φ be the function given by

$$\varphi \begin{pmatrix} 1 & a & b \\ 0 & 1 & c \\ 0 & 0 & 1 \end{pmatrix} = (a, c, e^{ib}).$$

Prove that φ is a surjective function onto $G = \mathbb{R} \times \mathbb{R} \times S^1$, and that if we define multiplication on this set by

$$(x_1, y_1, u_1) \cdot (x_2, y_2, u_2) = (x_1 + x_2, y_1 + y_2, e^{ix_1 y_2} u_1 u_2),$$

then G is a group and φ is a group homomorphism from H onto G.

(4) The *kernel* of a homomorphism $\varphi \colon G_1 \to G_2$ is defined as

$$\mathrm{Ker}\,(\varphi) = \{ a \in G_1 \mid \varphi(a) = e_2 \}.$$

Find explicitly the kernel of φ and show that it is a subgroup of H.

Problem 2.2. For any $m \in \mathbb{Z}$ with $m > 0$, the subset $m\mathbb{Z} = \{ mk \mid k \in \mathbb{Z} \}$ is an abelian subgroup of $\mathbb{Z}$. Check this.

(1) Give a group isomorphism (an invertible homomorphism) from $m\mathbb{Z}$ to $\mathbb{Z}$.

(2) Check that the inclusion map $i\colon m\mathbb{Z} \to \mathbb{Z}$ given by $i(mk) = mk$ is a group homomorphism. Prove that if $m \geq 2$ then there is no group homomorphism $p\colon \mathbb{Z} \to m\mathbb{Z}$ such that $p \circ i = \mathrm{id}$.

Remark: The above shows that abelian groups fail to have some of the properties of vector spaces. We will show later that a linear map satisfying the condition $p \circ i = \mathrm{id}$ always exists.

Problem 2.3. Let $E = \mathbb{R} \times \mathbb{R}$, and define the addition operation

$$(x_1, y_1) + (x_2, y_2) = (x_1 + x_2, y_1, +y_2), \quad x_1, x_2, y_1, y_2 \in \mathbb{R},$$

and the multiplication operation $\cdot\colon \mathbb{R} \times E \to E$ by

$$\lambda \cdot (x, y) = (\lambda x, y), \quad \lambda, x, y \in \mathbb{R}.$$

Show that E with the above operations $+$ and $\cdot$ is not a vector space. Which of the axioms is violated?

Problem 2.4. (1) Prove that the axioms of vector spaces imply that

$$\alpha \cdot 0 = 0$$
$$0 \cdot v = 0$$
$$\alpha \cdot (-v) = -(\alpha \cdot v)$$
$$(-\alpha) \cdot v = -(\alpha \cdot v),$$

for all $v \in E$ and all $\alpha \in K$, where E is a vector space over K.

(2) For every $\lambda \in \mathbb{R}$ and every $x = (x_1, \ldots, x_n) \in \mathbb{R}^n$, define λx by

$$\lambda x = \lambda(x_1, \ldots, x_n) = (\lambda x_1, \ldots, \lambda x_n).$$

Recall that every vector $x = (x_1, \ldots, x_n) \in \mathbb{R}^n$ can be written uniquely as

$$x = x_1 e_1 + \cdots + x_n e_n,$$

where $e_i = (0, \ldots, 0, 1, 0, \ldots, 0)$, with a single 1 in position i. For any operation $\cdot\colon \mathbb{R} \times \mathbb{R}^n \to \mathbb{R}^n$, if $\cdot$ satisfies the Axiom (V1) of a vector space, then prove that for any $\alpha \in \mathbb{R}$, we have

$$\alpha \cdot x = \alpha \cdot (x_1 e_1 + \cdots + x_n e_n) = \alpha \cdot (x_1 e_1) + \cdots + \alpha \cdot (x_n e_n).$$

Conclude that $\cdot$ is completely determined by its action on the one-dimensional subspaces of $\mathbb{R}^n$ spanned by $e_1, \ldots, e_n$.

(3) Use (2) to define operations $\cdot\colon \mathbb{R} \times \mathbb{R}^n \to \mathbb{R}^n$ that satisfy the Axioms (V1–V3), but for which Axiom V4 fails.

(4) For any operation $\cdot : \mathbb{R} \times \mathbb{R}^n \to \mathbb{R}^n$, prove that if $\cdot$ satisfies the Axioms (V2–V3), then for every rational number $r \in \mathbb{Q}$ and every vector $x \in \mathbb{R}^n$, we have

$$r \cdot x = r(1 \cdot x).$$

In the above equation, $1 \cdot x$ is some vector $(y_1, \ldots, y_n) \in \mathbb{R}^n$ not necessarily equal to $x = (x_1, \ldots, x_n)$, and

$$r(1 \cdot x) = (ry_1, \ldots, ry_n),$$

as in Part (2).

Use (4) to conclude that any operation $\cdot : \mathbb{Q} \times \mathbb{R}^n \to \mathbb{R}^n$ that satisfies the Axioms (V1–V3) is completely determined by the action of 1 on the one-dimensional subspaces of $\mathbb{R}^n$ spanned by $e_1, \ldots, e_n$.

Problem 2.5. Let A_1 be the following matrix:

$$A_1 = \begin{pmatrix} 2 & 3 & 1 \\ 1 & 2 & -1 \\ -3 & -5 & 1 \end{pmatrix}.$$

Prove that the columns of A_1 are linearly independent. Find the coordinates of the vector $x = (6, 2, -7)$ over the basis consisting of the column vectors of A_1.

Problem 2.6. Let A_2 be the following matrix:

$$A_2 = \begin{pmatrix} 1 & 2 & 1 & 1 \\ 2 & 3 & 2 & 3 \\ -1 & 0 & 1 & -1 \\ -2 & -1 & 3 & 0 \end{pmatrix}.$$

Express the fourth column of A_2 as a linear combination of the first three columns of A_2. Is the vector $x = (7, 14, -1, 2)$ a linear combination of the columns of A_2?

Problem 2.7. Let A_3 be the following matrix:

$$A_3 = \begin{pmatrix} 1 & 1 & 1 \\ 1 & 1 & 2 \\ 1 & 2 & 3 \end{pmatrix}.$$

Prove that the columns of A_1 are linearly independent. Find the coordinates of the vector $x = (6, 9, 14)$ over the basis consisting of the column vectors of A_3.

Problem 2.8. Let A_4 be the following matrix:

$$A_4 = \begin{pmatrix} 1 & 2 & 1 & 1 \\ 2 & 3 & 2 & 3 \\ -1 & 0 & 1 & -1 \\ -2 & -1 & 4 & 0 \end{pmatrix}.$$

Prove that the columns of A_4 are linearly independent. Find the coordinates of the vector $x = (7, 14, -1, 2)$ over the basis consisting of the column vectors of A_4.

Problem 2.9. Consider the following Haar matrix

$$H = \begin{pmatrix} 1 & 1 & 1 & 0 \\ 1 & 1 & -1 & 0 \\ 1 & -1 & 0 & 1 \\ 1 & -1 & 0 & -1 \end{pmatrix}.$$

Prove that the columns of H are linearly independent.
Hint. Compute the product $H^\top H$.

Problem 2.10. Consider the following Hadamard matrix

$$H_4 = \begin{pmatrix} 1 & 1 & 1 & 1 \\ 1 & -1 & 1 & -1 \\ 1 & 1 & -1 & -1 \\ 1 & -1 & -1 & 1 \end{pmatrix}.$$

Prove that the columns of H_4 are linearly independent.
Hint. Compute the product $H_4^\top H_4$.

Problem 2.11. In solving this problem, **do not use determinants**.

(1) Let $(u_1, \ldots, u_m)$ and $(v_1, \ldots, v_m)$ be two families of vectors in some vector space E. Assume that each v_i is a linear combination of the u_js, so that

$$v_i = a_{i\,1} u_1 + \cdots + a_{i\,m} u_m, \quad 1 \leq i \leq m,$$

and that the matrix $A = (a_{i\,j})$ is an upper-triangular matrix, which means that if $1 \leq j < i \leq m$, then $a_{i\,j} = 0$. Prove that if $(u_1, \ldots, u_m)$ are linearly independent and if all the diagonal entries of A are nonzero, then $(v_1, \ldots, v_m)$ are also linearly independent.
Hint. Use induction on m.

(2) Let $A = (a_{i\,j})$ be an upper-triangular matrix. Prove that if all the diagonal entries of A are nonzero, then A is invertible and the inverse A^{-1} of A is also upper-triangular.

Hint. Use induction on m.

Prove that if A is invertible, then all the diagonal entries of A are nonzero.

(3) Prove that if the families $(u_1, \ldots, u_m)$ and $(v_1, \ldots, v_m)$ are related as in (1), then $(u_1, \ldots, u_m)$ are linearly independent iff $(v_1, \ldots, v_m)$ are linearly independent.

Problem 2.12. In solving this problem, **do not use determinants**. Consider the $n \times n$ matrix

$$A = \begin{pmatrix} 1 & 2 & 0 & 0 & \ldots & 0 & 0 \\ 0 & 1 & 2 & 0 & \ldots & 0 & 0 \\ 0 & 0 & 1 & 2 & \ldots & 0 & 0 \\ \vdots & \vdots & \ddots & \ddots & \ddots & \vdots & \vdots \\ 0 & 0 & \ldots & 0 & 1 & 2 & 0 \\ 0 & 0 & \ldots & 0 & 0 & 1 & 2 \\ 0 & 0 & \ldots & 0 & 0 & 0 & 1 \end{pmatrix}.$$

(1) Find the solution $x = (x_1, \ldots, x_n)$ of the linear system

$$Ax = b,$$

for

$$b = \begin{pmatrix} b_1 \\ b_2 \\ \vdots \\ b_n \end{pmatrix}.$$

(2) Prove that the matrix A is invertible and find its inverse A^{-1}. Given that the number of atoms in the universe is estimated to be $\leq 10^{82}$, compare the size of the coefficients the inverse of A to 10^{82}, if $n \geq 300$.

(3) Assume b is perturbed by a small amount Δb (note that Δb is a vector). Find the new solution of the system

$$A(x + \Delta x) = b + \Delta b,$$

where Δx is also a vector. In the case where $b = (0, \ldots, 0, 1)$, and $\Delta b = (0, \ldots, 0, \epsilon)$, show that

$$|(\Delta x)_1| = 2^{n-1}|\epsilon|$$

(where $(\Delta x)_1$ is the first component of Δx).

(4) Prove that $(A - I)^n = 0$.

Problem 2.13. An $n \times n$ matrix N is *nilpotent* if there is some integer $r \geq 1$ such that $N^r = 0$.

(1) Prove that if N is a nilpotent matrix, then the matrix $I - N$ is invertible and

$$(I - N)^{-1} = I + N + N^2 + \cdots + N^{r-1}.$$

(2) Compute the inverse of the following matrix A using (1):

$$A = \begin{pmatrix} 1 & 2 & 3 & 4 & 5 \\ 0 & 1 & 2 & 3 & 4 \\ 0 & 0 & 1 & 2 & 3 \\ 0 & 0 & 0 & 1 & 2 \\ 0 & 0 & 0 & 0 & 1 \end{pmatrix}.$$

Problem 2.14. (1) Let A be an $n \times n$ matrix. If A is invertible, prove that for any $x \in \mathbb{R}^n$, if $Ax = 0$, then $x = 0$.

(2) Let A be an $m \times n$ matrix and let B be an $n \times m$ matrix. Prove that $I_m - AB$ is invertible iff $I_n - BA$ is invertible.

Hint. If for all $x \in \mathbb{R}^n$, $Mx = 0$ implies that $x = 0$, then M is invertible.

Problem 2.15. Consider the following $n \times n$ matrix, for $n \geq 3$:

$$B = \begin{pmatrix} 1 & -1 & -1 & -1 & \cdots & -1 & -1 \\ 1 & -1 & 1 & 1 & \cdots & 1 & 1 \\ 1 & 1 & -1 & 1 & \cdots & 1 & 1 \\ 1 & 1 & 1 & -1 & \cdots & 1 & 1 \\ \vdots & \vdots & \vdots & \vdots & \vdots & \vdots & \vdots \\ 1 & 1 & 1 & 1 & \cdots & -1 & 1 \\ 1 & 1 & 1 & 1 & \cdots & 1 & -1 \end{pmatrix}.$$

(1) If we denote the columns of B by $b_1, \ldots, b_n$, prove that

$$(n - 3)b_1 - (b_2 + \cdots + b_n) = 2(n - 2)e_1$$
$$b_1 - b_2 = 2(e_1 + e_2)$$
$$b_1 - b_3 = 2(e_1 + e_3)$$
$$\vdots \qquad\qquad \vdots$$
$$b_1 - b_n = 2(e_1 + e_n),$$

where $e_1, \ldots, e_n$ are the canonical basis vectors of $\mathbb{R}^n$.

(2) Prove that B is invertible and that its inverse $A = (a_{ij})$ is given by

$$a_{11} = \frac{(n-3)}{2(n-2)}, \quad a_{i1} = -\frac{1}{2(n-2)} \quad 2 \leq i \leq n$$

and

$$a_{ii} = -\frac{(n-3)}{2(n-2)}, \quad 2 \le i \le n$$

$$a_{ji} = \frac{1}{2(n-2)}, \quad 2 \le i \le n, j \ne i.$$

(3) Show that the n diagonal $n \times n$ matrices D_i defined such that the diagonal entries of D_i are equal the entries (from top down) of the ith column of B form a basis of the space of $n \times n$ diagonal matrices (matrices with zeros everywhere except possibly on the diagonal). For example, when $n = 4$, we have

$$D_1 = \begin{pmatrix} 1 & 0 & 0 & 0 \\ 0 & 1 & 0 & 0 \\ 0 & 0 & 1 & 0 \\ 0 & 0 & 0 & 1 \end{pmatrix} \qquad D_2 = \begin{pmatrix} -1 & 0 & 0 & 0 \\ 0 & -1 & 0 & 0 \\ 0 & 0 & 1 & 0 \\ 0 & 0 & 0 & 1 \end{pmatrix},$$

$$D_3 = \begin{pmatrix} -1 & 0 & 0 & 0 \\ 0 & 1 & 0 & 0 \\ 0 & 0 & -1 & 0 \\ 0 & 0 & 0 & 1 \end{pmatrix}, \qquad D_4 = \begin{pmatrix} -1 & 0 & 0 & 0 \\ 0 & 1 & 0 & 0 \\ 0 & 0 & 1 & 0 \\ 0 & 0 & 0 & -1 \end{pmatrix}.$$

Problem 2.16. Given any $m \times n$ matrix A and any $n \times p$ matrix B, if we denote the columns of A by $A^1, \ldots, A^n$ and the rows of B by $B_1, \ldots, B_n$, prove that

$$AB = A^1 B_1 + \cdots + A^n B_n.$$

Problem 2.17. Let $f \colon E \to F$ be a linear map which is also a bijection (it is injective and surjective). Prove that the inverse function $f^{-1} \colon F \to E$ is linear.

Problem 2.18. Given two vectors spaces E and F, let $(u_i)_{i \in I}$ be any basis of E and let $(v_i)_{i \in I}$ be any family of vectors in F. Prove that the unique linear map $f \colon E \to F$ such that $f(u_i) = v_i$ for all $i \in I$ is surjective iff $(v_i)_{i \in I}$ spans F.

Problem 2.19. Let $f \colon E \to F$ be a linear map with $\dim(E) = n$ and $\dim(F) = m$. Prove that f has rank 1 iff f is represented by an $m \times n$ matrix of the form

$$A = uv^\top$$

with u a nonzero column vector of dimension m and v a nonzero column vector of dimension n.

Problem 2.20. Find a nontrivial linear dependence among the linear forms

$$\varphi_1(x,y,z) = 2x{-}y{+}3z, \quad \varphi_2(x,y,z) = 3x{-}5y{+}z, \quad \varphi_3(x,y,z) = 4x{-}7y{+}z.$$

Problem 2.21. Prove that the linear forms

$$\varphi_1(x,y,z) = x{+}2y{+}z, \quad \varphi_2(x,y,z) = 2x{+}3y{+}3z, \quad \varphi_3(x,y,z) = 3x{+}7y{+}z$$

are linearly independent. Express the linear form $\varphi(x,y,z) = x + y + z$ as a linear combination of $\varphi_1, \varphi_2, \varphi_3$.

Chapter 3

Matrices and Linear Maps

In this chapter, all vector spaces are defined over an arbitrary field K. For the sake of concreteness, the reader may safely assume that $K = \mathbb{R}$.

3.1 Representation of Linear Maps by Matrices

Proposition 2.14 shows that given two vector spaces E and F and a basis $(u_j)_{j \in J}$ of E, every linear map $f \colon E \to F$ is uniquely determined by the family $(f(u_j))_{j \in J}$ of the images under f of the vectors in the basis $(u_j)_{j \in J}$.

If we also have a basis $(v_i)_{i \in I}$ of F, then every vector $f(u_j)$ can be written in a unique way as

$$f(u_j) = \sum_{i \in I} a_{i\,j} v_i,$$

where $j \in J$, for a family of scalars $(a_{i\,j})_{i \in I}$. Thus, with respect to the two bases $(u_j)_{j \in J}$ of E and $(v_i)_{i \in I}$ of F, the linear map f is completely determined by a "$I \times J$-matrix" $M(f) = (a_{i\,j})_{i \in I,\ j \in J}$.

Remark: Note that we intentionally assigned the index set J to the basis $(u_j)_{j \in J}$ of E, and the index set I to the basis $(v_i)_{i \in I}$ of F, so that the rows of the matrix $M(f)$ associated with $f \colon E \to F$ are indexed by I, and the columns of the matrix $M(f)$ are indexed by J. Obviously, this causes a mildly unpleasant reversal. If we had considered the bases $(u_i)_{i \in I}$ of E and $(v_j)_{j \in J}$ of F, we would obtain a $J \times I$-matrix $M(f) = (a_{j\,i})_{j \in J,\ i \in I}$. No matter what we do, there will be a reversal! We decided to stick to the bases $(u_j)_{j \in J}$ of E and $(v_i)_{i \in I}$ of F, so that we get an $I \times J$-matrix $M(f)$, knowing that we may occasionally suffer from this decision!

When I and J are finite, and say, when $|I| = m$ and $|J| = n$, the linear map f is determined by the matrix $M(f)$ whose entries in the j-th column are the components of the vector $f(u_j)$ over the basis $(v_1, \ldots, v_m)$, that is, the matrix

$$
M(f) = \begin{pmatrix}
a_{11} & a_{12} & \cdots & a_{1n} \\
a_{21} & a_{22} & \cdots & a_{2n} \\
\vdots & \vdots & \ddots & \vdots \\
a_{m1} & a_{m2} & \cdots & a_{mn}
\end{pmatrix}
$$

whose entry on Row i and Column j is a_{ij} $(1 \le i \le m, 1 \le j \le n)$.

We will now show that when *E and F have finite dimension, linear maps can be very conveniently represented by matrices, and that composition of linear maps corresponds to matrix multiplication.* We will follow rather closely an elegant presentation method due to Emil Artin.

Let E and F be two vector spaces, and assume that E has a finite basis $(u_1, \ldots, u_n)$ and that F has a finite basis $(v_1, \ldots, v_m)$. Recall that we have shown that every vector $x \in E$ can be written in a unique way as

$$
x = x_1 u_1 + \cdots + x_n u_n,
$$

and similarly every vector $y \in F$ can be written in a unique way as

$$
y = y_1 v_1 + \cdots + y_m v_m.
$$

Let $f: E \to F$ be a linear map between E and F. Then for every $x = x_1 u_1 + \cdots + x_n u_n$ in E, by linearity, we have

$$
f(x) = x_1 f(u_1) + \cdots + x_n f(u_n).
$$

Let

$$
f(u_j) = a_{1j} v_1 + \cdots + a_{mj} v_m,
$$

or more concisely,

$$
f(u_j) = \sum_{i=1}^{m} a_{ij} v_i,
$$

for every j, $1 \le j \le n$. This can be expressed by writing the coefficients $a_{1j}, a_{2j}, \ldots, a_{mj}$ of $f(u_j)$ over the basis $(v_1, \ldots, v_m)$, as the jth column of a matrix, as shown below:

$$
\begin{array}{c}
 \quad f(u_1)\ f(u_2)\ \cdots\ f(u_n) \\
\begin{array}{c}
v_1 \\ v_2 \\ \vdots \\ v_m
\end{array}
\begin{pmatrix}
a_{11} & a_{12} & \cdots & a_{1n} \\
a_{21} & a_{22} & \cdots & a_{2n} \\
\vdots & \vdots & \ddots & \vdots \\
a_{m1} & a_{m2} & \cdots & a_{mn}
\end{pmatrix}.
\end{array}
$$

Then substituting the right-hand side of each $f(u_j)$ into the expression for $f(x)$, we get

$$f(x) = x_1 \left(\sum_{i=1}^{m} a_{i\,1} v_i \right) + \cdots + x_n \left(\sum_{i=1}^{m} a_{i\,n} v_i \right),$$

which, by regrouping terms to obtain a linear combination of the v_i, yields

$$f(x) = \left(\sum_{j=1}^{n} a_{1\,j} x_j \right) v_1 + \cdots + \left(\sum_{j=1}^{n} a_{m\,j} x_j \right) v_m.$$

Thus, letting $f(x) = y = y_1 v_1 + \cdots + y_m v_m$, we have

$$y_i = \sum_{j=1}^{n} a_{i\,j} x_j \tag{3.1}$$

for all i, $1 \le i \le m$.

To make things more concrete, let us treat the case where $n = 3$ and $m = 2$. In this case,

$$f(u_1) = a_{11} v_1 + a_{21} v_2$$
$$f(u_2) = a_{12} v_1 + a_{22} v_2$$
$$f(u_3) = a_{13} v_1 + a_{23} v_2,$$

which in matrix form is expressed by

$$
\begin{array}{c}
f(u_1) \; f(u_2) \; f(u_3) \\
\begin{array}{c} v_1 \\ v_2 \end{array}
\left(\begin{array}{ccc}
a_{11} & a_{12} & a_{13} \\
a_{21} & a_{22} & a_{23}
\end{array} \right),
\end{array}
$$

and for any $x = x_1 u_1 + x_2 u_2 + x_3 u_3$, we have

$$
\begin{aligned}
f(x) &= f(x_1 u_1 + x_2 u_2 + x_3 u_3) \\
&= x_1 f(u_1) + x_2 f(u_2) + x_3 f(u_3) \\
&= x_1 (a_{11} v_1 + a_{21} v_2) + x_2 (a_{12} v_1 + a_{22} v_2) + x_3 (a_{13} v_1 + a_{23} v_2) \\
&= (a_{11} x_1 + a_{12} x_2 + a_{13} x_3) v_1 + (a_{21} x_1 + a_{22} x_2 + a_{23} x_3) v_2.
\end{aligned}
$$

Consequently, since

$$y = y_1 v_1 + y_2 v_2,$$

we have

$$y_1 = a_{11} x_1 + a_{12} x_2 + a_{13} x_3$$
$$y_2 = a_{21} x_1 + a_{22} x_2 + a_{23} x_3.$$

This agrees with the matrix equation

$$\begin{pmatrix} y_1 \\ y_2 \end{pmatrix} = \begin{pmatrix} a_{11} & a_{12} & a_{13} \\ a_{21} & a_{22} & a_{23} \end{pmatrix} \begin{pmatrix} x_1 \\ x_2 \\ x_3 \end{pmatrix}.$$

We now formalize the representation of linear maps by matrices.

Definition 3.1. Let E and F be two vector spaces, and let $(u_1, \ldots, u_n)$ be a basis for E, and $(v_1, \ldots, v_m)$ be a basis for F. Each vector $x \in E$ expressed in the basis $(u_1, \ldots, u_n)$ as $x = x_1 u_1 + \cdots + x_n u_n$ is represented by the column matrix

$$M(x) = \begin{pmatrix} x_1 \\ \vdots \\ x_n \end{pmatrix}$$

and similarly for each vector $y \in F$ expressed in the basis $(v_1, \ldots, v_m)$.

Every linear map $f \colon E \to F$ is represented by the matrix $M(f) = (a_{ij})$, where a_{ij} is the i-th component of the vector $f(u_j)$ over the basis $(v_1, \ldots, v_m)$, i.e., where

$$f(u_j) = \sum_{i=1}^{m} a_{ij} v_i, \quad \text{for every } j, \ 1 \le j \le n.$$

The coefficients $a_{1j}, a_{2j}, \ldots, a_{mj}$ of $f(u_j)$ over the basis $(v_1, \ldots, v_m)$ form the jth column of the matrix $M(f)$ shown below:

$$
\begin{array}{c}
 \quad f(u_1) \ \ f(u_2) \ \ldots \ f(u_n) \\
\begin{array}{c}
v_1 \\
v_2 \\
\vdots \\
v_m
\end{array}
\begin{pmatrix}
a_{11} & a_{12} & \cdots & a_{1n} \\
a_{21} & a_{22} & \cdots & a_{2n} \\
\vdots & \vdots & \ddots & \vdots \\
a_{m1} & a_{m2} & \cdots & a_{mn}
\end{pmatrix}.
\end{array}
$$

The matrix $M(f)$ associated with the linear map $f \colon E \to F$ is called the *matrix of f with respect to the bases* $(u_1, \ldots, u_n)$ *and* $(v_1, \ldots, v_m)$. When $E = F$ and the basis $(v_1, \ldots, v_m)$ is identical to the basis $(u_1, \ldots, u_n)$ of E, the matrix $M(f)$ associated with $f \colon E \to E$ (as above) is called the *matrix of f with respect to the basis* $(u_1, \ldots, u_n)$.

Remark: As in the remark after Definition 2.14, there is no reason to assume that the vectors in the bases $(u_1, \ldots, u_n)$ and $(v_1, \ldots, v_m)$ are ordered in any particular way. However, it is often convenient to assume the natural ordering. When this is so, authors sometimes refer to the matrix $M(f)$ as the matrix of f with respect to the *ordered bases* $(u_1, \ldots, u_n)$ and $(v_1, \ldots, v_m)$.

Let us illustrate the representation of a linear map by a matrix in a concrete situation. Let E be the vector space $\mathbb{R}[X]_4$ of polynomials of degree at most 4, let F be the vector space $\mathbb{R}[X]_3$ of polynomials of degree at most 3, and let the linear map be the derivative map d: that is,

$$d(P + Q) = dP + dQ$$
$$d(\lambda P) = \lambda dP,$$

with $\lambda \in \mathbb{R}$. We choose $(1, x, x^2, x^3, x^4)$ as a basis of E and $(1, x, x^2, x^3)$ as a basis of F. Then the 4×5 matrix D associated with d is obtained by expressing the derivative dx^i of each basis vector x^i for $i = 0, 1, 2, 3, 4$ over the basis $(1, x, x^2, x^3)$. We find

$$D = \begin{pmatrix} 0 & 1 & 0 & 0 & 0 \\ 0 & 0 & 2 & 0 & 0 \\ 0 & 0 & 0 & 3 & 0 \\ 0 & 0 & 0 & 0 & 4 \end{pmatrix}.$$

If P denotes the polynomial

$$P = 3x^4 - 5x^3 + x^2 - 7x + 5,$$

we have

$$dP = 12x^3 - 15x^2 + 2x - 7.$$

The polynomial P is represented by the vector $(5, -7, 1, -5, 3)$, the polynomial dP is represented by the vector $(-7, 2, -15, 12)$, and we have

$$\begin{pmatrix} 0 & 1 & 0 & 0 & 0 \\ 0 & 0 & 2 & 0 & 0 \\ 0 & 0 & 0 & 3 & 0 \\ 0 & 0 & 0 & 0 & 4 \end{pmatrix} \begin{pmatrix} 5 \\ -7 \\ 1 \\ -5 \\ 3 \end{pmatrix} = \begin{pmatrix} -7 \\ 2 \\ -15 \\ 12 \end{pmatrix},$$

as expected! The kernel (nullspace) of d consists of the polynomials of degree 0, that is, the constant polynomials. Therefore $\dim(\text{Ker }d) = 1$, and from

$$\dim(E) = \dim(\text{Ker }d) + \dim(\text{Im }d)$$

(see Theorem 5.1), we get $\dim(\text{Im }d) = 4$ (since $\dim(E) = 5$).

For fun, let us figure out the linear map from the vector space $\mathbb{R}[X]_3$ to the vector space $\mathbb{R}[X]_4$ given by integration (finding the primitive, or anti-derivative) of x^i, for $i = 0, 1, 2, 3$. The 5×4 matrix S representing $\int$ with respect to the same bases as before is

$$S = \begin{pmatrix} 0 & 0 & 0 & 0 \\ 1 & 0 & 0 & 0 \\ 0 & 1/2 & 0 & 0 \\ 0 & 0 & 1/3 & 0 \\ 0 & 0 & 0 & 1/4 \end{pmatrix}.$$

We verify that $DS = I_4$,

$$\begin{pmatrix} 0 & 1 & 0 & 0 & 0 \\ 0 & 0 & 2 & 0 & 0 \\ 0 & 0 & 0 & 3 & 0 \\ 0 & 0 & 0 & 0 & 4 \end{pmatrix} \begin{pmatrix} 0 & 0 & 0 & 0 \\ 1 & 0 & 0 & 0 \\ 0 & 1/2 & 0 & 0 \\ 0 & 0 & 1/3 & 0 \\ 0 & 0 & 0 & 1/4 \end{pmatrix} = \begin{pmatrix} 1 & 0 & 0 & 0 \\ 0 & 1 & 0 & 0 \\ 0 & 0 & 1 & 0 \\ 0 & 0 & 0 & 1 \end{pmatrix}.$$

This is to be expected by the fundamental theorem of calculus since the derivative of an integral returns the function. As we will shortly see, the above matrix product corresponds to this functional composition. The equation $DS = I_4$ shows that S is injective and has D as a left inverse. However, $SD \neq I_5$, and instead

$$\begin{pmatrix} 0 & 0 & 0 & 0 \\ 1 & 0 & 0 & 0 \\ 0 & 1/2 & 0 & 0 \\ 0 & 0 & 1/3 & 0 \\ 0 & 0 & 0 & 1/4 \end{pmatrix} \begin{pmatrix} 0 & 1 & 0 & 0 & 0 \\ 0 & 0 & 2 & 0 & 0 \\ 0 & 0 & 0 & 3 & 0 \\ 0 & 0 & 0 & 0 & 4 \end{pmatrix} = \begin{pmatrix} 0 & 0 & 0 & 0 & 0 \\ 0 & 1 & 0 & 0 & 0 \\ 0 & 0 & 1 & 0 & 0 \\ 0 & 0 & 0 & 1 & 0 \\ 0 & 0 & 0 & 0 & 1 \end{pmatrix},$$

because constant polynomials (polynomials of degree 0) belong to the kernel of D.

3.2   Composition of Linear Maps and Matrix Multiplication

Let us now consider how the composition of linear maps is expressed in terms of bases.

Let E, F, and G, be three vectors spaces with respective bases $(u_1, \ldots, u_p)$ for E, $(v_1, \ldots, v_n)$ for F, and $(w_1, \ldots, w_m)$ for G. Let $g \colon E \to F$ and $f \colon F \to G$ be linear maps. As explained earlier, $g \colon E \to F$ is determined by the images of the basis vectors u_j, and $f \colon F \to G$ is determined by the images of the basis vectors v_k. We would like to understand how $f \circ g \colon E \to G$ is determined by the images of the basis vectors u_j.

Remark: Note that we are considering linear maps $g \colon E \to F$ and $f \colon F \to G$, instead of $f \colon E \to F$ and $g \colon F \to G$, which yields the composition $f \circ g \colon E \to G$ instead of $g \circ f \colon E \to G$. Our perhaps unusual choice is motivated by the fact that if f is represented by a matrix $M(f) = (a_{i\,k})$ and g is represented by a matrix $M(g) = (b_{k\,j})$, then $f \circ g \colon E \to G$ is represented by the product AB of the matrices A and B. If we had adopted the other choice where $f \colon E \to F$ and $g \colon F \to G$, then $g \circ f \colon E \to G$ would be represented by the product BA. Personally, we find it easier to remember the formula for the entry in Row i and Column j of the product of two matrices when this product is written by AB, rather than BA. Obviously, this is a matter of taste! We will have to live with our perhaps unorthodox choice.

Thus, let

$$f(v_k) = \sum_{i=1}^{m} a_{i\,k} w_i,$$

for every k, $1 \le k \le n$, and let

$$g(u_j) = \sum_{k=1}^{n} b_{k\,j} v_k,$$

for every j, $1 \le j \le p$; in matrix form, we have

$$
\begin{array}{c}
f(v_1)\; f(v_2)\; \ldots\; f(v_n) \\[4pt]
\begin{array}{c} w_1 \\ w_2 \\ \vdots \\ w_m \end{array}
\begin{pmatrix}
a_{11} & a_{12} & \cdots & a_{1n} \\
a_{21} & a_{22} & \cdots & a_{2n} \\
\vdots & \vdots & \ddots & \vdots \\
a_{m1} & a_{m2} & \cdots & a_{mn}
\end{pmatrix}
\end{array}
$$

and

$$
\begin{array}{c}
g(u_1)\; g(u_2)\; \ldots\; g(u_p) \\[4pt]
\begin{array}{c} v_1 \\ v_2 \\ \vdots \\ v_n \end{array}
\begin{pmatrix}
b_{11} & b_{12} & \cdots & b_{1p} \\
b_{21} & b_{22} & \cdots & b_{2p} \\
\vdots & \vdots & \ddots & \vdots \\
b_{n1} & b_{n2} & \cdots & b_{np}
\end{pmatrix}
\end{array}.
$$

By previous considerations, for every

$$x = x_1 u_1 + \cdots + x_p u_p,$$

letting $g(x) = y = y_1 v_1 + \cdots + y_n v_n$, we have

$$y_k = \sum_{j=1}^{p} b_{k\,j} x_j \tag{3.2}$$

for all k, $1 \leq k \leq n$, and for every

$$y = y_1 v_1 + \cdots + y_n v_n,$$

letting $f(y) = z = z_1 w_1 + \cdots + z_m w_m$, we have

$$z_i = \sum_{k=1}^{n} a_{i\,k} y_k \tag{3.3}$$

for all i, $1 \leq i \leq m$. Then if $y = g(x)$ and $z = f(y)$, we have $z = f(g(x))$, and in view of (3.2) and (3.3), we have

$$z_i = \sum_{k=1}^{n} a_{i\,k} \left(\sum_{j=1}^{p} b_{k\,j} x_j \right)$$

$$= \sum_{k=1}^{n} \sum_{j=1}^{p} a_{i\,k} b_{k\,j} x_j$$

$$= \sum_{j=1}^{p} \sum_{k=1}^{n} a_{i\,k} b_{k\,j} x_j$$

$$= \sum_{j=1}^{p} \left(\sum_{k=1}^{n} a_{i\,k} b_{k\,j} \right) x_j.$$

Thus, defining $c_{i\,j}$ such that

$$c_{i\,j} = \sum_{k=1}^{n} a_{i\,k} b_{k\,j},$$

for $1 \leq i \leq m$, and $1 \leq j \leq p$, we have

$$z_i = \sum_{j=1}^{p} c_{i\,j} x_j. \tag{3.4}$$

Identity (3.4) shows that the composition of linear maps corresponds to the product of matrices.

Then given a linear map $f \colon E \to F$ represented by the matrix $M(f) = (a_{i\,j})$ w.r.t. the bases $(u_1, \ldots, u_n)$ and $(v_1, \ldots, v_m)$, by equation (3.1), namely

$$y_i = \sum_{j=1}^{n} a_{i\,j} x_j \quad 1 \leq i \leq m,$$

and the definition of matrix multiplication, the equation $y = f(x)$ corresponds to the matrix equation $M(y) = M(f)M(x)$, that is,

$$\begin{pmatrix} y_1 \\ \vdots \\ y_m \end{pmatrix} = \begin{pmatrix} a_{11} & \cdots & a_{1n} \\ \vdots & \ddots & \vdots \\ a_{m1} & \cdots & a_{mn} \end{pmatrix} \begin{pmatrix} x_1 \\ \vdots \\ x_n \end{pmatrix}.$$

Recall that

$$\begin{pmatrix} a_{11} & a_{12} & \cdots & a_{1n} \\ a_{21} & a_{22} & \cdots & a_{2n} \\ \vdots & \vdots & \ddots & \vdots \\ a_{m1} & a_{m2} & \cdots & a_{mn} \end{pmatrix} \begin{pmatrix} x_1 \\ x_2 \\ \vdots \\ x_n \end{pmatrix} = x_1 \begin{pmatrix} a_{11} \\ a_{21} \\ \vdots \\ a_{m1} \end{pmatrix} + x_2 \begin{pmatrix} a_{12} \\ a_{22} \\ \vdots \\ a_{m2} \end{pmatrix} + \cdots + x_n \begin{pmatrix} a_{1n} \\ a_{2n} \\ \vdots \\ a_{mn} \end{pmatrix}.$$

Sometimes, it is necessary to incorporate the bases $(u_1, \ldots, u_n)$ and $(v_1, \ldots, v_m)$ in the notation for the matrix $M(f)$ expressing f with respect to these bases. This turns out to be a messy enterprise!

We propose the following course of action:

Definition 3.2. Write $\mathcal{U} = (u_1, \ldots, u_n)$ and $\mathcal{V} = (v_1, \ldots, v_m)$ for the bases of E and F, and denote by $M_{\mathcal{U},\mathcal{V}}(f)$ the *matrix of f with respect to the bases $\mathcal{U}$ and $\mathcal{V}$*. Furthermore, write $x_{\mathcal{U}}$ for the coordinates $M(x) = (x_1, \ldots, x_n)$ of $x \in E$ w.r.t. the basis $\mathcal{U}$ and write $y_{\mathcal{V}}$ for the coordinates $M(y) = (y_1, \ldots, y_m)$ of $y \in F$ w.r.t. the basis $\mathcal{V}$. Then

$$y = f(x)$$

is expressed in matrix form by

$$y_{\mathcal{V}} = M_{\mathcal{U},\mathcal{V}}(f)\, x_{\mathcal{U}}.$$

When $\mathcal{U} = \mathcal{V}$, we abbreviate $M_{\mathcal{U},\mathcal{V}}(f)$ as $M_{\mathcal{U}}(f)$.

The above notation seems reasonable, but it has the slight disadvantage that in the expression $M_{\mathcal{U},\mathcal{V}}(f)x_{\mathcal{U}}$, the input argument $x_{\mathcal{U}}$ which is fed to the matrix $M_{\mathcal{U},\mathcal{V}}(f)$ does not appear next to the subscript $\mathcal{U}$ in $M_{\mathcal{U},\mathcal{V}}(f)$. We could have used the notation $M_{\mathcal{V},\mathcal{U}}(f)$, and some people do that. But then, we find a bit confusing that $\mathcal{V}$ comes before $\mathcal{U}$ when f maps from the space E with the basis $\mathcal{U}$ to the space F with the basis $\mathcal{V}$. So, we prefer to use the notation $M_{\mathcal{U},\mathcal{V}}(f)$.

Be aware that other authors such as Meyer [Meyer (2000)] use the notation $[f]_{\mathcal{U},\mathcal{V}}$, and others such as Dummit and Foote [Dummit and Foote (1999)] use the notation $M_{\mathcal{U}}^{\mathcal{V}}(f)$, instead of $M_{\mathcal{U},\mathcal{V}}(f)$. This gets worse! You may find the notation $M_{\mathcal{V}}^{\mathcal{U}}(f)$ (as in Lang [Lang (1993)]), or $_{\mathcal{U}}[f]_{\mathcal{V}}$, or other strange notations.

Definition 3.2 shows that the function which associates to a linear map $f\colon E \to F$ the matrix $M(f)$ w.r.t. the bases $(u_1, \ldots, u_n)$ and $(v_1, \ldots, v_m)$ has the property that matrix multiplication corresponds to composition of linear maps. This allows us to transfer properties of linear maps to matrices. Here is an illustration of this technique:

Proposition 3.1. *(1) Given any matrices $A \in \mathrm{M}_{m,n}(K)$, $B \in \mathrm{M}_{n,p}(K)$, and $C \in \mathrm{M}_{p,q}(K)$, we have*

$$(AB)C = A(BC);$$

that is, matrix multiplication is associative.

(2) Given any matrices $A, B \in \mathrm{M}_{m,n}(K)$, and $C, D \in \mathrm{M}_{n,p}(K)$, for all $\lambda \in K$, we have

$$(A + B)C = AC + BC$$
$$A(C + D) = AC + AD$$
$$(\lambda A)C = \lambda(AC)$$
$$A(\lambda C) = \lambda(AC),$$

so that matrix multiplication $\cdot\colon \mathrm{M}_{m,n}(K) \times \mathrm{M}_{n,p}(K) \to \mathrm{M}_{m,p}(K)$ is bilinear.

Proof. (1) Every $m \times n$ matrix $A = (a_{ij})$ defines the function $f_A\colon K^n \to K^m$ given by

$$f_A(x) = Ax,$$

for all $x \in K^n$. It is immediately verified that f_A is linear and that the matrix $M(f_A)$ representing f_A over the canonical bases in K^n and K^m is equal to A. Then Formula (4) proves that

$$M(f_A \circ f_B) = M(f_A)M(f_B) = AB,$$

so we get

$$M((f_A \circ f_B) \circ f_C) = M(f_A \circ f_B)M(f_C) = (AB)C$$

and

$$M(f_A \circ (f_B \circ f_C)) = M(f_A)M(f_B \circ f_C) = A(BC),$$

and since composition of functions is associative, we have $(f_A \circ f_B) \circ f_C = f_A \circ (f_B \circ f_C)$, which implies that

$$(AB)C = A(BC).$$

(2) It is immediately verified that if $f_1, f_2 \in \mathrm{Hom}_K(E, F)$, $A, B \in \mathrm{M}_{m,n}(K)$, $(u_1, \ldots, u_n)$ is any basis of E, and $(v_1, \ldots, v_m)$ is any basis of F, then

$$M(f_1 + f_2) = M(f_1) + M(f_2)$$
$$f_{A+B} = f_A + f_B.$$

Then we have

$$
\begin{aligned}
(A + B)C &= M(f_{A+B})M(f_C) \\
&= M(f_{A+B} \circ f_C) \\
&= M((f_A + f_B) \circ f_C)) \\
&= M((f_A \circ f_C) + (f_B \circ f_C)) \\
&= M(f_A \circ f_C) + M(f_B \circ f_C) \\
&= M(f_A)M(f_C) + M(f_B)M(f_C) \\
&= AC + BC.
\end{aligned}
$$

The equation $A(C + D) = AC + AD$ is proven in a similar fashion, and the last two equations are easily verified. We could also have verified all the identities by making matrix computations. $\qquad\square$

Note that Proposition 3.1 implies that the vector space $\mathrm{M}_n(K)$ of square matrices is a (noncommutative) ring with unit I_n. (It even shows that $\mathrm{M}_n(K)$ is an associative *algebra*.)

The following proposition states the main properties of the mapping $f \mapsto M(f)$ between $\mathrm{Hom}(E, F)$ and $\mathrm{M}_{m,n}$. In short, it is an isomorphism of vector spaces.

Proposition 3.2. *Given three vector spaces E, F, G, with respective bases $(u_1, \ldots, u_p)$, $(v_1, \ldots, v_n)$, and $(w_1, \ldots, w_m)$, the mapping $M \colon \mathrm{Hom}(E, F) \to \mathrm{M}_{n,p}$ that associates the matrix $M(g)$ to a linear map $g \colon E \to F$ satisfies the following properties for all $x \in E$, all $g, h \colon E \to F$, and all $f \colon F \to G$:*

$$M(g(x)) = M(g)M(x)$$
$$M(g + h) = M(g) + M(h)$$
$$M(\lambda g) = \lambda M(g)$$
$$M(f \circ g) = M(f)M(g),$$

where $M(x)$ is the column vector associated with the vector x and $M(g(x))$ is the column vector associated with $g(x)$, as explained in Definition 3.1.

Thus, $M \colon \operatorname{Hom}(E, F) \to \mathrm{M}_{n,p}$ is an isomorphism of vector spaces, and when $p = n$ and the basis $(v_1, \ldots, v_n)$ is identical to the basis $(u_1, \ldots, u_p)$, $M \colon \operatorname{Hom}(E, E) \to \mathrm{M}_n$ is an isomorphism of rings.

Proof. That $M(g(x)) = M(g)M(x)$ was shown by Definition 3.2 or equivalently by Formula (1). The identities $M(g + h) = M(g) + M(h)$ and $M(\lambda g) = \lambda M(g)$ are straightforward, and $M(f \circ g) = M(f)M(g)$ follows from Identity (4) and the definition of matrix multiplication. The mapping $M \colon \operatorname{Hom}(E, F) \to \mathrm{M}_{n,p}$ is clearly injective, and since every matrix defines a linear map (see Proposition 3.1), it is also surjective, and thus bijective. In view of the above identities, it is an isomorphism (and similarly for $M \colon \operatorname{Hom}(E, E) \to \mathrm{M}_n$, where Proposition 3.1 is used to show that M_n is a ring). $\qquad\square$

In view of Proposition 3.2, it seems preferable to represent vectors from a vector space of finite dimension as column vectors rather than row vectors. *Thus, from now on, we will denote vectors of $\mathbb{R}^n$ (or more generally, of K^n) as column vectors.*

3.3 Change of Basis Matrix

It is important to observe that the isomorphism $M \colon \operatorname{Hom}(E, F) \to \mathrm{M}_{n,p}$ given by Proposition 3.2 depends on the choice of the bases $(u_1, \ldots, u_p)$ and $(v_1, \ldots, v_n)$, and similarly for the isomorphism $M \colon \operatorname{Hom}(E, E) \to \mathrm{M}_n$, which depends on the choice of the basis $(u_1, \ldots, u_n)$. Thus, it would be useful to know how a change of basis affects the representation of a linear map $f \colon E \to F$ as a matrix. The following simple proposition is needed.

Proposition 3.3. *Let E be a vector space, and let $(u_1, \ldots, u_n)$ be a basis of E. For every family $(v_1, \ldots, v_n)$, let $P = (a_{ij})$ be the matrix defined such that $v_j = \sum_{i=1}^{n} a_{ij} u_i$. The matrix P is invertible iff $(v_1, \ldots, v_n)$ is a basis of E.*

Proof. Note that we have $P = M(f)$, the matrix associated with the unique linear map $f \colon E \to E$ such that $f(u_i) = v_i$. By Proposition 2.14, f is bijective iff $(v_1, \ldots, v_n)$ is a basis of E. Furthermore, it is obvious that the identity matrix I_n is the matrix associated with the identity id$\colon E \to E$ w.r.t. any basis. If f is an isomorphism, then $f \circ f^{-1} = f^{-1} \circ f = $ id, and by Proposition 3.2, we get $M(f)M(f^{-1}) = M(f^{-1})M(f) = I_n$, showing that P is invertible and that $M(f^{-1}) = P^{-1}$. $\qquad\square$

Proposition 3.3 suggests the following definition.

Definition 3.3. Given a vector space E of dimension n, for any two bases $(u_1, \ldots, u_n)$ and $(v_1, \ldots, v_n)$ of E, let $P = (a_{ij})$ be the invertible matrix defined such that

$$v_j = \sum_{i=1}^{n} a_{ij} u_i,$$

which is also the matrix of the identity id: $E \to E$ with respect to the bases $(v_1, \ldots, v_n)$ and $(u_1, \ldots, u_n)$, *in that order*. Indeed, we express each id$(v_j) = v_j$ over the basis $(u_1, \ldots, u_n)$. The coefficients $a_{1j}, a_{2j}, \ldots, a_{nj}$ of v_j over the basis $(u_1, \ldots, u_n)$ form the jth column of the matrix P shown below:

$$
\begin{array}{c}
\\
u_1 \\
u_2 \\
\vdots \\
u_n
\end{array}
\begin{array}{cccc}
v_1 & v_2 & \ldots & v_n \\
\end{array}
\left(
\begin{array}{cccc}
a_{11} & a_{12} & \ldots & a_{1n} \\
a_{21} & a_{22} & \ldots & a_{2n} \\
\vdots & \vdots & \ddots & \vdots \\
a_{n1} & a_{n2} & \ldots & a_{nn}
\end{array}
\right).
$$

The matrix P is called the *change of basis matrix from* $(u_1, \ldots, u_n)$ *to* $(v_1, \ldots, v_n)$.

Clearly, the change of basis matrix from $(v_1, \ldots, v_n)$ to $(u_1, \ldots, u_n)$ is P^{-1}. Since $P = (a_{ij})$ is the matrix of the identity id: $E \to E$ with respect to the bases $(v_1, \ldots, v_n)$ and $(u_1, \ldots, u_n)$, given any vector $x \in E$, if $x = x_1 u_1 + \cdots + x_n u_n$ over the basis $(u_1, \ldots, u_n)$ and $x = x_1' v_1 + \cdots + x_n' v_n$ over the basis $(v_1, \ldots, v_n)$, from Proposition 3.2, we have

$$
\begin{pmatrix} x_1 \\ \vdots \\ x_n \end{pmatrix} = \begin{pmatrix} a_{11} & \ldots & a_{1n} \\ \vdots & \ddots & \vdots \\ a_{n1} & \ldots & a_{nn} \end{pmatrix} \begin{pmatrix} x_1' \\ \vdots \\ x_n' \end{pmatrix},
$$

showing that the *old* coordinates (x_i) of x (over $(u_1, \ldots, u_n)$) are expressed in terms of the *new* coordinates (x_i') of x (over $(v_1, \ldots, v_n)$).

Now we face the painful task of assigning a "good" notation incorporating the bases $\mathcal{U} = (u_1, \ldots, u_n)$ and $\mathcal{V} = (v_1, \ldots, v_n)$ into the notation for the change of basis matrix from $\mathcal{U}$ to $\mathcal{V}$. Because the change of basis matrix from $\mathcal{U}$ to $\mathcal{V}$ is the matrix of the identity map id$_E$ *with respect to the bases* $\mathcal{V}$ *and* $\mathcal{U}$ *in that order*, we could denote it by $M_{\mathcal{V}, \mathcal{U}}(\text{id})$ (Meyer

[Meyer (2000)] uses the notation $[I]_{\mathcal{V},\mathcal{U}})$. We prefer to use an abbreviation for $M_{\mathcal{V},\mathcal{U}}(\text{id})$.

Definition 3.4. The *change of basis matrix from* $\mathcal{U}$ *to* $\mathcal{V}$ is denoted

$$P_{\mathcal{V},\mathcal{U}}.$$

Note that

$$P_{\mathcal{U},\mathcal{V}} = P_{\mathcal{V},\mathcal{U}}^{-1}.$$

Then, if we write $x_{\mathcal{U}} = (x_1, \ldots, x_n)$ for the *old* coordinates of x with respect to the basis $\mathcal{U}$ and $x_{\mathcal{V}} = (x_1', \ldots, x_n')$ for the *new* coordinates of x with respect to the basis $\mathcal{V}$, we have

$$x_{\mathcal{U}} = P_{\mathcal{V},\mathcal{U}}\, x_{\mathcal{V}}, \quad x_{\mathcal{V}} = P_{\mathcal{V},\mathcal{U}}^{-1}\, x_{\mathcal{U}}.$$

The above may look backward, but remember that the matrix $M_{\mathcal{U},\mathcal{V}}(f)$ takes input expressed over the basis $\mathcal{U}$ to output expressed over the basis $\mathcal{V}$. Consequently, $P_{\mathcal{V},\mathcal{U}}$ takes input expressed over the basis $\mathcal{V}$ to output expressed over the basis $\mathcal{U}$, and $x_{\mathcal{U}} = P_{\mathcal{V},\mathcal{U}}\, x_{\mathcal{V}}$ matches this point of view!

Beware that some authors (such as Artin [Artin (1991)]) define the change of basis matrix from $\mathcal{U}$ to $\mathcal{V}$ as $P_{\mathcal{U},\mathcal{V}} = P_{\mathcal{V},\mathcal{U}}^{-1}$. Under this point of view, the old basis $\mathcal{U}$ is expressed in terms of the new basis $\mathcal{V}$. We find this a bit unnatural. Also, in practice, it seems that the new basis is often expressed in terms of the old basis, rather than the other way around.

Since the matrix $P = P_{\mathcal{V},\mathcal{U}}$ expresses the *new* basis $(v_1, \ldots, v_n)$ in terms of the *old* basis $(u_1, \ldots, u_n)$, we observe that the coordinates (x_i) of a vector x vary in the *opposite direction* of the change of basis. For this reason, vectors are sometimes said to be *contravariant*. However, this expression does not make sense! Indeed, a vector in an intrinsic quantity that does not depend on a specific basis. What makes sense is that the *coordinates* of a vector vary in a contravariant fashion.

Let us consider some concrete examples of change of bases.

Example 3.1. Let $E = F = \mathbb{R}^2$, with $u_1 = (1,0)$, $u_2 = (0,1)$, $v_1 = (1,1)$ and $v_2 = (-1,1)$. The change of basis matrix P from the basis $\mathcal{U} = (u_1, u_2)$ to the basis $\mathcal{V} = (v_1, v_2)$ is

$$P = \begin{pmatrix} 1 & -1 \\ 1 & 1 \end{pmatrix}$$

and its inverse is

$$P^{-1} = \begin{pmatrix} 1/2 & 1/2 \\ -1/2 & 1/2 \end{pmatrix}.$$

The old coordinates (x_1, x_2) with respect to (u_1, u_2) are expressed in terms of the new coordinates (x_1', x_2') with respect to (v_1, v_2) by

$$\begin{pmatrix} x_1 \\ x_2 \end{pmatrix} = \begin{pmatrix} 1 & -1 \\ 1 & 1 \end{pmatrix} \begin{pmatrix} x_1' \\ x_2' \end{pmatrix},$$

and the new coordinates (x_1', x_2') with respect to (v_1, v_2) are expressed in terms of the old coordinates (x_1, x_2) with respect to (u_1, u_2) by

$$\begin{pmatrix} x_1' \\ x_2' \end{pmatrix} = \begin{pmatrix} 1/2 & 1/2 \\ -1/2 & 1/2 \end{pmatrix} \begin{pmatrix} x_1 \\ x_2 \end{pmatrix}.$$

Example 3.2. Let $E = F = \mathbb{R}[X]_3$ be the set of polynomials of degree at most 3, and consider the bases $\mathcal{U} = (1, x, x^2, x^3)$ and $\mathcal{V} = (B_0^3(x), B_1^3(x), B_2^3(x), B_3^3(x))$, where $B_0^3(x)$, $B_1^3(x)$, $B_2^3(x)$, $B_3^3(x)$ are the *Bernstein polynomials* of degree 3, given by

$$B_0^3(x) = (1-x)^3 \quad B_1^3(x) = 3(1-x)^2 x \quad B_2^3(x) = 3(1-x)x^2 \quad B_3^3(x) = x^3.$$

By expanding the Bernstein polynomials, we find that the change of basis matrix $P_{\mathcal{V}, \mathcal{U}}$ is given by

$$P_{\mathcal{V}, \mathcal{U}} = \begin{pmatrix} 1 & 0 & 0 & 0 \\ -3 & 3 & 0 & 0 \\ 3 & -6 & 3 & 0 \\ -1 & 3 & -3 & 1 \end{pmatrix}.$$

We also find that the inverse of $P_{\mathcal{V}, \mathcal{U}}$ is

$$P_{\mathcal{V}, \mathcal{U}}^{-1} = \begin{pmatrix} 1 & 0 & 0 & 0 \\ 1 & 1/3 & 0 & 0 \\ 1 & 2/3 & 1/3 & 0 \\ 1 & 1 & 1 & 1 \end{pmatrix}.$$

Therefore, the coordinates of the polynomial $2x^3 - x + 1$ over the basis $\mathcal{V}$ are

$$\begin{pmatrix} 1 \\ 2/3 \\ 1/3 \\ 2 \end{pmatrix} = \begin{pmatrix} 1 & 0 & 0 & 0 \\ 1 & 1/3 & 0 & 0 \\ 1 & 2/3 & 1/3 & 0 \\ 1 & 1 & 1 & 1 \end{pmatrix} \begin{pmatrix} 1 \\ -1 \\ 0 \\ 2 \end{pmatrix},$$

and so

$$2x^3 - x + 1 = B_0^3(x) + \frac{2}{3}B_1^3(x) + \frac{1}{3}B_2^3(x) + 2B_3^3(x).$$

3.4   The Effect of a Change of Bases on Matrices

The effect of a change of bases on the representation of a linear map is described in the following proposition.

Proposition 3.4. *Let E and F be vector spaces, let $\mathcal{U} = (u_1, \ldots, u_n)$ and $\mathcal{U}' = (u'_1, \ldots, u'_n)$ be two bases of E, and let $\mathcal{V} = (v_1, \ldots, v_m)$ and $\mathcal{V}' = (v'_1, \ldots, v'_m)$ be two bases of F. Let $P = P_{\mathcal{U}',\mathcal{U}}$ be the change of basis matrix from $\mathcal{U}$ to $\mathcal{U}'$, and let $Q = P_{\mathcal{V}',\mathcal{V}}$ be the change of basis matrix from $\mathcal{V}$ to $\mathcal{V}'$. For any linear map $f\colon E \to F$, let $M(f) = M_{\mathcal{U},\mathcal{V}}(f)$ be the matrix associated to f w.r.t. the bases $\mathcal{U}$ and $\mathcal{V}$, and let $M'(f) = M_{\mathcal{U}',\mathcal{V}'}(f)$ be the matrix associated to f w.r.t. the bases $\mathcal{U}'$ and $\mathcal{V}'$. We have*

$$M'(f) = Q^{-1} M(f) P,$$

or more explicitly

$$M_{\mathcal{U}',\mathcal{V}'}(f) = P_{\mathcal{V}',\mathcal{V}}^{-1} M_{\mathcal{U},\mathcal{V}}(f) P_{\mathcal{U}',\mathcal{U}} = P_{\mathcal{V},\mathcal{V}'} M_{\mathcal{U},\mathcal{V}}(f) P_{\mathcal{U}',\mathcal{U}}.$$

Proof. Since $f\colon E \to F$ can be written as $f = \mathrm{id}_F \circ f \circ \mathrm{id}_E$, since P is the matrix of id_E w.r.t. the bases $(u'_1, \ldots, u'_n)$ and $(u_1, \ldots, u_n)$, and Q^{-1} is the matrix of id_F w.r.t. the bases $(v_1, \ldots, v_m)$ and $(v'_1, \ldots, v'_m)$, by Proposition 3.2, we have $M'(f) = Q^{-1} M(f) P$. $\qquad\square$

As a corollary, we get the following result.

Corollary 3.1. *Let E be a vector space, and let $\mathcal{U} = (u_1, \ldots, u_n)$ and $\mathcal{U}' = (u'_1, \ldots, u'_n)$ be two bases of E. Let $P = P_{\mathcal{U}',\mathcal{U}}$ be the change of basis matrix from $\mathcal{U}$ to $\mathcal{U}'$. For any linear map $f\colon E \to E$, let $M(f) = M_{\mathcal{U}}(f)$ be the matrix associated to f w.r.t. the basis $\mathcal{U}$, and let $M'(f) = M_{\mathcal{U}'}(f)$ be the matrix associated to f w.r.t. the basis $\mathcal{U}'$. We have*

$$M'(f) = P^{-1} M(f) P,$$

or more explicitly,

$$M_{\mathcal{U}'}(f) = P_{\mathcal{U}',\mathcal{U}}^{-1} M_{\mathcal{U}}(f) P_{\mathcal{U}',\mathcal{U}} = P_{\mathcal{U},\mathcal{U}'} M_{\mathcal{U}}(f) P_{\mathcal{U}',\mathcal{U}}.$$

Example 3.3. Let $E = \mathbb{R}^2$, $\mathcal{U} = (e_1, e_2)$ where $e_1 = (1,0)$ and $e_2 = (0,1)$ are the canonical basis vectors, let $\mathcal{V} = (v_1, v_2) = (e_1, e_1 - e_2)$, and let

$$A = \begin{pmatrix} 2 & 1 \\ 0 & 1 \end{pmatrix}.$$

The change of basis matrix $P = P_{\mathcal{V},\mathcal{U}}$ from $\mathcal{U}$ to $\mathcal{V}$ is

$$P = \begin{pmatrix} 1 & 1 \\ 0 & -1 \end{pmatrix},$$

and we check that

$$P^{-1} = P.$$

Therefore, in the basis $\mathcal{V}$, the matrix representing the linear map f defined by A is

$$A' = P^{-1}AP = PAP = \begin{pmatrix} 1 & 1 \\ 0 & -1 \end{pmatrix} \begin{pmatrix} 2 & 1 \\ 0 & 1 \end{pmatrix} \begin{pmatrix} 1 & 1 \\ 0 & -1 \end{pmatrix} = \begin{pmatrix} 2 & 0 \\ 0 & 1 \end{pmatrix} = D,$$

a diagonal matrix. In the basis $\mathcal{V}$, it is clear what the action of f is: it is a stretch by a factor of 2 in the v_1 direction and it is the identity in the v_2 direction. Observe that v_1 and v_2 are not orthogonal.

What happened is that we *diagonalized* the matrix A. The diagonal entries 2 and 1 are the *eigenvalues* of A (and f), and v_1 and v_2 are corresponding *eigenvectors*. We will come back to eigenvalues and eigenvectors later on.

The above example showed that the same linear map can be represented by different matrices. This suggests making the following definition:

Definition 3.5. Two $n \times n$ matrices A and B are said to be *similar* iff there is some invertible matrix P such that

$$B = P^{-1}AP.$$

It is easily checked that similarity is an equivalence relation. From our previous considerations, *two $n \times n$ matrices A and B are similar iff they represent the same linear map with respect to two different bases.* The following surprising fact can be shown: Every square matrix A is similar to its transpose $A^\top$. The proof requires advanced concepts (the Jordan form or similarity invariants).

If $\mathcal{U} = (u_1, \ldots, u_n)$ and $\mathcal{V} = (v_1, \ldots, v_n)$ are two bases of E, the change of basis matrix

$$P = P_{\mathcal{V},\mathcal{U}} = \begin{pmatrix} a_{11} & a_{12} & \cdots & a_{1n} \\ a_{21} & a_{22} & \cdots & a_{2n} \\ \vdots & \vdots & \ddots & \vdots \\ a_{n1} & a_{n2} & \cdots & a_{nn} \end{pmatrix}$$

from $(u_1, \ldots, u_n)$ to $(v_1, \ldots, v_n)$ is the matrix whose jth column consists of the coordinates of v_j over the basis $(u_1, \ldots, u_n)$, which means that

$$v_j = \sum_{i=1}^{n} a_{ij} u_i.$$

It is natural to extend the matrix notation and to express the vector $\begin{pmatrix} v_1 \\ \vdots \\ v_n \end{pmatrix}$

in E^n as the product of a matrix times the vector $\begin{pmatrix} u_1 \\ \vdots \\ u_n \end{pmatrix}$ in E^n, namely as

$$
\begin{pmatrix} v_1 \\ v_2 \\ \vdots \\ v_n \end{pmatrix} = \begin{pmatrix} a_{11} & a_{21} & \cdots & a_{n1} \\ a_{12} & a_{22} & \cdots & a_{n2} \\ \vdots & \vdots & \ddots & \vdots \\ a_{1n} & a_{2n} & \cdots & a_{nn} \end{pmatrix} \begin{pmatrix} u_1 \\ u_2 \\ \vdots \\ u_n \end{pmatrix},
$$

but notice that the matrix involved is not P, but its *transpose* $P^\top$.

This observation has the following consequence: if $\mathcal{U} = (u_1, \ldots, u_n)$ and $\mathcal{V} = (v_1, \ldots, v_n)$ are two bases of E and if

$$
\begin{pmatrix} v_1 \\ \vdots \\ v_n \end{pmatrix} = A \begin{pmatrix} u_1 \\ \vdots \\ u_n \end{pmatrix},
$$

that is,

$$
v_i = \sum_{j=1}^{n} a_{ij} u_j,
$$

for any vector $w \in E$, if

$$
w = \sum_{i=1}^{n} x_i u_i = \sum_{k=1}^{n} y_k v_k,
$$

then

$$
\begin{pmatrix} x_1 \\ \vdots \\ x_n \end{pmatrix} = A^\top \begin{pmatrix} y_1 \\ \vdots \\ y_n \end{pmatrix},
$$

and so

$$
\begin{pmatrix} y_1 \\ \vdots \\ y_n \end{pmatrix} = (A^\top)^{-1} \begin{pmatrix} x_1 \\ \vdots \\ x_n \end{pmatrix}.
$$

It is easy to see that $(A^\top)^{-1} = (A^{-1})^\top$. Also, if $\mathcal{U} = (u_1, \ldots, u_n)$, $\mathcal{V} = (v_1, \ldots, v_n)$, and $\mathcal{W} = (w_1, \ldots, w_n)$ are three bases of E, and if the change of basis matrix from $\mathcal{U}$ to $\mathcal{V}$ is $P = P_{\mathcal{V},\mathcal{U}}$ and the change of basis matrix from $\mathcal{V}$ to $\mathcal{W}$ is $Q = P_{\mathcal{W},\mathcal{V}}$, then

$$\begin{pmatrix} v_1 \\ \vdots \\ v_n \end{pmatrix} = P^\top \begin{pmatrix} u_1 \\ \vdots \\ u_n \end{pmatrix}, \quad \begin{pmatrix} w_1 \\ \vdots \\ w_n \end{pmatrix} = Q^\top \begin{pmatrix} v_1 \\ \vdots \\ v_n \end{pmatrix},$$

so

$$\begin{pmatrix} w_1 \\ \vdots \\ w_n \end{pmatrix} = Q^\top P^\top \begin{pmatrix} u_1 \\ \vdots \\ u_n \end{pmatrix} = (PQ)^\top \begin{pmatrix} u_1 \\ \vdots \\ u_n \end{pmatrix},$$

which means that the change of basis matrix $P_{\mathcal{W},\mathcal{U}}$ from $\mathcal{U}$ to $\mathcal{W}$ is PQ. This proves that

$$P_{\mathcal{W},\mathcal{U}} = P_{\mathcal{V},\mathcal{U}} P_{\mathcal{W},\mathcal{V}}.$$

Even though matrices are indispensable since they are *the* major tool in applications of linear algebra, one should not lose track of the fact that

linear maps are more fundamental because they are intrinsic objects that do not depend on the choice of bases. Consequently, we advise the reader to try to think in terms of linear maps rather than reduce everything to matrices.

In our experience, this is particularly effective when it comes to proving results about linear maps and matrices, where proofs involving linear maps are often more "conceptual." These proofs are usually more general because they do not depend on the fact that the dimension is finite. Also, instead of thinking of a matrix decomposition as a purely algebraic operation, it is often illuminating to view it as a *geometric decomposition*. This is the case of the SVD, which in geometric terms says that every linear map can be factored as a rotation, followed by a rescaling along orthogonal axes and then another rotation.

After all,

a matrix is a representation of a linear map,

and most decompositions of a matrix reflect the fact that with a *suitable choice of a basis (or bases)*, the linear map is a represented by a matrix having a special shape. The problem is then to find such bases.

Still, for the beginner, matrices have a certain irresistible appeal, and we confess that it takes a certain amount of practice to reach the point where it becomes more natural to deal with linear maps. We still recommend it! For example, try to translate a result stated in terms of matrices into a result stated in terms of linear maps. Whenever we tried this exercise, we learned something.

Also, always try to keep in mind that

linear maps are geometric in nature; they act on space.

3.5 Summary

The main concepts and results of this chapter are listed below:

- The representation of linear maps by *matrices*.
- The *matrix representation mapping* $M : \text{Hom}(E, F) \to \text{M}_{n,p}$ and the representation isomorphism (Proposition 3.2).
- *Change of basis matrix* and Proposition 3.4.

3.6 Problems

Problem 3.1. Prove that the column vectors of the matrix A_1 given by

$$A_1 = \begin{pmatrix} 1 & 2 & 3 \\ 2 & 3 & 7 \\ 1 & 3 & 1 \end{pmatrix}$$

are linearly independent.

Prove that the coordinates of the column vectors of the matrix B_1 over the basis consisting of the column vectors of A_1 given by

$$B_1 = \begin{pmatrix} 3 & 5 & 1 \\ 1 & 2 & 1 \\ 4 & 3 & -6 \end{pmatrix}$$

are the columns of the matrix P_1 given by

$$P_1 = \begin{pmatrix} -27 & -61 & -41 \\ 9 & 18 & 9 \\ 4 & 10 & 8 \end{pmatrix}.$$

Give a nontrivial linear dependence of the columns of P_1. Check that $B_1 = A_1 P_1$. Is the matrix B_1 invertible?

Problem 3.2. Prove that the column vectors of the matrix A_2 given by

$$A_2 = \begin{pmatrix} 1\ 1\ 1\ 1 \\ 1\ 2\ 1\ 3 \\ 1\ 1\ 2\ 2 \\ 1\ 1\ 1\ 3 \end{pmatrix}$$

are linearly independent.

Prove that the column vectors of the matrix B_2 given by

$$B_2 = \begin{pmatrix} 1\ -2\ 2\ -2 \\ 0\ -3\ 2\ -3 \\ 3\ -5\ 5\ -4 \\ 3\ -4\ 4\ -4 \end{pmatrix}$$

are linearly independent.

Prove that the coordinates of the column vectors of the matrix B_2 over the basis consisting of the column vectors of A_2 are the columns of the matrix P_2 given by

$$P_2 = \begin{pmatrix} 2\ \ 0\ \ 1\ -1 \\ -3\ \ 1\ -2\ \ 1 \\ 1\ -2\ \ 2\ -1 \\ 1\ -1\ \ 1\ -1 \end{pmatrix}.$$

Check that $A_2 P_2 = B_2$. Prove that

$$P_2^{-1} = \begin{pmatrix} -1\ -1\ -1\ \ 1 \\ 2\ \ 1\ \ 1\ -2 \\ 2\ \ 1\ \ 2\ -3 \\ -1\ -1\ \ 0\ -1 \end{pmatrix}.$$

What are the coordinates over the basis consisting of the column vectors of B_2 of the vector whose coordinates over the basis consisting of the column vectors of A_1 are $(2, -3, 0, 0)$?

Problem 3.3. Consider the polynomials

$$B_0^2(t) = (1-t)^2 \quad B_1^2(t) = 2(1-t)t \quad B_2^2(t) = t^2$$
$$B_0^3(t) = (1-t)^3 \quad B_1^3(t) = 3(1-t)^2 t \quad B_2^3(t) = 3(1-t)t^2 \quad B_3^3(t) = t^3,$$

known as the *Bernstein polynomials* of degree 2 and 3.

(1) Show that the Bernstein polynomials $B_0^2(t), B_1^2(t), B_2^2(t)$ are expressed as linear combinations of the basis $(1, t, t^2)$ of the vector space of polynomials of degree at most 2 as follows:

$$\begin{pmatrix} B_0^2(t) \\ B_1^2(t) \\ B_2^2(t) \end{pmatrix} = \begin{pmatrix} 1 & -2 & 1 \\ 0 & 2 & -2 \\ 0 & 0 & 1 \end{pmatrix} \begin{pmatrix} 1 \\ t \\ t^2 \end{pmatrix}.$$

Prove that

$$B_0^2(t) + B_1^2(t) + B_2^2(t) = 1.$$

(2) Show that the Bernstein polynomials $B_0^3(t), B_1^3(t), B_2^3(t), B_3^3(t)$ are expressed as linear combinations of the basis $(1, t, t^2, t^3)$ of the vector space of polynomials of degree at most 3 as follows:

$$\begin{pmatrix} B_0^3(t) \\ B_1^3(t) \\ B_2^3(t) \\ B_3^3(t) \end{pmatrix} = \begin{pmatrix} 1 & -3 & 3 & -1 \\ 0 & 3 & -6 & 3 \\ 0 & 0 & 3 & -3 \\ 0 & 0 & 0 & 1 \end{pmatrix} \begin{pmatrix} 1 \\ t \\ t^2 \\ t^3 \end{pmatrix}.$$

Prove that

$$B_0^3(t) + B_1^3(t) + B_2^3(t) + B_3^3(t) = 1.$$

(3) Prove that the Bernstein polynomials of degree 2 are linearly independent, and that the Bernstein polynomials of degree 3 are linearly independent.

Problem 3.4. Recall that the *binomial coefficient* $\binom{m}{k}$ is given by

$$\binom{m}{k} = \frac{m!}{k!(m-k)!},$$

with $0 \leq k \leq m$.

For any $m \geq 1$, we have the $m + 1$ *Bernstein polynomials* of degree m given by

$$B_k^m(t) = \binom{m}{k}(1-t)^{m-k}t^k, \quad 0 \leq k \leq m.$$

(1) Prove that

$$B_k^m(t) = \sum_{j=k}^{m} (-1)^{j-k} \binom{m}{j}\binom{j}{k} t^j. \tag{3.5}$$

Use the above to prove that $B_0^m(t), \ldots, B_m^m(t)$ are linearly independent.

(2) Prove that

$$B_0^m(t) + \cdots + B_m^m(t) = 1.$$

(3) What can you say about the symmetries of the $(m+1) \times (m+1)$ matrix expressing $B_0^m, \ldots, B_m^m$ in terms of the basis $1, t, \ldots, t^m$?

Prove your claim (beware that in equation (3.5) the coefficient of t^j in B_k^m is the entry on the $(k+1)$th row of the $(j+1)$th column, since $0 \le k, j \le m$. Make appropriate modifications to the indices).

What can you say about the sum of the entries on each row of the above matrix? What about the sum of the entries on each column?

(4) The purpose of this question is to express the t^i in terms of the Bernstein polynomials $B_0^m(t), \ldots, B_m^m(t)$, with $0 \le i \le m$.

First, prove that

$$t^i = \sum_{j=0}^{m-i} t^i B_j^{m-i}(t), \quad 0 \le i \le m.$$

Then prove that

$$\binom{m}{i}\binom{m-i}{j} = \binom{m}{i+j}\binom{i+j}{i}.$$

Use the above facts to prove that

$$t^i = \sum_{j=0}^{m-i} \frac{\binom{i+j}{i}}{\binom{m}{i}} B_{i+j}^m(t).$$

Conclude that the Bernstein polynomials $B_0^m(t), \ldots, B_m^m(t)$ form a basis of the vector space of polynomials of degree $\le m$.

Compute the matrix expressing $1, t, t^2$ in terms of $B_0^2(t), B_1^2(t), B_2^2(t)$, and the matrix expressing $1, t, t^2, t^3$ in terms of $B_0^3(t), B_1^3(t), B_2^3(t), B_3^3(t)$.

You should find

$$\begin{pmatrix} 1 & 1 & 1 \\ 0 & 1/2 & 1 \\ 0 & 0 & 1 \end{pmatrix}$$

and

$$\begin{pmatrix} 1 & 1 & 1 & 1 \\ 0 & 1/3 & 2/3 & 1 \\ 0 & 0 & 1/3 & 1 \\ 0 & 0 & 0 & 1 \end{pmatrix}.$$

(5) A *polynomial curve* $C(t)$ *of degree* m in the plane is the set of points
$C(t) = \begin{pmatrix} x(t) \\ y(t) \end{pmatrix}$ given by two polynomials of degree $\leq m$,

$$x(t) = \alpha_0 t^{m_1} + \alpha_1 t^{m_1-1} + \cdots + \alpha_{m_1}$$
$$y(t) = \beta_0 t^{m_2} + \beta_1 t^{m_2-1} + \cdots + \beta_{m_2},$$

with $1 \leq m_1, m_2 \leq m$ and $\alpha_0, \beta_0 \neq 0$.

Prove that there exist $m + 1$ points $b_0, \ldots, b_m \in \mathbb{R}^2$ so that

$$C(t) = \begin{pmatrix} x(t) \\ y(t) \end{pmatrix} = B_0^m(t) b_0 + B_1^m(t) b_1 + \cdots + B_m^m(t) b_m$$

for all $t \in \mathbb{R}$, with $C(0) = b_0$ and $C(1) = b_m$. Are the points $b_1, \ldots, b_{m-1}$ generally on the curve?

We say that the curve C is a *Bézier curve* and $(b_0, \ldots, b_m)$ is the list of *control points* of the curve (control points need not be distinct).

Remark: Because $B_0^m(t) + \cdots + B_m^m(t) = 1$ and $B_i^m(t) \geq 0$ when $t \in [0, 1]$, the curve segment $C[0, 1]$ corresponding to $t \in [0, 1]$ belongs to the convex hull of the control points. This is an important property of Bézier curves which is used in geometric modeling to find the intersection of curve segments. Bézier curves play an important role in computer graphics and geometric modeling, but also in robotics because they can be used to model the trajectories of moving objects.

Problem 3.5. Consider the $n \times n$ matrix

$$A = \begin{pmatrix} 0 & 0 & 0 & \cdots & 0 & -a_n \\ 1 & 0 & 0 & \cdots & 0 & -a_{n-1} \\ 0 & 1 & 0 & \cdots & 0 & -a_{n-2} \\ \vdots & \ddots & \ddots & \ddots & \vdots & \vdots \\ 0 & 0 & 0 & \ddots & 0 & -a_2 \\ 0 & 0 & 0 & \cdots & 1 & -a_1 \end{pmatrix},$$

with $a_n \neq 0$.

(1) Find a matrix P such that
$$A^\top = P^{-1} A P.$$

What happens when $a_n = 0$?

Hint. First, try $n = 3, 4, 5$. Such a matrix must have zeros above the "antidiagonal," and identical entries p_{ij} for all $i, j \geq 0$ such that $i + j = n + k$, where $k = 1, \ldots, n$.

(2) Prove that if $a_n = 1$ and if $a_1, \ldots, a_{n-1}$ are integers, then P can be chosen so that the entries in P^{-1} are also integers.

Problem 3.6. For any matrix $A \in M_n(\mathbb{C})$, let R_A and L_A be the maps from $M_n(\mathbb{C})$ to itself defined so that

$$L_A(B) = AB, \quad R_A(B) = BA, \quad \text{for all } B \in M_n(\mathbb{C}).$$

(1) Check that L_A and R_A are linear, and that L_A and R_B commute for all A, B.

Let $\mathrm{ad}_A : M_n(\mathbb{C}) \to M_n(\mathbb{C})$ be the linear map given by

$$\mathrm{ad}_A(B) = L_A(B) - R_A(B) = AB - BA = [A, B], \quad \text{for all } B \in M_n(\mathbb{C}).$$

Note that $[A, B]$ is the Lie bracket.

(2) Prove that if A is invertible, then L_A and R_A are invertible; in fact, $(L_A)^{-1} = L_{A^{-1}}$ and $(R_A)^{-1} = R_{A^{-1}}$. Prove that if $A = PBP^{-1}$ for some invertible matrix P, then

$$L_A = L_P \circ L_B \circ L_P^{-1}, \quad R_A = R_P^{-1} \circ R_B \circ R_P.$$

(3) Recall that the n^2 matrices E_{ij} defined such that all entries in E_{ij} are zero except the (i, j)th entry, which is equal to 1, form a basis of the vector space $M_n(\mathbb{C})$. Consider the partial ordering of the E_{ij} defined such that for $i = 1, \ldots, n$, if $n \geq j > k \geq 1$, then then E_{ij} precedes E_{ik}, and for $j = 1, \ldots, n$, if $1 \leq i < h \leq n$, then E_{ij} precedes E_{hj}.

Draw the Hasse diagram of the partial order defined above when $n = 3$.

There are total orderings extending this partial ordering. How would you find them algorithmically? Check that the following is such a total order:

$$(1, 3), \ (1, 2), \ (1, 1), \ (2, 3), \ (2, 2), \ (2, 1), \ (3, 3), \ (3, 2), \ (3, 1).$$

(4) Let the total order of the basis (E_{ij}) extending the partial ordering defined in (2) be given by

$$(i, j) < (h, k) \quad \text{iff} \quad \begin{cases} i = h \text{ and } j > k \\ \text{or } i < h. \end{cases}$$

Let R be the $n \times n$ permutation matrix given by

$$R = \begin{pmatrix} 0 & 0 & \ldots & 0 & 1 \\ 0 & 0 & \ldots & 1 & 0 \\ \vdots & \vdots & \ddots & \vdots & \vdots \\ 0 & 1 & \ldots & 0 & 0 \\ 1 & 0 & \ldots & 0 & 0 \end{pmatrix}.$$

Observe that $R^{-1} = R$. Prove that for any $n \geq 1$, the matrix of L_A is given by $A \otimes I_n$, and the matrix of R_A is given by $I_n \otimes RA^\top R$ (over the basis

(E_{ij}) ordered as specified above), where $\otimes$ is the *Kronecker product* (also called *tensor product*) of matrices defined in Definition 4.4.

Hint. Figure out what are $R_B(E_{ij}) = E_{ij}B$ and $L_B(E_{ij}) = BE_{ij}$.

(5) Prove that if A is upper triangular, then the matrices representing L_A and R_A are also upper triangular.

Note that if instead of the ordering

$$E_{1n}, E_{1n-1}, \ldots, E_{11}, E_{2n}, \ldots, E_{21}, \ldots, E_{nn}, \ldots, E_{n1},$$

that I proposed you use the standard lexicographic ordering

$$E_{11}, E_{12}, \ldots, E_{1n}, E_{21}, \ldots, E_{2n}, \ldots, E_{n1}, \ldots, E_{nn},$$

then the matrix representing L_A is still $A \otimes I_n$, but the matrix representing R_A is $I_n \otimes A^\top$. In this case, if A is upper-triangular, then the matrix of R_A is *lower triangular*. This is the motivation for using the first basis (avoid upper becoming lower).

Chapter 4

Haar Bases, Haar Wavelets, Hadamard Matrices

In this chapter, we discuss two types of matrices that have applications in computer science and engineering:

(1) Haar matrices and the corresponding Haar wavelets, a fundamental tool in signal processing and computer graphics.

(2) Hadamard matrices which have applications in error correcting codes, signal processing, and low rank approximation.

4.1 Introduction to Signal Compression Using Haar Wavelets

We begin by considering *Haar wavelets* in $\mathbb{R}^4$. Wavelets play an important role in audio and video signal processing, especially for *compressing* long signals into much smaller ones that still retain enough information so that when they are played, we can't see or hear any difference.

Consider the four vectors w_1, w_2, w_3, w_4 given by

$$
w_1 = \begin{pmatrix} 1 \\ 1 \\ 1 \\ 1 \end{pmatrix} \qquad
w_2 = \begin{pmatrix} 1 \\ 1 \\ -1 \\ -1 \end{pmatrix} \qquad
w_3 = \begin{pmatrix} 1 \\ -1 \\ 0 \\ 0 \end{pmatrix} \qquad
w_4 = \begin{pmatrix} 0 \\ 0 \\ 1 \\ -1 \end{pmatrix}.
$$

Note that these vectors are pairwise orthogonal, so they are indeed linearly independent (we will see this in a later chapter). Let $\mathcal{W} = \{w_1, w_2, w_3, w_4\}$ be the *Haar basis*, and let $\mathcal{U} = \{e_1, e_2, e_3, e_4\}$ be the canonical basis of $\mathbb{R}^4$. The change of basis matrix $W = P_{\mathcal{W},\mathcal{U}}$ from $\mathcal{U}$ to $\mathcal{W}$ is given by

$$
W = \begin{pmatrix} 1 & 1 & 1 & 0 \\ 1 & 1 & -1 & 0 \\ 1 & -1 & 0 & 1 \\ 1 & -1 & 0 & -1 \end{pmatrix},
$$

and we easily find that the inverse of W is given by

$$W^{-1} = \begin{pmatrix} 1/4 & 0 & 0 & 0 \\ 0 & 1/4 & 0 & 0 \\ 0 & 0 & 1/2 & 0 \\ 0 & 0 & 0 & 1/2 \end{pmatrix} \begin{pmatrix} 1 & 1 & 1 & 1 \\ 1 & 1 & -1 & -1 \\ 1 & -1 & 0 & 0 \\ 0 & 0 & 1 & -1 \end{pmatrix}.$$

So the vector $v = (6, 4, 5, 1)$ over the basis $\mathcal{U}$ becomes $c = (c_1, c_2, c_3, c_4)$ over the Haar basis $\mathcal{W}$, with

$$\begin{pmatrix} c_1 \\ c_2 \\ c_3 \\ c_4 \end{pmatrix} = \begin{pmatrix} 1/4 & 0 & 0 & 0 \\ 0 & 1/4 & 0 & 0 \\ 0 & 0 & 1/2 & 0 \\ 0 & 0 & 0 & 1/2 \end{pmatrix} \begin{pmatrix} 1 & 1 & 1 & 1 \\ 1 & 1 & -1 & -1 \\ 1 & -1 & 0 & 0 \\ 0 & 0 & 1 & -1 \end{pmatrix} \begin{pmatrix} 6 \\ 4 \\ 5 \\ 1 \end{pmatrix} = \begin{pmatrix} 4 \\ 1 \\ 1 \\ 2 \end{pmatrix}.$$

Given a signal $v = (v_1, v_2, v_3, v_4)$, we first *transform* v into its coefficients $c = (c_1, c_2, c_3, c_4)$ over the Haar basis by computing $c = W^{-1}v$. Observe that

$$c_1 = \frac{v_1 + v_2 + v_3 + v_4}{4}$$

is the overall *average* value of the signal v. The coefficient c_1 corresponds to the background of the image (or of the sound). Then, c_2 gives the coarse details of v, whereas, c_3 gives the details in the first part of v, and c_4 gives the details in the second half of v.

Reconstruction of the signal consists in computing $v = Wc$. The trick for good *compression* is to throw away some of the coefficients of c (set them to zero), obtaining a *compressed signal* $\widehat{c}$, and still retain enough crucial information so that the reconstructed signal $\widehat{v} = W\widehat{c}$ looks almost as good as the original signal v. Thus, the steps are:

$$\text{input } v \longrightarrow \text{ coefficients } c$$
$$= W^{-1}v \longrightarrow \text{ compressed } \widehat{c} \longrightarrow \text{ compressed } \widehat{v} = W\widehat{c}.$$

This kind of compression scheme makes modern video conferencing possible.

It turns out that there is a faster way to find $c = W^{-1}v$, without actually using W^{-1}. This has to do with the multiscale nature of Haar wavelets.

Given the original signal $v = (6, 4, 5, 1)$ shown in Figure 4.1, we compute averages and half differences obtaining Figure 4.2. We get the coefficients $c_3 = 1$ and $c_4 = 2$. Then again we compute averages and half differences obtaining Figure 4.3. We get the coefficients $c_1 = 4$ and $c_2 = 1$. Note that

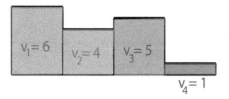

Fig. 4.1   The original signal v.

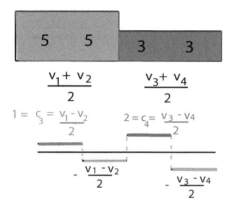

Fig. 4.2   First averages and first half differences.

the original signal v can be reconstructed from the two signals in Figure 4.2, and the signal on the left of Figure 4.2 can be reconstructed from the two signals in Figure 4.3. In particular, the data from Figure 4.2 gives us

$$5 + 1 = \frac{v_1 + v_2}{2} + \frac{v_1 - v_2}{2} = v_1$$

$$5 - 1 = \frac{v_1 + v_2}{2} - \frac{v_1 - v_2}{2} = v_2$$

$$3 + 2 = \frac{v_3 + v_4}{2} + \frac{v_3 - v_4}{2} = v_3$$

$$3 - 2 = \frac{v_3 + v_4}{2} - \frac{v_3 - v_4}{2} = v_4.$$

4.2   Haar Bases and Haar Matrices, Scaling Properties of Haar Wavelets

The method discussed in Section 4.2 can be generalized to signals of any length 2^n. The previous case corresponds to $n = 2$. Let us consider the

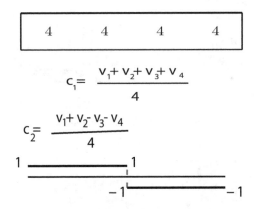

Fig. 4.3   Second averages and second half differences.

case $n = 3$. The *Haar basis* $(w_1, w_2, w_3, w_4, w_5, w_6, w_7, w_8)$ is given by the matrix

$$W = \begin{pmatrix} 1 & 1 & 1 & 0 & 1 & 0 & 0 & 0 \\ 1 & 1 & 1 & 0 & -1 & 0 & 0 & 0 \\ 1 & 1 & -1 & 0 & 0 & 1 & 0 & 0 \\ 1 & 1 & -1 & 0 & 0 & -1 & 0 & 0 \\ 1 & -1 & 0 & 1 & 0 & 0 & 1 & 0 \\ 1 & -1 & 0 & 1 & 0 & 0 & -1 & 0 \\ 1 & -1 & 0 & -1 & 0 & 0 & 0 & 1 \\ 1 & -1 & 0 & -1 & 0 & 0 & 0 & -1 \end{pmatrix}.$$

The columns of this matrix are orthogonal, and it is easy to see that

$$W^{-1} = \mathrm{diag}(1/8, 1/8, 1/4, 1/4, 1/2, 1/2, 1/2, 1/2)W^\top.$$

A pattern is beginning to emerge. It looks like the second Haar basis vector w_2 is the "mother" of all the other basis vectors, except the first, whose purpose is to perform averaging. Indeed, in general, given

$$w_2 = \underbrace{(1, \ldots, 1, -1, \ldots, -1)}_{2^n},$$

the other Haar basis vectors are obtained by a "scaling and shifting process." Starting from w_2, the scaling process generates the vectors

$$w_3, w_5, w_9, \ldots, w_{2^j+1}, \ldots, w_{2^{n-1}+1},$$

such that $w_{2^{j+1}+1}$ is obtained from w_{2^j+1} by forming two consecutive blocks of 1 and -1 of half the size of the blocks in w_{2^j+1}, and setting all other entries to zero. Observe that w_{2^j+1} has 2^j blocks of 2^{n-j} elements. The shifting process consists in shifting the blocks of 1 and -1 in w_{2^j+1} to the right by inserting a block of $(k-1)2^{n-j}$ zeros from the left, with $0 \leq j \leq n-1$ and $1 \leq k \leq 2^j$. Note that our convention is to use j as the scaling index and k as the shifting index. Thus, we obtain the following formula for w_{2^j+k}:

$$
w_{2^j+k}(i) = \begin{cases} 0 & 1 \leq i \leq (k-1)2^{n-j} \\ 1 & (k-1)2^{n-j}+1 \leq i \leq (k-1)2^{n-j}+2^{n-j-1} \\ -1 & (k-1)2^{n-j}+2^{n-j-1}+1 \leq i \leq k2^{n-j} \\ 0 & k2^{n-j}+1 \leq i \leq 2^n, \end{cases}
$$

with $0 \leq j \leq n-1$ and $1 \leq k \leq 2^j$. Of course

$$
w_1 = \underbrace{(1,\ldots,1)}_{2^n}.
$$

The above formulae look a little better if we change our indexing slightly by letting k vary from 0 to $2^j - 1$, and using the index j instead of 2^j.

Definition 4.1. The vectors of the *Haar basis* of dimension 2^n are denoted by

$$
w_1, h_0^0, h_0^1, h_1^1, h_0^2, h_1^2, h_2^2, h_3^2, \ldots, h_k^j, \ldots, h_{2^{n-1}-1}^{n-1},
$$

where

$$
h_k^j(i) = \begin{cases} 0 & 1 \leq i \leq k2^{n-j} \\ 1 & k2^{n-j}+1 \leq i \leq k2^{n-j}+2^{n-j-1} \\ -1 & k2^{n-j}+2^{n-j-1}+1 \leq i \leq (k+1)2^{n-j} \\ 0 & (k+1)2^{n-j}+1 \leq i \leq 2^n, \end{cases}
$$

with $0 \leq j \leq n-1$ and $0 \leq k \leq 2^j - 1$. The $2^n \times 2^n$ matrix whose columns are the vectors

$$
w_1, h_0^0, h_0^1, h_1^1, h_0^2, h_1^2, h_2^2, h_3^2, \ldots, h_k^j, \ldots, h_{2^{n-1}-1}^{n-1},
$$

(in that order), is called the *Haar matrix* of dimension 2^n, and is denoted by W_n.

It turns out that there is a way to understand these formulae better if we interpret a vector $u = (u_1, \ldots, u_m)$ as a piecewise linear function over the interval $[0, 1)$.

Definition 4.2. Given a vector $u = (u_1, \ldots, u_m)$, the *piecewise linear function* $\mathrm{plf}(u)$ is defined such that

$$
\mathrm{plf}(u)(x) = u_i, \qquad \frac{i-1}{m} \leq x < \frac{i}{m}, \ 1 \leq i \leq m.
$$

In words, the function plf(u) has the value u_1 on the interval $[0, 1/m)$, the value u_2 on $[1/m, 2/m)$, *etc.*, and the value u_m on the interval $[(m-1)/m, 1)$.

For example, the piecewise linear function associated with the vector
$$u = (2.4, 2.2, 2.15, 2.05, 6.8, 2.8, -1.1, -1.3)$$
is shown in Figure 4.4.

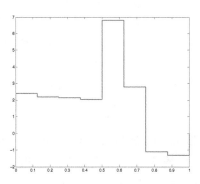

Fig. 4.4 The piecewise linear function plf(u).

Then each basis vector h_k^j corresponds to the function
$$\psi_k^j = \text{plf}(h_k^j).$$
In particular, for all n, the Haar basis vectors
$$h_0^0 = w_2 = \underbrace{(1, \ldots, 1, -1, \ldots, -1)}_{2^n}$$
yield the same piecewise linear function ψ given by
$$\psi(x) = \begin{cases} 1 & \text{if } 0 \le x < 1/2 \\ -1 & \text{if } 1/2 \le x < 1 \\ 0 & \text{otherwise,} \end{cases}$$
whose graph is shown in Figure 4.5. It is easy to see that ψ_k^j is given by the simple expression
$$\psi_k^j(x) = \psi(2^j x - k), \quad 0 \le j \le n - 1, \, 0 \le k \le 2^j - 1.$$
The above formula makes it clear that ψ_k^j is obtained from ψ by scaling and shifting.

Definition 4.3. The function $\phi_0^0 = \text{plf}(w_1)$ is the piecewise linear function with the constant value 1 on $[0, 1)$, and the functions $\psi_k^j = \text{plf}(h_k^j)$ together with ϕ_0^0 are known as the *Haar wavelets*.

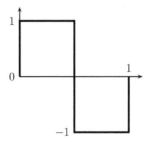

Fig. 4.5   The Haar wavelet ψ.

Rather than using W^{-1} to convert a vector u to a vector c of coefficients over the Haar basis, and the matrix W to reconstruct the vector u from its Haar coefficients c, we can use faster algorithms that use averaging and differencing.

If c is a vector of Haar coefficients of dimension 2^n, we compute the sequence of vectors $u^0, u^1, \ldots, u^n$ as follows:

$$u^0 = c$$
$$u^{j+1} = u^j$$
$$u^{j+1}(2i - 1) = u^j(i) + u^j(2^j + i)$$
$$u^{j+1}(2i) = u^j(i) - u^j(2^j + i),$$

for $j = 0, \ldots, n - 1$ and $i = 1, \ldots, 2^j$. The reconstructed vector (signal) is $u = u^n$.

If u is a vector of dimension 2^n, we compute the sequence of vectors $c^n, c^{n-1}, \ldots, c^0$ as follows:

$$c^n = u$$
$$c^j = c^{j+1}$$
$$c^j(i) = (c^{j+1}(2i - 1) + c^{j+1}(2i))/2$$
$$c^j(2^j + i) = (c^{j+1}(2i - 1) - c^{j+1}(2i))/2,$$

for $j = n - 1, \ldots, 0$ and $i = 1, \ldots, 2^j$. The vector over the Haar basis is $c = c^0$.

We leave it as an exercise to implement the above programs in `Matlab` using two variables u and c, and by building iteratively 2^j. Here is an example of the conversion of a vector to its Haar coefficients for $n = 3$.

Given the sequence $u = (31, 29, 23, 17, -6, -8, -2, -4)$, we get the sequence

$$c^3 = (31, 29, 23, 17, -6, -8, 2, -4)$$

$$c^2 = \left(\frac{31+29}{2}, \frac{23+17}{2}, \frac{-6-8}{2}, \frac{-2-4}{2}, \frac{31-29}{2}, \frac{23-17}{2}, \frac{-6-(-8)}{2}, \right.$$
$$\left. \frac{-2-(-4)}{2} \right)$$

$$= (30, 20, -7, -3, 1, 3, 1, 1)$$

$$c^1 = \left(\frac{30+20}{2}, \frac{-7-3}{2}, \frac{30-20}{2}, \frac{-7-(-3)}{2}, 1, 3, 1, 1 \right)$$

$$= (25, -5, 5, -2, 1, 3, 1, 1)$$

$$c^0 = \left(\frac{25-5}{2}, \frac{25-(-5)}{2}, 5, -2, 1, 3, 1, 1 \right)$$

$$= (10, 15, 5, -2, 1, 3, 1, 1)$$

so $c = (10, 15, 5, -2, 1, 3, 1, 1)$. Conversely, given $c = (10, 15, 5, -2, 1, 3, 1, 1)$, we get the sequence

$$u^0 = (10, 15, 5, -2, 1, 3, 1, 1)$$

$$u^1 = (10+15, 10-15, 5, -2, 1, 3, 1, 1) = (25, -5, 5, -2, 1, 3, 1, 1)$$

$$u^2 = (25+5, 25-5, -5+(-2), -5-(-2), 1, 3, 1, 1)$$

$$= (30, 20, -7, -3, 1, 3, 1, 1)$$

$$u^3 = (30+1, 30-1, 20+3, 20-3, -7+1, -7-1, -3+1, -3-1)$$

$$= (31, 29, 23, 17, -6, -8, -2, -4),$$

which gives back $u = (31, 29, 23, 17, -6, -8, -2, -4)$.

4.3   Kronecker Product Construction of Haar Matrices

There is another recursive method for constructing the Haar matrix W_n of dimension 2^n that makes it clearer why the columns of W_n are pairwise orthogonal, and why the above algorithms are indeed correct (which nobody seems to prove!). If we split W_n into two $2^n \times 2^{n-1}$ matrices, then the second matrix containing the last 2^{n-1} columns of W_n has a very simple structure: it consists of the vector

$$\underbrace{(1, -1, 0, \ldots, 0)}_{2^n}$$

and $2^{n-1} - 1$ shifted copies of it, as illustrated below for $n = 3$:

$$\begin{pmatrix} 1 & 0 & 0 & 0 \\ -1 & 0 & 0 & 0 \\ 0 & 1 & 0 & 0 \\ 0 & -1 & 0 & 0 \\ 0 & 0 & 1 & 0 \\ 0 & 0 & -1 & 0 \\ 0 & 0 & 0 & 1 \\ 0 & 0 & 0 & -1 \end{pmatrix}.$$

Observe that this matrix can be obtained from the identity matrix $I_{2^{n-1}}$, in our example

$$I_4 = \begin{pmatrix} 1 & 0 & 0 & 0 \\ 0 & 1 & 0 & 0 \\ 0 & 0 & 1 & 0 \\ 0 & 0 & 0 & 1 \end{pmatrix},$$

by forming the $2^n \times 2^{n-1}$ matrix obtained by replacing each 1 by the column vector

$$\begin{pmatrix} 1 \\ -1 \end{pmatrix}$$

and each zero by the column vector

$$\begin{pmatrix} 0 \\ 0 \end{pmatrix}.$$

Now the first half of W_n, that is the matrix consisting of the first 2^{n-1} columns of W_n, can be obtained from W_{n-1} by forming the $2^n \times 2^{n-1}$ matrix obtained by replacing each 1 by the column vector

$$\begin{pmatrix} 1 \\ 1 \end{pmatrix},$$

each -1 by the column vector

$$\begin{pmatrix} -1 \\ -1 \end{pmatrix},$$

and each zero by the column vector

$$\begin{pmatrix} 0 \\ 0 \end{pmatrix}.$$

For $n = 3$, the first half of W_3 is the matrix

$$\begin{pmatrix} 1 & 1 & 1 & 0 \\ 1 & 1 & 1 & 0 \\ 1 & 1 & -1 & 0 \\ 1 & 1 & -1 & 0 \\ 1 & -1 & 0 & 1 \\ 1 & -1 & 0 & 1 \\ 1 & -1 & 0 & -1 \\ 1 & -1 & 0 & -1 \end{pmatrix}$$

which is indeed obtained from

$$W_2 = \begin{pmatrix} 1 & 1 & 1 & 0 \\ 1 & 1 & -1 & 0 \\ 1 & -1 & 0 & 1 \\ 1 & -1 & 0 & -1 \end{pmatrix}$$

using the process that we just described.

These matrix manipulations can be described conveniently using a product operation on matrices known as the Kronecker product.

Definition 4.4. Given a $m \times n$ matrix $A = (a_{ij})$ and a $p \times q$ matrix $B = (b_{ij})$, the *Kronecker product* (or *tensor product*) $A \otimes B$ of A and B is the $mp \times nq$ matrix

$$A \otimes B = \begin{pmatrix} a_{11}B & a_{12}B & \cdots & a_{1n}B \\ a_{21}B & a_{22}B & \cdots & a_{2n}B \\ \vdots & \vdots & \ddots & \vdots \\ a_{m1}B & a_{m2}B & \cdots & a_{mn}B \end{pmatrix}.$$

It can be shown that $\otimes$ is associative and that

$$(A \otimes B)(C \otimes D) = AC \otimes BD$$
$$(A \otimes B)^{\top} = A^{\top} \otimes B^{\top},$$

whenever AC and BD are well defined. Then it is immediately verified that W_n is given by the following neat recursive equations:

$$W_n = \left(W_{n-1} \otimes \begin{pmatrix} 1 \\ 1 \end{pmatrix} I_{2^{n-1}} \otimes \begin{pmatrix} 1 \\ -1 \end{pmatrix} \right),$$

with $W_0 = (1)$. If we let

$$B_1 = 2 \begin{pmatrix} 1 & 0 \\ 0 & 1 \end{pmatrix} = \begin{pmatrix} 2 & 0 \\ 0 & 2 \end{pmatrix}$$

and for $n \geq 1$,

$$B_{n+1} = 2 \begin{pmatrix} B_n & 0 \\ 0 & I_{2^n} \end{pmatrix},$$

then it is not hard to use the Kronecker product formulation of W_n to obtain a rigorous proof of the equation

$$W_n^\top W_n = B_n, \quad \text{for all } n \geq 1.$$

The above equation offers a clean justification of the fact that the columns of W_n are pairwise orthogonal.

Observe that the right block (of size $2^n \times 2^{n-1}$) shows clearly how the detail coefficients in the second half of the vector c are added and subtracted to the entries in the first half of the partially reconstructed vector after $n-1$ steps.

4.4 Multiresolution Signal Analysis with Haar Bases

An important and attractive feature of the Haar basis is that it provides a *multiresolution analysis* of a signal. Indeed, given a signal u, if $c = (c_1, \ldots, c_{2^n})$ is the vector of its Haar coefficients, the coefficients with low index give coarse information about u, and the coefficients with high index represent fine information. For example, if u is an audio signal corresponding to a Mozart concerto played by an orchestra, c_1 corresponds to the "background noise," c_2 to the bass, c_3 to the first cello, c_4 to the second cello, c_5, c_6, c_7, c_7 to the violas, then the violins, *etc.* This multiresolution feature of wavelets can be exploited to compress a signal, that is, to use fewer coefficients to represent it. Here is an example.

Consider the signal

$$u = (2.4, 2.2, 2.15, 2.05, 6.8, 2.8, -1.1, -1.3),$$

whose Haar transform is

$$c = (2, 0.2, 0.1, 3, 0.1, 0.05, 2, 0.1).$$

The piecewise-linear curves corresponding to u and c are shown in Figure 4.6. Since some of the coefficients in c are small (smaller than or equal to 0.2) we can compress c by replacing them by 0. We get

$$c_2 = (2, 0, 0, 3, 0, 0, 2, 0),$$

and the reconstructed signal is

$$u_2 = (2, 2, 2, 2, 7, 3, -1, -1).$$

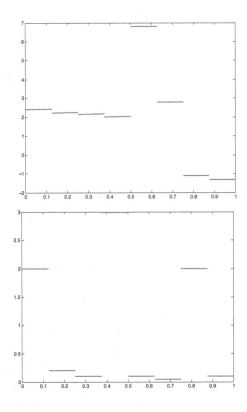

Fig. 4.6   A signal and its Haar transform.

The piecewise-linear curves corresponding to u_2 and c_2 are shown in Figure 4.7.

An interesting (and amusing) application of the Haar wavelets is to the compression of audio signals. It turns out that if your type **load handel** in **Matlab** an audio file will be loaded in a vector denoted by y, and if you type **sound(y)**, the computer will play this piece of music. You can convert y to its vector of Haar coefficients c. The length of y is 73113, so first truncate the tail of y to get a vector of length $65536 = 2^{16}$. A plot of the signals corresponding to y and c is shown in Figure 4.8. Then run a program that sets all coefficients of c whose absolute value is less that 0.05 to zero. This sets 37272 coefficients to 0. The resulting vector c_2 is converted to a signal y_2. A plot of the signals corresponding to y_2 and c_2 is shown in Figure 4.9. When you type **sound(y2)**, you find that the music doesn't differ much

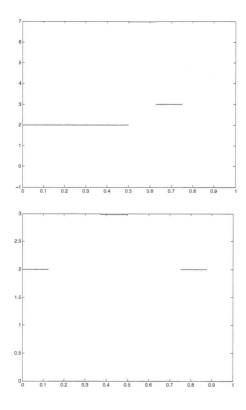

Fig. 4.7   A compressed signal and its compressed Haar transform.

from the original, although it sounds less crisp. You should play with other numbers greater than or less than 0.05. You should hear what happens when you type sound(c). It plays the music corresponding to the Haar transform c of y, and it is quite funny.

4.5   Haar Transform for Digital Images

Another neat property of the Haar transform is that it can be instantly generalized to matrices (even rectangular) without any extra effort! This allows for the compression of digital images. But first we address the issue of normalization of the Haar coefficients. As we observed earlier, the $2^n \times 2^n$ matrix W_n of Haar basis vectors has orthogonal columns, but its columns do not have unit length. As a consequence, $W_n^\top$ is not the inverse of W_n,

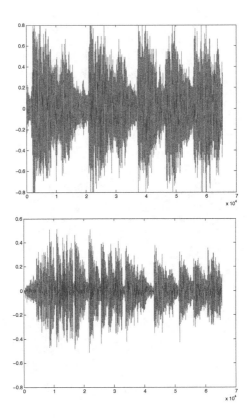

Fig. 4.8   The signal "handel" and its Haar transform.

but rather the matrix

$$W_n^{-1} = D_n W_n^\top$$

with

$$D_n = \operatorname{diag}\Big(2^{-n}, \underbrace{2^{-n}}_{2^0}, \underbrace{2^{-(n-1)}, 2^{-(n-1)}}_{2^1}, \underbrace{2^{-(n-2)}, \ldots, 2^{-(n-2)}}_{2^2}, \ldots,$$

$$\underbrace{2^{-1}, \ldots, 2^{-1}}_{2^{n-1}}\Big).$$

Definition 4.5. The orthogonal matrix

$$H_n = W_n D_n^{\frac{1}{2}}$$

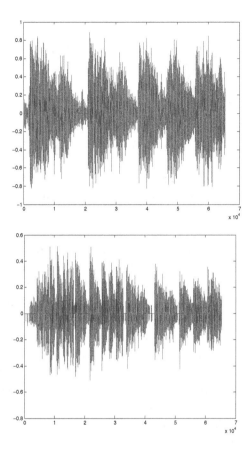

Fig. 4.9   The compressed signal "handel" and its Haar transform.

whose columns are the normalized Haar basis vectors, with

$$D_n^{\frac{1}{2}} = \mathrm{diag}\Big(2^{-\frac{n}{2}}, \underbrace{2^{-\frac{n}{2}}}_{2^0}, \underbrace{2^{-\frac{n-1}{2}}, 2^{-\frac{n-1}{2}}}_{2^1}, \underbrace{2^{-\frac{n-2}{2}}, \ldots, 2^{-\frac{n-2}{2}}}_{2^2}, \ldots, \underbrace{2^{-\frac{1}{2}}, \ldots, 2^{-\frac{1}{2}}}_{2^{n-1}}\Big)$$

is called the *normalized Haar transform matrix*. Given a vector (signal) u, we call $c = H_n^{\top} u$ the *normalized Haar coefficients* of u.

Because H_n is orthogonal, $H_n^{-1} = H_n^{\top}$.

Then a moment of reflection shows that we have to slightly modify the algorithms to compute $H_n^{\top} u$ and $H_n c$ as follows: When computing the sequence of u^js, use

$$u^{j+1}(2i - 1) = (u^j(i) + u^j(2^j + i))/\sqrt{2}$$
$$u^{j+1}(2i) = (u^j(i) - u^j(2^j + i))/\sqrt{2},$$

and when computing the sequence of c^js, use

$$c^j(i) = (c^{j+1}(2i - 1) + c^{j+1}(2i))/\sqrt{2}$$
$$c^j(2^j + i) = (c^{j+1}(2i - 1) - c^{j+1}(2i))/\sqrt{2}.$$

Note that things are now more symmetric, at the expense of a division by $\sqrt{2}$. However, for long vectors, it turns out that these algorithms are numerically more stable.

Remark: Some authors (for example, Stollnitz, Derose and Salesin [Stollnitz *et al.* (1996)]) rescale c by $1/\sqrt{2^n}$ and u by $\sqrt{2^n}$. This is because the norm of the basis functions ψ_k^j is not equal to 1 (under the inner product $\langle f, g \rangle = \int_0^1 f(t)g(t)dt$). The normalized basis functions are the functions $\sqrt{2^j}\psi_k^j$.

Let us now explain the 2D version of the Haar transform. We describe the version using the matrix W_n, the method using H_n being identical (except that $H_n^{-1} = H_n^{\top}$, but this does not hold for W_n^{-1}). Given a $2^m \times 2^n$ matrix A, we can first convert the *rows* of A to their Haar coefficients using the Haar transform W_n^{-1}, obtaining a matrix B, and then convert the *columns* of B to their Haar coefficients, using the matrix W_m^{-1}. Because columns and rows are exchanged in the first step,

$$B = A(W_n^{-1})^{\top},$$

and in the second step $C = W_m^{-1}B$, thus, we have

$$C = W_m^{-1}A(W_n^{-1})^{\top} = D_m W_m^{\top} A W_n D_n.$$

In the other direction, given a $2^m \times 2^n$ matrix C of Haar coefficients, we reconstruct the matrix A (the image) by first applying W_m to the columns of C, obtaining B, and then $W_n^{\top}$ to the rows of B. Therefore

$$A = W_m C W_n^{\top}.$$

Of course, we don't actually have to invert W_m and W_n and perform matrix multiplications. We just have to use our algorithms using averaging and differencing. Here is an example.

If the data matrix (the image) is the 8×8 matrix

$$A = \begin{pmatrix} 64 & 2 & 3 & 61 & 60 & 6 & 7 & 57 \\ 9 & 55 & 54 & 12 & 13 & 51 & 50 & 16 \\ 17 & 47 & 46 & 20 & 21 & 43 & 42 & 24 \\ 40 & 26 & 27 & 37 & 36 & 30 & 31 & 33 \\ 32 & 34 & 35 & 29 & 28 & 38 & 39 & 25 \\ 41 & 23 & 22 & 44 & 45 & 19 & 18 & 48 \\ 49 & 15 & 14 & 52 & 53 & 11 & 10 & 56 \\ 8 & 58 & 59 & 5 & 4 & 62 & 63 & 1 \end{pmatrix},$$

then applying our algorithms, we find that

$$C = \begin{pmatrix} 32.5 & 0 & 0 & 0 & 0 & 0 & 0 & 0 \\ 0 & 0 & 0 & 0 & 0 & 0 & 0 & 0 \\ 0 & 0 & 0 & 0 & 4 & -4 & 4 & -4 \\ 0 & 0 & 0 & 0 & 4 & -4 & 4 & -4 \\ 0 & 0 & 0.5 & 0.5 & 27 & -25 & 23 & -21 \\ 0 & 0 & -0.5 & -0.5 & -11 & 9 & -7 & 5 \\ 0 & 0 & 0.5 & 0.5 & -5 & 7 & -9 & 11 \\ 0 & 0 & -0.5 & -0.5 & 21 & -23 & 25 & -27 \end{pmatrix}.$$

As we can see, C has more zero entries than A; it is a compressed version of A. We can further compress C by setting to 0 all entries of absolute value at most 0.5. Then we get

$$C_2 = \begin{pmatrix} 32.5 & 0 & 0 & 0 & 0 & 0 & 0 & 0 \\ 0 & 0 & 0 & 0 & 0 & 0 & 0 & 0 \\ 0 & 0 & 0 & 0 & 4 & -4 & 4 & -4 \\ 0 & 0 & 0 & 0 & 4 & -4 & 4 & -4 \\ 0 & 0 & 0 & 0 & 27 & -25 & 23 & -21 \\ 0 & 0 & 0 & 0 & -11 & 9 & -7 & 5 \\ 0 & 0 & 0 & 0 & -5 & 7 & -9 & 11 \\ 0 & 0 & 0 & 0 & 21 & -23 & 25 & -27 \end{pmatrix}.$$

We find that the reconstructed image is

$$A_2 = \begin{pmatrix} 63.5 & 1.5 & 3.5 & 61.5 & 59.5 & 5.5 & 7.5 & 57.5 \\ 9.5 & 55.5 & 53.5 & 11.5 & 13.5 & 51.5 & 49.5 & 15.5 \\ 17.5 & 47.5 & 45.5 & 19.5 & 21.5 & 43.5 & 41.5 & 23.5 \\ 39.5 & 25.5 & 27.5 & 37.5 & 35.5 & 29.5 & 31.5 & 33.5 \\ 31.5 & 33.5 & 35.5 & 29.5 & 27.5 & 37.5 & 39.5 & 25.5 \\ 41.5 & 23.5 & 21.5 & 43.5 & 45.5 & 19.5 & 17.5 & 47.5 \\ 49.5 & 15.5 & 13.5 & 51.5 & 53.5 & 11.5 & 9.5 & 55.5 \\ 7.5 & 57.5 & 59.5 & 5.5 & 3.5 & 61.5 & 63.5 & 1.5 \end{pmatrix},$$

which is pretty close to the original image matrix A.

It turns out that Matlab has a wonderful command, image(X) (also imagesc(X), which often does a better job), which displays the matrix X has an image in which each entry is shown as a little square whose gray level is proportional to the numerical value of that entry (lighter if the value is higher, darker if the value is closer to zero; negative values are treated as zero). The images corresponding to A and C are shown in Figure 4.10. The compressed images corresponding to A_2 and C_2 are shown in Figure 4.11. The compressed versions appear to be indistinguishable from the originals!

If we use the normalized matrices H_m and H_n, then the equations

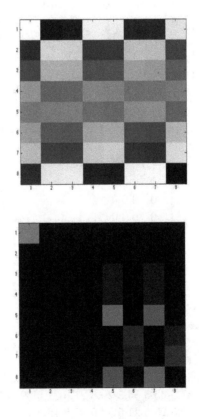

Fig. 4.10   An image and its Haar transform.

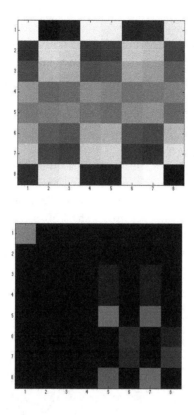

Fig. 4.11   Compressed image and its Haar transform.

relating the image matrix A and its normalized Haar transform C are

$$C = H_m^\top A H_n$$
$$A = H_m C H_n^\top.$$

The Haar transform can also be used to send large images progressively over the internet. Indeed, we can start sending the Haar coefficients of the matrix C starting from the coarsest coefficients (the first column from top down, then the second column, *etc.*), and at the receiving end we can start reconstructing the image as soon as we have received enough data.

Observe that instead of performing all rounds of averaging and differencing on each row and each column, we can perform partial encoding (and decoding). For example, we can perform a single round of averaging and differencing for each row and each column. The result is an image consisting of four subimages, where the top left quarter is a coarser version of the original, and the rest (consisting of three pieces) contain the finest detail coefficients. We can also perform two rounds of averaging and differencing, or three rounds, *etc.* The second round of averaging and differencing is applied to the top left quarter of the image. Generally, the kth round is applied to the $2^{m+1-k} \times 2^{n+1-k}$ submatrix consisting of the first 2^{m+1-k} rows and the first 2^{n+1-k} columns $(1 \leq k \leq n)$ of the matrix obtained at the end of the previous round. This process is illustrated on the image shown in Figure 4.12. The result of performing one round, two rounds, three rounds, and nine rounds of averaging is shown in Figure 4.13. Since our images have size 512×512, nine rounds of averaging yields the Haar transform, displayed as the image on the bottom right. The original image has completely disappeared! We leave it as a fun exercise to modify the algorithms involving averaging and differencing to perform k rounds of averaging/differencing. The reconstruction algorithm is a little tricky.

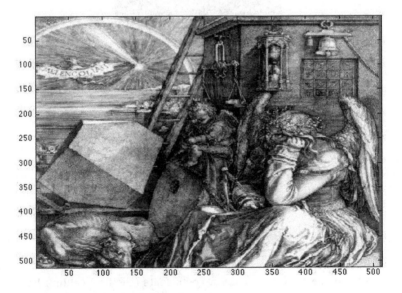

Fig. 4.12   Original drawing by Durer.

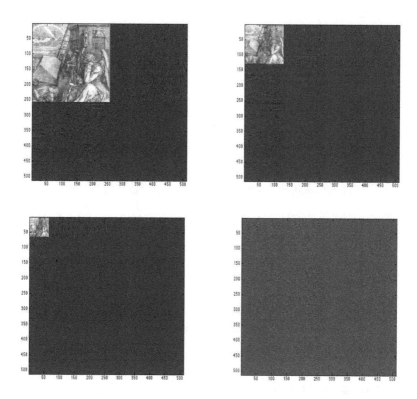

Fig. 4.13   Haar tranforms after one, two, three, and nine rounds of averaging.

A nice and easily accessible account of wavelets and their uses in image processing and computer graphics can be found in Stollnitz, Derose and Salesin [Stollnitz *et al.* (1996)]. A very detailed account is given in Strang and and Nguyen [Strang and Truong (1997)], but this book assumes a fair amount of background in signal processing.

We can find easily a basis of $2^n \times 2^n = 2^{2n}$ vectors w_{ij} $(2^n \times 2^n$ matrices) for the linear map that reconstructs an image from its Haar coefficients, in the sense that for any $2^n \times 2^n$ matrix C of Haar coefficients, the image matrix A is given by

$$A = \sum_{i=1}^{2^n} \sum_{j=1}^{2^n} c_{ij} w_{ij}.$$

Indeed, the matrix w_{ij} is given by the so-called outer product

$$w_{ij} = w_i (w_j)^\top.$$

Similarly, there is a basis of $2^n \times 2^n = 2^{2n}$ vectors h_{ij} ($2^n \times 2^n$ matrices) for the 2D Haar transform, in the sense that for any $2^n \times 2^n$ matrix A, its matrix C of Haar coefficients is given by

$$C = \sum_{i=1}^{2^n} \sum_{j=1}^{2^n} a_{ij} h_{ij}.$$

If the columns of W^{-1} are $w'_1, \ldots, w'_{2^n}$, then

$$h_{ij} = w'_i (w'_j)^\top.$$

We leave it as exercise to compute the bases (w_{ij}) and (h_{ij}) for $n = 2$, and to display the corresponding images using the command `imagesc`.

4.6   Hadamard Matrices

There is another famous family of matrices somewhat similar to Haar matrices, but these matrices have entries $+1$ and -1 (no zero entries).

Definition 4.6. A real $n \times n$ matrix H is a *Hadamard matrix* if $h_{ij} = \pm 1$ for all i, j such that $1 \le i, j \le n$ and if

$$H^\top H = n I_n.$$

Thus the columns of a Hadamard matrix are pairwise orthogonal. Because H is a square matrix, the equation $H^\top H = n I_n$ shows that H is invertible, so we also have $H H^\top = n I_n$. The following matrices are example of Hadamard matrices:

$$H_2 = \begin{pmatrix} 1 & 1 \\ 1 & -1 \end{pmatrix}, \quad H_4 = \begin{pmatrix} 1 & 1 & 1 & 1 \\ 1 & -1 & 1 & -1 \\ 1 & 1 & -1 & -1 \\ 1 & -1 & -1 & 1 \end{pmatrix},$$

and

$$H_8 = \begin{pmatrix} 1 & 1 & 1 & 1 & 1 & 1 & 1 & 1 \\ 1 & -1 & 1 & -1 & 1 & -1 & 1 & -1 \\ 1 & 1 & -1 & -1 & 1 & 1 & -1 & -1 \\ 1 & -1 & -1 & 1 & 1 & -1 & -1 & 1 \\ 1 & 1 & 1 & 1 & -1 & -1 & -1 & -1 \\ 1 & -1 & 1 & -1 & -1 & 1 & -1 & 1 \\ 1 & 1 & -1 & -1 & -1 & -1 & 1 & 1 \\ 1 & -1 & -1 & 1 & -1 & 1 & 1 & -1 \end{pmatrix}.$$

A natural question is to determine the positive integers n for which a Hadamard matrix of dimension n exists, but surprisingly this is an *open problem*. The *Hadamard conjecture* is that for every positive integer of the form $n = 4k$, there is a Hadamard matrix of dimension n.

What is known is a necessary condition and various sufficient conditions.

Theorem 4.1. *If H is an $n \times n$ Hadamard matrix, then either $n = 1, 2$, or $n = 4k$ for some positive integer k.*

Sylvester introduced a family of Hadamard matrices and proved that there are Hadamard matrices of dimension $n = 2^m$ for all $m \geq 1$ using the following construction.

Proposition 4.1. *(Sylvester, 1867) If H is a Hadamard matrix of dimension n, then the block matrix of dimension $2n$,*

$$\begin{pmatrix} H & H \\ H & -H \end{pmatrix},$$

is a Hadamard matrix.

If we start with

$$H_2 = \begin{pmatrix} 1 & 1 \\ 1 & -1 \end{pmatrix},$$

we obtain an infinite family of symmetric Hadamard matrices usually called *Sylvester–Hadamard* matrices and denoted by H_{2^m}. The Sylvester–Hadamard matrices H_2, H_4 and H_8 are shown on the previous page.

In 1893, Hadamard gave examples of Hadamard matrices for $n = 12$ and $n = 20$. At the present, Hadamard matrices are known for all $n = 4k \leq 1000$, *except for $n = 668, 716,$ and 892.*

Hadamard matrices have various applications to error correcting codes, signal processing, and numerical linear algebra; see Seberry, Wysocki and

Wysocki [Seberry *et al.* (2005)] and Tropp [Tropp (2011)]. For example, there is a code based on H_{32} that can correct 7 errors in any 32-bit encoded block, and can detect an eighth. This code was used on a Mariner spacecraft in 1969 to transmit pictures back to the earth.

For every $m \geq 0$, the piecewise affine functions $\text{plf}((H_{2^m})_i)$ associated with the 2^m rows of the Sylvester–Hadamard matrix H_{2^m} are functions on $[0,1]$ known as the *Walsh functions*. It is customary to index these 2^m functions by the integers $0, 1, \ldots, 2^m - 1$ in such a way that the Walsh function $\text{Wal}(k,t)$ is equal to the function $\text{plf}((H_{2^m})_i)$ associated with the Row i of H_{2^m} that contains k changes of signs between consecutive groups of $+1$ and consecutive groups of -1. For example, the fifth row of H_8, namely

$$\left(1\ -1\ -1\ 1\ 1\ -1\ -1\ 1\right),$$

has five consecutive blocks of $+1$s and -1s, four sign changes between these blocks, and thus is associated with $\text{Wal}(4,t)$. In particular, Walsh functions corresponding to the rows of H_8 (from top down) are:

$$\text{Wal}(0,t),\ \text{Wal}(7,t),\ \text{Wal}(3,t),\ \text{Wal}(4,t),$$
$$\text{Wal}(1,t),\ \text{Wal}(6,t),\ \text{Wal}(2,t),\ \text{Wal}(5,t).$$

Because of the connection between Sylvester–Hadamard matrices and Walsh functions, Sylvester–Hadamard matrices are called *Walsh–Hadamard matrices* by some authors. For every m, the 2^m Walsh functions are pairwise orthogonal. The countable set of Walsh functions $\text{Wal}(k,t)$ for all $m \geq 0$ and all k such that $0 \leq k \leq 2^m - 1$ can be ordered in such a way that it is an orthogonal Hilbert basis of the Hilbert space $\text{L}^2([0,1])$; see Seberry, Wysocki and Wysocki [Seberry *et al.* (2005)].

The Sylvester–Hadamard matrix H_{2^m} plays a role in various algorithms for dimension reduction and low-rank matrix approximation. There is a type of structured dimension-reduction map known as the *subsampled randomized Hadamard transform*, for short SRHT; see Tropp [Tropp (2011)] and Halko, Martinsson and Tropp [Halko *et al.* (2011)]. For $\ell \ll n = 2^m$, an *SRHT matrix* is an $\ell \times n$ matrix of the form

$$\Phi = \sqrt{\frac{n}{\ell}} RHD,$$

where

(1) D is a random $n \times n$ diagonal matrix whose entries are independent random signs.

(2) $H = n^{-1/2} H_n$, a normalized Sylvester–Hadamard matrix of dimension n.

(3) R is a random $\ell \times n$ matrix that restricts an n-dimensional vector to ℓ coordinates, chosen uniformly at random.

It is explained in Tropp [Tropp (2011)] that for any input x such that $\|x\|_2 = 1$, the probability that $|(HDx)_i| \geq \sqrt{n^{-1} \log(n)}$ for any i is quite small. Thus HD has the effect of "flattening" the input x. The main result about the SRHT is that it preserves the geometry of an entire subspace of vectors; see Tropp [Tropp (2011)] (Theorem 1.3).

4.7 Summary

The main concepts and results of this chapter are listed below:

- Haar basis vectors and a glimpse at *Haar wavelets*.
- *Kronecker product* (or *tensor product*) of matrices.
- Hadamard and Sylvester–Hadamard matrices.
- Walsh functions.

4.8 Problems

Problem 4.1. (Haar extravaganza) Consider the matrix

$$W_{3,3} = \begin{pmatrix} 1 & 0 & 0 & 0 & 1 & 0 & 0 & 0 \\ 1 & 0 & 0 & 0 & -1 & 0 & 0 & 0 \\ 0 & 1 & 0 & 0 & 0 & 1 & 0 & 0 \\ 0 & 1 & 0 & 0 & 0 & -1 & 0 & 0 \\ 0 & 0 & 1 & 0 & 0 & 0 & 1 & 0 \\ 0 & 0 & 1 & 0 & 0 & 0 & -1 & 0 \\ 0 & 0 & 0 & 1 & 0 & 0 & 0 & 1 \\ 0 & 0 & 0 & 1 & 0 & 0 & 0 & -1 \end{pmatrix}.$$

(1) Show that given any vector $c = (c_1, c_2, c_3, c_4, c_5, c_6, c_7, c_8)$, the result $W_{3,3}c$ of applying $W_{3,3}$ to c is

$$W_{3,3}c = (c_1 + c_5, c_1 - c_5, c_2 + c_6, c_2 - c_6, c_3 + c_7, c_3 - c_7, c_4 + c_8, c_4 - c_8),$$

the last step in reconstructing a vector from its Haar coefficients.

(2) Prove that the inverse of $W_{3,3}$ is $(1/2)W_{3,3}^\top$. Prove that the columns and the rows of $W_{3,3}$ are orthogonal.

(3) Let $W_{3,2}$ and $W_{3,1}$ be the following matrices:

$$
W_{3,2} = \begin{pmatrix}
1 & 0 & 1 & 0 & 0 & 0 & 0 & 0 \\
1 & 0 & -1 & 0 & 0 & 0 & 0 & 0 \\
0 & 1 & 0 & 1 & 0 & 0 & 0 & 0 \\
0 & 1 & 0 & -1 & 0 & 0 & 0 & 0 \\
0 & 0 & 0 & 0 & 1 & 0 & 0 & 0 \\
0 & 0 & 0 & 0 & 0 & 1 & 0 & 0 \\
0 & 0 & 0 & 0 & 0 & 0 & 1 & 0 \\
0 & 0 & 0 & 0 & 0 & 0 & 0 & 1
\end{pmatrix}, \quad
W_{3,1} = \begin{pmatrix}
1 & 1 & 0 & 0 & 0 & 0 & 0 & 0 \\
1 & -1 & 0 & 0 & 0 & 0 & 0 & 0 \\
0 & 0 & 1 & 0 & 0 & 0 & 0 & 0 \\
0 & 0 & 0 & 1 & 0 & 0 & 0 & 0 \\
0 & 0 & 0 & 0 & 1 & 0 & 0 & 0 \\
0 & 0 & 0 & 0 & 0 & 1 & 0 & 0 \\
0 & 0 & 0 & 0 & 0 & 0 & 1 & 0 \\
0 & 0 & 0 & 0 & 0 & 0 & 0 & 1
\end{pmatrix}.
$$

Show that given any vector $c = (c_1, c_2, c_3, c_4, c_5, c_6, c_7, c_8)$, the result $W_{3,2}c$ of applying $W_{3,2}$ to c is

$$W_{3,2}c = (c_1 + c_3, c_1 - c_3, c_2 + c_4, c_2 - c_4, c_5, c_6, c_7, c_8),$$

the second step in reconstructing a vector from its Haar coefficients, and the result $W_{3,1}c$ of applying $W_{3,1}$ to c is

$$W_{3,1}c = (c_1 + c_2, c_1 - c_2, c_3, c_4, c_5, c_6, c_7, c_8),$$

the first step in reconstructing a vector from its Haar coefficients.

Conclude that

$$W_{3,3}W_{3,2}W_{3,1} = W_3,$$

the Haar matrix

$$
W_3 = \begin{pmatrix}
1 & 1 & 1 & 0 & 1 & 0 & 0 & 0 \\
1 & 1 & 1 & 0 & -1 & 0 & 0 & 0 \\
1 & 1 & -1 & 0 & 0 & 1 & 0 & 0 \\
1 & 1 & -1 & 0 & 0 & -1 & 0 & 0 \\
1 & -1 & 0 & 1 & 0 & 0 & 1 & 0 \\
1 & -1 & 0 & 1 & 0 & 0 & -1 & 0 \\
1 & -1 & 0 & -1 & 0 & 0 & 0 & 1 \\
1 & -1 & 0 & -1 & 0 & 0 & 0 & -1
\end{pmatrix}.
$$

Hint. First check that

$$
W_{3,2}W_{3,1} = \begin{pmatrix} W_2 & 0_{4,4} \\ 0_{4,4} & I_4 \end{pmatrix},
$$

where

$$
W_2 = \begin{pmatrix}
1 & 1 & 1 & 0 \\
1 & 1 & -1 & 0 \\
1 & -1 & 0 & 1 \\
1 & -1 & 0 & -1
\end{pmatrix}.
$$

(4) Prove that the columns and the rows of $W_{3,2}$ and $W_{3,1}$ are orthogonal. Deduce from this that the columns of W_3 are orthogonal, and the rows of W_3^{-1} are orthogonal. Are the rows of W_3 orthogonal? Are the columns of W_3^{-1} orthogonal? Find the inverse of $W_{3,2}$ and the inverse of $W_{3,1}$.

Problem 4.2. This is a continuation of Problem 4.1.

(1) For any $n \geq 2$, the $2^n \times 2^n$ matrix $W_{n,n}$ is obtained form the two rows

$$\underbrace{1, 0, \ldots, 0}_{2^{n-1}}, \underbrace{1, 0, \ldots, 0}_{2^{n-1}}$$

$$\underbrace{1, 0, \ldots, 0}_{2^{n-1}}, \underbrace{-1, 0, \ldots, 0}_{2^{n-1}}$$

by shifting them $2^{n-1} - 1$ times over to the right by inserting a zero on the left each time.

Given any vector $c = (c_1, c_2, \ldots, c_{2^n})$, show that $W_{n,n}c$ is the result of the last step in the process of reconstructing a vector from its Haar coefficients c. Prove that $W_{n,n}^{-1} = (1/2)W_{n,n}^\top$, and that the columns and the rows of $W_{n,n}$ are orthogonal.

(2) Given a $m \times n$ matrix $A = (a_{ij})$ and a $p \times q$ matrix $B = (b_{ij})$, the *Kronecker product* (or *tensor product*) $A \otimes B$ of A and B is the $mp \times nq$ matrix

$$A \otimes B = \begin{pmatrix} a_{11}B & a_{12}B & \cdots & a_{1n}B \\ a_{21}B & a_{22}B & \cdots & a_{2n}B \\ \vdots & \vdots & \ddots & \vdots \\ a_{m1}B & a_{m2}B & \cdots & a_{mn}B \end{pmatrix}.$$

It can be shown (and you may use these facts without proof) that $\otimes$ is associative and that

$$(A \otimes B)(C \otimes D) = AC \otimes BD$$
$$(A \otimes B)^\top = A^\top \otimes B^\top,$$

whenever AC and BD are well defined.

Check that

$$W_{n,n} = \left(I_{2^{n-1}} \otimes \begin{pmatrix} 1 \\ 1 \end{pmatrix} \ I_{2^{n-1}} \otimes \begin{pmatrix} 1 \\ -1 \end{pmatrix} \right),$$

and that

$$W_n = \left(W_{n-1} \otimes \begin{pmatrix} 1 \\ 1 \end{pmatrix} \ I_{2^{n-1}} \otimes \begin{pmatrix} 1 \\ -1 \end{pmatrix} \right).$$

Use the above to reprove that

$$W_{n,n}W_{n,n}^\top = 2I_{2^n}.$$

Let

$$B_1 = 2\begin{pmatrix} 1 & 0 \\ 0 & 1 \end{pmatrix} = \begin{pmatrix} 2 & 0 \\ 0 & 2 \end{pmatrix}$$

and for $n \geq 1$,

$$B_{n+1} = 2\begin{pmatrix} B_n & 0 \\ 0 & I_{2^n} \end{pmatrix}.$$

Prove that

$$W_n^\top W_n = B_n, \quad \text{for all } n \geq 1.$$

(3) The matrix $W_{n,i}$ is obtained from the matrix $W_{i,i}$ $(1 \leq i \leq n-1)$ as follows:

$$W_{n,i} = \begin{pmatrix} W_{i,i} & 0_{2^i,2^n-2^i} \\ 0_{2^n-2^i,2^i} & I_{2^n-2^i} \end{pmatrix}.$$

It consists of four blocks, where $0_{2^i,2^n-2^i}$ and $0_{2^n-2^i,2^i}$ are matrices of zeros and $I_{2^n-2^i}$ is the identity matrix of dimension $2^n - 2^i$.

Explain what $W_{n,i}$ does to c and prove that

$$W_{n,n}W_{n,n-1}\cdots W_{n,1} = W_n,$$

where W_n is the Haar matrix of dimension 2^n.

Hint. Use induction on k, with the induction hypothesis

$$W_{n,k}W_{n,k-1}\cdots W_{n,1} = \begin{pmatrix} W_k & 0_{2^k,2^n-2^k} \\ 0_{2^n-2^k,2^k} & I_{2^n-2^k} \end{pmatrix}.$$

Prove that the columns and rows of $W_{n,k}$ are orthogonal, and use this to prove that the columns of W_n and the rows of W_n^{-1} are orthogonal. Are the rows of W_n orthogonal? Are the columns of W_n^{-1} orthogonal? Prove that

$$W_{n,k}^{-1} = \begin{pmatrix} \frac{1}{2}W_{k,k}^\top & 0_{2^k,2^n-2^k} \\ 0_{2^n-2^k,2^k} & I_{2^n-2^k} \end{pmatrix}.$$

Problem 4.3. Prove that if H is a Hadamard matrix of dimension n, then the block matrix of dimension $2n$,

$$\begin{pmatrix} H & H \\ H & -H \end{pmatrix},$$

is a Hadamard matrix.

Problem 4.4. Plot the graphs of the eight Walsh functions $\text{Wal}(k,t)$ for $k = 0, 1, \ldots, 7$.

Problem 4.5. Describe a recursive algorithm to compute the product $H_{2^m} x$ of the Sylvester–Hadamard matrix H_{2^m} by a vector $x \in \mathbb{R}^{2^m}$ that uses m recursive calls.

Chapter 5

Direct Sums, Rank-Nullity Theorem, Affine Maps

In this chapter all vector spaces are defined over an arbitrary field K. For the sake of concreteness, the reader may safely assume that $K = \mathbb{R}$.

5.1 Direct Products

There are some useful ways of forming new vector spaces from older ones.

Definition 5.1. Given $p \geq 2$ vector spaces $E_1, \ldots, E_p$, the product $F = E_1 \times \cdots \times E_p$ can be made into a vector space by defining addition and scalar multiplication as follows:

$$(u_1, \ldots, u_p) + (v_1, \ldots, v_p) = (u_1 + v_1, \ldots, u_p + v_p)$$
$$\lambda(u_1, \ldots, u_p) = (\lambda u_1, \ldots, \lambda u_p),$$

for all $u_i, v_i \in E_i$ and all $\lambda \in \mathbb{R}$. The zero vector of $E_1 \times \cdots \times E_p$ is the p-tuple

$$(\underbrace{0, \ldots, 0}_{p}),$$

where the ith zero is the zero vector of E_i.

With the above addition and multiplication, the vector space $F = E_1 \times \cdots \times E_p$ is called the *direct product* of the vector spaces $E_1, \ldots, E_p$.

As a special case, when $E_1 = \cdots = E_p = \mathbb{R}$, we find again the vector space $F = \mathbb{R}^p$. The *projection maps* $pr_i \colon E_1 \times \cdots \times E_p \to E_i$ given by

$$pr_i(u_1, \ldots, u_p) = u_i$$

are clearly linear. Similarly, the maps $in_i \colon E_i \to E_1 \times \cdots \times E_p$ given by

$$in_i(u_i) = (0, \ldots, 0, u_i, 0, \ldots, 0)$$

are injective and linear. If $\dim(E_i) = n_i$ and if $(e_1^i, \ldots, e_{n_i}^i)$ is a basis of E_i for $i = 1, \ldots, p$, then it is easy to see that the $n_1 + \cdots + n_p$ vectors

$$(e_1^1, 0, \ldots, 0), \qquad \ldots, \qquad (e_{n_1}^1, 0, \ldots, 0),$$
$$\vdots \qquad\qquad \vdots \qquad\qquad \vdots$$
$$(0, \ldots, 0, e_1^i, 0, \ldots, 0), \ldots, (0, \ldots, 0, e_{n_i}^i, 0, \ldots, 0),$$
$$\vdots \qquad\qquad \vdots \qquad\qquad \vdots$$
$$(0, \ldots, 0, e_1^p), \qquad \ldots, \qquad (0, \ldots, 0, e_{n_p}^p)$$

form a basis of $E_1 \times \cdots \times E_p$, and so

$$\dim(E_1 \times \cdots \times E_p) = \dim(E_1) + \cdots + \dim(E_p).$$

5.2 Sums and Direct Sums

Let us now consider a vector space E and p subspaces $U_1, \ldots, U_p$ of E. We have a map

$$a \colon U_1 \times \cdots \times U_p \to E$$

given by

$$a(u_1, \ldots, u_p) = u_1 + \cdots + u_p,$$

with $u_i \in U_i$ for $i = 1, \ldots, p$. It is clear that this map is linear, and so its image is a subspace of E denoted by

$$U_1 + \cdots + U_p$$

and called the *sum* of the subspaces $U_1, \ldots, U_p$. By definition,

$$U_1 + \cdots + U_p = \{u_1 + \cdots + u_p \mid u_i \in U_i,\ 1 \le i \le p\},$$

and it is immediately verified that $U_1 + \cdots + U_p$ is the smallest subspace of E containing $U_1, \ldots, U_p$. This also implies that $U_1 + \cdots + U_p$ does not depend on the order of the factors U_i; in particular,

$$U_1 + U_2 = U_2 + U_1.$$

Definition 5.2. For any vector space E and any $p \ge 2$ subspaces $U_1, \ldots, U_p$ of E, if the map $a \colon U_1 \times \cdots \times U_p \to E$ defined above is injective, then the sum $U_1 + \cdots + U_p$ is called a *direct sum* and it is denoted by

$$U_1 \oplus \cdots \oplus U_p.$$

The space E is the *direct sum* of the subspaces U_i if

$$E = U_1 \oplus \cdots \oplus U_p.$$

If the map a is injective, then by Proposition 2.13 we have $\operatorname{Ker} a = \{(\underbrace{0, \ldots, 0}_{p})\}$ where each 0 is the zero vector of E, which means that if $u_i \in U_i$ for $i = 1, \ldots, p$ and if

$$u_1 + \cdots + u_p = 0,$$

then $(u_1, \ldots, u_p) = (0, \ldots, 0)$, that is, $u_1 = 0, \ldots, u_p = 0$.

Proposition 5.1. *If the map $a \colon U_1 \times \cdots \times U_p \to E$ is injective, then every $u \in U_1 + \cdots + U_p$ has a unique expression as a sum*

$$u = u_1 + \cdots + u_p,$$

with $u_i \in U_i$, for $i = 1, \ldots, p$.

Proof. If

$$u = v_1 + \cdots + v_p = w_1 + \cdots + w_p,$$

with $v_i, w_i \in U_i$, for $i = 1, \ldots, p$, then we have

$$w_1 - v_1 + \cdots + w_p - v_p = 0,$$

and since $v_i, w_i \in U_i$ and each U_i is a subspace, $w_i - v_i \in U_i$. The injectivity of a implies that $w_i - v_i = 0$, that is, $w_i = v_i$ for $i = 1, \ldots, p$, which shows the uniqueness of the decomposition of u. $\qquad\square$

Proposition 5.2. *If the map $a \colon U_1 \times \cdots \times U_p \to E$ is injective, then any p nonzero vectors $u_1, \ldots, u_p$ with $u_i \in U_i$ are linearly independent.*

Proof. To see this, assume that

$$\lambda_1 u_1 + \cdots + \lambda_p u_p = 0$$

for some $\lambda_i \in \mathbb{R}$. Since $u_i \in U_i$ and U_i is a subspace, $\lambda_i u_i \in U_i$, and the injectivity of a implies that $\lambda_i u_i = 0$, for $i = 1, \ldots, p$. Since $u_i \neq 0$, we must have $\lambda_i = 0$ for $i = 1, \ldots, p$; that is, $u_1, \ldots, u_p$ with $u_i \in U_i$ and $u_i \neq 0$ are linearly independent. $\qquad\square$

Observe that if a is injective, then we must have $U_i \cap U_j = (0)$ whenever $i \neq j$. However, this condition is generally not sufficient if $p \geq 3$. For example, if $E = \mathbb{R}^2$ and U_1 the line spanned by $e_1 = (1, 0)$, U_2 is the line spanned by $d = (1, 1)$, and U_3 is the line spanned by $e_2 = (0, 1)$, then $U_1 \cap U_2 = U_1 \cap U_3 = U_2 \cap U_3 = \{(0, 0)\}$, but $U_1 + U_2 = U_1 + U_3 = U_2 + U_3 =$

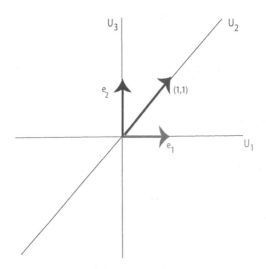

Fig. 5.1   The linear subspaces U_1, U_2, and U_3 illustrated as lines in $\mathbb{R}^2$.

$\mathbb{R}^2$, so $U_1 + U_2 + U_3$ is not a direct sum. For example, d is expressed in two different ways as

$$d = (1,1) = (1,0) + (0,1) = e_1 + e_2.$$

See Figure 5.1.

As in the case of a sum, $U_1 \oplus U_2 = U_2 \oplus U_1$. *Observe that when the map a is injective, then it is a linear isomorphism between $U_1 \times \cdots \times U_p$ and $U_1 \oplus \cdots \oplus U_p$.* The difference is that $U_1 \times \cdots \times U_p$ is defined even if the spaces U_i are not assumed to be subspaces of some common space.

If E is a direct sum $E = U_1 \oplus \cdots \oplus U_p$, since any p nonzero vectors $u_1, \ldots, u_p$ with $u_i \in U_i$ are linearly independent, if we pick a basis $(u_k)_{k \in I_j}$ in U_j for $j = 1, \ldots, p$, then $(u_i)_{i \in I}$ with $I = I_1 \cup \cdots \cup I_p$ is a basis of E. Intuitively, E is split into p independent subspaces.

Conversely, given a basis $(u_i)_{i \in I}$ of E, if we partition the index set I as $I = I_1 \cup \cdots \cup I_p$, then each subfamily $(u_k)_{k \in I_j}$ spans some subspace U_j of E, and it is immediately verified that we have a direct sum

$$E = U_1 \oplus \cdots \oplus U_p.$$

Definition 5.3. Let $f \colon E \to E$ be a linear map. For any subspace U of E, if $f(U) \subseteq U$ we say that U is *invariant under f*.

Assume that E is finite-dimensional, a direct sum $E = U_1 \oplus \cdots \oplus U_p$, and that each U_j is invariant under f. If we pick a basis $(u_i)_{i \in I}$ as above with $I = I_1 \cup \cdots \cup I_p$ and with each $(u_k)_{k \in I_j}$ a basis of U_j, since each U_j is invariant under f, the image $f(u_k)$ of every basis vector u_k with $k \in I_j$ belongs to U_j, so the matrix A representing f over the basis $(u_i)_{i \in I}$ is a *block diagonal* matrix of the form

$$
A = \begin{pmatrix} A_1 & & & \\ & A_2 & & \\ & & \ddots & \\ & & & A_p \end{pmatrix},
$$

with each block A_j a $d_j \times d_j$-matrix with $d_j = \dim(U_j)$ and all other entries equal to 0. If $d_j = 1$ for $j = 1, \ldots, p$, the matrix A is a diagonal matrix.

There are natural injections from each U_i to E denoted by $\mathrm{in}_i \colon U_i \to E$. Now, if $p = 2$, it is easy to determine the kernel of the map $a \colon U_1 \times U_2 \to E$. We have

$$
a(u_1, u_2) = u_1 + u_2 = 0 \quad \text{iff} \quad u_1 = -u_2, \, u_1 \in U_1, u_2 \in U_2,
$$

which implies that

$$
\mathrm{Ker}\, a = \{(u, -u) \mid u \in U_1 \cap U_2\}.
$$

Now, $U_1 \cap U_2$ is a subspace of E and the linear map $u \mapsto (u, -u)$ is clearly an isomorphism between $U_1 \cap U_2$ and $\mathrm{Ker}\, a$, so $\mathrm{Ker}\, a$ is isomorphic to $U_1 \cap U_2$. As a consequence, we get the following result:

Proposition 5.3. *Given any vector space E and any two subspaces U_1 and U_2, the sum $U_1 + U_2$ is a direct sum iff $U_1 \cap U_2 = (0)$.*

An interesting illustration of the notion of direct sum is the decomposition of a square matrix into its symmetric part and its skew-symmetric part. Recall that an $n \times n$ matrix $A \in \mathrm{M}_n$ is *symmetric* if $A^\top = A$, *skew-symmetric* if $A^\top = -A$. It is clear that

$$
\mathbf{S}(n) = \{A \in \mathrm{M}_n \mid A^\top = A\} \quad \text{and} \quad \mathbf{Skew}(n) = \{A \in \mathrm{M}_n \mid A^\top = -A\}
$$

are subspaces of M_n, and that $\mathbf{S}(n) \cap \mathbf{Skew}(n) = (0)$. Observe that for any matrix $A \in \mathrm{M}_n$, the matrix $H(A) = (A + A^\top)/2$ is symmetric and the matrix $S(A) = (A - A^\top)/2$ is skew-symmetric. Since

$$
A = H(A) + S(A) = \frac{A + A^\top}{2} + \frac{A - A^\top}{2},
$$

we see that $\mathrm{M}_n = \mathbf{S}(n) + \mathbf{Skew}(n)$, and since $\mathbf{S}(n) \cap \mathbf{Skew}(n) = (0)$, we have the direct sum

$$\mathrm{M}_n = \mathbf{S}(n) \oplus \mathbf{Skew}(n).$$

Remark: The vector space $\mathbf{Skew}(n)$ of skew-symmetric matrices is also denoted by $\mathfrak{so}(n)$. It is the *Lie algebra* of the group $\mathbf{SO}(n)$.

Proposition 5.3 can be generalized to any $p \geq 2$ subspaces at the expense of notation. The proof of the following proposition is left as an exercise.

Proposition 5.4. *Given any vector space E and any $p \geq 2$ subspaces $U_1, \ldots, U_p$, the following properties are equivalent:*

(1) The sum $U_1 + \cdots + U_p$ is a direct sum.

(2) We have

$$U_i \cap \left(\sum_{j=1, j \neq i}^{p} U_j \right) = (0), \quad i = 1, \ldots, p.$$

(3) We have

$$U_i \cap \left(\sum_{j=1}^{i-1} U_j \right) = (0), \quad i = 2, \ldots, p.$$

Because of the isomorphism

$$U_1 \times \cdots \times U_p \approx U_1 \oplus \cdots \oplus U_p,$$

we have

Proposition 5.5. *If E is any vector space, for any (finite-dimensional) subspaces $U_1, \ldots, U_p$ of E, we have*

$$\dim(U_1 \oplus \cdots \oplus U_p) = \dim(U_1) + \cdots + \dim(U_p).$$

If E is a direct sum

$$E = U_1 \oplus \cdots \oplus U_p,$$

since every $u \in E$ can be written in a unique way as

$$u = u_1 + \cdots + u_p$$

with $u_i \in U_i$ for $i = 1 \ldots, p$, we can define the maps $\pi_i \colon E \to U_i$, called *projections*, by

$$\pi_i(u) = \pi_i(u_1 + \cdots + u_p) = u_i.$$

It is easy to check that these maps are linear and satisfy the following properties:

$$\pi_j \circ \pi_i = \begin{cases} \pi_i & \text{if } i = j \\ 0 & \text{if } i \neq j, \end{cases}$$

$$\pi_1 + \cdots + \pi_p = \text{id}_E.$$

For example, in the case of the direct sum

$$\mathbf{M}_n = \mathbf{S}(n) \oplus \mathbf{Skew}(n),$$

the projection onto $\mathbf{S}(n)$ is given by

$$\pi_1(A) = H(A) = \frac{A + A^\top}{2},$$

and the projection onto $\mathbf{Skew}(n)$ is given by

$$\pi_2(A) = S(A) = \frac{A - A^\top}{2}.$$

Clearly, $H(A) + S(A) = A$, $H(H(A)) = H(A)$, $S(S(A)) = S(A)$, and $H(S(A)) = S(H(A)) = 0$.

A function f such that $f \circ f = f$ is said to be *idempotent*. Thus, the projections π_i are idempotent. Conversely, the following proposition can be shown:

Proposition 5.6. *Let E be a vector space. For any $p \geq 2$ linear maps $f_i \colon E \to E$, if*

$$f_j \circ f_i = \begin{cases} f_i & \text{if } i = j \\ 0 & \text{if } i \neq j, \end{cases}$$

$$f_1 + \cdots + f_p = \text{id}_E,$$

then if we let $U_i = f_i(E)$, we have a direct sum

$$E = U_1 \oplus \cdots \oplus U_p.$$

We also have the following proposition characterizing idempotent linear maps whose proof is also left as an exercise.

Proposition 5.7. *For every vector space E, if $f \colon E \to E$ is an idempotent linear map, i.e., $f \circ f = f$, then we have a direct sum*

$$E = \text{Ker } f \oplus \text{Im } f,$$

so that f is the projection onto its image $\text{Im } f$.

We are now ready to prove a very crucial result relating the rank and the dimension of the kernel of a linear map.

5.3   The Rank-Nullity Theorem; Grassmann's Relation

We begin with the following theorem which shows that given a linear map $f: E \to F$, its domain E is the direct sum of its kernel $\operatorname{Ker} f$ with some isomorphic copy of its image $\operatorname{Im} f$.

Theorem 5.1. *(Rank-nullity theorem) Let $f: E \to F$ be a linear map with finite image. For any choice of a basis $(f_1, \ldots, f_r)$ of $\operatorname{Im} f$, let $(u_1, \ldots, u_r)$ be any vectors in E such that $f_i = f(u_i)$, for $i = 1, \ldots, r$. If $s: \operatorname{Im} f \to E$ is the unique linear map defined by $s(f_i) = u_i$, for $i = 1, \ldots, r$, then s is injective, $f \circ s = \operatorname{id}$, and we have a direct sum*

$$E = \operatorname{Ker} f \oplus \operatorname{Im} s$$

as illustrated by the following diagram:

$$\operatorname{Ker} f \longrightarrow E = \operatorname{Ker} f \oplus \operatorname{Im} s \underset{s}{\overset{f}{\rightleftarrows}} \operatorname{Im} f \subseteq F.$$

See Figure 5.2. As a consequence, if E is finite-dimensional, then

$$\dim(E) = \dim(\operatorname{Ker} f) + \dim(\operatorname{Im} f) = \dim(\operatorname{Ker} f) + \operatorname{rk}(f).$$

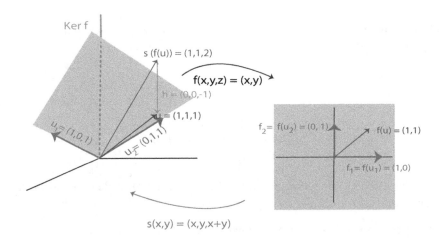

Fig. 5.2   Let $f: E \to F$ be the linear map from $\mathbb{R}^3$ to $\mathbb{R}^2$ given by $f(x, y, z) = (x, y)$. Then $s: \mathbb{R}^2 \to \mathbb{R}^3$ is given by $s(x, y) = (x, y, x+y)$ and maps the pink $\mathbb{R}^2$ isomorphically onto the slanted pink plane of $\mathbb{R}^3$ whose equation is $-x - y + z = 0$. Theorem 5.1 shows that $\mathbb{R}^3$ is the direct sum of the plane $-x - y + z = 0$ and the kernel of f which the orange z-axis.

Proof. The vectors $u_1, \ldots, u_r$ must be linearly independent since otherwise we would have a nontrivial linear dependence

$$\lambda_1 u_1 + \cdots + \lambda_r u_r = 0,$$

and by applying f, we would get the nontrivial linear dependence

$$0 = \lambda_1 f(u_1) + \cdots + \lambda_r f(u_r) = \lambda_1 f_1 + \cdots + \lambda_r f_r,$$

contradicting the fact that $(f_1, \ldots, f_r)$ is a basis. Therefore, the unique linear map s given by $s(f_i) = u_i$, for $i = 1, \ldots, r$, is a linear isomorphism between $\operatorname{Im} f$ and its image, the subspace spanned by $(u_1, \ldots, u_r)$. It is also clear by definition that $f \circ s = \operatorname{id}$. For any $u \in E$, let

$$h = u - (s \circ f)(u).$$

Since $f \circ s = \operatorname{id}$, we have

$$
\begin{aligned}
f(h) = f(u - (s \circ f)(u)) &= f(u) - (f \circ s \circ f)(u) \\
&= f(u) - (\operatorname{id} \circ f)(u) = f(u) - f(u) = 0,
\end{aligned}
$$

which shows that $h \in \operatorname{Ker} f$. Since $h = u - (s \circ f)(u)$, it follows that

$$u = h + s(f(u)),$$

with $h \in \operatorname{Ker} f$ and $s(f(u)) \in \operatorname{Im} s$, which proves that

$$E = \operatorname{Ker} f + \operatorname{Im} s.$$

Now if $u \in \operatorname{Ker} f \cap \operatorname{Im} s$, then $u = s(v)$ for some $v \in F$ and $f(u) = 0$ since $u \in \operatorname{Ker} f$. Since $u = s(v)$ and $f \circ s = \operatorname{id}$, we get

$$0 = f(u) = f(s(v)) = v,$$

and so $u = s(v) = s(0) = 0$. Thus, $\operatorname{Ker} f \cap \operatorname{Im} s = (0)$, which proves that we have a direct sum

$$E = \operatorname{Ker} f \oplus \operatorname{Im} s.$$

The equation

$$\dim(E) = \dim(\operatorname{Ker} f) + \dim(\operatorname{Im} f) = \dim(\operatorname{Ker} f) + \operatorname{rk}(f)$$

is an immediate consequence of the fact that the dimension is an additive property for direct sums, that by definition the rank of f is the dimension of the image of f, and that $\dim(\operatorname{Im} s) = \dim(\operatorname{Im} f)$, because s is an isomorphism between $\operatorname{Im} f$ and $\operatorname{Im} s$. $\qquad \square$

Remark: The statement $E = \operatorname{Ker} f \oplus \operatorname{Im} s$ holds if E has infinite dimension. It still holds if $\operatorname{Im}(f)$ also has infinite dimension.

Definition 5.4. The dimension $\dim(\operatorname{Ker} f)$ of the kernel of a linear map f is called the *nullity* of f.

We now derive some important results using Theorem 5.1.

Proposition 5.8. *Given a vector space E, if U and V are any two finite-dimensional subspaces of E, then*

$$\dim(U) + \dim(V) = \dim(U + V) + \dim(U \cap V),$$

an equation known as Grassmann's relation.

Proof. Recall that $U + V$ is the image of the linear map

$$a \colon U \times V \to E$$

given by

$$a(u, v) = u + v,$$

and that we proved earlier that the kernel $\operatorname{Ker} a$ of a is isomorphic to $U \cap V$. By Theorem 5.1,

$$\dim(U \times V) = \dim(\operatorname{Ker} a) + \dim(\operatorname{Im} a),$$

but $\dim(U \times V) = \dim(U) + \dim(V)$, $\dim(\operatorname{Ker} a) = \dim(U \cap V)$, and $\operatorname{Im} a = U + V$, so the Grassmann relation holds. $\qquad \square$

The Grassmann relation can be very useful to figure out whether two subspace have a nontrivial intersection in spaces of dimension > 3. For example, it is easy to see that in $\mathbb{R}^5$, there are subspaces U and V with $\dim(U) = 3$ and $\dim(V) = 2$ such that $U \cap V = (0)$; for example, let U be generated by the vectors $(1, 0, 0, 0, 0), (0, 1, 0, 0, 0), (0, 0, 1, 0, 0)$, and V be generated by the vectors $(0, 0, 0, 1, 0)$ and $(0, 0, 0, 0, 1)$. However, we claim that if $\dim(U) = 3$ and $\dim(V) = 3$, then $\dim(U \cap V) \geq 1$. Indeed, by the Grassmann relation, we have

$$\dim(U) + \dim(V) = \dim(U + V) + \dim(U \cap V),$$

namely

$$3 + 3 = 6 = \dim(U + V) + \dim(U \cap V),$$

and since $U + V$ is a subspace of $\mathbb{R}^5$, $\dim(U + V) \leq 5$, which implies

$$6 \leq 5 + \dim(U \cap V),$$

that is $1 \leq \dim(U \cap V)$.

As another consequence of Proposition 5.8, if U and V are two hyperplanes in a vector space of dimension n, so that $\dim(U) = n - 1$ and $\dim(V) = n - 1$, the reader should show that

$$\dim(U \cap V) \geq n - 2,$$

and so, if $U \neq V$, then

$$\dim(U \cap V) = n - 2.$$

Here is a characterization of direct sums that follows directly from Theorem 5.1.

Proposition 5.9. *If $U_1, \ldots, U_p$ are any subspaces of a finite dimensional vector space E, then*

$$\dim(U_1 + \cdots + U_p) \leq \dim(U_1) + \cdots + \dim(U_p),$$

and

$$\dim(U_1 + \cdots + U_p) = \dim(U_1) + \cdots + \dim(U_p)$$

iff the U_is form a direct sum $U_1 \oplus \cdots \oplus U_p$.

Proof. If we apply Theorem 5.1 to the linear map

$$a \colon U_1 \times \cdots \times U_p \to U_1 + \cdots + U_p$$

given by $a(u_1, \ldots, u_p) = u_1 + \cdots + u_p$, we get

$$\begin{aligned}
\dim(U_1 + \cdots + U_p) &= \dim(U_1 \times \cdots \times U_p) - \dim(\operatorname{Ker} a) \\
&= \dim(U_1) + \cdots + \dim(U_p) - \dim(\operatorname{Ker} a),
\end{aligned}$$

so the inequality follows. Since a is injective iff $\operatorname{Ker} a = (0)$, the U_is form a direct sum iff the second equation holds. $\qquad\square$

Another important corollary of Theorem 5.1 is the following result:

Proposition 5.10. *Let E and F be two vector spaces with the same finite dimension $\dim(E) = \dim(F) = n$. For every linear map $f \colon E \to F$, the following properties are equivalent:*

(a) f is bijective.
(b) f is surjective.
(c) f is injective.
(d) $\operatorname{Ker} f = (0)$.

Proof. Obviously, (a) implies (b).

If f is surjective, then $\operatorname{Im} f = F$, and so $\dim(\operatorname{Im} f) = n$. By Theorem 5.1,

$$\dim(E) = \dim(\operatorname{Ker} f) + \dim(\operatorname{Im} f),$$

and since $\dim(E) = n$ and $\dim(\operatorname{Im} f) = n$, we get $\dim(\operatorname{Ker} f) = 0$, which means that $\operatorname{Ker} f = (0)$, and so f is injective (see Proposition 2.13). This proves that (b) implies (c).

If f is injective, then by Proposition 2.13, $\operatorname{Ker} f = (0)$, so (c) implies (d).

Finally, assume that $\operatorname{Ker} f = (0)$, so that $\dim(\operatorname{Ker} f) = 0$ and f is injective (by Proposition 2.13). By Theorem 5.1,

$$\dim(E) = \dim(\operatorname{Ker} f) + \dim(\operatorname{Im} f),$$

and since $\dim(\operatorname{Ker} f) = 0$, we get

$$\dim(\operatorname{Im} f) = \dim(E) = \dim(F),$$

which proves that f is also surjective, and thus bijective. This proves that (d) implies (a) and concludes the proof. $\qquad\square$

One should be warned that Proposition 5.10 fails in infinite dimension. A linear map may be injective without being surjective and vice versa.

Here are a few applications of Proposition 5.10. Let A be an $n \times n$ matrix and assume that A some right inverse B, which means that B is an $n \times n$ matrix such that

$$AB = I.$$

The linear map associated with A is surjective, since for every $u \in \mathbb{R}^n$, we have $A(Bu) = u$. By Proposition 5.10, this map is bijective so B is actually the inverse of A; in particular $BA = I$.

Similarly, assume that A has a left inverse B, so that

$$BA = I.$$

This time the linear map associated with A is injective, because if $Au = 0$, then $BAu = B0 = 0$, and since $BA = I$ we get $u = 0$. Again, by Proposition 5.10, this map is bijective so B is actually the inverse of A; in particular $AB = I$.

Now assume that the linear system $Ax = b$ has some solution for every b. Then the linear map associated with A is surjective and by Proposition 5.10, A is invertible.

Finally assume that the linear system $Ax = b$ has at most one solution for every b. Then the linear map associated with A is injective and by Proposition 5.10, A is invertible.

We also have the following basic proposition about injective or surjective linear maps.

Proposition 5.11. *Let E and F be vector spaces, and let $f \colon E \to F$ be a linear map. If $f \colon E \to F$ is injective, then there is a surjective linear map $r \colon F \to E$ called a retraction, such that $r \circ f = \mathrm{id}_E$. See Figure 5.3. If $f \colon E \to F$ is surjective, then there is an injective linear map $s \colon F \to E$ called a section, such that $f \circ s = \mathrm{id}_F$. See Figure 5.2.*

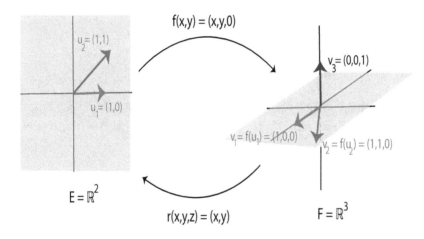

Fig. 5.3   Let $f \colon E \to F$ be the injective linear map from $\mathbb{R}^2$ to $\mathbb{R}^3$ given by $f(x, y) = (x, y, 0)$. Then a surjective retraction is given by $r \colon \mathbb{R}^3 \to \mathbb{R}^2$ is given by $r(x, y, z) = (x, y)$. Observe that $r(v_1) = u_1$, $r(v_2) = u_2$, and $r(v_3) = 0$.

Proof. Let $(u_i)_{i \in I}$ be a basis of E. Since $f \colon E \to F$ is an injective linear map, by Proposition 2.14, $(f(u_i))_{i \in I}$ is linearly independent in F. By Theorem 2.1, there is a basis $(v_j)_{j \in J}$ of F, where $I \subseteq J$, and where $v_i = f(u_i)$, for all $i \in I$. By Proposition 2.14, a linear map $r \colon F \to E$ can be defined such that $r(v_i) = u_i$, for all $i \in I$, and $r(v_j) = w$ for all $j \in (J - I)$, where w is any given vector in E, say $w = 0$. Since $r(f(u_i)) = u_i$ for all $i \in I$, by Proposition 2.14, we have $r \circ f = \mathrm{id}_E$.

Now assume that $f \colon E \to F$ is surjective. Let $(v_j)_{j \in J}$ be a basis of F. Since $f \colon E \to F$ is surjective, for every $v_j \in F$, there is some $u_j \in E$

such that $f(u_j) = v_j$. Since $(v_j)_{j \in J}$ is a basis of F, by Proposition 2.14, there is a unique linear map $s \colon F \to E$ such that $s(v_j) = u_j$. Also since $f(s(v_j)) = v_j$, by Proposition 2.14 (again), we must have $f \circ s = \mathrm{id}_F$. $\square$

Remark: Proposition 5.11 also holds if E or F has infinite dimension.

The converse of Proposition 5.11 is obvious.

The notion of rank of a linear map or of a matrix important, both theoretically and practically, since it is the key to the solvability of linear equations. We have the following simple proposition.

Proposition 5.12. *Given a linear map $f \colon E \to F$, the following properties hold:*

(i) $\mathrm{rk}(f) + \dim(\mathrm{Ker}\, f) = \dim(E)$.
(ii) $\mathrm{rk}(f) \leq \min(\dim(E), \dim(F))$.

Proof. Property (i) follows from Proposition 5.1. As for (ii), since $\mathrm{Im}\, f$ is a subspace of F, we have $\mathrm{rk}(f) \leq \dim(F)$, and since $\mathrm{rk}(f) + \dim(\mathrm{Ker}\, f) = \dim(E)$, we have $\mathrm{rk}(f) \leq \dim(E)$. $\square$

The rank of a matrix is defined as follows.

Definition 5.5. Given a $m \times n$-matrix $A = (a_{ij})$, the *rank* $\mathrm{rk}(A)$ of the matrix A is the maximum number of linearly independent columns of A (viewed as vectors in $\mathbb{R}^m$).

In view of Proposition 2.8, the rank of a matrix A is the dimension of the subspace of $\mathbb{R}^m$ generated by the columns of A. Let E and F be two vector spaces, and let $(u_1, \ldots, u_n)$ be a basis of E, and $(v_1, \ldots, v_m)$ a basis of F. Let $f \colon E \to F$ be a linear map, and let $M(f)$ be its matrix w.r.t. the bases $(u_1, \ldots, u_n)$ and $(v_1, \ldots, v_m)$. Since the rank $\mathrm{rk}(f)$ of f is the dimension of $\mathrm{Im}\, f$, which is generated by $(f(u_1), \ldots, f(u_n))$, the rank of f is the maximum number of linearly independent vectors in $(f(u_1), \ldots, f(u_n))$, which is equal to the number of linearly independent columns of $M(f)$, since F and $\mathbb{R}^m$ are isomorphic. *Thus, we have $\mathrm{rk}(f) = \mathrm{rk}(M(f))$, for every matrix representing f.*

We will see later, using duality, that the rank of a matrix A is also equal to the maximal number of linearly independent rows of A.

5.4 Affine Maps

We showed in Section 2.7 that every linear map f must send the zero vector to the zero vector; that is,

$$f(0) = 0.$$

Yet for any fixed nonzero vector $u \in E$ (where E is any vector space), the function t_u given by

$$t_u(x) = x + u, \quad \text{for all} \quad x \in E$$

shows up in practice (for example, in robotics). Functions of this type are called *translations*. They are *not* linear for $u \neq 0$, since $t_u(0) = 0 + u = u$.

More generally, functions combining linear maps and translations occur naturally in many applications (robotics, computer vision, *etc.*), so it is necessary to understand some basic properties of these functions. For this, the notion of affine combination turns out to play a key role.

Recall from Section 2.7 that for any vector space E, given any family $(u_i)_{i \in I}$ of vectors $u_i \in E$, an *affine combination* of the family $(u_i)_{i \in I}$ is an expression of the form

$$\sum_{i \in I} \lambda_i u_i \quad \text{with} \quad \sum_{i \in I} \lambda_i = 1,$$

where $(\lambda_i)_{i \in I}$ is a family of scalars.

A linear combination places no restriction on the scalars involved, but an affine combination is a linear combination *with the restriction that the scalars λ_i must add up to* 1. Nevertheless, a linear combination can always be viewed as an affine combination using the following trick involving 0. For any family $(u_i)_{i \in I}$ of vectors in E and for *any* family of scalars $(\lambda_i)_{i \in I}$, we can write the linear combination $\sum_{i \in I} \lambda_i u_i$ as an affine combination as follows:

$$\sum_{i \in I} \lambda_i u_i = \sum_{i \in I} \lambda_i u_i + \left(1 - \sum_{i \in I} \lambda_i\right) 0.$$

Affine combinations are also called *barycentric combinations*.

Although this is not obvious at first glance, the condition that the scalars λ_i add up to 1 ensures that affine combinations are preserved under translations. To make this precise, consider functions $f \colon E \to F$, where E and F are two vector spaces, such that there is some *linear map* $h \colon E \to F$ and some fixed vector $b \in F$ (a *translation vector*), such that

$$f(x) = h(x) + b, \quad \text{for all} \quad x \in E.$$

The map f given by

$$\begin{pmatrix} x_1 \\ x_2 \end{pmatrix} \mapsto \begin{pmatrix} 8/5 & -6/5 \\ 3/10 & 2/5 \end{pmatrix} \begin{pmatrix} x_1 \\ x_2 \end{pmatrix} + \begin{pmatrix} 1 \\ 1 \end{pmatrix}$$

is an example of the composition of a linear map with a translation.

We claim that functions of this type preserve affine combinations.

Proposition 5.13. *For any two vector spaces E and F, given any function $f: E \to F$ defined such that*

$$f(x) = h(x) + b, \quad \text{for all} \quad x \in E,$$

where $h: E \to F$ is a linear map and b is some fixed vector in F, for every affine combination $\sum_{i \in I} \lambda_i u_i$ (with $\sum_{i \in I} \lambda_i = 1$), we have

$$f\left(\sum_{i \in I} \lambda_i u_i\right) = \sum_{i \in I} \lambda_i f(u_i).$$

In other words, f preserves affine combinations.

Proof. By definition of f, using the fact that h is linear and the fact that $\sum_{i \in I} \lambda_i = 1$, we have

$$f\left(\sum_{i \in} \lambda_i u_i\right) = h\left(\sum_{i \in I} \lambda_i u_i\right) + b$$

$$= \sum_{i \in I} \lambda_i h(u_i) + 1b$$

$$= \sum_{i \in I} \lambda_i h(u_i) + \left(\sum_{i \in I} \lambda_i\right) b$$

$$= \sum_{i \in I} \lambda_i (h(u_i) + b)$$

$$= \sum_{i \in I} \lambda_i f(u_i),$$

as claimed. $\qquad\square$

Observe how the fact that $\sum_{i \in I} \lambda_i = 1$ was used in a crucial way in Line 3. Surprisingly, the converse of Proposition 5.13 also holds.

Proposition 5.14. *For any two vector spaces E and F, let $f: E \to F$ be any function that preserves affine combinations, i.e., for every affine combination $\sum_{i \in I} \lambda_i u_i$ (with $\sum_{i \in I} \lambda_i = 1$), we have*

$$f\left(\sum_{i \in I} \lambda_i u_i\right) = \sum_{i \in I} \lambda_i f(u_i).$$

Then for any $a \in E$, the function $h \colon E \to F$ given by

$$h(x) = f(a + x) - f(a)$$

is a linear map independent of a, and

$$f(a + x) = h(x) + f(a), \quad \text{for all} \quad x \in E.$$

In particular, for $a = 0$, if we let $c = f(0)$, then

$$f(x) = h(x) + c, \quad \text{for all} \quad x \in E.$$

Proof. First, let us check that h is linear. Since f preserves affine combinations and since $a + u + v = (a + u) + (a + v) - a$ is an affine combination $(1 + 1 - 1 = 1)$, we have

$$\begin{aligned}
h(u + v) &= f(a + u + v) - f(a) \\
&= f((a + u) + (a + v) - a) - f(a) \\
&= f(a + u) + f(a + v) - f(a) - f(a) \\
&= f(a + u) - f(a) + f(a + v) - f(a) \\
&= h(u) + h(v).
\end{aligned}$$

This proves that

$$h(u + v) = h(u) + h(v), \quad u, v \in E.$$

Observe that $a + \lambda u = \lambda(a + u) + (1 - \lambda)a$ is also an affine combination $(\lambda + 1 - \lambda = 1)$, so we have

$$\begin{aligned}
h(\lambda u) &= f(a + \lambda u) - f(a) \\
&= f(\lambda(a + u) + (1 - \lambda)a) - f(a) \\
&= \lambda f(a + u) + (1 - \lambda)f(a) - f(a) \\
&= \lambda(f(a + u) - f(a)) \\
&= \lambda h(u).
\end{aligned}$$

This proves that

$$h(\lambda u) = \lambda h(u), \quad u \in E, \ \lambda \in \mathbb{R}.$$

Therefore, h is indeed linear.

For any $b \in E$, since $b + u = (a + u) - a + b$ is an affine combination $(1 - 1 + 1 = 1)$, we have

$$\begin{aligned}
f(b + u) - f(b) &= f((a + u) - a + b) - f(b) \\
&= f(a + u) - f(a) + f(b) - f(b) \\
&= f(a + u) - f(a),
\end{aligned}$$

which proves that for all $a, b \in E$,

$$f(b + u) - f(b) = f(a + u) - f(a), \quad u \in E.$$

Therefore $h(x) = f(a + u) - f(a)$ does not depend on a, and it is obvious by the definition of h that

$$f(a + x) = h(x) + f(a), \quad \text{for all} \quad x \in E.$$

For $a = 0$, we obtain the last part of our proposition. $\qquad \square$

We should think of a as a *chosen origin* in E. The function f maps the origin a in E to the origin $f(a)$ in F. Proposition 5.14 shows that the definition of h does not depend on the origin chosen in E. Also, since

$$f(x) = h(x) + c, \quad \text{for all} \quad x \in E$$

for some fixed vector $c \in F$, we see that f is the composition of the linear map h with the translation t_c (in F).

The unique linear map h as above is called the *linear map associated with f*, and it is sometimes denoted by $\overrightarrow{f}$.

In view of Propositions 5.13 and 5.14, it is natural to make the following definition.

Definition 5.6. For any two vector spaces E and F, a function $f \colon E \to F$ is an *affine map* if f preserves affine combinations, *i.e.*, for every affine combination $\sum_{i \in I} \lambda_i u_i$ (with $\sum_{i \in I} \lambda_i = 1$), we have

$$f\left(\sum_{i \in I} \lambda_i u_i \right) = \sum_{i \in I} \lambda_i f(u_i).$$

Equivalently, a function $f \colon E \to F$ is an *affine map* if there is some linear map $h \colon E \to F$ (also denoted by $\overrightarrow{f}$) and some fixed vector $c \in F$ such that

$$f(x) = h(x) + c, \quad \text{for all} \quad x \in E.$$

Note that a linear map always maps the standard origin 0 in E to the standard origin 0 in F. However an affine map usually maps 0 to a nonzero vector $c = f(0)$. This is the "translation component" of the affine map.

When we deal with affine maps, it is often fruitful to think of the elements of E and F not only as vectors but also as *points*. In this point of view, *points can only be combined using affine combinations*, but vectors can be combined in an unrestricted fashion using linear combinations. We can also think of $u + v$ as the *result of translating the point u by the translation t_v*. These ideas lead to the definition of *affine spaces*.

The idea is that instead of a single space E, an affine space consists of two sets E and $\overrightarrow{E}$, where E is just an unstructured set of points, and $\overrightarrow{E}$ is a vector space. Furthermore, the vector space $\overrightarrow{E}$ acts on E. We can think of $\overrightarrow{E}$ as a set of *translations* specified by vectors, and given any point $a \in E$ and any vector (translation) $u \in \overrightarrow{E}$, the result of translating a by u is the point (not vector) $a + u$. Formally, we have the following definition.

Definition 5.7. An *affine space* is either the degenerate space reduced to the empty set, or a triple $\langle E, \overrightarrow{E}, + \rangle$ consisting of a nonempty set E (of *points*), a vector space $\overrightarrow{E}$ (of *translations*, or *free vectors*), and an action $+ \colon E \times \overrightarrow{E} \to E$, satisfying the following conditions.

(A1) $a + 0 = a$, for every $a \in E$.

(A2) $(a + u) + v = a + (u + v)$, for every $a \in E$, and every $u, v \in \overrightarrow{E}$.

(A3) For any two points $a, b \in E$, there is a unique $u \in \overrightarrow{E}$ such that $a + u = b$.

The unique vector $u \in \overrightarrow{E}$ such that $a + u = b$ is denoted by $\overrightarrow{ab}$, or sometimes by **ab**, or even by $b - a$. Thus, we also write

$$b = a + \overrightarrow{ab}$$

(or $b = a + \mathbf{ab}$, or even $b = a + (b - a)$).

It is important to note that *adding or rescaling points does not make sense*! However, using the fact that $\overrightarrow{E}$ acts on E is a special way (this action is transitive and faithful), it is possible to define rigorously the notion of *affine combinations* of points and to define affine spaces, affine maps, *etc.* However, this would lead us to far afield, and for our purposes it is enough to stick to vector spaces and we will not distinguish between vector addition $+$ and translation of a point by a vector $+$. Still, one should be aware that affine combinations really apply to points, and that points are not vectors!

If E and F are finite dimensional vector spaces with $\dim(E) = n$ and $\dim(F) = m$, then it is useful to represent an affine map with respect to bases in E in F. However, the translation part c of the affine map must be somehow incorporated. There is a standard trick to do this which amounts to viewing an affine map as a linear map between spaces of dimension $n + 1$ and $m + 1$. We also have the extra flexibility of choosing origins $a \in E$ and $b \in F$.

Let $(u_1, \ldots, u_n)$ be a basis of E, $(v_1, \ldots, v_m)$ be a basis of F, and let $a \in E$ and $b \in F$ be any two fixed vectors viewed as *origins*. Our affine map f has the property that if $v = f(u)$, then

$$v - b = f(a + u - a) - b = f(a) - b + h(u - a),$$

where the last equality made use of the fact that $h(x) = f(a + x) - f(a)$. If we let $y = v - b$, $x = u - a$, and $d = f(a) - b$, then

$$y = h(x) + d, \quad x \in E.$$

Over the basis $\mathcal{U} = (u_1, \ldots, u_n)$, we write

$$x = x_1 u_1 + \cdots + x_n u_n,$$

and over the basis $\mathcal{V} = (v_1, \ldots, v_m)$, we write

$$y = y_1 v_1 + \cdots + y_m v_m,$$
$$d = d_1 v_1 + \cdots + d_m v_m.$$

Then since

$$y = h(x) + d,$$

if we let A be the $m \times n$ matrix representing the linear map h, that is, the jth column of A consists of the coordinates of $h(u_j)$ over the basis $(v_1, \ldots, v_m)$, then we can write

$$y_{\mathcal{V}} = A x_{\mathcal{U}} + d_{\mathcal{V}}$$

where $x_{\mathcal{U}} = (x_1, \ldots, x_n)^\top$, $y_{\mathcal{V}} = (y_1, \ldots, y_m)^\top$, and $d_{\mathcal{V}} = (d_1, \ldots, d_m)^\top$. The above is the matrix representation of our affine map f with respect to $(a, (u_1, \ldots, u_n))$ and $(b, (v_1, \ldots, v_m))$.

The reason for using the origins a and b is that it gives us more flexibility. In particular, we can choose $b = f(a)$, and then f behaves like a linear map with respect to the origins a and $b = f(a)$.

When $E = F$, if there is some $a \in E$ such that $f(a) = a$ (a is a *fixed point* of f), then we can pick $b = a$. Then because $f(a) = a$, we get

$$v = f(u) = f(a + u - a) = f(a) + h(u - a) = a + h(u - a),$$

that is

$$v - a = h(u - a).$$

With respect to the new origin a, if we define x and y by

$$x = u - a$$
$$y = v - a,$$

then we get

$$y = h(x).$$

Therefore, f really behaves like a linear map, but *with respect to the new origin a* (not the *standard origin* 0). This is the case of a rotation around an axis that does not pass through the origin.

Remark: A pair $(a, (u_1, \ldots, u_n))$ where $(u_1, \ldots, u_n)$ is a basis of E and a is an origin chosen in E is called an *affine frame.*

We now describe the trick which allows us to incorporate the translation part d into the matrix A. We define the $(m + 1) \times (n + 1)$ matrix A' obtained by first adding d as the $(n+1)$th column and then $(\underbrace{0, \ldots, 0}_{n}, 1)$ as the $(m + 1)$th row:

$$A' = \begin{pmatrix} A & d \\ 0_n & 1 \end{pmatrix}.$$

It is clear that

$$\begin{pmatrix} y \\ 1 \end{pmatrix} = \begin{pmatrix} A & d \\ 0_n & 1 \end{pmatrix} \begin{pmatrix} x \\ 1 \end{pmatrix}$$

iff

$$y = Ax + d.$$

This amounts to considering a point $x \in \mathbb{R}^n$ as a point $(x, 1)$ *in the (affine) hyperplane H_{n+1} in $\mathbb{R}^{n+1}$ of equation $x_{n+1} = 1$.* Then an affine map is the restriction to the hyperplane H_{n+1} of the linear map $\widehat{f}$ from $\mathbb{R}^{n+1}$ to $\mathbb{R}^{m+1}$ corresponding to the matrix A' which maps H_{n+1} into H_{m+1} $(\widehat{f}(H_{n+1}) \subseteq H_{m+1})$. Figure 5.4 illustrates this process for $n = 2$.

For example, the map

$$\begin{pmatrix} x_1 \\ x_2 \end{pmatrix} \mapsto \begin{pmatrix} 1 & 1 \\ 1 & 3 \end{pmatrix} \begin{pmatrix} x_1 \\ x_2 \end{pmatrix} + \begin{pmatrix} 3 \\ 0 \end{pmatrix}$$

defines an affine map f which is represented in $\mathbb{R}^3$ by

$$\begin{pmatrix} x_1 \\ x_2 \\ 1 \end{pmatrix} \mapsto \begin{pmatrix} 1 & 1 & 3 \\ 1 & 3 & 0 \\ 0 & 0 & 1 \end{pmatrix} \begin{pmatrix} x_1 \\ x_2 \\ 1 \end{pmatrix}.$$

It is easy to check that the point $a = (6, -3)$ is fixed by f, which means that $f(a) = a$, so by translating the coordinate frame to the origin a, the affine map behaves like a linear map.

The idea of considering $\mathbb{R}^n$ as an hyperplane in $\mathbb{R}^{n+1}$ can be used to define *projective maps.*

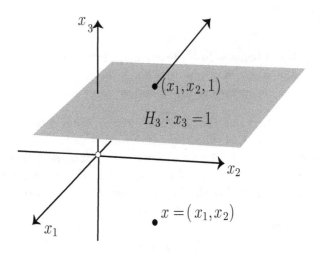

Fig. 5.4   Viewing $\mathbb{R}^n$ as a hyperplane in $\mathbb{R}^{n+1}$ $(n = 2)$.

5.5   Summary

The main concepts and results of this chapter are listed below:

- *Direct products, sums, direct sums.*
- *Projections.*
- The fundamental equation

$$\dim(E) = \dim(\operatorname{Ker} f) + \dim(\operatorname{Im} f) = \dim(\operatorname{Ker} f) + \operatorname{rk}(f)$$

(The rank-nullity theorem; Theorem 5.1).
- *Grassmann's relation*

$$\dim(U) + \dim(V) = \dim(U + V) + \dim(U \cap V).$$

- Characterizations of a bijective linear map $f \colon E \to F$.
- *Rank* of a matrix.
- *Affine Maps.*

5.6   Problems

Problem 5.1. Let V and W be two subspaces of a vector space E. Prove that if $V \cup W$ is a subspace of E, then either $V \subseteq W$ or $W \subseteq V$.

Problem 5.2. Prove that for every vector space E, if $f\colon E \to E$ is an idempotent linear map, i.e., $f \circ f = f$, then we have a direct sum

$$E = \operatorname{Ker} f \oplus \operatorname{Im} f,$$

so that f is the projection onto its image $\operatorname{Im} f$.

Problem 5.3. Let $U_1, \ldots, U_p$ be any $p \geq 2$ subspaces of some vector space E and recall that the linear map

$$a\colon U_1 \times \cdots \times U_p \to E$$

is given by

$$a(u_1, \ldots, u_p) = u_1 + \cdots + u_p,$$

with $u_i \in U_i$ for $i = 1, \ldots, p$.

(1) If we let $Z_i \subseteq U_1 \times \cdots \times U_p$ be given by

$$Z_i = \left\{ \left(u_1, \ldots, u_{i-1}, -\sum_{j=1, j \neq i}^{p} u_j, u_{i+1}, \ldots, u_p\right) \;\middle|\; \sum_{j=1, j \neq i}^{p} u_j \in U_i \cap \left(\sum_{j=1, j \neq i}^{p} U_j\right) \right\},$$

for $i = 1, \ldots, p$, then prove that

$$\operatorname{Ker} a = Z_1 = \cdots = Z_p.$$

In general, for any given i, the condition $U_i \cap \left(\sum_{j=1, j \neq i}^{p} U_j\right) = (0)$ does not necessarily imply that $Z_i = (0)$. Thus, let

$$Z = \left\{ \left(u_1, \ldots, u_{i-1}, u_i, u_{i+1}, \ldots, u_p\right) \;\middle|\; u_i = -\sum_{j=1, j \neq i}^{p} u_j, \; u_i \in U_i \cap \left(\sum_{j=1, j \neq i}^{p} U_j\right), \; 1 \leq i \leq p \right\}.$$

Since $\operatorname{Ker} a = Z_1 = \cdots = Z_p$, we have $Z = \operatorname{Ker} a$. Prove that if

$$U_i \cap \left(\sum_{j=1, j \neq i}^{p} U_j\right) = (0) \quad 1 \leq i \leq p,$$

then $Z = \operatorname{Ker} a = (0)$.

(2) Prove that $U_1 + \cdots + U_p$ is a direct sum iff

$$U_i \cap \left(\sum_{j=1, j \neq i}^{p} U_j\right) = (0) \quad 1 \leq i \leq p.$$

Problem 5.4. Assume that E is finite-dimensional, and let $f_i \colon E \to E$ be any $p \geq 2$ linear maps such that

$$f_1 + \cdots + f_p = \mathrm{id}_E.$$

Prove that the following properties are equivalent:

(1) $f_i^2 = f_i$, $1 \leq i \leq p$.
(2) $f_j \circ f_i = 0$, for all $i \neq j$, $1 \leq i, j \leq p$.

Hint. Use Problem 5.2.

Let $U_1, \ldots, U_p$ be any $p \geq 2$ subspaces of some vector space E. Prove that $U_1 + \cdots + U_p$ is a direct sum iff

$$U_i \cap \left(\sum_{j=1}^{i-1} U_j \right) = (0), \quad i = 2, \ldots, p.$$

Problem 5.5. Given any vector space E, a linear map $f \colon E \to E$ is an *involution* if $f \circ f = \mathrm{id}$.

(1) Prove that an involution f is invertible. What is its inverse?

(2) Let E_1 and E_{-1} be the subspaces of E defined as follows:

$$E_1 = \{ u \in E \mid f(u) = u \}$$
$$E_{-1} = \{ u \in E \mid f(u) = -u \}.$$

Prove that we have a direct sum

$$E = E_1 \oplus E_{-1}.$$

Hint. For every $u \in E$, write

$$u = \frac{u + f(u)}{2} + \frac{u - f(u)}{2}.$$

(3) If E is finite-dimensional and f is an involution, prove that there is some basis of E with respect to which the matrix of f is of the form

$$I_{k,n-k} = \begin{pmatrix} I_k & 0 \\ 0 & -I_{n-k} \end{pmatrix},$$

where I_k is the $k \times k$ identity matrix (similarly for I_{n-k}) and $k = \dim(E_1)$. Can you give a geometric interpretation of the action of f (especially when $k = n - 1$)?

Problem 5.6. An $n \times n$ matrix H is *upper Hessenberg* if $h_{jk} = 0$ for all (j, k) such that $j - k \geq 0$. An upper Hessenberg matrix is *unreduced* if $h_{i+1 i} \neq 0$ for $i = 1, \ldots, n - 1$.

Prove that if H is a singular unreduced upper Hessenberg matrix, then $\dim(\mathrm{Ker}\,(H)) = 1$.

Problem 5.7. Let A be any $n \times k$ matrix.

(1) Prove that the $k \times k$ matrix $A^\top A$ and the matrix A have the same nullspace. Use this to prove that $\operatorname{rank}(A^\top A) = \operatorname{rank}(A)$. Similarly, prove that the $n \times n$ matrix $AA^\top$ and the matrix $A^\top$ have the same nullspace, and conclude that $\operatorname{rank}(AA^\top) = \operatorname{rank}(A^\top)$.

We will prove later that $\operatorname{rank}(A^\top) = \operatorname{rank}(A)$.

(2) Let $a_1, \ldots, a_k$ be k linearly independent vectors in $\mathbb{R}^n$ $(1 \leq k \leq n)$, and let A be the $n \times k$ matrix whose ith column is a_i. Prove that $A^\top A$ has rank k, and that it is invertible. Let $P = A(A^\top A)^{-1}A^\top$ (an $n \times n$ matrix). Prove that

$$P^2 = P$$
$$P^\top = P.$$

What is the matrix P when $k = 1$?

(3) Prove that the image of P is the subspace V spanned by $a_1, \ldots, a_k$, or equivalently the set of all vectors in $\mathbb{R}^n$ of the form Ax, with $x \in \mathbb{R}^k$. Prove that the nullspace U of P is the set of vectors $u \in \mathbb{R}^n$ such that $A^\top u = 0$. Can you give a geometric interpretation of U?

Conclude that P is a projection of $\mathbb{R}^n$ onto the subspace V spanned by $a_1, \ldots, a_k$, and that

$$\mathbb{R}^n = U \oplus V.$$

Problem 5.8. A *rotation* R_θ in the plane $\mathbb{R}^2$ is given by the matrix

$$R_\theta = \begin{pmatrix} \cos\theta & -\sin\theta \\ \sin\theta & \cos\theta \end{pmatrix}.$$

(1) Use `Matlab` to show the action of a rotation R_θ on a simple figure such as a triangle or a rectangle, for various values of θ, including $\theta = \pi/6, \pi/4, \pi/3, \pi/2$.

(2) Prove that R_θ is invertible and that its inverse is $R_{-\theta}$.

(3) For any two rotations R_α and R_β, prove that

$$R_\beta \circ R_\alpha = R_\alpha \circ R_\beta = R_{\alpha+\beta}.$$

Use (2)–(3) to prove that the rotations in the plane form a commutative group denoted **SO**(2).

Problem 5.9. Consider the affine map $R_{\theta,(a_1,a_2)}$ in $\mathbb{R}^2$ given by

$$\begin{pmatrix} y_1 \\ y_2 \end{pmatrix} = \begin{pmatrix} \cos\theta & -\sin\theta \\ \sin\theta & \cos\theta \end{pmatrix} \begin{pmatrix} x_1 \\ x_2 \end{pmatrix} + \begin{pmatrix} a_1 \\ a_2 \end{pmatrix}.$$

(1) Prove that if $\theta \neq k2\pi$, with $k \in \mathbb{Z}$, then $R_{\theta,(a_1,a_2)}$ has a unique fixed point (c_1, c_2), that is, there is a unique point (c_1, c_2) such that

$$\begin{pmatrix} c_1 \\ c_2 \end{pmatrix} = R_{\theta,(a_1,a_2)} \begin{pmatrix} c_1 \\ c_2 \end{pmatrix},$$

and this fixed point is given by

$$\begin{pmatrix} c_1 \\ c_2 \end{pmatrix} = \frac{1}{2\sin(\theta/2)} \begin{pmatrix} \cos(\pi/2 - \theta/2) & -\sin(\pi/2 - \theta/2) \\ \sin(\pi/2 - \theta/2) & \cos(\pi/2 - \theta/2) \end{pmatrix} \begin{pmatrix} a_1 \\ a_2 \end{pmatrix}.$$

(2) In this question we still assume that $\theta \neq k2\pi$, with $k \in \mathbb{Z}$. By translating the coordinate system with origin $(0, 0)$ to the new coordinate system with origin (c_1, c_2), which means that if (x_1, x_2) are the coordinates with respect to the standard origin $(0, 0)$ and if (x_1', x_2') are the coordinates with respect to the new origin (c_1, c_2), we have

$$x_1 = x_1' + c_1$$
$$x_2 = x_2' + c_2$$

and similarly for (y_1, y_2) and (y_1', y_2'), then show that

$$\begin{pmatrix} y_1 \\ y_2 \end{pmatrix} = R_{\theta,(a_1,a_2)} \begin{pmatrix} x_1 \\ x_2 \end{pmatrix}$$

becomes

$$\begin{pmatrix} y_1' \\ y_2' \end{pmatrix} = R_\theta \begin{pmatrix} x_1' \\ x_2' \end{pmatrix}.$$

Conclude that with respect to the new origin (c_1, c_2), the affine map $R_{\theta,(a_1,a_2)}$ becomes the rotation R_θ. We say that $R_{\theta,(a_1,a_2)}$ is a *rotation of center* (c_1, c_2).

(3) Use `Matlab` to show the action of the affine map $R_{\theta,(a_1,a_2)}$ on a simple figure such as a triangle or a rectangle, for $\theta = \pi/3$ and various values of (a_1, a_2). Display the center (c_1, c_2) of the rotation.

What kind of transformations correspond to $\theta = k2\pi$, with $k \in \mathbb{Z}$?

(4) Prove that the inverse of $R_{\theta,(a_1,a_2)}$ is of the form $R_{-\theta,(b_1,b_2)}$, and find (b_1, b_2) in terms of θ and (a_1, a_2).

(5) Given two affine maps $R_{\alpha,(a_1,a_2)}$ and $R_{\beta,(b_1,b_2)}$, prove that

$$R_{\beta,(b_1,b_2)} \circ R_{\alpha,(a_1,a_2)} = R_{\alpha+\beta,(t_1,t_2)}$$

for some (t_1, t_2), and find (t_1, t_2) in terms of β, (a_1, a_2) and (b_1, b_2).

Even in the case where $(a_1, a_2) = (0, 0)$, prove that in general

$$R_{\beta,(b_1,b_2)} \circ R_\alpha \neq R_\alpha \circ R_{\beta,(b_1,b_2)}.$$

Use (4)–(5) to show that the affine maps of the plane defined in this problem form a nonabelian group denoted **SE**(2).

Prove that $R_{\beta,(b_1,b_2)} \circ R_{\alpha,(a_1,a_2)}$ is not a translation (possibly the identity) iff $\alpha + \beta \neq k2\pi$, for all $k \in \mathbb{Z}$. Find its center of rotation when $(a_1, a_2) = (0, 0)$.

If $\alpha + \beta = k2\pi$, then $R_{\beta,(b_1,b_2)} \circ R_{\alpha,(a_1,a_2)}$ is a pure translation. Find the translation vector of $R_{\beta,(b_1,b_2)} \circ R_{\alpha,(a_1,a_2)}$.

Problem 5.10. (Affine subspaces) A subset $\mathcal{A}$ of $\mathbb{R}^n$ is called an *affine subspace* if either $\mathcal{A} = \emptyset$, or there is some vector $a \in \mathbb{R}^n$ and some subspace U of $\mathbb{R}^n$ such that

$$\mathcal{A} = a + U = \{a + u \mid u \in U\}.$$

We define the dimension $\dim(\mathcal{A})$ of $\mathcal{A}$ as the dimension $\dim(U)$ of U.

(1) If $\mathcal{A} = a + U$, why is $a \in \mathcal{A}$?

What are affine subspaces of dimension 0? What are affine subspaces of dimension 1 (begin with $\mathbb{R}^2$)? What are affine subspaces of dimension 2 (begin with $\mathbb{R}^3$)?

Prove that any nonempty affine subspace is closed under affine combinations.

(2) Prove that if $\mathcal{A} = a + U$ is any nonempty affine subspace, then $\mathcal{A} = b + U$ for any $b \in \mathcal{A}$.

(3) Let $\mathcal{A}$ be any nonempty subset of $\mathbb{R}^n$ closed under affine combinations. For any $a \in \mathcal{A}$, prove that

$$U_a = \{x - a \in \mathbb{R}^n \mid x \in \mathcal{A}\}$$

is a (linear) subspace of $\mathbb{R}^n$ such that

$$\mathcal{A} = a + U_a.$$

Prove that U_a does not depend on the choice of $a \in \mathcal{A}$; that is, $U_a = U_b$ for all $a, b \in \mathcal{A}$. In fact, prove that

$$U_a = U = \{y - x \in \mathbb{R}^n \mid x, y \in \mathcal{A}\}, \quad \text{for all } a \in \mathcal{A},$$

and so

$$\mathcal{A} = a + U, \quad \text{for any } a \in \mathcal{A}.$$

Remark: The subspace U is called the *direction* of $\mathcal{A}$.

(4) Two nonempty affine subspaces $\mathcal{A}$ and $\mathcal{B}$ are said to be *parallel* iff they have the same direction. Prove that that if $\mathcal{A} \neq \mathcal{B}$ and $\mathcal{A}$ and $\mathcal{B}$ are parallel, then $\mathcal{A} \cap \mathcal{B} = \emptyset$.

Remark: The above shows that affine subspaces behave quite differently from linear subspaces.

Problem 5.11. (Affine frames and affine maps) For any vector $v = (v_1, \ldots, v_n) \in \mathbb{R}^n$, let $\widehat{v} \in \mathbb{R}^{n+1}$ be the vector $\widehat{v} = (v_1, \ldots, v_n, 1)$. Equivalently, $\widehat{v} = (\widehat{v}_1, \ldots, \widehat{v}_{n+1}) \in \mathbb{R}^{n+1}$ is the vector defined by

$$\widehat{v}_i = \begin{cases} v_i & \text{if } 1 \leq i \leq n, \\ 1 & \text{if } i = n+1. \end{cases}$$

(1) For any $m + 1$ vectors $(u_0, u_1, \ldots, u_m)$ with $u_i \in \mathbb{R}^n$ and $m \leq n$, prove that if the m vectors $(u_1 - u_0, \ldots, u_m - u_0)$ are linearly independent, then the $m + 1$ vectors $(\widehat{u}_0, \ldots, \widehat{u}_m)$ are linearly independent.

(2) Prove that if the $m+1$ vectors $(\widehat{u}_0, \ldots, \widehat{u}_m)$ are linearly independent, then for any choice of i, with $0 \leq i \leq m$, the m vectors $u_j - u_i$ for $j \in \{0, \ldots, m\}$ with $j - i \neq 0$ are linearly independent.

Any $m + 1$ vectors $(u_0, u_1, \ldots, u_m)$ such that the $m + 1$ vectors $(\widehat{u}_0, \ldots, \widehat{u}_m)$ are linearly independent are said to be *affinely independent*.

From (1) and (2), the vector $(u_0, u_1, \ldots, u_m)$ are affinely independent iff for any choice of i, with $0 \leq i \leq m$, the m vectors $u_j - u_i$ for $j \in \{0, \ldots, m\}$ with $j - i \neq 0$ are linearly independent. If $m = n$, we say that $n + 1$ affinely independent vectors $(u_0, u_1, \ldots, u_n)$ form an *affine frame* of $\mathbb{R}^n$.

(3) If $(u_0, u_1, \ldots, u_n)$ is an affine frame of $\mathbb{R}^n$, then prove that for every vector $v \in \mathbb{R}^n$, there is a unique $(n+1)$-tuple $(\lambda_0, \lambda_1, \ldots, \lambda_n) \in \mathbb{R}^{n+1}$, with $\lambda_0 + \lambda_1 + \cdots + \lambda_n = 1$, such that

$$v = \lambda_0 u_0 + \lambda_1 u_1 + \cdots + \lambda_n u_n.$$

The scalars $(\lambda_0, \lambda_1, \ldots, \lambda_n)$ are called the *barycentric* (or *affine*) *coordinates* of v w.r.t. the affine frame $(u_0, u_1, \ldots, u_n)$.

If we write $e_i = u_i - u_0$, for $i = 1, \ldots, n$, then prove that we have

$$v = u_0 + \lambda_1 e_1 + \cdots + \lambda_n e_n,$$

and since $(e_1, \ldots, e_n)$ is a basis of $\mathbb{R}^n$ (by (1) & (2)), the n-tuple $(\lambda_1, \ldots, \lambda_n)$ consists of the standard coordinates of $v - u_0$ over the basis $(e_1, \ldots, e_n)$.

Conversely, for any vector $u_0 \in \mathbb{R}^n$ and for any basis $(e_1, \ldots, e_n)$ of $\mathbb{R}^n$, let $u_i = u_0 + e_i$ for $i = 1, \ldots, n$. Prove that $(u_0, u_1, \ldots, u_n)$ is an affine frame of $\mathbb{R}^n$, and for any $v \in \mathbb{R}^n$, if

$$v = u_0 + x_1 e_1 + \cdots + x_n e_n,$$

with $(x_1, \ldots, x_n) \in \mathbb{R}^n$ (unique), then

$$v = (1 - (x_1 + \cdots + x_x))u_0 + x_1 u_1 + \cdots + x_n u_n,$$

so that $(1 - (x_1 + \cdots + x_x)), x_1, \cdots, x_n)$, are the barycentric coordinates of v w.r.t. the affine frame $(u_0, u_1, \ldots, u_n)$.

The above shows that there is a one-to-one correspondence between affine frames $(u_0, \ldots, u_n)$ and pairs $(u_0, (e_1, \ldots, e_n))$, with $(e_1, \ldots, e_n)$ a basis. Given an affine frame $(u_0, \ldots, u_n)$, we obtain the basis $(e_1, \ldots, e_n)$ with $e_i = u_i - u_0$, for $i = 1, \ldots, n$; given the pair $(u_0, (e_1, \ldots, e_n))$ where $(e_1, \ldots, e_n)$ is a basis, we obtain the affine frame $(u_0, \ldots, u_n)$, with $u_i = u_0 + e_i$, for $i = 1, \ldots, n$. There is also a one-to-one correspondence between barycentric coordinates w.r.t. the affine frame $(u_0, \ldots, u_n)$ and standard coordinates w.r.t. the basis $(e_1, \ldots, e_n)$. The barycentric cordinates $(\lambda_0, \lambda_1, \ldots, \lambda_n)$ of v (with $\lambda_0 + \lambda_1 + \cdots + \lambda_n = 1$) yield the standard coordinates $(\lambda_1, \ldots, \lambda_n)$ of $v - u_0$; the standard coordinates $(x_1, \ldots, x_n)$ of $v - u_0$ yield the barycentric coordinates $(1 - (x_1 + \cdots + x_n), x_1, \ldots, x_n)$ of v.

(4) Let $(u_0, \ldots, u_n)$ be any affine frame in $\mathbb{R}^n$ and let $(v_0, \ldots, v_n)$ be any vectors in $\mathbb{R}^m$. Prove that there is a *unique* affine map $f \colon \mathbb{R}^n \to \mathbb{R}^m$ such that

$$f(u_i) = v_i, \quad i = 0, \ldots, n.$$

(5) Let $(a_0, \ldots, a_n)$ be any affine frame in $\mathbb{R}^n$ and let $(b_0, \ldots, b_n)$ be any $n + 1$ points in $\mathbb{R}^n$. Prove that there is a unique $(n + 1) \times (n + 1)$ matrix

$$A = \begin{pmatrix} B & w \\ 0 & 1 \end{pmatrix}$$

corresponding to the unique affine map f such that

$$f(a_i) = b_i, \quad i = 0, \ldots, n,$$

in the sense that

$$A\widehat{a}_i = \widehat{b}_i, \quad i = 0, \ldots, n,$$

and that A is given by

$$A = \begin{pmatrix} \widehat{b}_0 & \widehat{b}_1 & \cdots & \widehat{b}_n \end{pmatrix} \begin{pmatrix} \widehat{a}_0 & \widehat{a}_1 & \cdots & \widehat{a}_n \end{pmatrix}^{-1}.$$

Make sure to prove that the bottom row of A is $(0, \ldots, 0, 1)$.

In the special case where $(a_0, \ldots, a_n)$ is the canonical affine frame with $a_i = e_{i+1}$ for $i = 0, \ldots, n - 1$ and $a_n = (0, \ldots, 0)$ (where e_i is the ith canonical basis vector), show that

$$\begin{pmatrix} \widehat{a}_0 & \widehat{a}_1 & \cdots & \widehat{a}_n \end{pmatrix} = \begin{pmatrix} 1 & 0 & \cdots & 0 & 0 \\ 0 & 1 & \cdots & 0 & 0 \\ \vdots & \vdots & \ddots & 0 & 0 \\ 0 & 0 & \cdots & 1 & 0 \\ 1 & 1 & \cdots & 1 & 1 \end{pmatrix}$$

and

$$\left(\widehat{a}_0\ \widehat{a}_1\ \cdots\ \widehat{a}_n\right)^{-1} = \begin{pmatrix} 1 & 0 & \cdots & 0 & 0 \\ 0 & 1 & \cdots & 0 & 0 \\ \vdots & \vdots & \ddots & 0 & 0 \\ 0 & 0 & \cdots & 1 & 0 \\ -1 & -1 & \cdots & -1 & 1 \end{pmatrix}.$$

For example, when $n = 2$, if we write $b_i = (x_i, y_i)$, then we have

$$A = \begin{pmatrix} x_1 & x_2 & x_3 \\ y_1 & y_2 & y_3 \\ 1 & 1 & 1 \end{pmatrix} \begin{pmatrix} 1 & 0 & 0 \\ 0 & 1 & 0 \\ -1 & -1 & 1 \end{pmatrix} = \begin{pmatrix} x_1 - x_3 & x_2 - x_3 & x_3 \\ y_1 - y_3 & y_2 - y_3 & y_3 \\ 0 & 0 & 1 \end{pmatrix}.$$

(6) Recall that a nonempty affine subspace $\mathcal{A}$ of $\mathbb{R}^n$ is any nonempty subset of $\mathbb{R}^n$ closed under affine combinations. For any affine map $f \colon \mathbb{R}^n \to \mathbb{R}^m$, for any affine subspace $\mathcal{A}$ of $\mathbb{R}^n$, and any affine subspace $\mathcal{B}$ of $\mathbb{R}^m$, prove that $f(\mathcal{A})$ is an affine subspace of $\mathbb{R}^m$, and that $f^{-1}(\mathcal{B})$ is an affine subspace of $\mathbb{R}^n$.

Chapter 6

Determinants

In this chapter all vector spaces are defined over an arbitrary field K. For the sake of concreteness, the reader may safely assume that $K = \mathbb{R}$.

6.1 Permutations, Signature of a Permutation

This chapter contains a review of determinants and their use in linear algebra. We begin with permutations and the signature of a permutation. Next we define multilinear maps and alternating multilinear maps. Determinants are introduced as alternating multilinear maps taking the value 1 on the unit matrix (following Emil Artin). It is then shown how to compute a determinant using the Laplace expansion formula, and the connection with the usual definition is made. It is shown how determinants can be used to invert matrices and to solve (at least in theory!) systems of linear equations (the Cramer formulae). The determinant of a linear map is defined. We conclude by defining the characteristic polynomial of a matrix (and of a linear map) and by proving the celebrated Cayley–Hamilton theorem which states that every matrix is a "zero" of its characteristic polynomial (we give two proofs; one computational, the other one more conceptual).

Determinants can be defined in several ways. For example, determinants can be defined in a fancy way in terms of the exterior algebra (or alternating algebra) of a vector space. We will follow a more algorithmic approach due to Emil Artin. No matter which approach is followed, we need a few preliminaries about permutations on a finite set. We need to show that every permutation on n elements is a product of transpositions and that the parity of the number of transpositions involved is an invariant of the permutation. Let $[n] = \{1, 2 \ldots, n\}$, where $n \in \mathbb{N}$, and $n > 0$.

Definition 6.1. A *permutation on n elements* is a bijection $\pi \colon [n] \to [n]$.

When $n = 1$, the only function from $[1]$ to $[1]$ is the constant map: $1 \mapsto 1$. Thus, we will assume that $n \geq 2$. A *transposition* is a permutation $\tau \colon [n] \to [n]$ such that, for some $i < j$ (with $1 \leq i < j \leq n$), $\tau(i) = j$, $\tau(j) = i$, and $\tau(k) = k$, for all $k \in [n] - \{i, j\}$. In other words, a transposition exchanges two distinct elements $i, j \in [n]$.

If τ is a transposition, clearly, $\tau \circ \tau = \mathrm{id}$. We will also use the terminology product of permutations (or transpositions) as a synonym for composition of permutations.

A permutation σ on n elements, say $\sigma(i) = k_i$ for $i = 1, \ldots, n$, can be represented in functional notation by the $2 \times n$ array

$$\begin{pmatrix} 1 & \cdots & i & \cdots & n \\ k_1 & \cdots & k_i & \cdots & k_n \end{pmatrix}$$

known as *Cauchy two-line notation*. For example, we have the permutation σ denoted by

$$\begin{pmatrix} 1\ 2\ 3\ 4\ 5\ 6 \\ 2\ 4\ 3\ 6\ 5\ 1 \end{pmatrix}.$$

A more concise notation often used in computer science and in combinatorics is to represent a permutation by its image, namely by the sequence

$$\sigma(1)\ \sigma(2)\ \cdots\ \sigma(n)$$

written as a row vector without commas separating the entries. The above is known as the *one-line notation*. For example, in the one-line notation, our previous permutation σ is represented by

$$2\ 4\ 3\ 6\ 5\ 1.$$

The reason for not enclosing the above sequence within parentheses is avoid confusion with the notation for cycles, for which is it customary to include parentheses.

Clearly, the composition of two permutations is a permutation and every permutation has an inverse which is also a permutation. Therefore, the set of permutations on $[n]$ is a *group* often denoted $\mathfrak{S}_n$ and called the *symmetric group* on n elements.

It is easy to show by induction that the group $\mathfrak{S}_n$ has $n!$ elements. The following proposition shows the importance of transpositions.

Proposition 6.1. *For every $n \geq 2$, every permutation $\pi \colon [n] \to [n]$ can be written as a nonempty composition of transpositions.*

Proof. We proceed by induction on n. If $n = 2$, there are exactly two permutations on [2], the transposition τ exchanging 1 and 2, and the identity. However, $\text{id}_2 = \tau^2$. Now let $n \geq 3$. If $\pi(n) = n$, since by the induction hypothesis, the restriction of π to $[n-1]$ can be written as a product of transpositions, π itself can be written as a product of transpositions. If $\pi(n) = k \neq n$, letting τ be the transposition such that $\tau(n) = k$ and $\tau(k) = n$, it is clear that $\tau \circ \pi$ leaves n invariant, and by the induction hypothesis, we have $\tau \circ \pi = \tau_m \circ \ldots \circ \tau_1$ for some transpositions, and thus

$$\pi = \tau \circ \tau_m \circ \ldots \circ \tau_1,$$

a product of transpositions (since $\tau \circ \tau = \text{id}_n$). $\qquad\square$

Remark: When $\pi = \text{id}_n$ is the identity permutation, we can agree that the composition of 0 transpositions is the identity. Proposition 6.1 shows that the transpositions generate the group of permutations $\mathfrak{S}_n$.

A transposition τ that exchanges two consecutive elements k and $k+1$ of $[n]$ $(1 \leq k \leq n-1)$ may be called a *basic* transposition. We leave it as a simple exercise to prove that every transposition can be written as a product of basic transpositions. In fact, the transposition that exchanges k and $k+p$ $(1 \leq p \leq n-k)$ can be realized using $2p-1$ basic transpositions. Therefore, the group of permutations $\mathfrak{S}_n$ is also generated by the basic transpositions.

Given a permutation written as a product of transpositions, we now show that the parity of the number of transpositions is an invariant. For this, we introduce the following function.

Definition 6.2. For every $n \geq 2$, let $\Delta \colon \mathbb{Z}^n \to \mathbb{Z}$ be the function given by

$$\Delta(x_1, \ldots, x_n) = \prod_{1 \leq i < j \leq n} (x_i - x_j).$$

More generally, for any permutation $\sigma \in \mathfrak{S}_n$, define $\Delta(x_{\sigma(1)}, \ldots, x_{\sigma(n)})$ by

$$\Delta(x_{\sigma(1)}, \ldots, x_{\sigma(n)}) = \prod_{1 \leq i < j \leq n} (x_{\sigma(i)} - x_{\sigma(j)}).$$

The expression $\Delta(x_1, \ldots, x_n)$ is often called the *discriminant* of $(x_1, \ldots, x_n)$.

$\Delta(x_1, \ldots, x_n) \neq 0$. The discriminant consists of $\binom{n}{2}$ factors. When $n = 3$,

$$\Delta(x_1, x_2, x_3) = (x_1 - x_2)(x_1 - x_3)(x_2 - x_3).$$

If σ is the permutation

$$\begin{pmatrix} 1\ 2\ 3 \\ 2\ 3\ 2 \end{pmatrix},$$

then

$$\Delta(x_{\sigma(1)}, x_{\sigma(2)}, x_{\sigma(3)}) = (x_{\sigma(1)} - x_{\sigma(2)})(x_{\sigma(1)} - x_{\sigma(3)})(x_{\sigma(2)} - x_{\sigma(3)})$$
$$= (x_2 - x_3)(x_2 - x_1)(x_3 - x_1).$$

Observe that

$$\Delta(x_{\sigma(1)}, x_{\sigma(2)}, x_{\sigma(3)}) = (-1)^2 \Delta(x_1, x_2, x_3),$$

since two transpositions applied to the identity permutation 1 2 3 (written in one-line notation) give rise to 2 3 1. This result regarding the parity of $\Delta(x_{\sigma(1)}, \dots, x_{\sigma(n)})$ is generalized by the following proposition.

Proposition 6.2. *For every basic transposition τ of $[n]$ $(n \geq 2)$, we have*

$$\Delta(x_{\tau(1)}, \dots, x_{\tau(n)}) = -\Delta(x_1, \dots, x_n).$$

The above also holds for every transposition, and more generally, for every composition of transpositions $\sigma = \tau_p \circ \cdots \circ \tau_1$, we have

$$\Delta(x_{\sigma(1)}, \dots, x_{\sigma(n)}) = (-1)^p \Delta(x_1, \dots, x_n).$$

Consequently, for every permutation σ of $[n]$, the parity of the number p of transpositions involved in any decomposition of σ as $\sigma = \tau_p \circ \cdots \circ \tau_1$ is an invariant (only depends on σ).

Proof. Suppose τ exchanges x_k and x_{k+1}. The terms $x_i - x_j$ that are affected correspond to $i = k$, or $i = k+1$, or $j = k$, or $j = k+1$. The contribution of these terms in $\Delta(x_1, \dots, x_n)$ is

$$(x_k - x_{k+1})[(x_k - x_{k+2}) \cdots (x_k - x_n)][(x_{k+1} - x_{k+2}) \cdots (x_{k+1} - x_n)]$$
$$[(x_1 - x_k) \cdots (x_{k-1} - x_k)][(x_1 - x_{k+1}) \cdots (x_{k-1} - x_{k+1})].$$

When we exchange x_k and x_{k+1}, the first factor is multiplied by -1, the second and the third factor are exchanged, and the fourth and the fifth factor are exchanged, so the whole product $\Delta(x_1, \dots, x_n)$ is is indeed multiplied by -1, that is,

$$\Delta(x_{\tau(1)}, \dots, x_{\tau(n)}) = -\Delta(x_1, \dots, x_n).$$

For the second statement, first we observe that since every transposition τ can be written as the composition of an odd number of basic transpositions (see the the remark following Proposition 6.1), we also have

$$\Delta(x_{\tau(1)}, \dots, x_{\tau(n)}) = -\Delta(x_1, \dots, x_n).$$

Next we proceed by induction on the number p of transpositions involved in the decomposition of a permutation σ.

The base case $p = 1$ has just been proven. If $p \geq 2$, if we write $\omega = \tau_{p-1} \circ \cdots \circ \tau_1$, then $\sigma = \tau_p \circ \omega$ and

$$
\begin{aligned}
\Delta(x_{\sigma(1)}, \ldots, x_{\sigma(n)}) &= \Delta(x_{\tau_p(\omega(1))}, \ldots, x_{\tau_p(\omega(n))}) \\
&= -\Delta(x_{\omega(1)}, \ldots, x_{\omega(n)}) \\
&= -(-1)^{p-1}\Delta(x_1, \ldots, x_n) \\
&= (-1)^{p}\Delta(x_1, \ldots, x_n),
\end{aligned}
$$

where we used the induction hypothesis from the second to the third line, establishing the induction hypothesis. Since $\Delta(x_{\sigma(1)}, \ldots, x_{\sigma(n)})$ only depends on σ, the equation

$$
\Delta(x_{\sigma(1)}, \ldots, x_{\sigma(n)}) = (-1)^{p}\Delta(x_1, \ldots, x_n).
$$

shows that the parity $(-1)^p$ of the number of transpositions in any decomposition of σ is an invariant. $\square$

In view of Proposition 6.2, the following definition makes sense:

Definition 6.3. For every permutation σ of $[n]$, the parity $\epsilon(\sigma)$ (or $\mathrm{sgn}(\sigma)$) of the number of transpositions involved in any decomposition of σ is called the *signature* (or *sign*) of σ.

Obviously $\epsilon(\tau) = -1$ for every transposition τ (since $(-1)^1 = -1$).

A simple way to compute the signature of a permutation is to count its number of inversions.

Definition 6.4. Given any permutation σ on n elements, we say that a pair (i, j) of indices $i, j \in \{1, \ldots, n\}$ such that $i < j$ and $\sigma(i) > \sigma(j)$ is an *inversion* of the permutation σ.

For example, the permutation σ given by

$$
\begin{pmatrix} 1\ 2\ 3\ 4\ 5\ 6 \\ 2\ 4\ 3\ 6\ 5\ 1 \end{pmatrix}
$$

has seven inversions

$$
(1, 6), \ (2, 3), \ (2, 6), \ (3, 6), \ (4, 5), \ (4, 6), \ (5, 6).
$$

Proposition 6.3. *The signature $\epsilon(\sigma)$ of any permutation σ is equal to the parity $(-1)^{I(\sigma)}$ of the number $I(\sigma)$ of inversions in σ.*

Proof. In the product
$$\Delta(x_{\sigma(1)}, \ldots, x_{\sigma(n)}) = \prod_{1 \leq i < j \leq n} (x_{\sigma(i)} - x_{\sigma(j)}),$$
the terms $x_{\sigma(i)} - x_{\sigma(j)}$ for which $\sigma(i) < \sigma(j)$ occur in $\Delta(x_1, \ldots, x_n)$, whereas the terms $x_{\sigma(i)} - x_{\sigma(j)}$ for which $\sigma(i) > \sigma(j)$ occur in $\Delta(x_1, \ldots, x_n)$ with a minus sign. Therefore, the number ν of terms in $\Delta(x_{\sigma(1)}, \ldots, x_{\sigma(n)})$ whose sign is the opposite of a term in $\Delta(x_1, \ldots, x_n)$, is equal to the number $I(\sigma)$ of inversions in σ, which implies that
$$\Delta(x_{\sigma(1)}, \ldots, x_{\sigma(n)}) = (-1)^{I(\sigma)} \Delta(x_1, \ldots, x_n).$$
By Proposition 6.2, the sign of $(-1)^{I(\sigma)}$ is equal to the signature of σ. $\quad\square$

For example, the permutation
$$\begin{pmatrix} 1\ 2\ 3\ 4\ 5\ 6 \\ 2\ 4\ 3\ 6\ 5\ 1 \end{pmatrix}$$
has odd signature since it has seven inversions and $(-1)^7 = -1$.

Remark: When $\pi = \mathrm{id}_n$ is the identity permutation, since we agreed that the composition of 0 transpositions is the identity, it it still correct that $(-1)^0 = \epsilon(\mathrm{id}) = +1$. From Proposition 6.2, it is immediate that $\epsilon(\pi' \circ \pi) = \epsilon(\pi')\epsilon(\pi)$. In particular, since $\pi^{-1} \circ \pi = \mathrm{id}_n$, we get $\epsilon(\pi^{-1}) = \epsilon(\pi)$.

We can now proceed with the definition of determinants.

6.2   Alternating Multilinear Maps

First we define multilinear maps, symmetric multilinear maps, and alternating multilinear maps.

Remark: Most of the definitions and results presented in this section also hold when K is a commutative ring and when we consider modules over K (free modules, when bases are needed).

Let $E_1, \ldots, E_n$, and F, be vector spaces over a field K, where $n \geq 1$.

Definition 6.5. A function $f \colon E_1 \times \ldots \times E_n \to F$ is a *multilinear map (or an n-linear map)* if it is linear in each argument, holding the others fixed. More explicitly, for every i, $1 \leq i \leq n$, for all $x_1 \in E_1, \ldots, x_{i-1} \in E_{i-1}$, $x_{i+1} \in E_{i+1}, \ldots, x_n \in E_n$, for all $x, y \in E_i$, for all $\lambda \in K$,
$$f(x_1, \ldots, x_{i-1}, x + y, x_{i+1}, \ldots, x_n) = f(x_1, \ldots, x_{i-1}, x, x_{i+1}, \ldots, x_n)$$
$$+ f(x_1, \ldots, x_{i-1}, y, x_{i+1}, \ldots, x_n),$$
$$f(x_1, \ldots, x_{i-1}, \lambda x, x_{i+1}, \ldots, x_n) = \lambda f(x_1, \ldots, x_{i-1}, x, x_{i+1}, \ldots, x_n).$$

When $F = K$, we call f an *n-linear form (or multilinear form)*. If $n \geq 2$ and $E_1 = E_2 = \ldots = E_n$, an n-linear map $f \colon E \times \ldots \times E \to F$ is called *symmetric*, if $f(x_1, \ldots, x_n) = f(x_{\pi(1)}, \ldots, x_{\pi(n)})$ for every permutation π on $\{1, \ldots, n\}$. An n-linear map $f \colon E \times \ldots \times E \to F$ is called *alternating*, if $f(x_1, \ldots, x_n) = 0$ whenever $x_i = x_{i+1}$ for some i, $1 \leq i \leq n - 1$ (in other words, when two adjacent arguments are equal). It does no harm to agree that when $n = 1$, a linear map is considered to be both symmetric and alternating, and we will do so.

When $n = 2$, a 2-linear map $f \colon E_1 \times E_2 \to F$ is called a *bilinear map*. We have already seen several examples of bilinear maps. Multiplication $\cdot \colon K \times K \to K$ is a bilinear map, treating K as a vector space over itself.

The operation $\langle -, - \rangle \colon E^* \times E \to K$ applying a linear form to a vector is a bilinear map.

Symmetric bilinear maps (and multilinear maps) play an important role in geometry (inner products, quadratic forms) and in differential calculus (partial derivatives).

A bilinear map is symmetric if $f(u, v) = f(v, u)$, for all $u, v \in E$.

Alternating multilinear maps satisfy the following simple but crucial properties.

Proposition 6.4. *Let* $f \colon E \times \ldots \times E \to F$ *be an n-linear alternating map, with* $n \geq 2$. *The following properties hold:*

(1)

$$f(\ldots, x_i, x_{i+1}, \ldots) = -f(\ldots, x_{i+1}, x_i, \ldots)$$

(2)

$$f(\ldots, x_i, \ldots, x_j, \ldots) = 0,$$

where $x_i = x_j$, *and* $1 \leq i < j \leq n$.
(3)

$$f(\ldots, x_i, \ldots, x_j, \ldots) = -f(\ldots, x_j, \ldots, x_i, \ldots),$$

where $1 \leq i < j \leq n$.
(4)

$$f(\ldots, x_i, \ldots) = f(\ldots, x_i + \lambda x_j, \ldots),$$

for any $\lambda \in K$, *and where* $i \neq j$.

Proof. (1) By multilinearity applied twice, we have

$$f(\ldots, x_i + x_{i+1}, x_i + x_{i+1}, \ldots) = f(\ldots, x_i, x_i, \ldots) + f(\ldots, x_i, x_{i+1}, \ldots)$$
$$+ f(\ldots, x_{i+1}, x_i, \ldots) + f(\ldots, x_{i+1}, x_{i+1}, \ldots),$$

and since f is alternating, this yields

$$0 = f(\ldots, x_i, x_{i+1}, \ldots) + f(\ldots, x_{i+1}, x_i, \ldots),$$

that is, $f(\ldots, x_i, x_{i+1}, \ldots) = -f(\ldots, x_{i+1}, x_i, \ldots)$.

(2) If $x_i = x_j$ and i and j are not adjacent, we can interchange x_i and x_{i+1}, and then x_i and x_{i+2}, *etc.*, until x_i and x_j become adjacent. By (1),

$$f(\ldots, x_i, \ldots, x_j, \ldots) = \epsilon f(\ldots, x_i, x_j, \ldots),$$

where $\epsilon = +1$ or -1, but $f(\ldots, x_i, x_j, \ldots) = 0$, since $x_i = x_j$, and (2) holds.

(3) follows from (2) as in (1). (4) is an immediate consequence of (2). $\square$

Proposition 6.4 will now be used to show a fundamental property of alternating multilinear maps. First we need to extend the matrix notation a little bit. Let E be a vector space over K. Given an $n \times n$ matrix $A = (a_{ij})$ over K, we can define a map $L(A) \colon E^n \to E^n$ as follows:

$$L(A)_1(u) = a_{11}u_1 + \cdots + a_{1n}u_n,$$

$$\cdots$$

$$L(A)_n(u) = a_{n1}u_1 + \cdots + a_{nn}u_n,$$

for all $u_1, \ldots, u_n \in E$ and with $u = (u_1, \ldots, u_n)$. It is immediately verified that $L(A)$ is linear. Then given two $n \times n$ matrices $A = (a_{ij})$ and $B = (b_{ij})$, by repeating the calculations establishing the product of matrices (just before Definition 2.14), we can show that

$$L(AB) = L(A) \circ L(B).$$

It is then convenient to use the matrix notation to describe the effect of the linear map $L(A)$, as

$$\begin{pmatrix} L(A)_1(u) \\ L(A)_2(u) \\ \vdots \\ L(A)_n(u) \end{pmatrix} = \begin{pmatrix} a_{11} & a_{12} & \ldots & a_{1n} \\ a_{21} & a_{22} & \ldots & a_{2n} \\ \vdots & \vdots & \ddots & \vdots \\ a_{n1} & a_{n2} & \ldots & a_{nn} \end{pmatrix} \begin{pmatrix} u_1 \\ u_2 \\ \vdots \\ u_n \end{pmatrix}.$$

Lemma 6.1. *Let* $f \colon E \times \ldots \times E \to F$ *be an n-linear alternating map. Let* $(u_1, \ldots, u_n)$ *and* $(v_1, \ldots, v_n)$ *be two families of n vectors, such that,*

$$v_1 = a_{11}u_1 + \cdots + a_{n1}u_n,$$

$$\cdots$$

$$v_n = a_{1n}u_1 + \cdots + a_{nn}u_n.$$

Equivalently, letting

$$A = \begin{pmatrix} a_{11} & a_{12} & \dots & a_{1n} \\ a_{21} & a_{22} & \dots & a_{2n} \\ \vdots & \vdots & \ddots & \vdots \\ a_{n1} & a_{n2} & \dots & a_{nn} \end{pmatrix},$$

assume that we have

$$\begin{pmatrix} v_1 \\ v_2 \\ \vdots \\ v_n \end{pmatrix} = A^\top \begin{pmatrix} u_1 \\ u_2 \\ \vdots \\ u_n \end{pmatrix}.$$

Then,

$$f(v_1, \dots, v_n) = \Big(\sum_{\pi \in \mathfrak{S}_n} \epsilon(\pi) a_{\pi(1)\,1} \cdots a_{\pi(n)\,n} \Big) f(u_1, \dots, u_n),$$

where the sum ranges over all permutations π on $\{1, \dots, n\}$.

Proof. Expanding $f(v_1, \dots, v_n)$ by multilinearity, we get a sum of terms of the form

$$a_{\pi(1)\,1} \cdots a_{\pi(n)\,n} f(u_{\pi(1)}, \dots, u_{\pi(n)}),$$

for all possible functions $\pi \colon \{1, \dots, n\} \to \{1, \dots, n\}$. However, because f is alternating, only the terms for which π is a permutation are nonzero. By Proposition 6.1, every permutation π is a product of transpositions, and by Proposition 6.2, the parity $\epsilon(\pi)$ of the number of transpositions only depends on π. Then applying Proposition 6.4 (3) to each transposition in π, we get

$$a_{\pi(1)\,1} \cdots a_{\pi(n)\,n} f(u_{\pi(1)}, \dots, u_{\pi(n)}) = \epsilon(\pi) a_{\pi(1)\,1} \cdots a_{\pi(n)\,n} f(u_1, \dots, u_n).$$

Thus, we get the expression of the lemma. $\qquad \square$

For the case of $n = 2$, the proof details of Lemma 6.1 become

$$\begin{aligned}
f(v_1, v_2) &= f(a_{11}u_1 + a_{21}u_2, a_{12}u_1 + a_{22}u_2) \\
&= f(a_{11}u_1 + a_{21}u_2, a_{12}u_1) + f(a_{11}u_1 + a_{21}u_2, a_{22}u_2) \\
&= f(a_{11}u_1, a_{12}u_1) + f(a_{21}u_2, a_{12}u_1) \\
&\quad + f(a_{11}u_a, a_{22}u_2) + f(a_{21}u_2, a_{22}u_2) \\
&= a_{11}a_{12}f(u_1, u_1) + a_{21}a_{12}f(u_2, u_1) + a_{11}a_{22}f(u_1, u_2) \\
&\quad + a_{21}a_{22}f(u_2, u_2) \\
&= a_{21}a_{12}f(u_2, u_1)a_{11}a_{22}f(u_1, u_2) \\
&= (a_{11}a_{22} - a_{12}a_{21}) f(u_1, u_2).
\end{aligned}$$

Hopefully the reader will recognize the quantity $a_{11}a_{22} - a_{12}a_{22}$. It is the determinant of the 2×2 matrix

$$A = \begin{pmatrix} a_{11} & a_{12} \\ a_{21} & a_{22} \end{pmatrix}.$$

This is no accident. The quantity

$$\det(A) = \sum_{\pi \in \mathfrak{S}_n} \epsilon(\pi) a_{\pi(1)\,1} \cdots a_{\pi(n)\,n}$$

is in fact the value of the determinant of A (which, as we shall see shortly, is also equal to the determinant of $A^\top$). However, working directly with the above definition is quite awkward, and we will proceed via a slightly indirect route

Remark: The reader might have been puzzled by the fact that it is the transpose matrix $A^\top$ rather than A itself that appears in Lemma 6.1. The reason is that if we want the generic term in the determinant to be

$$\epsilon(\pi) a_{\pi(1)\,1} \cdots a_{\pi(n)\,n},$$

where the permutation applies to the first index, then we have to express the v_js in terms of the u_is in terms of $A^\top$ as we did. Furthermore, since

$$v_j = a_{1\,j} u_1 + \cdots + a_{i\,j} u_i + \cdots + a_{n\,j} u_n,$$

we see that v_j corresponds to the jth column of the matrix A, and so the determinant is viewed as a function of the *columns* of A.

The literature is split on this point. Some authors prefer to define a determinant as we did. Others use A itself, which amounts to viewing det as a function of the rows, in which case we get the expression

$$\sum_{\sigma \in \mathfrak{S}_n} \epsilon(\sigma) a_{1\,\sigma(1)} \cdots a_{n\,\sigma(n)}.$$

Corollary 6.1 show that these two expressions are equal, so it doesn't matter which is chosen. This is a matter of taste.

6.3   Definition of a Determinant

Recall that the set of all square $n \times n$-matrices with coefficients in a field K is denoted by $\mathrm{M}_n(K)$.

Definition 6.6. A *determinant* is defined as any map

$$D \colon \mathrm{M}_n(K) \to K,$$

which, when viewed as a map on $(K^n)^n$, i.e., a map of the n columns of a matrix, is n-linear alternating and such that $D(I_n) = 1$ for the identity matrix I_n. Equivalently, we can consider a vector space E of dimension n, some fixed basis $(e_1, \ldots, e_n)$, and define

$$D \colon E^n \to K$$

as an n-linear alternating map such that $D(e_1, \ldots, e_n) = 1$.

First we will show that such maps D exist, using an inductive definition that also gives a recursive method for computing determinants. Actually, we will define a family $(\mathcal{D}_n)_{n \geq 1}$ of (finite) sets of maps $D \colon M_n(K) \to K$. Second we will show that determinants are in fact uniquely defined, that is, we will show that each $\mathcal{D}_n$ consists of a *single map*. This will show the equivalence of the direct definition $\det(A)$ of Lemma 6.1 with the inductive definition $D(A)$. Finally, we will prove some basic properties of determinants, using the uniqueness theorem.

Given a matrix $A \in M_n(K)$, we denote its n columns by $A^1, \ldots, A^n$. In order to describe the recursive process to define a determinant we need the notion of a minor.

Definition 6.7. Given any $n \times n$ matrix with $n \geq 2$, for any two indices i, j with $1 \leq i, j \leq n$, let A_{ij} be the $(n-1) \times (n-1)$ matrix obtained by deleting Row i and Column j from A and called a *minor*:

$$A_{ij} = \begin{pmatrix} & & & \times & & & \\ & & & \times & & & \\ \times & \times & \times & \times & \times & \times & \times \\ & & & \times & & & \\ & & & \times & & & \\ & & & \times & & & \\ & & & \times & & & \end{pmatrix}.$$

For example, if

$$A = \begin{pmatrix} 2 & -1 & 0 & 0 & 0 \\ -1 & 2 & -1 & 0 & 0 \\ 0 & -1 & 2 & -1 & 0 \\ 0 & 0 & -1 & 2 & -1 \\ 0 & 0 & 0 & -1 & 2 \end{pmatrix}$$

then

$$A_{23} = \begin{pmatrix} 2 & -1 & 0 & 0 \\ 0 & -1 & -1 & 0 \\ 0 & 0 & 2 & -1 \\ 0 & 0 & -1 & 2 \end{pmatrix}.$$

Definition 6.8. For every $n \geq 1$, we define a finite set $\mathcal{D}_n$ of maps $D \colon \mathrm{M}_n(K) \to K$ inductively as follows:

When $n = 1$, $\mathcal{D}_1$ consists of the single map D such that, $D(A) = a$, where $A = (a)$, with $a \in K$.

Assume that $\mathcal{D}_{n-1}$ has been defined, where $n \geq 2$. Then $\mathcal{D}_n$ consists of all the maps D such that, for some i, $1 \leq i \leq n$,

$$D(A) = (-1)^{i+1} a_{i1} D(A_{i1}) + \cdots + (-1)^{i+n} a_{in} D(A_{in}),$$

where for every j, $1 \leq j \leq n$, $D(A_{ij})$ is the result of applying any D in $\mathcal{D}_{n-1}$ to the minor A_{ij}.

We confess that the use of the same letter D for the member of $\mathcal{D}_n$ being defined, and for members of $\mathcal{D}_{n-1}$, may be slightly confusing. We considered using subscripts to distinguish, but this seems to complicate things unnecessarily. One should not worry too much anyway, since it will turn out that each $\mathcal{D}_n$ contains just one map.

Each $(-1)^{i+j} D(A_{ij})$ is called the *cofactor* of a_{ij}, and the inductive expression for $D(A)$ is called a *Laplace expansion of D according to the i-th Row*. Given a matrix $A \in \mathrm{M}_n(K)$, each $D(A)$ is called a *determinant* of A.

We can think of each member of $\mathcal{D}_n$ as an *algorithm* to evaluate "the" determinant of A. The main point is that these algorithms, which recursively evaluate a determinant using all possible Laplace row expansions, all yield the *same result*, $\det(A)$.

We will prove shortly that $D(A)$ is uniquely defined (at the moment, it is not clear that $\mathcal{D}_n$ consists of a single map). Assuming this fact, given a $n \times n$-matrix $A = (a_{ij})$,

$$A = \begin{pmatrix} a_{11} & a_{12} & \ldots & a_{1n} \\ a_{21} & a_{22} & \ldots & a_{2n} \\ \vdots & \vdots & \ddots & \vdots \\ a_{n1} & a_{n2} & \ldots & a_{nn} \end{pmatrix},$$

its determinant is denoted by $D(A)$ or $\det(A)$, or more explicitly by

$$\det(A) = \begin{vmatrix} a_{11} & a_{12} & \ldots & a_{1n} \\ a_{21} & a_{22} & \ldots & a_{2n} \\ \vdots & \vdots & \ddots & \vdots \\ a_{n1} & a_{n2} & \ldots & a_{nn} \end{vmatrix}.$$

Let us first consider some examples.

Example 6.1.

(1) When $n = 2$, if

$$A = \begin{pmatrix} a & b \\ c & d \end{pmatrix},$$

then by expanding according to any row, we have

$$D(A) = ad - bc.$$

(2) When $n = 3$, if

$$A = \begin{pmatrix} a_{11} & a_{12} & a_{13} \\ a_{21} & a_{22} & a_{23} \\ a_{31} & a_{32} & a_{33} \end{pmatrix},$$

then by expanding according to the first row, we have

$$D(A) = a_{11} \begin{vmatrix} a_{22} & a_{23} \\ a_{32} & a_{33} \end{vmatrix} - a_{12} \begin{vmatrix} a_{21} & a_{23} \\ a_{31} & a_{33} \end{vmatrix} + a_{13} \begin{vmatrix} a_{21} & a_{22} \\ a_{31} & a_{32} \end{vmatrix},$$

that is,

$$D(A) = a_{11}(a_{22}a_{33} - a_{32}a_{23}) - a_{12}(a_{21}a_{33} - a_{31}a_{23})$$
$$+ a_{13}(a_{21}a_{32} - a_{31}a_{22}),$$

which gives the explicit formula

$$D(A) = a_{11}a_{22}a_{33} + a_{21}a_{32}a_{13} + a_{31}a_{12}a_{23}$$
$$- a_{11}a_{32}a_{23} - a_{21}a_{12}a_{33} - a_{31}a_{22}a_{13}.$$

We now show that each $D \in \mathcal{D}_n$ is a determinant (map).

Lemma 6.2. *For every $n \geq 1$, for every $D \in \mathcal{D}_n$ as defined in Definition 6.8, D is an alternating multilinear map such that $D(I_n) = 1$.*

Proof. By induction on n, it is obvious that $D(I_n) = 1$. Let us now prove that D is multilinear. Let us show that D is linear in each column. Consider any Column k. Since

$$D(A) = (-1)^{i+1}a_{i1}D(A_{i1}) + \cdots + (-1)^{i+j}a_{ij}D(A_{ij}) + \cdots$$
$$+ (-1)^{i+n}a_{in}D(A_{in}),$$

if $j \neq k$, then by induction, $D(A_{ij})$ is linear in Column k, and a_{ij} does not belong to Column k, so $(-1)^{i+j}a_{ij}D(A_{ij})$ is linear in Column k. If $j = k$, then $D(A_{ij})$ does not depend on Column $k = j$, since A_{ij} is obtained from A by deleting Row i and Column $j = k$, and a_{ij} belongs to Column $j = k$. Thus, $(-1)^{i+j}a_{ij}D(A_{ij})$ is linear in Column k. Consequently, in all cases, $(-1)^{i+j}a_{ij}D(A_{ij})$ is linear in Column k, and thus, $D(A)$ is linear in Column k.

Let us now prove that D is alternating. Assume that two adjacent columns of A are equal, say $A^k = A^{k+1}$. Assume that $j \neq k$ and $j \neq k+1$. Then the matrix A_{ij} has two identical adjacent columns, and by the induction hypothesis, $D(A_{ij}) = 0$. The remaining terms of $D(A)$ are

$$(-1)^{i+k}a_{ik}D(A_{ik}) + (-1)^{i+k+1}a_{ik+1}D(A_{ik+1}).$$

However, the two matrices A_{ik} and A_{ik+1} are equal, since we are assuming that Columns k and $k+1$ of A are identical and A_{ik} is obtained from A by deleting Row i and Column k while A_{ik+1} is obtained from A by deleting Row i and Column $k+1$. Similarly, $a_{ik} = a_{ik+1}$, since Columns k and $k+1$ of A are equal. But then,

$$(-1)^{i+k}a_{ik}D(A_{ik}) + (-1)^{i+k+1}a_{ik+1}D(A_{ik+1})$$
$$= (-1)^{i+k}a_{ik}D(A_{ik}) - (-1)^{i+k}a_{ik}D(A_{ik}) = 0.$$

This shows that D is alternating and completes the proof. $\square$

Lemma 6.2 shows the existence of determinants. We now prove their uniqueness.

Theorem 6.1. *For every $n \geq 1$, for every $D \in \mathcal{D}_n$, for every matrix $A \in \mathrm{M}_n(K)$, we have*

$$D(A) = \sum_{\pi \in \mathfrak{S}_n} \epsilon(\pi)a_{\pi(1)\,1} \cdots a_{\pi(n)\,n},$$

where the sum ranges over all permutations π on $\{1,\ldots,n\}$. As a consequence, $\mathcal{D}_n$ consists of a single map for every $n \geq 1$, and this map is given by the above explicit formula.

Proof. Consider the standard basis $(e_1,\ldots,e_n)$ of K^n, where $(e_i)_i = 1$ and $(e_i)_j = 0$, for $j \neq i$. Then each column A^j of A corresponds to a vector v_j whose coordinates over the basis $(e_1,\ldots,e_n)$ are the components of A^j, that is, we can write

$$v_1 = a_{1\,1}e_1 + \cdots + a_{n\,1}e_n,$$

$$\cdots$$

$$v_n = a_{1\,n}e_1 + \cdots + a_{n\,n}e_n.$$

Since by Lemma 6.2, each D is a multilinear alternating map, by applying Lemma 6.1, we get

$$D(A) = D(v_1,\ldots,v_n) = \Big(\sum_{\pi \in \mathfrak{S}_n} \epsilon(\pi)a_{\pi(1)\,1} \cdots a_{\pi(n)\,n} \Big) D(e_1,\ldots,e_n),$$

where the sum ranges over all permutations π on $\{1, \ldots, n\}$. But $D(e_1, \ldots, e_n) = D(I_n)$, and by Lemma 6.2, we have $D(I_n) = 1$. Thus,

$$D(A) = \sum_{\pi \in \mathfrak{S}_n} \epsilon(\pi) a_{\pi(1)\,1} \cdots a_{\pi(n)\,n},$$

where the sum ranges over all permutations π on $\{1, \ldots, n\}$. $\square$

From now on we will favor the notation $\det(A)$ over $D(A)$ for the determinant of a square matrix.

Remark: There is a geometric interpretation of determinants which we find quite illuminating. Given n linearly independent vectors $(u_1, \ldots, u_n)$ in $\mathbb{R}^n$, the set

$$P_n = \{\lambda_1 u_1 + \cdots + \lambda_n u_n \mid 0 \leq \lambda_i \leq 1, \, 1 \leq i \leq n\}$$

is called a *parallelotope*. If $n = 2$, then P_2 is a *parallelogram* and if $n = 3$, then P_3 is a *parallelepiped*, a skew box having u_1, u_2, u_3 as three of its corner sides. See Figures 6.1 and 6.2.

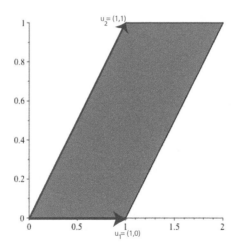

Fig. 6.1   The parallelogram in $\mathbb{R}^w$ spanned by the vectors $u_1 = (1, 0)$ and $u_2 = (1, 1)$.

Then it turns out that $\det(u_1, \ldots, u_n)$ is the *signed volume* of the parallelotope P_n (where volume means n-dimensional volume). The sign of this volume accounts for the orientation of P_n in $\mathbb{R}^n$.

We can now prove some properties of determinants.

Corollary 6.1. *For every matrix $A \in \mathrm{M}_n(K)$, we have $\det(A) = \det(A^\top)$.*

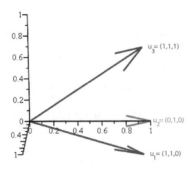

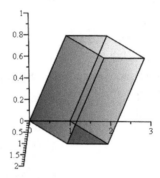

Fig. 6.2 The parallelepiped in $\mathbb{R}^3$ spanned by the vectors $u_1 = (1, 1, 0)$, $u_2 = (0, 1, 0)$, and $u_3 = (0, 0, 1)$.

Proof. By Theorem 6.1, we have

$$\det(A) = \sum_{\pi \in \mathfrak{S}_n} \epsilon(\pi) a_{\pi(1)\,1} \cdots a_{\pi(n)\,n},$$

where the sum ranges over all permutations π on $\{1, \ldots, n\}$. Since a permutation is invertible, every product

$$a_{\pi(1)\,1} \cdots a_{\pi(n)\,n}$$

can be rewritten as

$$a_{1\,\pi^{-1}(1)} \cdots a_{n\,\pi^{-1}(n)},$$

and since $\epsilon(\pi^{-1}) = \epsilon(\pi)$ and the sum is taken over all permutations on $\{1, \ldots, n\}$, we have

$$\sum_{\pi \in \mathfrak{S}_n} \epsilon(\pi) a_{\pi(1)\,1} \cdots a_{\pi(n)\,n} = \sum_{\sigma \in \mathfrak{S}_n} \epsilon(\sigma) a_{1\,\sigma(1)} \cdots a_{n\,\sigma(n)},$$

where π and σ range over all permutations. But it is immediately verified that

$$\det(A^\top) = \sum_{\sigma \in \mathfrak{S}_n} \epsilon(\sigma) a_{1\,\sigma(1)} \cdots a_{n\,\sigma(n)}. \qquad \square$$

A useful consequence of Corollary 6.1 is that the determinant of a matrix is also a multilinear alternating map of its *rows*. This fact, combined with the fact that the determinant of a matrix is a multilinear alternating map of its columns, is often useful for finding short-cuts in computing determinants. We illustrate this point on the following example which shows up in polynomial interpolation.

Example 6.2. Consider the so-called *Vandermonde determinant*

$$V(x_1, \ldots, x_n) = \begin{vmatrix} 1 & 1 & \ldots & 1 \\ x_1 & x_2 & \ldots & x_n \\ x_1^2 & x_2^2 & \ldots & x_n^2 \\ \vdots & \vdots & \ddots & \vdots \\ x_1^{n-1} & x_2^{n-1} & \ldots & x_n^{n-1} \end{vmatrix}.$$

We claim that

$$V(x_1, \ldots, x_n) = \prod_{1 \le i < j \le n} (x_j - x_i),$$

with $V(x_1, \ldots, x_n) = 1$, when $n = 1$. We prove it by induction on $n \ge 1$. The case $n = 1$ is obvious. Assume $n \ge 2$. We proceed as follows: multiply Row $n - 1$ by x_1 and subtract it from Row n (the last row), then multiply Row $n - 2$ by x_1 and subtract it from Row $n - 1$, *etc.*, multiply Row $i - 1$ by x_1 and subtract it from row i, until we reach Row 1. We obtain the following determinant:

$$V(x_1, \ldots, x_n) = \begin{vmatrix} 1 & 1 & \ldots & 1 \\ 0 & x_2 - x_1 & \ldots & x_n - x_1 \\ 0 & x_2(x_2 - x_1) & \ldots & x_n(x_n - x_1) \\ \vdots & \vdots & \ddots & \vdots \\ 0 & x_2^{n-2}(x_2 - x_1) & \ldots & x_n^{n-2}(x_n - x_1) \end{vmatrix}.$$

Now expanding this determinant according to the first column and using multilinearity, we can factor $(x_i - x_1)$ from the column of index $i - 1$ of the matrix obtained by deleting the first row and the first column, and thus

$$V(x_1, \ldots, x_n) = (x_2 - x_1)(x_3 - x_1) \cdots (x_n - x_1)V(x_2, \ldots, x_n),$$

which establishes the induction step.

Remark: Observe that

$$\Delta(x_1, \ldots, x_n) = V(x_n, \ldots, x_1) = (-1)^{\binom{n}{2}} V(x_1, \ldots x_n),$$

where $\Delta(x_1, \ldots, x_n)$ is the discriminant of $(x_1, \ldots, x_n)$ introduced in Definition 6.2.

Lemma 6.1 can be reformulated nicely as follows.

Proposition 6.5. *Let* $f : E \times \ldots \times E \to F$ *be an n-linear alternating map. Let* $(u_1, \ldots, u_n)$ *and* $(v_1, \ldots, v_n)$ *be two families of n vectors, such that*

$$v_1 = a_{11} u_1 + \cdots + a_{1n} u_n,$$

$$\cdots$$

$$v_n = a_{n1} u_1 + \cdots + a_{nn} u_n.$$

Equivalently, letting

$$A = \begin{pmatrix} a_{11} & a_{12} & \cdots & a_{1n} \\ a_{21} & a_{22} & \cdots & a_{2n} \\ \vdots & \vdots & \ddots & \vdots \\ a_{n1} & a_{n2} & \cdots & a_{nn} \end{pmatrix},$$

assume that we have

$$\begin{pmatrix} v_1 \\ v_2 \\ \vdots \\ v_n \end{pmatrix} = A \begin{pmatrix} u_1 \\ u_2 \\ \vdots \\ u_n \end{pmatrix}.$$

Then,

$$f(v_1, \ldots, v_n) = \det(A) f(u_1, \ldots, u_n).$$

Proof. The only difference with Lemma 6.1 is that here we are using $A^\top$ instead of A. Thus, by Lemma 6.1 and Corollary 6.1, we get the desired result. $\quad\square$

As a consequence, we get the very useful property that the determinant of a product of matrices is the product of the determinants of these matrices.

Proposition 6.6. *For any two $n \times n$-matrices A and B, we have $\det(AB) = \det(A)\det(B)$.*

Proof. We use Proposition 6.5 as follows: let $(e_1, \ldots, e_n)$ be the standard basis of K^n, and let

$$\begin{pmatrix} w_1 \\ w_2 \\ \vdots \\ w_n \end{pmatrix} = AB \begin{pmatrix} e_1 \\ e_2 \\ \vdots \\ e_n \end{pmatrix}.$$

Then we get

$$\det(w_1, \ldots, w_n) = \det(AB) \det(e_1, \ldots, e_n) = \det(AB),$$

since $\det(e_1, \ldots, e_n) = 1$. Now letting

$$\begin{pmatrix} v_1 \\ v_2 \\ \vdots \\ v_n \end{pmatrix} = B \begin{pmatrix} e_1 \\ e_2 \\ \vdots \\ e_n \end{pmatrix},$$

we get

$$\det(v_1, \ldots, v_n) = \det(B),$$

and since

$$\begin{pmatrix} w_1 \\ w_2 \\ \vdots \\ w_n \end{pmatrix} = A \begin{pmatrix} v_1 \\ v_2 \\ \vdots \\ v_n \end{pmatrix},$$

we get

$$\det(w_1, \ldots, w_n) = \det(A) \det(v_1, \ldots, v_n) = \det(A) \det(B). \qquad \square$$

It should be noted that all the results of this section, up to now, also hold when K is a commutative ring and not necessarily a field. We can now characterize when an $n \times n$-matrix A is invertible in terms of its determinant $\det(A)$.

6.4   Inverse Matrices and Determinants

In the next two sections, K is a commutative ring and when needed a field.

Definition 6.9. Let K be a commutative ring. Given a matrix $A \in \mathrm{M}_n(K)$, let $\widetilde{A} = (b_{ij})$ be the matrix defined such that

$$b_{ij} = (-1)^{i+j} \det(A_{ji}),$$

the cofactor of a_{ji}. The matrix $\widetilde{A}$ is called the *adjugate* of A, and each matrix A_{ji} is called a *minor* of the matrix A.

For example, if

$$A = \begin{pmatrix} 1 & 1 & 1 \\ 2 & -2 & -2 \\ 3 & 3 & -3 \end{pmatrix},$$

we have

$$b_{11} = \det(A_{11}) = \begin{vmatrix} -2 & -2 \\ 3 & -3 \end{vmatrix} = 12 \qquad b_{12} = -\det(A_{21}) = -\begin{vmatrix} 1 & 1 \\ 3 & -3 \end{vmatrix} = 6$$

$$b_{13} = \det(A_{31}) = \begin{vmatrix} 1 & 1 \\ -2 & -2 \end{vmatrix} = 0 \qquad b_{21} = -\det(A_{12}) = -\begin{vmatrix} 2 & -2 \\ 3 & -3 \end{vmatrix} = 0$$

$$b_{22} = \det(A_{22}) = \begin{vmatrix} 1 & 1 \\ 3 & -3 \end{vmatrix} = -6 \qquad b_{23} = -\det(A_{32}) = -\begin{vmatrix} 1 & 1 \\ 2 & -2 \end{vmatrix} = 4$$

$$b_{31} = \det(A_{13}) = \begin{vmatrix} 2 & -2 \\ 3 & 3 \end{vmatrix} = 12 \qquad b_{32} = -\det(A_{23}) = -\begin{vmatrix} 1 & 1 \\ 3 & 3 \end{vmatrix} = 0$$

$$b_{33} = \det(A_{33}) = \begin{vmatrix} 1 & 1 \\ 2 & -2 \end{vmatrix} = -4,$$

we find that

$$\widetilde{A} = \begin{pmatrix} 12 & 6 & 0 \\ 0 & -6 & 4 \\ 12 & 0 & -4 \end{pmatrix}.$$

 Note the reversal of the indices in

$$b_{ij} = (-1)^{i+j} \det(A_{ji}).$$

Thus, $\widetilde{A}$ is the *transpose* of the matrix of cofactors of elements of A.

We have the following proposition.

Proposition 6.7. *Let K be a commutative ring. For every matrix $A \in M_n(K)$, we have*

$$A\widetilde{A} = \widetilde{A}A = \det(A)I_n.$$

As a consequence, A is invertible iff $\det(A)$ is invertible, and if so, $A^{-1} = (\det(A))^{-1}\widetilde{A}$.

Proof. If $\widetilde{A} = (b_{ij})$ and $A\widetilde{A} = (c_{ij})$, we know that the entry c_{ij} in row i and column j of $A\widetilde{A}$ is

$$c_{ij} = a_{i1}b_{1j} + \cdots + a_{ik}b_{kj} + \cdots + a_{in}b_{nj},$$

which is equal to

$$a_{i\,1}(-1)^{j+1}\det(A_{j\,1}) + \cdots + a_{i\,n}(-1)^{j+n}\det(A_{j\,n}).$$

If $j = i$, then we recognize the expression of the expansion of $\det(A)$ according to the i-th row:

$$c_{i\,i} = \det(A) = a_{i\,1}(-1)^{i+1}\det(A_{i\,1}) + \cdots + a_{i\,n}(-1)^{i+n}\det(A_{i\,n}).$$

If $j \neq i$, we can form the matrix A' by replacing the j-th row of A by the i-th row of A. Now the matrix $A_{j\,k}$ obtained by deleting row j and column k from A is equal to the matrix $A'_{j\,k}$ obtained by deleting row j and column k from A', since A and A' only differ by the j-th row. Thus,

$$\det(A_{j\,k}) = \det(A'_{j\,k}),$$

and we have

$$c_{i\,j} = a_{i\,1}(-1)^{j+1}\det(A'_{j\,1}) + \cdots + a_{i\,n}(-1)^{j+n}\det(A'_{j\,n}).$$

However, this is the expansion of $\det(A')$ according to the j-th row, since the j-th row of A' is equal to the i-th row of A. Furthermore, since A' has two identical rows i and j, because det is an alternating map of the rows (see an earlier remark), we have $\det(A') = 0$. Thus, we have shown that $c_{i\,i} = \det(A)$, and $c_{i\,j} = 0$, when $j \neq i$, and so

$$A\widetilde{A} = \det(A)I_n.$$

It is also obvious from the definition of $\widetilde{A}$, that

$$\widetilde{A}^\top = \widetilde{A^\top}.$$

Then applying the first part of the argument to $A^\top$, we have

$$A^\top\widetilde{A^\top} = \det(A^\top)I_n,$$

and since $\det(A^\top) = \det(A)$, $\widetilde{A}^\top = \widetilde{A^\top}$, and $(\widetilde{A}A)^\top = A^\top\widetilde{A}^\top$, we get

$$\det(A)I_n = A^\top\widetilde{A^\top} = A^\top\widetilde{A}^\top = (\widetilde{A}A)^\top,$$

that is,

$$(\widetilde{A}A)^\top = \det(A)I_n,$$

which yields

$$\widetilde{A}A = \det(A)I_n,$$

since $I_n^\top = I_n$. This proves that

$$A\widetilde{A} = \widetilde{A}A = \det(A)I_n.$$

As a consequence, if $\det(A)$ is invertible, we have $A^{-1} = (\det(A))^{-1}\widetilde{A}$. Conversely, if A is invertible, from $AA^{-1} = I_n$, by Proposition 6.6, we have $\det(A)\det(A^{-1}) = 1$, and $\det(A)$ is invertible. $\qquad\square$

For example, we saw earlier that

$$A = \begin{pmatrix} 1 & 1 & 1 \\ 2 & -2 & -2 \\ 3 & 3 & -3 \end{pmatrix} \quad \text{and} \quad \widetilde{A} = \begin{pmatrix} 12 & 6 & 0 \\ 0 & -6 & 4 \\ 12 & 0 & -4 \end{pmatrix},$$

and we have

$$\begin{pmatrix} 1 & 1 & 1 \\ 2 & -2 & -2 \\ 3 & 3 & -3 \end{pmatrix} \begin{pmatrix} 12 & 6 & 0 \\ 0 & -6 & 4 \\ 12 & 0 & -4 \end{pmatrix} = 24 \begin{pmatrix} 1 & 0 & 0 \\ 0 & 1 & 0 \\ 0 & 0 & 1 \end{pmatrix}$$

with $\det(A) = 24$.

When K is a field, an element $a \in K$ is invertible iff $a \neq 0$. In this case, the second part of the proposition can be stated as A is invertible iff $\det(A) \neq 0$. Note in passing that this method of computing the inverse of a matrix is usually not practical.

6.5   Systems of Linear Equations and Determinants

We now consider some applications of determinants to linear independence and to solving systems of linear equations. Although these results hold for matrices over certain rings, their proofs require more sophisticated methods. Therefore, we assume again that K is a field (usually, $K = \mathbb{R}$ or $K = \mathbb{C}$).

Let A be an $n \times n$-matrix, x a column vectors of variables, and b another column vector, and let $A^1, \ldots, A^n$ denote the columns of A. Observe that the system of equations $Ax = b$,

$$\begin{pmatrix} a_{11} & a_{12} & \ldots & a_{1n} \\ a_{21} & a_{22} & \ldots & a_{2n} \\ \vdots & \vdots & \ddots & \vdots \\ a_{n1} & a_{n2} & \ldots & a_{nn} \end{pmatrix} \begin{pmatrix} x_1 \\ x_2 \\ \vdots \\ x_n \end{pmatrix} = \begin{pmatrix} b_1 \\ b_2 \\ \vdots \\ b_n \end{pmatrix}$$

is equivalent to

$$x_1 A^1 + \cdots + x_j A^j + \cdots + x_n A^n = b,$$

since the equation corresponding to the i-th row is in both cases

$$a_{i1} x_1 + \cdots + a_{ij} x_j + \cdots + a_{in} x_n = b_i.$$

First we characterize linear independence of the column vectors of a matrix A in terms of its determinant.

Proposition 6.8. *Given an $n \times n$-matrix A over a field K, the columns $A^1, \ldots, A^n$ of A are linearly dependent iff $\det(A) = \det(A^1, \ldots, A^n) = 0$. Equivalently, A has rank n iff $\det(A) \neq 0$.*

Proof. First assume that the columns $A^1, \ldots, A^n$ of A are linearly dependent. Then there are $x_1, \ldots, x_n \in K$, such that

$$x_1 A^1 + \cdots + x_j A^j + \cdots + x_n A^n = 0,$$

where $x_j \neq 0$ for some j. If we compute

$$\det(A^1, \ldots, x_1 A^1 + \cdots + x_j A^j + \cdots + x_n A^n, \ldots, A^n)$$
$$= \det(A^1, \ldots, 0, \ldots, A^n) = 0,$$

where 0 occurs in the j-th position. By multilinearity, all terms containing two identical columns A^k for $k \neq j$ vanish, and we get

$$\det(A^1, \ldots, x_1 A^1 + \cdots + x_j A^j + \cdots + x_n A^n, \ldots, A^n) = x_j \det(A^1, \ldots, A^n) = 0.$$

Since $x_j \neq 0$ and K is a field, we must have $\det(A^1, \ldots, A^n) = 0$.

Conversely, we show that if the columns $A^1, \ldots, A^n$ of A are linearly independent, then $\det(A^1, \ldots, A^n) \neq 0$. If the columns $A^1, \ldots, A^n$ of A are linearly independent, then they form a basis of K^n, and we can express the standard basis $(e_1, \ldots, e_n)$ of K^n in terms of $A^1, \ldots, A^n$. Thus, we have

$$\begin{pmatrix} e_1 \\ e_2 \\ \vdots \\ e_n \end{pmatrix} = \begin{pmatrix} b_{11} & b_{12} & \ldots & b_{1n} \\ b_{21} & b_{22} & \ldots & b_{2n} \\ \vdots & \vdots & \ddots & \vdots \\ b_{n1} & b_{n2} & \ldots & b_{nn} \end{pmatrix} \begin{pmatrix} A^1 \\ A^2 \\ \vdots \\ A^n \end{pmatrix},$$

for some matrix $B = (b_{ij})$, and by Proposition 6.5, we get

$$\det(e_1, \ldots, e_n) = \det(B) \det(A^1, \ldots, A^n),$$

and since $\det(e_1, \ldots, e_n) = 1$, this implies that $\det(A^1, \ldots, A^n) \neq 0$ (and $\det(B) \neq 0$). For the second assertion, recall that the rank of a matrix is equal to the maximum number of linearly independent columns, and the conclusion is clear. $\square$

We now characterize when a system of linear equations of the form $Ax = b$ has a unique solution.

Proposition 6.9. *Given an $n \times n$-matrix A over a field K, the following properties hold:*

(1) For every column vector b, there is a unique column vector x such that $Ax = b$ iff the only solution to $Ax = 0$ is the trivial vector $x = 0$, iff $\det(A) \neq 0$.

(2) If $\det(A) \neq 0$, *the unique solution of* $Ax = b$ *is given by the expressions*

$$x_j = \frac{\det(A^1, \ldots, A^{j-1}, b, A^{j+1}, \ldots, A^n)}{\det(A^1, \ldots, A^{j-1}, A^j, A^{j+1}, \ldots, A^n)},$$

known as Cramer's rules.

(3) The system of linear equations $Ax = 0$ *has a nonzero solution iff* $\det(A) = 0$.

Proof. (1) Assume that $Ax = b$ has a single solution x_0, and assume that $Ay = 0$ with $y \neq 0$. Then,

$$A(x_0 + y) = Ax_0 + Ay = Ax_0 + 0 = b,$$

and $x_0 + y \neq x_0$ is another solution of $Ax = b$, contradicting the hypothesis that $Ax = b$ has a single solution x_0. Thus, $Ax = 0$ only has the trivial solution. Now assume that $Ax = 0$ only has the trivial solution. This means that the columns $A^1, \ldots, A^n$ of A are linearly independent, and by Proposition 6.8, we have $\det(A) \neq 0$. Finally, if $\det(A) \neq 0$, by Proposition 6.7, this means that A is invertible, and then for every b, $Ax = b$ is equivalent to $x = A^{-1}b$, which shows that $Ax = b$ has a single solution.

(2) Assume that $Ax = b$. If we compute

$$\det(A^1, \ldots, x_1 A^1 + \cdots + x_j A^j + \cdots + x_n A^n, \ldots, A^n) = \det(A^1, \ldots, b, \ldots, A^n),$$

where b occurs in the j-th position, by multilinearity, all terms containing two identical columns A^k for $k \neq j$ vanish, and we get

$$x_j \det(A^1, \ldots, A^n) = \det(A^1, \ldots, A^{j-1}, b, A^{j+1}, \ldots, A^n),$$

for every j, $1 \leq j \leq n$. Since we assumed that $\det(A) = \det(A^1, \ldots, A^n) \neq 0$, we get the desired expression.

(3) Note that $Ax = 0$ has a nonzero solution iff $A^1, \ldots, A^n$ are linearly dependent (as observed in the proof of Proposition 6.8), which, by Proposition 6.8, is equivalent to $\det(A) = 0$. $\qquad\square$

As pleasing as Cramer's rules are, it is usually impractical to solve systems of linear equations using the above expressions. However, these formula imply an interesting fact, which is that the solution of the system $Ax = b$ are continuous in A and b. If we assume that the entries in A are continuous functions $a_{ij}(t)$ and the entries in b are are also continuous functions $b_j(t)$ of a real parameter t, since determinants are polynomial functions of their entries, the expressions

$$x_j(t) = \frac{\det(A^1, \ldots, A^{j-1}, b, A^{j+1}, \ldots, A^n)}{\det(A^1, \ldots, A^{j-1}, A^j, A^{j+1}, \ldots, A^n)}$$

are ratios of polynomials, and thus are also continuous as long as $\det(A(t))$ is nonzero. Similarly, if the functions $a_{ij}(t)$ and $b_j(t)$ are differentiable, so are the $x_j(t)$.

6.6 Determinant of a Linear Map

Given a vector space E of finite dimension n, given a basis $(u_1, \ldots, u_n)$ of E, for every linear map $f \colon E \to E$, if $M(f)$ is the matrix of f w.r.t. the basis $(u_1, \ldots, u_n)$, we can define $\det(f) = \det(M(f))$. If $(v_1, \ldots, v_n)$ is any other basis of E, and if P is the change of basis matrix, by Corollary 3.1, the matrix of f with respect to the basis $(v_1, \ldots, v_n)$ is $P^{-1}M(f)P$. By Proposition 6.6, we have

$$\det(P^{-1}M(f)P) = \det(P^{-1})\det(M(f))\det(P) =$$
$$\det(P^{-1})\det(P)\det(M(f)) = \det(M(f)).$$

Thus, $\det(f)$ is indeed independent of the basis of E.

Definition 6.10. Given a vector space E of finite dimension, for any linear map $f \colon E \to E$, we define the *determinant* $\det(f)$ *of* f as the determinant $\det(M(f))$ of the matrix of f in any basis (since, from the discussion just before this definition, this determinant does not depend on the basis).

Then we have the following proposition.

Proposition 6.10. *Given any vector space E of finite dimension n, a linear map $f \colon E \to E$ is invertible iff $\det(f) \neq 0$.*

Proof. The linear map $f \colon E \to E$ is invertible iff its matrix $M(f)$ in any basis is invertible (by Proposition 3.2), iff $\det(M(f)) \neq 0$, by Proposition 6.7. $\square$

Given a vector space of finite dimension n, it is easily seen that the set of bijective linear maps $f \colon E \to E$ such that $\det(f) = 1$ is a group under composition. This group is a subgroup of the general linear group $\mathbf{GL}(E)$. It is called the *special linear group (of E)*, and it is denoted by $\mathbf{SL}(E)$, or when $E = K^n$, by $\mathbf{SL}(n, K)$, or even by $\mathbf{SL}(n)$.

6.7 The Cayley–Hamilton Theorem

We next discuss an interesting and important application of Proposition 6.7, the *Cayley–Hamilton theorem*. The results of this section apply to matrices over any commutative ring K. First we need the concept of the characteristic polynomial of a matrix.

Definition 6.11. If K is any commutative ring, for every $n \times n$ matrix $A \in \mathrm{M}_n(K)$, the *characteristic polynomial* $P_A(X)$ of A is the determinant

$$P_A(X) = \det(XI - A).$$

The characteristic polynomial $P_A(X)$ is a polynomial in $K[X]$, the ring of polynomials in the indeterminate X with coefficients in the ring K. For example, when $n = 2$, if

$$A = \begin{pmatrix} a & b \\ c & d \end{pmatrix},$$

then

$$P_A(X) = \begin{vmatrix} X - a & -b \\ -c & X - d \end{vmatrix} = X^2 - (a + d)X + ad - bc.$$

We can substitute the matrix A for the variable X in the polynomial $P_A(X)$, obtaining a *matrix* P_A. If we write

$$P_A(X) = X^n + c_1 X^{n-1} + \cdots + c_n,$$

then

$$P_A = A^n + c_1 A^{n-1} + \cdots + c_n I.$$

We have the following remarkable theorem.

Theorem 6.2. *(Cayley–Hamilton) If K is any commutative ring, for every $n \times n$ matrix $A \in \mathrm{M}_n(K)$, if we let*

$$P_A(X) = X^n + c_1 X^{n-1} + \cdots + c_n$$

be the characteristic polynomial of A, then

$$P_A = A^n + c_1 A^{n-1} + \cdots + c_n I = 0.$$

Proof. We can view the matrix $B = XI - A$ as a matrix with coefficients in the polynomial ring $K[X]$, and then we can form the matrix $\widetilde{B}$ which is the transpose of the matrix of cofactors of elements of B. Each entry in $\widetilde{B}$ is an $(n-1) \times (n-1)$ determinant, and thus a polynomial of degree a most $n - 1$, so we can write $\widetilde{B}$ as

$$\widetilde{B} = X^{n-1} B_0 + X^{n-2} B_1 + \cdots + B_{n-1},$$

for some $n \times n$ matrices $B_0, \ldots, B_{n-1}$ with coefficients in K. For example, when $n = 2$, we have

$$B = \begin{pmatrix} X - a & -b \\ -c & X - d \end{pmatrix}, \quad \widetilde{B} = \begin{pmatrix} X - d & b \\ c & X - a \end{pmatrix} = X \begin{pmatrix} 1 & 0 \\ 0 & 1 \end{pmatrix} + \begin{pmatrix} -d & b \\ c & -a \end{pmatrix}.$$

By Proposition 6.7, we have

$$B\widetilde{B} = \det(B)I = P_A(X)I.$$

On the other hand, we have

$$B\widetilde{B} = (XI - A)(X^{n-1}B_0 + X^{n-2}B_1 + \cdots + X^{n-j-1}B_j + \cdots + B_{n-1}),$$

and by multiplying out the right-hand side, we get

$$B\widetilde{B} = X^n D_0 + X^{n-1}D_1 + \cdots + X^{n-j}D_j + \cdots + D_n,$$

with

$$D_0 = B_0$$
$$D_1 = B_1 - AB_0$$
$$\vdots$$
$$D_j = B_j - AB_{j-1}$$
$$\vdots$$
$$D_{n-1} = B_{n-1} - AB_{n-2}$$
$$D_n = -AB_{n-1}.$$

Since

$$P_A(X)I = (X^n + c_1 X^{n-1} + \cdots + c_n)I,$$

the equality

$$X^n D_0 + X^{n-1}D_1 + \cdots + D_n = (X^n + c_1 X^{n-1} + \cdots + c_n)I$$

is an equality between two matrices, so it requires that all corresponding entries are equal, and since these are polynomials, the coefficients of these polynomials must be identical, which is equivalent to the set of equations

$$I = B_0$$
$$c_1 I = B_1 - AB_0$$
$$\vdots$$
$$c_j I = B_j - AB_{j-1}$$
$$\vdots$$
$$c_{n-1} I = B_{n-1} - AB_{n-2}$$
$$c_n I = -AB_{n-1},$$

for all j, with $1 \leq j \leq n - 1$. If, as in the table below,

$$A^n = A^n B_0$$
$$c_1 A^{n-1} = A^{n-1}(B_1 - AB_0)$$
$$\vdots$$
$$c_j A^{n-j} = A^{n-j}(B_j - AB_{j-1})$$
$$\vdots$$
$$c_{n-1} A = A(B_{n-1} - AB_{n-2})$$
$$c_n I = -AB_{n-1},$$

we multiply the first equation by A^n, the last by I, and generally the $(j + 1)$th by A^{n-j}, when we add up all these new equations, we see that the right-hand side adds up to 0, and we get our desired equation

$$A^n + c_1 A^{n-1} + \cdots + c_n I = 0,$$

as claimed. $\square$

As a concrete example, when $n = 2$, the matrix

$$A = \begin{pmatrix} a & b \\ c & d \end{pmatrix}$$

satisfies the equation

$$A^2 - (a + d)A + (ad - bc)I = 0.$$

Most readers will probably find the proof of Theorem 6.2 rather clever but very mysterious and unmotivated. The conceptual difficulty is that we really need to understand how polynomials in one variable "act" on vectors in terms of the matrix A. This can be done and yields a more "natural" proof. Actually, the reasoning is simpler and more general if we free ourselves from matrices and instead consider a finite-dimensional vector space E and some given linear map $f \colon E \to E$. Given any polynomial $p(X) = a_0 X^n + a_1 X^{n-1} + \cdots + a_n$ with coefficients in the field K, we define the *linear map* $p(f) \colon E \to E$ by

$$p(f) = a_0 f^n + a_1 f^{n-1} + \cdots + a_n \mathrm{id},$$

where $f^k = f \circ \cdots \circ f$, the k-fold composition of f with itself. Note that

$$p(f)(u) = a_0 f^n(u) + a_1 f^{n-1}(u) + \cdots + a_n u,$$

for every vector $u \in E$. Then we define a new kind of scalar multiplication $\cdot\colon K[X] \times E \to E$ by polynomials as follows: for every polynomial $p(X) \in K[X]$, for every $u \in E$,

$$p(X) \cdot u = p(f)(u).$$

It is easy to verify that this is a "good action," which means that

$$p \cdot (u + v) = p \cdot u + p \cdot v$$
$$(p + q) \cdot u = p \cdot u + q \cdot u$$
$$(pq) \cdot u = p \cdot (q \cdot u)$$
$$1 \cdot u = u,$$

for all $p, q \in K[X]$ and all $u, v \in E$. With this new scalar multiplication, E is a $K[X]$-module.

If $p = \lambda$ is just a scalar in K (a polynomial of degree 0), then

$$\lambda \cdot u = (\lambda \mathrm{id})(u) = \lambda u,$$

which means that K acts on E by scalar multiplication as before. If $p(X) = X$ (the monomial X), then

$$X \cdot u = f(u).$$

Now if we pick a basis $(e_1, \ldots, e_n)$ of E, if a polynomial $p(X) \in K[X]$ has the property that

$$p(X) \cdot e_i = 0, \quad i = 1, \ldots, n,$$

then this means that $p(f)(e_i) = 0$ for $i = 1, \ldots, n$, which means that the linear map $p(f)$ vanishes on E. We can also check, as we did in Section 6.2, that if A and B are two $n \times n$ matrices and if $(u_1, \ldots, u_n)$ are any n vectors, then

$$A \cdot \left(B \cdot \begin{pmatrix} u_1 \\ \vdots \\ u_n \end{pmatrix} \right) = (AB) \cdot \begin{pmatrix} u_1 \\ \vdots \\ u_n \end{pmatrix}.$$

This suggests the plan of attack for our second proof of the Cayley–Hamilton theorem. For simplicity, we prove the theorem for vector spaces over a field. The proof goes through for a free module over a commutative ring.

Theorem 6.3. *(Cayley–Hamilton) For every finite-dimensional vector space over a field K, for every linear map $f\colon E \to E$, for every basis $(e_1, \ldots, e_n)$, if A is the matrix over f over the basis $(e_1, \ldots, e_n)$ and if*

$$P_A(X) = X^n + c_1 X^{n-1} + \cdots + c_n$$

is the characteristic polynomial of A, then

$$P_A(f) = f^n + c_1 f^{n-1} + \cdots + c_n \mathrm{id} = 0.$$

Proof. Since the columns of A consist of the vector $f(e_j)$ expressed over the basis $(e_1, \ldots, e_n)$, we have

$$f(e_j) = \sum_{i=1}^{n} a_{ij} e_i, \quad 1 \le j \le n.$$

Using our action of $K[X]$ on E, the above equations can be expressed as

$$X \cdot e_j = \sum_{i=1}^{n} a_{ij} \cdot e_i, \quad 1 \le j \le n,$$

which yields

$$\sum_{i=1}^{j-1} -a_{ij} \cdot e_i + (X - a_{jj}) \cdot e_j + \sum_{i=j+1}^{n} -a_{ij} \cdot e_i = 0, \qquad 1 \le j \le n.$$

Observe that the transpose of the characteristic polynomial shows up, so the above system can be written as

$$\begin{pmatrix} X - a_{11} & -a_{21} & \cdots & -a_{n1} \\ -a_{12} & X - a_{22} & \cdots & -a_{n2} \\ \vdots & \vdots & \vdots & \vdots \\ -a_{1n} & -a_{2n} & \cdots & X - a_{nn} \end{pmatrix} \cdot \begin{pmatrix} e_1 \\ e_2 \\ \vdots \\ e_n \end{pmatrix} = \begin{pmatrix} 0 \\ 0 \\ \vdots \\ 0 \end{pmatrix}.$$

If we let $B = XI - A^\top$, then as in the previous proof, if $\widetilde{B}$ is the transpose of the matrix of cofactors of B, we have

$$\widetilde{B}B = \det(B)I = \det(XI - A^\top)I = \det(XI - A)I = P_A I.$$

But since

$$B \cdot \begin{pmatrix} e_1 \\ e_2 \\ \vdots \\ e_n \end{pmatrix} = \begin{pmatrix} 0 \\ 0 \\ \vdots \\ 0 \end{pmatrix},$$

and since $\widetilde{B}$ is matrix whose entries are polynomials in $K[X]$, it makes sense to multiply on the left by $\widetilde{B}$ and we get

$$\widetilde{B} \cdot B \cdot \begin{pmatrix} e_1 \\ e_2 \\ \vdots \\ e_n \end{pmatrix} = (\widetilde{B}B) \cdot \begin{pmatrix} e_1 \\ e_2 \\ \vdots \\ e_n \end{pmatrix} = P_A I \cdot \begin{pmatrix} e_1 \\ e_2 \\ \vdots \\ e_n \end{pmatrix} = \widetilde{B} \cdot \begin{pmatrix} 0 \\ 0 \\ \vdots \\ 0 \end{pmatrix} = \begin{pmatrix} 0 \\ 0 \\ \vdots \\ 0 \end{pmatrix};$$

that is,

$$P_A \cdot e_j = 0, \quad j = 1, \ldots, n,$$

which proves that $P_A(f) = 0$, as claimed. $\square$

If K is a field, then the characteristic polynomial of a linear map $f\colon E \to E$ is independent of the basis $(e_1, \ldots, e_n)$ chosen in E. To prove this, observe that the matrix of f over another basis will be of the form $P^{-1}AP$, for some inverible matrix P, and then

$$
\begin{aligned}
\det(XI - P^{-1}AP) &= \det(XP^{-1}IP - P^{-1}AP) \\
&= \det(P^{-1}(XI - A)P) \\
&= \det(P^{-1})\det(XI - A)\det(P) \\
&= \det(XI - A).
\end{aligned}
$$

Therefore, the characteristic polynomial of a linear map is intrinsic to f, and it is denoted by P_f.

The zeros (roots) of the characteristic polynomial of a linear map f are called the *eigenvalues* of f. They play an important role in theory and applications. We will come back to this topic later on.

6.8   Permanents

Recall that the explicit formula for the determinant of an $n \times n$ matrix is

$$
\det(A) = \sum_{\pi \in \mathfrak{S}_n} \epsilon(\pi) a_{\pi(1)\,1} \cdots a_{\pi(n)\,n}.
$$

If we drop the sign $\epsilon(\pi)$ of every permutation from the above formula, we obtain a quantity known as the *permanent*:

$$
\mathrm{per}(A) = \sum_{\pi \in \mathfrak{S}_n} a_{\pi(1)\,1} \cdots a_{\pi(n)\,n}.
$$

Permanents and determinants were investigated as early as 1812 by Cauchy. It is clear from the above definition that the permanent is a multilinear symmetric form. We also have

$$
\mathrm{per}(A) = \mathrm{per}(A^\top),
$$

and the following unsigned version of the Laplace expansion formula:

$$
\mathrm{per}(A) = a_{i\,1}\mathrm{per}(A_{i\,1}) + \cdots + a_{i\,j}\mathrm{per}(A_{i\,j}) + \cdots + a_{i\,n}\mathrm{per}(A_{i\,n}),
$$

for $i = 1, \ldots, n$. However, unlike determinants which have a clear geometric interpretation as signed volumes, permanents do not have any natural geometric interpretation. Furthermore, determinants can be evaluated efficiently, for example using the conversion to row reduced echelon form, but computing the permanent is hard.

Permanents turn out to have various combinatorial interpretations. One of these is in terms of perfect matchings of bipartite graphs which we now discuss.

See Definition 18.5 for the definition of an undirected graph. A *bipartite* (undirected) graph $G = (V, E)$ is a graph whose set of nodes V can be partitioned into two nonempty disjoint subsets V_1 and V_2, such that every edge $e \in E$ has one endpoint in V_1 and one endpoint in V_2.

An example of a bipartite graph with 14 nodes is shown in Figure 6.3; its nodes are partitioned into the two sets $\{x_1, x_2, x_3, x_4, x_5, x_6, x_7\}$ and $\{y_1, y_2, y_3, y_4, y_5, y_6, y_7\}$.

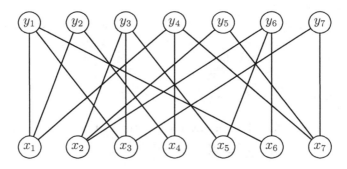

Fig. 6.3   A bipartite graph G.

A *matching* in a graph $G = (V, E)$ (bipartite or not) is a set M of pairwise non-adjacent edges, which means that no two edges in M share a common vertex. A *perfect matching* is a matching such that every node in V is incident to some edge in the matching M (every node in V is an endpoint of some edge in M). Figure 6.4 shows a perfect matching (in red) in the bipartite graph G.

Obviously, a perfect matching in a bipartite graph can exist only if its set of nodes has a partition in two blocks of equal size, say $\{x_1, \ldots, x_m\}$ and $\{y_1, \ldots, y_m\}$. Then there is a bijection between perfect matchings and bijections $\pi \colon \{x_1, \ldots, x_m\} \to \{y_1, \ldots, y_m\}$ such that $\pi(x_i) = y_j$ iff there is an edge between x_i and y_j.

Now every bipartite graph G with a partition of its nodes into two sets of equal size as above is represented by an $m \times m$ matrix $A = (a_{ij})$ such that $a_{ij} = 1$ iff there is an edge between x_i and y_j, and $a_{ij} = 0$ otherwise. Using the interpretation of perfect matchings as bijections

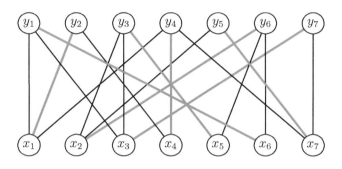

Fig. 6.4   A perfect matching in the bipartite graph G.

$\pi\colon \{x_1,\ldots,x_m\} \to \{y_1,\ldots,y_m\}$, we see that *the permanent* per(A) *of the* $(0,1)$-*matrix* A *representing the bipartite graph* G *counts the number of perfect matchings in* G.

In a famous paper published in 1979, Leslie Valiant proves that computing the permanent is a #P-complete problem. Such problems are suspected to be intractable. It is known that if a polynomial-time algorithm existed to solve a #P-complete problem, then we would have $P = NP$, which is believed to be very unlikely.

Another combinatorial interpretation of the permanent can be given in terms of systems of distinct representatives. Given a finite set S, let $(A_1,\ldots,A_n)$ be any sequence of nonempty subsets of S (not necessarily distinct). A *system of distinct representatives* (for short *SDR*) of the sets $A_1,\ldots,A_n$ is a sequence of n distinct elements $(a_1,\ldots,a_n)$, with $a_i \in A_i$ for $i = 1,\ldots,n$. The number of SDR's of a sequence of sets plays an important role in combinatorics. Now, if $S = \{1,2,\ldots,n\}$ and if we associate to any sequence $(A_1,\ldots,A_n)$ of nonempty subsets of S the matrix $A = (a_{ij})$ defined such that $a_{ij} = 1$ if $j \in A_i$ and $a_{ij} = 0$ otherwise, then *the permanent* per(A) *counts the number of SDR's of the sets* $A_1,\ldots,A_n$.

This interpretation of permanents in terms of SDR's can be used to prove bounds for the permanents of various classes of matrices. Interested readers are referred to van Lint and Wilson [van Lint and Wilson (2001)] (Chapters 11 and 12). In particular, a proof of a theorem known as *Van der Waerden conjecture* is given in Chapter 12. This theorem states that for any $n \times n$ matrix A with nonnegative entries in which all row-sums and column-sums are 1 (doubly stochastic matrices), we have

$$\mathrm{per}(A) \geq \frac{n!}{n^n},$$

with equality for the matrix in which all entries are equal to $1/n$.

6.9   Summary

The main concepts and results of this chapter are listed below:

- *Permutations, transpositions, basics transpositions.*
- Every permutation can be written as a composition of permutations.
- The *parity* of the number of transpositions involved in any decomposition of a permutation σ is an invariant; it is the *signature* $\epsilon(\sigma)$ of the permutation σ.
- *Multilinear maps* (also called *n-linear maps*); *bilinear maps*.
- *Symmetric* and *alternating* multilinear maps.
- A basic property of alternating multilinear maps (Lemma 6.1) and the introduction of the formula expressing a determinant.
- Definition of a *determinant* as a multlinear alternating map $D \colon \mathrm{M}_n(K) \to K$ such that $D(I) = 1$.
- We define the set of algorithms $\mathcal{D}_n$, to compute the determinant of an $n \times n$ matrix.
- *Laplace expansion according to the ith row; cofactors.*
- We prove that the algorithms in $\mathcal{D}_n$ compute determinants (Lemma 6.2).
- We prove that all algorithms in $\mathcal{D}_n$ compute the same determinant (Theorem 6.1).
- We give an interpretation of determinants as *signed volumes*.
- We prove that $\det(A) = \det(A^\top)$.
- We prove that $\det(AB) = \det(A)\det(B)$.
- The *adjugate* $\widetilde{A}$ of a matrix A.
- Formula for the inverse in terms of the adjugate.
- A matrix A is invertible iff $\det(A) \neq 0$.
- Solving linear equations using *Cramer's rules*.
- Determinant of a linear map.
- The *characteristic polynomial* of a matrix.
- The *Cayley–Hamilton theorem*.
- The action of the polynomial ring induced by a linear map on a vector space.

- *Permanents.*
- Permanents count the number of perfect matchings in bipartite graphs.
- Computing the permanent is a #P-perfect problem (L. Valiant).
- Permanents count the number of SDRs of sequences of subsets of a given set.

6.10 Further Readings

Thorough expositions of the material covered in Chapters 2–5 and 6 can be found in Strang [Strang (1988, 1986)], Lax [Lax (2007)], Lang [Lang (1993)], Artin [Artin (1991)], Mac Lane and Birkhoff [Mac Lane and Birkhoff (1967)], Hoffman and Kunze [Kenneth and Ray (1971)], Dummit and Foote [Dummit and Foote (1999)], Bourbaki [Bourbaki (1970, 1981a)], Van Der Waerden [Van Der Waerden (1973)], Serre [Serre (2010)], Horn and Johnson [Horn and Johnson (1990)], and Bertin [Bertin (1981)]. These notions of linear algebra are nicely put to use in classical geometry, see Berger [Berger (1990a,b)], Tisseron [Tisseron (1994)] and Dieudonné [Dieudonné (1965)].

6.11 Problems

Problem 6.1. Prove that every transposition can be written as a product of basic transpositions.

Problem 6.2. (1) Given two vectors in $\mathbb{R}^2$ of coordinates $(c_1 - a_1, c_2 - a_2)$ and $(b_1 - a_1, b_2 - a_2)$, prove that they are linearly dependent iff

$$\begin{vmatrix} a_1 & b_1 & c_1 \\ a_2 & b_2 & c_2 \\ 1 & 1 & 1 \end{vmatrix} = 0.$$

(2) Given three vectors in $\mathbb{R}^3$ of coordinates $(d_1 - a_1, d_2 - a_2, d_3 - a_3)$, $(c_1 - a_1, c_2 - a_2, c_3 - a_3)$, and $(b_1 - a_1, b_2 - a_2, b_3 - a_3)$, prove that they are linearly dependent iff

$$\begin{vmatrix} a_1 & b_1 & c_1 & d_1 \\ a_2 & b_2 & c_2 & d_2 \\ a_3 & b_3 & c_3 & d_3 \\ 1 & 1 & 1 & 1 \end{vmatrix} = 0.$$

Problem 6.3. Let A be the $(m+n) \times (m+n)$ block matrix (over any field K) given by

$$A = \begin{pmatrix} A_1 & A_2 \\ 0 & A_4 \end{pmatrix},$$

where A_1 is an $m \times m$ matrix, A_2 is an $m \times n$ matrix, and A_4 is an $n \times n$ matrix. Prove that $\det(A) = \det(A_1)\det(A_4)$.

Use the above result to prove that if A is an upper triangular $n \times n$ matrix, then $\det(A) = a_{11}a_{22}\cdots a_{nn}$.

Problem 6.4. Prove that if $n \geq 3$, then

$$
\det \begin{pmatrix}
1 + x_1y_1 & 1 + x_1y_2 & \cdots & 1 + x_1y_n \\
1 + x_2y_1 & 1 + x_2y_2 & \cdots & 1 + x_2y_n \\
\vdots & \vdots & \vdots & \vdots \\
1 + x_ny_1 & 1 + x_ny_2 & \cdots & 1 + x_ny_n
\end{pmatrix} = 0.
$$

Problem 6.5. Prove that

$$
\begin{vmatrix}
1 & 4 & 9 & 16 \\
4 & 9 & 16 & 25 \\
9 & 16 & 25 & 36 \\
16 & 25 & 36 & 49
\end{vmatrix} = 0.
$$

Problem 6.6. Consider the $n \times n$ symmetric matrix

$$
A = \begin{pmatrix}
1 & 2 & 0 & 0 & \cdots & 0 & 0 \\
2 & 5 & 2 & 0 & \cdots & 0 & 0 \\
0 & 2 & 5 & 2 & \cdots & 0 & 0 \\
\vdots & \vdots & \ddots & \ddots & \ddots & \vdots & \vdots \\
0 & 0 & \cdots & 2 & 5 & 2 & 0 \\
0 & 0 & \cdots & 0 & 2 & 5 & 2 \\
0 & 0 & \cdots & 0 & 0 & 2 & 5
\end{pmatrix}.
$$

(1) Find an upper-triangular matrix R such that $A = R^\top R$.

(2) Prove that $\det(A) = 1$.

(3) Consider the sequence

$$
p_0(\lambda) = 1
$$
$$
p_1(\lambda) = 1 - \lambda
$$
$$
p_k(\lambda) = (5 - \lambda)p_{k-1}(\lambda) - 4p_{k-2}(\lambda) \quad 2 \leq k \leq n.
$$

Prove that

$$
\det(A - \lambda I) = p_n(\lambda).
$$

Remark: It can be shown that $p_n(\lambda)$ has n distinct (real) roots and that the roots of $p_k(\lambda)$ separate the roots of $p_{k+1}(\lambda)$.

Problem 6.7. Let B be the $n \times n$ matrix ($n \geq 3$) given by

$$B = \begin{pmatrix} 1 & -1 & -1 & -1 & \cdots & -1 & -1 \\ 1 & -1 & 1 & 1 & \cdots & 1 & 1 \\ 1 & 1 & -1 & 1 & \cdots & 1 & 1 \\ 1 & 1 & 1 & -1 & \cdots & 1 & 1 \\ \vdots & \vdots & \vdots & \vdots & \vdots & \vdots & \vdots \\ 1 & 1 & 1 & 1 & \cdots & -1 & 1 \\ 1 & 1 & 1 & 1 & \cdots & 1 & -1 \end{pmatrix}.$$

Prove that

$$\det(B) = (-1)^n (n-2) 2^{n-1}.$$

Problem 6.8. Given a field K (say $K = \mathbb{R}$ or $K = \mathbb{C}$), given any two polynomials $p(X), q(X) \in K[X]$, we says that $q(X)$ *divides* $p(X)$ (and that $p(X)$ *is a multiple of* $q(X)$) iff there is some polynomial $s(X) \in K[X]$ such that

$$p(X) = q(X)s(X).$$

In this case we say that $q(X)$ *is a factor of* $p(X)$, and if $q(X)$ has degree at least one, we say that $q(X)$ *is a nontrivial factor of* $p(X)$.

Let $f(X)$ and $g(X)$ be two polynomials in $K[X]$ with

$$f(X) = a_0 X^m + a_1 X^{m-1} + \cdots + a_m$$

of degree $m \geq 1$ and

$$g(X) = b_0 X^n + b_1 X^{n-1} + \cdots + b_n$$

of degree $n \geq 1$ (with $a_0, b_0 \neq 0$).

You will need the following result which you need not prove:

Two polynomials $f(X)$ and $g(X)$ with $\deg(f) = m \geq 1$ and $\deg(g) = n \geq 1$ have some common nontrivial factor iff there exist two nonzero polynomials $p(X)$ and $q(X)$ such that

$$fp = gq,$$

with $\deg(p) \leq n - 1$ and $\deg(q) \leq m - 1$.

(1) Let $\mathcal{P}_m$ denote the vector space of all polynomials in $K[X]$ of degree at most $m - 1$, and let $T \colon \mathcal{P}_n \times \mathcal{P}_m \to \mathcal{P}_{m+n}$ be the map given by

$$T(p, q) = fp + gq, \quad p \in \mathcal{P}_n, \ q \in \mathcal{P}_m,$$

where f and g are some fixed polynomials of degree $m \geq 1$ and $n \geq 1$.

Prove that the map T is linear.

(2) Prove that T is not injective iff f and g have a common nontrivial factor.

(3) Prove that f and g have a nontrivial common factor iff $R(f, g) = 0$, where $R(f, g)$ is the determinant given by

$$R(f,g) = \begin{vmatrix} a_0 & a_1 & \cdots & \cdots & a_m & 0 & \cdots & \cdots & \cdots & \cdots & 0 \\ 0 & a_0 & a_1 & \cdots & \cdots & a_m & 0 & \cdots & \cdots & \cdots & 0 \\ \cdots & \cdots & \cdots & \cdots & \cdots & \cdots & \cdots & \cdots & \cdots & \cdots & \cdots \\ \cdots & \cdots & \cdots & \cdots & \cdots & \cdots & \cdots & \cdots & \cdots & \cdots & \cdots \\ \cdots & \cdots & \cdots & \cdots & \cdots & \cdots & \cdots & \cdots & \cdots & \cdots & \cdots \\ \cdots & \cdots & \cdots & \cdots & \cdots & \cdots & \cdots & \cdots & \cdots & \cdots & \cdots \\ 0 & \cdots & \cdots & \cdots & \cdots & 0 & a_0 & a_1 & \cdots & \cdots & a_m \\ b_0 & b_1 & \cdots & \cdots & \cdots & \cdots & b_n & 0 & \cdots & 0 \\ 0 & b_0 & b_1 & \cdots & \cdots & \cdots & \cdots & b_n & 0 & \cdots \\ \cdots & \cdots & \cdots & \cdots & \cdots & \cdots & \cdots & \cdots & \cdots & \cdots & \cdots \\ 0 & \cdots & 0 & b_0 & b_1 & \cdots & \cdots & \cdots & \cdots & \cdots & b_n \end{vmatrix}.$$

The above determinant is called the _resultant of f and g_.

Note that the matrix of the resultant is an $(n + m) \times (n + m)$ matrix, with the first row (involving the a_is) occurring n times, each time shifted over to the right by one column, and the $(n + 1)$th row (involving the b_js) occurring m times, each time shifted over to the right by one column.

Hint. Express the matrix of T over some suitable basis.

(4) Compute the resultant in the following three cases:

(a) $m = n = 1$, and write $f(X) = aX + b$ and $g(X) = cX + d$.
(b) $m = 1$ and $n \geq 2$ arbitrary.
(c) $f(X) = aX^2 + bX + c$ and $g(X) = 2aX + b$.

(5) Compute the resultant of $f(X) = X^3 + pX + q$ and $g(X) = 3X^2 + p$, and

$$f(X) = a_0 X^2 + a_1 X + a_2$$
$$g(X) = b_0 X^2 + b_1 X + b_2.$$

In the second case, you should get

$$4R(f,g) = (2a_0 b_2 - a_1 b_1 + 2a_2 b_0)^2 - (4a_0 a_2 - a_1^2)(4b_0 b_2 - b_1^2).$$

Problem 6.9. Let A, B, C, D be $n \times n$ real or complex matrices.

(1) Prove that if A is invertible and if $AC = CA$, then

$$\det \begin{pmatrix} A & B \\ C & D \end{pmatrix} = \det(AD - CB).$$

(2) Prove that if H is an $n \times n$ Hadamard matrix $(n \geq 2)$, then $|\det(H)| = n^{n/2}$.

(3) Prove that if H is an $n \times n$ Hadamard matrix $(n \geq 2)$, then

$$\det \begin{pmatrix} H & H \\ H & -H \end{pmatrix} = (2n)^n.$$

Problem 6.10. Compute the product of the following determinants

$$\begin{vmatrix} a & -b & -c & -d \\ b & a & -d & c \\ c & d & a & -b \\ d & -c & b & a \end{vmatrix} \begin{vmatrix} x & -y & -z & -t \\ y & x & -t & z \\ z & t & x & -y \\ t & -z & y & x \end{vmatrix}$$

to prove the following identity (due to Euler):

$$(a^2 + b^2 + c^2 + d^2)(x^2 + y^2 + z^2 + t^2)$$
$$= (ax + by + cz + dt)^2 + (ay - bx + ct - dz)^2$$
$$+ (az - bt - cx + dy)^2 + (at + bz - cy + dx)^2.$$

Problem 6.11. Let A be an $n \times n$ matrix with integer entries. Prove that A^{-1} exists and has integer entries if and only if $\det(A) = \pm 1$.

Problem 6.12. Let A be an $n \times n$ real or complex matrix.

(1) Prove that if $A^\top = -A$ (A is *skew-symmetric*) and if n is odd, then $\det(A) = 0$.

(2) Prove that

$$\begin{vmatrix} 0 & a & b & c \\ -a & 0 & d & e \\ -b & -d & 0 & f \\ -c & -e & -f & 0 \end{vmatrix} = (af - be + dc)^2.$$

Problem 6.13. A *Cauchy matrix* is a matrix of the form

$$\begin{pmatrix} \dfrac{1}{\lambda_1 - \sigma_1} & \dfrac{1}{\lambda_1 - \sigma_2} & \cdots & \dfrac{1}{\lambda_1 - \sigma_n} \\ \dfrac{1}{\lambda_2 - \sigma_1} & \dfrac{1}{\lambda_2 - \sigma_2} & \cdots & \dfrac{1}{\lambda_2 - \sigma_n} \\ \vdots & \vdots & \vdots & \vdots \\ \dfrac{1}{\lambda_n - \sigma_1} & \dfrac{1}{\lambda_n - \sigma_2} & \cdots & \dfrac{1}{\lambda_n - \sigma_n} \end{pmatrix}$$

where $\lambda_i \neq \sigma_j$, for all i, j, with $1 \leq i, j \leq n$. Prove that the determinant C_n of a Cauchy matrix as above is given by

$$C_n = \frac{\prod_{i=2}^{n} \prod_{j=1}^{i-1} (\lambda_i - \lambda_j)(\sigma_j - \sigma_i)}{\prod_{i=1}^{n} \prod_{j=1}^{n} (\lambda_i - \sigma_j)}.$$

Problem 6.14. Let $(\alpha_1, \ldots, \alpha_{m+1})$ be a sequence of pairwise distinct scalars in $\mathbb{R}$ and let $(\beta_1, \ldots, \beta_{m+1})$ be any sequence of scalars in $\mathbb{R}$, not necessarily distinct.

(1) Prove that there is a unique polynomial P of degree at most m such that

$$P(\alpha_i) = \beta_i, \quad 1 \leq i \leq m + 1.$$

Hint. Remember Vandermonde!

(2) Let $L_i(X)$ be the polynomial of degree m given by

$$L_i(X) = \frac{(X - \alpha_1) \cdots (X - \alpha_{i-1})(X - \alpha_{i+1}) \cdots (X - \alpha_{m+1})}{(\alpha_i - \alpha_1) \cdots (\alpha_i - \alpha_{i-1})(\alpha_i - \alpha_{i+1}) \cdots (\alpha_i - \alpha_{m+1})},$$
$$1 \leq i \leq m + 1.$$

The polynomials $L_i(X)$ are known as *Lagrange polynomial interpolants*. Prove that

$$L_i(\alpha_j) = \delta_{ij} \quad 1 \leq i, j \leq m + 1.$$

Prove that

$$P(X) = \beta_1 L_1(X) + \cdots + \beta_{m+1} L_{m+1}(X)$$

is the unique polynomial of degree at most m such that

$$P(\alpha_i) = \beta_i, \quad 1 \leq i \leq m + 1.$$

(3) Prove that $L_1(X), \ldots, L_{m+1}(X)$ are linearly independent, and that they form a basis of all polynomials of degree at most m.

How is 1 (the constant polynomial 1) expressed over the basis $(L_1(X), \ldots, L_{m+1}(X))$?

Give the expression of every polynomial $P(X)$ of degree at most m over the basis $(L_1(X), \ldots, L_{m+1}(X))$.

(4) Prove that the dual basis $(L_1^*, \ldots, L_{m+1}^*)$ of the basis $(L_1(X), \ldots, L_{m+1}(X))$ consists of the linear forms L_i^* given by

$$L_i^*(P) = P(\alpha_i),$$

for every polynomial P of degree at most m; this is simply *evaluation at* α_i.

Chapter 7

Gaussian Elimination, LU-Factorization, Cholesky Factorization, Reduced Row Echelon Form

In this chapter we assume that all vector spaces are over the field $\mathbb{R}$. All results that do not rely on the ordering on $\mathbb{R}$ or on taking square roots hold for arbitrary fields.

7.1 Motivating Example: Curve Interpolation

Curve interpolation is a problem that arises frequently in computer graphics and in robotics (path planning). There are many ways of tackling this problem and in this section we will describe a solution using *cubic splines.* Such splines consist of cubic Bézier curves. They are often used because they are cheap to implement and give more flexibility than quadratic Bézier curves.

A *cubic Bézier curve* $C(t)$ (in $\mathbb{R}^2$ or $\mathbb{R}^3$) is specified by a list of four *control points* (b_0, b_1, b_2, b_3) and is given parametrically by the equation

$$C(t) = (1 - t)^3 \, b_0 + 3(1 - t)^2 t \, b_1 + 3(1 - t) t^2 \, b_2 + t^3 \, b_3.$$

Clearly, $C(0) = b_0$, $C(1) = b_3$, and for $t \in [0, 1]$, the point $C(t)$ belongs to the convex hull of the control points b_0, b_1, b_2, b_3. The polynomials

$$(1 - t)^3, \quad 3(1 - t)^2 t, \quad 3(1 - t)t^2, \quad t^3$$

are the *Bernstein polynomials* of degree 3.

Typically, we are only interested in the curve segment corresponding to the values of t in the interval $[0, 1]$. Still, the placement of the control points drastically affects the shape of the curve segment, which can even have a self-intersection; See Figures 7.1, 7.2, 7.3 illustrating various configurations.

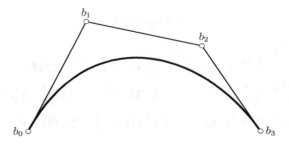

Fig. 7.1   A "standard" Bézier curve.

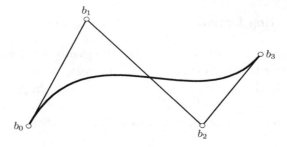

Fig. 7.2   A Bézier curve with an inflection point.

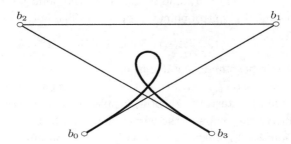

Fig. 7.3   A self-intersecting Bézier curve.

Interpolation problems require finding curves passing through some given data points and possibly satisfying some extra constraints.

A *Bézier spline curve* F is a curve which is made up of curve segments which are Bézier curves, say $C_1, \ldots, C_m$ ($m \geq 2$). We will assume that F defined on $[0, m]$, so that for $i = 1, \ldots, m$,

$$F(t) = C_i(t - i + 1), \quad i - 1 \leq t \leq i.$$

Typically, some smoothness is required between any two junction points, that is, between any two points $C_i(1)$ and $C_{i+1}(0)$, for $i = 1, \ldots, m - 1$. We require that $C_i(1) = C_{i+1}(0)$ (C^0-*continuity*), and typically that the derivatives of C_i at 1 and of C_{i+1} at 0 agree up to second order derivatives. This is called C^2-*continuity*, and it ensures that the tangents agree as well as the curvatures.

There are a number of interpolation problems, and we consider one of the most common problems which can be stated as follows:

Problem: Given $N + 1$ data points $x_0, \ldots, x_N$, find a C^2 cubic spline curve F such that $F(i) = x_i$ for all i, $0 \leq i \leq N$ ($N \geq 2$).

A way to solve this problem is to find $N + 3$ auxiliary points $d_{-1}, \ldots, d_{N+1}$, called *de Boor control points*, from which N Bézier curves can be found. Actually,

$$d_{-1} = x_0 \quad \text{and} \quad d_{N+1} = x_N$$

so we only need to find $N + 1$ points $d_0, \ldots, d_N$.

It turns out that the C^2-continuity constraints on the N Bézier curves yield only $N - 1$ equations, so d_0 and d_N can be chosen arbitrarily. In practice, d_0 and d_N are chosen according to various *end conditions*, such as prescribed velocities at x_0 and x_N. For the time being, we will assume that d_0 and d_N are given.

Figure 7.4 illustrates an interpolation problem involving $N+1 = 7+1 = 8$ data points. The control points d_0 and d_7 were chosen arbitrarily.

It can be shown that $d_1, \ldots, d_{N-1}$ are given by the linear system

$$\begin{pmatrix} \frac{7}{2} & 1 & & & \\ 1 & 4 & 1 & & 0 \\ & \ddots & \ddots & \ddots & \\ 0 & & 1 & 4 & 1 \\ & & & 1 & \frac{7}{2} \end{pmatrix} \begin{pmatrix} d_1 \\ d_2 \\ \vdots \\ d_{N-2} \\ d_{N-1} \end{pmatrix} = \begin{pmatrix} 6x_1 - \frac{3}{2}d_0 \\ 6x_2 \\ \vdots \\ 6x_{N-2} \\ 6x_{N-1} - \frac{3}{2}d_N \end{pmatrix}.$$

We will show later that the above matrix is invertible because it is strictly diagonally dominant.

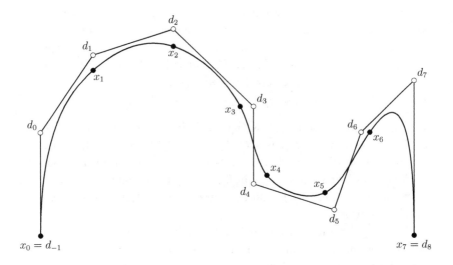

Fig. 7.4 A C^2 cubic interpolation spline curve passing through the points $x_0, x_1, x_2, x_3,$ x_4, x_5, x_6, x_7.

Once the above system is solved, the Bézier cubics $C_1, \ldots, C_N$ are determined as follows (we assume $N \geq 2$): For $2 \leq i \leq N - 1$, the control points $(b_0^i, b_1^i, b_2^i, b_3^i)$ of C_i are given by

$$b_0^i = x_{i-1}$$
$$b_1^i = \frac{2}{3} d_{i-1} + \frac{1}{3} d_i$$
$$b_2^i = \frac{1}{3} d_{i-1} + \frac{2}{3} d_i$$
$$b_3^i = x_i.$$

The control points $(b_0^1, b_1^1, b_2^1, b_3^1)$ of C_1 are given by

$$b_0^1 = x_0$$
$$b_1^1 = d_0$$
$$b_2^1 = \frac{1}{2} d_0 + \frac{1}{2} d_1$$
$$b_3^1 = x_1,$$

and the control points $(b_0^N, b_1^N, b_2^N, b_3^N)$ of C_N are given by

$$b_0^N = x_{N-1}$$
$$b_1^N = \frac{1}{2}d_{N-1} + \frac{1}{2}d_N$$
$$b_2^N = d_N$$
$$b_3^N = x_N.$$

Figure 7.5 illustrates this process spline interpolation for $N = 7$.

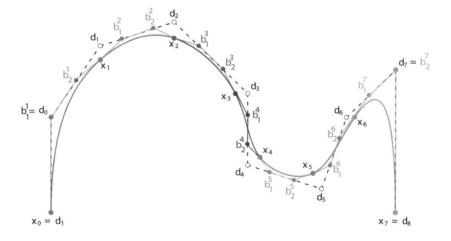

Fig. 7.5 A C^2 cubic interpolation of $x_0, x_1, x_2, x_3, x_4, x_5, x_6, x_7$ with associated color coded Bézier cubics.

We will now describe various methods for solving linear systems. Since the matrix of the above system is tridiagonal, there are specialized methods which are more efficient than the general methods. We will discuss a few of these methods.

7.2 Gaussian Elimination

Let A be an $n \times n$ matrix, let $b \in \mathbb{R}^n$ be an n-dimensional vector and assume that A is invertible. Our goal is to solve the system $Ax = b$. Since A is assumed to be invertible, we know that this system has a unique solution $x = A^{-1}b$. Experience shows that two counter-intuitive facts are revealed:

(1) One should avoid computing the inverse A^{-1} of A explicitly. This is inefficient since it would amount to solving the n linear systems $Au^{(j)} = e_j$ for $j = 1, \ldots, n$, where $e_j = (0, \ldots, 1, \ldots, 0)$ is the jth canonical basis vector of $\mathbb{R}^n$ (with a 1 is the jth slot). By doing so, we would replace the resolution of a single system by the resolution of n systems, and we would still have to multiply A^{-1} by b.

(2) One does not solve (large) linear systems by computing determinants (using Cramer's formulae) since this method requires a number of additions (resp. multiplications) proportional to $(n+1)!$ (resp. $(n+2)!$).

The key idea on which most direct methods (as opposed to iterative methods, that look for an approximation of the solution) are based is that if A is an upper-triangular matrix, which means that $a_{ij} = 0$ for $1 \leq j < i \leq n$ (resp. lower-triangular, which means that $a_{ij} = 0$ for $1 \leq i < j \leq n$), then computing the solution x is trivial. Indeed, say A is an upper-triangular matrix

$$A = \begin{pmatrix} a_{11} & a_{12} & \cdots & a_{1\,n-2} & a_{1\,n-1} & a_{1\,n} \\ 0 & a_{22} & \cdots & a_{2\,n-2} & a_{2\,n-1} & a_{2\,n} \\ 0 & 0 & \ddots & \vdots & \vdots & \vdots \\ & & & \ddots & \vdots & \vdots \\ 0 & 0 & \cdots & 0 & a_{n-1\,n-1} & a_{n-1\,n} \\ 0 & 0 & \cdots & 0 & 0 & a_{n\,n} \end{pmatrix}.$$

Then $\det(A) = a_{11}a_{22}\cdots a_{nn} \neq 0$, which implies that $a_{ii} \neq 0$ for $i = 1, \ldots, n$, and we can solve the system $Ax = b$ from bottom-up by *back-substitution*. That is, first we compute x_n from the last equation, next plug this value of x_n into the next to the last equation and compute x_{n-1} from it, *etc*. This yields

$$x_n = a_{nn}^{-1} b_n$$
$$x_{n-1} = a_{n-1\,n-1}^{-1}(b_{n-1} - a_{n-1\,n}x_n)$$
$$\vdots$$
$$x_1 = a_{11}^{-1}(b_1 - a_{12}x_2 - \cdots - a_{1\,n}x_n).$$

Note that the use of determinants can be avoided to prove that if A is invertible then $a_{ii} \neq 0$ for $i = 1, \ldots, n$. Indeed, it can be shown directly (by induction) that an upper (or lower) triangular matrix is invertible iff all its diagonal entries are nonzero.

If A is lower-triangular, we solve the system from top-down by *forward-substitution*.

Thus, what we need is a method for transforming a matrix to an equivalent one in upper-triangular form. This can be done by *elimination*. Let us illustrate this method on the following example:

$$
\begin{aligned}
2x &+ y + z = 5 \\
4x &- 6y = -2 \\
-2x &+ 7y + 2z = 9.
\end{aligned}
$$

We can eliminate the variable x from the second and the third equation as follows: Subtract twice the first equation from the second and add the first equation to the third. We get the new system

$$
\begin{aligned}
2x + y &+ z = 5 \\
- 8y &- 2z = -12 \\
8y &+ 3z = 14.
\end{aligned}
$$

This time we can eliminate the variable y from the third equation by adding the second equation to the third:

$$
\begin{aligned}
2x + y &+ z = 5 \\
- 8y &- 2z = -12 \\
&z = 2.
\end{aligned}
$$

This last system is upper-triangular. Using back-substitution, we find the solution: $z = 2$, $y = 1$, $x = 1$.

Observe that we have performed only *row operations*. The general method is to iteratively eliminate variables using simple row operations (namely, adding or subtracting a multiple of a row to another row of the matrix) while simultaneously applying these operations to the vector b, to obtain a system, $MAx = Mb$, where MA is upper-triangular. Such a method is called *Gaussian elimination*. However, one extra twist is needed for the method to work in all cases: It may be necessary to permute rows, as illustrated by the following example:

$$
\begin{aligned}
x &+ y + z = 1 \\
x &+ y + 3z = 1 \\
2x &+ 5y + 8z = 1.
\end{aligned}
$$

In order to eliminate x from the second and third row, we subtract the first row from the second and we subtract twice the first row from the third:

$$
\begin{aligned}
x + y &+ z = 1 \\
2z &= 0 \\
3y &+ 6z = -1.
\end{aligned}
$$

Now the trouble is that y does not occur in the second row; so, we can't eliminate y from the third row by adding or subtracting a multiple of the second row to it. The remedy is simple: Permute the second and the third row! We get the system:

$$\begin{aligned} x + y + z &= 1 \\ 3y + 6z &= -1 \\ 2z &= 0, \end{aligned}$$

which is already in triangular form. Another example where some permutations are needed is:

$$\begin{aligned} z &= 1 \\ -2x + 7y + 2z &= 1 \\ 4x - 6y &= -1. \end{aligned}$$

First we permute the first and the second row, obtaining

$$\begin{aligned} -2x + 7y + 2z &= 1 \\ z &= 1 \\ 4x - 6y &= -1, \end{aligned}$$

and then we add twice the first row to the third, obtaining:

$$\begin{aligned} -2x + 7y + 2z &= 1 \\ z &= 1 \\ 8y + 4z &= 1. \end{aligned}$$

Again we permute the second and the third row, getting

$$\begin{aligned} -2x + 7y + 2z &= 1 \\ 8y + 4z &= 1 \\ z &= 1, \end{aligned}$$

an upper-triangular system. Of course, in this example, z is already solved and we could have eliminated it first, but for the general method, we need to proceed in a systematic fashion.

We now describe the method of *Gaussian elimination* applied to a linear system $Ax = b$, where A is assumed to be invertible. We use the variable k to keep track of the stages of elimination. Initially, $k = 1$.

(1) The first step is to pick some nonzero entry $a_{i\,1}$ in the first column of A. Such an entry must exist, since A is invertible (otherwise, the first column of A would be the zero vector, and the columns of A would not be linearly independent. Equivalently, we would have $\det(A) = 0$). The actual choice of such an element has some impact on the numerical

stability of the method, but this will be examined later. For the time
being, we assume that some arbitrary choice is made. This chosen
element is called the *pivot* of the elimination step and is denoted π_1
(so, in this first step, $\pi_1 = a_{i\,1}$).

(2) Next we permute the row (i) corresponding to the pivot with the first
row. Such a step is called *pivoting*. So after this permutation, the first
element of the first row is nonzero.

(3) We now eliminate the variable x_1 from all rows except the first by
adding suitable multiples of the first row to these rows. More precisely
we add $-a_{i\,1}/\pi_1$ times the first row to the ith row for $i = 2, \ldots, n$. At
the end of this step, all entries in the first column are zero except the
first.

(4) Increment k by 1. If $k = n$, stop. Otherwise, $k < n$, and then iteratively
repeat Steps (1), (2), (3) on the $(n - k + 1) \times (n - k + 1)$ subsystem
obtained by deleting the first $k - 1$ rows and $k - 1$ columns from the
current system.

If we let $A_1 = A$ and $A_k = (a_{i\,j}^{(k)})$ be the matrix obtained after $k - 1$
elimination steps $(2 \le k \le n)$, then the kth elimination step is applied to
the matrix A_k of the form

$$A_k = \begin{pmatrix} a_{1\,1}^{(k)} & a_{1\,2}^{(k)} & \cdots & \cdots & \cdots & a_{1\,n}^{(k)} \\ 0 & a_{2\,2}^{(k)} & \cdots & \cdots & \cdots & a_{2\,n}^{(k)} \\ \vdots & \ddots & \ddots & & & \vdots \\ 0 & 0 & 0 & a_{k\,k}^{(k)} & \cdots & a_{k\,n}^{(k)} \\ \vdots & \vdots & \vdots & \vdots & & \vdots \\ 0 & 0 & 0 & a_{n\,k}^{(k)} & \cdots & a_{n\,n}^{(k)} \end{pmatrix}.$$

Actually, note that

$$a_{i\,j}^{(k)} = a_{i\,j}^{(i)}$$

for all i, j with $1 \le i \le k - 2$ and $i \le j \le n$, since the first $k - 1$ rows
remain unchanged after the $(k - 1)$th step.

We will prove later that $\det(A_k) = \pm \det(A)$. Consequently, A_k is
invertible. The fact that A_k is invertible iff A is invertible can also be
shown without determinants from the fact that there is some invertible
matrix M_k such that $A_k = M_k A$, as we will see shortly.

Since A_k is invertible, some entry $a_{i\,k}^{(k)}$ with $k \le i \le n$ is nonzero.
Otherwise, the last $n - k + 1$ entries in the first k columns of A_k would be

zero, and the first k columns of A_k would yield k vectors in $\mathbb{R}^{k-1}$. But then the first k columns of A_k would be linearly dependent and A_k would not be invertible, a contradiction. This situation is illustrated by the following matrix for $n = 5$ and $k = 3$:

$$\begin{pmatrix} a_{11}^{(3)} & a_{12}^{(3)} & a_{13}^{(3)} & a_{13}^{(3)} & a_{15}^{(3)} \\ 0 & a_{22}^{(3)} & a_{23}^{(3)} & a_{24}^{(3)} & a_{25}^{(3)} \\ 0 & 0 & 0 & a_{34}^{(3)} & a_{35}^{(3)} \\ 0 & 0 & 0 & a_{44}^{(3)} & a_{4n}^{(3)} \\ 0 & 0 & 0 & a_{54}^{(3)} & a_{55}^{(3)} \end{pmatrix}.$$

The first three columns of the above matrix are linearly dependent.

So one of the entries $a_{ik}^{(k)}$ with $k \le i \le n$ can be chosen as pivot, and we permute the kth row with the ith row, obtaining the matrix $\alpha^{(k)} = (\alpha_{jl}^{(k)})$. The new pivot is $\pi_k = \alpha_{kk}^{(k)}$, and we zero the entries $i = k+1, \dots, n$ in column k by adding $-\alpha_{ik}^{(k)}/\pi_k$ times row k to row i. At the end of this step, we have A_{k+1}. Observe that the first $k-1$ rows of A_k are identical to the first $k-1$ rows of A_{k+1}.

The process of Gaussian elimination is illustrated in schematic form below:

$$\begin{pmatrix} \times & \times & \times & \times \\ \times & \times & \times & \times \\ \times & \times & \times & \times \\ \times & \times & \times & \times \end{pmatrix} \Longrightarrow \begin{pmatrix} \times & \times & \times & \times \\ 0 & \times & \times & \times \\ 0 & \times & \times & \times \\ 0 & \times & \times & \times \end{pmatrix} \Longrightarrow \begin{pmatrix} \times & \times & \times & \times \\ 0 & \times & \times & \times \\ 0 & 0 & \times & \times \\ 0 & 0 & \times & \times \end{pmatrix} \Longrightarrow \begin{pmatrix} \times & \times & \times & \times \\ 0 & \times & \times & \times \\ 0 & 0 & \times & \times \\ 0 & 0 & 0 & \times \end{pmatrix}.$$

7.3 Elementary Matrices and Row Operations

It is easy to figure out what kind of matrices perform the elementary row operations used during Gaussian elimination. The key point is that if $A = PB$, where A, B are $m \times n$ matrices and P is a square matrix of dimension m, if (as usual) we denote the rows of A and B by $A_1, \dots, A_m$ and $B_1, \dots, B_m$, then the formula

$$a_{ij} = \sum_{k=1}^{m} p_{ik} b_{kj}$$

giving the (i, j)th entry in A shows that the ith row of A is a *linear combination* of the rows of B:

$$A_i = p_{i1} B_1 + \cdots + p_{im} B_m.$$

Therefore, *multiplication of a matrix on the left by a square matrix performs row operations*. Similarly, multiplication of a matrix on the right by a square matrix performs column operations

The permutation of the kth row with the ith row is achieved by multiplying A on the left by the *transposition matrix* $P(i, k)$, which is the matrix obtained from the identity matrix by permuting rows i and k, *i.e.*,

$$P(i,k) = \begin{pmatrix} 1 & & & & & & & & \\ & 1 & & & & & & & \\ & & 0 & & 1 & & & & \\ & & & 1 & & & & & \\ & & & & \ddots & & & & \\ & & & & & 1 & & & \\ & & 1 & & 0 & & & & \\ & & & & & & 1 & & \\ & & & & & & & 1 \end{pmatrix}.$$

For example, if $m = 3$,

$$P(1,3) = \begin{pmatrix} 0 & 0 & 1 \\ 0 & 1 & 0 \\ 1 & 0 & 0 \end{pmatrix},$$

then

$$P(1,3)B = \begin{pmatrix} 0 & 0 & 1 \\ 0 & 1 & 0 \\ 1 & 0 & 0 \end{pmatrix} \begin{pmatrix} b_{11} & b_{12} & \cdots & \cdots & \cdots & b_{1n} \\ b_{21} & b_{22} & \cdots & \cdots & \cdots & b_{2n} \\ b_{31} & b_{32} & \cdots & \cdots & \cdots & b_{3n} \end{pmatrix} = \begin{pmatrix} b_{31} & b_{32} & \cdots & \cdots & \cdots & b_{3n} \\ b_{21} & b_{22} & \cdots & \cdots & \cdots & b_{2n} \\ b_{11} & b_{12} & \cdots & \cdots & \cdots & b_{1n} \end{pmatrix}.$$

Observe that $\det(P(i, k)) = -1$. Furthermore, $P(i, k)$ is *symmetric* $(P(i, k)^\top = P(i, k))$, and

$$P(i, k)^{-1} = P(i, k).$$

During the permutation Step (2), if row k and row i need to be permuted, the matrix A is multiplied on the left by the matrix P_k such that $P_k = P(i, k)$, else we set $P_k = I$.

Adding β times row j to row i (with $i \neq j$) is achieved by multiplying A on the left by the *elementary matrix*,

$$E_{i,j;\beta} = I + \beta e_{ij},$$

where

$$(e_{ij})_{kl} = \begin{cases} 1 & \text{if } k = i \text{ and } l = j \\ 0 & \text{if } k \neq i \text{ or } l \neq j, \end{cases}$$

i.e.,

$$
E_{i,j;\beta} = \begin{pmatrix} 1 & & & & & & & \\ & 1 & & & & & & \\ & & 1 & & & & & \\ & & & 1 & & & & \\ & & & & \ddots & & & \\ & & & & & 1 & & \\ & \beta & & & & & 1 & \\ & & & & & & & 1 \\ & & & & & & & & 1 \end{pmatrix}
\quad \text{or} \quad
E_{i,j;\beta} = \begin{pmatrix} 1 & & & & & & & \\ & 1 & & & & & & \\ & & 1 & & \beta & & & \\ & & & 1 & & & & \\ & & & & \ddots & & & \\ & & & & & 1 & & \\ & & & & & & 1 & \\ & & & & & & & 1 \\ & & & & & & & & 1 \end{pmatrix},
$$

on the left, $i > j$, and on the right, $i < j$. The index i is the index of the row that is *changed* by the multiplication. For example, if $m = 3$ and we want to add twice row 1 to row 3, since $\beta = 2$, $j = 1$ and $i = 3$, we form

$$
E_{3,1;2} = I + 2e_{31} = \begin{pmatrix} 1 & 0 & 0 \\ 0 & 1 & 0 \\ 0 & 0 & 1 \end{pmatrix} + \begin{pmatrix} 0 & 0 & 0 \\ 0 & 0 & 0 \\ 2 & 0 & 0 \end{pmatrix} = \begin{pmatrix} 1 & 0 & 0 \\ 0 & 1 & 0 \\ 2 & 0 & 1 \end{pmatrix},
$$

and calculate

$$
E_{3,1;2} B = \begin{pmatrix} 1 & 0 & 0 \\ 0 & 1 & 0 \\ 2 & 0 & 1 \end{pmatrix} \begin{pmatrix} b_{11} & b_{12} & \cdots & \cdots & \cdots & b_{1n} \\ b_{21} & b_{22} & \cdots & \cdots & \cdots & b_{2n} \\ b_{31} & b_{32} & \cdots & \cdots & \cdots & b_{3n} \end{pmatrix}
$$

$$
= \begin{pmatrix} b_{11} & b_{12} & \cdots & \cdots & \cdots b_{1n} \\ b_{21} & b_{22} & \cdots & \cdots & \cdots b_{2n} \\ 2b_{11} + b_{31} & 2b_{12} + b_{32} & \cdots & \cdots & \cdots 2b_{1n} + b_{3n} \end{pmatrix}.
$$

Observe that the inverse of $E_{i,j;\beta} = I + \beta e_{ij}$ is $E_{i,j;-\beta} = I - \beta e_{ij}$ and that $\det(E_{i,j;\beta}) = 1$. Therefore, during Step 3 (the elimination step), the matrix A is multiplied on the left by a product E_k of matrices of the form $E_{i,k;\beta_{i,k}}$, with $i > k$.

Consequently, we see that

$$
A_{k+1} = E_k P_k A_k,
$$

and then

$$
A_k = E_{k-1} P_{k-1} \cdots E_1 P_1 A.
$$

This justifies the claim made earlier that $A_k = M_k A$ for some invertible matrix M_k; we can pick

$$
M_k = E_{k-1} P_{k-1} \cdots E_1 P_1,
$$

a product of invertible matrices.

The fact that $\det(P(i,k)) = -1$ and that $\det(E_{i,j;\beta}) = 1$ implies immediately the fact claimed above: We always have

$$\det(A_k) = \pm \det(A).$$

Furthermore, since

$$A_k = E_{k-1} P_{k-1} \cdots E_1 P_1 A$$

and since Gaussian elimination stops for $k = n$, the matrix

$$A_n = E_{n-1} P_{n-1} \cdots E_2 P_2 E_1 P_1 A$$

is upper-triangular. Also note that if we let $M = E_{n-1} P_{n-1} \cdots E_2 P_2 E_1 P_1$, then $\det(M) = \pm 1$, and

$$\det(A) = \pm \det(A_n).$$

The matrices $P(i,k)$ and $E_{i,j;\beta}$ are called *elementary matrices*. We can summarize the above discussion in the following theorem:

Theorem 7.1. *(Gaussian elimination) Let A be an $n \times n$ matrix (invertible or not). Then there is some invertible matrix M so that $U = MA$ is upper-triangular. The pivots are all nonzero iff A is invertible.*

Proof. We already proved the theorem when A is invertible, as well as the last assertion. Now A is singular iff some pivot is zero, say at Stage k of the elimination. If so, we must have $a_{ik}^{(k)} = 0$ for $i = k, \ldots, n$; but in this case, $A_{k+1} = A_k$ and we may pick $P_k = E_k = I$. □

Remark: Obviously, the matrix M can be computed as

$$M = E_{n-1} P_{n-1} \cdots E_2 P_2 E_1 P_1,$$

but this expression is of no use. Indeed, what we need is M^{-1}; when no permutations are needed, it turns out that M^{-1} can be obtained immediately from the matrices E_k's, in fact, from their inverses, and no multiplications are necessary.

Remark: Instead of looking for an invertible matrix M so that MA is upper-triangular, we can look for an invertible matrix M so that MA is a diagonal matrix. Only a simple change to Gaussian elimination is needed. At every Stage k, after the pivot has been found and pivoting been performed, if necessary, in addition to adding suitable multiples of the kth row to the rows *below* row k in order to zero the entries in column k for $i = k+1, \ldots, n$, also add suitable multiples of the kth row to the rows *above* row k in order to zero the entries in column k for $i = 1, \ldots, k - 1$. Such steps are also achieved by multiplying on the left by elementary matrices $E_{i,k;\beta_{i,k}}$, except that $i < k$, so that these matrices are not lower-triangular matrices. Nevertheless, at the end of the process, we find that $A_n = MA$, is a diagonal matrix.

This method is called the *Gauss–Jordan factorization*. Because it is more expensive than Gaussian elimination, this method is not used much in practice. However, Gauss–Jordan factorization can be used to compute the inverse of a matrix A. Indeed, we find the jth column of A^{-1} by solving the system $Ax^{(j)} = e_j$ (where e_j is the jth canonical basis vector of $\mathbb{R}^n$). By applying Gauss–Jordan, we are led to a system of the form $D_j x^{(j)} = M_j e_j$, where D_j is a diagonal matrix, and we can immediately compute $x^{(j)}$.

It remains to discuss the choice of the pivot, and also conditions that guarantee that no permutations are needed during the Gaussian elimination process. We begin by stating a necessary and sufficient condition for an invertible matrix to have an LU-factorization (*i.e.*, Gaussian elimination does not require pivoting).

7.4   *LU*-Factorization

Definition 7.1. We say that an invertible matrix A has an *LU-factorization* if it can be written as $A = LU$, where U is upper-triangular invertible and L is lower-triangular, with $L_{ii} = 1$ for $i = 1, \ldots, n$.

A lower-triangular matrix with diagonal entries equal to 1 is called a *unit lower-triangular* matrix. Given an $n \times n$ matrix $A = (a_{ij})$, for any k with $1 \leq k \leq n$, let $A(1:k, 1:k)$ denote the submatrix of A whose entries are a_{ij}, where $1 \leq i, j \leq k$.[1] For example, if A is the 5×5 matrix

$$A = \begin{pmatrix} a_{11} & a_{12} & a_{13} & a_{14} & a_{15} \\ a_{21} & a_{22} & a_{23} & a_{24} & a_{25} \\ a_{31} & a_{32} & a_{33} & a_{34} & a_{35} \\ a_{41} & a_{42} & a_{43} & a_{44} & a_{45} \\ a_{51} & a_{52} & a_{53} & a_{54} & a_{55} \end{pmatrix},$$

---

[1] We are using `Matlab`'s notation.

then

$$A(1:3, 1:3) = \begin{pmatrix} a_{11} & a_{12} & a_{13} \\ a_{21} & a_{22} & a_{23} \\ a_{31} & a_{32} & a_{33} \end{pmatrix}.$$

Proposition 7.1. *Let A be an invertible $n \times n$-matrix. Then A has an LU-factorization $A = LU$ iff every matrix $A(1:k, 1:k)$ is invertible for $k = 1, \ldots, n$. Furthermore, when A has an LU-factorization, we have*

$$\det(A(1:k, 1:k)) = \pi_1 \pi_2 \cdots \pi_k, \quad k = 1, \ldots, n,$$

where π_k is the pivot obtained after $k - 1$ elimination steps. Therefore, the kth pivot is given by

$$\pi_k = \begin{cases} a_{11} = \det(A(1:1, 1:1)) & \text{if } k = 1 \\ \dfrac{\det(A(1:k, 1:k))}{\det(A(1:k-1, 1:k-1))} & \text{if } k = 2, \ldots, n. \end{cases}$$

Proof. First assume that $A = LU$ is an LU-factorization of A. We can write

$$A = \begin{pmatrix} A(1:k, 1:k) & A_2 \\ A_3 & A_4 \end{pmatrix} = \begin{pmatrix} L_1 & 0 \\ L_3 & L_4 \end{pmatrix} \begin{pmatrix} U_1 & U_2 \\ 0 & U_4 \end{pmatrix} = \begin{pmatrix} L_1 U_1 & L_1 U_2 \\ L_3 U_1 & L_3 U_2 + L_4 U_4 \end{pmatrix},$$

where L_1, L_4 are unit lower-triangular and U_1, U_4 are upper-triangular. (Note, $A(1:k, 1:k)$, L_1, and U_1 are $k \times k$ matrices; A_2 and U_2 are $k \times (n-k)$ matrices; A_3 and L_3 are $(n-k) \times k$ matrices; A_4, L_4, and U_4 are $(n-k) \times (n-k)$ matrices.) Thus,

$$A(1:k, 1:k) = L_1 U_1,$$

and since U is invertible, U_1 is also invertible (the determinant of U is the product of the diagonal entries in U, which is the product of the diagonal entries in U_1 and U_4). As L_1 is invertible (since its diagonal entries are equal to 1), we see that $A(1:k, 1:k)$ is invertible for $k = 1, \ldots, n$.

Conversely, assume that $A(1:k, 1:k)$ is invertible for $k = 1, \ldots, n$. We just need to show that Gaussian elimination does not need pivoting. We prove by induction on k that the kth step does not need pivoting.

This holds for $k = 1$, since $A(1:1, 1:1) = (a_{11})$, so $a_{11} \neq 0$. Assume that no pivoting was necessary for the first $k - 1$ steps ($2 \leq k \leq n - 1$). In this case, we have

$$E_{k-1} \cdots E_2 E_1 A = A_k,$$

where $L = E_{k-1} \cdots E_2 E_1$ is a unit lower-triangular matrix and $A_k(1 : k, 1 : k)$ is upper-triangular, so that $LA = A_k$ can be written as

$$\begin{pmatrix} L_1 & 0 \\ L_3 & L_4 \end{pmatrix} \begin{pmatrix} A(1 : k, 1 : k) & A_2 \\ A_3 & A_4 \end{pmatrix} = \begin{pmatrix} U_1 & B_2 \\ 0 & B_4 \end{pmatrix},$$

where L_1 is unit lower-triangular and U_1 is upper-triangular. (Once again $A(1 : k, 1 : k)$, L_1, and U_1 are $k \times k$ matrices; A_2 and B_2 are $k \times (n - k)$ matrices; A_3 and L_3 are $(n - k) \times k$ matrices; A_4, L_4, and B_4 are $(n - k) \times (n - k)$ matrices.) But then,

$$L_1 A(1 : k, 1 : k)) = U_1,$$

where L_1 is invertible (in fact, $\det(L_1) = 1$), and since by hypothesis $A(1 : k, 1 : k)$ is invertible, U_1 is also invertible, which implies that $(U_1)_{kk} \neq 0$, since U_1 is upper-triangular. Therefore, no pivoting is needed in Step k, establishing the induction step. Since $\det(L_1) = 1$, we also have

$$\det(U_1) = \det(L_1 A(1 : k, 1 : k)) = \det(L_1) \det(A(1 : k, 1 : k))$$
$$= \det(A(1 : k, 1 : k)),$$

and since U_1 is upper-triangular and has the pivots $\pi_1, \ldots, \pi_k$ on its diagonal, we get

$$\det(A(1 : k, 1 : k)) = \pi_1 \pi_2 \cdots \pi_k, \quad k = 1, \ldots, n,$$

as claimed. □

Remark: The use of determinants in the first part of the proof of Proposition 7.1 can be avoided if we use the fact that a triangular matrix is invertible iff all its diagonal entries are nonzero.

Corollary 7.1. *(LU-Factorization) Let A be an invertible $n \times n$-matrix. If every matrix $A(1 : k, 1 : k)$ is invertible for $k = 1, \ldots, n$, then Gaussian elimination requires no pivoting and yields an LU-factorization $A = LU$.*

Proof. We proved in Proposition 7.1 that in this case Gaussian elimination requires no pivoting. Then since every elementary matrix $E_{i,k;\beta}$ is lower-triangular (since we always arrange that the pivot π_k occurs above the rows that it operates on), since $E_{i,k;\beta}^{-1} = E_{i,k;-\beta}$ and the E_ks are products of $E_{i,k;\beta_{i,k}}$s, from

$$E_{n-1} \cdots E_2 E_1 A = U,$$

where U is an upper-triangular matrix, we get

$$A = LU,$$

where $L = E_1^{-1} E_2^{-1} \cdots E_{n-1}^{-1}$ is a lower-triangular matrix. Furthermore, as the diagonal entries of each $E_{i,k;\beta}$ are 1, the diagonal entries of each E_k are also 1.   □

Example 7.1. The reader should verify that

$$\begin{pmatrix} 2 & 1 & 1 & 0 \\ 4 & 3 & 3 & 1 \\ 8 & 7 & 9 & 5 \\ 6 & 7 & 9 & 8 \end{pmatrix} = \begin{pmatrix} 1 & 0 & 0 & 0 \\ 2 & 1 & 0 & 0 \\ 4 & 3 & 1 & 0 \\ 3 & 4 & 1 & 1 \end{pmatrix} \begin{pmatrix} 2 & 1 & 1 & 0 \\ 0 & 1 & 1 & 1 \\ 0 & 0 & 2 & 2 \\ 0 & 0 & 0 & 2 \end{pmatrix}$$

is an LU-factorization.

One of the main reasons why the existence of an LU-factorization for a matrix A is interesting is that if we need to solve *several* linear systems $Ax = b$ corresponding to the same matrix A, we can do this cheaply by solving the two triangular systems

$$Lw = b, \quad \text{and} \quad Ux = w.$$

There is a certain asymmetry in the LU-decomposition $A = LU$ of an invertible matrix A. Indeed, the diagonal entries of L are all 1, but this is generally false for U. This asymmetry can be eliminated as follows: if

$$D = \text{diag}(u_{11}, u_{22}, \ldots, u_{nn})$$

is the diagonal matrix consisting of the diagonal entries in U (the pivots), then we if let $U' = D^{-1}U$, we can write

$$A = LDU',$$

where L is lower-triangular, U' is upper-triangular, all diagonal entries of both L and U' are 1, and D is a diagonal matrix of pivots. Such a decomposition leads to the following definition.

Definition 7.2. We say that an invertible $n \times n$ matrix A has an LDU-factorization if it can be written as $A = LDU'$, where L is lower-triangular, U' is upper-triangular, all diagonal entries of both L and U' are 1, and D is a diagonal matrix.

We will see shortly than if A is real symmetric, then $U' = L^\top$.

As we will see a bit later, real symmetric positive definite matrices satisfy the condition of Proposition 7.1. *Therefore, linear systems involving real symmetric positive definite matrices can be solved by Gaussian elimination without pivoting.* Actually, it is possible to do better: this is the Cholesky factorization.

If a square invertible matrix A has an LU-factorization, then it is possible to find L and U while performing Gaussian elimination. Recall that at Step k, we pick a pivot $\pi_k = a_{ik}^{(k)} \neq 0$ in the portion consisting of the entries of index $j \geq k$ of the k-th column of the matrix A_k obtained so far, we swap rows i and k if necessary (the pivoting step), and then we zero the entries of index $j = k + 1, \ldots, n$ in column k. Schematically, we have the following steps:

$$
\begin{pmatrix}
\times & \times & \times & \times & \times \\
0 & \times & \times & \times & \times \\
0 & \times & \times & \times & \times \\
0 & a_{ik}^{(k)} & \times & \times & \times \\
0 & \times & \times & \times & \times
\end{pmatrix}
\overset{\text{pivot}}{\Longrightarrow}
\begin{pmatrix}
\times & \times & \times & \times & \times \\
0 & a_{ik}^{(k)} & \times & \times & \times \\
0 & \times & \times & \times & \times \\
0 & \times & \times & \times & \times \\
0 & \times & \times & \times & \times
\end{pmatrix}
\overset{\text{elim}}{\Longrightarrow}
\begin{pmatrix}
\times & \times & \times & \times & \times \\
0 & \times & \times & \times & \times \\
0 & \mathbf{0} & \times & \times & \times \\
0 & \mathbf{0} & \times & \times & \times \\
0 & \mathbf{0} & \times & \times & \times
\end{pmatrix}.
$$

More precisely, after permuting row k and row i (the pivoting step), if the entries in column k below row k are $\alpha_{k+1k}, \ldots, \alpha_{nk}$, then we add $-\alpha_{jk}/\pi_k$ times row k to row j; this process is illustrated below:

$$
\begin{pmatrix}
a_{kk}^{(k)} \\
a_{k+1k}^{(k)} \\
\vdots \\
a_{ik}^{(k)} \\
\vdots \\
a_{nk}^{(k)}
\end{pmatrix}
\overset{\text{pivot}}{\Longrightarrow}
\begin{pmatrix}
a_{ik}^{(k)} \\
a_{k+1k}^{(k)} \\
\vdots \\
a_{kk}^{(k)} \\
\vdots \\
a_{nk}^{(k)}
\end{pmatrix}
=
\begin{pmatrix}
\pi_k \\
\alpha_{k+1k} \\
\vdots \\
\alpha_{ik} \\
\vdots \\
\alpha_{nk}
\end{pmatrix}
\overset{\text{elim}}{\Longrightarrow}
\begin{pmatrix}
\pi_k \\
\mathbf{0} \\
\vdots \\
\mathbf{0} \\
\vdots \\
\mathbf{0}
\end{pmatrix}.
$$

Then if we write $\ell_{jk} = \alpha_{jk}/\pi_k$ for $j = k + 1, \ldots, n$, the kth column of L is

$$
\begin{pmatrix}
0 \\
\vdots \\
0 \\
1 \\
\ell_{k+1k} \\
\vdots \\
\ell_{nk}
\end{pmatrix}.
$$

Observe that the signs of the multipliers $-\alpha_{jk}/\pi_k$ have been flipped. Thus, we obtain the unit lower triangular matrix

$$L = \begin{pmatrix} 1 & 0 & 0 & \cdots & 0 \\ \ell_{21} & 1 & 0 & \cdots & 0 \\ \ell_{31} & \ell_{32} & 1 & \cdots & 0 \\ \vdots & \vdots & \vdots & \ddots & 0 \\ \ell_{n1} & \ell_{n2} & \ell_{n3} & \cdots & 1 \end{pmatrix}.$$

It is easy to see (and this is proven in Theorem 7.2) that the inverse of L is obtained from L by flipping the signs of the ℓ_{ij}:

$$L^{-1} = \begin{pmatrix} 1 & 0 & 0 & \cdots & 0 \\ -\ell_{21} & 1 & 0 & \cdots & 0 \\ -\ell_{31} & -\ell_{32} & 1 & \cdots & 0 \\ \vdots & \vdots & \vdots & \ddots & 0 \\ -\ell_{n1} & -\ell_{n2} & -\ell_{n3} & \cdots & 1 \end{pmatrix}.$$

Furthermore, if the result of Gaussian elimination (without pivoting) is $U = E_{n-1} \cdots E_1 A$, then

$$E_k = \begin{pmatrix} 1 & \cdots & & 0 & 0 & \cdots & 0 \\ \vdots & \ddots & & \vdots & \vdots & \vdots & \vdots \\ 0 & \cdots & & 1 & 0 & \cdots & 0 \\ 0 & \cdots & & -\ell_{k+1k} & 1 & \cdots & 0 \\ \vdots & \vdots & & \vdots & \vdots & \ddots & \vdots \\ 0 & \cdots & & -\ell_{nk} & 0 & \cdots & 1 \end{pmatrix} \quad \text{and} \quad E_k^{-1} = \begin{pmatrix} 1 & \cdots & & 0 & 0 & \cdots & 0 \\ \vdots & \ddots & & \vdots & \vdots & \vdots & \vdots \\ 0 & \cdots & & 1 & 0 & \cdots & 0 \\ 0 & \cdots & & \ell_{k+1k} & 1 & \cdots & 0 \\ \vdots & \vdots & & \vdots & \vdots & \ddots & \vdots \\ 0 & \cdots & & \ell_{nk} & 0 & \cdots & 1 \end{pmatrix},$$

so the kth column of E_k is the kth column of L^{-1}.

Here is an example illustrating the method.

Example 7.2. Given

$$A = A_1 = \begin{pmatrix} 1 & 1 & 1 & 0 \\ 1 & -1 & 0 & 1 \\ 1 & 1 & -1 & 0 \\ 1 & -1 & 0 & -1 \end{pmatrix},$$

we have the following sequence of steps: The first pivot is $\pi_1 = 1$ in row 1, and we subtract row 1 from rows 2, 3, and 4. We get

$$A_2 = \begin{pmatrix} 1 & 1 & 1 & 0 \\ 0 & -2 & -1 & 1 \\ 0 & 0 & -2 & 0 \\ 0 & -2 & -1 & -1 \end{pmatrix} \quad L_1 = \begin{pmatrix} 1 & 0 & 0 & 0 \\ 1 & 1 & 0 & 0 \\ 1 & 0 & 1 & 0 \\ 1 & 0 & 0 & 1 \end{pmatrix}.$$

The next pivot is $\pi_2 = -2$ in row 2, and we subtract row 2 from row 4 (and add 0 times row 2 to row 3). We get

$$A_3 = \begin{pmatrix} 1 & 1 & 1 & 0 \\ 0 & -2 & -1 & 1 \\ 0 & 0 & -2 & 0 \\ 0 & 0 & 0 & -2 \end{pmatrix} \quad L_2 = \begin{pmatrix} 1 & 0 & 0 & 0 \\ 1 & 1 & 0 & 0 \\ 1 & 0 & 1 & 0 \\ 1 & 1 & 0 & 1 \end{pmatrix}.$$

The next pivot is $\pi_3 = -2$ in row 3, and since the fourth entry in column 3 is already a zero, we add 0 times row 3 to row 4. We get

$$A_4 = \begin{pmatrix} 1 & 1 & 1 & 0 \\ 0 & -2 & -1 & 1 \\ 0 & 0 & -2 & 0 \\ 0 & 0 & 0 & -2 \end{pmatrix} \quad L_3 = \begin{pmatrix} 1 & 0 & 0 & 0 \\ 1 & 1 & 0 & 0 \\ 1 & 0 & 1 & 0 \\ 1 & 1 & 0 & 1 \end{pmatrix}.$$

The procedure is finished, and we have

$$L = L_3 = \begin{pmatrix} 1 & 0 & 0 & 0 \\ 1 & 1 & 0 & 0 \\ 1 & 0 & 1 & 0 \\ 1 & 1 & 0 & 1 \end{pmatrix} \quad U = A_4 = \begin{pmatrix} 1 & 1 & 1 & 0 \\ 0 & -2 & -1 & 1 \\ 0 & 0 & -2 & 0 \\ 0 & 0 & 0 & -2 \end{pmatrix}.$$

It is easy to check that indeed

$$LU = \begin{pmatrix} 1 & 0 & 0 & 0 \\ 1 & 1 & 0 & 0 \\ 1 & 0 & 1 & 0 \\ 1 & 1 & 0 & 1 \end{pmatrix} \begin{pmatrix} 1 & 1 & 1 & 0 \\ 0 & -2 & -1 & 1 \\ 0 & 0 & -2 & 0 \\ 0 & 0 & 0 & -2 \end{pmatrix} = \begin{pmatrix} 1 & 1 & 1 & 0 \\ 1 & -1 & 0 & 1 \\ 1 & 1 & -1 & 0 \\ 1 & -1 & 0 & -1 \end{pmatrix} = A.$$

We now show how to extend the above method to deal with pivoting efficiently. This is the $PA = LU$ factorization.

7.5 $PA = LU$ Factorization

The following easy proposition shows that, in principle, A can be premultiplied by some permutation matrix P, so that PA can be converted to upper-triangular form without using any pivoting. Permutations are discussed in some detail in Section 6.1, but for now we just need this definition. For the precise connection between the notion of permutation (as discussed in Section 6.1) and permutation matrices, see Problem 7.16.

Definition 7.3. A *permutation matrix* is a square matrix that has a single 1 in every row and every column and zeros everywhere else.

It is shown in Section 6.1 that every permutation matrix is a product of transposition matrices (the $P(i, k)$s), and that P is invertible with inverse $P^\top$.

Proposition 7.2. *Let A be an invertible $n \times n$-matrix. There is some permutation matrix P so that $(PA)(1 : k, 1 : k)$ is invertible for $k = 1, \ldots, n$.*

Proof. The case $n = 1$ is trivial, and so is the case $n = 2$ (we swap the rows if necessary). If $n \geq 3$, we proceed by induction. Since A is invertible, its columns are linearly independent; in particular, its first $n - 1$ columns are also linearly independent. Delete the last column of A. Since the remaining $n - 1$ columns are linearly independent, there are also $n - 1$ linearly independent rows in the corresponding $n \times (n - 1)$ matrix. Thus, there is a permutation of these n rows so that the $(n - 1) \times (n - 1)$ matrix consisting of the first $n - 1$ rows is invertible. But then there is a corresponding permutation matrix P_1, so that the first $n - 1$ rows and columns of $P_1 A$ form an invertible matrix A'. Applying the induction hypothesis to the $(n - 1) \times (n - 1)$ matrix A', we see that there some permutation matrix P_2 (leaving the nth row fixed), so that $(P_2 P_1 A)(1 : k, 1 : k)$ is invertible, for $k = 1, \ldots, n - 1$. Since A is invertible in the first place and P_1 and P_2 are invertible, $P_1 P_2 A$ is also invertible, and we are done. $\square$

Remark: One can also prove Proposition 7.2 using a clever reordering of the Gaussian elimination steps suggested by Trefethen and Bau [Trefethen and Bau III (1997)] (Lecture 21). Indeed, we know that if A is invertible, then there are permutation matrices P_i and products of elementary matrices E_i, so that

$$A_n = E_{n-1} P_{n-1} \cdots E_2 P_2 E_1 P_1 A,$$

where $U = A_n$ is upper-triangular. For example, when $n = 4$, we have $E_3 P_3 E_2 P_2 E_1 P_1 A = U$. We can define new matrices E_1', E_2', E_3' which are still products of elementary matrices so that we have

$$E_3' E_2' E_1' P_3 P_2 P_1 A = U.$$

Indeed, if we let $E_3' = E_3$, $E_2' = P_3 E_2 P_3^{-1}$, and $E_1' = P_3 P_2 E_1 P_2^{-1} P_3^{-1}$, we easily verify that each E_k' is a product of elementary matrices and that

$$E_3' E_2' E_1' P_3 P_2 P_1 = E_3 (P_3 E_2 P_3^{-1})(P_3 P_2 E_1 P_2^{-1} P_3^{-1}) P_3 P_2 P_1$$
$$= E_3 P_3 E_2 P_2 E_1 P_1.$$

It can also be proven that E_1', E_2', E_3' are lower triangular (see Theorem 7.2).

In general, we let

$$E_k' = P_{n-1} \cdots P_{k+1} E_k P_{k+1}^{-1} \cdots P_{n-1}^{-1},$$

and we have

$$E_{n-1}' \cdots E_1' P_{n-1} \cdots P_1 A = U,$$

where each E_j' is a lower triangular matrix (see Theorem 7.2).

It is remarkable that if pivoting steps are necessary during Gaussian elimination, a very simple modification of the algorithm for finding an *LU*-factorization yields the matrices L, U, and P, such that $PA = LU$. To describe this new method, since the diagonal entries of L are 1s, it is convenient to write

$$L = I + \Lambda.$$

Then in assembling the matrix Λ while performing Gaussian elimination with pivoting, we make the same transposition on the rows of Λ (really Λ_{k-1}) that we make on the rows of A (really A_k) during a pivoting step involving row k and row i. We also assemble P by starting with the identity matrix and applying to P the same row transpositions that we apply to A and Λ. Here is an example illustrating this method.

Example 7.3. Given

$$A = A_1 = \begin{pmatrix} 1 & 1 & 1 & 0 \\ 1 & 1 & -1 & 0 \\ 1 & -1 & 0 & 1 \\ 1 & -1 & 0 & -1 \end{pmatrix},$$

we have the following sequence of steps: We initialize $\Lambda_0 = 0$ and $P_0 = I_4$. The first pivot is $\pi_1 = 1$ in row 1, and we subtract row 1 from rows 2, 3, and 4. We get

$$A_2 = \begin{pmatrix} 1 & 1 & 1 & 0 \\ 0 & 0 & -2 & 0 \\ 0 & -2 & -1 & 1 \\ 0 & -2 & -1 & -1 \end{pmatrix} \quad \Lambda_1 = \begin{pmatrix} 0 & 0 & 0 & 0 \\ 1 & 0 & 0 & 0 \\ 1 & 0 & 0 & 0 \\ 1 & 0 & 0 & 0 \end{pmatrix} \quad P_1 = \begin{pmatrix} 1 & 0 & 0 & 0 \\ 0 & 1 & 0 & 0 \\ 0 & 0 & 1 & 0 \\ 0 & 0 & 0 & 1 \end{pmatrix}.$$

The next pivot is $\pi_2 = -2$ in row 3, so we permute row 2 and 3; we also apply this permutation to Λ and P:

$$A_3' = \begin{pmatrix} 1 & 1 & 1 & 0 \\ 0 & -2 & -1 & 1 \\ 0 & 0 & -2 & 0 \\ 0 & -2 & -1 & -1 \end{pmatrix} \quad \Lambda_2' = \begin{pmatrix} 0 & 0 & 0 & 0 \\ 1 & 0 & 0 & 0 \\ 1 & 0 & 0 & 0 \\ 1 & 0 & 0 & 0 \end{pmatrix} \quad P_2 = \begin{pmatrix} 1 & 0 & 0 & 0 \\ 0 & 0 & 1 & 0 \\ 0 & 1 & 0 & 0 \\ 0 & 0 & 0 & 1 \end{pmatrix}.$$

Next we subtract row 2 from row 4 (and add 0 times row 2 to row 3). We get

$$A_3 = \begin{pmatrix} 1 & 1 & 1 & 0 \\ 0 & -2 & -1 & 1 \\ 0 & 0 & -2 & 0 \\ 0 & 0 & 0 & -2 \end{pmatrix} \quad \Lambda_2 = \begin{pmatrix} 0 & 0 & 0 & 0 \\ 1 & 0 & 0 & 0 \\ 1 & 0 & 0 & 0 \\ 1 & 1 & 0 & 0 \end{pmatrix} \quad P_2 = \begin{pmatrix} 1 & 0 & 0 & 0 \\ 0 & 0 & 1 & 0 \\ 0 & 1 & 0 & 0 \\ 0 & 0 & 0 & 1 \end{pmatrix}.$$

The next pivot is $\pi_3 = -2$ in row 3, and since the fourth entry in column 3 is already a zero, we add 0 times row 3 to row 4. We get

$$A_4 = \begin{pmatrix} 1 & 1 & 1 & 0 \\ 0 & -2 & -1 & 1 \\ 0 & 0 & -2 & 0 \\ 0 & 0 & 0 & -2 \end{pmatrix} \quad \Lambda_3 = \begin{pmatrix} 0 & 0 & 0 & 0 \\ 1 & 0 & 0 & 0 \\ 1 & 0 & 0 & 0 \\ 1 & 1 & 0 & 0 \end{pmatrix} \quad P_3 = \begin{pmatrix} 1 & 0 & 0 & 0 \\ 0 & 0 & 1 & 0 \\ 0 & 1 & 0 & 0 \\ 0 & 0 & 0 & 1 \end{pmatrix}.$$

The procedure is finished, and we have

$$L = \Lambda_3 + I = \begin{pmatrix} 1 & 0 & 0 & 0 \\ 1 & 1 & 0 & 0 \\ 1 & 0 & 1 & 0 \\ 1 & 1 & 0 & 1 \end{pmatrix} \quad U = A_4 = \begin{pmatrix} 1 & 1 & 1 & 0 \\ 0 & -2 & -1 & 1 \\ 0 & 0 & -2 & 0 \\ 0 & 0 & 0 & -2 \end{pmatrix}$$

$$P = P_3 = \begin{pmatrix} 1 & 0 & 0 & 0 \\ 0 & 0 & 1 & 0 \\ 0 & 1 & 0 & 0 \\ 0 & 0 & 0 & 1 \end{pmatrix}.$$

It is easy to check that indeed

$$LU = \begin{pmatrix} 1 & 0 & 0 & 0 \\ 1 & 1 & 0 & 0 \\ 1 & 0 & 1 & 0 \\ 1 & 1 & 0 & 1 \end{pmatrix} \begin{pmatrix} 1 & 1 & 1 & 0 \\ 0 & -2 & -1 & 1 \\ 0 & 0 & -2 & 0 \\ 0 & 0 & 0 & -2 \end{pmatrix} = \begin{pmatrix} 1 & 1 & 1 & 0 \\ 1 & -1 & 0 & 1 \\ 1 & 1 & -1 & 0 \\ 1 & -1 & 0 & -1 \end{pmatrix}$$

and

$$PA = \begin{pmatrix} 1 & 0 & 0 & 0 \\ 0 & 0 & 1 & 0 \\ 0 & 1 & 0 & 0 \\ 0 & 0 & 0 & 1 \end{pmatrix} \begin{pmatrix} 1 & 1 & 1 & 0 \\ 1 & 1 & -1 & 0 \\ 1 & -1 & 0 & 1 \\ 1 & -1 & 0 & -1 \end{pmatrix} = \begin{pmatrix} 1 & 1 & 1 & 0 \\ 1 & -1 & 0 & 1 \\ 1 & 1 & -1 & 0 \\ 1 & -1 & 0 & -1 \end{pmatrix}.$$

Using the idea in the remark before the above example, we can prove the theorem below which shows the correctness of the algorithm for computing P, L and U using a simple adaptation of Gaussian elimination.

We are not aware of a detailed proof of Theorem 7.2 in the standard texts. Although Golub and Van Loan [Golub and Van Loan (1996)] state a version of this theorem as their Theorem 3.1.4, they say that "The proof is a messy subscripting argument." Meyer [Meyer (2000)] also provides a sketch of proof (see the end of Section 3.10). In view of this situation, we offer a complete proof. It does involve a lot of subscripts and superscripts, but in our opinion, it contains some techniques that go far beyond symbol manipulation.

Theorem 7.2. *For every invertible $n \times n$-matrix A, the following hold:*

(1) There is some permutation matrix P, some upper-triangular matrix U, and some unit lower-triangular matrix L, so that $PA = LU$ (recall, $L_{ii} = 1$ for $i = 1, \ldots, n$). Furthermore, if $P = I$, then L and U are unique and they are produced as a result of Gaussian elimination without pivoting.

(2) If $E_{n-1} \ldots E_1 A = U$ is the result of Gaussian elimination without pivoting, write as usual $A_k = E_{k-1} \ldots E_1 A$ (with $A_k = (a_{ij}^{(k)})$), and let $\ell_{ik} = a_{ik}^{(k)}/a_{kk}^{(k)}$, with $1 \le k \le n - 1$ and $k + 1 \le i \le n$. Then

$$L = \begin{pmatrix} 1 & 0 & 0 & \cdots & 0 \\ \ell_{21} & 1 & 0 & \cdots & 0 \\ \ell_{31} & \ell_{32} & 1 & \cdots & 0 \\ \vdots & \vdots & \vdots & \ddots & 0 \\ \ell_{n1} & \ell_{n2} & \ell_{n3} & \cdots & 1 \end{pmatrix},$$

where the kth column of L is the kth column of E_k^{-1}, for $k = 1, \ldots, n-1$.

(3) If $E_{n-1}P_{n-1} \cdots E_1 P_1 A = U$ is the result of Gaussian elimination with some pivoting, write $A_k = E_{k-1}P_{k-1} \cdots E_1 P_1 A$, and define E_j^k, with $1 \le j \le n - 1$ and $j \le k \le n - 1$, such that, for $j = 1, \ldots, n - 2$,

$$E_j^j = E_j$$
$$E_j^k = P_k E_j^{k-1} P_k, \quad for\ k = j + 1, \ldots, n - 1,$$

and

$$E_{n-1}^{n-1} = E_{n-1}.$$

Then,

$$E_j^k = P_k P_{k-1} \cdots P_{j+1} E_j P_{j+1} \cdots P_{k-1} P_k$$
$$U = E_{n-1}^{n-1} \cdots E_1^{n-1} P_{n-1} \cdots P_1 A,$$

and if we set

$$P = P_{n-1} \cdots P_1$$
$$L = (E_1^{n-1})^{-1} \cdots (E_{n-1}^{n-1})^{-1},$$

then

$$PA = LU. \tag{7.1}$$

Furthermore,

$$(E_j^k)^{-1} = I + \mathcal{E}_j^k, \quad 1 \le j \le n-1, \; j \le k \le n-1,$$

where $\mathcal{E}_j^k$ is a lower triangular matrix of the form

$$\mathcal{E}_j^k = \begin{pmatrix} 0 & \cdots & 0 & 0 & \cdots & 0 \\ \vdots & \ddots & \vdots & \vdots & \vdots & \vdots \\ 0 & \cdots & 0 & 0 & \cdots & 0 \\ 0 & \cdots & \ell_{j+1j}^{(k)} & 0 & \cdots & 0 \\ \vdots & \vdots & \vdots & \vdots & \ddots & \vdots \\ 0 & \cdots & \ell_{nj}^{(k)} & 0 & \cdots & 0 \end{pmatrix},$$

we have

$$E_j^k = I - \mathcal{E}_j^k,$$

and

$$\mathcal{E}_j^k = P_k \mathcal{E}_j^{k-1}, \quad 1 \le j \le n-2, \; j+1 \le k \le n-1,$$

where $P_k = I$ or else $P_k = P(k,i)$ for some i such that $k + 1 \le i \le n$; if $P_k \ne I$, this means that $(E_j^k)^{-1}$ is obtained from $(E_j^{k-1})^{-1}$ by permuting the entries on rows i and k in column j. Because the matrices $(E_j^k)^{-1}$ are all lower triangular, the matrix L is also lower triangular.

In order to find L, define lower triangular $n \times n$ matrices Λ_k of the form

$$\Lambda_k = \begin{pmatrix} 0 & 0 & 0 & 0 & 0 & \cdots & \cdots & 0 \\ \lambda_{21}^{(k)} & 0 & 0 & 0 & 0 & \vdots & \vdots & 0 \\ \lambda_{31}^{(k)} & \lambda_{32}^{(k)} & \ddots & 0 & 0 & \vdots & \vdots & 0 \\ \vdots & \vdots & \ddots & 0 & 0 & \vdots & \vdots & \vdots \\ \lambda_{k+11}^{(k)} & \lambda_{k+12}^{(k)} & \cdots & \lambda_{k+1k}^{(k)} & 0 & \cdots & \cdots & 0 \\ \lambda_{k+21}^{(k)} & \lambda_{k+22}^{(k)} & \cdots & \lambda_{k+2k}^{(k)} & 0 & \ddots & \cdots & 0 \\ \vdots & \vdots & \ddots & \vdots & \vdots & \vdots & \ddots & \vdots \\ \lambda_{n1}^{(k)} & \lambda_{n2}^{(k)} & \cdots & \lambda_{nk}^{(k)} & 0 & \cdots & \cdots & 0 \end{pmatrix}$$

to assemble the columns of L iteratively as follows: let

$$(-\ell_{k+1\,k}^{(k)}, \ldots, -\ell_{nk}^{(k)})$$

be the last $n - k$ elements of the kth column of E_k, and define Λ_k inductively by setting

$$\Lambda_1 = \begin{pmatrix} 0 & 0 & \cdots & 0 \\ \ell_{21}^{(1)} & 0 & \cdots & 0 \\ \vdots & \vdots & \ddots & \vdots \\ \ell_{n1}^{(1)} & 0 & \cdots & 0 \end{pmatrix},$$

then for $k = 2, \ldots, n - 1$, define

$$\Lambda_k' = P_k \Lambda_{k-1}, \tag{7.2}$$

and $\Lambda_k = (I + \Lambda_k')E_k^{-1} - I$, with

$$\Lambda_k = \begin{pmatrix} 0 & 0 & 0 & 0 & 0 & \cdots & \cdots & 0 \\ \lambda_{21}^{'(k-1)} & 0 & 0 & 0 & 0 & \vdots & \vdots & 0 \\ \lambda_{31}^{'(k-1)} & \lambda_{32}^{'(k-1)} & \ddots & 0 & 0 & \vdots & \vdots & 0 \\ \vdots & \vdots & \ddots & 0 & 0 & \vdots & \vdots & \vdots \\ \lambda_{k1}^{'(k-1)} & \lambda_{k2}^{'(k-1)} & \cdots & \lambda_{k\,(k-1)}^{'(k-1)} & 0 & \cdots & \cdots & 0 \\ \lambda_{k+1\,1}^{'(k-1)} & \lambda_{k+1\,2}^{'(k-1)} & \cdots & \lambda_{k+1\,(k-1)}^{'(k-1)} & \ell_{k+1\,k}^{(k)} & \ddots & \cdots & 0 \\ \vdots & \vdots & \ddots & \vdots & \vdots & \vdots & \ddots & \vdots \\ \lambda_{n1}^{'(k-1)} & \lambda_{n2}^{'(k-1)} & \cdots & \lambda_{n\,k-1}^{'(k-1)} & \ell_{nk}^{(k)} & \cdots & \cdots & 0 \end{pmatrix},$$

with $P_k = I$ or $P_k = P(k, i)$ for some $i > k$. This means that in assembling L, row k and row i of Λ_{k-1} need to be permuted when a pivoting step permuting row k and row i of A_k is required. Then

$$I + \Lambda_k = (E_1^k)^{-1} \cdots (E_k^k)^{-1}$$
$$\Lambda_k = \mathcal{E}_1^k + \cdots + \mathcal{E}_k^k,$$

for $k = 1, \ldots, n - 1$, and therefore

$$L = I + \Lambda_{n-1}.$$

The proof of Theorem 7.2, which is very technical, is given in Section 7.6. We emphasize again that Part (3) of Theorem 7.2 shows the remarkable fact that in assembling the matrix L while performing Gaussian elimination with pivoting, the only change to the algorithm is to make the same transposition on the rows of Λ_{k-1} that we make on the rows of A (really A_k) during a pivoting step involving row k and row i. We can also assemble P by starting with the identity matrix and applying to P the same row transpositions that we apply to A and Λ. Here is an example illustrating this method.

Example 7.4. Consider the matrix

$$A = \begin{pmatrix} 1 & 2 & -3 & 4 \\ 4 & 8 & 12 & -8 \\ 2 & 3 & 2 & 1 \\ -3 & -1 & 1 & -4 \end{pmatrix}.$$

We set $P_0 = I_4$, and we can also set $\Lambda_0 = 0$. The first step is to permute row 1 and row 2, using the pivot 4. We also apply this permutation to P_0:

$$A'_1 = \begin{pmatrix} 4 & 8 & 12 & -8 \\ 1 & 2 & -3 & 4 \\ 2 & 3 & 2 & 1 \\ -3 & -1 & 1 & -4 \end{pmatrix} \quad P_1 = \begin{pmatrix} 0 & 1 & 0 & 0 \\ 1 & 0 & 0 & 0 \\ 0 & 0 & 1 & 0 \\ 0 & 0 & 0 & 1 \end{pmatrix}.$$

Next we subtract $1/4$ times row 1 from row 2, $1/2$ times row 1 from row 3, and add $3/4$ times row 1 to row 4, and start assembling Λ:

$$A_2 = \begin{pmatrix} 4 & 8 & 12 & -8 \\ 0 & 0 & -6 & 6 \\ 0 & -1 & -4 & 5 \\ 0 & 5 & 10 & -10 \end{pmatrix} \quad \Lambda_1 = \begin{pmatrix} 0 & 0 & 0 & 0 \\ 1/4 & 0 & 0 & 0 \\ 1/2 & 0 & 0 & 0 \\ -3/4 & 0 & 0 & 0 \end{pmatrix} \quad P_1 = \begin{pmatrix} 0 & 1 & 0 & 0 \\ 1 & 0 & 0 & 0 \\ 0 & 0 & 1 & 0 \\ 0 & 0 & 0 & 1 \end{pmatrix}.$$

Next we permute row 2 and row 4, using the pivot 5. We also apply this permutation to Λ and P:

$$A'_3 = \begin{pmatrix} 4 & 8 & 12 & -8 \\ 0 & 5 & 10 & -10 \\ 0 & -1 & -4 & 5 \\ 0 & 0 & -6 & 6 \end{pmatrix} \quad \Lambda'_2 = \begin{pmatrix} 0 & 0 & 0 & 0 \\ -3/4 & 0 & 0 & 0 \\ 1/2 & 0 & 0 & 0 \\ 1/4 & 0 & 0 & 0 \end{pmatrix} \quad P_2 = \begin{pmatrix} 0 & 1 & 0 & 0 \\ 0 & 0 & 0 & 1 \\ 0 & 0 & 1 & 0 \\ 1 & 0 & 0 & 0 \end{pmatrix}.$$

Next we add $1/5$ times row 2 to row 3, and update Λ'_2:

$$A_3 = \begin{pmatrix} 4 & 8 & 12 & -8 \\ 0 & 5 & 10 & -10 \\ 0 & 0 & -2 & 3 \\ 0 & 0 & -6 & 6 \end{pmatrix} \quad \Lambda_2 = \begin{pmatrix} 0 & 0 & 0 & 0 \\ -3/4 & 0 & 0 & 0 \\ 1/2 & -1/5 & 0 & 0 \\ 1/4 & 0 & 0 & 0 \end{pmatrix} \quad P_2 = \begin{pmatrix} 0 & 1 & 0 & 0 \\ 0 & 0 & 0 & 1 \\ 0 & 0 & 1 & 0 \\ 1 & 0 & 0 & 0 \end{pmatrix}.$$

Next we permute row 3 and row 4, using the pivot -6. We also apply this permutation to Λ and P:

$$A'_4 = \begin{pmatrix} 4 & 8 & 12 & -8 \\ 0 & 5 & 10 & -10 \\ 0 & 0 & -6 & 6 \\ 0 & 0 & -2 & 3 \end{pmatrix} \quad \Lambda'_3 = \begin{pmatrix} 0 & 0 & 0 & 0 \\ -3/4 & 0 & 0 & 0 \\ 1/4 & 0 & 0 & 0 \\ 1/2 & -1/5 & 0 & 0 \end{pmatrix} \quad P_3 = \begin{pmatrix} 0 & 1 & 0 & 0 \\ 0 & 0 & 0 & 1 \\ 1 & 0 & 0 & 0 \\ 0 & 0 & 1 & 0 \end{pmatrix}.$$

Finally we subtract $1/3$ times row 3 from row 4, and update Λ'_3:

$$A_4 = \begin{pmatrix} 4 & 8 & 12 & -8 \\ 0 & 5 & 10 & -10 \\ 0 & 0 & -6 & 6 \\ 0 & 0 & 0 & 1 \end{pmatrix} \quad \Lambda_3 = \begin{pmatrix} 0 & 0 & 0 & 0 \\ -3/4 & 0 & 0 & 0 \\ 1/4 & 0 & 0 & 0 \\ 1/2 & -1/5 & 1/3 & 0 \end{pmatrix} \quad P_3 = \begin{pmatrix} 0 & 1 & 0 & 0 \\ 0 & 0 & 0 & 1 \\ 1 & 0 & 0 & 0 \\ 0 & 0 & 1 & 0 \end{pmatrix}.$$

Consequently, adding the identity to Λ_3, we obtain

$$L = \begin{pmatrix} 1 & 0 & 0 & 0 \\ -3/4 & 1 & 0 & 0 \\ 1/4 & 0 & 1 & 0 \\ 1/2 & -1/5 & 1/3 & 1 \end{pmatrix}, \quad U = \begin{pmatrix} 4 & 8 & 12 & -8 \\ 0 & 5 & 10 & -10 \\ 0 & 0 & -6 & 6 \\ 0 & 0 & 0 & 1 \end{pmatrix}, \quad P = \begin{pmatrix} 0 & 1 & 0 & 0 \\ 0 & 0 & 0 & 1 \\ 1 & 0 & 0 & 0 \\ 0 & 0 & 1 & 0 \end{pmatrix}.$$

We check that

$$PA = \begin{pmatrix} 0 & 1 & 0 & 0 \\ 0 & 0 & 0 & 1 \\ 1 & 0 & 0 & 0 \\ 0 & 0 & 1 & 0 \end{pmatrix} \begin{pmatrix} 1 & 2 & -3 & 4 \\ 4 & 8 & 12 & -8 \\ 2 & 3 & 2 & 1 \\ -3 & -1 & 1 & -4 \end{pmatrix} = \begin{pmatrix} 4 & 8 & 12 & -8 \\ -3 & -1 & 1 & -4 \\ 1 & 2 & -3 & 4 \\ 2 & 3 & 2 & 1 \end{pmatrix},$$

and that

$$LU = \begin{pmatrix} 1 & 0 & 0 & 0 \\ -3/4 & 1 & 0 & 0 \\ 1/4 & 0 & 1 & 0 \\ 1/2 & -1/5 & 1/3 & 1 \end{pmatrix} \begin{pmatrix} 4 & 8 & 12 & -8 \\ 0 & 5 & 10 & -10 \\ 0 & 0 & -6 & 6 \\ 0 & 0 & 0 & 1 \end{pmatrix} = \begin{pmatrix} 4 & 8 & 12 & -8 \\ -3 & -1 & 1 & -4 \\ 1 & 2 & -3 & 4 \\ 2 & 3 & 2 & 1 \end{pmatrix} = PA.$$

Note that if one willing to overwrite the lower triangular part of the evolving matrix A, one can store the evolving Λ there, since these entries will eventually be zero anyway! There is also no need to save explicitly the permutation matrix P. One could instead record the permutation steps in an extra column (record the vector $(\pi(1), \ldots, \pi(n))$ corresponding to the permutation π applied to the rows). We let the reader write such a bold and space-efficient version of LU-decomposition!

Remark: In `Matlab` the function `lu` returns the matrices P, L, U involved in the $PA = LU$ factorization using the call $[L, U, P] = \text{lu}(A)$.

As a corollary of Theorem 7.2(1), we can show the following result.

Proposition 7.3. *If an invertible real symmetric matrix A has an LU-decomposition, then A has a factorization of the form*

$$A = LDL^\top,$$

where L is a lower-triangular matrix whose diagonal entries are equal to 1, and where D consists of the pivots. Furthermore, such a decomposition is unique.

Proof. If A has an LU-factorization, then it has an LDU factorization

$$A = LDU,$$

where L is lower-triangular, U is upper-triangular, and the diagonal entries of both L and U are equal to 1. Since A is symmetric, we have

$$LDU = A = A^\top = U^\top DL^\top,$$

with $U^\top$ lower-triangular and $DL^\top$ upper-triangular. By the uniqueness of LU-factorization (Part (1) of Theorem 7.2), we must have $L = U^\top$ (and $DU = DL^\top$), thus $U = L^\top$, as claimed.   □

Remark: It can be shown that Gaussian elimination plus back-substitution requires $n^3/3 + O(n^2)$ additions, $n^3/3 + O(n^2)$ multiplications and $n^2/2 + O(n)$ divisions.

7.6   Proof of Theorem 7.2 ⊛

Proof. (1) The only part that has not been proven is the uniqueness part (when $P = I$). Assume that A is invertible and that $A = L_1 U_1 = L_2 U_2$, with L_1, L_2 unit lower-triangular and U_1, U_2 upper-triangular. Then we have

$$L_2^{-1} L_1 = U_2 U_1^{-1}.$$

However, it is obvious that L_2^{-1} is lower-triangular and that U_1^{-1} is upper-triangular, and so $L_2^{-1} L_1$ is lower-triangular and $U_2 U_1^{-1}$ is upper-triangular. Since the diagonal entries of L_1 and L_2 are 1, the above equality is only possible if $U_2 U_1^{-1} = I$, that is, $U_1 = U_2$, and so $L_1 = L_2$.

(2) When $P = I$, we have $L = E_1^{-1} E_2^{-1} \cdots E_{n-1}^{-1}$, where E_k is the product of $n - k$ elementary matrices of the form $E_{i,k;-\ell_i}$, where $E_{i,k;-\ell_i}$

subtracts ℓ_i times row k from row i, with $\ell_{ik} = a_{ik}^{(k)}/a_{kk}^{(k)}$, $1 \leq k \leq n-1$, and $k+1 \leq i \leq n$. Then it is immediately verified that

$$E_k = \begin{pmatrix} 1 & \cdots & 0 & 0 & \cdots & 0 \\ \vdots & \ddots & \vdots & \vdots & \vdots & \vdots \\ 0 & \cdots & 1 & 0 & \cdots & 0 \\ 0 & \cdots & -\ell_{k+1k} & 1 & \cdots & 0 \\ \vdots & \vdots & \vdots & \vdots & \ddots & \vdots \\ 0 & \cdots & -\ell_{nk} & 0 & \cdots & 1 \end{pmatrix},$$

and that

$$E_k^{-1} = \begin{pmatrix} 1 & \cdots & 0 & 0 & \cdots & 0 \\ \vdots & \ddots & \vdots & \vdots & \vdots & \vdots \\ 0 & \cdots & 1 & 0 & \cdots & 0 \\ 0 & \cdots & \ell_{k+1k} & 1 & \cdots & 0 \\ \vdots & \vdots & \vdots & \vdots & \ddots & \vdots \\ 0 & \cdots & \ell_{nk} & 0 & \cdots & 1 \end{pmatrix}.$$

If we define L_k by

$$L_k = \begin{pmatrix} 1 & 0 & 0 & 0 & 0 & \vdots & 0 \\ \ell_{21} & 1 & 0 & 0 & 0 & \vdots & 0 \\ \ell_{31} & \ell_{32} & \ddots & 0 & 0 & \vdots & 0 \\ \vdots & \vdots & \ddots & 1 & 0 & \vdots & 0 \\ \ell_{k+11} & \ell_{k+12} & \cdots & \ell_{k+1k} & 1 & \cdots & 0 \\ \vdots & \vdots & \ddots & \vdots & 0 & \vdots & 0 \\ \ell_{n1} & \ell_{n2} & \cdots & \ell_{nk} & 0 & \cdots & 1 \end{pmatrix}$$

for $k = 1, \ldots, n-1$, we easily check that $L_1 = E_1^{-1}$, and that

$$L_k = L_{k-1} E_k^{-1}, \quad 2 \leq k \leq n-1,$$

because multiplication on the right by E_k^{-1} adds ℓ_i times column i to column k (of the matrix L_{k-1}) with $i > k$, and column i of L_{k-1} has only the nonzero entry 1 as its ith element. Since

$$L_k = E_1^{-1} \cdots E_k^{-1}, \quad 1 \leq k \leq n-1,$$

we conclude that $L = L_{n-1}$, proving our claim about the shape of L.

(3)

Step 1. Prove (7.1).

First we prove by induction on k that

$$A_{k+1} = E_k^k \cdots E_1^k P_k \cdots P_1 A, \quad k = 1, \ldots, n-2.$$

For $k = 1$, we have $A_2 = E_1 P_1 A = E_1^1 P_1 A$, since $E_1^1 = E_1$, so our assertion holds trivially.

Now if $k \geq 2$,

$$A_{k+1} = E_k P_k A_k,$$

and by the induction hypothesis,

$$A_k = E_{k-1}^{k-1} \cdots E_2^{k-1} E_1^{k-1} P_{k-1} \cdots P_1 A.$$

Because P_k is either the identity or a transposition, $P_k^2 = I$, so by inserting occurrences of $P_k P_k$ as indicated below we can write

$$
\begin{aligned}
A_{k+1} &= E_k P_k A_k \\
&= E_k P_k E_{k-1}^{k-1} \cdots E_2^{k-1} E_1^{k-1} P_{k-1} \cdots P_1 A \\
&= E_k P_k E_{k-1}^{k-1} (P_k P_k) \cdots (P_k P_k) E_2^{k-1} (P_k P_k) E_1^{k-1} (P_k P_k) P_{k-1} \cdots P_1 A \\
&= E_k (P_k E_{k-1}^{k-1} P_k) \cdots (P_k E_2^{k-1} P_k)(P_k E_1^{k-1} P_k) P_k P_{k-1} \cdots P_1 A.
\end{aligned}
$$

Observe that P_k has been "moved" to the right of the elimination steps. However, by definition,

$$
\begin{aligned}
E_j^k &= P_k E_j^{k-1} P_k, \quad j = 1, \ldots, k-1 \\
E_k^k &= E_k,
\end{aligned}
$$

so we get

$$A_{k+1} = E_k^k E_{k-1}^k \cdots E_2^k E_1^k P_k \cdots P_1 A,$$

establishing the induction hypothesis. For $k = n - 2$, we get

$$U = A_{n-1} = E_{n-1}^{n-1} \cdots E_1^{n-1} P_{n-1} \cdots P_1 A,$$

as claimed, and the factorization $PA = LU$ with

$$
\begin{aligned}
P &= P_{n-1} \cdots P_1 \\
L &= (E_1^{n-1})^{-1} \cdots (E_{n-1}^{n-1})^{-1}
\end{aligned}
$$

is clear.

Step 2. Prove that the matrices $(E_j^k)^{-1}$ are lower-triangular. To achieve this, we prove that the matrices $\mathcal{E}_j^k$ are strictly lower triangular matrices of a very special form.

Since for $j = 1, \ldots, n-2$, we have $E_j^j = E_j$,

$$E_j^k = P_k E_j^{k-1} P_k, \quad k = j+1, \ldots, n-1,$$

since $E_{n-1}^{n-1} = E_{n-1}$ and $P_k^{-1} = P_k$, we get $(E_j^j)^{-1} = E_j^{-1}$ for $j = 1, \ldots,$
$n-1$, and for $j = 1, \ldots, n-2$, we have

$$(E_j^k)^{-1} = P_k (E_j^{k-1})^{-1} P_k, \quad k = j+1, \ldots, n-1.$$

Since

$$(E_j^{k-1})^{-1} = I + \mathcal{E}_j^{k-1}$$

and $P_k = P(k, i)$ is a transposition or $P_k = I$, so $P_k^2 = I$, and we get

$$(E_j^k)^{-1} = P_k (E_j^{k-1})^{-1} P_k = P_k (I + \mathcal{E}_j^{k-1}) P_k = P_k^2 + P_k \mathcal{E}_j^{k-1} P_k$$
$$= I + P_k \mathcal{E}_j^{k-1} P_k.$$

Therefore, we have

$$(E_j^k)^{-1} = I + P_k \mathcal{E}_j^{k-1} P_k, \quad 1 \le j \le n-2, \ j+1 \le k \le n-1.$$

We prove for $j = 1, \ldots, n-1$, that for $k = j, \ldots, n-1$, each $\mathcal{E}_j^k$ is a lower
triangular matrix of the form

$$\mathcal{E}_j^k = \begin{pmatrix} 0 & \cdots & 0 & 0 & \cdots & 0 \\ \vdots & \ddots & \vdots & \vdots & \vdots & \vdots \\ 0 & \cdots & 0 & 0 & \cdots & 0 \\ 0 & \cdots & \ell_{j+1\,j}^{(k)} & 0 & \cdots & 0 \\ \vdots & \vdots & \vdots & \vdots & \ddots & \vdots \\ 0 & \cdots & \ell_{n\,j}^{(k)} & 0 & \cdots & 0 \end{pmatrix},$$

and that

$$\mathcal{E}_j^k = P_k \mathcal{E}_j^{k-1}, \quad 1 \le j \le n-2, \ j+1 \le k \le n-1,$$

with $P_k = I$ or $P_k = P(k, i)$ for some i such that $k+1 \le i \le n$.

For each j ($1 \le j \le n-1$) we proceed by induction on $k = j, \ldots, n-1$.
Since $(E_j^j)^{-1} = E_j^{-1}$ and since E_j^{-1} is of the above form, the base case
holds.

For the induction step, we only need to consider the case where $P_k = P(k, i)$ is a transposition, since the case where $P_k = I$ is trivial. We have
to figure out what $P_k \mathcal{E}_j^{k-1} P_k = P(k, i) \mathcal{E}_j^{k-1} P(k, i)$ is. However, since

$$\mathcal{E}_j^{k-1} = \begin{pmatrix} 0 & \cdots & 0 & 0 & \cdots & 0 \\ \vdots & \ddots & \vdots & \vdots & \vdots & \vdots \\ 0 & \cdots & 0 & 0 & \cdots & 0 \\ 0 & \cdots & \ell_{j+1\,j}^{(k-1)} & 0 & \cdots & 0 \\ \vdots & \vdots & \vdots & \vdots & \ddots & \vdots \\ 0 & \cdots & \ell_{n\,j}^{(k-1)} & 0 & \cdots & 0 \end{pmatrix},$$

and because $k+1 \le i \le n$ and $j \le k-1$, multiplying $\mathcal{E}_j^{k-1}$ on the right by $P(k,i)$ will permute *columns* i and k, which are columns of zeros, so

$$P(k,i)\,\mathcal{E}_j^{k-1}\,P(k,i) = P(k,i)\,\mathcal{E}_j^{k-1},$$

and thus,

$$(E_j^k)^{-1} = I + P(k,i)\,\mathcal{E}_j^{k-1}.$$

But since

$$(E_j^k)^{-1} = I + \mathcal{E}_j^k,$$

we deduce that

$$\mathcal{E}_j^k = P(k,i)\,\mathcal{E}_j^{k-1}.$$

We also know that multiplying $\mathcal{E}_j^{k-1}$ on the left by $P(k,i)$ will permute *rows* i and k, which shows that $\mathcal{E}_j^k$ has the desired form, as claimed. Since all $\mathcal{E}_j^k$ are strictly lower triangular, all $(E_j^k)^{-1} = I + \mathcal{E}_j^k$ are lower triangular, so the product

$$L = (E_1^{n-1})^{-1} \cdots (E_{n-1}^{n-1})^{-1}$$

is also lower triangular.

Step 3. Express L as $L = I + \Lambda_{n-1}$, with $\Lambda_{n-1} = \mathcal{E}_1^1 + \cdots + \mathcal{E}_{n-1}^{n-1}$. From Step 1 of Part (3), we know that

$$L = (E_1^{n-1})^{-1} \cdots (E_{n-1}^{n-1})^{-1}.$$

We prove by induction on k that

$$I + \Lambda_k = (E_1^k)^{-1} \cdots (E_k^k)^{-1}$$
$$\Lambda_k = \mathcal{E}_1^k + \cdots + \mathcal{E}_k^k,$$

for $k = 1, \ldots, n-1$.

If $k = 1$, we have $E_1^1 = E_1$ and

$$E_1 = \begin{pmatrix} 1 & 0 & \cdots & 0 \\ -\ell_{21}^{(1)} & 1 & \cdots & 0 \\ \vdots & \vdots & \ddots & \vdots \\ -\ell_{n1}^{(1)} & 0 & \cdots & 1 \end{pmatrix}.$$

We also get

$$(E_1^{-1})^{-1} = \begin{pmatrix} 1 & 0 & \cdots & 0 \\ \ell_{21}^{(1)} & 1 & \cdots & 0 \\ \vdots & \vdots & \ddots & \vdots \\ \ell_{n1}^{(1)} & 0 & \cdots & 1 \end{pmatrix} = I + \Lambda_1.$$

Since $(E_1^{-1})^{-1} = I + \mathcal{E}_1^1$, we find that we get $\Lambda_1 = \mathcal{E}_1^1$, and the base step holds.

Since $(E_j^k)^{-1} = I + \mathcal{E}_j^k$ with

$$
\mathcal{E}_j^k = \begin{pmatrix}
0 & \cdots & 0 & 0 & \cdots & 0 \\
\vdots & \ddots & \vdots & \vdots & \vdots & \vdots \\
0 & \cdots & 0 & 0 & \cdots & 0 \\
0 & \cdots & \ell_{j+1\,j}^{(k)} & 0 & \cdots & 0 \\
\vdots & \vdots & \vdots & \vdots & \ddots & \vdots \\
0 & \cdots & \ell_{nj}^{(k)} & 0 & \cdots & 0
\end{pmatrix}
$$

and $\mathcal{E}_i^k \mathcal{E}_j^k = 0$ if $i < j$, as in part (2) for the computation involving the products of L_k's, we get

$$(E_1^{k-1})^{-1} \cdots (E_{k-1}^{k-1})^{-1} = I + \mathcal{E}_1^{k-1} + \cdots + \mathcal{E}_{k-1}^{k-1}, \quad 2 \le k \le n. \quad (7.3)$$

Similarly, from the fact that $\mathcal{E}_j^{k-1} P(k, i) = \mathcal{E}_j^{k-1}$ if $i \ge k+1$ and $j \le k-1$ and since

$$(E_j^k)^{-1} = I + P_k \mathcal{E}_j^{k-1}, \quad 1 \le j \le n-2,\ j+1 \le k \le n-1,$$

we get

$$(E_1^k)^{-1} \cdots (E_{k-1}^k)^{-1} = I + P_k(\mathcal{E}_1^{k-1} + \cdots + \mathcal{E}_{k-1}^{k-1}), \quad 2 \le k \le n-1. \quad (7.4)$$

By the induction hypothesis,

$$I + \Lambda_{k-1} = (E_1^{k-1})^{-1} \cdots (E_{k-1}^{k-1})^{-1},$$

and from (7.3), we get

$$\Lambda_{k-1} = \mathcal{E}_1^{k-1} + \cdots + \mathcal{E}_{k-1}^{k-1}.$$

Using (7.4), we deduce that

$$(E_1^k)^{-1} \cdots (E_{k-1}^k)^{-1} = I + P_k \Lambda_{k-1}.$$

Since $E_k^k = E_k$, we obtain

$$(E_1^k)^{-1} \cdots (E_{k-1}^k)^{-1}(E_k^k)^{-1} = (I + P_k \Lambda_{k-1})E_k^{-1}.$$

However, by definition

$$I + \Lambda_k = (I + P_k \Lambda_{k-1})E_k^{-1},$$

which proves that

$$I + \Lambda_k = (E_1^k)^{-1} \cdots (E_{k-1}^k)^{-1}(E_k^k)^{-1}, \quad (7.5)$$

and finishes the induction step for the proof of this formula.

If we apply equation (7.3) again with $k+1$ in place of k, we have

$$(E_1^k)^{-1} \cdots (E_k^k)^{-1} = I + \mathcal{E}_1^k + \cdots + \mathcal{E}_k^k,$$

and together with (7.5), we obtain,

$$\Lambda_k = \mathcal{E}_1^k + \cdots + \mathcal{E}_k^k,$$

also finishing the induction step for the proof of this formula. For $k = n-1$ in (7.5), we obtain the desired equation: $L = I + \Lambda_{n-1}$. $\square$

7.7 Dealing with Roundoff Errors; Pivoting Strategies

Let us now briefly comment on the choice of a pivot. Although theoretically, any pivot can be chosen, the possibility of roundoff errors implies that it is not a good idea to pick very small pivots. The following example illustrates this point. Consider the linear system

$$10^{-4}x + y = 1$$
$$x \ + y = 2.$$

Since 10^{-4} is nonzero, it can be taken as pivot, and we get

$$10^{-4}x + \quad y \quad = \quad 1$$
$$(1 - 10^4)y = 2 - 10^4.$$

Thus, the exact solution is

$$x = \frac{10^4}{10^4 - 1}, \quad y = \frac{10^4 - 2}{10^4 - 1}.$$

However, if roundoff takes place on the fourth digit, then $10^4 - 1 = 9999$ and $10^4 - 2 = 9998$ will be rounded off both to 9990, and then the solution is $x = 0$ and $y = 1$, very far from the exact solution where $x \approx 1$ and $y \approx 1$. The problem is that we picked a very small pivot. If instead we permute the equations, the pivot is 1, and after elimination we get the system

$$x + \quad y \quad = \quad 2$$
$$(1 - 10^{-4})y = 1 - 2 \times 10^{-4}.$$

This time, $1 - 10^{-4} = 0.9999$ and $1 - 2 \times 10^{-4} = 0.9998$ are rounded off to 0.999 and the solution is $x = 1, y = 1$, much closer to the exact solution.

To remedy this problem, one may use the strategy of *partial pivoting*. This consists of choosing during Step k ($1 \leq k \leq n - 1$) one of the entries $a_{ik}^{(k)}$ such that

$$|a_{ik}^{(k)}| = \max_{k \leq p \leq n} |a_{pk}^{(k)}|.$$

By maximizing the value of the pivot, we avoid dividing by undesirably small pivots.

Remark: A matrix, A, is called *strictly column diagonally dominant* iff

$$|a_{jj}| > \sum_{i=1, \, i \neq j}^{n} |a_{ij}|, \quad \text{for } j = 1, \ldots, n$$

(resp. *strictly row diagonally dominant* iff

$$|a_{ii}| > \sum_{j=1,\, j \neq i}^{n} |a_{ij}|, \quad \text{for } i = 1, \ldots, n.)$$

For example, the matrix

$$\begin{pmatrix} \frac{7}{2} & 1 & & & \\ 1 & 4 & 1 & & 0 \\ & \ddots & \ddots & \ddots & \\ 0 & & 1 & 4 & 1 \\ & & & 1 & \frac{7}{2} \end{pmatrix}$$

of the curve interpolation problem discussed in Section 7.1 is strictly column (and row) diagonally dominant.

It has been known for a long time (before 1900, say by Hadamard) that if a matrix A is strictly column diagonally dominant (resp. strictly row diagonally dominant), then it is invertible. It can also be shown that if A is strictly column diagonally dominant, then Gaussian elimination with partial pivoting does not actually require pivoting (see Problem 7.12).

Another strategy, called *complete pivoting*, consists in choosing some entry $a_{ij}^{(k)}$, where $k \leq i, j \leq n$, such that

$$|a_{ij}^{(k)}| = \max_{k \leq p, q \leq n} |a_{pq}^{(k)}|.$$

However, in this method, if the chosen pivot is not in column k, it is also necessary to permute columns. This is achieved by multiplying on the right by a permutation matrix. However, complete pivoting tends to be too expensive in practice, and partial pivoting is the method of choice.

A special case where the LU-factorization is particularly efficient is the case of tridiagonal matrices, which we now consider.

7.8 Gaussian Elimination of Tridiagonal Matrices

Consider the tridiagonal matrix

$$A = \begin{pmatrix} b_1 & c_1 & & & & & \\ a_2 & b_2 & c_2 & & & & \\ & a_3 & b_3 & c_3 & & & \\ & & \ddots & \ddots & \ddots & & \\ & & & a_{n-2} & b_{n-2} & c_{n-2} & \\ & & & & a_{n-1} & b_{n-1} & c_{n-1} \\ & & & & & a_n & b_n \end{pmatrix}.$$

Define the sequence

$$\delta_0 = 1, \quad \delta_1 = b_1, \quad \delta_k = b_k \delta_{k-1} - a_k c_{k-1} \delta_{k-2}, \quad 2 \le k \le n.$$

Proposition 7.4. *If A is the tridiagonal matrix above, then $\delta_k = \det(A(1 : k, 1 : k))$ for $k = 1, \ldots, n$.*

Proof. By expanding $\det(A(1 : k, 1 : k))$ with respect to its last row, the proposition follows by induction on k. $\square$

Theorem 7.3. *If A is the tridiagonal matrix above and $\delta_k \ne 0$ for $k = 1, \ldots, n$, then A has the following LU-factorization:*

$$
A =
\begin{pmatrix}
1 & & & & & \\
a_2 \dfrac{\delta_0}{\delta_1} & 1 & & & & \\
& a_3 \dfrac{\delta_1}{\delta_2} & 1 & & & \\
& & \ddots & \ddots & & \\
& & & a_{n-1} \dfrac{\delta_{n-3}}{\delta_{n-2}} & 1 & \\
& & & & a_n \dfrac{\delta_{n-2}}{\delta_{n-1}} & 1
\end{pmatrix}
\begin{pmatrix}
\dfrac{\delta_1}{\delta_0} & c_1 & & & & \\
& \dfrac{\delta_2}{\delta_1} & c_2 & & & \\
& & \dfrac{\delta_3}{\delta_2} & c_3 & & \\
& & & \ddots & \ddots & \\
& & & & \dfrac{\delta_{n-1}}{\delta_{n-2}} & c_{n-1} \\
& & & & & \dfrac{\delta_n}{\delta_{n-1}}
\end{pmatrix}.
$$

Proof. Since $\delta_k = \det(A(1 : k, 1 : k)) \ne 0$ for $k = 1, \ldots, n$, by Theorem 7.2 (and Proposition 7.1), we know that A has a unique LU-factorization. Therefore, it suffices to check that the proposed factorization works. We easily check that

$$
\begin{aligned}
(LU)_{k\,k+1} &= c_k, \quad 1 \le k \le n-1 \\
(LU)_{k\,k-1} &= a_k, \quad 2 \le k \le n \\
(LU)_{k\,l} &= 0, \quad |k - l| \ge 2 \\
(LU)_{11} &= \frac{\delta_1}{\delta_0} = b_1 \\
(LU)_{k\,k} &= \frac{a_k c_{k-1} \delta_{k-2} + \delta_k}{\delta_{k-1}} = b_k, \quad 2 \le k \le n,
\end{aligned}
$$

since $\delta_k = b_k \delta_{k-1} - a_k c_{k-1} \delta_{k-2}$. $\square$

It follows that there is a simple method to solve a linear system $Ax = d$ where A is tridiagonal (and $\delta_k \ne 0$ for $k = 1, \ldots, n$). For this, it is convenient to "squeeze" the diagonal matrix Δ defined such that $\Delta_{kk} = \delta_k/\delta_{k-1}$ into the factorization so that $A = (L\Delta)(\Delta^{-1}U)$, and if we let

$$z_1 = \frac{c_1}{b_1}, \quad z_k = c_k\frac{\delta_{k-1}}{\delta_k}, \quad 2 \le k \le n-1, \quad z_n = \frac{\delta_n}{\delta_{n-1}} = b_n - a_n z_{n-1},$$

$A = (L\Delta)(\Delta^{-1}U)$ is written as

$$A = \begin{pmatrix} \frac{c_1}{z_1} & & & & & \\ a_2 & \frac{c_2}{z_2} & & & & \\ & a_3 & \frac{c_3}{z_3} & & & \\ & & \ddots & \ddots & & \\ & & & a_{n-1} & \frac{c_{n-1}}{z_{n-1}} & \\ & & & & a_n & z_n \end{pmatrix} \begin{pmatrix} 1 & z_1 & & & & \\ & 1 & z_2 & & & \\ & & 1 & z_3 & & \\ & & & \ddots & \ddots & \\ & & & & 1 & z_{n-2} \\ & & & & & 1 & z_{n-1} \\ & & & & & & 1 \end{pmatrix}.$$

As a consequence, the system $Ax = d$ can be solved by constructing three sequences: First, the sequence

$$z_1 = \frac{c_1}{b_1}, \quad z_k = \frac{c_k}{b_k - a_k z_{k-1}}, \quad k = 2,\ldots,n-1, \quad z_n = b_n - a_n z_{n-1},$$

corresponding to the recurrence $\delta_k = b_k\delta_{k-1} - a_k c_{k-1}\delta_{k-2}$ and obtained by dividing both sides of this equation by δ_{k-1}, next

$$w_1 = \frac{d_1}{b_1}, \quad w_k = \frac{d_k - a_k w_{k-1}}{b_k - a_k z_{k-1}}, \quad k = 2,\ldots,n,$$

corresponding to solving the system $L\Delta w = d$, and finally

$$x_n = w_n, \quad x_k = w_k - z_k x_{k+1}, \quad k = n-1, n-2, \ldots, 1,$$

corresponding to solving the system $\Delta^{-1}Ux = w$.

Remark: It can be verified that this requires $3(n-1)$ additions, $3(n-1)$ multiplications, and $2n$ divisions, a total of $8n-6$ operations, which is much less that the $O(2n^3/3)$ required by Gaussian elimination in general.

We now consider the special case of symmetric positive definite matrices (SPD matrices).

7.9   SPD Matrices and the Cholesky Decomposition

Recall that an $n \times n$ real symmetric matrix A is *positive definite* iff

$$x^\top A x > 0 \quad \text{for all } x \in \mathbb{R}^n \text{ with } x \neq 0.$$

Equivalently, A is symmetric positive definite iff all its eigenvalues are strictly positive. The following facts about a symmetric positive definite matrix A are easily established (some left as an exercise):

(1) The matrix A is invertible. (Indeed, if $Ax = 0$, then $x^\top A x = 0$, which implies $x = 0$.)

(2) We have $a_{ii} > 0$ for $i = 1, \ldots, n$. (Just observe that for $x = e_i$, the ith canonical basis vector of $\mathbb{R}^n$, we have $e_i^\top A e_i = a_{ii} > 0$.)

(3) For every $n \times n$ real invertible matrix Z, the matrix $Z^\top A Z$ is real symmetric positive definite iff A is real symmetric positive definite.

(4) The set of $n \times n$ real symmetric positive definite matrices is convex. This means that if A and B are two $n \times n$ symmetric positive definite matrices, then for any $\lambda \in \mathbb{R}$ such that $0 \leq \lambda \leq 1$, the matrix $(1 - \lambda)A + \lambda B$ is also symmetric positive definite. Clearly since A and B are symmetric, $(1 - \lambda)A + \lambda B$ is also symmetric. For any nonzero $x \in \mathbb{R}^n$, we have $x^\top A x > 0$ and $x^\top B x > 0$, so

$$x^\top ((1 - \lambda)A + \lambda B)x = (1 - \lambda)x^\top A x + \lambda x^\top B x > 0,$$

because $0 \leq \lambda \leq 1$, so $1 - \lambda \geq 0$ and $\lambda \geq 0$, and $1 - \lambda$ and λ can't be zero simultaneously.

(5) The set of $n \times n$ real symmetric positive definite matrices is a cone. This means that if A is symmetric positive definite and if $\lambda > 0$ is any real, then λA is symmetric positive definite. Clearly λA is symmetric, and for nonzero $x \in \mathbb{R}^n$, we have $x^\top A x > 0$, and since $\lambda > 0$, we have $x^\top \lambda A x = \lambda x^\top A x > 0$.

Remark: Given a complex $m \times n$ matrix A, we define the matrix $\overline{A}$ as the $m \times n$ matrix $\overline{A} = (\overline{a_{ij}})$. Then we define A^* as the $n \times m$ matrix $A^* = (\overline{A})^\top = \overline{(A^\top)}$. The $n \times n$ complex matrix A is *Hermitian* if $A^* = A$. This is the complex analog of the notion of a real symmetric matrix. A Hermitian matrix A is *positive definite* if

$$z^* A z > 0 \quad \text{for all } z \in \mathbb{C}^n \text{ with } z \neq 0.$$

It is easily verified that Properties (1)–(5) hold for Hermitian positive definite matrices; replace $\top$ by $*$.

It is instructive to characterize when a 2×2 real symmetric matrix A is positive definite. Write

$$A = \begin{pmatrix} a & c \\ c & b \end{pmatrix}.$$

Then we have

$$\begin{pmatrix} x & y \end{pmatrix} \begin{pmatrix} a & c \\ c & b \end{pmatrix} \begin{pmatrix} x \\ y \end{pmatrix} = ax^2 + 2cxy + by^2.$$

If the above expression is strictly positive for all nonzero vectors $\begin{pmatrix} x \\ y \end{pmatrix}$, then for $x = 1, y = 0$ we get $a > 0$ and for $x = 0, y = 1$ we get $b > 0$. Then we can write

$$ax^2 + 2cxy + by^2 = \left(\sqrt{a}x + \frac{c}{\sqrt{a}}y \right)^2 + by^2 - \frac{c^2}{a}y^2$$

$$= \left(\sqrt{a}x + \frac{c}{\sqrt{a}}y \right)^2 + \frac{1}{a}\left(ab - c^2\right)y^2. \qquad (7.6)$$

Since $a > 0$, if $ab - c^2 \leq 0$, then we can choose $y > 0$ so that the second term is negative or zero, and we can set $x = -(c/a)y$ to make the first term zero, in which case $ax^2 + 2cxy + by^2 \leq 0$, so we must have $ab - c^2 > 0$.

Conversely, if $a > 0, b > 0$ and $ab > c^2$, then for any $(x, y) \neq (0, 0)$, if $y = 0$, then $x \neq 0$ and the first term of (7.6) is positive, and if $y \neq 0$, then the second term of (7.6) is positive. Therefore, the symmetric matrix A is positive definite iff

$$a > 0, \ b > 0, \ ab > c^2. \qquad (7.7)$$

Note that $ab - c^2 = \det(A)$, so the third condition says that $\det(A) > 0$.

Observe that the condition $b > 0$ is redundant, since if $a > 0$ and $ab > c^2$, then we must have $b > 0$ (and similarly $b > 0$ and $ab > c^2$ implies that $a > 0$).

We can try to visualize the space of 2×2 real symmetric positive definite matrices in $\mathbb{R}^3$, by viewing (a, b, c) as the coordinates along the x, y, z axes. Then the locus determined by the strict inequalities in (7.7) corresponds to the region on the side of the cone of equation $xy = z^2$ that does not contain the origin and for which $x > 0$ and $y > 0$. For $z = \delta$ fixed, the equation $xy = \delta^2$ define a hyperbola in the plane $z = \delta$. The cone of equation $xy = z^2$ consists of the lines through the origin that touch the hyperbola $xy = 1$ in the plane $z = 1$. We only consider the branch of this hyperbola for which $x > 0$ and $y > 0$. See Figure 7.6.

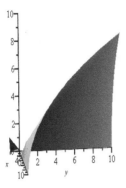

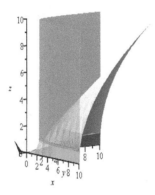

Fig. 7.6 Two views of the surface $xy = z^2$ in $\mathbb{R}^3$. The intersection of the surface with a constant z plane results in a hyperbola. The region associated with the 2×2 symmetric positive definite matrices lies in "front" of the green side.

It is not hard to show that the inverse of a real symmetric positive definite matrix is also real symmetric positive definite, but the product of two real symmetric positive definite matrices may *not* be symmetric positive definite, as the following example shows:

$$\begin{pmatrix} 1 & 1 \\ 1 & 2 \end{pmatrix} \begin{pmatrix} 1/\sqrt{2} & -1\sqrt{2} \\ -1/\sqrt{2} & 3/\sqrt{2} \end{pmatrix} = \begin{pmatrix} 0 & 2/\sqrt{2} \\ -1/\sqrt{2} & 5/\sqrt{2} \end{pmatrix}.$$

According to the above criterion, the two matrices on the left-hand side are real symmetric positive definite, but the matrix on the right-hand side is not even symmetric, and

$$(-6\ 1) \begin{pmatrix} 0 & 2/\sqrt{2} \\ -1/\sqrt{2} & 5/\sqrt{2} \end{pmatrix} \begin{pmatrix} -6 \\ 1 \end{pmatrix} = (-6\ 1) \begin{pmatrix} 2/\sqrt{2} \\ 11/\sqrt{2} \end{pmatrix} = -1/\sqrt{5},$$

even though its eigenvalues are both real and positive.

Next we show that a real symmetric positive definite matrix has a special LU-factorization of the form $A = BB^\top$, where B is a lower-triangular matrix whose diagonal elements are strictly positive. This is the *Cholesky factorization*.

First we note that a symmetric positive definite matrix satisfies the condition of Proposition 7.1.

Proposition 7.5. *If A is a real symmetric positive definite matrix, then $A(1 : k, 1 : k)$ is symmetric positive definite and thus invertible for $k = 1, \ldots, n$.*

Proof. Since A is symmetric, each $A(1 : k, 1 : k)$ is also symmetric. If $w \in \mathbb{R}^k$, with $1 \leq k \leq n$, we let $x \in \mathbb{R}^n$ be the vector with $x_i = w_i$ for $i = 1, \ldots, k$ and $x_i = 0$ for $i = k + 1, \ldots, n$. Now since A is symmetric positive definite, we have $x^\top A x > 0$ for all $x \in \mathbb{R}^n$ with $x \neq 0$. This holds in particular for all vectors x obtained from nonzero vectors $w \in \mathbb{R}^k$ as defined earlier, and clearly

$$x^\top A x = w^\top A(1 : k, 1 : k)\, w,$$

which implies that $A(1 : k, 1 : k)$ is positive definite. Thus, by Fact 1 above, $A(1 : k, 1 : k)$ is also invertible. $\qquad\square$

Proposition 7.5 also holds for a complex Hermitian positive definite matrix. Proposition 7.5 can be strengthened as follows: *A real symmetric (or complex Hermitian) matrix A is positive definite iff* $\det(A(1 : k, 1 : k)) > 0$ *for* $k = 1, \ldots, n$.

The above fact is known as *Sylvester's criterion*. We will prove it after establishing the Cholesky factorization.

Let A be an $n \times n$ real symmetric positive definite matrix and write

$$A = \begin{pmatrix} a_{11} & W^\top \\ W & C \end{pmatrix},$$

where C is an $(n - 1) \times (n - 1)$ symmetric matrix and W is an $(n - 1) \times 1$ matrix. Since A is symmetric positive definite, $a_{11} > 0$, and we can compute $\alpha = \sqrt{a_{11}}$. The trick is that we can factor A uniquely as

$$A = \begin{pmatrix} a_{11} & W^\top \\ W & C \end{pmatrix} = \begin{pmatrix} \alpha & 0 \\ W/\alpha & I \end{pmatrix} \begin{pmatrix} 1 & 0 \\ 0 & C - WW^\top/a_{11} \end{pmatrix} \begin{pmatrix} \alpha & W^\top/\alpha \\ 0 & I \end{pmatrix},$$

i.e., as $A = B_1 A_1 B_1^\top$, where B_1 is lower-triangular with positive diagonal entries. Thus, B_1 is invertible, and by Fact (3) above, A_1 is also symmetric positive definite.

Remark: The matrix $C - WW^\top/a_{11}$ is known as the *Schur complement* of the matrix (a_{11}).

Theorem 7.4. *(Cholesky factorization) Let A be a real symmetric positive definite matrix. Then there is some real lower-triangular matrix B so that $A = BB^\top$. Furthermore, B can be chosen so that its diagonal elements are strictly positive, in which case B is unique.*

Proof. We proceed by induction on the dimension n of A. For $n = 1$, we must have $a_{11} > 0$, and if we let $\alpha = \sqrt{a_{11}}$ and $B = (\alpha)$, the theorem holds trivially. If $n \geq 2$, as we explained above, again we must have $a_{11} > 0$, and we can write

$$A = \begin{pmatrix} a_{11} & W^\top \\ W & C \end{pmatrix} = \begin{pmatrix} \alpha & 0 \\ W/\alpha & I \end{pmatrix} \begin{pmatrix} 1 & 0 \\ 0 & C - WW^\top/a_{11} \end{pmatrix} \begin{pmatrix} \alpha & W^\top/\alpha \\ 0 & I \end{pmatrix} = B_1 A_1 B_1^\top,$$

where $\alpha = \sqrt{a_{11}}$, the matrix B_1 is invertible and

$$A_1 = \begin{pmatrix} 1 & 0 \\ 0 & C - WW^\top/a_{11} \end{pmatrix}$$

is symmetric positive definite. However, this implies that $C - WW^\top/a_{11}$ is also symmetric positive definite (consider $x^\top A_1 x$ for every $x \in \mathbb{R}^n$ with $x \neq 0$ and $x_1 = 0$). Thus, we can apply the induction hypothesis to $C - WW^\top/a_{11}$ (which is an $(n-1) \times (n-1)$ matrix), and we find a unique lower-triangular matrix L with positive diagonal entries so that

$$C - WW^\top/a_{11} = LL^\top.$$

But then we get

$$\begin{aligned}
A &= \begin{pmatrix} \alpha & 0 \\ W/\alpha & I \end{pmatrix} \begin{pmatrix} 1 & 0 \\ 0 & C - WW^\top/a_{11} \end{pmatrix} \begin{pmatrix} \alpha & W^\top/\alpha \\ 0 & I \end{pmatrix} \\
&= \begin{pmatrix} \alpha & 0 \\ W/\alpha & I \end{pmatrix} \begin{pmatrix} 1 & 0 \\ 0 & LL^\top \end{pmatrix} \begin{pmatrix} \alpha & W^\top/\alpha \\ 0 & I \end{pmatrix} \\
&= \begin{pmatrix} \alpha & 0 \\ W/\alpha & I \end{pmatrix} \begin{pmatrix} 1 & 0 \\ 0 & L \end{pmatrix} \begin{pmatrix} 1 & 0 \\ 0 & L^\top \end{pmatrix} \begin{pmatrix} \alpha & W^\top/\alpha \\ 0 & I \end{pmatrix} \\
&= \begin{pmatrix} \alpha & 0 \\ W/\alpha & L \end{pmatrix} \begin{pmatrix} \alpha & W^\top/\alpha \\ 0 & L^\top \end{pmatrix}.
\end{aligned}$$

Therefore, if we let

$$B = \begin{pmatrix} \alpha & 0 \\ W/\alpha & L \end{pmatrix},$$

we have a unique lower-triangular matrix with positive diagonal entries and $A = BB^\top$. $\qquad\square$

Remark: The uniqueness of the Cholesky decomposition can also be established using the uniqueness of an LU-decomposition. Indeed, if $A = B_1 B_1^\top = B_2 B_2^\top$ where B_1 and B_2 are lower triangular with positive diagonal entries, if we let Δ_1 (resp. Δ_2) be the diagonal matrix consisting of the diagonal entries of B_1 (resp. B_2) so that $(\Delta_k)_{ii} = (B_k)_{ii}$ for $k = 1, 2$, then we have two LU-decompositions

$$A = (B_1 \Delta_1^{-1})(\Delta_1 B_1^\top) = (B_2 \Delta_2^{-1})(\Delta_2 B_2^\top)$$

with $B_1 \Delta_1^{-1}, B_2 \Delta_2^{-1}$ unit lower triangular, and $\Delta_1 B_1^\top, \Delta_2 B_2^\top$ upper triangular. By uniqueness of LU-factorization (Theorem 7.2(1)), we have

$$B_1 \Delta_1^{-1} = B_2 \Delta_2^{-1}, \quad \Delta_1 B_1^\top = \Delta_2 B_2^\top,$$

and the second equation yields

$$B_1 \Delta_1 = B_2 \Delta_2. \tag{7.8}$$

The diagonal entries of $B_1 \Delta_1$ are $(B_1)_{ii}^2$ and similarly the diagonal entries of $B_2 \Delta_2$ are $(B_2)_{ii}^2$, so the above equation implies that

$$(B_1)_{ii}^2 = (B_2)_{ii}^2, \quad i = 1, \dots, n.$$

Since the diagonal entries of both B_1 and B_2 are assumed to be positive, we must have

$$(B_1)_{ii} = (B_2)_{ii}, \quad i = 1, \dots, n;$$

that is, $\Delta_1 = \Delta_2$, and since both are invertible, we conclude from (7.8) that $B_1 = B_2$.

Theorem 7.4 also holds for complex Hermitian positive definite matrices. In this case, we have $A = BB^*$ for some unique lower triangular matrix B with positive diagonal entries.

The proof of Theorem 7.4 immediately yields an algorithm to compute B from A by solving for a lower triangular matrix B such that $A = BB^\top$ (where both A and B are real matrices). For $j = 1, \dots, n$,

$$b_{jj} = \left(a_{jj} - \sum_{k=1}^{j-1} b_{jk}^2 \right)^{1/2},$$

and for $i = j + 1, \dots, n$ (and $j = 1, \dots, n-1$)

$$b_{ij} = \left(a_{ij} - \sum_{k=1}^{j-1} b_{ik} b_{jk} \right) / b_{jj}.$$

The above formulae are used to compute the jth column of B from top-down, using the first $j - 1$ columns of B previously computed, and the matrix A. In the case of $n = 3$, $A = BB^\top$ yields

$$\begin{pmatrix} a_{11} & a_{12} & a_{31} \\ a_{21} & a_{22} & a_{32} \\ a_{31} & a_{32} & a_{33} \end{pmatrix} = \begin{pmatrix} b_{11} & 0 & 0 \\ b_{21} & b_{22} & 0 \\ b_{31} & b_{32} & b_{33} \end{pmatrix} \begin{pmatrix} b_{11} & b_{21} & b_{31} \\ 0 & b_{22} & b_{32} \\ 0 & 0 & b_{33} \end{pmatrix}$$

$$= \begin{pmatrix} b_{11}^2 & b_{11}b_{21} & b_{11}b_{31} \\ b_{11}b_{21} & b_{21}^2 + b_{22}^2 & b_{21}b_{31} + b_{22}b_{32} \\ b_{11}b_{31} & b_{21}b_{31} + b_{22}b_{32} & b_{31}^2 + b_{32}^2 + b_{33}^2 \end{pmatrix}.$$

We work down the first column of A, compare entries, and discover that

$$a_{11} = b_{11}^2 \qquad\qquad b_{11} = \sqrt{a_{11}}$$

$$a_{21} = b_{11}b_{21} \qquad\qquad b_{21} = \frac{a_{21}}{b_{11}}$$

$$a_{31} = b_{11}b_{31} \qquad\qquad b_{31} = \frac{a_{31}}{b_{11}}.$$

Next we work down the second column of A using previously calculated expressions for b_{21} and b_{31} to find that

$$a_{22} = b_{21}^2 + b_{22}^2 \qquad\qquad b_{22} = \left(a_{22} - b_{21}^2\right)^{\frac{1}{2}}$$

$$a_{32} = b_{21}b_{31} + b_{22}b_{32} \qquad\qquad b_{32} = \frac{a_{32} - b_{21}b_{31}}{b_{22}}.$$

Finally, we use the third column of A and the previously calculated expressions for b_{31} and b_{32} to determine b_{33} as

$$a_{33} = b_{31}^2 + b_{32}^2 + b_{33}^2 \qquad\qquad b_{33} = \left(a_{33} - b_{31}^2 - b_{32}^2\right)^{\frac{1}{2}}.$$

For another example, if

$$A = \begin{pmatrix} 1 & 1 & 1 & 1 & 1 & 1 \\ 1 & 2 & 2 & 2 & 2 & 2 \\ 1 & 2 & 3 & 3 & 3 & 3 \\ 1 & 2 & 3 & 4 & 4 & 4 \\ 1 & 2 & 3 & 4 & 5 & 5 \\ 1 & 2 & 3 & 4 & 5 & 6 \end{pmatrix},$$

we find that

$$B = \begin{pmatrix} 1 & 0 & 0 & 0 & 0 & 0 \\ 1 & 1 & 0 & 0 & 0 & 0 \\ 1 & 1 & 1 & 0 & 0 & 0 \\ 1 & 1 & 1 & 1 & 0 & 0 \\ 1 & 1 & 1 & 1 & 1 & 0 \\ 1 & 1 & 1 & 1 & 1 & 1 \end{pmatrix}.$$

We leave it as an exercise to find similar formulae (involving conjugation) to factor a complex Hermitian positive definite matrix A as $A = BB^*$. The following `Matlab` program implements the Cholesky factorization.

```
function B = Cholesky(A)
n = size(A,1);
B = zeros(n,n);
for j = 1:n-1;
   if j == 1
      B(1,1) = sqrt(A(1,1));
      for i = 2:n
         B(i,1) = A(i,1)/B(1,1);
      end
   else
      B(j,j) = sqrt(A(j,j) - B(j,1:j-1)*B(j,1:j-1)');
      for i = j+1:n
         B(i,j) = (A(i,j) - B(i,1:j-1)*B(j,1:j-1)')/B(j,j);
      end
   end
end
B(n,n) = sqrt(A(n,n) - B(n,1:n-1)*B(n,1:n-1)');
end
```

If we run the above algorithm on the following matrix

$$
A = \begin{pmatrix} 4 & 1 & 0 & 0 & 0 \\ 1 & 4 & 1 & 0 & 0 \\ 0 & 1 & 4 & 1 & 0 \\ 0 & 0 & 1 & 4 & 1 \\ 0 & 0 & 0 & 1 & 4 \end{pmatrix},
$$

we obtain

$$
B = \begin{pmatrix} 2.0000 & 0 & 0 & 0 & 0 \\ 0.5000 & 1.9365 & 0 & 0 & 0 \\ 0 & 0.5164 & 1.9322 & 0 & 0 \\ 0 & 0 & 0.5175 & 1.9319 & 0 \\ 0 & 0 & 0 & 0.5176 & 1.9319 \end{pmatrix}.
$$

The Cholesky factorization can be used to solve linear systems $Ax = b$ where A is symmetric positive definite: Solve the two systems $Bw = b$ and $B^\top x = w$.

Remark: It can be shown that this methods requires $n^3/6 + O(n^2)$ additions, $n^3/6 + O(n^2)$ multiplications, $n^2/2 + O(n)$ divisions, and $O(n)$ square root extractions. Thus, the Cholesky method requires half of the number of operations required by Gaussian elimination (since Gaussian elimination requires $n^3/3 + O(n^2)$ additions, $n^3/3 + O(n^2)$ multiplications, and $n^2/2 + O(n)$ divisions). It also requires half of the space (only B is needed, as opposed to both L and U). Furthermore, it can be shown that Cholesky's method is numerically stable (see Trefethen and Bau [Trefethen and Bau III (1997)], Lecture 23). In `Matlab` the function `chol` returns the lower-triangular matrix B such that $A = BB^\top$ using the call $B = \mathrm{chol}(A,$ 'lower').

Remark: If $A = BB^\top$, where B is any invertible matrix, then A is symmetric positive definite.

Proof. Obviously, $BB^\top$ is symmetric, and since B is invertible, $B^\top$ is invertible, and from

$$x^\top A x = x^\top B B^\top x = (B^\top x)^\top B^\top x,$$

it is clear that $x^\top A x > 0$ if $x \neq 0$. □

We now give three more criteria for a symmetric matrix to be positive definite.

Proposition 7.6. *Let A be any $n \times n$ real symmetric matrix. The following conditions are equivalent:*

(a) A is positive definite.
(b) All principal minors of A are positive; that is: $\det(A(1:k,1:k)) > 0$ for $k = 1, \ldots, n$ (Sylvester's criterion).
(c) A has an LU-factorization and all pivots are positive.
(d) A has an $LDL^\top$-factorization and all pivots in D are positive.

Proof. By Proposition 7.5, if A is symmetric positive definite, then each matrix $A(1:k,1:k)$ is symmetric positive definite for $k = 1, \ldots, n$. By the Cholsesky decomposition, $A(1:k,1:k) = Q^\top Q$ for some invertible matrix Q, so $\det(A(1:k,1:k)) = \det(Q)^2 > 0$. This shows that (a) implies (b).

If $\det(A(1:k,1:k)) > 0$ for $k = 1, \ldots, n$, then each $A(1:k,1:k)$ is invertible. By Proposition 7.1, the matrix A has an LU-factorization, and since the pivots π_k are given by

$$\pi_k = \begin{cases} a_{11} = \det(A(1:1,1:1)) & \text{if } k = 1 \\ \dfrac{\det(A(1:k,1:k))}{\det(A(1:k-1,1:k-1))} & \text{if } k = 2,\dots,n, \end{cases}$$

we see that $\pi_k > 0$ for $k = 1,\dots,n$. Thus (b) implies (c).

Assume A has an LU-factorization and that the pivots are all positive. Since A is symmetric, this implies that A has a factorization of the form

$$A = LDL^\top,$$

with L lower-triangular with 1s on its diagonal, and where D is a diagonal matrix with positive entries on the diagonal (the pivots). This shows that (c) implies (d).

Given a factorization $A = LDL^\top$ with all pivots in D positive, if we form the diagonal matrix

$$\sqrt{D} = \operatorname{diag}(\sqrt{\pi_1},\dots,\sqrt{\pi_n})$$

and if we let $B = L\sqrt{D}$, then we have

$$A = BB^\top,$$

with B lower-triangular and invertible. By the remark before Proposition 7.6, A is positive definite. Hence, (d) implies (a).   □

Criterion (c) yields a simple computational test to check whether a symmetric matrix is positive definite. There is one more criterion for a symmetric matrix to be positive definite: its eigenvalues must be positive. We will have to learn about the spectral theorem for symmetric matrices to establish this criterion.

Proposition 7.6 also holds for complex Hermitian positive definite matrices, where in (d), the factorization $LDL^\top$ is replaced by LDL^*.

For more on the stability analysis and efficient implementation methods of Gaussian elimination, LU-factoring and Cholesky factoring, see Demmel [Demmel (1997)], Trefethen and Bau [Trefethen and Bau III (1997)], Ciarlet [Ciarlet (1989)], Golub and Van Loan [Golub and Van Loan (1996)], Meyer [Meyer (2000)], Strang [Strang (1986, 1988)], and Kincaid and Cheney [Kincaid and Cheney (1996)].

7.10 Reduced Row Echelon Form (RREF)

Gaussian elimination described in Section 7.2 can also be applied to rectangular matrices. This yields a method for determining whether a system $Ax = b$ is solvable and a description of all the solutions when the system is solvable, for any rectangular $m \times n$ matrix A.

It turns out that the discussion is simpler if we rescale all pivots to be 1, and for this we need a third kind of elementary matrix. For any $\lambda \neq 0$, let $E_{i,\lambda}$ be the $n \times n$ diagonal matrix

$$
E_{i,\lambda} =
\begin{pmatrix}
1 & & & & & & \\
& \ddots & & & & & \\
& & 1 & & & & \\
& & & \lambda & & & \\
& & & & 1 & & \\
& & & & & \ddots & \\
& & & & & & 1
\end{pmatrix},
$$

with $(E_{i,\lambda})_{ii} = \lambda$ $(1 \leq i \leq n)$. Note that $E_{i,\lambda}$ is also given by

$$
E_{i,\lambda} = I + (\lambda - 1)e_{ii},
$$

and that $E_{i,\lambda}$ is invertible with

$$
E_{i,\lambda}^{-1} = E_{i,\lambda^{-1}}.
$$

Now after $k - 1$ elimination steps, if the bottom portion

$$
(a_{kk}^{(k)}, a_{k+1k}^{(k)}, \ldots, a_{mk}^{(k)})
$$

of the kth column of the current matrix A_k is nonzero so that a pivot π_k can be chosen, after a permutation of rows if necessary, we also divide row k by π_k to obtain the pivot 1, and not only do we zero all the entries $i = k+1, \ldots, m$ in column k, but also all the entries $i = 1, \ldots, k-1$, so that the only nonzero entry in column k is a 1 in row k. These row operations are achieved by multiplication on the left by elementary matrices. If $a_{kk}^{(k)} = a_{k+1k}^{(k)} = \cdots = a_{mk}^{(k)} = 0$, we move on to column $k + 1$.

When the kth column contains a pivot, the kth stage of the procedure for converting a matrix to *rref* consists of the following three steps illustrated below:

$$\begin{pmatrix} 1 & \times & 0 & \times & \times & \times & \times \\ 0 & 0 & 1 & \times & \times & \times & \times \\ 0 & 0 & 0 & \times & \times & \times & \times \\ 0 & 0 & 0 & \times & \times & \times & \times \\ 0 & 0 & 0 & a_{ik}^{(k)} & \times & \times & \times \\ 0 & 0 & 0 & \times & \times & \times & \times \end{pmatrix} \overset{\text{pivot}}{\Longrightarrow} \begin{pmatrix} 1 & \times & 0 & \times & \times & \times & \times \\ 0 & 0 & 1 & \times & \times & \times & \times \\ 0 & 0 & 0 & a_{ik}^{(k)} & \times & \times & \times \\ 0 & 0 & 0 & \times & \times & \times & \times \\ 0 & 0 & 0 & \times & \times & \times & \times \\ 0 & 0 & 0 & \times & \times & \times & \times \end{pmatrix} \overset{\text{rescale}}{\Longrightarrow}$$

$$\begin{pmatrix} 1 & \times & 0 & \times & \times & \times & \times \\ 0 & 0 & 1 & \times & \times & \times & \times \\ 0 & 0 & 0 & 1 & \times & \times & \times \\ 0 & 0 & 0 & \times & \times & \times & \times \\ 0 & 0 & 0 & \times & \times & \times & \times \\ 0 & 0 & 0 & \times & \times & \times & \times \end{pmatrix} \overset{\text{elim}}{\Longrightarrow} \begin{pmatrix} 1 & \times & 0 & \mathbf{0} & \times & \times & \times \\ 0 & 0 & 1 & \mathbf{0} & \times & \times & \times \\ 0 & 0 & 0 & 1 & \times & \times & \times \\ 0 & 0 & 0 & \mathbf{0} & \times & \times & \times \\ 0 & 0 & 0 & \mathbf{0} & \times & \times & \times \\ 0 & 0 & 0 & \mathbf{0} & \times & \times & \times \end{pmatrix} .$$

If the kth column does not contain a pivot, we simply move on to the next column.

The result is that after performing such elimination steps, we obtain a matrix that has a special shape known as a *reduced row echelon matrix*, for short *rref*.

Here is an example illustrating this process: Starting from the matrix

$$A_1 = \begin{pmatrix} 1 & 0 & 2 & 1 & 5 \\ 1 & 1 & 5 & 2 & 7 \\ 1 & 2 & 8 & 4 & 12 \end{pmatrix},$$

we perform the following steps

$$A_1 \longrightarrow A_2 = \begin{pmatrix} 1 & 0 & 2 & 1 & 5 \\ 0 & 1 & 3 & 1 & 2 \\ 0 & 2 & 6 & 3 & 7 \end{pmatrix},$$

by subtracting row 1 from row 2 and row 3;

$$A_2 \longrightarrow \begin{pmatrix} 1 & 0 & 2 & 1 & 5 \\ 0 & 2 & 6 & 3 & 7 \\ 0 & 1 & 3 & 1 & 2 \end{pmatrix} \longrightarrow \begin{pmatrix} 1 & 0 & 2 & 1 & 5 \\ 0 & 1 & 3 & 3/2 & 7/2 \\ 0 & 1 & 3 & 1 & 2 \end{pmatrix} \longrightarrow A_3 = \begin{pmatrix} 1 & 0 & 2 & 1 & 5 \\ 0 & 1 & 3 & 3/2 & 7/2 \\ 0 & 0 & 0 & -1/2 & -3/2 \end{pmatrix},$$

after choosing the pivot 2 and permuting row 2 and row 3, dividing row 2 by 2, and subtracting row 2 from row 3;

$$A_3 \longrightarrow \begin{pmatrix} 1 & 0 & 2 & 1 & 5 \\ 0 & 1 & 3 & 3/2 & 7/2 \\ 0 & 0 & 0 & 1 & 3 \end{pmatrix} \longrightarrow A_4 = \begin{pmatrix} 1 & 0 & 2 & 0 & 2 \\ 0 & 1 & 3 & 0 & -1 \\ 0 & 0 & 0 & 1 & 3 \end{pmatrix},$$

after dividing row 3 by $-1/2$, subtracting row 3 from row 1, and subtracting $(3/2) \times$ row 3 from row 2.

It is clear that columns $1, 2$ and 4 are linearly independent, that column 3 is a linear combination of columns 1 and 2, and that column 5 is a linear combination of columns $1, 2, 4$.

In general, the sequence of steps leading to a reduced echelon matrix is not unique. For example, we could have chosen 1 instead of 2 as the second pivot in matrix A_2. Nevertheless, *the reduced row echelon matrix obtained from any given matrix is unique*; that is, it does not depend on the the sequence of steps that are followed during the reduction process. This fact is not so easy to prove rigorously, but we will do it later.

If we want to solve a linear system of equations of the form $Ax = b$, we apply elementary row operations to both the matrix A and the right-hand side b. To do this conveniently, we form the *augmented matrix* (A, b), which is the $m \times (n + 1)$ matrix obtained by adding b as an extra column to the matrix A. For example if

$$A = \begin{pmatrix} 1 & 0 & 2 & 1 \\ 1 & 1 & 5 & 2 \\ 1 & 2 & 8 & 4 \end{pmatrix} \quad \text{and} \quad b = \begin{pmatrix} 5 \\ 7 \\ 12 \end{pmatrix},$$

then the augmented matrix is

$$(A, b) = \begin{pmatrix} 1 & 0 & 2 & 1 & 5 \\ 1 & 1 & 5 & 2 & 7 \\ 1 & 2 & 8 & 4 & 12 \end{pmatrix}.$$

Now for any matrix M, since

$$M(A, b) = (MA, Mb),$$

performing elementary row operations on (A, b) is equivalent to simultaneously performing operations on both A and b. For example, consider the system

$$\begin{aligned} x_1 \quad\quad + 2x_3 + x_4 &= 5 \\ x_1 + x_2 + 5x_3 + 2x_4 &= 7 \\ x_1 + 2x_2 + 8x_3 + 4x_4 &= 12. \end{aligned}$$

Its augmented matrix is the matrix

$$(A, b) = \begin{pmatrix} 1 & 0 & 2 & 1 & 5 \\ 1 & 1 & 5 & 2 & 7 \\ 1 & 2 & 8 & 4 & 12 \end{pmatrix}$$

considered above, so the reduction steps applied to this matrix yield the system

$$\begin{aligned}
x_1 \quad + 2x_3 \quad &= \ 2 \\
x_2 + 3x_3 \quad &= -1 \\
x_4 &= \ 3.
\end{aligned}$$

This reduced system has the same set of solutions as the original, and obviously x_3 can be chosen arbitrarily. Therefore, our system has infinitely many solutions given by

$$x_1 = 2 - 2x_3, \quad x_2 = -1 - 3x_3, \quad x_4 = 3,$$

where x_3 is arbitrary.

The following proposition shows that the set of solutions of a system $Ax = b$ is preserved by any sequence of row operations.

Proposition 7.7. *Given any $m \times n$ matrix A and any vector $b \in \mathbb{R}^m$, for any sequence of elementary row operations $E_1, \ldots, E_k$, if $P = E_k \cdots E_1$ and $(A', b') = P(A, b)$, then the solutions of $Ax = b$ are the same as the solutions of $A'x = b'$.*

Proof. Since each elementary row operation E_i is invertible, so is P, and since $(A', b') = P(A, b)$, then $A' = PA$ and $b' = Pb$. If x is a solution of the original system $Ax = b$, then multiplying both sides by P we get $PAx = Pb$; that is, $A'x = b'$, so x is a solution of the new system. Conversely, assume that x is a solution of the new system, that is $A'x = b'$. Then because $A' = PA$, $b' = Pb$, and P is invertible, we get

$$Ax = P^{-1}A'x = P^{-1}b' = b,$$

so x is a solution of the original system $Ax = b$. $\square$

Another important fact is this:

Proposition 7.8. *Given an $m \times n$ matrix A, for any sequence of row operations $E_1, \ldots, E_k$, if $P = E_k \cdots E_1$ and $B = PA$, then the subspaces spanned by the rows of A and the rows of B are identical. Therefore, A and B have the same row rank. Furthermore, the matrices A and B also have the same (column) rank.*

Proof. Since $B = PA$, from a previous observation, the rows of B are linear combinations of the rows of A, so the span of the rows of B is a subspace of the span of the rows of A. Since P is invertible, $A = P^{-1}B$, so by the same reasoning the span of the rows of A is a subspace of the span of the rows of B. Therefore, the subspaces spanned by the rows of A and the rows of B are identical, which implies that A and B have the same row rank.

Proposition 7.7 implies that the systems $Ax = 0$ and $Bx = 0$ have the same solutions. Since Ax is a linear combinations of the columns of A and Bx is a linear combinations of the columns of B, the maximum number of linearly independent columns in A is equal to the maximum number of linearly independent columns in B; that is, A and B have the same rank. $\qquad\square$

Remark: The subspaces spanned by the columns of A and B can be different! However, their dimension must be the same.

We will show in Section 7.14 that the row rank is equal to the column rank. This will also be proven in Proposition 10.11 Let us now define precisely what is a reduced row echelon matrix.

Definition 7.4. An $m \times n$ matrix A is a *reduced row echelon matrix* iff the following conditions hold:

(a) The first nonzero entry in every row is 1. This entry is called a *pivot*.
(b) The first nonzero entry of row $i + 1$ is to the right of the first nonzero entry of row i.
(c) The entries above a pivot are zero.

If a matrix satisfies the above conditions, we also say that it is in *reduced row echelon form*, for short *rref*.

Note that Condition (b) implies that the entries below a pivot are also zero. For example, the matrix

$$A = \begin{pmatrix} 1 & 6 & 0 & 1 \\ 0 & 0 & 1 & 2 \\ 0 & 0 & 0 & 0 \end{pmatrix}$$

is a reduced row echelon matrix. In general, a matrix in *rref* has the following shape:

$$\begin{pmatrix} 1 & 0 & 0 & \times & \times & 0 & 0 & \times \\ 0 & 1 & 0 & \times & \times & 0 & 0 & \times \\ 0 & 0 & 1 & \times & \times & 0 & 0 & \times \\ 0 & 0 & 0 & 0 & 0 & 1 & 0 & \times \\ 0 & 0 & 0 & 0 & 0 & 0 & 1 & \times \\ 0 & 0 & 0 & 0 & 0 & 0 & 0 & 0 \\ 0 & 0 & 0 & 0 & 0 & 0 & 0 & 0 \end{pmatrix}$$

if the last row consists of zeros, or

$$\begin{pmatrix} 1 & 0 & 0 & \times & \times & 0 & 0 & \times & 0 & \times \\ 0 & 1 & 0 & \times & \times & 0 & 0 & \times & 0 & \times \\ 0 & 0 & 1 & \times & \times & 0 & 0 & \times & 0 & \times \\ 0 & 0 & 0 & 0 & 0 & 1 & 0 & \times & 0 & \times \\ 0 & 0 & 0 & 0 & 0 & 0 & 1 & \times & \times & 0 \\ 0 & 0 & 0 & 0 & 0 & 0 & 0 & 0 & 1 & \times \end{pmatrix}$$

if the last row contains a pivot.

The following proposition shows that every matrix can be converted to a reduced row echelon form using row operations.

Proposition 7.9. *Given any $m \times n$ matrix A, there is a sequence of row operations $E_1, \ldots, E_k$ such that if $P = E_k \cdots E_1$, then $U = PA$ is a reduced row echelon matrix.*

Proof. We proceed by induction on m. If $m = 1$, then either all entries on this row are zero, so $A = 0$, or if a_j is the first nonzero entry in A, let $P = (a_j^{-1})$ (a 1×1 matrix); clearly, PA is a reduced row echelon matrix.

Let us now assume that $m \geq 2$. If $A = 0$, we are done, so let us assume that $A \neq 0$. Since $A \neq 0$, there is a leftmost column j which is nonzero, so pick any pivot $\pi = a_{ij}$ in the jth column, permute row i and row 1 if necessary, multiply the new first row by π^{-1}, and clear out the other entries in column j by subtracting suitable multiples of row 1. At the end of this process, we have a matrix A_1 that has the following shape:

$$A_1 = \begin{pmatrix} 0 & \cdots & 0 & 1 & * & \cdots & * \\ 0 & \cdots & 0 & 0 & * & \cdots & * \\ \vdots & & \vdots & \vdots & \vdots & & \vdots \\ 0 & \cdots & 0 & 0 & * & \cdots & * \end{pmatrix},$$

where $*$ stands for an arbitrary scalar, or more concisely

$$A_1 = \begin{pmatrix} 0 & 1 & B \\ 0 & 0 & D \end{pmatrix},$$

where D is a $(m-1) \times (n-j)$ matrix (and B is a $1 \times n - j$ matrix). If $j = n$, we are done. Otherwise, by the induction hypothesis applied to D, there is a sequence of row operations that converts D to a reduced row echelon matrix R', and these row operations do not affect the first row of A_1, which means that A_1 is reduced to a matrix of the form

$$R = \begin{pmatrix} 0 & 1 & B \\ 0 & 0 & R' \end{pmatrix}.$$

Because R' is a reduced row echelon matrix, the matrix R satisfies Conditions (a) and (b) of the reduced row echelon form. Finally, the entries above all pivots in R' can be cleared out by subtracting suitable multiples of the rows of R' containing a pivot. The resulting matrix also satisfies Condition (c), and the induction step is complete. $\quad\square$

Remark: There is a `Matlab` function named `rref` that converts any matrix to its reduced row echelon form.

If A is any matrix and if R is a reduced row echelon form of A, the second part of Proposition 7.8 can be sharpened a little, since the structure of a reduced row echelon matrix makes it clear that its rank is equal to the number of pivots.

Proposition 7.10. *The rank of a matrix A is equal to the number of pivots in its rref R.*

7.11 RREF, Free Variables, and Homogenous Linear Systems

Given a system of the form $Ax = b$, we can apply the reduction procedure to the augmented matrix (A, b) to obtain a reduced row echelon matrix (A', b') such that the system $A'x = b'$ has the same solutions as the original system $Ax = b$. The advantage of the reduced system $A'x = b'$ is that there is a simple test to check whether this system is solvable, and to find its solutions if it is solvable.

Indeed, if any row of the matrix A' is zero and if the corresponding entry in b' is nonzero, then it is a pivot and we have the "equation"

$$0 = 1,$$

which means that the system $A'x = b'$ has no solution. On the other hand, if there is no pivot in b', then for every row i in which $b'_i \neq 0$, there is some column j in A' where the entry on row i is 1 (a pivot). Consequently, we can assign arbitrary values to the variable x_k if column k does not contain a pivot, and then solve for the pivot variables.

For example, if we consider the reduced row echelon matrix

$$(A', b') = \begin{pmatrix} 1 & 6 & 0 & 1 & 0 \\ 0 & 0 & 1 & 2 & 0 \\ 0 & 0 & 0 & 0 & 1 \end{pmatrix},$$

there is no solution to $A'x = b'$ because the third equation is $0 = 1$. On the other hand, the reduced system

$$(A', b') = \begin{pmatrix} 1 & 6 & 0 & 1 & 1 \\ 0 & 0 & 1 & 2 & 3 \\ 0 & 0 & 0 & 0 & 0 \end{pmatrix}$$

has solutions. We can pick the variables x_2, x_4 corresponding to nonpivot columns arbitrarily, and then solve for x_3 (using the second equation) and x_1 (using the first equation).

The above reasoning proves the following theorem:

Theorem 7.5. *Given any system $Ax = b$ where A is a $m \times n$ matrix, if the augmented matrix (A, b) is a reduced row echelon matrix, then the system $Ax = b$ has a solution iff there is no pivot in b. In that case, an arbitrary value can be assigned to the variable x_j if column j does not contain a pivot.*

Definition 7.5. Nonpivot variables are often called *free variables*.

Putting Proposition 7.9 and Theorem 7.5 together we obtain a criterion to decide whether a system $Ax = b$ has a solution: Convert the augmented system (A, b) to a row reduced echelon matrix (A', b') and check whether b' has no pivot.

Remark: When writing a program implementing row reduction, we may stop when the last column of the matrix A is reached. In this case, the test whether the system $Ax = b$ is solvable is that the row-reduced matrix A' has no zero row of index $i > r$ such that $b'_i \neq 0$ (where r is the number of pivots, and b' is the row-reduced right-hand side).

If we have a *homogeneous system* $Ax = 0$, which means that $b = 0$, of course $x = 0$ is always a solution, but Theorem 7.5 implies that if the system $Ax = 0$ has more variables than equations, then it has some nonzero solution (we call it a *nontrivial solution*).

Proposition 7.11. *Given any homogeneous system $Ax = 0$ of m equations* in n variables, if $m < n$, then there is a nonzero vector $x \in \mathbb{R}^n$ such that $Ax = 0$.

Proof. Convert the matrix A to a reduced row echelon matrix A'. We know that $Ax = 0$ iff $A'x = 0$. If r is the number of pivots of A', we must have $r \leq m$, so by Theorem 7.5 we may assign arbitrary values to $n - r > 0$ nonpivot variables and we get nontrivial solutions. □

Theorem 7.5 can also be used to characterize when a square matrix is invertible. First, note the following simple but important fact:

If a square $n \times n$ matrix A is a row reduced echelon matrix, then either A is the identity or the bottom row of A is zero.

Proposition 7.12. *Let A be a square matrix of dimension n. The following conditions are equivalent:*

(a) The matrix A can be reduced to the identity by a sequence of elementary row operations.

(b) The matrix A is a product of elementary matrices.

(c) The matrix A is invertible.

(d) The system of homogeneous equations $Ax = 0$ has only the trivial solution $x = 0$.

Proof. First we prove that (a) implies (b). If (a) can be reduced to the identity by a sequence of row operations $E_1, \ldots, E_p$, this means that $E_p \cdots E_1 A = I$. Since each E_i is invertible, we get

$$A = E_1^{-1} \cdots E_p^{-1},$$

where each E_i^{-1} is also an elementary row operation, so (b) holds. Now if (b) holds, since elementary row operations are invertible, A is invertible and (c) holds. If A is invertible, we already observed that the homogeneous system $Ax = 0$ has only the trivial solution $x = 0$, because from $Ax = 0$, we get $A^{-1}Ax = A^{-1}0$; that is, $x = 0$. It remains to prove that (d) implies (a) and for this we prove the contrapositive: if (a) does not hold, then (d) does not hold.

Using our basic observation about reducing square matrices, if A does not reduce to the identity, then A reduces to a row echelon matrix A' whose bottom row is zero. Say $A' = PA$, where P is a product of elementary row operations. Because the bottom row of A' is zero, the system $A'x = 0$ has at most $n - 1$ nontrivial equations, and by Proposition 7.11, this system has a nontrivial solution x. But then, $Ax = P^{-1}A'x = 0$ with $x \neq 0$, contradicting the fact that the system $Ax = 0$ is assumed to have only the trivial solution. Therefore, (d) implies (a) and the proof is complete. $\quad\square$

Proposition 7.12 yields a method for computing the inverse of an invertible matrix A: reduce A to the identity using elementary row operations, obtaining

$$E_p \cdots E_1 A = I.$$

Multiplying both sides by A^{-1} we get

$$A^{-1} = E_p \cdots E_1.$$

From a practical point of view, we can build up the product $E_p \cdots E_1$ by reducing to row echelon form the augmented $n \times 2n$ matrix (A, I_n) obtained by adding the n columns of the identity matrix to A. This is just another way of performing the Gauss–Jordan procedure.

Here is an example: let us find the inverse of the matrix

$$A = \begin{pmatrix} 5 & 4 \\ 6 & 5 \end{pmatrix}.$$

We form the 2×4 block matrix

$$(A, I) = \begin{pmatrix} 5 & 4 & 1 & 0 \\ 6 & 5 & 0 & 1 \end{pmatrix}$$

and apply elementary row operations to reduce A to the identity. For example:

$$(A, I) = \begin{pmatrix} 5 & 4 & 1 & 0 \\ 6 & 5 & 0 & 1 \end{pmatrix} \longrightarrow \begin{pmatrix} 5 & 4 & 1 & 0 \\ 1 & 1 & -1 & 1 \end{pmatrix}$$

by subtracting row 1 from row 2,

$$\begin{pmatrix} 5 & 4 & 1 & 0 \\ 1 & 1 & -1 & 1 \end{pmatrix} \longrightarrow \begin{pmatrix} 1 & 0 & 5 & -4 \\ 1 & 1 & -1 & 1 \end{pmatrix}$$

by subtracting $4 \times$ row 2 from row 1,

$$\begin{pmatrix} 1 & 0 & 5 & -4 \\ 1 & 1 & -1 & 1 \end{pmatrix} \longrightarrow \begin{pmatrix} 1 & 0 & 5 & -4 \\ 0 & 1 & -6 & 5 \end{pmatrix} = (I, A^{-1}),$$

by subtracting row 1 from row 2. Thus

$$A^{-1} = \begin{pmatrix} 5 & -4 \\ -6 & 5 \end{pmatrix}.$$

Proposition 7.12 can also be used to give an elementary proof of the fact that if a square matrix A has a left inverse B (resp. a right inverse B), so that $BA = I$ (resp. $AB = I$), then A is invertible and $A^{-1} = B$. This is an interesting exercise, try it!

7.12 Uniqueness of RREF Form

For the sake of completeness, we prove that the reduced row echelon form of a matrix is unique. The neat proof given below is borrowed and adapted from W. Kahan.

Proposition 7.13. *Let A be any $m \times n$ matrix. If U and V are two reduced row echelon matrices obtained from A by applying two sequences of elementary row operations $E_1, \ldots, E_p$ and $F_1, \ldots, F_q$, so that*

$$U = E_p \cdots E_1 A \quad and \quad V = F_q \cdots F_1 A,$$

then $U = V$ and $E_p \cdots E_1 = F_q \cdots F_1$. In other words, the reduced row echelon form of any matrix is unique.

Proof. Let

$$C = E_p \cdots E_1 F_1^{-1} \cdots F_q^{-1}$$

so that

$$U = CV \quad and \quad V = C^{-1}U.$$

We prove by induction on n that $U = V$ (and $C = I$).

Let ℓ_j denote the jth column of the identity matrix I_n, and let $u_j = U\ell_j$, $v_j = V\ell_j$, $c_j = C\ell_j$, and $a_j = A\ell_j$, be the jth column of U, V, C, and A respectively.

First I claim that $u_j = 0$ iff $v_j = 0$ iff $a_j = 0$.

Indeed, if $v_j = 0$, then (because $U = CV$) $u_j = Cv_j = 0$, and if $u_j = 0$, then $v_j = C^{-1}u_j = 0$. Since $U = E_p \cdots E_1 A$, we also get $a_j = 0$ iff $u_j = 0$.

Therefore, we may simplify our task by striking out columns of zeros from U, V, and A, since they will have corresponding indices. We still use n to denote the number of columns of A. Observe that because U and V are reduced row echelon matrices with no zero columns, we must have $u_1 = v_1 = \ell_1$.

Claim. If U and V are reduced row echelon matrices without zero columns such that $U = CV$, for all $k \geq 1$, if $k \leq n$, then ℓ_k occurs in U iff ℓ_k occurs in V, and if ℓ_k does occur in U, then

(1) ℓ_k occurs for the same column index j_k in both U and V;
(2) the first j_k columns of U and V match;
(3) the subsequent columns in U and V (of column index $> j_k$) whose coordinates of index $k+1$ through m are all equal to 0 also match. Let n_k be the rightmost index of such a column, with $n_k = j_k$ if there is none.

(4) the first n_k columns of C match the first n_k columns of I_n.

We prove this claim by induction on k.

For the base case $k = 1$, we already know that $u_1 = v_1 = \ell_1$. We also have

$$c_1 = C\ell_1 = Cv_1 = u_1 = \ell_1.$$

If $v_j = \lambda\ell_1$ for some $\lambda \in \mathbb{R}$, then

$$u_j = U\ell_j = CV\ell_j = Cv_j = \lambda C\ell_1 = \lambda c_1 = \lambda\ell_1 = v_j.$$

A similar argument using C^{-1} shows that if $u_j = \lambda\ell_1$, then $v_j = u_j$. Therefore, all the columns of U and V proportional to ℓ_1 match, which establishes the base case. Observe that if ℓ_2 appears in U, then it must appear in both U and V for the same index, and if not then $n_1 = n$ and $U = V$.

Next us now prove the induction step. If $n_k = n$, then $U = V$ and we are done. Otherwise, ℓ_{k+1} appears in both U and V, in which case, by (2) and (3) of the induction hypothesis, it appears in both U and V for the same index, say j_{k+1}. Thus, $u_{j_{k+1}} = v_{j_{k+1}} = \ell_{k+1}$. It follows that

$$c_{k+1} = C\ell_{k+1} = Cv_{j_{k+1}} = u_{j_{k+1}} = \ell_{k+1},$$

so the first j_{k+1} columns of C match the first j_{k+1} columns of I_n.

Consider any subsequent column v_j (with $j > j_{k+1}$) whose elements beyond the $(k+1)$th all vanish. Then v_j is a linear combination of columns of V to the left of v_j, so

$$u_j = Cv_j = v_j$$

because the first $k+1$ columns of C match the first column of I_n. Similarly, any subsequent column u_j (with $j > j_{k+1}$) whose elements beyond the $(k+1)$th all vanish is equal to v_j. Therefore, all the subsequent columns in U and V (of index $> j_{k+1}$) whose elements beyond the $(k+1)$th all vanish also match, so the first n_{k+1} columns of C match the first n_{k+1} columns of C, which completes the induction hypothesis.

We can now prove that $U = V$ (recall that we may assume that U and V have no zero columns). We noted earlier that $u_1 = v_1 = \ell_1$, so there is a largest $k \leq n$ such that ℓ_k occurs in U. Then the previous claim implies that all the columns of U and V match, which means that $U = V$. $\square$

The reduction to row echelon form also provides a method to describe the set of solutions of a linear system of the form $Ax = b$.

7.13   Solving Linear Systems Using RREF

First we have the following simple result.

Proposition 7.14. *Let A be any $m \times n$ matrix and let $b \in \mathbb{R}^m$ be any vector. If the system $Ax = b$ has a solution, then the set Z of all solutions of this system is the set*

$$Z = x_0 + \mathrm{Ker}\,(A) = \{x_0 + x \mid Ax = 0\},$$

where $x_0 \in \mathbb{R}^n$ is any solution of the system $Ax = b$, which means that $Ax_0 = b$ (x_0 is called a special solution), and where $\mathrm{Ker}\,(A) = \{x \in \mathbb{R}^n \mid Ax = 0\}$, the set of solutions of the homogeneous system associated with $Ax = b$.

Proof. Assume that the system $Ax = b$ is solvable and let x_0 and x_1 be any two solutions so that $Ax_0 = b$ and $Ax_1 = b$. Subtracting the first equation from the second, we get

$$A(x_1 - x_0) = 0,$$

which means that $x_1 - x_0 \in \mathrm{Ker}\,(A)$. Therefore, $Z \subseteq x_0 + \mathrm{Ker}\,(A)$, where x_0 is a special solution of $Ax = b$. Conversely, if $Ax_0 = b$, then for any $z \in \mathrm{Ker}\,(A)$, we have $Az = 0$, and so

$$A(x_0 + z) = Ax_0 + Az = b + 0 = b,$$

which shows that $x_0 + \mathrm{Ker}\,(A) \subseteq Z$. Therefore, $Z = x_0 + \mathrm{Ker}\,(A)$. $\square$

Given a linear system $Ax = b$, reduce the augmented matrix (A, b) to its row echelon form (A', b'). As we showed before, the system $Ax = b$ has a solution iff b' contains no pivot. Assume that this is the case. Then, if (A', b') has r pivots, which means that A' has r pivots since b' has no pivot, we know that the first r columns of I_m appear in A'.

We can permute the columns of A' and renumber the variables in x correspondingly so that the first r columns of I_m match the first r columns of A', and then our reduced echelon matrix is of the form (R, b') with

$$R = \begin{pmatrix} I_r & F \\ 0_{m-r,r} & 0_{m-r,n-r} \end{pmatrix}$$

and

$$b' = \begin{pmatrix} d \\ 0_{m-r} \end{pmatrix},$$

where F is a $r \times (n - r)$ matrix and $d \in \mathbb{R}^r$. Note that R has $m - r$ zero rows.

Then because

$$\begin{pmatrix} I_r & F \\ 0_{m-r,r} & 0_{m-r,n-r} \end{pmatrix} \begin{pmatrix} d \\ 0_{n-r} \end{pmatrix} = \begin{pmatrix} d \\ 0_{m-r} \end{pmatrix} = b',$$

we see that

$$x_0 = \begin{pmatrix} d \\ 0_{n-r} \end{pmatrix}$$

is a special solution of $Rx = b'$, and thus to $Ax = b$. In other words, we get a special solution by assigning the first r components of b' to the pivot variables and setting the nonpivot variables (the *free variables*) to zero.

Here is an example of the preceding construction taken from Kumpel and Thorpe [Kumpel and Thorpe (1983)]. The linear system

$$x_1 - x_2 + x_3 + x_4 - 2x_5 = -1$$
$$-2x_1 + 2x_2 - x_3 + x_5 = 2$$
$$x_1 - x_2 + 2x_3 + 3x_4 - 5x_5 = -1,$$

is represented by the augmented matrix

$$(A, b) = \begin{pmatrix} 1 & -1 & 1 & 1 & -2 & -1 \\ -2 & 2 & -1 & 0 & 1 & 2 \\ 1 & -1 & 2 & 3 & -5 & -1 \end{pmatrix},$$

where A is a 3×5 matrix. The reader should find that the row echelon form of this system is

$$(A', b') = \begin{pmatrix} 1 & -1 & 0 & -1 & 1 & -1 \\ 0 & 0 & 1 & 2 & -3 & 0 \\ 0 & 0 & 0 & 0 & 0 & 0 \end{pmatrix}.$$

The 3×5 matrix A' has rank 2. We permute the second and third columns (which is equivalent to interchanging variables x_2 and x_3) to form

$$R = \begin{pmatrix} I_2 & F \\ 0_{1,2} & 0_{1,3} \end{pmatrix}, \qquad F = \begin{pmatrix} -1 & -1 & 1 \\ 0 & 2 & -3 \end{pmatrix}.$$

Then a special solution to this linear system is given by

$$x_0 = \begin{pmatrix} d \\ 0_3 \end{pmatrix} = \begin{pmatrix} -1 \\ 0 \\ 0_3 \end{pmatrix}.$$

We can also find a basis of the kernel (nullspace) of A using F. If $x = (u, v)$ is in the kernel of A, with $u \in \mathbb{R}^r$ and $v \in \mathbb{R}^{n-r}$, then x is also in the kernel of R, which means that $Rx = 0$; that is,

$$\begin{pmatrix} I_r & F \\ 0_{m-r,r} & 0_{m-r,n-r} \end{pmatrix} \begin{pmatrix} u \\ v \end{pmatrix} = \begin{pmatrix} u + Fv \\ 0_{m-r} \end{pmatrix} = \begin{pmatrix} 0_r \\ 0_{m-r} \end{pmatrix}.$$

Therefore, $u = -Fv$, and $\text{Ker}(A)$ consists of all vectors of the form

$$\begin{pmatrix} -Fv \\ v \end{pmatrix} = \begin{pmatrix} -F \\ I_{n-r} \end{pmatrix} v,$$

for any arbitrary $v \in \mathbb{R}^{n-r}$. It follows that the $n - r$ columns of the matrix

$$N = \begin{pmatrix} -F \\ I_{n-r} \end{pmatrix}$$

form a basis of the kernel of A. This is because N contains the identity matrix I_{n-r} as a submatrix, so the columns of N are linearly independent. In summary, if $N^1, \ldots, N^{n-r}$ are the columns of N, then the general solution of the equation $Ax = b$ is given by

$$x = \begin{pmatrix} d \\ 0_{n-r} \end{pmatrix} + x_{r+1} N^1 + \cdots + x_n N^{n-r},$$

where $x_{r+1}, \ldots, x_n$ are the free variables; that is, the nonpivot variables.

Going back to our example from Kumpel and Thorpe [Kumpel and Thorpe (1983)], we see that

$$N = \begin{pmatrix} -F \\ I_3 \end{pmatrix} = \begin{pmatrix} 1 & 1 & -1 \\ 0 & -2 & -3 \\ 1 & 0 & 0 \\ 0 & 1 & 0 \\ 0 & 0 & 1 \end{pmatrix},$$

and that the general solution is given by

$$x = \begin{pmatrix} -1 \\ 0 \\ 0 \\ 0 \\ 0 \end{pmatrix} + x_3 \begin{pmatrix} 1 \\ 0 \\ 1 \\ 0 \\ 0 \end{pmatrix} + x_4 \begin{pmatrix} 1 \\ -2 \\ 0 \\ 1 \\ 0 \end{pmatrix} + x_5 \begin{pmatrix} -1 \\ -3 \\ 0 \\ 0 \\ 1 \end{pmatrix}.$$

In the general case where the columns corresponding to pivots are mixed with the columns corresponding to free variables, we find the special solution as follows. Let $i_1 < \cdots < i_r$ be the indices of the columns corresponding to pivots. Assign b'_k to the pivot variable x_{i_k} for $k = 1, \ldots, r$, and set all other variables to 0. To find a basis of the kernel, we form the $n - r$ vectors N^k obtained as follows. Let $j_1 < \cdots < j_{n-r}$ be the indices of the columns corresponding to free variables. For every column j_k corresponding to a free variable ($1 \leq k \leq n-r$), form the vector N^k defined so that the entries $N^k_{i_1}, \ldots, N^k_{i_r}$ are equal to the negatives of the first r entries in column j_k (flip the sign of these entries); let $N^k_{j_k} = 1$, and set all other entries to zero. Schematically, if the column of index j_k (corresponding to the free variable x_{j_k}) is

$$
\begin{pmatrix}
\alpha_1 \\
\vdots \\
\alpha_r \\
0 \\
\vdots \\
0
\end{pmatrix},
$$

then the vector N^k is given by

$$
\begin{matrix}
1 \\
\vdots \\
i_1 - 1 \\
i_1 \\
i_1 + 1 \\
\vdots \\
i_r - 1 \\
i_r \\
i_r + 1 \\
\vdots \\
j_k - 1 \\
j_k \\
j_k + 1 \\
\vdots \\
n
\end{matrix}
\begin{pmatrix}
0 \\
\vdots \\
0 \\
-\alpha_1 \\
0 \\
\vdots \\
0 \\
-\alpha_r \\
0 \\
\vdots \\
0 \\
1 \\
0 \\
\vdots \\
0
\end{pmatrix}.
$$

The presence of the 1 in position j_k guarantees that $N^1, \ldots, N^{n-r}$ are linearly independent.

As an illustration of the above method, consider the problem of finding a basis of the subspace V of $n \times n$ matrices $A \in \mathrm{M}_n(\mathbb{R})$ satisfying the following properties:

(1) The sum of the entries in every row has the same value (say c_1);
(2) The sum of the entries in every column has the same value (say c_2).

It turns out that $c_1 = c_2$ and that the $2n - 2$ equations corresponding to the above conditions are linearly independent. We leave the proof of these facts as an interesting exercise. It can be shown using the duality theorem (Theorem 10.1) that the dimension of the space V of matrices satisfying the above equations is $n^2 - (2n - 2)$. Let us consider the case $n = 4$. There are 6 equations, and the space V has dimension 10. The equations are

$$a_{11} + a_{12} + a_{13} + a_{14} - a_{21} - a_{22} - a_{23} - a_{24} = 0$$
$$a_{21} + a_{22} + a_{23} + a_{24} - a_{31} - a_{32} - a_{33} - a_{34} = 0$$
$$a_{31} + a_{32} + a_{33} + a_{34} - a_{41} - a_{42} - a_{43} - a_{44} = 0$$
$$a_{11} + a_{21} + a_{31} + a_{41} - a_{12} - a_{22} - a_{32} - a_{42} = 0$$
$$a_{12} + a_{22} + a_{32} + a_{42} - a_{13} - a_{23} - a_{33} - a_{43} = 0$$
$$a_{13} + a_{23} + a_{33} + a_{43} - a_{14} - a_{24} - a_{34} - a_{44} = 0,$$

and the corresponding matrix is

$$A = \begin{pmatrix} 1 & 1 & 1 & 1 & -1 & -1 & -1 & -1 & 0 & 0 & 0 & 0 & 0 & 0 & 0 & 0 \\ 0 & 0 & 0 & 0 & 1 & 1 & 1 & 1 & -1 & -1 & -1 & -1 & 0 & 0 & 0 & 0 \\ 0 & 0 & 0 & 0 & 0 & 0 & 0 & 0 & 1 & 1 & 1 & 1 & -1 & -1 & -1 & -1 \\ 1 & -1 & 0 & 0 & 1 & -1 & 0 & 0 & 1 & -1 & 0 & 0 & 1 & -1 & 0 & 0 \\ 0 & 1 & -1 & 0 & 0 & 1 & -1 & 0 & 0 & 1 & -1 & 0 & 0 & 1 & -1 & 0 \\ 0 & 0 & 1 & -1 & 0 & 0 & 1 & -1 & 0 & 0 & 1 & -1 & 0 & 0 & 1 & -1 \end{pmatrix}.$$

The result of performing the reduction to row echelon form yields the following matrix in rref:

$$U = \begin{pmatrix} 1 & 0 & 0 & 0 & 0 & -1 & -1 & -1 & 0 & -1 & -1 & -1 & 2 & 1 & 1 & 1 \\ 0 & 1 & 0 & 0 & 0 & 1 & 0 & 0 & 0 & 1 & 0 & 0 & -1 & 0 & -1 & -1 \\ 0 & 0 & 1 & 0 & 0 & 0 & 1 & 0 & 0 & 0 & 1 & 0 & -1 & -1 & 0 & -1 \\ 0 & 0 & 0 & 1 & 0 & 0 & 0 & 1 & 0 & 0 & 0 & 1 & -1 & -1 & -1 & 0 \\ 0 & 0 & 0 & 0 & 1 & 1 & 1 & 1 & 0 & 0 & 0 & 0 & -1 & -1 & -1 & -1 \\ 0 & 0 & 0 & 0 & 0 & 0 & 0 & 0 & 1 & 1 & 1 & 1 & -1 & -1 & -1 & -1 \end{pmatrix}$$

The list *pivlist* of indices of the pivot variables and the list *freelist* of indices of the free variables is given by

$$pivlist = (1, 2, 3, 4, 5, 9),$$
$$freelist = (\mathbf{6, 7, 8, 10, 11, 12, 13, 14, 15, 16}).$$

After applying the algorithm to find a basis of the kernel of U, we find the following 16×10 matrix

$$
BK = \begin{pmatrix}
1 & 1 & 1 & 1 & 1 & 1 & -2 & -1 & -1 & -1 \\
-1 & 0 & 0 & -1 & 0 & 0 & 1 & 0 & 1 & 1 \\
0 & -1 & 0 & 0 & -1 & 0 & 1 & 1 & 0 & 1 \\
0 & 0 & -1 & 0 & 0 & -1 & 1 & 1 & 1 & 0 \\
-1 & -1 & -1 & 0 & 0 & 0 & 1 & 1 & 1 & 1 \\
1 & 0 & 0 & 0 & 0 & 0 & 0 & 0 & 0 & 0 \\
0 & 1 & 0 & 0 & 0 & 0 & 0 & 0 & 0 & 0 \\
0 & 0 & 1 & 0 & 0 & 0 & 0 & 0 & 0 & 0 \\
0 & 0 & 0 & -1 & -1 & -1 & 1 & 1 & 1 & 1 \\
0 & 0 & 0 & 1 & 0 & 0 & 0 & 0 & 0 & 0 \\
0 & 0 & 0 & 0 & 1 & 0 & 0 & 0 & 0 & 0 \\
0 & 0 & 0 & 0 & 0 & 1 & 0 & 0 & 0 & 0 \\
0 & 0 & 0 & 0 & 0 & 0 & 1 & 0 & 0 & 0 \\
0 & 0 & 0 & 0 & 0 & 0 & 0 & 1 & 0 & 0 \\
0 & 0 & 0 & 0 & 0 & 0 & 0 & 0 & 1 & 0 \\
0 & 0 & 0 & 0 & 0 & 0 & 0 & 0 & 0 & 1
\end{pmatrix}.
$$

The reader should check that that in each column j of BK, the lowest bold 1 belongs to the row whose index is the jth element in *freelist*, and that in each column j of BK, the signs of the entries whose indices belong to *pivlist* are the flipped signs of the 6 entries in the column U corresponding to the jth index in *freelist*. We can now read off from BK the 4×4 matrices that form a basis of V: every column of BK corresponds to a matrix whose rows have been concatenated. We get the following 10 matrices:

$$
M_1 = \begin{pmatrix}
1 & -1 & 0 & 0 \\
-1 & 1 & 0 & 0 \\
0 & 0 & 0 & 0 \\
0 & 0 & 0 & 0
\end{pmatrix}, \quad
M_2 = \begin{pmatrix}
1 & 0 & -1 & 0 \\
-1 & 0 & 1 & 0 \\
0 & 0 & 0 & 0 \\
0 & 0 & 0 & 0
\end{pmatrix}, \quad
M_3 = \begin{pmatrix}
1 & 0 & 0 & -1 \\
-1 & 0 & 0 & 1 \\
0 & 0 & 0 & 0 \\
0 & 0 & 0 & 0
\end{pmatrix},
$$

$$
M_4 = \begin{pmatrix}
1 & -1 & 0 & 0 \\
0 & 0 & 0 & 0 \\
-1 & 1 & 0 & 0 \\
0 & 0 & 0 & 0
\end{pmatrix}, \quad
M_5 = \begin{pmatrix}
1 & 0 & -1 & 0 \\
0 & 0 & 0 & 0 \\
-1 & 0 & 1 & 0 \\
0 & 0 & 0 & 0
\end{pmatrix}, \quad
M_6 = \begin{pmatrix}
1 & 0 & 0 & -1 \\
0 & 0 & 0 & 0 \\
-1 & 0 & 0 & 1 \\
0 & 0 & 0 & 0
\end{pmatrix},
$$

$$M_7 = \begin{pmatrix} -2 & 1 & 1 & 1 \\ 1 & 0 & 0 & 0 \\ 1 & 0 & 0 & 0 \\ 1 & 0 & 0 & 0 \end{pmatrix}, \quad M_8 = \begin{pmatrix} -1 & 0 & 1 & 1 \\ 1 & 0 & 0 & 0 \\ 1 & 0 & 0 & 0 \\ 0 & 1 & 0 & 0 \end{pmatrix}, \quad M_9 = \begin{pmatrix} -1 & 1 & 0 & 1 \\ 1 & 0 & 0 & 0 \\ 1 & 0 & 0 & 0 \\ 0 & 0 & 1 & 0 \end{pmatrix},$$

$$M_{10} = \begin{pmatrix} -1 & 1 & 1 & 0 \\ 1 & 0 & 0 & 0 \\ 1 & 0 & 0 & 0 \\ 0 & 0 & 0 & 1 \end{pmatrix}.$$

Recall that a *magic square* is a square matrix that satisfies the two conditions about the sum of the entries in each row and in each column to be the same number, and also the additional two constraints that the main descending and the main ascending diagonals add up to this common number. Furthermore, the entries are also required to be positive integers. For $n = 4$, the additional two equations are

$$a_{22} + a_{33} + a_{44} - a_{12} - a_{13} - a_{14} = 0$$
$$a_{41} + a_{32} + a_{23} - a_{11} - a_{12} - a_{13} = 0,$$

and the 8 equations stating that a matrix is a magic square are linearly independent. Again, by running row elimination, we get a basis of the "generalized magic squares" whose entries are not restricted to be positive integers. We find a basis of 8 matrices. For $n = 3$, we find a basis of 3 matrices.

A magic square is said to be *normal* if its entries are precisely the integers $1, 2 \ldots, n^2$. Then since the sum of these entries is

$$1 + 2 + 3 + \cdots + n^2 = \frac{n^2(n^2 + 1)}{2},$$

and since each row (and column) sums to the same number, this common value (the *magic sum*) is

$$\frac{n(n^2 + 1)}{2}.$$

It is easy to see that there are no normal magic squares for $n = 2$. For $n = 3$, the magic sum is 15, for $n = 4$, it is 34, and for $n = 5$, it is 65.

In the case $n = 3$, we have the additional condition that the rows and columns add up to 15, so we end up with a solution parametrized by two numbers x_1, x_2; namely,

$$\begin{pmatrix} x_1 + x_2 - 5 & 10 - x_2 & 10 - x_1 \\ 20 - 2x_1 - x_2 & 5 & 2x_1 + x_2 - 10 \\ x_1 & x_2 & 15 - x_1 - x_2 \end{pmatrix}.$$

Thus, in order to find a normal magic square, we have the additional inequality constraints

$$x_1 + x_2 > 5$$
$$x_1 < 10$$
$$x_2 < 10$$
$$2x_1 + x_2 < 20$$
$$2x_1 + x_2 > 10$$
$$x_1 > 0$$
$$x_2 > 0$$
$$x_1 + x_2 < 15,$$

and all 9 entries in the matrix must be distinct. After a tedious case analysis, we discover the remarkable fact that there is a unique normal magic square (up to rotations and reflections):

$$\begin{pmatrix} 2 & 7 & 6 \\ 9 & 5 & 1 \\ 4 & 3 & 8 \end{pmatrix}.$$

It turns out that there are 880 different normal magic squares for $n = 4$, and $275{,}305{,}224$ normal magic squares for $n = 5$ (up to rotations and reflections). Even for $n = 4$, it takes a fair amount of work to enumerate them all! Finding the number of magic squares for $n > 5$ is an open problem!

7.14   Elementary Matrices and Columns Operations

Instead of performing elementary row operations on a matrix A, we can perform elementary columns operations, which means that we multiply A by elementary matrices on the *right*. As elementary row and column operations, $P(i, k)$, $E_{i,j;\beta}$, $E_{i,\lambda}$ perform the following actions:

(1) As a row operation, $P(i, k)$ permutes row i and row k.
(2) As a column operation, $P(i, k)$ permutes column i and column k.
(3) The inverse of $P(i, k)$ is $P(i, k)$ itself.
(4) As a row operation, $E_{i,j;\beta}$ adds β times row j to row i.
(5) As a column operation, $E_{i,j;\beta}$ adds β times column i to column j (note the switch in the indices).
(6) The inverse of $E_{i,j;\beta}$ is $E_{i,j;-\beta}$.

(7) As a row operation, $E_{i,\lambda}$ multiplies row i by λ.

(8) As a column operation, $E_{i,\lambda}$ multiplies column i by λ.

(9) The inverse of $E_{i,\lambda}$ is $E_{i,\lambda^{-1}}$.

We can define the notion of a reduced column echelon matrix and show that every matrix can be reduced to a unique reduced column echelon form. Now given any $m \times n$ matrix A, if we first convert A to its reduced row echelon form R, it is easy to see that we can apply elementary column operations that will reduce R to a matrix of the form

$$\begin{pmatrix} I_r & 0_{r,n-r} \\ 0_{m-r,r} & 0_{m-r,n-r} \end{pmatrix},$$

where r is the number of pivots (obtained during the row reduction). Therefore, for every $m \times n$ matrix A, there exist two sequences of elementary matrices $E_1, \ldots, E_p$ and $F_1, \ldots, F_q$, such that

$$E_p \cdots E_1 A F_1 \cdots F_q = \begin{pmatrix} I_r & 0_{r,n-r} \\ 0_{m-r,r} & 0_{m-r,n-r} \end{pmatrix}.$$

The matrix on the right-hand side is called the *rank normal form* of A. Clearly, r is the rank of A. As a corollary we obtain the following important result whose proof is immediate.

Proposition 7.15. *A matrix A and its transpose $A^\top$ have the same rank.*

7.15 Transvections and Dilatations ⊛

In this section we characterize the linear isomorphisms of a vector space E that leave every vector in some hyperplane fixed. These maps turn out to be the linear maps that are represented in some suitable basis by elementary matrices of the form $E_{i,j;\beta}$ (transvections) or $E_{i,\lambda}$ (dilatations). Furthermore, the transvections generate the group $\mathbf{SL}(E)$, and the dilatations generate the group $\mathbf{GL}(E)$.

Let H be any hyperplane in E, and pick some (nonzero) vector $v \in E$ such that $v \notin H$, so that

$$E = H \oplus Kv.$$

Assume that $f \colon E \to E$ is a linear isomorphism such that $f(u) = u$ for all $u \in H$, and that f is not the identity. We have

$$f(v) = h + \alpha v, \quad \text{for some } h \in H \text{ and some } \alpha \in K,$$

with $\alpha \neq 0$, because otherwise we would have $f(v) = h = f(h)$ since $h \in H$, contradicting the injectivity of f ($v \neq h$ since $v \notin H$). For any $x \in E$, if we write

$$x = y + tv, \quad \text{for some } y \in H \text{ and some } t \in K,$$

then

$$f(x) = f(y) + f(tv) = y + tf(v) = y + th + t\alpha v,$$

and since $\alpha x = \alpha y + t\alpha v$, we get

$$f(x) - \alpha x = (1 - \alpha)y + th$$
$$f(x) - x = t(h + (\alpha - 1)v).$$

Observe that if E is finite-dimensional, by picking a basis of E consisting of v and basis vectors of H, then the matrix of f is a lower triangular matrix whose diagonal entries are all 1 except the first entry which is equal to α. Therefore, $\det(f) = \alpha$.

Case 1. $\alpha \neq 1$.

We have $f(x) = \alpha x$ iff $(1 - \alpha)y + th = 0$ iff

$$y = \frac{t}{\alpha - 1} h.$$

Then if we let $w = h + (\alpha - 1)v$, for $y = (t/(\alpha - 1))h$, we have

$$x = y + tv = \frac{t}{\alpha - 1}h + tv = \frac{t}{\alpha - 1}(h + (\alpha - 1)v) = \frac{t}{\alpha - 1}w,$$

which shows that $f(x) = \alpha x$ iff $x \in Kw$. Note that $w \notin H$, since $\alpha \neq 1$ and $v \notin H$. Therefore,

$$E = H \oplus Kw,$$

and f is the identity on H and a magnification by α on the line $D = Kw$.

Definition 7.6. Given a vector space E, for any hyperplane H in E, any nonzero vector $u \in E$ such that $u \notin H$, and any scalar $\alpha \neq 0, 1$, a linear map f such that $f(x) = x$ for all $x \in H$ and $f(x) = \alpha x$ for every $x \in D = Ku$ is called a *dilatation of hyperplane* H, *direction* D, *and scale factor* α.

If π_H and π_D are the projections of E onto H and D, then we have

$$f(x) = \pi_H(x) + \alpha\pi_D(x).$$

The inverse of f is given by

$$f^{-1}(x) = \pi_H(x) + \alpha^{-1}\pi_D(x).$$

When $\alpha = -1$, we have $f^2 = \mathrm{id}$, and f is a symmetry about the hyperplane H in the direction D. This situation includes orthogonal reflections about H.

Case 2. $\alpha = 1$.

In this case,

$$f(x) - x = th,$$

that is, $f(x) - x \in Kh$ for all $x \in E$. Assume that the hyperplane H is given as the kernel of some linear form φ, and let $a = \varphi(v)$. We have $a \neq 0$, since $v \notin H$. For any $x \in E$, we have

$$\varphi(x - a^{-1}\varphi(x)v) = \varphi(x) - a^{-1}\varphi(x)\varphi(v) = \varphi(x) - \varphi(x) = 0,$$

which shows that $x - a^{-1}\varphi(x)v \in H$ for all $x \in E$. Since every vector in H is fixed by f, we get

$$\begin{aligned} x - a^{-1}\varphi(x)v &= f(x - a^{-1}\varphi(x)v) \\ &= f(x) - a^{-1}\varphi(x)f(v), \end{aligned}$$

so

$$f(x) = x + \varphi(x)(f(a^{-1}v) - a^{-1}v).$$

Since $f(z) - z \in Kh$ for all $z \in E$, we conclude that $u = f(a^{-1}v) - a^{-1}v = \beta h$ for some $\beta \in K$, so $\varphi(u) = 0$, and we have

$$f(x) = x + \varphi(x)u, \quad \varphi(u) = 0. \tag{7.9}$$

A linear map defined as above is denoted by $\tau_{\varphi,u}$.

Conversely for any linear map $f = \tau_{\varphi,u}$ given by equation (7.9), where φ is a nonzero linear form and u is some vector $u \in E$ such that $\varphi(u) = 0$, if $u = 0$, then f is the identity, so assume that $u \neq 0$. If so, we have $f(x) = x$ iff $\varphi(x) = 0$, that is, iff $x \in H$. We also claim that the inverse of f is obtained by changing u to $-u$. Actually, we check the slightly more general fact that

$$\tau_{\varphi,u} \circ \tau_{\varphi,w} = \tau_{\varphi,u+w}.$$

Indeed, using the fact that $\varphi(w) = 0$, we have

$$\begin{aligned} \tau_{\varphi,u}(\tau_{\varphi,w}(x)) &= \tau_{\varphi,w}(x) + \varphi(\tau_{\varphi,w}(x))u \\ &= \tau_{\varphi,w}(x) + (\varphi(x) + \varphi(x)\varphi(w))u \\ &= \tau_{\varphi,w}(x) + \varphi(x)u \\ &= x + \varphi(x)w + \varphi(x)u \\ &= x + \varphi(x)(u + w). \end{aligned}$$

For $v = -u$, we have $\tau_{\varphi,u+v} = \varphi_{\varphi,0} = \mathrm{id}$, so $\tau_{\varphi,u}^{-1} = \tau_{\varphi,-u}$, as claimed.

Therefore, we proved that every linear isomorphism of E that leaves every vector in some hyperplane H fixed and has the property that $f(x) - x \in H$ for all $x \in E$ is given by a map $\tau_{\varphi,u}$ as defined by Equation $(*)$, where φ is some nonzero linear form defining H and u is some vector in H. We have $\tau_{\varphi,u} = \mathrm{id}$ iff $u = 0$.

Definition 7.7. Given any hyperplane H in E, for any nonzero nonlinear form $\varphi \in E^*$ defining H (which means that $H = \mathrm{Ker}\,(\varphi)$) and any nonzero vector $u \in H$, the linear map $f = \tau_{\varphi,u}$ given by

$$\tau_{\varphi,u}(x) = x + \varphi(x)u, \quad \varphi(u) = 0,$$

for all $x \in E$ is called a *transvection of hyperplane H and direction u*. The map $f = \tau_{\varphi,u}$ leaves every vector in H fixed, and $f(x) - x \in Ku$ for all $x \in E$.

The above arguments show the following result.

Proposition 7.16. *Let $f \colon E \to E$ be a bijective linear map and assume that $f \neq \mathrm{id}$ and that $f(x) = x$ for all $x \in H$, where H is some hyperplane in E. If there is some nonzero vector $u \in E$ such that $u \notin H$ and $f(u) - u \in H$, then f is a transvection of hyperplane H; otherwise, f is a dilatation of hyperplane H.*

Proof. Using the notation as above, for some $v \notin H$, we have $f(v) = h + \alpha v$ with $\alpha \neq 0$, and write $u = y + tv$ with $y \in H$ and $t \neq 0$ since $u \notin H$. If $f(u) - u \in H$, from

$$f(u) - u = t(h + (\alpha - 1)v),$$

we get $(\alpha - 1)v \in H$, and since $v \notin H$, we must have $\alpha = 1$, and we proved that f is a transvection. Otherwise, $\alpha \neq 0, 1$, and we proved that f is a dilatation. $\qquad\square$

If E is finite-dimensional, then $\alpha = \det(f)$, so we also have the following result.

Proposition 7.17. *Let $f \colon E \to E$ be a bijective linear map of a finite-dimensional vector space E and assume that $f \neq \mathrm{id}$ and that $f(x) = x$ for all $x \in H$, where H is some hyperplane in E. If $\det(f) = 1$, then f is a transvection of hyperplane H; otherwise, f is a dilatation of hyperplane H.*

Suppose that f is a dilatation of hyperplane H and direction u, and say $\det(f) = \alpha \neq 0, 1$. Pick a basis $(u, e_2, \ldots, e_n)$ of E where $(e_2, \ldots, e_n)$ is a basis of H. Then the matrix of f is of the form

$$\begin{pmatrix} \alpha & 0 & \cdots & 0 \\ 0 & 1 & & 0 \\ \vdots & & \ddots & \vdots \\ 0 & 0 & \cdots & 1 \end{pmatrix},$$

which is an elementary matrix of the form $E_{1,\alpha}$. Conversely, it is clear that every elementary matrix of the form $E_{i,\alpha}$ with $\alpha \neq 0, 1$ is a dilatation.

Now, assume that f is a transvection of hyperplane H and direction $u \in H$. Pick some $v \notin H$, and pick some basis $(u, e_3, \ldots, e_n)$ of H, so that $(v, u, e_3, \ldots, e_n)$ is a basis of E. Since $f(v) - v \in Ku$, the matrix of f is of the form

$$\begin{pmatrix} 1 & 0 & \cdots & 0 \\ \alpha & 1 & & 0 \\ \vdots & & \ddots & \vdots \\ 0 & 0 & \cdots & 1 \end{pmatrix},$$

which is an elementary matrix of the form $E_{2,1;\alpha}$. Conversely, it is clear that every elementary matrix of the form $E_{i,j;\alpha}$ $(\alpha \neq 0)$ is a transvection.

The following proposition is an interesting exercise that requires good mastery of the elementary row operations $E_{i,j;\beta}$; see Problems 7.10 and 7.11.

Proposition 7.18. *Given any invertible $n \times n$ matrix A, there is a matrix S such that*

$$SA = \begin{pmatrix} I_{n-1} & 0 \\ 0 & \alpha \end{pmatrix} = E_{n,\alpha},$$

with $\alpha = \det(A)$, and where S is a product of elementary matrices of the form $E_{i,j;\beta}$; that is, S is a composition of transvections.

Surprisingly, every transvection is the composition of two dilatations!

Proposition 7.19. *If the field K is not of characteristic 2, then every transvection f of hyperplane H can be written as $f = d_2 \circ d_1$, where d_1, d_2 are dilatations of hyperplane H, where the direction of d_1 can be chosen arbitrarily.*

Proof. Pick some dilatation d_1 of hyperplane H and scale factor $\alpha \neq 0, 1$. Then, $d_2 = f \circ d_1^{-1}$ leaves every vector in H fixed, and $\det(d_2) = \alpha^{-1} \neq 1$. By Proposition 7.17, the linear map d_2 is a dilatation of hyperplane H, and we have $f = d_2 \circ d_1$, as claimed. $\qquad\square$

Observe that in Proposition 7.19, we can pick $\alpha = -1$; that is, every transvection of hyperplane H is the compositions of two symmetries about the hyperplane H, one of which can be picked arbitrarily.

Remark: Proposition 7.19 holds as long as $K \neq \{0, 1\}$.

The following important result is now obtained.

Theorem 7.6. *Let E be any finite-dimensional vector space over a field K of characteristic not equal to 2. Then the group $\mathbf{SL}(E)$ is generated by the transvections, and the group $\mathbf{GL}(E)$ is generated by the dilatations.*

Proof. Consider any $f \in \mathbf{SL}(E)$, and let A be its matrix in any basis. By Proposition 7.18, there is a matrix S such that

$$SA = \begin{pmatrix} I_{n-1} & 0 \\ 0 & \alpha \end{pmatrix} = E_{n,\alpha},$$

with $\alpha = \det(A)$, and where S is a product of elementary matrices of the form $E_{i,j;\beta}$. Since $\det(A) = 1$, we have $\alpha = 1$, and the result is proven. Otherwise, if f is invertible but $f \notin \mathbf{SL}(E)$, the above equation shows $E_{n,\alpha}$ is a dilatation, S is a product of transvections, and by Proposition 7.19, every transvection is the composition of two dilatations. Thus, the second result is also proven. $\qquad\square$

We conclude this section by proving that any two transvections are conjugate in $\mathbf{GL}(E)$. Let $\tau_{\varphi,u}$ $(u \neq 0)$ be a transvection and let $g \in \mathbf{GL}(E)$ be any invertible linear map. We have

$$(g \circ \tau_{\varphi,u} \circ g^{-1})(x) = g(g^{-1}(x) + \varphi(g^{-1}(x))u)$$
$$= x + \varphi(g^{-1}(x))g(u).$$

Let us find the hyperplane determined by the linear form $x \mapsto \varphi(g^{-1}(x))$. This is the set of vectors $x \in E$ such that $\varphi(g^{-1}(x)) = 0$, which holds iff $g^{-1}(x) \in H$ iff $x \in g(H)$. Therefore, $\operatorname{Ker}(\varphi \circ g^{-1}) = g(H) = H'$, and we have $g(u) \in g(H) = H'$, so $g \circ \tau_{\varphi,u} \circ g^{-1}$ is the transvection of hyperplane $H' = g(H)$ and direction $u' = g(u)$ (with $u' \in H'$).

Conversely, let $\tau_{\psi,u'}$ be some transvection $(u' \neq 0)$. Pick some vectors v, v' such that $\varphi(v) = \psi(v') = 1$, so that

$$E = H \oplus Kv = H' \oplus Kv'.$$

There is a linear map $g \in \mathbf{GL}(E)$ such that $g(u) = u'$, $g(v) = v'$, and $g(H) = H'$. To define g, pick a basis $(v, u, e_2, \ldots, e_{n-1})$ where $(u, e_2, \ldots, e_{n-1})$ is a basis of H and pick a basis $(v', u', e'_2, \ldots, e'_{n-1})$ where $(u', e'_2, \ldots, e'_{n-1})$ is a basis of H'; then g is defined so that $g(v) = v'$, $g(u) = u'$, and $g(e_i) = g(e'_i)$, for $i = 2, \ldots, n - 1$. If $n = 2$, then e_i and e'_i are missing. Then, we have

$$(g \circ \tau_{\varphi,u} \circ g^{-1})(x) = x + \varphi(g^{-1}(x))u'.$$

Now $\varphi \circ g^{-1}$ also determines the hyperplane $H' = g(H)$, so we have $\varphi \circ g^{-1} = \lambda\psi$ for some nonzero λ in K. Since $v' = g(v)$, we get

$$\varphi(v) = \varphi \circ g^{-1}(v') = \lambda\psi(v'),$$

and since $\varphi(v) = \psi(v') = 1$, we must have $\lambda = 1$. It follows that

$$(g \circ \tau_{\varphi,u} \circ g^{-1})(x) = x + \psi(x)u' = \tau_{\psi,u'}(x).$$

In summary, we proved almost all parts the following result.

Proposition 7.20. *Let E be any finite-dimensional vector space. For every transvection $\tau_{\varphi,u}$ $(u \neq 0)$ and every linear map $g \in \mathbf{GL}(E)$, the map $g \circ \tau_{\varphi,u} \circ g^{-1}$ is the transvection of hyperplane $g(H)$ and direction $g(u)$ (that is, $g \circ \tau_{\varphi,u} \circ g^{-1} = \tau_{\varphi \circ g^{-1}, g(u)}$). For every other transvection $\tau_{\psi,u'}$ $(u' \neq 0)$, there is some $g \in \mathbf{GL}(E)$ such $\tau_{\psi,u'} = g \circ \tau_{\varphi,u} \circ g^{-1}$; in other words any two transvections $(\neq \mathrm{id})$ are conjugate in $\mathbf{GL}(E)$. Moreover, if $n \geq 3$, then the linear isomorphism g as above can be chosen so that $g \in \mathbf{SL}(E)$.*

Proof. We just need to prove that if $n \geq 3$, then for any two transvections $\tau_{\varphi,u}$ and $\tau_{\psi,u'}$ $(u, u' \neq 0)$, there is some $g \in \mathbf{SL}(E)$ such that $\tau_{\psi,u'} = g \circ \tau_{\varphi,u} \circ g^{-1}$. As before, we pick a basis $(v, u, e_2, \ldots, e_{n-1})$ where $(u, e_2, \ldots, e_{n-1})$ is a basis of H, we pick a basis $(v', u', e'_2, \ldots, e'_{n-1})$ where $(u', e'_2, \ldots, e'_{n-1})$ is a basis of H', and we define g as the unique linear map such that $g(v) = v'$, $g(u) = u'$, and $g(e_i) = e'_i$, for $i = 1, \ldots, n - 1$. But in this case, both H and $H' = g(H)$ have dimension at least 2, so in any basis of H' including u', there is some basis vector e'_2 independent of u', and we can rescale e'_2 in such a way that the matrix of g over the two bases has determinant $+1$. $\square$

7.16   Summary

The main concepts and results of this chapter are listed below:

- One does not solve (large) linear systems by computing determinants.
- *Upper-triangular* (*lower-triangular*) matrices.
- Solving by *back-substitution* (*forward-substitution*).
- *Gaussian elimination.*
- Permuting rows.
- The *pivot* of an elimination step; *pivoting*.
- *Transposition matrix*; *elementary matrix*.
- The *Gaussian elimination theorem* (Theorem 7.1).
- *Gauss–Jordan factorization.*
- *LU-factorization*; Necessary and sufficient condition for the existence of an
  LU-factorization (Proposition 7.1).
- *LDU-factorization.*
- "$PA = LU$ theorem" (Theorem 7.2).
- $LDL^\top$-*factorization* of a symmetric matrix.
- Avoiding small pivots: *partial pivoting*; *complete pivoting*.
- Gaussian elimination of tridiagonal matrices.
- *LU*-factorization of tridiagonal matrices.
- *Symmetric positive definite* matrices (SPD matrices).
- *Cholesky factorization* (Theorem 7.4).
- Criteria for a symmetric matrix to be positive definite; *Sylvester's criterion*.
- *Reduced row echelon form.*
- Reduction of a rectangular matrix to its row echelon form.
- Using the reduction to row echelon form to decide whether a system $Ax = b$ is solvable, and to find its solutions, using a *special* solution and a basis of the *homogeneous system* $Ax = 0$.
- *Magic squares.*
- *Transvections and dilatations.*

7.17   Problems

Problem 7.1. Solve the following linear systems by Gaussian elimination:

$$\begin{pmatrix} 2 & 3 & 1 \\ 1 & 2 & -1 \\ -3 & -5 & 1 \end{pmatrix} \begin{pmatrix} x \\ y \\ z \end{pmatrix} = \begin{pmatrix} 6 \\ 2 \\ -7 \end{pmatrix}, \quad \begin{pmatrix} 1 & 1 & 1 \\ 1 & 1 & 2 \\ 1 & 2 & 3 \end{pmatrix} \begin{pmatrix} x \\ y \\ z \end{pmatrix} = \begin{pmatrix} 6 \\ 9 \\ 14 \end{pmatrix}.$$

Problem 7.2. Solve the following linear system by Gaussian elimination:

$$\begin{pmatrix} 1 & 2 & 1 & 1 \\ 2 & 3 & 2 & 3 \\ -1 & 0 & 1 & -1 \\ -2 & -1 & 4 & 0 \end{pmatrix} \begin{pmatrix} x_1 \\ x_2 \\ x_3 \\ x_4 \end{pmatrix} = \begin{pmatrix} 7 \\ 14 \\ -1 \\ 2 \end{pmatrix}.$$

Problem 7.3. Consider the matrix

$$A = \begin{pmatrix} 1 & c & 0 \\ 2 & 4 & 1 \\ 3 & 5 & 1 \end{pmatrix}.$$

When applying Gaussian elimination, which value of c yields zero in the second pivot position? Which value of c yields zero in the third pivot position? In this case, what can your say about the matrix A?

Problem 7.4. Solve the system

$$\begin{pmatrix} 2 & 1 & 1 & 0 \\ 4 & 3 & 3 & 1 \\ 8 & 7 & 9 & 5 \\ 6 & 7 & 9 & 8 \end{pmatrix} \begin{pmatrix} x_1 \\ x_2 \\ x_3 \\ x_4 \end{pmatrix} = \begin{pmatrix} 1 \\ -1 \\ -1 \\ 1 \end{pmatrix}$$

using the LU-factorization of Example 7.1.

Problem 7.5. Apply **rref** to the matrix

$$A_2 = \begin{pmatrix} 1 & 2 & 1 & 1 \\ 2 & 3 & 2 & 3 \\ -1 & 0 & 1 & -1 \\ -2 & -1 & 3 & 0 \end{pmatrix}.$$

Problem 7.6. Apply **rref** to the matrix

$$\begin{pmatrix} 1 & 4 & 9 & 16 \\ 4 & 9 & 16 & 25 \\ 9 & 16 & 25 & 36 \\ 16 & 25 & 36 & 49 \end{pmatrix}.$$

Problem 7.7. (1) Prove that the dimension of the subspace of 2×2 matrices A, such that the sum of the entries of every row is the same (say c_1) and the sum of entries of every column is the same (say c_2) is 2.

(2) Prove that the dimension of the subspace of 2×2 matrices A, such that the sum of the entries of every row is the same (say c_1), the sum of entries of every column is the same (say c_2), and $c_1 = c_2$ is also 2. Prove that every such matrix is of the form

$$\begin{pmatrix} a & b \\ b & a \end{pmatrix},$$

and give a basis for this subspace.

(3) Prove that the dimension of the subspace of 3×3 matrices A, such that the sum of the entries of every row is the same (say c_1), the sum of entries of every column is the same (say c_2), and $c_1 = c_2$ is 5. Begin by showing that the above constraints are given by the set of equations

$$\begin{pmatrix} 1 & 1 & 1 & -1 & -1 & -1 & 0 & 0 & 0 \\ 0 & 0 & 0 & 1 & 1 & 1 & -1 & -1 & -1 \\ 1 & -1 & 0 & 1 & -1 & 0 & 1 & -1 & 0 \\ 0 & 1 & -1 & 0 & 1 & -1 & 0 & 1 & -1 \\ 0 & 1 & 1 & -1 & 0 & 0 & -1 & 0 & 0 \end{pmatrix} \begin{pmatrix} a_{11} \\ a_{12} \\ a_{13} \\ a_{21} \\ a_{22} \\ a_{23} \\ a_{31} \\ a_{32} \\ a_{33} \end{pmatrix} = \begin{pmatrix} 0 \\ 0 \\ 0 \\ 0 \\ 0 \end{pmatrix}.$$

Prove that every matrix satisfying the above constraints is of the form

$$\begin{pmatrix} a+b-c & -a+c+e & -b+c+d \\ -a-b+c+d+e & a & b \\ c & d & e \end{pmatrix},$$

with $a, b, c, d, e \in \mathbb{R}$. Find a basis for this subspace. (Use the method to find a basis for the kernel of a matrix.)

Problem 7.8. If A is an $n \times n$ symmetric matrix and B is any $n \times n$ invertible matrix, prove that A is positive definite iff $B^\top A B$ is positive definite.

Problem 7.9. (1) Consider the matrix

$$A_4 = \begin{pmatrix} 2 & -1 & 0 & 0 \\ -1 & 2 & -1 & 0 \\ 0 & -1 & 2 & -1 \\ 0 & 0 & -1 & 2 \end{pmatrix}.$$

Find three matrices of the form $E_{2,1;\beta_1}, E_{3,2;\beta_2}, E_{4,3;\beta_3}$, such that
$$E_{4,3;\beta_3} E_{3,2;\beta_2} E_{2,1;\beta_1} A_4 = U_4$$
where U_4 is an upper triangular matrix. Compute
$$M = E_{4,3;\beta_3} E_{3,2;\beta_2} E_{2,1;\beta_1}$$
and check that
$$MA_4 = U_4 = \begin{pmatrix} 2 & -1 & 0 & 0 \\ 0 & 3/2 & -1 & 0 \\ 0 & 0 & 4/3 & -1 \\ 0 & 0 & 0 & 5/4 \end{pmatrix}.$$

(2) Now consider the matrix
$$A_5 = \begin{pmatrix} 2 & -1 & 0 & 0 & 0 \\ -1 & 2 & -1 & 0 & 0 \\ 0 & -1 & 2 & -1 & 0 \\ 0 & 0 & -1 & 2 & -1 \\ 0 & 0 & 0 & -1 & 2 \end{pmatrix}.$$

Find four matrices of the form $E_{2,1;\beta_1}, E_{3,2;\beta_2}, E_{4,3;\beta_3}, E_{5,4;\beta_4}$, such that
$$E_{5,4;\beta_4} E_{4,3;\beta_3} E_{3,2;\beta_2} E_{2,1;\beta_1} A_5 = U_5$$
where U_5 is an upper triangular matrix. Compute
$$M = E_{5,4;\beta_4} E_{4,3;\beta_3} E_{3,2;\beta_2} E_{2,1;\beta_1}$$
and check that
$$MA_5 = U_5 = \begin{pmatrix} 2 & -1 & 0 & 0 & 0 \\ 0 & 3/2 & -1 & 0 & 0 \\ 0 & 0 & 4/3 & -1 & 0 \\ 0 & 0 & 0 & 5/4 & -1 \\ 0 & 0 & 0 & 0 & 6/5 \end{pmatrix}.$$

(3) Write a `Matlab` program defining the function Ematrix(n, i, j, b) which is the $n \times n$ matrix that adds b times row j to row i. Also write some `Matlab` code that produces an $n \times n$ matrix A_n generalizing the matrices A_4 and A_5.

Use your program to figure out which five matrices $E_{i,j;\beta}$ reduce A_6 to the upper triangular matrix
$$U_6 = \begin{pmatrix} 2 & -1 & 0 & 0 & 0 & 0 \\ 0 & 3/2 & -1 & 0 & 0 & 0 \\ 0 & 0 & 4/3 & -1 & 0 & 0 \\ 0 & 0 & 0 & 5/4 & -1 & 0 \\ 0 & 0 & 0 & 0 & 6/5 & -1 \\ 0 & 0 & 0 & 0 & 0 & 7/6 \end{pmatrix}.$$

Also use your program to figure out which six matrices $E_{i,j;\beta}$ reduce A_7 to the upper triangular matrix

$$U_7 = \begin{pmatrix} 2 & -1 & 0 & 0 & 0 & 0 & 0 \\ 0 & 3/2 & -1 & 0 & 0 & 0 & 0 \\ 0 & 0 & 4/3 & -1 & 0 & 0 & 0 \\ 0 & 0 & 0 & 5/4 & -1 & 0 & 0 \\ 0 & 0 & 0 & 0 & 6/5 & -1 & 0 \\ 0 & 0 & 0 & 0 & 0 & 7/6 & -1 \\ 0 & 0 & 0 & 0 & 0 & 0 & 8/7 \end{pmatrix}.$$

(4) Find the lower triangular matrices L_6 and L_7 such that

$$L_6 U_6 = A_6$$

and

$$L_7 U_7 = A_7.$$

(5) It is natural to conjecture that there are $n-1$ matrices of the form $E_{i,j;\beta}$ that reduce A_n to the upper triangular matrix

$$U_n = \begin{pmatrix} 2 & -1 & 0 & 0 & 0 & 0 & 0 \\ 0 & 3/2 & -1 & 0 & 0 & 0 & 0 \\ 0 & 0 & 4/3 & -1 & 0 & 0 & 0 \\ 0 & 0 & 0 & 5/4 & -1 & 0 & 0 \\ 0 & 0 & 0 & 0 & 6/5 & \ddots & \vdots \\ \vdots & \vdots & \vdots & \vdots & \ddots & \ddots & -1 \\ 0 & 0 & 0 & 0 & \cdots & 0 & (n+1)/n \end{pmatrix},$$

namely,

$$E_{2,1;1/2}, E_{3,2;2/3}, E_{4,3;3/4}, \cdots, E_{n,n-1;(n-1)/n}.$$

It is also natural to conjecture that the lower triangular matrix L_n such that

$$L_n U_n = A_n$$

is given by

$$L_n = E_{2,1;-1/2} E_{3,2;-2/3} E_{4,3;-3/4} \cdots E_{n,n-1;-(n-1)/n},$$

that is,

$$L_n = \begin{pmatrix} 1 & 0 & 0 & 0 & 0 & 0 & 0 \\ -1/2 & 1 & 0 & 0 & 0 & 0 & 0 \\ 0 & -2/3 & 1 & 0 & 0 & 0 & 0 \\ 0 & 0 & -3/4 & 1 & 0 & 0 & 0 \\ 0 & 0 & 0 & -4/5 & 1 & \ddots & \vdots \\ \vdots & \vdots & \vdots & \vdots & \ddots & \ddots & 0 \\ 0 & 0 & 0 & 0 & \cdots & -(n-1)/n & 1 \end{pmatrix}.$$

Prove the above conjectures.

(6) Prove that the last column of A_n^{-1} is

$$\begin{pmatrix} 1/(n+1) \\ 2/(n+1) \\ \vdots \\ n/(n+1) \end{pmatrix}.$$

Problem 7.10. (1) Let A be any invertible 2×2 matrix

$$A = \begin{pmatrix} a & b \\ c & d \end{pmatrix}.$$

Prove that there is an invertible matrix S such that

$$SA = \begin{pmatrix} 1 & 0 \\ 0 & ad - bc \end{pmatrix},$$

where S is the product of at most four elementary matrices of the form $E_{i,j;\beta}$.

Conclude that every matrix A in $\mathbf{SL}(2)$ (the group of invertible 2×2 matrices A with $\det(A) = +1$) is the product of at most four elementary matrices of the form $E_{i,j;\beta}$.

For any $a \neq 0, 1$, give an explicit factorization as above for

$$A = \begin{pmatrix} a & 0 \\ 0 & a^{-1} \end{pmatrix}.$$

What is this decomposition for $a = -1$?

(2) Recall that a rotation matrix R (a member of the group $\mathbf{SO}(2)$) is a matrix of the form

$$R = \begin{pmatrix} \cos\theta & -\sin\theta \\ \sin\theta & \cos\theta \end{pmatrix}.$$

Prove that if $\theta \neq k\pi$ (with $k \in \mathbb{Z}$), any rotation matrix can be written as a product

$$R = ULU,$$

where U is upper triangular and L is lower triangular of the form

$$U = \begin{pmatrix} 1 & u \\ 0 & 1 \end{pmatrix}, \quad L = \begin{pmatrix} 1 & 0 \\ v & 1 \end{pmatrix}.$$

Therefore, every plane rotation (except a flip about the origin when $\theta = \pi$) can be written as the composition of three shear transformations!

Problem 7.11. (1) Recall that $E_{i,d}$ is the diagonal matrix

$$E_{i,d} = \text{diag}(1, \ldots, 1, d, 1, \ldots, 1),$$

whose diagonal entries are all $+1$, except the (i, i)th entry which is equal to d.

Given any $n \times n$ matrix A, for any pair (i, j) of distinct row indices $(1 \leq i, j \leq n)$, prove that there exist two elementary matrices $E_1(i, j)$ and $E_2(i, j)$ of the form $E_{k,\ell;\beta}$, such that

$$E_{j,-1} E_1(i, j) E_2(i, j) E_1(i, j) A = P(i, j) A,$$

the matrix obtained from the matrix A by permuting row i and row j. Equivalently, we have

$$E_1(i, j) E_2(i, j) E_1(i, j) A = E_{j,-1} P(i, j) A,$$

the matrix obtained from A by permuting row i and row j and multiplying row j by -1.

Prove that for every $i = 2, \ldots, n$, there exist four elementary matrices $E_3(i, d), E_4(i, d), E_5(i, d), E_6(i, d)$ of the form $E_{k,\ell;\beta}$, such that

$$E_6(i, d) E_5(i, d) E_4(i, d) E_3(i, d) E_{n,d} = E_{i,d}.$$

What happens when $d = -1$, that is, what kind of simplifications occur?

Prove that all permutation matrices can be written as products of elementary operations of the form $E_{k,\ell;\beta}$ and the operation $E_{n,-1}$.

(2) Prove that for every invertible $n \times n$ matrix A, there is a matrix S such that

$$SA = \begin{pmatrix} I_{n-1} & 0 \\ 0 & d \end{pmatrix} = E_{n,d},$$

with $d = \det(A)$, and where S is a product of elementary matrices of the form $E_{k,\ell;\beta}$.

In particular, every matrix in $\mathbf{SL}(n)$ (the group of invertible $n \times n$ matrices A with $\det(A) = +1$) can be written as a product of elementary matrices of the form $E_{k,\ell;\beta}$. Prove that at most $n(n+1) - 2$ such transformations are needed.

(3) Prove that every matrix in $\mathbf{SL}(n)$ can be written as a product of at most $(n-1)(\max\{n,3\}+1)$ elementary matrices of the form $E_{k,\ell;\beta}$.

Problem 7.12. A matrix A is called *strictly column diagonally dominant* iff

$$|a_{jj}| > \sum_{i=1,\, i \neq j}^{n} |a_{ij}|, \quad \text{for } j = 1, \ldots, n$$

Prove that if A is strictly column diagonally dominant, then Gaussian elimination with partial pivoting does not require pivoting, and A is invertible.

Problem 7.13. (1) Find a lower triangular matrix E such that

$$E \begin{pmatrix} 1\,0\,0\,0 \\ 1\,1\,0\,0 \\ 1\,2\,1\,0 \\ 1\,3\,3\,1 \end{pmatrix} = \begin{pmatrix} 1\,0\,0\,0 \\ 0\,1\,0\,0 \\ 0\,1\,1\,0 \\ 0\,1\,2\,1 \end{pmatrix}.$$

(2) What is the effect of the product (on the left) with

$$E_{4,3;-1} E_{3,2;-1} E_{4,3;-1} E_{2,1;-1} E_{3,2;-1} E_{4,3;-1}$$

on the matrix

$$Pa_3 = \begin{pmatrix} 1\,0\,0\,0 \\ 1\,1\,0\,0 \\ 1\,2\,1\,0 \\ 1\,3\,3\,1 \end{pmatrix}.$$

(3) Find the inverse of the matrix Pa_3.

(4) Consider the $(n+1) \times (n+1)$ Pascal matrix Pa_n whose ith row is given by the binomial coefficients

$$\binom{i-1}{j-1},$$

with $1 \le i \le n+1$, $1 \le j \le n+1$, and with the usual convention that

$$\binom{0}{0} = 1, \quad \binom{i}{j} = 0 \quad \text{if } j > i.$$

The matrix Pa_3 is shown in Question (c) and Pa_4 is shown below:

$$Pa_4 = \begin{pmatrix} 1 & 0 & 0 & 0 & 0 \\ 1 & 1 & 0 & 0 & 0 \\ 1 & 2 & 1 & 0 & 0 \\ 1 & 3 & 3 & 1 & 0 \\ 1 & 4 & 6 & 4 & 1 \end{pmatrix}.$$

Find n elementary matrices $E_{i_k,j_k;\beta_k}$ such that

$$E_{i_n,j_n;\beta_n} \cdots E_{i_1,j_1;\beta_1} Pa_n = \begin{pmatrix} 1 & 0 \\ 0 & Pa_{n-1} \end{pmatrix}.$$

Use the above to prove that the inverse of Pa_n is the lower triangular matrix whose ith row is given by the signed binomial coefficients

$$(-1)^{i+j-2}\binom{i-1}{j-1},$$

with $1 \le i \le n+1$, $1 \le j \le n+1$. For example,

$$Pa_4^{-1} = \begin{pmatrix} 1 & 0 & 0 & 0 & 0 \\ -1 & 1 & 0 & 0 & 0 \\ 1 & -2 & 1 & 0 & 0 \\ -1 & 3 & -3 & 1 & 0 \\ 1 & -4 & 6 & -4 & 1 \end{pmatrix}.$$

Hint. Given any $n \times n$ matrix A, multiplying A by the elementary matrix $E_{i,j;\beta}$ *on the right* yields the matrix $AE_{i,j;\beta}$ in which β times the ith column is added to the jth column.

Problem 7.14. (1) Implement the method for converting a rectangular matrix to reduced row echelon form in `Matlab`.

(2) Use the above method to find the inverse of an invertible $n \times n$ matrix A by applying it to the the $n \times 2n$ matrix $[A\,I]$ obtained by adding the n columns of the identity matrix to A.

(3) Consider the matrix

$$A = \begin{pmatrix} 1 & 2 & 3 & 4 & \cdots & n \\ 2 & 3 & 4 & 5 & \cdots & n+1 \\ 3 & 4 & 5 & 6 & \cdots & n+2 \\ \vdots & \vdots & \vdots & & \ddots & \vdots \\ n & n+1 & n+2 & n+3 & \cdots & 2n-1 \end{pmatrix}.$$

Using your program, find the row reduced echelon form of A for $n = 4, \ldots, 20$.

Also run the `Matlab` `rref` function and compare results. Your program probably disagrees with `rref` even for small values of n. The problem is that some pivots are very small and the normalization step (to make the pivot 1) causes roundoff errors. Use a tolerance parameter to fix this problem.

What can you conjecture about the rank of A?

(4) Prove that the matrix A has the following row reduced form:

$$R = \begin{pmatrix} 1 & 0 & -1 & -2 & \cdots & -(n-2) \\ 0 & 1 & 2 & 3 & \cdots & n-1 \\ 0 & 0 & 0 & 0 & \cdots & 0 \\ \vdots & \vdots & \vdots & & \ddots & \vdots \\ 0 & 0 & 0 & 0 & \cdots & 0 \end{pmatrix}.$$

Deduce from the above that A has rank 2.

Hint. Some well chosen sequence of row operations.

(5) Use your program to show that if you add any number greater than or equal to $(2/25)n^2$ to every diagonal entry of A you get an invertible matrix! In fact, running the `Matlab` fuction `chol` should tell you that these matrices are SPD (symmetric, positive definite).

Problem 7.15. Let A be an $n \times n$ complex Hermitian positive definite matrix. Prove that the lower-triangular matrix B with positive diagonal entries such that $A = BB^*$ is given by the following formulae: For $j = 1, \ldots, n$,

$$b_{jj} = \left(a_{jj} - \sum_{k=1}^{j-1} |b_{jk}|^2 \right)^{1/2},$$

and for $i = j+1, \ldots, n$ (and $j = 1, \ldots, n-1$)

$$b_{ij} = \left(a_{ij} - \sum_{k=1}^{j-1} b_{ik} b_{jk} \right) / b_{jj}.$$

Problem 7.16. (Permutations and permutation matrices) A permutation can be viewed as an operation permuting the rows of a matrix. For example, the permutation

$$\begin{pmatrix} 1 & 2 & 3 & 4 \\ 3 & 4 & 2 & 1 \end{pmatrix}$$

corresponds to the matrix

$$P_\pi = \begin{pmatrix} 0\ 0\ 0\ 1 \\ 0\ 0\ 1\ 0 \\ 1\ 0\ 0\ 0 \\ 0\ 1\ 0\ 0 \end{pmatrix}.$$

Observe that the matrix P_π has a single 1 on every row and every column, all other entries being zero, and that if we multiply any 4×4 matrix A by P_π on the left, then the rows of A are permuted according to the permutation π; that is, the $\pi(i)$th row of $P_\pi A$ is the ith row of A. For example,

$$P_\pi A = \begin{pmatrix} 0\ 0\ 0\ 1 \\ 0\ 0\ 1\ 0 \\ 1\ 0\ 0\ 0 \\ 0\ 1\ 0\ 0 \end{pmatrix} \begin{pmatrix} a_{11}\ a_{12}\ a_{13}\ a_{14} \\ a_{21}\ a_{22}\ a_{23}\ a_{24} \\ a_{31}\ a_{32}\ a_{33}\ a_{34} \\ a_{41}\ a_{42}\ a_{43}\ a_{44} \end{pmatrix} = \begin{pmatrix} a_{41}\ a_{42}\ a_{43}\ a_{44} \\ a_{31}\ a_{32}\ a_{33}\ a_{34} \\ a_{11}\ a_{12}\ a_{13}\ a_{14} \\ a_{21}\ a_{22}\ a_{23}\ a_{24} \end{pmatrix}.$$

Equivalently, the ith row of $P_\pi A$ is the $\pi^{-1}(i)$th row of A. In order for the matrix P_π to move the ith row of A to the $\pi(i)$th row, the $\pi(i)$th row of P_π must have a 1 in column i and zeros everywhere else; this means that the ith column of P_π contains the basis vector $e_{\pi(i)}$, the vector that has a 1 in position $\pi(i)$ and zeros everywhere else.

This is the general situation and it leads to the following definition.

Definition 7.8. Given any permutation $\pi \colon [n] \to [n]$, the *permutation matrix* $P_\pi = (p_{ij})$ representing π is the matrix given by

$$p_{ij} = \begin{cases} 1 & \text{if } i = \pi(j) \\ 0 & \text{if } i \neq \pi(j); \end{cases}$$

equivalently, the jth column of P_π is the basis vector $e_{\pi(j)}$. A *permutation matrix* P is any matrix of the form P_π (where P is an $n \times n$ matrix, and $\pi \colon [n] \to [n]$ is a permutation, for some $n \geq 1$).

Remark: There is a confusing point about the notation for permutation matrices. A permutation matrix P acts on a matrix A by multiplication on the left by permuting the rows of A. As we said before, this means that the $\pi(i)$th row of $P_\pi A$ is the ith row of A, or equivalently that the ith row of $P_\pi A$ is the $\pi^{-1}(i)$th row of A. But then observe that the row index of the entries of the ith row of PA is $\pi^{-1}(i)$, and not $\pi(i)$! See the following example:

$$\begin{pmatrix} 0 & 0 & 0 & 1 \\ 0 & 0 & 1 & 0 \\ 1 & 0 & 0 & 0 \\ 0 & 1 & 0 & 0 \end{pmatrix} \begin{pmatrix} a_{11} & a_{12} & a_{13} & a_{14} \\ a_{21} & a_{22} & a_{23} & a_{24} \\ a_{31} & a_{32} & a_{33} & a_{34} \\ a_{41} & a_{42} & a_{43} & a_{44} \end{pmatrix} = \begin{pmatrix} a_{41} & a_{42} & a_{43} & a_{44} \\ a_{31} & a_{32} & a_{33} & a_{34} \\ a_{11} & a_{12} & a_{13} & a_{14} \\ a_{21} & a_{22} & a_{23} & a_{24} \end{pmatrix},$$

where

$$\pi^{-1}(1) = 4$$
$$\pi^{-1}(2) = 3$$
$$\pi^{-1}(3) = 1$$
$$\pi^{-1}(4) = 2.$$

Prove the following results

(1) Given any two permutations $\pi_1, \pi_2 \colon [n] \to [n]$, the permutation matrix $P_{\pi_2 \circ \pi_1}$ representing the composition of π_1 and π_2 is equal to the product $P_{\pi_2} P_{\pi_1}$ of the permutation matrices P_{π_1} and P_{π_2} representing π_1 and π_2; that is,

$$P_{\pi_2 \circ \pi_1} = P_{\pi_2} P_{\pi_1}.$$

(2) The matrix $P_{\pi_1^{-1}}$ representing the inverse of the permutation π_1 is the inverse $P_{\pi_1}^{-1}$ of the matrix P_{π_1} representing the permutation π_1; that is,

$$P_{\pi_1^{-1}} = P_{\pi_1}^{-1}.$$

Furthermore,

$$P_{\pi_1}^{-1} = (P_{\pi_1})^{\top}.$$

(3) Prove that if P is the matrix associated with a transposition, then $\det(P) = -1$.

(4) Prove that if P is a permutation matrix, then $\det(P) = \pm 1$.

(5) Use permutation matrices to give another proof of the fact that the parity of the number of transpositions used to express a permutation π depends only on π.

Chapter 8

Vector Norms and Matrix Norms

8.1 Normed Vector Spaces

In order to define how close two vectors or two matrices are, and in order to define the convergence of sequences of vectors or matrices, we can use the notion of a norm. Recall that $\mathbb{R}_+ = \{x \in \mathbb{R} \mid x \geq 0\}$. Also recall that if $z = a + ib \in \mathbb{C}$ is a complex number, with $a, b \in \mathbb{R}$, then $\bar{z} = a - ib$ and $|z| = \sqrt{z\bar{z}} = \sqrt{a^2 + b^2}$ ($|z|$ is the *modulus* of z).

Definition 8.1. Let E be a vector space over a field K, where K is either the field $\mathbb{R}$ of reals, or the field $\mathbb{C}$ of complex numbers. A *norm* on E is a function $\| \ \| \colon E \to \mathbb{R}_+$, assigning a nonnegative real number $\|u\|$ to any vector $u \in E$, and satisfying the following conditions for all $x, y, z \in E$ and $\lambda \in K$:

(N1) $\|x\| \geq 0$, and $\|x\| = 0$ iff $x = 0$.   (positivity)
(N2) $\|\lambda x\| = |\lambda| \, \|x\|$.   (homogeneity (or scaling))
(N3) $\|x + y\| \leq \|x\| + \|y\|$.   (triangle inequality)

A vector space E together with a norm $\| \ \|$ is called a *normed vector space*.

By (N2), setting $\lambda = -1$, we obtain

$$\|-x\| = \|(-1)x\| = |-1| \, \|x\| = \|x\| \, ;$$

that is, $\|-x\| = \|x\|$. From (N3), we have

$$\|x\| = \|x - y + y\| \leq \|x - y\| + \|y\| \, ,$$

which implies that

$$\|x\| - \|y\| \leq \|x - y\| \, .$$

By exchanging x and y and using the fact that by (N2),

$$\|y - x\| = \|-(x - y)\| = \|x - y\|,$$

we also have

$$\|y\| - \|x\| \le \|x - y\|.$$

Therefore,

$$\big|\|x\| - \|y\|\big| \le \|x - y\|, \quad \text{for all } x, y \in E. \tag{8.1}$$

Observe that setting $\lambda = 0$ in (N2), we deduce that $\|0\| = 0$ without assuming (N1). Then by setting $y = 0$ in (8.1), we obtain

$$\big|\|x\|\big| \le \|x\|, \quad \text{for all } x \in E.$$

Therefore, the condition $\|x\| \ge 0$ in (N1) follows from (N2) and (N3), and (N1) can be replaced by the weaker condition

(N1') For all $x \in E$, if $\|x\| = 0$, then $x = 0$,

A function $\| \ \| : E \to \mathbb{R}$ satisfying Axioms (N2) and (N3) is called a *seminorm*. From the above discussion, a seminorm also has the properties
$\|x\| \ge 0$ for all $x \in E$, and $\|0\| = 0$.
However, there may be nonzero vectors $x \in E$ such that $\|x\| = 0$.
Let us give some examples of normed vector spaces.

Example 8.1.

(1) Let $E = \mathbb{R}$, and $\|x\| = |x|$, the absolute value of x.
(2) Let $E = \mathbb{C}$, and $\|z\| = |z|$, the modulus of z.
(3) Let $E = \mathbb{R}^n$ (or $E = \mathbb{C}^n$). There are three standard norms. For every $(x_1, \ldots, x_n) \in E$, we have the norm $\|x\|_1$, defined such that,

$$\|x\|_1 = |x_1| + \cdots + |x_n|,$$

we have the *Euclidean norm* $\|x\|_2$, defined such that,

$$\|x\|_2 = \left(|x_1|^2 + \cdots + |x_n|^2\right)^{\frac{1}{2}},$$

and the *sup*-norm $\|x\|_\infty$, defined such that,

$$\|x\|_\infty = \max\{|x_i| \mid 1 \le i \le n\}.$$

More generally, we define the ℓ^p-*norm* (for $p \ge 1$) by

$$\|x\|_p = \left(|x_1|^p + \cdots + |x_n|^p\right)^{1/p}.$$

See Figures 8.1 through 8.4.

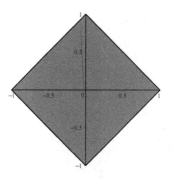

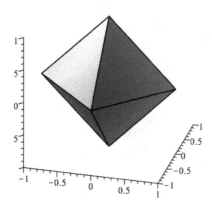

Fig. 8.1 The top figure is $\{x \in \mathbb{R}^2 \mid \|x\|_1 \leq 1\}$, while the bottom figure is $\{x \in \mathbb{R}^3 \mid \|x\|_1 \leq 1\}$.

There are other norms besides the ℓ^p-norms. Here are some examples.

(1) For $E = \mathbb{R}^2$,
$$\|(u_1, u_2)\| = |u_1| + 2|u_2|.$$
See Figure 8.5.

(2) For $E = \mathbb{R}^2$,
$$\|(u_1, u_2)\| = \left((u_1 + u_2)^2 + u_1^2\right)^{1/2}.$$
See Figure 8.6.

(3) For $E = \mathbb{C}^2$,
$$\|(u_1, u_2)\| = |u_1 + iu_2| + |u_1 - iu_2|.$$

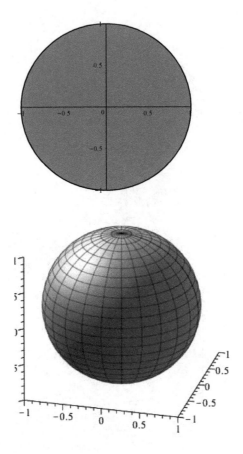

Fig. 8.2   The top figure is $\{x \in \mathbb{R}^2 \mid \|x\|_2 \leq 1\}$, while the bottom figure is $\{x \in \mathbb{R}^3 \mid \|x\|_2 \leq 1\}$.

The reader should check that they satisfy all the axioms of a norm.

Some work is required to show the triangle inequality for the ℓ^p-norm.

Proposition 8.1. *If $E = \mathbb{C}^n$ or $E = \mathbb{R}^n$, for every real number $p \geq 1$, the ℓ^p-norm is indeed a norm.*

Proof. The cases $p = 1$ and $p = \infty$ are easy and left to the reader. If $p > 1$, then let $q > 1$ such that

$$\frac{1}{p} + \frac{1}{q} = 1.$$

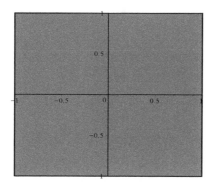

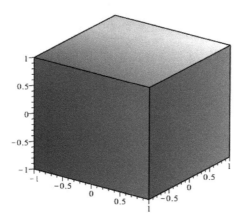

Fig. 8.3 The top figure is $\{x \in \mathbb{R}^2 \mid \|x\|_\infty \le 1\}$, while the bottom figure is $\{x \in \mathbb{R}^3 \mid \|x\|_\infty \le 1\}$.

We will make use of the following fact: for all $\alpha, \beta \in \mathbb{R}$, if $\alpha, \beta \ge 0$, then

$$\alpha\beta \le \frac{\alpha^p}{p} + \frac{\beta^q}{q}. \tag{8.2}$$

To prove the above inequality, we use the fact that the exponential function $t \mapsto e^t$ satisfies the following convexity inequality:

$$e^{\theta x + (1-\theta)y} \le \theta e^x + (1-\theta)e^y,$$

for all $x, y \in \mathbb{R}$ and all θ with $0 \le \theta \le 1$.

Since the case $\alpha\beta = 0$ is trivial, let us assume that $\alpha > 0$ and $\beta > 0$. If we replace θ by $1/p$, x by $p \log \alpha$ and y by $q \log \beta$, then we get

$$e^{\frac{1}{p}p\log\alpha + \frac{1}{q}q\log\beta} \le \frac{1}{p}e^{p\log\alpha} + \frac{1}{q}e^{q\log\beta},$$

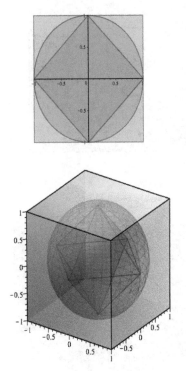

Fig. 8.4   The relationships between the closed unit balls from the ℓ^1-norm, the Euclidean norm, and the sup-norm.

which simplifies to

$$\alpha\beta \le \frac{\alpha^p}{p} + \frac{\beta^q}{q},$$

as claimed.

We will now prove that for any two vectors $u, v \in E$, (where E is of dimension n), we have

$$\sum_{i=1}^{n} |u_i v_i| \le \|u\|_p \|v\|_q. \tag{8.3}$$

Since the above is trivial if $u = 0$ or $v = 0$, let us assume that $u \neq 0$ and

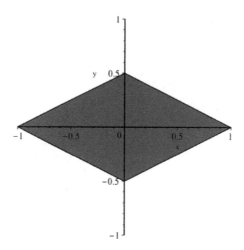

Fig. 8.5 The unit closed unit ball $\{(u_1, u_2) \in \mathbb{R}^2 \mid \|(u_1, u_2)\| \leq 1\}$, where $\|(u_1, u_2)\| = |u_1| + 2|u_2|$.

$v \neq 0$. Then Inequality (8.2) with $\alpha = |u_i| / \|u\|_p$ and $\beta = |v_i| / \|v\|_q$ yields

$$\frac{|u_i v_i|}{\|u\|_p \|v\|_q} \leq \frac{|u_i|^p}{p \|u\|_p^p} + \frac{|v_i|^q}{q \|u\|_q^q},$$

for $i = 1, \ldots, n$, and by summing up these inequalities, we get

$$\sum_{i=1}^{n} |u_i v_i| \leq \|u\|_p \|v\|_q,$$

as claimed. To finish the proof, we simply have to prove that property (N3) holds, since (N1) and (N2) are clear. For $i = 1, \ldots, n$, we can write

$$(|u_i| + |v_i|)^p = |u_i|(|u_i| + |v_i|)^{p-1} + |v_i|(|u_i| + |v_i|)^{p-1},$$

so that by summing up these equations we get

$$\sum_{i=1}^{n}(|u_i| + |v_i|)^p = \sum_{i=1}^{n} |u_i|(|u_i| + |v_i|)^{p-1} + \sum_{i=1}^{n} |v_i|(|u_i| + |v_i|)^{p-1},$$

and using Inequality (8.3), with $V \in E$ where $V_i = (|u_i| + |v_i|)^{p-1}$, we get

$$\sum_{i=1}^{n}(|u_i| + |v_i|)^p \leq \|u\|_p \|V\|_q + \|v\|_p \|V\|_q$$

$$= (\|u\|_p + \|v\|_p)\left(\sum_{i=1}^{n}(|u_i| + |v_i|)^{(p-1)q}\right)^{1/q}.$$

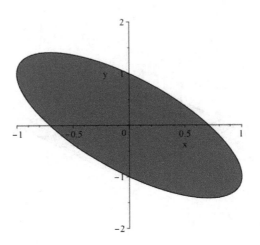

Fig. 8.6   The unit closed unit ball $\{(u_1, u_2) \in \mathbb{R}^2 \mid \|(u_1, u_2)\| \leq 1\}$, where $\|(u_1, u_2)\| = ((u_1 + u_2)^2 + u_1^2)^{1/2}$.

However, $1/p + 1/q = 1$ implies $pq = p + q$, that is, $(p-1)q = p$, so we have

$$\sum_{i=1}^{n}(|u_i| + |v_i|)^p \leq (\|u\|_p + \|v\|_p)\left(\sum_{i=1}^{n}(|u_i| + |v_i|)^p\right)^{1/q},$$

which yields

$$\left(\sum_{i=1}^{n}(|u_i| + |v_i|)^p\right)^{1-1/q} = \left(\sum_{i=1}^{n}(|u_i| + |v_i|)^p\right)^{1/p} \leq \|u\|_p + \|v\|_p.$$

Since $|u_i + v_i| \leq |u_i| + |v_i|$, the above implies the triangle inequality $\|u + v\|_p \leq \|u\|_p + \|v\|_p$, as claimed. $\square$

For $p > 1$ and $1/p + 1/q = 1$, the inequality

$$\sum_{i=1}^{n}|u_i v_i| \leq \left(\sum_{i=1}^{n}|u_i|^p\right)^{1/p}\left(\sum_{i=1}^{n}|v_i|^q\right)^{1/q}$$

is known as *Hölder's inequality*. For $p = 2$, it is the *Cauchy–Schwarz inequality*.

Actually, if we define the *Hermitian inner product* $\langle -, - \rangle$ on $\mathbb{C}^n$ by

$$\langle u, v \rangle = \sum_{i=1}^{n} u_i \overline{v}_i,$$

where $u = (u_1, \ldots, u_n)$ and $v = (v_1, \ldots, v_n)$, then

$$|\langle u, v \rangle| \leq \sum_{i=1}^{n} |u_i \overline{v}_i| = \sum_{i=1}^{n} |u_i v_i|,$$

so Hölder's inequality implies the following inequalities.

Corollary 8.1. *(Hölder's inequalities) For any real numbers p, q, such that $p, q \geq 1$ and*

$$\frac{1}{p} + \frac{1}{q} = 1,$$

(with $q = +\infty$ if $p = 1$ and $p = +\infty$ if $q = 1$), we have the inequalities

$$\sum_{i=1}^{n} |u_i v_i| \leq \left(\sum_{i=1}^{n} |u_i|^p \right)^{1/p} \left(\sum_{i=1}^{n} |v_i|^q \right)^{1/q}$$

and

$$|\langle u, v \rangle| \leq \|u\|_p \|v\|_q, \qquad u, v \in \mathbb{C}^n.$$

For $p = 2$, this is the standard Cauchy–Schwarz inequality. The triangle inequality for the ℓ^p-norm,

$$\left(\sum_{i=1}^{n} (|u_i + v_i|)^p \right)^{1/p} \leq \left(\sum_{i=1}^{n} |u_i|^p \right)^{1/p} + \left(\sum_{i=1}^{n} |v_i|^q \right)^{1/q},$$

is known as *Minkowski's inequality*.

When we restrict the Hermitian inner product to real vectors, $u, v \in \mathbb{R}^n$, we get the *Euclidean inner product*

$$\langle u, v \rangle = \sum_{i=1}^{n} u_i v_i.$$

It is very useful to observe that if we represent (as usual) $u = (u_1, \ldots, u_n)$ and $v = (v_1, \ldots, v_n)$ (in $\mathbb{R}^n$) by column vectors, then their Euclidean inner product is given by

$$\langle u, v \rangle = u^\top v = v^\top u,$$

and when $u, v \in \mathbb{C}^n$, their Hermitian inner product is given by

$$\langle u, v \rangle = v^* u = \overline{u^* v}.$$

In particular, when $u = v$, in the complex case we get

$$\|u\|_2^2 = u^* u,$$

and in the real case this becomes

$$\|u\|_2^2 = u^\top u.$$

As convenient as these notations are, we still recommend that you do not abuse them; the notation $\langle u, v \rangle$ is more intrinsic and still "works" when our vector space is infinite dimensional.

Remark: If $0 < p < 1$, then $x \mapsto \|x\|_p$ is not a norm because the triangle inequality *fails*. For example, consider $x = (2, 0)$ and $y = (0, 2)$. Then $x + y = (2, 2)$, and we have $\|x\|_p = (2^p + 0^p)^{1/p} = 2$, $\|y\|_p = (0^p + 2^p)^{1/p} = 2$, and $\|x + y\|_p = (2^p + 2^p)^{1/p} = 2^{(p+1)/p}$. Thus

$$\|x + y\|_p = 2^{(p+1)/p}, \quad \|x\|_p + \|y\|_p = 4 = 2^2.$$

Since $0 < p < 1$, we have $2p < p + 1$, that is, $(p+1)/p > 2$, so $2^{(p+1)/p} > 2^2 = 4$, and the triangle inequality $\|x + y\|_p \leq \|x\|_p + \|y\|_p$ fails.

Observe that

$$\|(1/2)x\|_p = (1/2) \|x\|_p = \|(1/2)y\|_p = (1/2) \|y\|_p = 1,$$

$$\|(1/2)(x + y)\|_p = 2^{1/p},$$

and since $p < 1$, we have $2^{1/p} > 2$, so

$$\|(1/2)(x + y)\|_p = 2^{1/p} > 2 = (1/2) \|x\|_p + (1/2) \|y\|_p,$$

and the map $x \mapsto \|x\|_p$ is not convex.

For $p = 0$, for any $x \in \mathbb{R}^n$, we have

$$\|x\|_0 = |\{i \in \{1, \ldots, n\} \mid x_i \neq 0\}|,$$

the number of nonzero components of x. The map $x \mapsto \|x\|_0$ is not a norm this time because Axiom (N2) fails. For example,

$$\|(1, 0)\|_0 = \|(10, 0)\|_0 = 1 \neq 10 = 10 \|(1, 0)\|_0.$$

The map $x \mapsto \|x\|_0$ is also not convex. For example,

$$\|(1/2)(2, 2)\|_0 = \|(1, 1)\|_0 = 2,$$

and

$$\|(2, 0)\|_0 = \|(0, 2)\|_0 = 1,$$

but

$$\|(1/2)(2, 2)\|_0 = 2 > 1 = (1/2) \|(2, 0)\|_0 + (1/2) \|(0, 2)\|_0.$$

Nevertheless, the "zero-norm" $x \mapsto \|x\|_0$ is used in machine learning as a regularizing term which encourages sparsity, namely increases the number of zero components of the vector x.

The following proposition is easy to show.

Proposition 8.2. *The following inequalities hold for all* $x \in \mathbb{R}^n$ *(or* $x \in \mathbb{C}^n$*):*

$$\|x\|_\infty \leq \|x\|_1 \leq n\|x\|_\infty,$$
$$\|x\|_\infty \leq \|x\|_2 \leq \sqrt{n}\|x\|_\infty,$$
$$\|x\|_2 \leq \|x\|_1 \leq \sqrt{n}\|x\|_2.$$

Proposition 8.2 is actually a special case of a very important result: *in a finite-dimensional vector space, any two norms are equivalent.*

Definition 8.2. Given any (real or complex) vector space E, two norms $\| \ \|_a$ and $\| \ \|_b$ are *equivalent* iff there exists some positive reals $C_1, C_2 > 0$, such that

$$\|u\|_a \leq C_1 \|u\|_b \quad \text{and} \quad \|u\|_b \leq C_2 \|u\|_a, \text{ for all } u \in E.$$

Given any norm $\| \ \|$ on a vector space of dimension n, for any basis $(e_1, \dots, e_n)$ of E, observe that for any vector $x = x_1 e_1 + \cdots + x_n e_n$, we have

$$\|x\| = \|x_1 e_1 + \cdots + x_n e_n\| \leq |x_1| \|e_1\| + \cdots + |x_n| \|e_n\|$$
$$\leq C(|x_1| + \cdots + |x_n|) = C \|x\|_1,$$

with $C = \max_{1 \leq i \leq n} \|e_i\|$ and with the norm $\|x\|_1$ defined as

$$\|x\|_1 = \|x_1 e_1 + \cdots + x_n e_n\| = |x_1| + \cdots + |x_n|.$$

The above implies that

$$| \|u\| - \|v\| | \leq \|u - v\| \leq C \|u - v\|_1,$$

and this implies the following corollary.

Corollary 8.2. *For any norm* $u \mapsto \|u\|$ *on a finite-dimensional (complex or real) vector space E, the map* $u \mapsto \|u\|$ *is continuous with respect to the norm* $\| \ \|_1$.

Let S_1^{n-1} be the unit sphere with respect to the norm $\| \ \|_1$, namely

$$S_1^{n-1} = \{x \in E \mid \|x\|_1 = 1\}.$$

Now S_1^{n-1} is a closed and bounded subset of a finite-dimensional vector space, so by Heine–Borel (or equivalently, by Bolzano–Weiertrass), S_1^{n-1} is compact. On the other hand, it is a well known result of analysis that any continuous real-valued function on a nonempty compact set has a minimum and a maximum, and that they are achieved. Using these facts, we can prove the following important theorem:

Theorem 8.1. *If E is any real or complex vector space of finite dimension, then any two norms on E are equivalent.*

Proof. It is enough to prove that any norm $\| \ \|$ is equivalent to the 1-norm. We already proved that the function $x \mapsto \|x\|$ is continuous with respect to the norm $\| \ \|_1$, and we observed that the unit sphere S_1^{n-1} is compact. Now we just recalled that because the function $f \colon x \mapsto \|x\|$ is continuous and because S_1^{n-1} is compact, the function f has a minimum m and a maximum M, and because $\|x\|$ is never zero on S_1^{n-1}, we must have $m > 0$. Consequently, we just proved that if $\|x\|_1 = 1$, then

$$0 < m \leq \|x\| \leq M,$$

so for any $x \in E$ with $x \neq 0$, we get

$$m \leq \|x/\|x\|_1\| \leq M,$$

which implies

$$m \|x\|_1 \leq \|x\| \leq M \|x\|_1 \,.$$

Since the above inequality holds trivially if $x = 0$, we just proved that $\| \ \|$ and $\| \ \|_1$ are equivalent, as claimed. $\qquad\square$

Remark: Let P be a $n \times n$ symmetric positive definite matrix. It is immediately verified that the map $x \mapsto \|x\|_P$ given by

$$\|x\|_P = (x^\top P x)^{1/2}$$

is a norm on $\mathbb{R}^n$ called a *quadratic norm*. Using some convex analysis (the Löwner–John ellipsoid), it can be shown that *any* norm $\| \ \|$ on $\mathbb{R}^n$ can be approximated by a quadratic norm in the sense that there is a quadratic norm $\| \ \|_P$ such that

$$\|x\|_P \leq \|x\| \leq \sqrt{n} \, \|x\|_P \qquad \text{for all } x \in \mathbb{R}^n;$$

see Boyd and Vandenberghe [Boyd and Vandenberghe (2004)], Section 8.4.1. Next we will consider norms on matrices.

8.2 Matrix Norms

For simplicity of exposition, we will consider the vector spaces $M_n(\mathbb{R})$ and $M_n(\mathbb{C})$ of square $n \times n$ matrices. Most results also hold for the spaces $M_{m,n}(\mathbb{R})$ and $M_{m,n}(\mathbb{C})$ of rectangular $m \times n$ matrices. Since $n \times n$ matrices can be multiplied, the idea behind matrix norms is that they should behave "well" with respect to matrix multiplication.

Definition 8.3. A *matrix norm* $\|\ \|$ on the space of square $n \times n$ matrices in $M_n(K)$, with $K = \mathbb{R}$ or $K = \mathbb{C}$, is a norm on the vector space $M_n(K)$, with the additional property called *submultiplicativity* that

$$\|AB\| \leq \|A\| \, \|B\| \,,$$

for all $A, B \in M_n(K)$. A norm on matrices satisfying the above property is often called a *submultiplicative* matrix norm.

Since $I^2 = I$, from $\|I\| = \|I^2\| \leq \|I\|^2$, we get $\|I\| \geq 1$, for every matrix norm.

Before giving examples of matrix norms, we need to review some basic definitions about matrices. Given any matrix $A = (a_{ij}) \in M_{m,n}(\mathbb{C})$, the *conjugate* $\overline{A}$ of A is the matrix such that

$$\overline{A}_{ij} = \overline{a}_{ij}, \quad 1 \leq i \leq m, 1 \leq j \leq n.$$

The *transpose* of A is the $n \times m$ matrix $A^\top$ such that

$$A^\top_{ij} = a_{ji}, \quad 1 \leq i \leq m, 1 \leq j \leq n.$$

The *adjoint* of A is the $n \times m$ matrix A^* such that

$$A^* = \overline{(A^\top)} = (\overline{A})^\top.$$

When A is a real matrix, $A^* = A^\top$. A matrix $A \in M_n(\mathbb{C})$ is *Hermitian* if

$$A^* = A.$$

If A is a real matrix $(A \in M_n(\mathbb{R}))$, we say that A is *symmetric* if

$$A^\top = A.$$

A matrix $A \in M_n(\mathbb{C})$ is *normal* if

$$AA^* = A^*A,$$

and if A is a real matrix, it is *normal* if

$$AA^\top = A^\top A.$$

A matrix $U \in M_n(\mathbb{C})$ is *unitary* if

$$UU^* = U^*U = I.$$

A real matrix $Q \in M_n(\mathbb{R})$ is *orthogonal* if

$$QQ^\top = Q^\top Q = I.$$

Given any matrix $A = (a_{ij}) \in M_n(\mathbb{C})$, the *trace* $\mathrm{tr}(A)$ of A is the sum of its diagonal elements

$$\mathrm{tr}(A) = a_{11} + \cdots + a_{nn}.$$

It is easy to show that the trace is a linear map, so that

$$\mathrm{tr}(\lambda A) = \lambda \mathrm{tr}(A)$$

and

$$\mathrm{tr}(A + B) = \mathrm{tr}(A) + \mathrm{tr}(B).$$

Moreover, if A is an $m \times n$ matrix and B is an $n \times m$ matrix, it is not hard to show that

$$\mathrm{tr}(AB) = \mathrm{tr}(BA).$$

We also review eigenvalues and eigenvectors. We content ourselves with definition involving matrices. A more general treatment will be given later on (see Chapter 14).

Definition 8.4. Given any square matrix $A \in M_n(\mathbb{C})$, a complex number $\lambda \in \mathbb{C}$ is an *eigenvalue* of A if there is some *nonzero* vector $u \in \mathbb{C}^n$, such that

$$Au = \lambda u.$$

If λ is an eigenvalue of A, then the *nonzero* vectors $u \in \mathbb{C}^n$ such that $Au = \lambda u$ are called *eigenvectors of A associated with* λ; together with the zero vector, these eigenvectors form a subspace of $\mathbb{C}^n$ denoted by $E_\lambda(A)$, and called the *eigenspace associated with* λ.

Remark: Note that Definition 8.4 *requires an eigenvector to be nonzero*. A somewhat unfortunate consequence of this requirement is that the set of eigenvectors is *not* a subspace, since the zero vector is missing! On the positive side, whenever eigenvectors are involved, there is no need to say that they are nonzero. The fact that eigenvectors are nonzero is implicitly used in all the arguments involving them, so it seems safer (but perhaps not as elegant) to stipulate that eigenvectors should be nonzero.

If A is a square real matrix $A \in M_n(\mathbb{R})$, then we restrict Definition 8.4 to real eigenvalues $\lambda \in \mathbb{R}$ and real eigenvectors. However, it should be noted that although every complex matrix always has at least some complex eigenvalue, a real matrix may not have any real eigenvalues. For example, the matrix

$$A = \begin{pmatrix} 0 & -1 \\ 1 & 0 \end{pmatrix}$$

has the complex eigenvalues i and $-i$, but no real eigenvalues. Thus, typically even for real matrices, we consider complex eigenvalues.

Observe that $\lambda \in \mathbb{C}$ is an eigenvalue of A

- iff $Au = \lambda u$ for some nonzero vector $u \in \mathbb{C}^n$
- iff $(\lambda I - A)u = 0$
- iff the matrix $\lambda I - A$ defines a linear map which has a nonzero kernel, that is,
- iff $\lambda I - A$ not invertible.

However, from Proposition 6.7, $\lambda I - A$ is not invertible iff

$$\det(\lambda I - A) = 0.$$

Now $\det(\lambda I - A)$ is a polynomial of degree n in the indeterminate λ, in fact, of the form

$$\lambda^n - \mathrm{tr}(A)\lambda^{n-1} + \cdots + (-1)^n \det(A).$$

Thus we see that the eigenvalues of A are the zeros (also called *roots*) of the above polynomial. Since every complex polynomial of degree n has exactly n roots, counted with their multiplicity, we have the following definition:

Definition 8.5. Given any square $n \times n$ matrix $A \in M_n(\mathbb{C})$, the polynomial

$$\det(\lambda I - A) = \lambda^n - \mathrm{tr}(A)\lambda^{n-1} + \cdots + (-1)^n \det(A)$$

is called the *characteristic polynomial* of A. The n (not necessarily distinct) roots $\lambda_1, \ldots, \lambda_n$ of the characteristic polynomial are all the *eigenvalues* of A and constitute the *spectrum* of A. We let

$$\rho(A) = \max_{1 \leq i \leq n} |\lambda_i|$$

be the largest modulus of the eigenvalues of A, called the *spectral radius* of A.

Since the eigenvalue $\lambda_1, \ldots, \lambda_n$ of A are the zeros of the polynomial

$$\det(\lambda I - A) = \lambda^n - \mathrm{tr}(A)\lambda^{n-1} + \cdots + (-1)^n \det(A),$$

we deduce (see Section 14.1 for details) that

$$\mathrm{tr}(A) = \lambda_1 + \cdots + \lambda_n$$
$$\det(A) = \lambda_1 \cdots \lambda_n.$$

Proposition 8.3. *For any matrix norm* $\| \ \|$ *on* $\mathrm{M}_n(\mathbb{C})$ *and for any square* $n \times n$ *matrix* $A \in \mathrm{M}_n(\mathbb{C})$, *we have*

$$\rho(A) \leq \|A\|.$$

Proof. Let λ be some eigenvalue of A for which $|\lambda|$ is maximum, that is, such that $|\lambda| = \rho(A)$. If $u \ (\neq 0)$ is any eigenvector associated with λ and if U is the $n \times n$ matrix whose columns are all u, then $Au = \lambda u$ implies

$$AU = \lambda U,$$

and since

$$|\lambda| \, \|U\| = \|\lambda U\| = \|AU\| \leq \|A\| \, \|U\|$$

and $U \neq 0$, we have $\|U\| \neq 0$, and get

$$\rho(A) = |\lambda| \leq \|A\|,$$

as claimed. $\qquad\qquad\qquad\qquad\qquad\qquad\qquad\qquad\qquad\qquad\qquad\square$

Proposition 8.3 also holds for any real matrix norm $\| \ \|$ on $\mathrm{M}_n(\mathbb{R})$ but the proof is more subtle and requires the notion of induced norm. We prove it after giving Definition 8.7.

It turns out that if A is a real $n \times n$ symmetric matrix, then the eigenvalues of A are all real and there is some orthogonal matrix Q such that

$$A = Q\mathrm{diag}(\lambda_1, \ldots, \lambda_n)Q^\top,$$

where $\mathrm{diag}(\lambda_1, \ldots, \lambda_n)$ denotes the matrix whose only nonzero entries (if any) are its diagonal entries, which are the (real) eigenvalues of A. Similarly, if A is a complex $n \times n$ Hermitian matrix, then the eigenvalues of A are all real and there is some unitary matrix U such that

$$A = U\mathrm{diag}(\lambda_1, \ldots, \lambda_n)U^*,$$

where $\mathrm{diag}(\lambda_1, \ldots, \lambda_n)$ denotes the matrix whose only nonzero entries (if any) are its diagonal entries, which are the (real) eigenvalues of A. See Chapter 16 for the proof of these results.

We now return to matrix norms. We begin with the so-called *Frobenius norm*, which is just the norm $\| \ \|_2$ on $\mathbb{C}^{n^2}$, where the $n \times n$ matrix A is viewed as the vector obtained by concatenating together the rows (or the columns) of A. The reader should check that for any $n \times n$ complex matrix $A = (a_{ij})$,

$$\left(\sum_{i,j=1}^{n} |a_{ij}|^2 \right)^{1/2} = \sqrt{\mathrm{tr}(A^*A)} = \sqrt{\mathrm{tr}(AA^*)}.$$

Definition 8.6. The *Frobenius norm* $\| \ \|_F$ is defined so that for every square $n \times n$ matrix $A \in \mathrm{M}_n(\mathbb{C})$,

$$\|A\|_F = \left(\sum_{i,j=1}^{n} |a_{ij}|^2 \right)^{1/2} = \sqrt{\mathrm{tr}(AA^*)} = \sqrt{\mathrm{tr}(A^*A)}.$$

The following proposition show that the Frobenius norm is a matrix norm satisfying other nice properties.

Proposition 8.4. *The Frobenius norm $\| \ \|_F$ on $\mathrm{M}_n(\mathbb{C})$ satisfies the following properties:*

(1) It is a matrix norm; that is, $\|AB\|_F \leq \|A\|_F \|B\|_F$, for all $A, B \in \mathrm{M}_n(\mathbb{C})$.

(2) It is unitarily invariant, which means that for all unitary matrices U, V, we have

$$\|A\|_F = \|UA\|_F = \|AV\|_F = \|UAV\|_F.$$

(3) $\sqrt{\rho(A^*A)} \leq \|A\|_F \leq \sqrt{n}\sqrt{\rho(A^*A)}$, for all $A \in \mathrm{M}_n(\mathbb{C})$.

Proof. (1) The only property that requires a proof is the fact $\|AB\|_F \leq \|A\|_F \|B\|_F$. This follows from the Cauchy–Schwarz inequality:

$$\|AB\|_F^2 = \sum_{i,j=1}^{n} \left| \sum_{k=1}^{n} a_{ik}b_{kj} \right|^2$$

$$\leq \sum_{i,j=1}^{n} \left(\sum_{h=1}^{n} |a_{ih}|^2 \right) \left(\sum_{k=1}^{n} |b_{kj}|^2 \right)$$

$$= \left(\sum_{i,h=1}^{n} |a_{ih}|^2 \right) \left(\sum_{k,j=1}^{n} |b_{kj}|^2 \right) = \|A\|_F^2 \|B\|_F^2.$$

(2) We have

$$\|A\|_F^2 = \mathrm{tr}(A^*A) = \mathrm{tr}(VV^*A^*A) = \mathrm{tr}(V^*A^*AV) = \|AV\|_F^2,$$

and

$$\|A\|_F^2 = \text{tr}(A^*A) = \text{tr}(A^*U^*UA) = \|UA\|_F^2.$$

The identity

$$\|A\|_F = \|UAV\|_F$$

follows from the previous two.

(3) It is well known that the trace of a matrix is equal to the sum of its eigenvalues. Furthermore, A^*A is symmetric positive semidefinite (which means that its eigenvalues are nonnegative), so $\rho(A^*A)$ is the largest eigenvalue of A^*A and

$$\rho(A^*A) \leq \text{tr}(A^*A) \leq n\rho(A^*A),$$

which yields (3) by taking square roots. $\qquad\square$

Remark: The Frobenius norm is also known as the *Hilbert–Schmidt norm* or the *Schur norm*. So many famous names associated with such a simple thing!

8.3 Subordinate Norms

We now give another method for obtaining matrix norms using subordinate norms. First we need a proposition that shows that in a finite-dimensional space, the linear map induced by a matrix is bounded, and thus continuous.

Proposition 8.5. *For every norm $\|\ \|$ on $\mathbb{C}^n$ (or $\mathbb{R}^n$), for every matrix $A \in \text{M}_n(\mathbb{C})$ (or $A \in \text{M}_n(\mathbb{R})$), there is a real constant $C_A \geq 0$, such that*

$$\|Au\| \leq C_A \|u\|,$$

for every vector $u \in \mathbb{C}^n$ (or $u \in \mathbb{R}^n$ if A is real).

Proof. For every basis $(e_1, \ldots, e_n)$ of $\mathbb{C}^n$ (or $\mathbb{R}^n$), for every vector $u = u_1e_1 + \cdots + u_ne_n$, we have

$$\begin{aligned}
\|Au\| &= \|u_1A(e_1) + \cdots + u_nA(e_n)\| \\
&\leq |u_1| \|A(e_1)\| + \cdots + |u_n| \|A(e_n)\| \\
&\leq C_1(|u_1| + \cdots + |u_n|) = C_1 \|u\|_1,
\end{aligned}$$

where $C_1 = \max_{1 \leq i \leq n} \|A(e_i)\|$. By Theorem 8.1, the norms $\|\ \|$ and $\|\ \|_1$ are equivalent, so there is some constant $C_2 > 0$ so that $\|u\|_1 \leq C_2 \|u\|$ for all u, which implies that

$$\|Au\| \leq C_A \|u\|,$$

where $C_A = C_1C_2$. $\qquad\square$

Proposition 8.5 says that every linear map on a finite-dimensional space is *bounded*. This implies that every linear map on a finite-dimensional space is continuous. Actually, it is not hard to show that a linear map on a normed vector space E is bounded iff it is continuous, regardless of the dimension of E.

Proposition 8.5 implies that for every matrix $A \in M_n(\mathbb{C})$ (or $A \in M_n(\mathbb{R})$),

$$\sup_{\substack{x \in \mathbb{C}^n \\ x \neq 0}} \frac{\|Ax\|}{\|x\|} \leq C_A.$$

Since $\|\lambda u\| = |\lambda| \|u\|$, for every nonzero vector x, we have

$$\frac{\|Ax\|}{\|x\|} = \frac{\|x\| \, \|A(x/\|x\|)\|}{\|x\|} = \|A(x/\|x\|)\|,$$

which implies that

$$\sup_{\substack{x \in \mathbb{C}^n \\ x \neq 0}} \frac{\|Ax\|}{\|x\|} = \sup_{\substack{x \in \mathbb{C}^n \\ \|x\| = 1}} \|Ax\|.$$

Similarly

$$\sup_{\substack{x \in \mathbb{R}^n \\ x \neq 0}} \frac{\|Ax\|}{\|x\|} = \sup_{\substack{x \in \mathbb{R}^n \\ \|x\| = 1}} \|Ax\|.$$

The above considerations justify the following definition.

Definition 8.7. If $\| \ \|$ is any norm on $\mathbb{C}^n$, we define the function $\| \ \|_{\mathrm{op}}$ on $M_n(\mathbb{C})$ by

$$\|A\|_{\mathrm{op}} = \sup_{\substack{x \in \mathbb{C}^n \\ x \neq 0}} \frac{\|Ax\|}{\|x\|} = \sup_{\substack{x \in \mathbb{C}^n \\ \|x\| = 1}} \|Ax\|.$$

The function $A \mapsto \|A\|_{\mathrm{op}}$ is called the *subordinate matrix norm* or *operator norm* induced by the norm $\| \ \|$.

Another notation for the operator norm of a matrix A (in particular, used by Horn and Johnson [Horn and Johnson (1990)]), is $\|A\|$.

It is easy to check that the function $A \mapsto \|A\|_{\mathrm{op}}$ is indeed a norm, and by definition, it satisfies the property

$$\|Ax\| \leq \|A\|_{\mathrm{op}} \|x\|, \quad \text{for all } x \in \mathbb{C}^n.$$

A norm $\| \ \|_{op}$ on $M_n(\mathbb{C})$ satisfying the above property is said to be *subordinate* to the vector norm $\| \ \|$ on $\mathbb{C}^n$. As a consequence of the above inequality, we have

$$\|ABx\| \leq \|A\|_{op} \|Bx\| \leq \|A\|_{op} \|B\|_{op} \|x\|,$$

for all $x \in \mathbb{C}^n$, which implies that

$$\|AB\|_{op} \leq \|A\|_{op} \|B\|_{op} \quad \text{for all } A, B \in M_n(\mathbb{C}),$$

showing that $A \mapsto \|A\|_{op}$ is a matrix norm (it is submultiplicative).

Observe that the operator norm is also defined by

$$\|A\|_{op} = \inf\{\lambda \in \mathbb{R} \mid \|Ax\| \leq \lambda \|x\|, \text{ for all } x \in \mathbb{C}^n\}.$$

Since the function $x \mapsto \|Ax\|$ is continuous (because $\big| \|Ay\| - \|Ax\| \big| \leq \|Ay - Ax\| \leq C_A \|x - y\|$) and the unit sphere $S^{n-1} = \{x \in \mathbb{C}^n \mid \|x\| = 1\}$ is compact, there is some $x \in \mathbb{C}^n$ such that $\|x\| = 1$ and

$$\|Ax\| = \|A\|_{op}.$$

Equivalently, there is some $x \in \mathbb{C}^n$ such that $x \neq 0$ and

$$\|Ax\| = \|A\|_{op} \|x\|.$$

The definition of an operator norm also implies that

$$\|I\|_{op} = 1.$$

The above shows that the Frobenius norm is not a subordinate matrix norm (why?).

If $\| \ \|$ is a vector norm on $\mathbb{C}^n$, the operator norm $\| \ \|_{op}$ that it induces applies to matrices in $M_n(\mathbb{C})$. If we are careful to denote vectors and matrices so that no confusion arises, for example, by using lower case letters for vectors and upper case letters for matrices, it should be clear that $\|A\|_{op}$ is the operator norm of the matrix A and that $\|x\|$ is the vector norm of x. Consequently, following common practice to alleviate notation, we will drop the subscript "op" and simply write $\|A\|$ instead of $\|A\|_{op}$.

The notion of subordinate norm can be slightly generalized.

Definition 8.8. If $K = \mathbb{R}$ or $K = \mathbb{C}$, for any norm $\| \ \|$ on $M_{m,n}(K)$, and for any two norms $\| \ \|_a$ on K^n and $\| \ \|_b$ on K^m, we say that the norm $\| \ \|$ is *subordinate* to the norms $\| \ \|_a$ and $\| \ \|_b$ if

$$\|Ax\|_b \leq \|A\| \|x\|_a \quad \text{for all } A \in M_{m,n}(K) \text{ and all } x \in K^n.$$

Remark: For any norm $\| \; \|$ on $\mathbb{C}^n$, we can define the function $\| \; \|_{\mathbb{R}}$ on $\mathrm{M}_n(\mathbb{R})$ by

$$\|A\|_{\mathbb{R}} = \sup_{\substack{x \in \mathbb{R}^n \\ x \neq 0}} \frac{\|Ax\|}{\|x\|} = \sup_{\substack{x \in \mathbb{R}^n \\ \|x\|=1}} \|Ax\| \, .$$

The function $A \mapsto \|A\|_{\mathbb{R}}$ is a matrix norm on $\mathrm{M}_n(\mathbb{R})$, and

$$\|A\|_{\mathbb{R}} \leq \|A\| \, ,$$

for all real matrices $A \in \mathrm{M}_n(\mathbb{R})$. However, it is possible to construct vector norms $\| \; \|$ on $\mathbb{C}^n$ and *real* matrices A such that

$$\|A\|_{\mathbb{R}} < \|A\| \, .$$

In order to avoid this kind of difficulties, we define subordinate matrix norms over $\mathrm{M}_n(\mathbb{C})$. Luckily, it turns out that $\|A\|_{\mathbb{R}} = \|A\|$ for the vector norms, $\| \; \|_1, \| \; \|_2$, and $\| \; \|_\infty$.

We now prove Proposition 8.3 for real matrix norms.

Proposition 8.6. *For any matrix norm $\| \; \|$ on $\mathrm{M}_n(\mathbb{R})$ and for any square $n \times n$ matrix $A \in \mathrm{M}_n(\mathbb{R})$, we have*

$$\rho(A) \leq \|A\| \, .$$

Proof. We follow the proof in Denis Serre's book [Serre (2010)]. If A is a real matrix, the problem is that the eigenvectors associated with the eigenvalue of maximum modulus may be complex. We use a trick based on the fact that for every matrix A (real or complex),

$$\rho(A^k) = (\rho(A))^k,$$

which is left as an exercise (use Proposition 14.4 which shows that if $(\lambda_1, \ldots, \lambda_n)$ are the (not necessarily distinct) eigenvalues of A, then $(\lambda_1^k, \ldots, \lambda_n^k)$ are the eigenvalues of A^k, for $k \geq 1$).

Pick any complex matrix norm $\| \; \|_c$ on $\mathbb{C}^n$ (for example, the Frobenius norm, or any subordinate matrix norm induced by a norm on $\mathbb{C}^n$). The restriction of $\| \; \|_c$ to real matrices is a real norm that we also denote by $\| \; \|_c$. Now by Theorem 8.1, since $\mathrm{M}_n(\mathbb{R})$ has finite dimension n^2, there is some constant $C > 0$ so that

$$\|B\|_c \leq C \|B\| \, , \quad \text{for all} \quad B \in \mathrm{M}_n(\mathbb{R}).$$

Furthermore, for every $k \geq 1$ and for every real $n \times n$ matrix A, by Proposition 8.3, $\rho(A^k) \leq \left\|A^k\right\|_c$, and because $\| \; \|$ is a matrix norm, $\left\|A^k\right\| \leq \|A\|^k$, so we have

$$(\rho(A))^k = \rho(A^k) \leq \left\|A^k\right\|_c \leq C \left\|A^k\right\| \leq C \|A\|^k \, ,$$

for all $k \geq 1$. It follows that

$$\rho(A) \leq C^{1/k} \|A\|, \quad \text{for all} \quad k \geq 1.$$

However because $C > 0$, we have $\lim_{k \mapsto \infty} C^{1/k} = 1$ (we have $\lim_{k \mapsto \infty} \frac{1}{k} \log(C) = 0$). Therefore, we conclude that

$$\rho(A) \leq \|A\|,$$

as desired. $\square$

We now determine explicitly what are the subordinate matrix norms associated with the vector norms $\| \ \|_1$, $\| \ \|_2$, and $\| \ \|_\infty$.

Proposition 8.7. *For every square matrix $A = (a_{ij}) \in \mathrm{M}_n(\mathbb{C})$, we have*

$$\|A\|_1 = \sup_{\substack{x \in \mathbb{C}^n \\ \|x\|_1 = 1}} \|Ax\|_1 = \max_j \sum_{i=1}^{n} |a_{ij}|$$

$$\|A\|_\infty = \sup_{\substack{x \in \mathbb{C}^n \\ \|x\|_\infty = 1}} \|Ax\|_\infty = \max_i \sum_{j=1}^{n} |a_{ij}|$$

$$\|A\|_2 = \sup_{\substack{x \in \mathbb{C}^n \\ \|x\|_2 = 1}} \|Ax\|_2 = \sqrt{\rho(A^*A)} = \sqrt{\rho(AA^*)}.$$

Note that $\|A\|_1$ is the maximum of the ℓ^1-norms of the columns of A and $\|A\|_\infty$ is the maximum of the ℓ^1-norms of the rows of A. Furthermore, $\|A^*\|_2 = \|A\|_2$, the norm $\| \ \|_2$ is unitarily invariant, which means that

$$\|A\|_2 = \|UAV\|_2$$

for all unitary matrices U, V, and if A is a normal matrix, then $\|A\|_2 = \rho(A)$.

Proof. For every vector u, we have

$$\|Au\|_1 = \sum_i \left| \sum_j a_{ij} u_j \right| \leq \sum_j |u_j| \sum_i |a_{ij}| \leq \left(\max_j \sum_i |a_{ij}| \right) \|u\|_1,$$

which implies that

$$\|A\|_1 \leq \max_j \sum_{i=1}^{n} |a_{ij}|.$$

It remains to show that equality can be achieved. For this let j_0 be some index such that

$$\max_j \sum_i |a_{ij}| = \sum_i |a_{ij_0}|,$$

and let $u_i = 0$ for all $i \neq j_0$ and $u_{j_0} = 1$.

In a similar way, we have

$$\|Au\|_\infty = \max_i \left| \sum_j a_{ij} u_j \right| \leq \left(\max_i \sum_j |a_{ij}| \right) \|u\|_\infty,$$

which implies that

$$\|A\|_\infty \leq \max_i \sum_{j=1}^n |a_{ij}|.$$

To achieve equality, let i_0 be some index such that

$$\max_i \sum_j |a_{ij}| = \sum_j |a_{i_0 j}|.$$

The reader should check that the vector given by

$$u_j = \begin{cases} \dfrac{\overline{a_{i_0 j}}}{|a_{i_0 j}|} & \text{if } a_{i_0 j} \neq 0 \\ 1 & \text{if } a_{i_0 j} = 0 \end{cases}$$

works.

We have

$$\|A\|_2^2 = \sup_{\substack{x \in \mathbb{C}^n \\ x^* x = 1}} \|Ax\|_2^2 = \sup_{\substack{x \in \mathbb{C}^n \\ x^* x = 1}} x^* A^* A x.$$

Since the matrix $A^* A$ is symmetric, it has real eigenvalues and it can be diagonalized with respect to a unitary matrix. These facts can be used to prove that the function $x \mapsto x^* A^* A x$ has a maximum on the sphere $x^* x = 1$ equal to the largest eigenvalue of $A^* A$, namely, $\rho(A^* A)$. We postpone the proof until we discuss optimizing quadratic functions. Therefore,

$$\|A\|_2 = \sqrt{\rho(A^* A)}.$$

Let us now prove that $\rho(A^* A) = \rho(A A^*)$. First assume that $\rho(A^* A) > 0$. In this case, there is some eigenvector $u \, (\neq 0)$ such that

$$A^* A u = \rho(A^* A) u,$$

and since $\rho(A^* A) > 0$, we must have $Au \neq 0$. Since $Au \neq 0$,

$$A A^* (A u) = A(A^* A u) = \rho(A^* A) A u$$

which means that $\rho(A^* A)$ is an eigenvalue of $A A^*$, and thus

$$\rho(A^* A) \leq \rho(A A^*).$$

Because $(A^*)^* = A$, by replacing A by A^*, we get

$$\rho(A A^*) \leq \rho(A^* A),$$

and so $\rho(A^*A) = \rho(AA^*)$.

If $\rho(A^*A) = 0$, then we must have $\rho(AA^*) = 0$, since otherwise by the previous reasoning we would have $\rho(A^*A) = \rho(AA^*) > 0$. Hence, in all case

$$\|A\|_2^2 = \rho(A^*A) = \rho(AA^*) = \|A^*\|_2^2.$$

For any unitary matrices U and V, it is an easy exercise to prove that V^*A^*AV and A^*A have the same eigenvalues, so

$$\|A\|_2^2 = \rho(A^*A) = \rho(V^*A^*AV) = \|AV\|_2^2,$$

and also

$$\|A\|_2^2 = \rho(A^*A) = \rho(A^*U^*UA) = \|UA\|_2^2.$$

Finally, if A is a normal matrix $(AA^* = A^*A)$, it can be shown that there is some unitary matrix U so that

$$A = UDU^*,$$

where $D = \mathrm{diag}(\lambda_1, \ldots, \lambda_n)$ is a diagonal matrix consisting of the eigenvalues of A, and thus

$$A^*A = (UDU^*)^*UDU^* = UD^*U^*UDU^* = UD^*DU^*.$$

However, $D^*D = \mathrm{diag}(|\lambda_1|^2, \ldots, |\lambda_n|^2)$, which proves that

$$\rho(A^*A) = \rho(D^*D) = \max_i |\lambda_i|^2 = (\rho(A))^2,$$

so that $\|A\|_2 = \rho(A)$. $\square$

Definition 8.9. For $A = (a_{ij}) \in \mathrm{M}_n(\mathbb{C})$, the norm $\|A\|_2 =$ is often called the *spectral norm*.

Observe that Property (3) of Proposition 8.4 says that

$$\|A\|_2 \le \|A\|_F \le \sqrt{n}\,\|A\|_2,$$

which shows that the Frobenius norm is an upper bound on the spectral norm. The Frobenius norm is much easier to compute than the spectral norm.

The reader will check that the above proof still holds if the matrix A is real (change unitary to orthogonal), confirming the fact that $\|A\|_{\mathbb{R}} = \|A\|$ for the vector norms $\| \|_1, \| \|_2$, and $\| \|_\infty$. It is also easy to verify that the proof goes through for *rectangular* $m \times n$ matrices, with the same formulae. Similarly, the Frobenius norm given by

$$\|A\|_F = \left(\sum_{i=1}^{m} \sum_{j=1}^{n} |a_{ij}|^2 \right)^{1/2} = \sqrt{\mathrm{tr}(A^*A)} = \sqrt{\mathrm{tr}(AA^*)}$$

is also a norm on rectangular matrices. For these norms, whenever AB makes sense, we have

$$\|AB\| \le \|A\| \|B\|.$$

Remark: It can be shown that for any two real numbers $p, q \ge 1$ such that $\dfrac{1}{p} + \dfrac{1}{q} = 1$, we have

$$\|A^*\|_q = \|A\|_p = \sup\{\Re(y^* Ax) \mid \|x\|_p = 1, \|y\|_q = 1\}$$
$$= \sup\{|\langle Ax, y\rangle| \mid \|x\|_p = 1, \|y\|_q = 1\},$$

where $\|A^*\|_q$ and $\|A\|_p$ are the operator norms.

Remark: Let $(E, \|\ \|)$ and $(F, \|\ \|)$ be two normed vector spaces (for simplicity of notation, we use the same symbol $\|\ \|$ for the norms on E and F; this should not cause any confusion). Recall that a function $f\colon E \to F$ is *continuous* if for every $a \in E$, for every $\epsilon > 0$, there is some $\eta > 0$ such that for all $x \in E$,

$$\text{if} \quad \|x - a\| \le \eta \quad \text{then} \quad \|f(x) - f(a)\| \le \epsilon.$$

It is not hard to show that a *linear map* $f\colon E \to F$ is continuous iff there is some constant $C \ge 0$ such that

$$\|f(x)\| \le C \|x\| \text{ for all } x \in E.$$

If so, we say that f is *bounded* (or a *linear bounded operator*). We let $\mathcal{L}(E; F)$ denote the set of all continuous (equivalently, bounded) linear maps from E to F. Then we can define the *operator norm* (or *subordinate norm*) $\|\ \|$ on $\mathcal{L}(E; F)$ as follows: for every $f \in \mathcal{L}(E; F)$,

$$\|f\| = \sup_{\substack{x \in E \\ x \ne 0}} \frac{\|f(x)\|}{\|x\|} = \sup_{\substack{x \in E \\ \|x\| = 1}} \|f(x)\|,$$

or equivalently by

$$\|f\| = \inf\{\lambda \in \mathbb{R} \mid \|f(x)\| \le \lambda \|x\|, \text{ for all } x \in E\}.$$

It is not hard to show that the map $f \mapsto \|f\|$ is a norm on $\mathcal{L}(E; F)$ satisfying the property

$$\|f(x)\| \le \|f\| \|x\|$$

for all $x \in E$, and that if $f \in \mathcal{L}(E; F)$ and $g \in \mathcal{L}(F; G)$, then

$$\|g \circ f\| \le \|g\| \|f\|.$$

Operator norms play an important role in functional analysis, especially when the spaces E and F are *complete*.

8.4 Inequalities Involving Subordinate Norms

In this section we discuss two technical inequalities which will be needed
for certain proofs in the last three sections of this chapter. First we prove a
proposition which will be needed when we deal with the condition number
of a matrix.

Proposition 8.8. *Let $\| \ \|$ be any matrix norm, and let $B \in M_n(\mathbb{C})$ such
that $\|B\| < 1$.*

*(1) If $\| \ \|$ is a subordinate matrix norm, then the matrix $I + B$ is invertible
and*

$$\left\|(I + B)^{-1}\right\| \leq \frac{1}{1 - \|B\|}.$$

*(2) If a matrix of the form $I+B$ is singular, then $\|B\| \geq 1$ for every matrix
norm (not necessarily subordinate).*

Proof. (1) Observe that $(I + B)u = 0$ implies $Bu = -u$, so

$$\|u\| = \|Bu\|.$$

Recall that

$$\|Bu\| \leq \|B\| \, \|u\|$$

for every subordinate norm. Since $\|B\| < 1$, if $u \neq 0$, then

$$\|Bu\| < \|u\|,$$

which contradicts $\|u\| = \|Bu\|$. Therefore, we must have $u = 0$, which
proves that $I + B$ is injective, and thus bijective, i.e., invertible. Then we
have

$$(I + B)^{-1} + B(I + B)^{-1} = (I + B)(I + B)^{-1} = I,$$

so we get

$$(I + B)^{-1} = I - B(I + B)^{-1},$$

which yields

$$\left\|(I + B)^{-1}\right\| \leq 1 + \|B\| \left\|(I + B)^{-1}\right\|,$$

and finally,

$$\left\|(I + B)^{-1}\right\| \leq \frac{1}{1 - \|B\|}.$$

(2) If $I+B$ is singular, then -1 is an eigenvalue of B, and by Proposition 8.3,
we get $\rho(B) \leq \|B\|$, which implies $1 \leq \rho(B) \leq \|B\|$. $\square$

The second inequality is a result is that is needed to deal with the convergence of sequences of powers of matrices.

Proposition 8.9. *For every matrix $A \in M_n(\mathbb{C})$ and for every $\epsilon > 0$, there is some subordinate matrix norm $\| \ \|$ such that*

$$\|A\| \leq \rho(A) + \epsilon.$$

Proof. By Theorem 14.1, there exists some invertible matrix U and some upper triangular matrix T such that

$$A = UTU^{-1},$$

and say that

$$T = \begin{pmatrix} \lambda_1 & t_{12} & t_{13} & \cdots & t_{1n} \\ 0 & \lambda_2 & t_{23} & \cdots & t_{2n} \\ \vdots & \vdots & \ddots & \vdots & \vdots \\ 0 & 0 & \cdots & \lambda_{n-1} & t_{n-1\,n} \\ 0 & 0 & \cdots & 0 & \lambda_n \end{pmatrix},$$

where $\lambda_1, \ldots, \lambda_n$ are the eigenvalues of A. For every $\delta \neq 0$, define the diagonal matrix

$$D_\delta = \mathrm{diag}(1, \delta, \delta^2, \ldots, \delta^{n-1}),$$

and consider the matrix

$$(UD_\delta)^{-1} A (UD_\delta) = D_\delta^{-1} T D_\delta = \begin{pmatrix} \lambda_1 & \delta t_{12} & \delta^2 t_{13} & \cdots & \delta^{n-1} t_{1n} \\ 0 & \lambda_2 & \delta t_{23} & \cdots & \delta^{n-2} t_{2n} \\ \vdots & \vdots & \ddots & \vdots & \vdots \\ 0 & 0 & \cdots & \lambda_{n-1} & \delta t_{n-1\,n} \\ 0 & 0 & \cdots & 0 & \lambda_n \end{pmatrix}.$$

Now define the function $\| \ \| : M_n(\mathbb{C}) \to \mathbb{R}$ by

$$\|B\| = \left\| (UD_\delta)^{-1} B (UD_\delta) \right\|_\infty,$$

for every $B \in M_n(\mathbb{C})$. Then it is easy to verify that the above function is the matrix norm subordinate to the vector norm

$$v \mapsto \left\| (UD_\delta)^{-1} v \right\|_\infty.$$

Furthermore, for every $\epsilon > 0$, we can pick δ so that

$$\sum_{j=i+1}^{n} |\delta^{j-i} t_{ij}| \leq \epsilon, \quad 1 \leq i \leq n-1,$$

and by definition of the norm $\| \ \|_\infty$, we get

$$\|A\| \leq \rho(A) + \epsilon,$$

which shows that the norm that we have constructed satisfies the required properties. $\qquad \square$

Note that equality is generally not possible; consider the matrix

$$A = \begin{pmatrix} 0 & 1 \\ 0 & 0 \end{pmatrix},$$

for which $\rho(A) = 0 < \|A\|$, since $A \neq 0$.

8.5   Condition Numbers of Matrices

Unfortunately, there exist linear systems $Ax = b$ whose solutions are not stable under small perturbations of either b or A. For example, consider the system

$$\begin{pmatrix} 10 & 7 & 8 & 7 \\ 7 & 5 & 6 & 5 \\ 8 & 6 & 10 & 9 \\ 7 & 5 & 9 & 10 \end{pmatrix} \begin{pmatrix} x_1 \\ x_2 \\ x_3 \\ x_4 \end{pmatrix} = \begin{pmatrix} 32 \\ 23 \\ 33 \\ 31 \end{pmatrix}.$$

The reader should check that it has the solution $x = (1, 1, 1, 1)$. If we perturb slightly the right-hand side as $b + \Delta b$, where

$$\Delta b = \begin{pmatrix} 0.1 \\ -0.1 \\ 0.1 \\ -0.1 \end{pmatrix},$$

we obtain the new system

$$\begin{pmatrix} 10 & 7 & 8 & 7 \\ 7 & 5 & 6 & 5 \\ 8 & 6 & 10 & 9 \\ 7 & 5 & 9 & 10 \end{pmatrix} \begin{pmatrix} x_1 + \Delta x_1 \\ x_2 + \Delta x_2 \\ x_3 + \Delta x_3 \\ x_4 + \Delta x_4 \end{pmatrix} = \begin{pmatrix} 32.1 \\ 22.9 \\ 33.1 \\ 30.9 \end{pmatrix}.$$

The new solution turns out to be $x + \Delta x = (9.2, -12.6, 4.5, -1.1)$, where

$$\Delta x = (9.2, -12.6, 4.5, -1.1) - (1, 1, 1, 1) = (8.2, -13.6, 3.5, -2.1).$$

Then a relative error of the data in terms of the one-norm,

$$\frac{\|\Delta b\|_1}{\|b\|_1} = \frac{0.4}{119} = \frac{4}{1190} \approx \frac{1}{300},$$

produces a relative error in the input

$$\frac{\|\Delta x\|_1}{\|x\|_1} = \frac{27.4}{4} \approx 7.$$

So a relative order of the order $1/300$ in the data produces a relative error of the order $7/1$ in the solution, which represents an amplification of the relative error of the order 2100.

Now let us perturb the matrix slightly, obtaining the new system

$$\begin{pmatrix} 10 & 7 & 8.1 & 7.2 \\ 7.08 & 5.04 & 6 & 5 \\ 8 & 5.98 & 9.98 & 9 \\ 6.99 & 4.99 & 9 & 9.98 \end{pmatrix} \begin{pmatrix} x_1 + \Delta x_1 \\ x_2 + \Delta x_2 \\ x_3 + \Delta x_3 \\ x_4 + \Delta x_4 \end{pmatrix} = \begin{pmatrix} 32 \\ 23 \\ 33 \\ 31 \end{pmatrix}.$$

This time the solution is $x + \Delta x = (-81, 137, -34, 22)$. Again a small change in the data alters the result rather drastically. Yet the original system is symmetric, has determinant 1, and has integer entries. The problem is that the matrix of the system is badly conditioned, a concept that we will now explain.

Given an invertible matrix A, first assume that we perturb b to $b + \Delta b$, and let us analyze the change between the two exact solutions x and $x + \Delta x$ of the two systems

$$Ax = b$$
$$A(x + \Delta x) = b + \Delta b.$$

We also assume that we have some norm $\| \ \|$ and we use the *subordinate* matrix norm on matrices. From

$$Ax = b$$
$$Ax + A\Delta x = b + \Delta b,$$

we get

$$\Delta x = A^{-1}\Delta b,$$

and we conclude that

$$\|\Delta x\| \leq \|A^{-1}\| \|\Delta b\|$$
$$\|b\| \leq \|A\| \|x\|.$$

Consequently, the relative error in the result $\|\Delta x\| / \|x\|$ is bounded in terms of the relative error $\|\Delta b\| / \|b\|$ in the data as follows:

$$\frac{\|\Delta x\|}{\|x\|} \leq (\|A\| \|A^{-1}\|) \frac{\|\Delta b\|}{\|b\|}.$$

Now let us assume that A is perturbed to $A + \Delta A$, and let us analyze the change between the exact solutions of the two systems

$$Ax = b$$
$$(A + \Delta A)(x + \Delta x) = b.$$

The second equation yields $Ax + A\Delta x + \Delta A(x + \Delta x) = b$, and by subtracting the first equation we get

$$\Delta x = -A^{-1}\Delta A(x + \Delta x).$$

It follows that

$$\|\Delta x\| \le \left\|A^{-1}\right\| \|\Delta A\| \|x + \Delta x\|,$$

which can be rewritten as

$$\frac{\|\Delta x\|}{\|x + \Delta x\|} \le \left(\|A\| \left\|A^{-1}\right\| \right)\frac{\|\Delta A\|}{\|A\|}.$$

Observe that the above reasoning is valid even if the matrix $A + \Delta A$ is singular, as long as $x + \Delta x$ is a solution of the second system. Furthermore, if $\|\Delta A\|$ is small enough, it is not unreasonable to expect that the ratio $\|\Delta x\| / \|x + \Delta x\|$ is close to $\|\Delta x\| / \|x\|$. This will be made more precise later.

In summary, for each of the two perturbations, we see that the relative error in the result is bounded by the relative error in the data, *multiplied the number* $\|A\| \left\|A^{-1}\right\|$. In fact, this factor turns out to be optimal and this suggests the following definition:

Definition 8.10. For any subordinate matrix norm $\| \ \|$, for any invertible matrix A, the number

$$\mathrm{cond}(A) = \|A\| \left\|A^{-1}\right\|$$

is called the *condition number* of A relative to $\| \ \|$.

The condition number $\mathrm{cond}(A)$ measures the sensitivity of the linear system $Ax = b$ to variations in the data b and A; a feature referred to as the *condition* of the system. Thus, when we says that a linear system is *ill-conditioned*, we mean that the condition number of its matrix is large. We can sharpen the preceding analysis as follows:

Proposition 8.10. *Let A be an invertible matrix and let x and $x + \Delta x$ be the solutions of the linear systems*

$$Ax = b$$

$$A(x + \Delta x) = b + \Delta b.$$

If $b \neq 0$, then the inequality

$$\frac{\|\Delta x\|}{\|x\|} \le \mathrm{cond}(A)\frac{\|\Delta b\|}{\|b\|}$$

holds and is the best possible. This means that for a given matrix A, there exist some vectors $b \neq 0$ and $\Delta b \neq 0$ for which equality holds.

Proof. We already proved the inequality. Now, because $\| \; \|$ is a subordinate matrix norm, there exist some vectors $x \neq 0$ and $\Delta b \neq 0$ for which

$$\left\| A^{-1}\Delta b \right\| = \left\| A^{-1} \right\| \left\| \Delta b \right\| \quad \text{and} \quad \left\| Ax \right\| = \left\| A \right\| \left\| x \right\|.$$

$\square$

Proposition 8.11. *Let A be an invertible matrix and let x and $x + \Delta x$ be the solutions of the two systems*

$$Ax = b$$
$$(A + \Delta A)(x + \Delta x) = b.$$

If $b \neq 0$, then the inequality

$$\frac{\| \Delta x \|}{\| x + \Delta x \|} \leq \operatorname{cond}(A)\frac{\| \Delta A \|}{\| A \|}$$

holds and is the best possible. This means that given a matrix A, there exist a vector $b \neq 0$ and a matrix $\Delta A \neq 0$ for which equality holds. Furthermore, if $\| \Delta A \|$ is small enough (for instance, if $\| \Delta A \| < 1/ \left\| A^{-1} \right\|$), we have

$$\frac{\| \Delta x \|}{\| x \|} \leq \operatorname{cond}(A)\frac{\| \Delta A \|}{\| A \|}(1 + O(\| \Delta A \|));$$

in fact, we have

$$\frac{\| \Delta x \|}{\| x \|} \leq \operatorname{cond}(A)\frac{\| \Delta A \|}{\| A \|} \left(\frac{1}{1 - \left\| A^{-1} \right\| \| \Delta A \|} \right).$$

Proof. The first inequality has already been proven. To show that equality can be achieved, let w be any vector such that $w \neq 0$ and

$$\left\| A^{-1}w \right\| = \left\| A^{-1} \right\| \| w \|,$$

and let $\beta \neq 0$ be any real number. Now the vectors

$$\Delta x = -\beta A^{-1}w$$
$$x + \Delta x = w$$
$$b = (A + \beta I)w$$

and the matrix

$$\Delta A = \beta I$$

satisfy the equations

$$Ax = b$$
$$(A + \Delta A)(x + \Delta x) = b$$
$$\| \Delta x \| = |\beta| \left\| A^{-1}w \right\| = \| \Delta A \| \left\| A^{-1} \right\| \| x + \Delta x \|.$$

Finally we can pick β so that $-\beta$ is not equal to any of the eigenvalues of A, so that $A + \Delta A = A + \beta I$ is invertible and b is is nonzero.

If $\|\Delta A\| < 1/\|A^{-1}\|$, then

$$\|A^{-1}\Delta A\| \le \|A^{-1}\| \, \|\Delta A\| < 1,$$

so by Proposition 8.8, the matrix $I + A^{-1}\Delta A$ is invertible and

$$\|(I + A^{-1}\Delta A)^{-1}\| \le \frac{1}{1 - \|A^{-1}\Delta A\|} \le \frac{1}{1 - \|A^{-1}\| \, \|\Delta A\|}.$$

Recall that we proved earlier that

$$\Delta x = -A^{-1}\Delta A(x + \Delta x),$$

and by adding x to both sides and moving the right-hand side to the left-hand side yields

$$(I + A^{-1}\Delta A)(x + \Delta x) = x,$$

and thus

$$x + \Delta x = (I + A^{-1}\Delta A)^{-1}x,$$

which yields

$$\Delta x = ((I + A^{-1}\Delta A)^{-1} - I)x = (I + A^{-1}\Delta A)^{-1}(I - (I + A^{-1}\Delta A))x$$
$$= -(I + A^{-1}\Delta A)^{-1}A^{-1}(\Delta A)x.$$

From this and

$$\|(I + A^{-1}\Delta A)^{-1}\| \le \frac{1}{1 - \|A^{-1}\| \, \|\Delta A\|},$$

we get

$$\|\Delta x\| \le \frac{\|A^{-1}\| \, \|\Delta A\|}{1 - \|A^{-1}\| \, \|\Delta A\|} \, \|x\|,$$

which can be written as

$$\frac{\|\Delta x\|}{\|x\|} \le \mathrm{cond}(A)\frac{\|\Delta A\|}{\|A\|}\left(\frac{1}{1 - \|A^{-1}\| \, \|\Delta A\|}\right),$$

which is the kind of inequality that we were seeking. $\qquad\square$

Remark: If A and b are perturbed simultaneously, so that we get the "perturbed" system

$$(A + \Delta A)(x + \Delta x) = b + \Delta b,$$

it can be shown that if $\|\Delta A\| < 1/\|A^{-1}\|$ (and $b \neq 0$), then

$$\frac{\|\Delta x\|}{\|x\|} \leq \frac{\mathrm{cond}(A)}{1 - \|A^{-1}\|\,\|\Delta A\|} \left(\frac{\|\Delta A\|}{\|A\|} + \frac{\|\Delta b\|}{\|b\|} \right);$$

see Demmel [Demmel (1997)], Section 2.2 and Horn and Johnson [Horn and Johnson (1990)], Section 5.8.

We now list some properties of condition numbers and figure out what $\mathrm{cond}(A)$ is in the case of the spectral norm (the matrix norm induced by $\|\ \|_2$). First, we need to introduce a very important factorization of matrices, the *singular value decomposition*, for short, *SVD*.

It can be shown (see Section 20.2) that given any $n \times n$ matrix $A \in M_n(\mathbb{C})$, there exist two unitary matrices U and V, and a *real* diagonal matrix $\Sigma = \mathrm{diag}(\sigma_1, \ldots, \sigma_n)$, with $\sigma_1 \geq \sigma_2 \geq \cdots \geq \sigma_n \geq 0$, such that

$$A = V\Sigma U^*.$$

Definition 8.11. Given a complex $n \times n$ matrix A, a triple (U, V, Σ) such that $A = V\Sigma U^\top$, where U and V are $n \times n$ unitary matrices and $\Sigma = \mathrm{diag}(\sigma_1, \ldots, \sigma_n)$ is a diagonal matrix of real numbers $\sigma_1 \geq \sigma_2 \geq \cdots \geq \sigma_n \geq 0$, is called a *singular decomposition* (for short *SVD*) of A. If A is a real matrix, then U and V are orthogonal matrices. The nonnegative numbers $\sigma_1, \ldots, \sigma_n$ are called the *singular values* of A.

The factorization $A = V\Sigma U^*$ implies that

$$A^*A = U\Sigma^2 U^* \quad \text{and} \quad AA^* = V\Sigma^2 V^*,$$

which shows that $\sigma_1^2, \ldots, \sigma_n^2$ are the eigenvalues of *both* A^*A and AA^*, that the columns of U are corresponding eigenvectors for A^*A, and that the columns of V are corresponding eigenvectors for AA^*.

Since σ_1^2 is the largest eigenvalue of A^*A (and AA^*), note that $\sqrt{\rho(A^*A)} = \sqrt{\rho(AA^*)} = \sigma_1$.

Corollary 8.3. *The spectral norm $\|A\|_2$ of a matrix A is equal to the largest singular value of A. Equivalently, the spectral norm $\|A\|_2$ of a matrix A is equal to the ℓ^∞-norm of its vector of singular values,*

$$\|A\|_2 = \max_{1 \leq i \leq n} \sigma_i = \|(\sigma_1, \ldots, \sigma_n)\|_\infty.$$

Since the Frobenius norm of a matrix A is defined by $\|A\|_F = \sqrt{\mathrm{tr}(A^*A)}$ and since

$$\mathrm{tr}(A^*A) = \sigma_1^2 + \cdots + \sigma_n^2$$

where $\sigma_1^2, \ldots, \sigma_n^2$ are the eigenvalues of A^*A, we see that

$$\|A\|_F = (\sigma_1^2 + \cdots + \sigma_n^2)^{1/2} = \|(\sigma_1, \ldots, \sigma_n)\|_2.$$

Corollary 8.4. *The Frobenius norm of a matrix is given by the ℓ^2-norm of its vector of singular values; $\|A\|_F = \|(\sigma_1, \ldots, \sigma_n)\|_2$.*

In the case of a normal matrix if $\lambda_1, \ldots, \lambda_n$ are the (complex) eigenvalues of A, then

$$\sigma_i = |\lambda_i|, \quad 1 \le i \le n.$$

Proposition 8.12. *For every invertible matrix $A \in \mathrm{M}_n(\mathbb{C})$, the following properties hold:*

(1)

$$\mathrm{cond}(A) \ge 1,$$
$$\mathrm{cond}(A) = \mathrm{cond}(A^{-1})$$
$$\mathrm{cond}(\alpha A) = \mathrm{cond}(A) \quad \text{for all } \alpha \in \mathbb{C} - \{0\}.$$

(2) If $\mathrm{cond}_2(A)$ denotes the condition number of A with respect to the spectral norm, then

$$\mathrm{cond}_2(A) = \frac{\sigma_1}{\sigma_n},$$

where $\sigma_1 \ge \cdots \ge \sigma_n$ are the singular values of A.

(3) If the matrix A is normal, then

$$\mathrm{cond}_2(A) = \frac{|\lambda_1|}{|\lambda_n|},$$

where $\lambda_1, \ldots, \lambda_n$ are the eigenvalues of A sorted so that $|\lambda_1| \ge \cdots \ge |\lambda_n|$.

(4) If A is a unitary or an orthogonal matrix, then

$$\mathrm{cond}_2(A) = 1.$$

(5) The condition number $\mathrm{cond}_2(A)$ is invariant under unitary transformations, which means that

$$\mathrm{cond}_2(A) = \mathrm{cond}_2(UA) = \mathrm{cond}_2(AV),$$

for all unitary matrices U and V.

Proof. The properties in (1) are immediate consequences of the properties of subordinate matrix norms. In particular, $AA^{-1} = I$ implies

$$1 = \|I\| \leq \|A\| \, \|A^{-1}\| = \operatorname{cond}(A).$$

(2) We showed earlier that $\|A\|_2^2 = \rho(A^*A)$, which is the square of the modulus of the largest eigenvalue of A^*A. Since we just saw that the eigenvalues of A^*A are $\sigma_1^2 \geq \cdots \geq \sigma_n^2$, where $\sigma_1, \ldots, \sigma_n$ are the singular values of A, we have

$$\|A\|_2 = \sigma_1.$$

Now if A is invertible, then $\sigma_1 \geq \cdots \geq \sigma_n > 0$, and it is easy to show that the eigenvalues of $(A^*A)^{-1}$ are $\sigma_n^{-2} \geq \cdots \geq \sigma_1^{-2}$, which shows that

$$\|A^{-1}\|_2 = \sigma_n^{-1},$$

and thus

$$\operatorname{cond}_2(A) = \frac{\sigma_1}{\sigma_n}.$$

(3) This follows from the fact that $\|A\|_2 = \rho(A)$ for a normal matrix.

(4) If A is a unitary matrix, then $A^*A = AA^* = I$, so $\rho(A^*A) = 1$, and $\|A\|_2 = \sqrt{\rho(A^*A)} = 1$. We also have $\|A^{-1}\|_2 = \|A^*\|_2 = \sqrt{\rho(AA^*)} = 1$, and thus $\operatorname{cond}(A) = 1$.

(5) This follows immediately from the unitary invariance of the spectral norm. $\qquad\square$

Proposition 8.12 (4) shows that unitary and orthogonal transformations are very well-conditioned, and Part (5) shows that unitary transformations preserve the condition number.

In order to compute $\operatorname{cond}_2(A)$, we need to compute the top and bottom singular values of A, which may be hard. The inequality

$$\|A\|_2 \leq \|A\|_F \leq \sqrt{n} \, \|A\|_2,$$

may be useful in getting an approximation of $\operatorname{cond}_2(A) = \|A\|_2 \, \|A^{-1}\|_2$, if A^{-1} can be determined.

Remark: There is an interesting geometric characterization of $\operatorname{cond}_2(A)$. If $\theta(A)$ denotes the least angle between the vectors Au and Av as u and v range over all pairs of orthonormal vectors, then it can be shown that

$$\operatorname{cond}_2(A) = \cot(\theta(A)/2)).$$

Thus if A is nearly singular, then there will be some orthonormal pair u, v such that Au and Av are nearly parallel; the angle $\theta(A)$ will the be small and $\cot(\theta(A)/2))$ will be large. For more details, see Horn and Johnson [Horn and Johnson (1990)] (Section 5.8 and Section 7.4).

It should be noted that in general (if A is not a normal matrix) a matrix could have a very large condition number even if all its eigenvalues are identical! For example, if we consider the $n \times n$ matrix

$$A = \begin{pmatrix} 1 & 2 & 0 & 0 & \dots & 0 & 0 \\ 0 & 1 & 2 & 0 & \dots & 0 & 0 \\ 0 & 0 & 1 & 2 & \dots & 0 & 0 \\ \vdots & \vdots & \ddots & \ddots & \ddots & \vdots & \vdots \\ 0 & 0 & \dots & 0 & 1 & 2 & 0 \\ 0 & 0 & \dots & 0 & 0 & 1 & 2 \\ 0 & 0 & \dots & 0 & 0 & 0 & 1 \end{pmatrix},$$

it turns out that $\mathrm{cond}_2(A) \geq 2^{n-1}$.

A classical example of matrix with a very large condition number is the *Hilbert matrix* $H^{(n)}$, the $n \times n$ matrix with

$$H_{ij}^{(n)} = \left(\frac{1}{i+j-1} \right).$$

For example, when $n = 5$,

$$H^{(5)} = \begin{pmatrix} 1 & \frac{1}{2} & \frac{1}{3} & \frac{1}{4} & \frac{1}{5} \\ \frac{1}{2} & \frac{1}{3} & \frac{1}{4} & \frac{1}{5} & \frac{1}{6} \\ \frac{1}{3} & \frac{1}{4} & \frac{1}{5} & \frac{1}{6} & \frac{1}{7} \\ \frac{1}{4} & \frac{1}{5} & \frac{1}{6} & \frac{1}{7} & \frac{1}{8} \\ \frac{1}{5} & \frac{1}{6} & \frac{1}{7} & \frac{1}{8} & \frac{1}{9} \end{pmatrix}.$$

It can be shown that

$$\mathrm{cond}_2(H^{(5)}) \approx 4.77 \times 10^5.$$

Hilbert introduced these matrices in 1894 while studying a problem in approximation theory. The Hilbert matrix $H^{(n)}$ is symmetric positive definite. A closed-form formula can be given for its determinant (it is a special form of the so-called *Cauchy determinant*); see Problem 8.15. The inverse of $H^{(n)}$ can also be computed explicitly; see Problem 8.15. It can be shown that

$$\mathrm{cond}_2(H^{(n)}) = O((1 + \sqrt{2})^{4n}/\sqrt{n}).$$

Going back to our matrix

$$A = \begin{pmatrix} 10 & 7 & 8 & 7 \\ 7 & 5 & 6 & 5 \\ 8 & 6 & 10 & 9 \\ 7 & 5 & 9 & 10 \end{pmatrix},$$

which is a symmetric positive definite matrix, it can be shown that its eigenvalues, which in this case are also its singular values because A is SPD, are

$$\lambda_1 \approx 30.2887 > \lambda_2 \approx 3.858 > \lambda_3 \approx 0.8431 > \lambda_4 \approx 0.01015,$$

so that

$$\mathrm{cond}_2(A) = \frac{\lambda_1}{\lambda_4} \approx 2984.$$

The reader should check that for the perturbation of the right-hand side b used earlier, the relative errors $\|\Delta x\| / \|x\|$ and $\|\Delta x\| / \|x\|$ satisfy the inequality

$$\frac{\|\Delta x\|}{\|x\|} \leq \mathrm{cond}(A) \frac{\|\Delta b\|}{\|b\|}$$

and comes close to equality.

8.6 An Application of Norms: Solving Inconsistent Linear Systems

The problem of solving an inconsistent linear system $Ax = b$ often arises in practice. This is a system where b does not belong to the column space of A, usually with more equations than variables. Thus, such a system has no solution. Yet we would still like to "solve" such a system, at least approximately.

Such systems often arise when trying to fit some data. For example, we may have a set of 3D data points

$$\{p_1, \ldots, p_n\},$$

and we have reason to believe that these points are nearly coplanar. We would like to find a plane that best fits our data points. Recall that the equation of a plane is

$$\alpha x + \beta y + \gamma z + \delta = 0,$$

with $(\alpha, \beta, \gamma) \neq (0,0,0)$. Thus, every plane is either not parallel to the x-axis ($\alpha \neq 0$) or not parallel to the y-axis ($\beta \neq 0$) or not parallel to the z-axis ($\gamma \neq 0$).

Say we have reasons to believe that the plane we are looking for is not parallel to the z-axis. If we are wrong, in the least squares solution, one of the coefficients, α, β, will be very large. If $\gamma \neq 0$, then we may assume that our plane is given by an equation of the form

$$z = ax + by + d,$$

and we would like this equation to be satisfied for all the p_i's, which leads to a system of n equations in 3 unknowns a, b, d, with $p_i = (x_i, y_i, z_i)$;

$$ax_1 + by_1 + d = z_1$$

$$\vdots \qquad \vdots$$

$$ax_n + by_n + d = z_n.$$

However, if n is larger than 3, such a system generally has *no solution*. Since the above system can't be solved exactly, we can try to find a solution (a, b, d) that *minimizes the least-squares error*

$$\sum_{i=1}^{n} (ax_i + by_i + d - z_i)^2.$$

This is what Legendre and Gauss figured out in the early 1800's!

In general, given a linear system

$$Ax = b,$$

we solve the *least squares problem*: minimize $\|Ax - b\|_2^2$.

Fortunately, every $n \times m$-matrix A can be written as

$$A = VDU^\top$$

where U and V are orthogonal and D is a rectangular diagonal matrix with non-negative entries (*singular value decomposition, or SVD*); see Chapter 20.

The SVD can be used to solve an inconsistent system. It is shown in Chapter 21 that there is a vector x of smallest norm minimizing $\|Ax - b\|_2$. It is given by the (Penrose) *pseudo-inverse* of A (itself given by the SVD).

It has been observed that solving in the least-squares sense may give too much weight to "outliers," that is, points clearly outside the best-fit plane. In this case, it is preferable to minimize (the ℓ^1-norm)

$$\sum_{i=1}^{n} |ax_i + by_i + d - z_i|.$$

This does not appear to be a linear problem, but we can use a trick to convert this minimization problem into a linear program (which means a problem involving linear constraints).

Note that $|x| = \max\{x, -x\}$. So by introducing new variables $e_1, \ldots, e_n$, our minimization problem is equivalent to the linear program (LP):

$$\begin{aligned} \text{minimize} \quad & e_1 + \cdots + e_n \\ \text{subject to} \quad & ax_i + by_i + d - z_i \leq e_i \\ & -(ax_i + by_i + d - z_i) \leq e_i \\ & 1 \leq i \leq n. \end{aligned}$$

Observe that the constraints are equivalent to

$$e_i \geq |ax_i + by_i + d - z_i|, \qquad 1 \leq i \leq n.$$

For an optimal solution, we must have equality, since otherwise we could decrease some e_i and get an even better solution. Of course, we are no longer dealing with "pure" linear algebra, since our constraints are inequalities.

We prefer not getting into linear programming right now, but the above example provides a good reason to learn more about linear programming!

8.7 Limits of Sequences and Series

If $x \in \mathbb{R}$ or $x \in \mathbb{C}$ and if $|x| < 1$, it is well known that the sums $\sum_{k=0}^{n} x^k = 1 + x + x^2 + \cdots + x^n$ converge to the limit $1/(1-x)$ when n goes to infinity, and we write

$$\sum_{k=0}^{\infty} x^k = \frac{1}{1-x}.$$

For example,

$$\sum_{k=0}^{\infty} \frac{1}{2^k} = 2.$$

Similarly, the sums

$$S_n = \sum_{k=0}^{n} \frac{x^k}{k!}$$

converge to e^x when n goes to infinity, for every x (in $\mathbb{R}$ or $\mathbb{C}$). What if we replace x by a real of complex $n \times n$ matrix A?

The partial sums $\sum_{k=0}^{n} A^k$ and $\sum_{k=0}^{n} \frac{A^k}{k!}$ still make sense, but we have to define what is the limit of a sequence of matrices. This can be done in any normed vector space.

Definition 8.12. Let $(E, \|\|)$ be a normed vector space. A *sequence* $(u_n)_{n \in \mathbb{N}}$ in E is any function $u \colon \mathbb{N} \to E$. For any $v \in E$, the sequence (u_n) *converges to v* (and *v is the limit of the sequence (u_n)*) if for every $\epsilon > 0$, there is some integer $N > 0$ such that

$$\|u_n - v\| < \epsilon \quad \text{for all } n \geq N.$$

Often we assume that a sequence is indexed by $\mathbb{N} - \{0\}$, that is, its first term is u_1 rather than u_0.

If the sequence (u_n) converges to v, then since by the triangle inequality

$$\|u_m - u_n\| \leq \|u_m - v\| + \|v - u_n\|,$$

we see that for every $\epsilon > 0$, we can find $N > 0$ such that $\|u_m - v\| < \epsilon/2$ and $\|u_n - v\| < \epsilon/2$, and so

$$\|u_m - u_n\| < \epsilon \quad \text{for all } m, n \geq N.$$

The above property is *necessary* for a convergent sequence, but *not necessarily* sufficient. For example, if $E = \mathbb{Q}$, there are sequences of rationals satisfying the above condition, but whose limit is not a rational number. For example, the sequence $\sum_{k=1}^{n} \frac{1}{k!}$ converges to e, and the sequence $\sum_{k=0}^{n} (-1)^k \frac{1}{2k+1}$ converges to $\pi/4$, but e and $\pi/4$ are not rational (in fact, they are transcendental). However, $\mathbb{R}$ is constructed from $\mathbb{Q}$ to guarantee that sequences with the above property converge, and so is $\mathbb{C}$.

Definition 8.13. Given a normed vector space $(E, \| \|)$, a sequence (u_n) is a *Cauchy sequence* if for every $\epsilon > 0$, there is some $N > 0$ such that

$$\|u_m - u_n\| < \epsilon \quad \text{for all } m, n \geq N.$$

If every Cauchy sequence converges, then we say that E is *complete*. A complete normed vector spaces is also called a *Banach space*.

A fundamental property of $\mathbb{R}$ is that *it is complete*. It follows immediately that $\mathbb{C}$ is also complete. If E is a finite-dimensional real or complex vector space, since any two norms are equivalent, we can pick the ℓ^{∞} norm, and then by picking a basis in E, a sequence (u_n) of vectors in E converges iff the n sequences of coordinates (u_n^i) $(1 \leq i \leq n)$ converge, so *any finite-dimensional real or complex vector space is a Banach space*.

Let us now consider the convergence of series.

Definition 8.14. Given a normed vector space $(E, \| \ \|)$, a *series* is an infinite sum $\sum_{k=0}^{\infty} u_k$ of elements $u_k \in E$. We denote by S_n the partial sum of the first $n + 1$ elements,

$$S_n = \sum_{k=0}^{n} u_k.$$

Definition 8.15. We say that the series $\sum_{k=0}^{\infty} u_k$ *converges* to the limit $v \in E$ if the sequence (S_n) converges to v, i.e., given any $\epsilon > 0$, there exists a positive integer N such that for all $n \geq N$,

$$\|S_n - v\| < \epsilon.$$

In this case, we say that the series is *convergent*. We say that the series $\sum_{k=0}^{\infty} u_k$ *converges absolutely* if the series of norms $\sum_{k=0}^{\infty} \|u_k\|$ is convergent.

If the series $\sum_{k=0}^{\infty} u_k$ converges to v, since for all m, n with $m > n$ we have

$$\sum_{k=0}^{m} u_k - S_n = \sum_{k=0}^{m} u_k - \sum_{k=0}^{n} u_k = \sum_{k=n+1}^{m} u_k,$$

if we let m go to infinity (with n fixed), we see that the series $\sum_{k=n+1}^{\infty} u_k$ converges and that

$$v - S_n = \sum_{k=n+1}^{\infty} u_k.$$

There are series that are convergent but not absolutely convergent; for example, the series

$$\sum_{k=1}^{\infty} \frac{(-1)^{k-1}}{k}$$

converges to $\ln 2$, but $\sum_{k=1}^{\infty} \frac{1}{k}$ does not converge (this sum is infinite). If E is complete, the converse is an enormously useful result.

Proposition 8.13. *Assume $(E, \| \ \|)$ is a complete normed vector space. If a series $\sum_{k=0}^{\infty} u_k$ is absolutely convergent, then it is convergent.*

Proof. If $\sum_{k=0}^{\infty} u_k$ is absolutely convergent, then we prove that the sequence (S_m) is a Cauchy sequence; that is, for every $\epsilon > 0$, there is some $p > 0$ such that for all $n \geq m \geq p$,

$$\|S_n - S_m\| \leq \epsilon.$$

Observe that

$$\|S_n - S_m\| = \|u_{m+1} + \cdots + u_n\| \leq \|u_{m+1}\| + \cdots + \|u_n\|,$$

and since the sequence $\sum_{k=0}^{\infty} \|u_k\|$ converges, it satisfies Cauchy's criterion. Thus, the sequence (S_m) also satisfies Cauchy's criterion, and since E is a complete vector space, the sequence (S_m) converges. $\qquad\square$

Remark: It can be shown that if $(E, \|\ \|)$ is a normed vector space such that every absolutely convergent series is also convergent, then E must be complete (see Schwartz [Schwartz (1991)]).

An important corollary of absolute convergence is that if the terms in series $\sum_{k=0}^{\infty} u_k$ are rearranged, then the resulting series is still absolutely convergent and has the *same sum*. More precisely, let σ be any permutation (bijection) of the natural numbers. The series $\sum_{k=0}^{\infty} u_{\sigma(k)}$ is called a *rearrangement* of the original series. The following result can be shown (see Schwartz [Schwartz (1991)]).

Proposition 8.14. *Assume $(E, \|\ \|)$ is a normed vector space. If a series $\sum_{k=0}^{\infty} u_k$ is convergent as well as absolutely convergent, then for every permutation σ of $\mathbb{N}$, the series $\sum_{k=0}^{\infty} u_{\sigma(k)}$ is convergent and absolutely convergent, and its sum is equal to the sum of the original series:*

$$\sum_{k=0}^{\infty} u_{\sigma(k)} = \sum_{k=0}^{\infty} u_k.$$

In particular, if $(E, \|\ \|)$ is a complete normed vector space, then Proposition 8.14 holds.

We now apply Proposition 8.13 to the matrix exponential.

8.8 The Matrix Exponential

Proposition 8.15. *For any $n \times n$ real or complex matrix A, the series*

$$\sum_{k=0}^{\infty} \frac{A^k}{k!}$$

converges absolutely for any operator norm on $\mathrm{M}_n(\mathbb{C})$ (or $\mathrm{M}_n(\mathbb{R})$).

Proof. Pick any norm on $\mathbb{C}^n$ (or $\mathbb{R}^n$) and let $\|\|$ be the corresponding operator norm on $\mathrm{M}_n(\mathbb{C})$. Since $\mathrm{M}_n(\mathbb{C})$ has dimension n^2, it is complete. By Proposition 8.13, it suffices to show that the series of nonnegative reals $\sum_{k=0}^n \left\| \frac{A^k}{k!} \right\|$ converges. Since $\| \, \|$ is an operator norm, this a matrix norm, so we have

$$\sum_{k=0}^n \left\| \frac{A^k}{k!} \right\| \leq \sum_{k=0}^n \frac{\|A\|^k}{k!} \leq e^{\|A\|}.$$

Thus, the nondecreasing sequence of positive real numbers $\sum_{k=0}^n \left\| \frac{A^k}{k!} \right\|$ is bounded by $e^{\|A\|}$, and by a fundamental property of $\mathbb{R}$, it has a least upper bound which is its limit. $\square$

Definition 8.16. Let E be a finite-dimensional real of complex normed vector space. For any $n \times n$ matrix A, the limit of the series

$$\sum_{k=0}^{\infty} \frac{A^k}{k!}$$

is the *exponential of* A and is denoted e^A.

A basic property of the exponential $x \mapsto e^x$ with $x \in \mathbb{C}$ is

$$e^{x+y} = e^x e^y, \quad \text{for all } x, y \in \mathbb{C}.$$

As a consequence, e^x is always invertible and $(e^x)^{-1} = e^{-x}$. For matrices, because matrix multiplication is not commutative, in general,

$$e^{A+B} = e^A e^B$$

fails! This result is salvaged as follows.

Proposition 8.16. *For any two $n \times n$ complex matrices A and B, if A and B commute, that is, $AB = BA$, then*

$$e^{A+B} = e^A e^B.$$

A proof of Proposition 8.16 can be found in Gallier [Gallier (2011b)].

Since A and $-A$ commute, as a corollary of Proposition 8.16, we see that e^A is always invertible and that

$$(e^A)^{-1} = e^{-A}.$$

It is also easy to see that

$$(e^A)^{\top} = e^{A^{\top}}.$$

In general, there is no closed-form formula for the exponential e^A of a matrix A, but for skew symmetric matrices of dimension 2 and 3, there are explicit formulae. Everyone should enjoy computing the exponential e^A where

$$A = \begin{pmatrix} 0 & -\theta \\ \theta & 0 \end{pmatrix}.$$

If we write

$$J = \begin{pmatrix} 0 & -1 \\ 1 & 0 \end{pmatrix},$$

then

$$A = \theta J.$$

The key property is that

$$J^2 = -I.$$

Proposition 8.17. *If $A = \theta J$, then*

$$e^A = \cos\theta I + \sin\theta J = \begin{pmatrix} \cos\theta & -\sin\theta \\ \sin\theta & \cos\theta \end{pmatrix}.$$

Proof. We have

$$A^{4n} = \theta^{4n} I_2,$$
$$A^{4n+1} = \theta^{4n+1} J,$$
$$A^{4n+2} = -\theta^{4n+2} I_2,$$
$$A^{4n+3} = -\theta^{4n+3} J,$$

and so

$$e^A = I_2 + \frac{\theta}{1!}J - \frac{\theta^2}{2!}I_2 - \frac{\theta^3}{3!}J + \frac{\theta^4}{4!}I_2 + \frac{\theta^5}{5!}J - \frac{\theta^6}{6!}I_2 - \frac{\theta^7}{7!}J + \cdots.$$

Rearranging the order of the terms, we have

$$e^A = \left(1 - \frac{\theta^2}{2!} + \frac{\theta^4}{4!} - \frac{\theta^6}{6!} + \cdots\right) I_2 + \left(\frac{\theta}{1!} - \frac{\theta^3}{3!} + \frac{\theta^5}{5!} - \frac{\theta^7}{7!} + \cdots\right) J.$$

We recognize the power series for $\cos\theta$ and $\sin\theta$, and thus

$$e^A = \cos\theta I_2 + \sin\theta J,$$

that is

$$e^A = \begin{pmatrix} \cos\theta & -\sin\theta \\ \sin\theta & \cos\theta \end{pmatrix},$$

as claimed. $\square$

Thus, we see that the exponential of a 2×2 skew-symmetric matrix is a rotation matrix. This property generalizes to any dimension. An explicit formula when $n = 3$ (the Rodrigues' formula) is given in Section 11.7.

Proposition 8.18. *If B is an $n \times n$ (real) skew symmetric matrix, that is, $B^\top = -B$, then $Q = e^B$ is an orthogonal matrix, that is*

$$Q^\top Q = QQ^\top = I.$$

Proof. Since $B^\top = -B$, we have

$$Q^\top = (e^B)^\top = e^{B^\top} = e^{-B}.$$

Since B and $-B$ commute, we have

$$Q^\top Q = e^{-B}e^B = e^{-B+B} = e^0 = I.$$

Similarly,

$$QQ^\top = e^B e^{-B} = e^{B-B} = e^0 = I,$$

which concludes the proof. $\qquad \square$

It can also be shown that $\det(Q) = \det(e^B) = 1$, but this requires a better understanding of the eigenvalues of e^B (see Section 14.5). Furthermore, for every $n \times n$ rotation matrix Q (an orthogonal matrix Q such that $\det(Q) = 1$), there is a skew symmetric matrix B such that $Q = e^B$. This is a fundamental property which has applications in robotics for $n = 3$.

All familiar series have matrix analogs. For example, if $\|A\| < 1$ (where $\| \ \|$ is an operator norm), then the series $\sum_{k=0}^{\infty} A^k$ converges absolutely, and it can be shown that its limit is $(I - A)^{-1}$.

Another interesting series is the logarithm. For any $n \times n$ complex matrix A, if $\|A\| < 1$ (where $\| \ \|$ is an operator norm), then the series

$$\log(I + A) = \sum_{k=1}^{\infty}(-1)^{k+1}\frac{A^k}{k}$$

converges absolutely.

8.9 Summary

The main concepts and results of this chapter are listed below:

- *Norms* and *normed vector spaces*.
- The *triangle inequality*.
- The *Euclidean norm*; the *ℓ^p-norms*.

- *Hölder's inequality*; the *Cauchy–Schwarz inequality*; *Minkowski's inequality*.
- *Hermitian inner product* and *Euclidean inner product*.
- *Equivalent* norms.
- *All norms on a finite-dimensional vector space are equivalent* (Theorem 8.1).
- *Matrix norms*.
- *Hermitian, symmetric* and *normal* matrices. *Orthogonal* and *unitary* matrices.
- The *trace* of a matrix.
- *Eigenvalues* and *eigenvectors* of a matrix.
- The *characteristic polynomial* of a matrix.
- The *spectral radius* $\rho(A)$ of a matrix A.
- The *Frobenius norm*.
- The Frobenius norm is a *unitarily invariant* matrix norm.
- *Bounded* linear maps.
- *Subordinate matrix norms*.
- Characterization of the subordinate matrix norms for the vector norms $\|\ \|_1$, $\|\ \|_2$, and $\|\ \|_\infty$.
- The *spectral norm*.
- For every matrix $A \in M_n(\mathbb{C})$ and for every $\epsilon > 0$, there is some subordinate matrix norm $\|\ \|$ such that $\|A\| \le \rho(A) + \epsilon$.
- *Condition numbers* of matrices.
- Perturbation analysis of linear systems.
- The *singular value decomposition* (SVD).
- Properties of conditions numbers. Characterization of $\mathrm{cond}_2(A)$ in terms of the largest and smallest singular values of A.
- The *Hilbert matrix*: a very badly conditioned matrix.
- Solving inconsistent linear systems by the method of *least-squares*; *linear programming*.
- Convergence of sequences of vectors in a normed vector space.
- Cauchy sequences, complex normed vector spaces, Banach spaces.
- Convergence of series. Absolute convergence.
- The matrix exponential.
- Skew symmetric matrices and orthogonal matrices.

8.10 Problems

Problem 8.1. Let A be the following matrix:

$$B = \begin{pmatrix} 1 & 1/\sqrt{2} \\ 1/\sqrt{2} & 3/2 \end{pmatrix}.$$

Compute the operator 2-norm $\|A\|_2$ of A.

Problem 8.2. Prove Proposition 8.2, namely that the following inequalities hold for all $x \in \mathbb{R}^n$ (or $x \in \mathbb{C}^n$):

$$\|x\|_\infty \leq \|x\|_1 \leq n\|x\|_\infty,$$
$$\|x\|_\infty \leq \|x\|_2 \leq \sqrt{n}\|x\|_\infty,$$
$$\|x\|_2 \leq \|x\|_1 \leq \sqrt{n}\|x\|_2.$$

Problem 8.3. For any $p \geq 1$, prove that for all $x \in \mathbb{R}^n$,

$$\lim_{p \mapsto \infty} \|x\|_p = \|x\|_\infty.$$

Problem 8.4. Let A be an $n \times n$ matrix which is strictly row diagonally dominant, which means that

$$|a_{ii}| > \sum_{j \neq i} |a_{ij}|,$$

for $i = 1, \ldots, n$, and let

$$\delta = \min_i \left\{ |a_{ii}| - \sum_{j \neq i} |a_{ij}| \right\}.$$

The fact that A is strictly row diagonally dominant is equivalent to the condition $\delta > 0$.

(1) For any nonzero vector v, prove that

$$\|Av\|_\infty \geq \|v\|_\infty \, \delta.$$

Use the above to prove that A is invertible.

(2) Prove that

$$\|A^{-1}\|_\infty \leq \delta^{-1}.$$

Hint. Prove that

$$\sup_{v \neq 0} \frac{\|A^{-1}v\|_\infty}{\|v\|_\infty} = \sup_{w \neq 0} \frac{\|w\|_\infty}{\|Aw\|_\infty}.$$

Problem 8.5. Let A be any invertible complex $n \times n$ matrix.

(1) For any vector norm $\| \; \|$ on $\mathbb{C}^n$, prove that the function $\| \; \|_A : \mathbb{C}^n \to \mathbb{R}$ given by

$$\|x\|_A = \|Ax\| \quad \text{for all} \quad x \in \mathbb{C}^n,$$

is a vector norm.

(2) Prove that the operator norm induced by $\| \; \|_A$, also denoted by $\| \; \|_A$, is given by

$$\|B\|_A = \|ABA^{-1}\| \quad \text{for every } n \times n \text{ matrix} \quad B,$$

where $\|ABA^{-1}\|$ uses the operator norm induced by $\| \; \|$.

Problem 8.6. Give an example of a norm on $\mathbb{C}^n$ and of a *real* matrix A such that

$$\|A\|_{\mathbb{R}} < \|A\| ,$$

where $\|-\|_{\mathbb{R}}$ and $\|-\|$ are the operator norms associated with the vector norm $\|-\|$.

Hint. This can already be done for $n = 2$.

Problem 8.7. Let $\| \; \|$ be any operator norm. Given an invertible $n \times n$ matrix A, if $c = 1/(2\,\|A^{-1}\|)$, then for every $n \times n$ matrix H, if $\|H\| \le c$, then $A + H$ is invertible. Furthermore, show that if $\|H\| \le c$, then $\|(A + H)^{-1}\| \le 1/c$.

Problem 8.8. Let A be any $m \times n$ matrix and let $\lambda \in \mathbb{R}$ be any positive real number $\lambda > 0$.

(1) Prove that $A^\top A + \lambda I_n$ and $AA^\top + \lambda I_m$ are invertible.

(2) Prove that

$$A^\top (AA^\top + \lambda I_m)^{-1} = (A^\top A + \lambda I_n)^{-1} A^\top.$$

Remark: The expressions above correspond to the matrix for which the function

$$\Phi(x) = (Ax - b)^\top (Ax - b) + \lambda x^\top x$$

achieves a minimum. It shows up in machine learning (kernel methods).

Problem 8.9. Let Z be a $q \times p$ real matrix. Prove that if $I_p - Z^\top Z$ is positive definite, then the $(p + q) \times (p + q)$ matrix

$$S = \begin{pmatrix} I_p & Z^\top \\ Z & I_q \end{pmatrix}$$

is symmetric positive definite.

Problem 8.10. Prove that for any real or complex square matrix A, we have

$$\|A\|_2^2 \le \|A\|_1 \|A\|_\infty \,,$$

where the above norms are operator norms.
Hint. Use Proposition 8.7 (among other things, it shows that $\|A\|_1 = \|A^\top\|_\infty$).

Problem 8.11. Show that the map $A \mapsto \rho(A)$ (where $\rho(A)$ is the spectral radius of A) is neither a norm nor a matrix norm. In particular, find two 2×2 matrices A and B such that

$$\rho(A + B) > \rho(A) + \rho(B) = 0 \quad \text{and} \quad \rho(AB) > \rho(A)\rho(B) = 0.$$

Problem 8.12. Define the map $A \mapsto M(A)$ (defined on $n \times n$ real or complex $n \times n$ matrices) by

$$M(A) = \max\{|a_{ij}| \mid 1 \le i, j \le n\}.$$

(1) Prove that

$$M(AB) \le nM(A)M(B)$$

for all $n \times n$ matrices A and B.

(2) Give a counter-example of the inequality

$$M(AB) \le M(A)M(B).$$

(3) Prove that the map $A \mapsto \|A\|_M$ given by

$$\|A\|_M = nM(A) = n\max\{|a_{ij}| \mid 1 \le i, j \le n\}$$

is a matrix norm.

Problem 8.13. Let S be a real symmetric positive definite matrix.

(1) Use the Cholesky factorization to prove that there is some upper-triangular matrix C, unique if its diagonal elements are strictly positive, such that $S = C^\top C$.

(2) For any $x \in \mathbb{R}^n$, define

$$\|x\|_S = (x^\top S x)^{1/2}.$$

Prove that

$$\|x\|_S = \|Cx\|_2 \,,$$

and that the map $x \mapsto \|x\|_S$ is a norm.

Problem 8.14. Let A be a real 2×2 matrix

$$A = \begin{pmatrix} a_{11} & a_{12} \\ a_{21} & a_{22} \end{pmatrix}.$$

(1) Prove that the squares of the singular values $\sigma_1 \geq \sigma_2$ of A are the roots of the quadratic equation

$$X^2 - \mathrm{tr}(A^\top A)X + |\det(A)|^2 = 0.$$

(2) If we let

$$\mu(A) = \frac{a_{11}^2 + a_{12}^2 + a_{21}^2 + a_{22}^2}{2|a_{11}a_{22} - a_{12}a_{21}|},$$

prove that

$$\mathrm{cond}_2(A) = \frac{\sigma_1}{\sigma_2} = \mu(A) + (\mu(A)^2 - 1)^{1/2}.$$

(3) Consider the subset $\mathcal{S}$ of 2×2 invertible matrices whose entries a_{ij} are integers such that $0 \leq a_{ij} \leq 100$.

Prove that the functions $\mathrm{cond}_2(A)$ and $\mu(A)$ reach a maximum on the set $\mathcal{S}$ for the same values of A.

Check that for the matrix

$$A_m = \begin{pmatrix} 100 & 99 \\ 99 & 98 \end{pmatrix}$$

we have

$$\mu(A_m) = 19,603 \quad \det(A_m) = -1$$

and

$$\mathrm{cond}_2(A_m) \approx 39,206.$$

(4) Prove that for all $A \in \mathcal{S}$, if $|\det(A)| \geq 2$ then $\mu(A) \leq 10,000$. Conclude that the maximum of $\mu(A)$ on $\mathcal{S}$ is achieved for matrices such that $\det(A) = \pm 1$. Prove that finding matrices that maximize μ on $\mathcal{S}$ is equivalent to finding some integers n_1, n_2, n_3, n_4 such that

$$0 \leq n_4 \leq n_3 \leq n_2 \leq n_1 \leq 100$$
$$n_1^2 + n_2^2 + n_3^2 + n_4^2 \geq 100^2 + 99^2 + 99^2 + 98^2 = 39,206$$
$$|n_1 n_4 - n_2 n_3| = 1.$$

You may use without proof that the fact that the only solution to the above constraints is the multiset

$$\{100, 99, 99, 98\}.$$

(5) Deduce from part (4) that the matrices in $\mathcal{S}$ for which μ has a maximum value are

$$A_m = \begin{pmatrix} 100 & 99 \\ 99 & 98 \end{pmatrix} \quad \begin{pmatrix} 98 & 99 \\ 99 & 100 \end{pmatrix} \quad \begin{pmatrix} 99 & 100 \\ 98 & 99 \end{pmatrix} \quad \begin{pmatrix} 99 & 98 \\ 100 & 99 \end{pmatrix}$$

and check that μ has the same value for these matrices. Conclude that

$$\max_{A \in \mathcal{S}} \mathrm{cond}_2(A) = \mathrm{cond}_2(A_m).$$

(6) Solve the system

$$\begin{pmatrix} 100 & 99 \\ 99 & 98 \end{pmatrix} \begin{pmatrix} x_1 \\ x_2 \end{pmatrix} = \begin{pmatrix} 199 \\ 197 \end{pmatrix}.$$

Perturb the right-hand side b by

$$\Delta b = \begin{pmatrix} -0.0097 \\ 0.0106 \end{pmatrix}$$

and solve the new system

$$A_m y = b + \Delta b$$

where $y = (y_1, y_2)$. Check that

$$\Delta x = y - x = \begin{pmatrix} 2 \\ -2.0203 \end{pmatrix}.$$

Compute $\|x\|_2$, $\|\Delta x\|_2$, $\|b\|_2$, $\|\Delta b\|_2$, and estimate

$$c = \frac{\|\Delta x\|_2}{\|x\|_2} \left(\frac{\|\Delta b\|_2}{\|b\|_2} \right)^{-1}.$$

Check that

$$c \approx \mathrm{cond}_2(A_m) = 39,206.$$

Problem 8.15. Consider a real 2×2 matrix with zero trace of the form

$$A = \begin{pmatrix} a & b \\ c & -a \end{pmatrix}.$$

(1) Prove that

$$A^2 = (a^2 + bc)I_2 = -\det(A)I_2.$$

If $a^2 + bc = 0$, prove that

$$e^A = I_2 + A.$$

(2) If $a^2 + bc < 0$, let $\omega > 0$ be such that $\omega^2 = -(a^2 + bc)$. Prove that

$$e^A = \cos \omega\, I_2 + \frac{\sin \omega}{\omega} A.$$

(3) If $a^2 + bc > 0$, let $\omega > 0$ be such that $\omega^2 = a^2 + bc$. Prove that

$$e^A = \cosh \omega\, I_2 + \frac{\sinh \omega}{\omega} A.$$

(3) Prove that in all cases

$$\det \left(e^A\right) = 1 \quad \text{and} \quad \operatorname{tr}(A) \geq -2.$$

(4) Prove that there exist some real 2×2 matrix B with $\det(B) = 1$ such that there is no real 2×2 matrix A with zero trace such that $e^A = B$.

Problem 8.16. Recall that the Hilbert matrix is given by

$$H_{ij}^{(n)} = \left(\frac{1}{i + j - 1} \right).$$

(1) Prove that

$$\det(H^{(n)}) = \frac{(1!2! \cdots (n-1)!)^4}{1!2! \cdots (2n-1)!},$$

thus the reciprocal of an integer.

Hint. Use Problem 6.13.

(2) Amazingly, the entries of the inverse of $H^{(n)}$ are integers. Prove that $(H^{(n)})^{-1} = (\alpha_{ij})$, with

$$\alpha_{ij} = (-1)^{i+j}(i+j-1)\binom{n+i-1}{n-j}\binom{n+j-1}{n-i}\binom{i+j-2}{i-1}^2.$$

Chapter 9

Iterative Methods for Solving Linear Systems

9.1 Convergence of Sequences of Vectors and Matrices

In Chapter 7 we discussed some of the main methods for solving systems of linear equations. These methods are *direct methods*, in the sense that they yield exact solutions (assuming infinite precision!).

Another class of methods for solving linear systems consists in approximating solutions using *iterative methods*. The basic idea is this: Given a linear system $Ax = b$ (with A a square invertible matrix in $M_n(\mathbb{C})$), find another matrix $B \in M_n(\mathbb{C})$ and a vector $c \in \mathbb{C}^n$, such that

(1) The matrix $I - B$ is invertible.
(2) The unique solution $\widetilde{x}$ of the system $Ax = b$ is *identical* to the unique solution $\widetilde{u}$ of the system

$$u = Bu + c,$$

and then starting from any vector u_0, compute the sequence (u_k) given by

$$u_{k+1} = Bu_k + c, \quad k \in \mathbb{N}.$$

Under certain conditions (to be clarified soon), the sequence (u_k) converges to a limit $\widetilde{u}$ which is the unique solution of $u = Bu + c$, and thus of $Ax = b$.

Consequently, it is important to find conditions that ensure the convergence of the above sequences and to have tools to compare the "rate" of convergence of these sequences. Thus, we begin with some general results about the convergence of sequences of vectors and matrices.

Let $(E, \| \ \|)$ be a normed vector space. Recall from Section 8.7 that a sequence (u_k) of vectors $u_k \in E$ *converges to a limit* $u \in E$, if for every $\epsilon > 0$, there some natural number N such that

$$\|u_k - u\| \leq \epsilon, \quad \text{for all } k \geq N.$$

We write

$$u = \lim_{k \mapsto \infty} u_k.$$

If E is a finite-dimensional vector space and $\dim(E) = n$, we know from Theorem 8.1 that any two norms are equivalent, and if we choose the norm $\| \ \|_\infty$, we see that the convergence of the sequence of vectors u_k is equivalent to the convergence of the n sequences of scalars formed by the components of these vectors (over any basis). The same property applies to the finite-dimensional vector space $\mathrm{M}_{m,n}(K)$ of $m \times n$ matrices (with $K = \mathbb{R}$ or $K = \mathbb{C}$), which means that the convergence of a sequence of matrices $A_k = (a_{ij}^{(k)})$ is equivalent to the convergence of the $m \times n$ sequences of scalars $(a_{ij}^{(k)})$, with i, j fixed ($1 \le i \le m$, $1 \le j \le n$).

The first theorem below gives a necessary and sufficient condition for the sequence (B^k) of powers of a matrix B to converge to the zero matrix. Recall that the spectral radius $\rho(B)$ of a matrix B is the maximum of the moduli $|\lambda_i|$ of the eigenvalues of B.

Theorem 9.1. *For any square matrix B, the following conditions are equivalent:*

(1) $\lim_{k \mapsto \infty} B^k = 0$,
(2) $\lim_{k \mapsto \infty} B^k v = 0$, *for all vectors v,*
(3) $\rho(B) < 1$,
(4) $\|B\| < 1$, *for some subordinate matrix norm $\| \ \|$.*

Proof. Assume (1) and let $\| \ \|$ be a vector norm on E and $\| \ \|$ be the corresponding matrix norm. For every vector $v \in E$, because $\| \ \|$ is a matrix norm, we have

$$\|B^k v\| \le \|B^k\| \|v\|,$$

and since $\lim_{k \mapsto \infty} B^k = 0$ means that $\lim_{k \mapsto \infty} \|B^k\| = 0$, we conclude that $\lim_{k \mapsto \infty} \|B^k v\| = 0$, that is, $\lim_{k \mapsto \infty} B^k v = 0$. This proves that (1) implies (2).

Assume (2). If we had $\rho(B) \ge 1$, then there would be some eigenvector $u \ (\ne 0)$ and some eigenvalue λ such that

$$Bu = \lambda u, \quad |\lambda| = \rho(B) \ge 1,$$

but then the sequence $(B^k u)$ would not converge to 0, because $B^k u = \lambda^k u$ and $|\lambda^k| = |\lambda|^k \ge 1$. It follows that (2) implies (3).

Assume that (3) holds, that is, $\rho(B) < 1$. By Proposition 8.9, we can find $\epsilon > 0$ small enough that $\rho(B) + \epsilon < 1$, and a subordinate matrix norm $\| \ \|$ such that

$$\|B\| \leq \rho(B) + \epsilon,$$

which is (4).

Finally, assume (4). Because $\| \ \|$ is a matrix norm,

$$\|B^k\| \leq \|B\|^k,$$

and since $\|B\| < 1$, we deduce that (1) holds. $\qquad\square$

The following proposition is needed to study the rate of convergence of iterative methods.

Proposition 9.1. *For every square matrix $B \in M_n(\mathbb{C})$ and every matrix norm $\| \ \|$, we have*

$$\lim_{k \mapsto \infty} \|B^k\|^{1/k} = \rho(B).$$

Proof. We know from Proposition 8.3 that $\rho(B) \leq \|B\|$, and since $\rho(B) = (\rho(B^k))^{1/k}$, we deduce that

$$\rho(B) \leq \|B^k\|^{1/k} \quad \text{for all } k \geq 1,$$

and so

$$\rho(B) \leq \lim_{k \mapsto \infty} \|B^k\|^{1/k}.$$

Now let us prove that for every $\epsilon > 0$, there is some integer $N(\epsilon)$ such that

$$\|B^k\|^{1/k} \leq \rho(B) + \epsilon \quad \text{for all } k \geq N(\epsilon),$$

which proves that

$$\lim_{k \mapsto \infty} \|B^k\|^{1/k} \leq \rho(B),$$

and our proposition.

For any given $\epsilon > 0$, let B_ϵ be the matrix

$$B_\epsilon = \frac{B}{\rho(B) + \epsilon}.$$

Since $\|B_\epsilon\| < 1$, Theorem 9.1 implies that $\lim_{k \mapsto \infty} B_\epsilon^k = 0$. Consequently, there is some integer $N(\epsilon)$ such that for all $k \geq N(\epsilon)$, we have

$$\|B^k\| = \frac{\|B^k\|}{(\rho(B) + \epsilon)^k} \leq 1,$$

which implies that

$$\|B^k\|^{1/k} \leq \rho(B) + \epsilon,$$

as claimed. $\qquad\square$

We now apply the above results to the convergence of iterative methods.

9.2   Convergence of Iterative Methods

Recall that iterative methods for solving a linear system $Ax = b$ (with $A \in M_n(\mathbb{C})$ invertible) consists in finding some matrix B and some vector c, such that $I - B$ is invertible, and the unique solution $\widetilde{x}$ of $Ax = b$ is equal to the unique solution $\widetilde{u}$ of $u = Bu + c$. Then starting from *any* vector u_0, compute the sequence (u_k) given by

$$u_{k+1} = Bu_k + c, \quad k \in \mathbb{N},$$

and say that the iterative method is *convergent* iff

$$\lim_{k \mapsto \infty} u_k = \widetilde{u},$$

for *every* initial vector u_0.

Here is a fundamental criterion for the convergence of any iterative methods based on a matrix B, called the *matrix of the iterative method*.

Theorem 9.2. *Given a system* $u = Bu + c$ *as above, where* $I - B$ *is invertible, the following statements are equivalent:*

(1) The iterative method is convergent.
(2) $\rho(B) < 1$.
(3) $\|B\| < 1$, *for some subordinate matrix norm* $\| \; \|$.

Proof. Define the vector e_k (*error vector*) by

$$e_k = u_k - \widetilde{u},$$

where $\widetilde{u}$ is the unique solution of the system $u = Bu + c$. Clearly, the iterative method is convergent iff

$$\lim_{k \mapsto \infty} e_k = 0.$$

We claim that

$$e_k = B^k e_0, \quad k \geq 0,$$

where $e_0 = u_0 - \widetilde{u}$.

This is proven by induction on k. The base case $k = 0$ is trivial. By the induction hypothesis, $e_k = B^k e_0$, and since $u_{k+1} = Bu_k + c$, we get

$$u_{k+1} - \widetilde{u} = Bu_k + c - \widetilde{u},$$

and because $\widetilde{u} = B\widetilde{u} + c$ and $e_k = B^k e_0$ (by the induction hypothesis), we obtain

$$u_{k+1} - \widetilde{u} = Bu_k - B\widetilde{u} = B(u_k - \widetilde{u}) = Be_k = BB^k e_0 = B^{k+1} e_0,$$

proving the induction step. Thus, the iterative method converges iff

$$\lim_{k \mapsto \infty} B^k e_0 = 0.$$

Consequently, our theorem follows by Theorem 9.1. $\square$

The next proposition is needed to compare the rate of convergence of iterative methods. It shows that *asymptotically, the error vector* $e_k = B^k e_0$ *behaves at worst like* $(\rho(B))^k$.

Proposition 9.2. *Let* $\| \ \|$ *be any vector norm, let* $B \in M_n(\mathbb{C})$ *be a matrix such that* $I - B$ *is invertible, and let* $\widetilde{u}$ *be the unique solution of* $u = Bu + c$.
(1) If (u_k) *is any sequence defined iteratively by*

$$u_{k+1} = Bu_k + c, \quad k \in \mathbb{N},$$

then

$$\lim_{k \mapsto \infty} \left[\sup_{\|u_0 - \widetilde{u}\| = 1} \|u_k - \widetilde{u}\|^{1/k} \right] = \rho(B).$$

(2) Let B_1 *and* B_2 *be two matrices such that* $I - B_1$ *and* $I - B_2$ *are invertible, assume that both* $u = B_1 u + c_1$ *and* $u = B_2 u + c_2$ *have the same unique solution* $\widetilde{u}$, *and consider any two sequences* (u_k) *and* (v_k) *defined inductively by*

$$u_{k+1} = B_1 u_k + c_1$$
$$v_{k+1} = B_2 v_k + c_2,$$

with $u_0 = v_0$. *If* $\rho(B_1) < \rho(B_2)$, *then for any* $\epsilon > 0$, *there is some integer* $N(\epsilon)$, *such that for all* $k \geq N(\epsilon)$, *we have*

$$\sup_{\|u_0 - \widetilde{u}\| = 1} \left[\frac{\|v_k - \widetilde{u}\|}{\|u_k - \widetilde{u}\|} \right]^{1/k} \geq \frac{\rho(B_2)}{\rho(B_1) + \epsilon}.$$

Proof. Let $\| \ \|$ be the subordinate matrix norm. Recall that

$$u_k - \widetilde{u} = B^k e_0,$$

with $e_0 = u_0 - \widetilde{u}$. For every $k \in \mathbb{N}$, we have

$$(\rho(B_1))^k = \rho(B_1^k) \leq \|B_1^k\| = \sup_{\|e_0\| = 1} \|B_1^k e_0\|,$$

which implies

$$\rho(B_1) = \sup_{\|e_0\| = 1} \|B_1^k e_0\|^{1/k} = \|B_1^k\|^{1/k},$$

and Statement (1) follows from Proposition 9.1.
Because $u_0 = v_0$, we have

$$u_k - \widetilde{u} = B_1^k e_0$$
$$v_k - \widetilde{u} = B_2^k e_0,$$

with $e_0 = u_0 - \widetilde{u} = v_0 - \widetilde{u}$. Again, by Proposition 9.1, for every $\epsilon > 0$, there is some natural number $N(\epsilon)$ such that if $k \geq N(\epsilon)$, then

$$\sup_{\|e_0\|=1} \|B_1^k e_0\|^{1/k} \leq \rho(B_1) + \epsilon.$$

Furthermore, for all $k \geq N(\epsilon)$, there exists a vector $e_0 = e_0(k)$ such that

$$\|e_0\| = 1 \quad \text{and} \quad \|B_2^k e_0\|^{1/k} = \|B_2^k\|^{1/k} \geq \rho(B_2),$$

which implies Statement (2).   □

In light of the above, we see that when we investigate new iterative methods, we have to deal with the following two problems:

(1) Given an iterative method with matrix B, determine whether the method is convergent. This involves determining whether $\rho(B) < 1$, or equivalently whether there is a subordinate matrix norm such that $\|B\| < 1$. By Proposition 8.8, this implies that $I - B$ is invertible (since $\|-B\| = \|B\|$, Proposition 8.8 applies).

(2) Given two convergent iterative methods, compare them. The iterative method which is faster is that whose matrix has the smaller spectral radius.

We now discuss three iterative methods for solving linear systems:

(1) Jacobi's method
(2) Gauss–Seidel's method
(3) The relaxation method.

9.3   Description of the Methods of Jacobi, Gauss–Seidel, and Relaxation

The methods described in this section are instances of the following scheme: Given a linear system $Ax = b$, with A invertible, suppose we can write A in the form

$$A = M - N,$$

with M invertible, and "easy to invert," which means that M is close to being a diagonal or a triangular matrix (perhaps by blocks). Then $Au = b$ is equivalent to

$$Mu = Nu + b,$$

that is,
$$u = M^{-1}Nu + M^{-1}b.$$
Therefore, we are in the situation described in the previous sections with $B = M^{-1}N$ and $c = M^{-1}b$. In fact, since $A = M - N$, we have
$$B = M^{-1}N = M^{-1}(M - A) = I - M^{-1}A, \qquad (9.1)$$
which shows that $I - B = M^{-1}A$ is invertible. The iterative method associated with the matrix $B = M^{-1}N$ is given by
$$u_{k+1} = M^{-1}Nu_k + M^{-1}b, \quad k \geq 0, \qquad (9.2)$$
starting from any arbitrary vector u_0. From a practical point of view, we do not invert M, and instead we solve iteratively the systems
$$Mu_{k+1} = Nu_k + b, \quad k \geq 0.$$

Various methods correspond to various ways of choosing M and N from A. The first two methods choose M and N as disjoint submatrices of A, but the relaxation method allows some overlapping of M and N.

To describe the various choices of M and N, it is convenient to write A in terms of three submatrices D, E, F, as
$$A = D - E - F,$$
where the only nonzero entries in D are the diagonal entries in A, the only nonzero entries in E are entries in A below the the diagonal, and the only nonzero entries in F are entries in A above the diagonal. More explicitly, if

$$A = \begin{pmatrix} a_{11} & a_{12} & a_{13} & \cdots & a_{1n-1} & a_{1n} \\ a_{21} & a_{22} & a_{23} & \cdots & a_{2n-1} & a_{2n} \\ a_{31} & a_{32} & a_{33} & \cdots & a_{3n-1} & a_{3n} \\ \vdots & \vdots & \vdots & \ddots & \vdots & \vdots \\ a_{n-11} & a_{n-12} & a_{n-13} & \cdots & a_{n-1n-1} & a_{n-1n} \\ a_{n1} & a_{n2} & a_{n3} & \cdots & a_{nn-1} & a_{nn} \end{pmatrix},$$

then

$$D = \begin{pmatrix} a_{11} & 0 & 0 & \cdots & 0 & 0 \\ 0 & a_{22} & 0 & \cdots & 0 & 0 \\ 0 & 0 & a_{33} & \cdots & 0 & 0 \\ \vdots & \vdots & \vdots & \ddots & \vdots & \vdots \\ 0 & 0 & 0 & \cdots & a_{n-1n-1} & 0 \\ 0 & 0 & 0 & \cdots & 0 & a_{nn} \end{pmatrix},$$

$$-E = \begin{pmatrix} 0 & 0 & 0 & \cdots & 0 & 0 \\ a_{21} & 0 & 0 & \cdots & 0 & 0 \\ a_{31} & a_{32} & 0 & \cdots & 0 & 0 \\ \vdots & \vdots & \ddots & \ddots & \vdots & \vdots \\ a_{n-1\,1} & a_{n-1\,2} & a_{n-1\,3} & \ddots & 0 & 0 \\ a_{n\,1} & a_{n\,2} & a_{n\,3} & \cdots & a_{n\,n-1} & 0 \end{pmatrix},$$

$$-F = \begin{pmatrix} 0 & a_{12} & a_{13} & \cdots & a_{1n-1} & a_{1n} \\ 0 & 0 & a_{23} & \cdots & a_{2n-1} & a_{2n} \\ 0 & 0 & 0 & \ddots & a_{3n-1} & a_{3n} \\ \vdots & \vdots & \vdots & \ddots & \ddots & \vdots \\ 0 & 0 & 0 & \cdots & 0 & a_{n-1\,n} \\ 0 & 0 & 0 & \cdots & 0 & 0 \end{pmatrix}.$$

In *Jacobi's method*, we assume that *all* diagonal entries in A are nonzero, and we pick

$$M = D$$
$$N = E + F,$$

so that by (9.1),

$$B = M^{-1}N = D^{-1}(E + F) = I - D^{-1}A.$$

As a matter of notation, we let

$$J = I - D^{-1}A = D^{-1}(E + F),$$

which is called *Jacobi's matrix*. The corresponding method, *Jacobi's iterative method*, computes the sequence (u_k) using the recurrence

$$u_{k+1} = D^{-1}(E + F)u_k + D^{-1}b, \quad k \geq 0.$$

In practice, we iteratively solve the systems

$$Du_{k+1} = (E + F)u_k + b, \quad k \geq 0.$$

If we write $u_k = (u_1^k, \ldots, u_n^k)$, we solve iteratively the following system:

$$
\begin{array}{rcllll}
a_{11}u_1^{k+1} & = & & -a_{12}u_2^k & \cdots & -a_{1n}u_n^k & +b_1 \\
a_{22}u_2^{k+1} & = & -a_{21}u_1^k & & \cdots & -a_{2n}u_n^k & +b_2 \\
\vdots & & \vdots & \vdots & & & \\
a_{n-1\,n-1}u_{n-1}^{k+1} & = & -a_{n-1\,1}u_1^k & \cdots & & -a_{n-1\,n}u_n^k & +b_{n-1} \\
a_{n\,n}u_n^{k+1} & = & -a_{n\,1}u_1^k & -a_{n\,2}u_2^k & -a_{n\,n-1}u_{n-1}^k & & +b_n
\end{array}
$$

In Matlab one step of Jacobi iteration is achieved by the following function:

```
function v = Jacobi2(A,b,u)
n = size(A,1);
v = zeros(n,1);
    for i = 1:n
        v(i,1)  = u(i,1) + (-A(i,:)*u + b(i))/A(i,i);
    end
end
```

In order to run m iteration steps, run the following function:

```
function u = jacobi(A,b,u0,m)
  u = u0;
  for j = 1:m
    u = Jacobi2(A,b,u);
  end
end
```

Example 9.1. Consider the linear system

$$
\begin{pmatrix}
2 & -1 & 0 & 0 \\
-1 & 2 & -1 & 0 \\
0 & -1 & 2 & -1 \\
0 & 0 & -1 & 2
\end{pmatrix}
\begin{pmatrix}
x_1 \\ x_2 \\ x_3 \\ x_4
\end{pmatrix}
=
\begin{pmatrix}
25 \\ -24 \\ 21 \\ -15
\end{pmatrix}.
$$

We check immediately that the solution is

$$x_1 = 11, \; x_2 = -3, \; x_3 = 7, \; x_4 = -4.$$

It is easy to see that the Jacobi matrix is

$$
J = \frac{1}{2}
\begin{pmatrix}
0 & 1 & 0 & 0 \\
1 & 0 & 1 & 0 \\
0 & 1 & 0 & 1 \\
0 & 0 & 1 & 0
\end{pmatrix}.
$$

After 10 Jacobi iterations, we find the approximate solution

$$x_1 = 10.2588, \ x_2 = -2.5244, \ x_3 = 5.8008, \ x_4 = -3.7061.$$

After 20 iterations, we find the approximate solution

$$x_1 = 10.9110, \ x_2 = -2.9429, \ x_3 = 6.8560, \ x_4 = -3.9647.$$

After 50 iterations, we find the approximate solution

$$x_1 = 10.9998, \ x_2 = -2.9999, \ x_3 = 6.9998, \ x_4 = -3.9999,$$

and after 60 iterations, we find the approximate solution

$$x_1 = 11.0000, \ x_2 = -3.0000, \ x_3 = 7.0000, \ x_4 = -4.0000,$$

correct up to at least four decimals.

It can be shown (see Problem 9.6) that the eigenvalues of J are

$$\cos\left(\frac{\pi}{5}\right), \ \cos\left(\frac{2\pi}{5}\right), \ \cos\left(\frac{3\pi}{5}\right), \ \cos\left(\frac{4\pi}{5}\right),$$

so the spectral radius of $J = B$ is

$$\rho(J) = \cos\left(\frac{\pi}{5}\right) = 0.8090 < 1.$$

By Theorem 9.2, Jacobi's method converges for the matrix of this example.

Observe that we can try to "speed up" the method by using the new value u_1^{k+1} instead of u_1^k in solving for u_2^{k+2} using the second equations, and more generally, use $u_1^{k+1}, \ldots, u_{i-1}^{k+1}$ instead of $u_1^k, \ldots, u_{i-1}^k$ in solving for u_i^{k+1} in the ith equation. This observation leads to the system

$$
\begin{array}{lllll}
a_{11}u_1^{k+1} & = & -a_{12}u_2^k & \cdots & -a_{1n}u_n^k & +b_1 \\
a_{22}u_2^{k+1} & = & -a_{21}u_1^{k+1} & \cdots & -a_{2n}u_n^k & +b_2 \\
\vdots & \vdots & \vdots & & & \\
a_{n-1\,n-1}u_{n-1}^{k+1} & = & -a_{n-1\,1}u_1^{k+1} & \cdots & -a_{n-1\,n}u_n^k & +b_{n-1} \\
a_{n\,n}u_n^{k+1} & = & -a_{n\,1}u_1^{k+1} & -a_{n\,2}u_2^{k+1} & -a_{n\,n-1}u_{n-1}^{k+1} & +b_n
\end{array}
$$

which, in matrix form, is written

$$Du_{k+1} = Eu_{k+1} + Fu_k + b.$$

Because D is invertible and E is lower triangular, the matrix $D - E$ is invertible, so the above equation is equivalent to

$$u_{k+1} = (D - E)^{-1}Fu_k + (D - E)^{-1}b, \quad k \geq 0.$$

The above corresponds to choosing M and N to be

$$M = D - E$$
$$N = F,$$

and the matrix B is given by

$$B = M^{-1}N = (D - E)^{-1}F.$$

Since $M = D - E$ is invertible, we know that $I - B = M^{-1}A$ is also invertible.

The method that we just described is the *iterative method of Gauss–Seidel*, and the matrix B is called the *matrix of Gauss–Seidel* and denoted by $\mathcal{L}_1$, with

$$\mathcal{L}_1 = (D - E)^{-1}F.$$

One of the advantages of the method of Gauss–Seidel is that is requires only half of the memory used by Jacobi's method, since we only need

$$u_1^{k+1}, \dots, u_{i-1}^{k+1}, u_{i+1}^k, \dots, u_n^k$$

to compute u_i^{k+1}. We also show that in certain important cases (for example, if A is a tridiagonal matrix), the method of Gauss–Seidel converges faster than Jacobi's method (in this case, they both converge or diverge simultaneously).

In `Matlab` one step of Gauss–Seidel iteration is achieved by the following function:

```
function u = GaussSeidel3(A,b,u)
n = size(A,1);
for i = 1:n
  u(i,1)  = u(i,1) + (-A(i,:)*u + b(i))/A(i,i);
end
end
```

It is remarkable that the only difference with `Jacobi2` is that the same variable u is used on both sides of the assignment. In order to run m iteration steps, run the following function:

```
function u = GaussSeidel1(A,b,u0,m)
  u = u0;
  for j = 1:m
    u = GaussSeidel3(A,b,u);
  end
end
```

Example 9.2. Consider the same linear system

$$\begin{pmatrix} 2 & -1 & 0 & 0 \\ -1 & 2 & -1 & 0 \\ 0 & -1 & 2 & -1 \\ 0 & 0 & -1 & 2 \end{pmatrix} \begin{pmatrix} x_1 \\ x_2 \\ x_3 \\ x_4 \end{pmatrix} = \begin{pmatrix} 25 \\ -24 \\ 21 \\ -15 \end{pmatrix}$$

as in Example 9.1, whose solution is

$$x_1 = 11, \ x_2 = -3, \ x_3 = 7, \ x_4 = -4.$$

After 10 Gauss–Seidel iterations, we find the approximate solution

$$x_1 = 10.9966, \ x_2 = -3.0044, \ x_3 = 6.9964, \ x_4 = -4.0018.$$

After 20 iterations, we find the approximate solution

$$x_1 = 11.0000, \ x_2 = -3.0001, \ x_3 = 6.9999, \ x_4 = -4.0000.$$

After 25 iterations, we find the approximate solution

$$x_1 = 11.0000, \ x_2 = -3.0000, \ x_3 = 7.0000, \ x_4 = -4.0000,$$

correct up to at least four decimals. We observe that for this example, Gauss–Seidel's method converges about twice as fast as Jacobi's method. It will be shown in Proposition 9.5 that for a tridiagonal matrix, the spectral radius of the Gauss–Seidel matrix $\mathcal{L}_1$ is given by

$$\rho(\mathcal{L}_1) = (\rho(J))^2,$$

so our observation is consistent with the theory.

The new ingredient in the *relaxation method* is to incorporate part of the matrix D into N: we define M and N by

$$M = \frac{D}{\omega} - E$$

$$N = \frac{1 - \omega}{\omega} D + F,$$

where $\omega \neq 0$ is a real parameter to be suitably chosen. Actually, we show in Section 9.4 that for the relaxation method to converge, we must have $\omega \in (0, 2)$. Note that the case $\omega = 1$ corresponds to the method of Gauss–Seidel.

If we assume that *all* diagonal entries of D are nonzero, the matrix M is invertible. The matrix B is denoted by $\mathcal{L}_\omega$ and called the *matrix of relaxation*, with

$$\mathcal{L}_\omega = \left(\frac{D}{\omega} - E\right)^{-1} \left(\frac{1 - \omega}{\omega} D + F\right) = (D - \omega E)^{-1}((1 - \omega)D + \omega F).$$

The number ω is called the *parameter of relaxation*.

When $\omega > 1$, the relaxation method is known as *successive overrelaxation*, abbreviated as *SOR*.

At first glance the relaxation matrix $\mathcal{L}_\omega$ seems at lot more complicated than the Gauss–Seidel matrix $\mathcal{L}_1$, but the iterative system associated with the relaxation method is very similar to the method of Gauss–Seidel, and is quite simple. Indeed, the system associated with the relaxation method is given by

$$\left(\frac{D}{\omega} - E\right) u_{k+1} = \left(\frac{1-\omega}{\omega}D + F\right) u_k + b,$$

which is equivalent to

$$(D - \omega E) u_{k+1} = ((1 - \omega)D + \omega F) u_k + \omega b,$$

and can be written

$$D u_{k+1} = D u_k - \omega (D u_k - E u_{k+1} - F u_k - b).$$

Explicitly, this is the system

$$a_{11} u_1^{k+1} = a_{11} u_1^k - \omega (a_{11} u_1^k + \cdots + a_{1n-1} u_{n-1}^k + a_{1n} u_n^k - b_1)$$
$$a_{22} u_2^{k+1} = a_{22} u_2^k - \omega (a_{21} u_1^{k+1} + \cdots + a_{2n-1} u_{n-1}^k + a_{2n} u_n^k - b_2)$$
$$\vdots$$
$$a_{nn} u_n^{k+1} = a_{nn} u_n^k - \omega (a_{n1} u_1^{k+1} + + \cdots + a_{nn-1} u_{n-1}^{k+1} + a_{nn} u_n^k - b_n).$$

In `Matlab` one step of relaxation iteration is achieved by the following function:

```
function u = relax3(A,b,u,omega)
n = size(A,1);
for i = 1:n
    u(i,1)  = u(i,1) + omega*(-A(i,:)*u + b(i))/A(i,i);
end
end
```

Observe that function `relax3` is obtained from the function `GaussSeidel3` by simply inserting ω in front of the expression $(-A(i,:) * u + b(i))/A(i,i)$. In order to run m iteration steps, run the following function:

```
function u = relax(A,b,u0,omega,m)
    u = u0;
    for j = 1:m
```

```
   u = relax3(A,b,u,omega);
 end
end
```

Example 9.3. Consider the same linear system as in Examples 9.1 and 9.2, whose solution is

$$x_1 = 11, \ x_2 = -3, \ x_3 = 7, \ x_4 = -4.$$

After 10 relaxation iterations with $\omega = 1.1$, we find the approximate solution

$$x_1 = 11.0026, \ x_2 = -2.9968, \ x_3 = 7.0024, \ x_4 = -3.9989.$$

After 10 iterations with $\omega = 1.2$, we find the approximate solution

$$x_1 = 11.0014, \ x_2 = -2.9985, \ x_3 = 7.0010, \ x_4 = -3.9996.$$

After 10 iterations with $\omega = 1.3$, we find the approximate solution

$$x_1 = 10.9996, \ x_2 = -3.0001, \ x_3 = 6.9999, \ x_4 = -4.0000.$$

After 10 iterations with $\omega = 1.27$, we find the approximate solution

$$x_1 = 11.0000, \ x_2 = -3.0000, \ x_3 = 7.0000, \ x_4 = -4.0000,$$

correct up to at least four decimals. We observe that for this example the method of relaxation with $\omega = 1.27$ converges faster than the method of Gauss–Seidel. This observation will be confirmed by Proposition 9.7.

What remains to be done is to find conditions that ensure the convergence of the relaxation method (and the Gauss–Seidel method), that is:

(1) Find conditions on ω, namely some interval $I \subseteq \mathbb{R}$ so that $\omega \in I$ implies $\rho(\mathcal{L}_\omega) < 1$; we will prove that $\omega \in (0, 2)$ is a necessary condition.
(2) Find if there exist some *optimal value* ω_0 of $\omega \in I$, so that

$$\rho(\mathcal{L}_{\omega_0}) = \inf_{\omega \in I} \rho(\mathcal{L}_\omega).$$

We will give partial answers to the above questions in the next section.

It is also possible to extend the methods of this section by using *block decompositions* of the form $A = D - E - F$, where $D, E,$ and F consist of blocks, and D is an invertible block-diagonal matrix. See Figure 9.1.

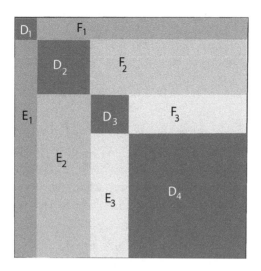

Fig. 9.1 A schematic representation of a block decomposition $A = D - E - F$, where $D = \cup_{i=1}^{4} D_i$, $E = \cup_{i=1}^{3} E_i$, and $F = \cup_{i=1}^{3} F_i$.

9.4 Convergence of the Methods of Gauss–Seidel and Relaxation

We begin with a general criterion for the convergence of an iterative method associated with a (complex) Hermitian positive definite matrix, $A = M - N$. Next we apply this result to the relaxation method.

Proposition 9.3. *Let A be any Hermitian positive definite matrix, written as*

$$A = M - N,$$

with M invertible. Then $M^* + N$ is Hermitian, and if it is positive definite, then

$$\rho(M^{-1}N) < 1,$$

so that the iterative method converges.

Proof. Since $M = A + N$ and A is Hermitian, $A^* = A$, so we get

$$M^* + N = A^* + N^* + N = A + N + N^* = M + N^* = (M^* + N)^*,$$

which shows that $M^* + N$ is indeed Hermitian.

Because A is Hermitian positive definite, the function

$$v \mapsto (v^* A v)^{1/2}$$

from $\mathbb{C}^n$ to $\mathbb{R}$ is a vector norm $\| \ \|$, and let $\| \ \|$ also denote its subordinate matrix norm. We prove that

$$\|M^{-1}N\| < 1,$$

which by Theorem 9.1 proves that $\rho(M^{-1}N) < 1$. By definition

$$\|M^{-1}N\| = \|I - M^{-1}A\| = \sup_{\|v\|=1} \|v - M^{-1}Av\|,$$

which leads us to evaluate $\|v - M^{-1}Av\|$ when $\|v\| = 1$. If we write $w = M^{-1}Av$, using the facts that $\|v\| = 1$, $v = A^{-1}Mw$, $A^* = A$, and $A = M - N$, we have

$$\begin{aligned}
\|v - w\|^2 &= (v-w)^* A(v-w) \\
&= \|v\|^2 - v^*Aw - w^*Av + w^*Aw \\
&= 1 - w^*M^*w - w^*Mw + w^*Aw \\
&= 1 - w^*(M^* + N)w.
\end{aligned}$$

Now since we assumed that $M^* + N$ is positive definite, if $w \neq 0$, then $w^*(M^* + N)w > 0$, and we conclude that

$$\text{if} \quad \|v\| = 1, \quad \text{then} \quad \|v - M^{-1}Av\| < 1.$$

Finally, the function

$$v \mapsto \|v - M^{-1}Av\|$$

is continuous as a composition of continuous functions, therefore it achieves its maximum on the compact subset $\{v \in \mathbb{C}^n \mid \|v\| = 1\}$, which proves that

$$\sup_{\|v\|=1} \|v - M^{-1}Av\| < 1,$$

and completes the proof. $\qquad\square$

Now as in the previous sections, we assume that A is written as $A = D - E - F$, with D invertible, possibly in block form. The next theorem provides a sufficient condition (which turns out to be also necessary) for the relaxation method to converge (and thus, for the method of Gauss–Seidel to converge). This theorem is known as the *Ostrowski–Reich theorem*.

Theorem 9.3. *If $A = D - E - F$ is Hermitian positive definite, and if $0 < \omega < 2$, then the relaxation method converges. This also holds for a block decomposition of A.*

Proof. Recall that for the relaxation method, $A = M - N$ with

$$M = \frac{D}{\omega} - E$$

$$N = \frac{1 - \omega}{\omega} D + F,$$

and because $D^* = D$, $E^* = F$ (since A is Hermitian) and $\omega \neq 0$ is real, we have

$$M^* + N = \frac{D^*}{\omega} - E^* + \frac{1 - \omega}{\omega} D + F = \frac{2 - \omega}{\omega} D.$$

If D consists of the diagonal entries of A, then we know from Section 7.8 that these entries are all positive, and since $\omega \in (0, 2)$, we see that the matrix $((2 - \omega)/\omega)D$ is positive definite. If D consists of diagonal blocks of A, because A is positive, definite, by choosing vectors z obtained by picking a nonzero vector for each block of D and padding with zeros, we see that each block of D is positive definite, and thus D itself is positive definite. Therefore, in all cases, $M^* + N$ is positive definite, and we conclude by using Proposition 9.3. $\qquad\square$

Remark: What if we allow the parameter ω to be a nonzero complex number $\omega \in \mathbb{C}$? In this case, we get

$$M^* + N = \frac{D^*}{\overline{\omega}} - E^* + \frac{1 - \omega}{\omega} D + F = \left(\frac{1}{\omega} + \frac{1}{\overline{\omega}} - 1 \right) D.$$

But,

$$\frac{1}{\omega} + \frac{1}{\overline{\omega}} - 1 = \frac{\omega + \overline{\omega} - \omega\overline{\omega}}{\omega\overline{\omega}} = \frac{1 - (\omega - 1)(\overline{\omega} - 1)}{|\omega|^2} = \frac{1 - |\omega - 1|^2}{|\omega|^2},$$

so the relaxation method also converges for $\omega \in \mathbb{C}$, provided that

$$|\omega - 1| < 1.$$

This condition reduces to $0 < \omega < 2$ if ω is real.

Unfortunately, Theorem 9.3 does not apply to Jacobi's method, but in special cases, Proposition 9.3 can be used to prove its convergence. On the positive side, if a matrix is strictly column (or row) diagonally dominant, then it can be shown that the method of Jacobi and the method of Gauss–Seidel both converge. The relaxation method also converges if $\omega \in (0, 1]$, but this is not a very useful result because the speed-up of convergence usually occurs for $\omega > 1$.

We now prove that, without *any* assumption on $A = D - E - F$, other than the fact that A and D are invertible, in order for the relaxation method to converge, we must have $\omega \in (0, 2)$.

Proposition 9.4. *Given any matrix $A = D - E - F$, with A and D invertible, for any $\omega \neq 0$, we have*

$$\rho(\mathcal{L}_\omega) \geq |\omega - 1|,$$

where $\mathcal{L}_\omega = \left(\dfrac{D}{\omega} - E \right)^{-1} \left(\dfrac{1-\omega}{\omega} D + F \right)$. Therefore, the relaxation method (possibly by blocks) does not converge unless $\omega \in (0, 2)$. If we allow ω to be complex, then we must have

$$|\omega - 1| < 1$$

for the relaxation method to converge.

Proof. Observe that the product $\lambda_1 \cdots \lambda_n$ of the eigenvalues of $\mathcal{L}_\omega$, which is equal to $\det(\mathcal{L}_\omega)$, is given by

$$\lambda_1 \cdots \lambda_n = \det(\mathcal{L}_\omega) = \frac{\det \left(\dfrac{1 - \omega}{\omega} D + F \right)}{\det \left(\dfrac{D}{\omega} - E \right)} = (1 - \omega)^n.$$

It follows that

$$\rho(\mathcal{L}_\omega) \geq |\lambda_1 \cdots \lambda_n|^{1/n} = |\omega - 1|.$$

The proof is the same if $\omega \in \mathbb{C}$. $\square$

9.5 Convergence of the Methods of Jacobi, Gauss–Seidel, and Relaxation for Tridiagonal Matrices

We now consider the case where A is a *tridiagonal matrix*, possibly by blocks. In this case, we obtain precise results about the spectral radius of J and $\mathcal{L}_\omega$, and as a consequence, about the convergence of these methods. We also obtain some information about the rate of convergence of these methods. We begin with the case $\omega = 1$, which is technically easier to deal with. The following proposition gives us the precise relationship between the spectral radii $\rho(J)$ and $\rho(\mathcal{L}_1)$ of the Jacobi matrix and the Gauss–Seidel matrix.

Proposition 9.5. *Let A be a tridiagonal matrix (possibly by blocks). If $\rho(J)$ is the spectral radius of the Jacobi matrix and $\rho(\mathcal{L}_1)$ is the spectral radius of the Gauss–Seidel matrix, then we have*

$$\rho(\mathcal{L}_1) = (\rho(J))^2.$$

Consequently, the method of Jacobi and the method of Gauss–Seidel both converge or both diverge simultaneously (even when A is tridiagonal by blocks); when they converge, the method of Gauss–Seidel converges faster than Jacobi's method.

Proof. We begin with a preliminary result. Let $A(\mu)$ with a tridiagonal matrix by block of the form

$$A(\mu) = \begin{pmatrix} A_1 & \mu^{-1}C_1 & 0 & 0 & \cdots & 0 \\ \mu B_1 & A_2 & \mu^{-1}C_2 & 0 & \cdots & 0 \\ 0 & \ddots & \ddots & \ddots & \cdots & \vdots \\ \vdots & \cdots & \ddots & \ddots & \ddots & 0 \\ 0 & \cdots & 0 & \mu B_{p-2} & A_{p-1} & \mu^{-1}C_{p-1} \\ 0 & \cdots & \cdots & 0 & \mu B_{p-1} & A_p \end{pmatrix},$$

then

$$\det(A(\mu)) = \det(A(1)), \quad \mu \neq 0.$$

To prove this fact, form the block diagonal matrix

$$P(\mu) = \mathrm{diag}(\mu I_1, \mu^2 I_2, \ldots, \mu^p I_p),$$

where I_j is the identity matrix of the same dimension as the block A_j. Then it is easy to see that

$$A(\mu) = P(\mu)A(1)P(\mu)^{-1},$$

and thus,

$$\det(A(\mu)) = \det(P(\mu)A(1)P(\mu)^{-1}) = \det(A(1)).$$

Since the Jacobi matrix is $J = D^{-1}(E + F)$, the eigenvalues of J are the zeros of the characteristic polynomial

$$p_J(\lambda) = \det(\lambda I - D^{-1}(E + F)),$$

and thus, they are also the zeros of the polynomial

$$q_J(\lambda) = \det(\lambda D - E - F) = \det(D)p_J(\lambda).$$

Similarly, since the Gauss–Seidel matrix is $\mathcal{L}_1 = (D - E)^{-1}F$, the zeros of the characteristic polynomial

$$p_{\mathcal{L}_1}(\lambda) = \det(\lambda I - (D - E)^{-1}F)$$

are also the zeros of the polynomial

$$q_{\mathcal{L}_1}(\lambda) = \det(\lambda D - \lambda E - F) = \det(D - E)p_{\mathcal{L}_1}(\lambda).$$

Since $A = D - E - F$ is tridiagonal (or tridiagonal by blocks), $\lambda^2 D - \lambda^2 E - F$ is also tridiagonal (or tridiagonal by blocks), and by using our preliminary result with $\mu = \lambda \neq 0$, we get

$$q_{\mathcal{L}_1}(\lambda^2) = \det(\lambda^2 D - \lambda^2 E - F) = \det(\lambda^2 D - \lambda E - \lambda F) = \lambda^n q_J(\lambda).$$

By continuity, the above equation also holds for $\lambda = 0$. But then we deduce that:

(1) For any $\beta \neq 0$, if β is an eigenvalue of $\mathcal{L}_1$, then $\beta^{1/2}$ and $-\beta^{1/2}$ are both eigenvalues of J, where $\beta^{1/2}$ is one of the complex square roots of β.

(2) For any $\alpha \neq 0$, if α and $-\alpha$ are both eigenvalues of J, then α^2 is an eigenvalue of $\mathcal{L}_1$.

The above immediately implies that $\rho(\mathcal{L}_1) = (\rho(J))^2$. $\qquad\square$

We now consider the more general situation where ω is any real in $(0, 2)$.

Proposition 9.6. *Let A be a tridiagonal matrix (possibly by blocks), and assume that the eigenvalues of the Jacobi matrix are all real. If $\omega \in (0, 2)$, then the method of Jacobi and the method of relaxation both converge or both diverge simultaneously (even when A is tridiagonal by blocks). When they converge, the function $\omega \mapsto \rho(\mathcal{L}_\omega)$ (for $\omega \in (0, 2)$) has a unique minimum equal to $\omega_0 - 1$ for*

$$\omega_0 = \frac{2}{1 + \sqrt{1 - (\rho(J))^2}},$$

where $1 < \omega_0 < 2$ if $\rho(J) > 0$. We also have $\rho(\mathcal{L}_1) = (\rho(J))^2$, as before.

Proof. The proof is very technical and can be found in Serre [Serre (2010)] and Ciarlet [Ciarlet (1989)]. As in the proof of the previous proposition, we begin by showing that the eigenvalues of the matrix $\mathcal{L}_\omega$ are the zeros of the polynomial

$$q_{\mathcal{L}_\omega}(\lambda) = \det\left(\frac{\lambda + \omega - 1}{\omega}D - \lambda E - F\right) = \det\left(\frac{D}{\omega} - E\right)p_{\mathcal{L}_\omega}(\lambda),$$

where $p_{\mathcal{L}_\omega}(\lambda)$ is the characteristic polynomial of $\mathcal{L}_\omega$. Then using the preliminary fact from Proposition 9.5, it is easy to show that

$$q_{\mathcal{L}_\omega}(\lambda^2) = \lambda^n q_J\left(\frac{\lambda^2 + \omega - 1}{\lambda\omega}\right),$$

for all $\lambda \in \mathbb{C}$, with $\lambda \neq 0$. This time we cannot extend the above equation to $\lambda = 0$. This leads us to consider the equation

$$\frac{\lambda^2 + \omega - 1}{\lambda\omega} = \alpha,$$

which is equivalent to

$$\lambda^2 - \alpha\omega\lambda + \omega - 1 = 0,$$

for all $\lambda \neq 0$. Since $\lambda \neq 0$, the above equivalence does not hold for $\omega = 1$, but this is not a problem since the case $\omega = 1$ has already been considered in the previous proposition. Then we can show the following:

(1) For any $\beta \neq 0$, if β is an eigenvalue of $\mathcal{L}_\omega$, then

$$\frac{\beta + \omega - 1}{\beta^{1/2}\omega}, \quad -\frac{\beta + \omega - 1}{\beta^{1/2}\omega}$$

are eigenvalues of J.
(2) For every $\alpha \neq 0$, if α and $-\alpha$ are eigenvalues of J, then $\mu_+(\alpha, \omega)$ and $\mu_-(\alpha, \omega)$ are eigenvalues of $\mathcal{L}_\omega$, where $\mu_+(\alpha, \omega)$ and $\mu_-(\alpha, \omega)$ are the squares of the roots of the equation

$$\lambda^2 - \alpha\omega\lambda + \omega - 1 = 0.$$

It follows that

$$\rho(\mathcal{L}_\omega) = \max_{\lambda \,|\, p_J(\lambda) = 0}\{\max(|\mu_+(\alpha, \omega)|, |\mu_-(\alpha, \omega)|)\},$$

and since we are assuming that J has real roots, we are led to study the function

$$M(\alpha, \omega) = \max\{|\mu_+(\alpha, \omega)|, |\mu_-(\alpha, \omega)|\},$$

where $\alpha \in \mathbb{R}$ and $\omega \in (0, 2)$. Actually, because $M(-\alpha, \omega) = M(\alpha, \omega)$, it is only necessary to consider the case where $\alpha \geq 0$.

Note that for $\alpha \neq 0$, the roots of the equation

$$\lambda^2 - \alpha\omega\lambda + \omega - 1 = 0.$$

are

$$\frac{\alpha\omega \pm \sqrt{\alpha^2\omega^2 - 4\omega + 4}}{2}.$$

In turn, this leads to consider the roots of the equation

$$\omega^2\alpha^2 - 4\omega + 4 = 0,$$

which are

$$\frac{2(1 \pm \sqrt{1 - \alpha^2})}{\alpha^2},$$

for $\alpha \neq 0$. Since we have

$$\frac{2(1 + \sqrt{1 - \alpha^2})}{\alpha^2} = \frac{2(1 + \sqrt{1 - \alpha^2})(1 - \sqrt{1 - \alpha^2})}{\alpha^2(1 - \sqrt{1 - \alpha^2})} = \frac{2}{1 - \sqrt{1 - \alpha^2}}$$

and

$$\frac{2(1 - \sqrt{1 - \alpha^2})}{\alpha^2} = \frac{2(1 + \sqrt{1 - \alpha^2})(1 - \sqrt{1 - \alpha^2})}{\alpha^2(1 + \sqrt{1 - \alpha^2})} = \frac{2}{1 + \sqrt{1 - \alpha^2}},$$

these roots are

$$\omega_0(\alpha) = \frac{2}{1 + \sqrt{1 - \alpha^2}}, \quad \omega_1(\alpha) = \frac{2}{1 - \sqrt{1 - \alpha^2}}.$$

Observe that the expression for $\omega_0(\alpha)$ is exactly the expression in the statement of our proposition! The rest of the proof consists in analyzing the variations of the function $M(\alpha, \omega)$ by considering various cases for α. In the end, we find that the minimum of $\rho(\mathcal{L}_\omega)$ is obtained for $\omega_0(\rho(J))$. The details are tedious and we omit them. The reader will find complete proofs in Serre [Serre (2010)] and Ciarlet [Ciarlet (1989)]. $\quad\square$

Combining the results of Theorem 9.3 and Proposition 9.6, we obtain the following result which gives precise information about the spectral radii of the matrices J, $\mathcal{L}_1$, and $\mathcal{L}_\omega$.

Proposition 9.7. *Let A be a tridiagonal matrix (possibly by blocks) which is Hermitian positive definite. Then the methods of Jacobi, Gauss–Seidel,* *and relaxation, all converge for* $\omega \in (0,2)$. *There is a unique optimal relaxation parameter*

$$\omega_0 = \frac{2}{1 + \sqrt{1 - (\rho(J))^2}},$$

such that

$$\rho(\mathcal{L}_{\omega_0}) = \inf_{0 < \omega < 2} \rho(\mathcal{L}_\omega) = \omega_0 - 1.$$

Furthermore, if $\rho(J) > 0$, *then*

$$\rho(\mathcal{L}_{\omega_0}) < \rho(\mathcal{L}_1) = (\rho(J))^2 < \rho(J),$$

and if $\rho(J) = 0$, *then* $\omega_0 = 1$ *and* $\rho(\mathcal{L}_1) = \rho(J) = 0$.

Proof. In order to apply Proposition 9.6, we have to check that $J = D^{-1}(E + F)$ has real eigenvalues. However, if α is any eigenvalue of J and if u is any corresponding eigenvector, then

$$D^{-1}(E + F)u = \alpha u$$

implies that

$$(E + F)u = \alpha Du,$$

and since $A = D - E - F$, the above shows that $(D - A)u = \alpha Du$, that is,

$$Au = (1 - \alpha)Du.$$

Consequently,

$$u^* Au = (1 - \alpha)u^* Du,$$

and since A and D are Hermitian positive definite, we have $u^* Au > 0$ and $u^* Du > 0$ if $u \neq 0$, which proves that $\alpha \in \mathbb{R}$. The rest follows from Theorem 9.3 and Proposition 9.6. $\qquad \square$

Remark: It is preferable to overestimate rather than underestimate the relaxation parameter when the optimum relaxation parameter is not known exactly.

9.6   Summary

The main concepts and results of this chapter are listed below:

- Iterative methods. Splitting A as $A = M - N$.
- *Convergence of a sequence of vectors or matrices.*
- A criterion for the convergence of the sequence (B^k) of powers of a matrix B to zero in terms of the spectral radius $\rho(B)$.
- A characterization of the spectral radius $\rho(B)$ as the limit of the sequence $(\|B^k\|^{1/k})$.
- A criterion of the convergence of iterative methods.
- Asymptotic behavior of iterative methods.
- Splitting A as $A = D - E - F$, and the methods of *Jacobi, Gauss–Seidel,* and *relaxation* (and *SOR*).
- The *Jacobi matrix*, $J = D^{-1}(E + F)$.
- The *Gauss–Seidel matrix*, $\mathcal{L}_1 = (D - E)^{-1}F$.
- The *matrix of relaxation*, $\mathcal{L}_\omega = (D - \omega E)^{-1}((1 - \omega)D + \omega F)$.
- Convergence of iterative methods: a general result when $A = M - N$ is Hermitian positive definite.
- A sufficient condition for the convergence of the methods of Jacobi, Gauss–Seidel, and relaxation. The *Ostrowski–Reich theorem*: A is Hermitian positive definite and $\omega \in (0, 2)$.
- A necessary condition for the convergence of the methods of Jacobi, Gauss–Seidel, and relaxation: $\omega \in (0, 2)$.
- The case of tridiagonal matrices (possibly by blocks). Simultaneous convergence or divergence of Jacobi's method and Gauss–Seidel's method, and comparison of the spectral radii of $\rho(J)$ and $\rho(\mathcal{L}_1)$: $\rho(\mathcal{L}_1) = (\rho(J))^2$.
- The case of tridiagonal Hermitian positive definite matrices (possibly by blocks). The methods of Jacobi, Gauss–Seidel, and relaxation, all converge.
- In the above case, there is a unique optimal relaxation parameter for which $\rho(\mathcal{L}_{\omega_0}) < \rho(\mathcal{L}_1) = (\rho(J))^2 < \rho(J)$ (if $\rho(J) \neq 0$).

9.7 Problems

Problem 9.1. Consider the matrix

$$A = \begin{pmatrix} 1 & 2 & -2 \\ 1 & 1 & 1 \\ 2 & 2 & 1 \end{pmatrix}.$$

Prove that $\rho(J) = 0$ and $\rho(\mathcal{L}_1) = 2$, so

$$\rho(J) < 1 < \rho(\mathcal{L}_1),$$

where J is Jacobi's matrix and $\mathcal{L}_1$ is the matrix of Gauss–Seidel.

Problem 9.2. Consider the matrix

$$A = \begin{pmatrix} 2 & -1 & 1 \\ 2 & 2 & 2 \\ -1 & -1 & 2 \end{pmatrix}.$$

Prove that $\rho(J) = \sqrt{5}/2$ and $\rho(\mathcal{L}_1) = 1/2$, so

$$\rho(\mathcal{L}_1) < \rho(J),$$

where where J is Jacobi's matrix and $\mathcal{L}_1$ is the matrix of Gauss–Seidel.

Problem 9.3. Consider the following linear system:

$$\begin{pmatrix} 2 & -1 & 0 & 0 \\ -1 & 2 & -1 & 0 \\ 0 & -1 & 2 & -1 \\ 0 & 0 & -1 & 2 \end{pmatrix} \begin{pmatrix} x_1 \\ x_2 \\ x_3 \\ x_4 \end{pmatrix} = \begin{pmatrix} 19 \\ 19 \\ -3 \\ -12 \end{pmatrix}.$$

(1) Solve the above system by Gaussian elimination.

(2) Compute the sequences of vectors $u_k = (u_1^k, u_2^k, u_3^k, u_4^k)$ for $k = 1, \ldots, 10$, using the methods of Jacobi, Gauss–Seidel, and relaxation for the following values of ω: $\omega = 1.1, 1.2, \ldots, 1.9$. In all cases, the initial vector is $u_0 = (0, 0, 0, 0)$.

Problem 9.4. Recall that a complex or real $n \times n$ matrix A is *strictly row diagonally dominant* if $|a_{ii}| > \sum_{j=1, j \neq i}^{n} |a_{ij}|$ for $i = 1, \ldots, n$.

(1) Prove that if A is strictly row diagonally dominant, then Jacobi's method converges.

(2) Prove that if A is strictly row diagonally dominant, then Gauss–Seidel's method converges.

Problem 9.5. Prove that the converse of Proposition 9.3 holds. That is, if A is a Hermitian positive definite matrix written as $A = M - N$ with M invertible, if the Hermitan matrix $M^* + N$ is positive definite, and if $\rho(M^{-1}N) < 1$, then A is positive definite.

Problem 9.6. Consider the following tridiagonal $n \times n$ matrix:

$$
A = \frac{1}{(n+1)^2}
\begin{pmatrix}
2 & -1 & 0 & & \\
-1 & 2 & -1 & & \\
& \ddots & \ddots & \ddots & \\
& & -1 & 2 & -1 \\
& & 0 & -1 & 2
\end{pmatrix}.
$$

(1) Prove that the eigenvalues of the Jacobi matrix J are given by

$$
\lambda_k = \cos\left(\frac{k\pi}{n+1}\right), \quad k = 1, \dots, n.
$$

Hint. First show that the Jacobi matrix is

$$
J = \frac{1}{2}
\begin{pmatrix}
0 & 1 & 0 & & \\
1 & 0 & 1 & & \\
& \ddots & \ddots & \ddots & \\
& & 1 & 0 & 1 \\
& & 0 & 1 & 0
\end{pmatrix}.
$$

Then the eigenvalues and the eigenvectors of J are solutions of the system of equations

$$
y_0 = 0
$$
$$
y_{k+1} + y_{k-1} = 2\lambda y_k, \quad k = 1, \dots, n
$$
$$
y_{n+1} = 0.
$$

It is well known that the general solution to the above recurrence is given by

$$
y_k = \alpha z_1^k + \beta z_2^k, \quad k = 0, \dots, n+1,
$$

(with $\alpha, \beta \neq 0$) where z_1 and z_2 are the zeros of the equation

$$
z^2 - 2\lambda z + 1 = 0.
$$

It follows that $z_2 = z_1^{-1}$ and $z_1 + z_2 = 2\lambda$. The boundary condition $y_0 = 0$ yields $\alpha + \beta = 0$, so $y_k = \alpha(z_1^k - z_1^{-k})$, and the boundary condition $y_{n+1} = 0$ yields

$$
z_1^{2(n+1)} = 1.
$$

Deduce that we may assume that the n possible values $(z_1)_k$ for z_1 are given by

$$(z_1)_k = e^{\frac{k\pi i}{n+1}}, \quad k = 1, \ldots, n,$$

and find

$$2\lambda_k = (z_1)_k + (z_1)_k^{-1}.$$

Show that an eigenvector $(y_1^{(k)}, \ldots, y_n^{(k)})$ associated with the eigenvalue λ_k is given by

$$y_j^{(k)} = \sin\left(\frac{kj\pi}{n+1}\right), \quad j = 1, \ldots, n.$$

(2) Find the spectral radius $\rho(J)$, $\rho(\mathcal{L}_1)$, and $\rho(\mathcal{L}_{\omega_0})$, as functions of $h = 1/(n+1)$.

Chapter 10

The Dual Space and Duality

In this chapter all vector spaces are defined over an arbitrary field K. For the sake of concreteness, the reader may safely assume that $K = \mathbb{R}$.

10.1 The Dual Space E^* and Linear Forms

In Section 2.8 we defined linear forms, the dual space $E^* = \mathrm{Hom}(E, K)$ of a vector space E, and showed the existence of dual bases for vector spaces of finite dimension.

In this chapter we take a deeper look at the connection between a space E and its dual space E^*. As we will see shortly, every linear map $f \colon E \to F$ gives rise to a linear map $f^\top \colon F^* \to E^*$, and it turns out that in a suitable basis, the matrix of $f^\top$ is the transpose of the matrix of f. *Thus, the notion of dual space provides a conceptual explanation of the phenomena associated with transposition.*

But it does more, because it allows us to view a linear equation as an element of the dual space E^*, and thus to view subspaces of E as solutions of sets of linear equations and vice-versa. The relationship between subspaces and sets of linear forms is the essence of *duality*, a term which is often used loosely, but can be made precise as a bijection between the set of subspaces of a given vector space E and the set of subspaces of its dual E^*. In this correspondence, a subspace V of E yields the subspace V^0 of E^* consisting of all linear forms that vanish on V (that is, have the value zero for all input in V).

Consider the following set of two "linear equations" in $\mathbb{R}^3$,

$$x - y + z = 0$$
$$x - y - z = 0,$$

and let us find out what is their set V of common solutions $(x, y, z) \in \mathbb{R}^3$. By subtracting the second equation from the first, we get $2z = 0$, and by adding the two equations, we find that $2(x-y) = 0$, so the set V of solutions is given by

$$y = x$$
$$z = 0.$$

This is a one dimensional subspace of $\mathbb{R}^3$. Geometrically, this is the line of equation $y = x$ in the plane $z = 0$ as illustrated by Figure 10.1.

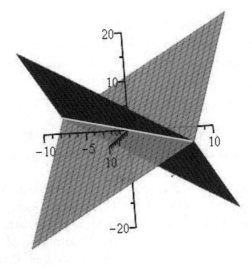

Fig. 10.1   The intersection of the magenta plane $x - y + z = 0$ with the blue-gray plane $x - y - z = 0$ is the pink line $y = x$.

Now why did we say that the above equations are linear? Because as functions of (x, y, z), both maps $f_1\colon (x, y, z) \mapsto x - y + z$ and $f_2\colon (x, y, z) \mapsto x - y - z$ are linear. The set of all such linear functions from $\mathbb{R}^3$ to $\mathbb{R}$ is a vector space; we used this fact to form linear combinations of the "equations" f_1 and f_2. Observe that the dimension of the subspace V is 1. The ambient space has dimension $n = 3$ and there are two "independent" equations f_1, f_2, so it appears that the dimension $\dim(V)$ of the subspace V defined by m independent equations is

$$\dim(V) = n - m,$$

which is indeed a general fact (proven in Theorem 10.1).

More generally, in $\mathbb{R}^n$, a linear equation is determined by an n-tuple $(a_1, \ldots, a_n) \in \mathbb{R}^n$, and the solutions of this linear equation are given by the n-tuples $(x_1, \ldots, x_n) \in \mathbb{R}^n$ such that

$$a_1 x_1 + \cdots + a_n x_n = 0;$$

these solutions constitute the kernel of the linear map $(x_1, \ldots, x_n) \mapsto a_1 x_1 + \cdots + a_n x_n$. The above considerations assume that we are working in the canonical basis $(e_1, \ldots, e_n)$ of $\mathbb{R}^n$, but we can define "linear equations" independently of bases and in any dimension, by viewing them as elements of the vector space $\mathrm{Hom}(E, K)$ of linear maps from E to the field K.

Definition 10.1. Given a vector space E, the vector space $\mathrm{Hom}(E, K)$ of linear maps from E to the field K is called the *dual space (or dual)* of E. The space $\mathrm{Hom}(E, K)$ is also denoted by E^*, and the linear maps in E^* are called *the linear forms*, or *covectors*. The dual space E^{**} of the space E^* is called the *bidual* of E.

As a matter of notation, linear forms $f \colon E \to K$ will also be denoted by starred symbol, such as u^*, x^*, etc.

Given a vector space E and any basis $(u_i)_{i \in I}$ for E, we can associate to each u_i a linear form $u_i^* \in E^*$, and the u_i^* have some remarkable properties.

Definition 10.2. Given a vector space E and any basis $(u_i)_{i \in I}$ for E, by Proposition 2.14, for every $i \in I$, there is a unique linear form u_i^* such that

$$u_i^*(u_j) = \begin{cases} 1 & \text{if } i = j \\ 0 & \text{if } i \neq j, \end{cases}$$

for every $j \in I$. The linear form u_i^* is called the *coordinate form* of index i w.r.t. the basis $(u_i)_{i \in I}$.

The reason for the terminology *coordinate form* was explained in Section 2.8.

We proved in Theorem 2.3 that if $(u_1, \ldots, u_n)$ is a basis of E, then $(u_1^*, \ldots, u_n^*)$ is a basis of E^* called the *dual basis*.

If $(u_1, \ldots, u_n)$ is a basis of $\mathbb{R}^n$ (more generally K^n), it is possible to find explicitly the dual basis $(u_1^*, \ldots, u_n^*)$, where each u_i^* is represented by a row vector.

Example 10.1. For example, consider the columns of the Bézier matrix

$$B_4 = \begin{pmatrix} 1 & -3 & 3 & -1 \\ 0 & 3 & -6 & 3 \\ 0 & 0 & 3 & -3 \\ 0 & 0 & 0 & 1 \end{pmatrix}.$$

In other words, we have the basis

$$u_1 = \begin{pmatrix} 1 \\ 0 \\ 0 \\ 0 \end{pmatrix} \qquad u_2 = \begin{pmatrix} -3 \\ 3 \\ 0 \\ 0 \end{pmatrix} \qquad u_3 = \begin{pmatrix} 3 \\ -6 \\ 3 \\ 0 \end{pmatrix} \qquad u_4 = \begin{pmatrix} -1 \\ 3 \\ -3 \\ 1 \end{pmatrix}.$$

Since the form u_1^* is defined by the conditions $u_1^*(u_1) = 1, u_1^*(u_2) = 0, u_1^*(u_3) = 0, u_1^*(u_4) = 0$, it is represented by a row vector $(\lambda_1 \ \lambda_2 \ \lambda_3 \ \lambda_4)$ such that

$$\begin{pmatrix} \lambda_1 & \lambda_2 & \lambda_3 & \lambda_4 \end{pmatrix} \begin{pmatrix} 1 & -3 & 3 & -1 \\ 0 & 3 & -6 & 3 \\ 0 & 0 & 3 & -3 \\ 0 & 0 & 0 & 1 \end{pmatrix} = \begin{pmatrix} 1 & 0 & 0 & 0 \end{pmatrix}.$$

This implies that u_1^* is the first row of the inverse of B_4. Since

$$B_4^{-1} = \begin{pmatrix} 1 & 1 & 1 & 1 \\ 0 & 1/3 & 2/3 & 1 \\ 0 & 0 & 1/3 & 1 \\ 0 & 0 & 0 & 1 \end{pmatrix},$$

the linear forms $(u_1^*, u_2^*, u_3^*, u_4^*)$ correspond to the rows of B_4^{-1}. In particular, u_1^* is represented by $(1 \ 1 \ 1 \ 1)$.

The above method works for any n. Given any basis $(u_1, \ldots, u_n)$ of $\mathbb{R}^n$, if P is the $n \times n$ matrix whose jth column is u_j, then the dual form u_i^* is given by the ith row of the matrix P^{-1}.

When E is of finite dimension n and $(u_1, \ldots, u_n)$ is a basis of E, by Theorem 10.1 (1), the family $(u_1^*, \ldots, u_n^*)$ is a basis of the dual space E^*. Let us see how the coordinates of a linear form $\varphi^* \in E^*$ over the dual basis $(u_1^*, \ldots, u_n^*)$ vary under a change of basis.

Let $(u_1, \ldots, u_n)$ and $(v_1, \ldots, v_n)$ be two bases of E, and let $P = (a_{ij})$ be the change of basis matrix from $(u_1, \ldots, u_n)$ to $(v_1, \ldots, v_n)$, so that

$$v_j = \sum_{i=1}^{n} a_{ij} u_i,$$

and let $P^{-1} = (b_{ij})$ be the inverse of P, so that

$$u_i = \sum_{j=1}^{n} b_{ji} v_j.$$

For fixed j, where $1 \le j \le n$, we want to find scalars $(c_i)_{i=1}^{n}$ such that

$$v_j^* = c_1 u_1^* + c_2 u_2^* + \cdots + c_n u_n^*.$$

To find each c_i, we evaluate the above expression at u_i. Since $u_i^*(u_j) = \delta_{ij}$ and $v_i^*(v_j) = \delta_{ij}$, we get

$$v_j^*(u_i) = (c_1 u_1^* + c_2 u_2^* + \cdots + c_n u_n^*)(u_i) = c_i$$

$$v_j^*(u_i) = v_j^* \left(\sum_{k=1}^{n} b_{ki} v_k \right) = b_{ji},$$

and thus

$$v_j^* = \sum_{i=1}^{n} b_{ji} u_i^*.$$

Similar calculations show that

$$u_i^* = \sum_{j=1}^{n} a_{ij} v_j^*.$$

This means that the change of basis from the dual basis $(u_1^*, \ldots, u_n^*)$ to the dual basis $(v_1^*, \ldots, v_n^*)$ is $(P^{-1})^\top$. Since

$$\varphi^* = \sum_{i=1}^{n} \varphi_i u_i^* = \sum_{i=1}^{n} \varphi_i \sum_{j=1}^{n} a_{ij} v_j^* = \sum_{j=1}^{n} \left(\sum_{i=1}^{n} a_{ij} \varphi_i \right) v_j = \sum_{i=1}^{n} \varphi_i' v_i^*,$$

we get

$$\varphi_j' = \sum_{i=1}^{n} a_{ij} \varphi_i,$$

so the new coordinates φ_j' are expressed in terms of the old coordinates φ_i using the matrix $P^\top$. If we use the row vectors $(\varphi_1, \ldots, \varphi_n)$ and $(\varphi_1', \ldots, \varphi_n')$, we have

$$(\varphi_1', \ldots, \varphi_n') = (\varphi_1, \ldots, \varphi_n) P.$$

These facts are summarized in the following proposition.

Proposition 10.1. *Let $(u_1, \ldots, u_n)$ and $(v_1, \ldots, v_n)$ be two bases of E, and let $P = (a_{ij})$ be the change of basis matrix from $(u_1, \ldots, u_n)$ to $(v_1, \ldots, v_n)$, so that*

$$v_j = \sum_{i=1}^{n} a_{ij} u_i.$$

Then the change of basis from the dual basis $(u_1^*, \ldots, u_n^*)$ to the dual basis $(v_1^*, \ldots, v_n^*)$ is $(P^{-1})^\top$, and for any linear form φ, the new coordinates φ_j' of φ are expressed in terms of the old coordinates φ_i of φ using the matrix $P^\top$; that is,

$$(\varphi_1', \ldots, \varphi_n') = (\varphi_1, \ldots, \varphi_n) P.$$

To best understand the preceding paragraph, recall Example 3.1, in which $E = \mathbb{R}^2$, $u_1 = (1,0)$, $u_2 = (0,1)$, and $v_1 = (1,1)$, $v_2 = (-1,1)$. Then P, the change of basis matrix from (u_1, u_2) to (v_1, v_2), is given by

$$P = \begin{pmatrix} 1 & -1 \\ 1 & 1 \end{pmatrix},$$

with $(v_1, v_2) = (u_1, u_2)P$, and $(u_1, u_2) = (v_1, v_2)P^{-1}$, where

$$P^{-1} = \begin{pmatrix} 1/2 & 1/2 \\ -1/2 & 1/2 \end{pmatrix}.$$

Let (u_1^*, u_2^*) be the dual basis for (u_1, u_2) and (v_1^*, v_2^*) be the dual basis for (v_1, v_2). We claim that

$$(v_1^*, v_2^*) = (u_1^*, u_2^*) \begin{pmatrix} 1/2 & -1/2 \\ 1/2 & 1/2 \end{pmatrix} = (u_1^*, u_2^*)(P^{-1})^\top.$$

Indeed, since $v_1^* = c_1 u_1^* + c_2 u_2^*$ and $v_2^* = C_1 u_1^* + C_2 u_2^*$ we find that

$$c_1 = v_1^*(u_1) = v_1^*(1/2v_1 - 1/2v_2) = 1/2$$
$$c_2 = v_1^*(u_2) = v_1^*(1/2v_1 + 1/2v_2) = 1/2$$
$$C_1 = v_2^*(u_1) = v_2^*(1/2v_1 - 1/2v_2) = -1/2$$
$$C_2 = v_2^*(u_2) = v_1^*(1/2v_1 + 1/2v_2) = 1/2.$$

Furthermore, since $(u_1^*, u_2^*) = (v_1^*, v_2^*)P^\top$ (since $(v_1^*, v_2^*) = (u_1^*, u_2^*)(P^\top)^{-1}$), we find that

$$\varphi^* = \varphi_1 u_1^* + \varphi_2 u_2^* = \varphi_1(v_1^* - v_2^*) + \varphi(v_1^* + v_2^*)$$
$$= (\varphi_1 + \varphi_2)v_1^* + (-\varphi_1 + \varphi_2)v_2^* = \varphi_1' v_1^* + \varphi_2' v_2'$$

Hence

$$\begin{pmatrix} 1 & 1 \\ -1 & 1 \end{pmatrix} \begin{pmatrix} \varphi_1 \\ \varphi_2 \end{pmatrix} = \begin{pmatrix} \varphi_1' \\ \varphi_2' \end{pmatrix},$$

where

$$P^\top = \begin{pmatrix} 1 & 1 \\ -1 & 1 \end{pmatrix}.$$

Comparing with the change of basis

$$v_j = \sum_{i=1}^n a_{ij} u_i,$$

we note that this time, the coordinates (φ_i) of the linear form φ^* change in the *same direction* as the change of basis. For this reason, we say that the coordinates of linear forms are *covariant*. By abuse of language, it is often said that linear forms are *covariant*, which explains why the term *covector* is also used for a linear form.

Observe that if $(e_1, \ldots, e_n)$ is a basis of the vector space E, then, as a linear map from E to K, every linear form $f \in E^*$ is represented by a $1 \times n$ matrix, that is, by a *row vector*

$$(\lambda_1 \; \cdots \; \lambda_n),$$

with respect to the basis $(e_1, \ldots, e_n)$ of E, and 1 of K, where $f(e_i) = \lambda_i$. A vector $u = \sum_{i=1}^n u_i e_i \in E$ is represented by a $n \times 1$ matrix, that is, by a *column vector*

$$\begin{pmatrix} u_1 \\ \vdots \\ u_n \end{pmatrix},$$

and the action of f on u, namely $f(u)$, is represented by the matrix product

$$(\lambda_1 \; \cdots \; \lambda_n) \begin{pmatrix} u_1 \\ \vdots \\ u_n \end{pmatrix} = \lambda_1 u_1 + \cdots + \lambda_n u_n.$$

On the other hand, with respect to the dual basis $(e_1^*, \ldots, e_n^*)$ of E^*, the linear form f is represented by the column vector

$$\begin{pmatrix} \lambda_1 \\ \vdots \\ \lambda_n \end{pmatrix}.$$

Remark: In many texts using tensors, vectors are often indexed with lower indices. If so, it is more convenient to write the coordinates of a vector x over the basis $(u_1, \ldots, u_n)$ as (x^i), using an upper index, so that

$$x = \sum_{i=1}^n x^i u_i,$$

and in a change of basis, we have

$$v_j = \sum_{i=1}^n a_j^i u_i$$

and

$$x^i = \sum_{j=1}^n a_j^i x'^j.$$

Dually, linear forms are indexed with upper indices. Then it is more convenient to write the coordinates of a covector φ^* over the dual basis $(u^{*1}, \ldots, u^{*n})$ as (φ_i), using a lower index, so that

$$\varphi^* = \sum_{i=1}^{n} \varphi_i u^{*i}$$

and in a change of basis, we have

$$u^{*i} = \sum_{j=1}^{n} a_j^i v^{*j}$$

and

$$\varphi_j' = \sum_{i=1}^{n} a_j^i \varphi_i.$$

With these conventions, the index of summation appears once in upper position and once in lower position, and the summation sign can be safely omitted, a trick due to *Einstein*. For example, we can write

$$\varphi_j' = a_j^i \varphi_i$$

as an abbreviation for

$$\varphi_j' = \sum_{i=1}^{n} a_j^i \varphi_i.$$

For another example of the use of Einstein's notation, if the vectors $(v_1, \ldots, v_n)$ are linear combinations of the vectors $(u_1, \ldots, u_n)$, with

$$v_i = \sum_{j=1}^{n} a_{ij} u_j, \quad 1 \le i \le n,$$

then the above equations are written as

$$v_i = a_i^j u_j, \quad 1 \le i \le n.$$

Thus, in Einstein's notation, the $n \times n$ matrix (a_{ij}) is denoted by (a_i^j), a $(1,1)$-*tensor*.

 Beware that some authors view a matrix as a mapping between *coordinates*, in which case the matrix (a_{ij}) is denoted by (a_j^i).

10.2 Pairing and Duality Between E and E^*

Given a linear form $u^* \in E^*$ and a vector $v \in E$, the result $u^*(v)$ of applying u^* to v is also denoted by $\langle u^*, v \rangle$. This defines a binary operation $\langle -, - \rangle \colon E^* \times E \to K$ satisfying the following properties:

$$\langle u_1^* + u_2^*, v \rangle = \langle u_1^*, v \rangle + \langle u_2^*, v \rangle$$

$$\langle u^*, v_1 + v_2 \rangle = \langle u^*, v_1 \rangle + \langle u^*, v_2 \rangle$$

$$\langle \lambda u^*, v \rangle = \lambda \langle u^*, v \rangle$$

$$\langle u^*, \lambda v \rangle = \lambda \langle u^*, v \rangle.$$

The above identities mean that $\langle -, - \rangle$ is a *bilinear map*, since it is linear in each argument. It is often called the *canonical pairing* between E^* and E. In view of the above identities, given any fixed vector $v \in E$, the map $\mathrm{eval}_v \colon E^* \to K$ (*evaluation at v*) defined such that

$$\mathrm{eval}_v(u^*) = \langle u^*, v \rangle = u^*(v) \quad \text{for every } u^* \in E^*$$

is a linear map from E^* to K, that is, eval_v is a linear form in E^{**}. Again, from the above identities, the map $\mathrm{eval}_E \colon E \to E^{**}$, defined such that

$$\mathrm{eval}_E(v) = \mathrm{eval}_v \quad \text{for every } v \in E,$$

is a linear map. Observe that

$$\mathrm{eval}_E(v)(u^*) = \mathrm{eval}_v(u^*) = \langle u^*, v \rangle = u^*(v), \quad \text{for all } v \in E \text{ and all } u^* \in E^*.$$

We shall see that the map eval_E is injective, and that it is an isomorphism when E has finite dimension.

We now formalize the notion of the set V^0 of linear equations vanishing on all vectors in a given subspace $V \subseteq E$, and the notion of the set U^0 of common solutions of a given set $U \subseteq E^*$ of linear equations. The duality theorem (Theorem 10.1) shows that the dimensions of V and V^0, and the dimensions of U and U^0, are related in a crucial way. It also shows that, in finite dimension, the maps $V \mapsto V^0$ and $U \mapsto U^0$ are inverse bijections from subspaces of E to subspaces of E^*.

Definition 10.3. Given a vector space E and its dual E^*, we say that a vector $v \in E$ and a linear form $u^* \in E^*$ are *orthogonal* iff $\langle u^*, v \rangle = 0$. Given a subspace V of E and a subspace U of E^*, we say that V *and U are orthogonal* iff $\langle u^*, v \rangle = 0$ for every $u^* \in U$ and every $v \in V$. Given a subset V of E (resp. a subset U of E^*), the *orthogonal V^0 of V* is the subspace V^0 of E^* defined such that

$$V^0 = \{u^* \in E^* \mid \langle u^*, v \rangle = 0, \text{ for every } v \in V\}$$

(resp. the *orthogonal U^0 of U* is the subspace U^0 of E defined such that

$$U^0 = \{v \in E \mid \langle u^*, v \rangle = 0, \text{ for every } u^* \in U\}).$$

The subspace $V^0 \subseteq E^*$ is also called the *annihilator* of V. The subspace $U^0 \subseteq E$ annihilated by $U \subseteq E^*$ does not have a special name. It seems reasonable to call it the *linear subspace (or linear variety) defined by U*.

Informally, V^0 is the *set of linear equations that vanish on V*, and U^0 is *the set of common zeros of all linear equations in U*. We can also define V^0 by

$$V^0 = \{u^* \in E^* \mid V \subseteq \operatorname{Ker} u^*\}$$

and U^0 by

$$U^0 = \bigcap_{u^* \in U} \operatorname{Ker} u^*.$$

Observe that $E^0 = \{0\} = (0)$, and $\{0\}^0 = E^*$.

Proposition 10.2. *If $V_1 \subseteq V_2 \subseteq E$, then $V_2^0 \subseteq V_1^0 \subseteq E^*$, and if $U_1 \subseteq U_2 \subseteq E^*$, then $U_2^0 \subseteq U_1^0 \subseteq E$. See Figure 10.2.*

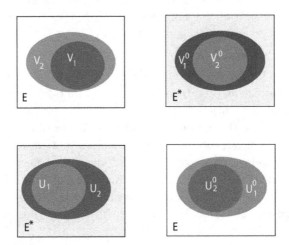

Fig. 10.2   The top pair of figures schematically illustrates the relation if $V_1 \subseteq V_2 \subseteq E$, then $V_2^0 \subseteq V_1^0 \subseteq E^*$, while the bottom pair of figures illustrates the relationship if $U_1 \subseteq U_2 \subseteq E^*$, then $U_2^0 \subseteq U_1^0 \subseteq E$.

Proof. Indeed, if $V_1 \subseteq V_2 \subseteq E$, then for any $f^* \in V_2^0$ we have $f^*(v) = 0$ for all $v \in V_2$, and thus $f^*(v) = 0$ for all $v \in V_1$, so $f^* \in V_1^0$. Similarly, if $U_1 \subseteq U_2 \subseteq E^*$, then for any $v \in U_2^0$, we have $f^*(v) = 0$ for all $f^* \in U_2$, so $f^*(v) = 0$ for all $f^* \in U_1$, which means that $v \in U_1^0$. $\square$

Here are some examples.

Example 10.2. Let $E = M_2(\mathbb{R})$, the space of real 2×2 matrices, and let V be the subspace of $M_2(\mathbb{R})$ spanned by the matrices

$$\begin{pmatrix} 0 & 1 \\ 1 & 0 \end{pmatrix}, \quad \begin{pmatrix} 1 & 0 \\ 0 & 0 \end{pmatrix}, \quad \begin{pmatrix} 0 & 0 \\ 0 & 1 \end{pmatrix}.$$

We check immediately that the subspace V consists of all matrices of the form

$$\begin{pmatrix} b & a \\ a & c \end{pmatrix},$$

that is, all symmetric matrices. The matrices

$$\begin{pmatrix} a_{11} & a_{12} \\ a_{21} & a_{22} \end{pmatrix}$$

in V satisfy the equation

$$a_{12} - a_{21} = 0,$$

and all scalar multiples of these equations, so V^0 is the subspace of E^* spanned by the linear form given by $u^*(a_{11}, a_{12}, a_{21}, a_{22}) = a_{12} - a_{21}$. By the duality theorem (Theorem 10.1) we have

$$\dim(V^0) = \dim(E) - \dim(V) = 4 - 3 = 1.$$

Example 10.3. The above example generalizes to $E = M_n(\mathbb{R})$ for any $n \geq 1$, but this time, consider the space U of linear forms asserting that a matrix A is symmetric; these are the linear forms spanned by the $n(n-1)/2$ equations

$$a_{ij} - a_{ji} = 0, \quad 1 \leq i < j \leq n.$$

Note there are no constraints on diagonal entries, and half of the equations

$$a_{ij} - a_{ji} = 0, \quad 1 \leq i \neq j \leq n$$

are redundant. It is easy to check that the equations (linear forms) for which $i < j$ are linearly independent. To be more precise, let U be the space of linear forms in E^* spanned by the linear forms

$$u_{ij}^*(a_{11}, \ldots, a_{1n}, a_{21}, \ldots, a_{2n}, \ldots, a_{n1}, \ldots, a_{nn}) = a_{ij} - a_{ji}, \quad 1 \leq i < j \leq n.$$

The dimension of U is $n(n-1)/2$. Then the set U^0 of common solutions of these equations is the space $\mathbf{S}(n)$ of symmetric matrices. By the duality theorem (Theorem 10.1), this space has dimension

$$\frac{n(n+1)}{2} = n^2 - \frac{n(n-1)}{2}.$$

We leave it as an exercise to find a basis of $\mathbf{S}(n)$.

Example 10.4. If $E = \mathrm{M}_n(\mathbb{R})$, consider the subspace U of linear forms in E^* spanned by the linear forms

$$u_{ij}^*(a_{11}, \ldots, a_{1n}, a_{21}, \ldots, a_{2n}, \ldots, a_{n1}, \ldots, a_{nn}) = a_{ij} + a_{ji}, \quad 1 \le i < j \le n$$

$$u_{ii}^*(a_{11}, \ldots, a_{1n}, a_{21}, \ldots, a_{2n}, \ldots, a_{n1}, \ldots, a_{nn}) = a_{ii}, \quad 1 \le i \le n.$$

It is easy to see that these linear forms are linearly independent, so $\dim(U) = n(n+1)/2$. The space U^0 of matrices $A \in \mathrm{M}_n(\mathbb{R})$ satisfying all of the above equations is clearly the space **Skew**(n) of skew-symmetric matrices. By the duality theorem (Theorem 10.1), the dimension of U^0 is

$$\frac{n(n-1)}{2} = n^2 - \frac{n(n+1)}{2}.$$

We leave it as an exercise to find a basis of **Skew**(n).

Example 10.5. For yet another example with $E = \mathrm{M}_n(\mathbb{R})$, for any $A \in \mathrm{M}_n(\mathbb{R})$, consider the linear form in E^* given by

$$\mathrm{tr}(A) = a_{11} + a_{22} + \cdots + a_{nn},$$

called the *trace* of A. The subspace U^0 of E consisting of all matrices A such that $\mathrm{tr}(A) = 0$ is a space of dimension $n^2 - 1$. We leave it as an exercise to find a basis of this space.

The dimension equations

$$\dim(V) + \dim(V^0) = \dim(E)$$

$$\dim(U) + \dim(U^0) = \dim(E)$$

are always true (if E is finite-dimensional). This is part of the duality theorem (Theorem 10.1).

Remark: In contrast with the previous examples, given a matrix $A \in \mathrm{M}_n(\mathbb{R})$, the equations asserting that $A^\top A = I$ are not linear constraints. For example, for $n = 2$, we have

$$a_{11}^2 + a_{21}^2 = 1$$

$$a_{21}^2 + a_{22}^2 = 1$$

$$a_{11}a_{12} + a_{21}a_{22} = 0.$$

Remarks:

(1) The notation V^0 (resp. U^0) for the orthogonal of a subspace V of E (resp. a subspace U of E^*) is not universal. Other authors use the notation $V^\perp$ (resp. $U^\perp$). However, the notation $V^\perp$ is also used to denote the orthogonal complement of a subspace V with respect to an inner product on a space E, in which case $V^\perp$ is a subspace of E and not a subspace of E^* (see Chapter 11). To avoid confusion, we prefer using the notation V^0.

(2) Since linear forms can be viewed as linear equations (at least in finite dimension), given a subspace (or even a subset) U of E^*, we can define the set $\mathcal{Z}(U)$ of *common zeros* of the equations in U by

$$\mathcal{Z}(U) = \{v \in E \mid u^*(v) = 0, \text{ for all } u^* \in U\}.$$

Of course $\mathcal{Z}(U) = U^0$, but the notion $\mathcal{Z}(U)$ can be generalized to more general kinds of equations, namely polynomial equations. In this more general setting, U is a set of *polynomials* in n variables with coefficients in a field K (where $n = \dim(E)$). Sets of the form $\mathcal{Z}(U)$ are called *algebraic varieties*. Linear forms correspond to the special case where homogeneous polynomials of degree 1 are considered.

If V is a subset of E, it is natural to associate with V the *set of polynomials in $K[X_1, \ldots, X_n]$ that vanish on V*. This set, usually denoted $\mathcal{I}(V)$, has some special properties that make it an *ideal*. If V is a linear subspace of E, it is natural to restrict our attention to the space V^0 of linear forms that vanish on V, and in this case we identify $\mathcal{I}(V)$ and V^0 (although technically, $\mathcal{I}(V)$ is no longer an ideal).

For any arbitrary set of polynomials $U \subseteq K[X_1, \ldots, X_n]$ (resp. subset $V \subseteq E$), the relationship between $\mathcal{I}(\mathcal{Z}(U))$ and U (resp. $\mathcal{Z}(\mathcal{I}(V))$ and V) is generally not simple, even though we always have

$$U \subseteq \mathcal{I}(\mathcal{Z}(U)) \quad (\text{resp. } V \subseteq \mathcal{Z}(\mathcal{I}(V))).$$

However, when the field K is algebraically closed, then $\mathcal{I}(\mathcal{Z}(U))$ is equal to the *radical* of the ideal U, a famous result due to Hilbert known as the *Nullstellensatz* (see Lang [Lang (1993)] or Dummit and Foote [Dummit and Foote (1999)]). The study of algebraic varieties is the main subject of *algebraic geometry*, a beautiful but formidable subject. For a taste of algebraic geometry, see Lang [Lang (1993)] or Dummit and Foote [Dummit and Foote (1999)].

The duality theorem (Theorem 10.1) shows that the situation is much simpler if we restrict our attention to linear subspaces; in this case

$$U = \mathcal{I}(\mathcal{Z}(U)) \quad \text{and} \quad V = \mathcal{Z}(\mathcal{I}(V)).$$

Proposition 10.3. *We have $V \subseteq V^{00}$ for every subspace V of E, and $U \subseteq U^{00}$ for every subspace U of E^*.*

Proof. Indeed, for any $v \in V$, to show that $v \in V^{00}$ we need to prove that $u^*(v) = 0$ for all $u^* \in V^0$. However, V^0 consists of all linear forms u^* such that $u^*(y) = 0$ for *all* $y \in V$; in particular, for a fixed $v \in V$, we have $u^*(v) = 0$ for all $u^* \in V^0$, as required.

Similarly, for any $u^* \in U$, to show that $u^* \in U^{00}$ we need to prove that $u^*(v) = 0$ for all $v \in U^0$. However, U^0 consists of all vectors v such that $f^*(v) = 0$ for *all* $f^* \in U$; in particular, for a fixed $u^* \in U$, we have $u^*(v) = 0$ for all $v \in U^0$, as required. □

We will see shortly that in finite dimension, we have $V = V^{00}$ and $U = U^{00}$.

10.3   The Duality Theorem and Some Consequences

Given a vector space E of dimension $n \geq 1$ and a subspace U of E, by Theorem 2.2, every basis $(u_1, \ldots, u_m)$ of U can be extended to a basis $(u_1, \ldots, u_n)$ of E. We have the following important theorem adapted from E. Artin [Artin (1957)] (Chapter 1).

Theorem 10.1. *(Duality theorem) Let E be a vector space of dimension n. The following properties hold:*

(a) *For every basis $(u_1, \ldots, u_n)$ of E, the family of coordinate forms $(u_1^*, \ldots, u_n^*)$ is a basis of E^* (called the dual basis of $(u_1, \ldots, u_n)$).*

(b) *For every subspace V of E, we have $V^{00} = V$.*

(c) *For every pair of subspaces V and W of E such that $E = V \oplus W$, with V of dimension m, for every basis $(u_1, \ldots, u_n)$ of E such that $(u_1, \ldots, u_m)$ is a basis of V and $(u_{m+1}, \ldots, u_n)$ is a basis of W, the family $(u_1^*, \ldots, u_m^*)$ is a basis of the orthogonal W^0 of W in E^*, so that*

$$\dim(W) + \dim(W^0) = \dim(E).$$

Furthermore, we have $W^{00} = W$.

(d) *For every subspace U of E^*, we have*

$$\dim(U) + \dim(U^0) = \dim(E),$$

where U^0 is the orthogonal of U in E, and $U^{00} = U$.

Proof. (a) This part was proven in Theorem 2.3.

(b) By Proposition 10.3 we have $V \subseteq V^{00}$. If $V \neq V^{00}$, then let $(u_1, \ldots, u_p)$ be a basis of V^{00} such that $(u_1, \ldots, u_m)$ is a basis of V, with $m < p$. Since $u_{m+1} \in V^{00}$, u_{m+1} is orthogonal to every linear form in V^0. By definition we have $u_{m+1}^*(u_i) = 0$ for all $i = 1, \ldots, m$, and thus $u_{m+1}^* \in V^0$. However, $u_{m+1}^*(u_{m+1}) = 1$, contradicting the fact that u_{m+1} is orthogonal to every linear form in V^0. Thus, $V = V^{00}$.

(c) Every linear form $f^* \in W^0$ is orthogonal to every u_j for $j = m + 1, \ldots, n$, and thus, $f^*(u_j) = 0$ for $j = m + 1, \ldots, n$. For such a linear form $f^* \in W^0$, let

$$g^* = f^*(u_1)u_1^* + \cdots + f^*(u_m)u_m^*.$$

We have $g^*(u_i) = f^*(u_i)$, for every i, $1 \leq i \leq m$. Furthermore, by definition, g^* vanishes on all u_j with $j = m + 1, \ldots, n$. Thus, f^* and g^* agree on the basis $(u_1, \ldots, u_n)$ of E, and so $g^* = f^*$. This shows that $(u_1^*, \ldots, u_m^*)$ generates W^0, and since it is also a linearly independent family, $(u_1^*, \ldots, u_m^*)$ is a basis of W^0. It is then obvious that $\dim(W) + \dim(W^0) = \dim(E)$, and by Part (b), we have $W^{00} = W$.

(d) The only remaining fact to prove is that $U^{00} = U$. Let $(f_1^*, \ldots, f_m^*)$ be a basis of U. Note that the map $h \colon E \to K^m$ defined such that

$$h(v) = (f_1^*(v), \ldots, f_m^*(v))$$

for every $v \in E$ is a linear map, and that its kernel $\operatorname{Ker} h$ is precisely U^0. Then by Proposition 5.1,

$$n = \dim(E) = \dim(\operatorname{Ker} h) + \dim(\operatorname{Im} h) \leq \dim(U^0) + m,$$

since $\dim(\operatorname{Im} h) \leq m$. Thus, $n - \dim(U^0) \leq m$. By (c), we have $\dim(U^0) + \dim(U^{00}) = \dim(E) = n$, so we get $\dim(U^{00}) \leq m$. However, by Proposition 10.3 it is clear that $U \subseteq U^{00}$, which implies $m = \dim(U) \leq \dim(U^{00})$, so $\dim(U) = \dim(U^{00}) = m$, and we must have $U = U^{00}$. $\square$

Part (a) of Theorem 10.1 shows that

$$\dim(E) = \dim(E^*),$$

and if $(u_1, \ldots, u_n)$ is a basis of E, then $(u_1^*, \ldots, u_n^*)$ is a basis of the dual space E^* called the *dual basis* of $(u_1, \ldots, u_n)$.

Define the function $\mathcal{E}$ ($\mathcal{E}$ for equations) from subspaces of E to subspaces of E^* and the function $\mathcal{Z}$ ($\mathcal{Z}$ for zeros) from subspaces of E^* to subspaces of E by

$$\mathcal{E}(V) = V^0, \quad V \subseteq E$$
$$\mathcal{Z}(U) = U^0, \quad U \subseteq E^*.$$

By Parts (c) and (d) of Theorem 10.1,

$$(\mathcal{Z} \circ \mathcal{E})(V) = V^{00} = V$$
$$(\mathcal{E} \circ \mathcal{Z})(U) = U^{00} = U,$$

so $\mathcal{Z} \circ \mathcal{E} = \mathrm{id}$ and $\mathcal{E} \circ \mathcal{Z} = \mathrm{id}$, and the maps $\mathcal{E}$ and $\mathcal{Z}$ are inverse bijections. These maps set up a *duality* between subspaces of E and subspaces of E^*. In particular, every subspace $V \subseteq E$ of dimension m is the set of common zeros of the space of linear forms (equations) V^0, which has dimension $n-m$. This confirms the claim we made about the dimension of the subspace defined by a set of linear equations.

One should be careful that this bijection does not hold if E has infinite dimension. Some restrictions on the dimensions of U and V are needed.

Remark: However, even if E is infinite-dimensional, the identity $V = V^{00}$ holds for every subspace V of E. The proof is basically the same but uses an infinite basis of V^{00} extending a basis of V.

We now discuss some applications of the duality theorem.

Problem 1. Suppose that V is a subspace of $\mathbb{R}^n$ of dimension m and that $(v_1, \ldots, v_m)$ is a basis of V. The problem is to find a basis of V^0.

We first extend $(v_1, \ldots, v_m)$ to a basis $(v_1, \ldots, v_n)$ of $\mathbb{R}^n$, and then by part (c) of Theorem 10.1, we know that $(v_{m+1}^*, \ldots, v_n^*)$ is a basis of V^0.

Example 10.6. For example, suppose that V is the subspace of $\mathbb{R}^4$ spanned by the two linearly independent vectors

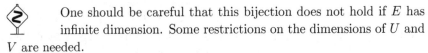

$$v_1 = \begin{pmatrix} 1 \\ 1 \\ 1 \\ 1 \end{pmatrix} \quad v_2 = \begin{pmatrix} 1 \\ 1 \\ -1 \\ -1 \end{pmatrix},$$

the first two vectors of the Haar basis in $\mathbb{R}^4$. The four columns of the Haar matrix

$$W = \begin{pmatrix} 1 & 1 & 1 & 0 \\ 1 & 1 & -1 & 0 \\ 1 & -1 & 0 & 1 \\ 1 & -1 & 0 & -1 \end{pmatrix}$$

form a basis of $\mathbb{R}^4$, and the inverse of W is given by

$$W^{-1} = \begin{pmatrix} 1/4 & 0 & 0 & 0 \\ 0 & 1/4 & 0 & 0 \\ 0 & 0 & 1/2 & 0 \\ 0 & 0 & 0 & 1/2 \end{pmatrix} \begin{pmatrix} 1 & 1 & 1 & 1 \\ 1 & 1 & -1 & -1 \\ 1 & -1 & 0 & 0 \\ 0 & 0 & 1 & -1 \end{pmatrix} = \begin{pmatrix} 1/4 & 1/4 & 1/4 & 1/4 \\ 1/4 & 1/4 & -1/4 & -1/4 \\ 1/2 & -1/2 & 0 & 0 \\ 0 & 0 & 1/2 & -1/2 \end{pmatrix}.$$

Since the dual basis $(v_1^*, v_2^*, v_3^*, v_4^*)$ is given by the rows of W^{-1}, the last two rows of W^{-1},

$$\begin{pmatrix} 1/2 & -1/2 & 0 & 0 \\ 0 & 0 & 1/2 & -1/2 \end{pmatrix},$$

form a basis of V^0. We also obtain a basis by rescaling by the factor $1/2$, so the linear forms given by the row vectors

$$\begin{pmatrix} 1 & -1 & 0 & 0 \\ 0 & 0 & 1 & -1 \end{pmatrix}$$

form a basis of V^0, the space of linear forms (linear equations) that vanish on the subspace V.

The method that we described to find V^0 requires first extending a basis of V and then inverting a matrix, but there is a more direct method. Indeed, let A be the $n \times m$ matrix whose columns are the basis vectors $(v_1, \ldots, v_m)$ of V. Then a linear form u represented by a row vector belongs to V^0 iff $uv_i = 0$ for $i = 1, \ldots, m$ iff

$$uA = 0$$

iff

$$A^\top u^\top = 0.$$

Therefore, all we need to do is to find a basis of the nullspace of $A^\top$. This can be done quite effectively using the reduction of a matrix to reduced row echelon form (rref); see Section 7.10.

Example 10.7. For example, if we reconsider the previous example, $A^\top u^\top = 0$ becomes

$$\begin{pmatrix} 1 & 1 & 1 & 1 \\ 1 & 1 & -1 & -1 \end{pmatrix} \begin{pmatrix} u_1 \\ u_2 \\ u_3 \\ u_4 \end{pmatrix} = \begin{pmatrix} 0 \\ 0 \end{pmatrix}.$$

Since the rref of $A^\top$ is

$$\begin{pmatrix} 1 & 1 & 0 & 0 \\ 0 & 0 & 1 & 1 \end{pmatrix},$$

the above system is equivalent to

$$\begin{pmatrix} 1 & 1 & 0 & 0 \\ 0 & 0 & 1 & 1 \end{pmatrix} \begin{pmatrix} u_1 \\ u_2 \\ u_3 \\ u_4 \end{pmatrix} = \begin{pmatrix} u_1 + u_2 \\ u_3 + u_4 \end{pmatrix} = \begin{pmatrix} 0 \\ 0 \end{pmatrix},$$

where the free variables are associated with u_2 and u_4. Thus to determine a basis for the kernel of $A^\top$, we set $u_2 = 1, u_4 = 0$ and $u_2 = 0, u_4 = 1$ and obtain a basis for V^0 as

$$\begin{pmatrix} 1 & -1 & 0 & 0 \end{pmatrix}, \qquad \begin{pmatrix} 0 & 0 & 1 & -1 \end{pmatrix}.$$

Problem 2. Let us now consider the problem of finding a basis of the hyperplane H in $\mathbb{R}^n$ defined by the equation

$$c_1 x_1 + \cdots + c_n x_n = 0.$$

More precisely, if $u^*(x_1, \ldots, x_n)$ is the linear form in $(\mathbb{R}^n)^*$ given by $u^*(x_1, \ldots, x_n) = c_1 x_1 + \cdots + c_n x_n$, then the hyperplane H is the kernel of u^*. Of course we assume that some c_j is nonzero, in which case the linear form u^* spans a one-dimensional subspace U of $(\mathbb{R}^n)^*$, and $U^0 = H$ has dimension $n - 1$.

Since u^* is not the linear form which is identically zero, there is a smallest positive index $j \le n$ such that $c_j \ne 0$, so our linear form is really $u^*(x_1, \ldots, x_n) = c_j x_j + \cdots + c_n x_n$. We claim that the following $n - 1$ vectors (in $\mathbb{R}^n$) form a basis of H:

$$
\begin{array}{c}
\begin{array}{cccccccc}
1 & 2 & \ldots & j-1 & j & j+1 & \ldots & n-1
\end{array} \\
\begin{array}{c}
1 \\ 2 \\ \vdots \\ j-1 \\ j \\ j+1 \\ j+2 \\ \vdots \\ n
\end{array}
\left(
\begin{array}{cccccccc}
1 & 0 & \ldots & 0 & 0 & 0 & \ldots & 0 \\
0 & 1 & \ldots & 0 & 0 & 0 & \ldots & 0 \\
\vdots & \vdots & \ddots & \vdots & \vdots & \vdots & \ddots & \vdots \\
0 & 0 & \ldots & 1 & 0 & 0 & \ldots & 0 \\
0 & 0 & \ldots & 0 & -c_{j+1}/c_j & -c_{j+2}/c_j & \ldots & -c_n/c_j \\
0 & 0 & \ldots & 0 & 1 & 0 & \ldots & 0 \\
0 & 0 & \ldots & 0 & 0 & 1 & \ldots & 0 \\
\vdots & \vdots & \ddots & \vdots & \vdots & \vdots & \ddots & \vdots \\
0 & 0 & \ldots & 0 & 0 & 0 & \ldots & 1
\end{array}
\right).
\end{array}
$$

Observe that the $(n - 1) \times (n - 1)$ matrix obtained by deleting row j is the identity matrix, so the columns of the above matrix are linearly independent. A simple calculation also shows that the linear form $u^*(x_1, \ldots, x_n) = c_j x_j + \cdots + c_n x_n$ vanishes on every column of the above matrix. For a concrete example in $\mathbb{R}^6$, if $u^*(x_1, \ldots, x_6) = x_3 + 2x_4 + 3x_5 + 4x_6$, we obtain the basis for the hyperplane H of equation

$$x_3 + 2x_4 + 3x_5 + 4x_6 = 0$$

given by the following matrix:

$$\begin{pmatrix} 1 & 0 & 0 & 0 & 0 \\ 0 & 1 & 0 & 0 & 0 \\ 0 & 0 & -2 & -3 & -4 \\ 0 & 0 & 1 & 0 & 0 \\ 0 & 0 & 0 & 1 & 0 \\ 0 & 0 & 0 & 0 & 1 \end{pmatrix}.$$

Problem 3. Conversely, given a hyperplane H in $\mathbb{R}^n$ given as the span of $n-1$ linearly vectors $(u_1, \ldots, u_{n-1})$, it is possible using determinants to find a linear form $(\lambda_1, \ldots, \lambda_n)$ that vanishes on H.

In the case $n = 3$, we are looking for a row vector $(\lambda_1, \lambda_2, \lambda_3)$ such that if

$$u = \begin{pmatrix} u_1 \\ u_2 \\ u_3 \end{pmatrix} \quad \text{and} \quad v = \begin{pmatrix} v_1 \\ v_2 \\ v_3 \end{pmatrix}$$

are two linearly independent vectors, then

$$\begin{pmatrix} u_1 & u_2 & u_2 \\ v_1 & v_2 & v_2 \end{pmatrix} \begin{pmatrix} \lambda_1 \\ \lambda_2 \\ \lambda_3 \end{pmatrix} = \begin{pmatrix} 0 \\ 0 \end{pmatrix},$$

and the cross-product $u \times v$ of u and v given by

$$u \times v = \begin{pmatrix} u_2 v_3 - u_3 v_2 \\ u_3 v_1 - u_1 v_3 \\ u_1 v_2 - u_2 v_1 \end{pmatrix}$$

is a solution. In other words, the equation of the plane spanned by u and v is

$$(u_2 v_3 - u_3 v_2)x + (u_3 v_1 - u_1 v_3)y + (u_1 v_2 - u_2 v_1)z = 0.$$

Problem 4. Here is another example illustrating the power of Theorem 10.1. Let $E = M_n(\mathbb{R})$, and consider the equations asserting that the sum of the entries in every row of a matrix $A \in M_n(\mathbb{R})$ is equal to the same number. We have $n-1$ equations

$$\sum_{j=1}^{n}(a_{ij} - a_{i+1j}) = 0, \quad 1 \leq i \leq n-1,$$

and it is easy to see that they are linearly independent. Therefore, the space U of linear forms in E^* spanned by the above linear forms (equations) has dimension $n-1$, and the space U^0 of matrices satisfying all these equations has dimension $n^2 - n + 1$. It is not so obvious to find a basis for this space.

We will now pin down the relationship between a vector space E and its bidual E^{**}.

10.4   The Bidual and Canonical Pairings

Proposition 10.4. *Let E be a vector space. The following properties hold:*

(a) The linear map $\mathrm{eval}_E \colon E \to E^{**}$ defined such that

$$\mathrm{eval}_E(v) = \mathrm{eval}_v \quad \textit{for all } v \in E,$$

that is, $\mathrm{eval}_E(v)(u^*) = \langle u^*, v \rangle = u^*(v)$ for every $u^* \in E^*$, is injective.

(b) When E is of finite dimension n, the linear map $\mathrm{eval}_E \colon E \to E^{**}$ is an isomorphism (called the canonical isomorphism).

Proof. (a) Let $(u_i)_{i \in I}$ be a basis of E, and let $v = \sum_{i \in I} v_i u_i$. If $\mathrm{eval}_E(v) = 0$, then in particular $\mathrm{eval}_E(v)(u_i^*) = 0$ for all u_i^*, and since

$$\mathrm{eval}_E(v)(u_i^*) = \langle u_i^*, v \rangle = v_i,$$

we have $v_i = 0$ for all $i \in I$, that is, $v = 0$, showing that $\mathrm{eval}_E \colon E \to E^{**}$ is injective.

If E is of finite dimension n, by Theorem 10.1, for every basis $(u_1, \ldots, u_n)$, the family $(u_1^*, \ldots, u_n^*)$ is a basis of the dual space E^*, and thus the family $(u_1^{**}, \ldots, u_n^{**})$ is a basis of the bidual E^{**}. This shows that $\dim(E) = \dim(E^{**}) = n$, and since by Part (a), we know that $\mathrm{eval}_E \colon E \to E^{**}$ is injective, in fact, $\mathrm{eval}_E \colon E \to E^{**}$ is bijective (by Proposition 5.10). $\square$

When E is of finite dimension and $(u_1, \ldots, u_n)$ is a basis of E, in view of the canonical isomorphism $\mathrm{eval}_E \colon E \to E^{**}$, the basis $(u_1^{**}, \ldots, u_n^{**})$ of the bidual is *identified* with $(u_1, \ldots, u_n)$.

Proposition 10.4 can be reformulated very fruitfully in terms of pairings, a remarkably useful concept discovered by Pontrjagin in 1931 (adapted from E. Artin [Artin (1957)], Chapter 1). Given two vector spaces E and F over a field K, we say that a function $\varphi \colon E \times F \to K$ is *bilinear* if for every $v \in V$, the map $u \mapsto \varphi(u, v)$ (from E to K) is linear, and for every $u \in E$, the map $v \mapsto \varphi(u, v)$ (from F to K) is linear.

Definition 10.4. Given two vector spaces E and F over K, a *pairing between E and F* is a bilinear map $\varphi \colon E \times F \to K$. Such a pairing is *nondegenerate* iff

(1) for every $u \in E$, if $\varphi(u, v) = 0$ for all $v \in F$, then $u = 0$, and
(2) for every $v \in F$, if $\varphi(u, v) = 0$ for all $u \in E$, then $v = 0$.

A pairing $\varphi\colon E \times F \to K$ is often denoted by $\langle -, - \rangle\colon E \times F \to K$. For example, the map $\langle -, - \rangle\colon E^* \times E \to K$ defined earlier is a nondegenerate pairing (use the proof of (a) in Proposition 10.4). If $E = F$ and $K = \mathbb{R}$, any inner product on E is a nondegenerate pairing (because an inner product is positive definite); see Chapter 11. Other interesting nondegenerate pairings arise in exterior algebra and differential geometry.

Given a pairing $\varphi\colon E \times F \to K$, we can define two maps $l_\varphi\colon E \to F^*$ and $r_\varphi\colon F \to E^*$ as follows: For every $u \in E$, we define the linear form $l_\varphi(u)$ in F^* such that

$$l_\varphi(u)(y) = \varphi(u, y) \quad \text{for every } y \in F,$$

and for every $v \in F$, we define the linear form $r_\varphi(v)$ in E^* such that

$$r_\varphi(v)(x) = \varphi(x, v) \quad \text{for every } x \in E.$$

We have the following useful proposition.

Proposition 10.5. *Given two vector spaces E and F over K, for every nondegenerate pairing $\varphi\colon E \times F \to K$ between E and F, the maps $l_\varphi\colon E \to F^*$ and $r_\varphi\colon F \to E^*$ are linear and injective. Furthermore, if E and F have finite dimension, then this dimension is the same and $l_\varphi\colon E \to F^*$ and $r_\varphi\colon F \to E^*$ are bijections.*

Proof. The maps $l_\varphi\colon E \to F^*$ and $r_\varphi\colon F \to E^*$ are linear because a pairing is bilinear. If $l_\varphi(u) = 0$ (the null form), then

$$l_\varphi(u)(v) = \varphi(u, v) = 0 \quad \text{for every } v \in F,$$

and since φ is nondegenerate, $u = 0$. Thus, $l_\varphi\colon E \to F^*$ is injective. Similarly, $r_\varphi\colon F \to E^*$ is injective. When F has finite dimension n, we have seen that F and F^* have the same dimension. Since $l_\varphi\colon E \to F^*$ is injective, we have $m = \dim(E) \leq \dim(F) = n$. The same argument applies to E, and thus $n = \dim(F) \leq \dim(E) = m$. But then, $\dim(E) = \dim(F)$, and $l_\varphi\colon E \to F^*$ and $r_\varphi\colon F \to E^*$ are bijections. $\qquad \square$

When E has finite dimension, the nondegenerate pairing $\langle -, - \rangle\colon E^* \times E \to K$ yields another proof of the existence of a natural isomorphism between E and E^{**}. When $E = F$, the nondegenerate pairing induced by an inner product on E yields a natural isomorphism between E and E^* (see Section 11.2).

We now show the relationship between hyperplanes and linear forms.

10.5 Hyperplanes and Linear Forms

Actually Proposition 10.6 below follows from Parts (c) and (d) of Theorem 10.1, but we feel that it is also interesting to give a more direct proof.

Proposition 10.6. *Let E be a vector space. The following properties hold:*

(a) *Given any nonnull linear form $f^* \in E^*$, its kernel $H = \operatorname{Ker} f^*$ is a hyperplane.*

(b) *For any hyperplane H in E, there is a (nonnull) linear form $f^* \in E^*$ such that $H = \operatorname{Ker} f^*$.*

(c) *Given any hyperplane H in E and any (nonnull) linear form $f^* \in E^*$ such that $H = \operatorname{Ker} f^*$, for every linear form $g^* \in E^*$, $H = \operatorname{Ker} g^*$ iff $g^* = \lambda f^*$ for some $\lambda \neq 0$ in K.*

Proof. (a) If $f^* \in E^*$ is nonnull, there is some vector $v_0 \in E$ such that $f^*(v_0) \neq 0$. Let $H = \operatorname{Ker} f^*$. For every $v \in E$, we have

$$f^* \left(v - \frac{f^*(v)}{f^*(v_0)} v_0 \right) = f^*(v) - \frac{f^*(v)}{f^*(v_0)} f^*(v_0) = f^*(v) - f^*(v) = 0.$$

Thus,

$$v - \frac{f^*(v)}{f^*(v_0)} v_0 = h \in H,$$

and

$$v = h + \frac{f^*(v)}{f^*(v_0)} v_0,$$

that is, $E = H + K v_0$. Also since $f^*(v_0) \neq 0$, we have $v_0 \notin H$, that is, $H \cap K v_0 = 0$. Thus, $E = H \oplus K v_0$, and H is a hyperplane.

(b) If H is a hyperplane, $E = H \oplus K v_0$ for some $v_0 \notin H$. Then every $v \in E$ can be written in a unique way as $v = h + \lambda v_0$. Thus there is a well-defined function $f^* \colon E \to K$, such that, $f^*(v) = \lambda$, for every $v = h + \lambda v_0$. We leave as a simple exercise the verification that f^* is a linear form. Since $f^*(v_0) = 1$, the linear form f^* is nonnull. Also, by definition, it is clear that $\lambda = 0$ iff $v \in H$, that is, $\operatorname{Ker} f^* = H$.

(c) Let H be a hyperplane in E, and let $f^* \in E^*$ be any (nonnull) linear form such that $H = \operatorname{Ker} f^*$. Clearly, if $g^* = \lambda f^*$ for some $\lambda \neq 0$, then $H = \operatorname{Ker} g^*$. Conversely, assume that $H = \operatorname{Ker} g^*$ for some nonnull linear form g^*. From (a), we have $E = H \oplus K v_0$, for some v_0 such that $f^*(v_0) \neq 0$ and $g^*(v_0) \neq 0$. Then observe that

$$g^* - \frac{g^*(v_0)}{f^*(v_0)} f^*$$

is a linear form that vanishes on H, since both f^* and g^* vanish on H, but also vanishes on Kv_0. Thus, $g^* = \lambda f^*$, with

$$\lambda = \frac{g^*(v_0)}{f^*(v_0)}.$$

$\square$

We leave as an exercise the fact that every subspace $V \neq E$ of a vector space E is the intersection of all hyperplanes that contain V. We now consider the notion of transpose of a linear map and of a matrix.

10.6 Transpose of a Linear Map and of a Matrix

Given a linear map $f \colon E \to F$, it is possible to define a map $f^\top \colon F^* \to E^*$ which has some interesting properties.

Definition 10.5. Given a linear map $f \colon E \to F$, the *transpose* $f^\top \colon F^* \to E^*$ *of* f is the linear map defined such that

$$f^\top(v^*) = v^* \circ f, \quad \text{for every } v^* \in F^*,$$

as shown in the diagram below:

$$
\begin{array}{ccc}
E & \xrightarrow{\ f\ } & F \\
& \underset{f^\top(v^*)}{\searrow} & \big\downarrow{\scriptstyle v^*} \\
& & K.
\end{array}
$$

Equivalently, the linear map $f^\top \colon F^* \to E^*$ is defined such that

$$\langle v^*, f(u) \rangle = \langle f^\top(v^*), u \rangle, \tag{10.1}$$

for all $u \in E$ and all $v^* \in F^*$.

It is easy to verify that the following properties hold:

$$(f + g)^\top = f^\top + g^\top$$
$$(g \circ f)^\top = f^\top \circ g^\top$$
$$\mathrm{id}_E^\top = \mathrm{id}_{E^*}.$$

Note the reversal of composition on the right-hand side of $(g \circ f)^\top = f^\top \circ g^\top$.

The equation $(g \circ f)^\top = f^\top \circ g^\top$ implies the following useful proposition.

Proposition 10.7. *If $f \colon E \to F$ is any linear map, then the following properties hold:*

(1) If f is injective, then $f^\top$ is surjective.
(2) If f is surjective, then $f^\top$ is injective.

Proof. If $f \colon E \to F$ is injective, then it has a retraction $r \colon F \to E$ such that $r \circ f = \mathrm{id}_E$, and if $f \colon E \to F$ is surjective, then it has a section $s \colon F \to E$ such that $f \circ s = \mathrm{id}_F$. Now if $f \colon E \to F$ is injective, then we have

$$(r \circ f)^\top = f^\top \circ r^\top = \mathrm{id}_{E^*},$$

which implies that $f^\top$ is surjective, and if f is surjective, then we have

$$(f \circ s)^\top = s^\top \circ f^\top = \mathrm{id}_{F^*},$$

which implies that $f^\top$ is injective. $\square$

The following proposition shows the relationship between orthogonality and transposition.

Proposition 10.8. *Given a linear map $f \colon E \to F$, for any subspace V of E, we have*

$$f(V)^0 = (f^\top)^{-1}(V^0) = \{w^* \in F^* \mid f^\top(w^*) \in V^0\}.$$

As a consequence,

$$\mathrm{Ker}\, f^\top = (\mathrm{Im}\, f)^0.$$

We also have

$$\mathrm{Ker}\, f = (\mathrm{Im}\, f^\top)^0.$$

Proof. We have

$$\langle w^*, f(v) \rangle = \langle f^\top(w^*), v \rangle,$$

for all $v \in E$ and all $w^* \in F^*$, and thus, we have $\langle w^*, f(v) \rangle = 0$ for every $v \in V$, i.e. $w^* \in f(V)^0$ iff $\langle f^\top(w^*), v \rangle = 0$ for every $v \in V$ iff $f^\top(w^*) \in V^0$, i.e. $w^* \in (f^\top)^{-1}(V^0)$, proving that

$$f(V)^0 = (f^\top)^{-1}(V^0).$$

Since we already observed that $E^0 = (0)$, letting $V = E$ in the above identity we obtain that

$$\mathrm{Ker}\, f^\top = (\mathrm{Im}\, f)^0.$$

From the equation

$$\langle w^*, f(v) \rangle = \langle f^\top(w^*), v \rangle,$$

we deduce that $v \in (\operatorname{Im} f^\top)^0$ iff $\langle f^\top(w^*), v \rangle = 0$ for all $w^* \in F^*$ iff $\langle w^*, f(v) \rangle = 0$ for all $w^* \in F^*$. Assume that $v \in (\operatorname{Im} f^\top)^0$. If we pick a basis $(w_i)_{i \in I}$ of F, then we have the linear forms $w_i^* \colon F \to K$ such that $w_i^*(w_j) = \delta_{ij}$, and since we must have $\langle w_i^*, f(v) \rangle = 0$ for all $i \in I$ and $(w_i)_{i \in I}$ is a basis of F, we conclude that $f(v) = 0$, and thus $v \in \operatorname{Ker} f$ (this is because $\langle w_i^*, f(v) \rangle$ is the coefficient of $f(v)$ associated with the basis vector w_i). Conversely, if $v \in \operatorname{Ker} f$, then $\langle w^*, f(v) \rangle = 0$ for all $w^* \in F^*$, so we conclude that $v \in (\operatorname{Im} f^\top)^0$. Therefore, $v \in (\operatorname{Im} f^\top)^0$ iff $v \in \operatorname{Ker} f$; that is,

$$\operatorname{Ker} f = (\operatorname{Im} f^\top)^0,$$

as claimed.   □

The following theorem shows the relationship between the rank of f and the rank of $f^\top$.

Theorem 10.2. *Given a linear map $f \colon E \to F$, the following properties hold.*

(a) The dual $(\operatorname{Im} f)^*$ of $\operatorname{Im} f$ is isomorphic to $\operatorname{Im} f^\top = f^\top(F^*)$; that is,

$$(\operatorname{Im} f)^* \cong \operatorname{Im} f^\top.$$

(b) If F is finite dimensional, then $\operatorname{rk}(f) = \operatorname{rk}(f^\top)$.

Proof. (a) Consider the linear maps

$$E \xrightarrow{\ p\ } \operatorname{Im} f \xrightarrow{\ j\ } F,$$

where $E \xrightarrow{\ p\ } \operatorname{Im} f$ is the surjective map induced by $E \xrightarrow{\ f\ } F$, and $\operatorname{Im} f \xrightarrow{\ j\ } F$ is the injective inclusion map of $\operatorname{Im} f$ into F. By definition, $f = j \circ p$. To simplify the notation, let $I = \operatorname{Im} f$. By Proposition 10.7, since $E \xrightarrow{\ p\ } I$ is surjective, $I^* \xrightarrow{\ p^\top\ } E^*$ is injective, and since $\operatorname{Im} f \xrightarrow{\ j\ } F$ is injective, $F^* \xrightarrow{\ j^\top\ } I^*$ is surjective. Since $f = j \circ p$, we also have

$$f^\top = (j \circ p)^\top = p^\top \circ j^\top,$$

and since $F^* \xrightarrow{\ j^\top\ } I^*$ is surjective, and $I^* \xrightarrow{\ p^\top\ } E^*$ is injective, we have an isomorphism between $(\operatorname{Im} f)^*$ and $f^\top(F^*)$.

(b) We already noted that Part (a) of Theorem 10.1 shows that $\dim(F) = \dim(F^*)$, for every vector space F of finite dimension. Consequently, $\dim(\operatorname{Im} f) = \dim((\operatorname{Im} f)^*)$, and thus, by Part (a) we have $\operatorname{rk}(f) = \operatorname{rk}(f^\top)$.

Remark: When both E and F are finite-dimensional, there is also a simple proof of (b) that doesn't use the result of Part (a). By Theorem 10.1(c)

$$\dim(\operatorname{Im} f) + \dim((\operatorname{Im} f)^0) = \dim(F),$$

and by Theorem 5.1

$$\dim(\operatorname{Ker} f^\top) + \dim(\operatorname{Im} f^\top) = \dim(F^*).$$

Furthermore, by Proposition 10.8, we have

$$\operatorname{Ker} f^\top = (\operatorname{Im} f)^0,$$

and since F is finite-dimensional $\dim(F) = \dim(F^*)$, so we deduce

$$\dim(\operatorname{Im} f) + \dim((\operatorname{Im} f)^0) = \dim((\operatorname{Im} f)^0) + \dim(\operatorname{Im} f^\top),$$

which yields $\dim(\operatorname{Im} f) = \dim(\operatorname{Im} f^\top)$; that is, $\operatorname{rk}(f) = \operatorname{rk}(f^\top)$. $\square$

The following proposition can be shown, but it requires a generalization of the duality theorem, so its proof is omitted.

Proposition 10.9. *If $f: E \to F$ is any linear map, then the following identities hold:*

$$\operatorname{Im} f^\top = (\operatorname{Ker}(f))^0$$
$$\operatorname{Ker}(f^\top) = (\operatorname{Im} f)^0$$
$$\operatorname{Im} f = (\operatorname{Ker}(f^\top))^0$$
$$\operatorname{Ker}(f) = (\operatorname{Im} f^\top)^0.$$

Observe that the second and the fourth equation have already be proven in Proposition 10.8. Since for any subspace $V \subseteq E$, even infinite-dimensional, we have $V^{00} = V$, the third equation follows from the second equation by taking orthogonals. Actually, the fourth equation follows from the first also by taking orthogonals. Thus the only equation to be proven is the first equation. We will give a proof later in the case where E is finite-dimensional (see Proposition 10.16).

The following proposition shows the relationship between the matrix representing a linear map $f \colon E \to F$ and the matrix representing its transpose $f^\top \colon F^* \to E^*$.

Proposition 10.10. *Let E and F be two vector spaces, and let $(u_1, \ldots, u_n)$ be a basis for E and $(v_1, \ldots, v_m)$ be a basis for F. Given any linear map $f \colon E \to F$, if $M(f)$ is the $m \times n$-matrix representing f w.r.t. the bases $(u_1, \ldots, u_n)$ and $(v_1, \ldots, v_m)$, then the $n \times m$-matrix $M(f^\top)$ representing $f^\top \colon F^* \to E^*$ w.r.t. the dual bases $(v_1^*, \ldots, v_m^*)$ and $(u_1^*, \ldots, u_n^*)$ is the transpose $M(f)^\top$ of $M(f)$.*

Proof. Recall that the entry a_{ij} in row i and column j of $M(f)$ is the i-th coordinate of $f(u_j)$ over the basis $(v_1, \ldots, v_m)$. By definition of v_i^*, we have $\langle v_i^*, f(u_j) \rangle = a_{ij}$. The entry $a_{ji}^\top$ in row j and column i of $M(f^\top)$ is the j-th coordinate of

$$f^\top(v_i^*) = a_{1i}^\top u_1^* + \cdots + a_{ji}^\top u_j^* + \cdots + a_{ni}^\top u_n^*$$

over the basis $(u_1^*, \ldots, u_n^*)$, which is just $a_{ji}^\top = f^\top(v_i^*)(u_j) = \langle f^\top(v_i^*), u_j \rangle$. Since

$$\langle v_i^*, f(u_j) \rangle = \langle f^\top(v_i^*), u_j \rangle,$$

we have $a_{ij} = a_{ji}^\top$, proving that $M(f^\top) = M(f)^\top$. $\qquad\square$

We now can give a very short proof of the fact that the rank of a matrix is equal to the rank of its transpose.

Proposition 10.11. *Given an $m \times n$ matrix A over a field K, we have $\mathrm{rk}(A) = \mathrm{rk}(A^\top)$.*

Proof. The matrix A corresponds to a linear map $f \colon K^n \to K^m$, and by Theorem 10.2, $\mathrm{rk}(f) = \mathrm{rk}(f^\top)$. By Proposition 10.10, the linear map $f^\top$ corresponds to $A^\top$. Since $\mathrm{rk}(A) = \mathrm{rk}(f)$, and $\mathrm{rk}(A^\top) = \mathrm{rk}(f^\top)$, we conclude that $\mathrm{rk}(A) = \mathrm{rk}(A^\top)$. $\qquad\square$

Thus, given an $m \times n$-matrix A, the maximum number of linearly independent columns is equal to the maximum number of linearly independent rows. There are other ways of proving this fact that do not involve the dual space, but instead some elementary transformations on rows and columns.

Proposition 10.11 immediately yields the following criterion for determining the rank of a matrix:

Proposition 10.12. *Given any $m \times n$ matrix A over a field K (typically $K = \mathbb{R}$ or $K = \mathbb{C}$), the rank of A is the maximum natural number r such* *that there is an invertible $r \times r$ submatrix of A obtained by selecting r rows and r columns of A.*

For example, the 3×2 matrix

$$A = \begin{pmatrix} a_{11} & a_{12} \\ a_{21} & a_{22} \\ a_{31} & a_{32} \end{pmatrix}$$

has rank 2 iff one of the three 2×2 matrices

$$\begin{pmatrix} a_{11} & a_{12} \\ a_{21} & a_{22} \end{pmatrix} \quad \begin{pmatrix} a_{11} & a_{12} \\ a_{31} & a_{32} \end{pmatrix} \quad \begin{pmatrix} a_{21} & a_{22} \\ a_{31} & a_{32} \end{pmatrix}$$

is invertible.

If we combine Proposition 6.8 with Proposition 10.12, we obtain the following criterion for finding the rank of a matrix.

Proposition 10.13. *Given any $m \times n$ matrix A over a field K (typically $K = \mathbb{R}$ or $K = \mathbb{C}$), the rank of A is the maximum natural number r such that there is an $r \times r$ submatrix B of A obtained by selecting r rows and r columns of A, such that $\det(B) \neq 0$.*

This is not a very efficient way of finding the rank of a matrix. We will see that there are better ways using various decompositions such as LU, QR, or SVD.

10.7   Properties of the Double Transpose

First we have the following property showing the naturality of the eval map.

Proposition 10.14. *For any linear map $f \colon E \to F$, we have*

$$f^{\top\top} \circ \mathrm{eval}_E = \mathrm{eval}_F \circ f,$$

or equivalently the following diagram commutes:

$$
\begin{array}{ccc}
E^{**} & \xrightarrow{\;f^{\top\top}\;} & F^{**} \\
{\scriptstyle \mathrm{eval}_E}\Big\uparrow & & \Big\uparrow{\scriptstyle \mathrm{eval}_F} \\
E & \xrightarrow[\;f\;]{} & F.
\end{array}
$$

Proof. For every $u \in E$ and every $\varphi \in F^*$, we have

$$
\begin{aligned}
(f^{\top\top} \circ \mathrm{eval}_E)(u)(\varphi) &= \langle f^{\top\top}(\mathrm{eval}_E(u)), \varphi \rangle \\
&= \langle \mathrm{eval}_E(u), f^{\top}(\varphi) \rangle \\
&= \langle f^{\top}(\varphi), u \rangle \\
&= \langle \varphi, f(u) \rangle \\
&= \langle \mathrm{eval}_F(f(u)), \varphi \rangle \\
&= \langle (\mathrm{eval}_F \circ f)(u), \varphi \rangle \\
&= (\mathrm{eval}_F \circ f)(u)(\varphi),
\end{aligned}
$$

which proves that $f^{\top\top} \circ \mathrm{eval}_E = \mathrm{eval}_F \circ f$, as claimed. $\qquad\square$

If E and F are finite-dimensional, then eval_E and eval_F are isomorphisms, so Proposition 10.14 shows that

$$
f^{\top\top} = \mathrm{eval}_F \circ f \circ \mathrm{eval}_E^{-1}. \tag{10.2}
$$

The above equation is often interpreted as follows: if we identify E with its bidual E^{**} and F with its bidual F^{**}, then $f^{\top\top} = f$. This is an abuse of notation; the rigorous statement is (10.2).

As a corollary of Proposition 10.14, we obtain the following result.

Proposition 10.15. *If* $\dim(E)$ *is finite, then we have*

$$
\mathrm{Ker}\,(f^{\top\top}) = \mathrm{eval}_E(\mathrm{Ker}\,(f)).
$$

Proof. Indeed, if E is finite-dimensional, the map $\mathrm{eval}_E \colon E \to E^{**}$ is an isomorphism, so every $\varphi \in E^{**}$ is of the form $\varphi = \mathrm{eval}_E(u)$ for some $u \in E$, the map $\mathrm{eval}_F \colon F \to F^{**}$ is injective, and we have

$$
\begin{aligned}
f^{\top\top}(\varphi) = 0 \quad &\text{iff} \quad f^{\top\top}(\mathrm{eval}_E(u)) = 0 \\
&\text{iff} \quad \mathrm{eval}_F(f(u)) = 0 \\
&\text{iff} \quad f(u) = 0 \\
&\text{iff} \quad u \in \mathrm{Ker}\,(f) \\
&\text{iff} \quad \varphi \in \mathrm{eval}_E(\mathrm{Ker}\,(f)),
\end{aligned}
$$

which proves that $\mathrm{Ker}\,(f^{\top\top}) = \mathrm{eval}_E(\mathrm{Ker}\,(f))$. $\qquad\square$

Remarks: If $\dim(E)$ is finite, following an argument of Dan Guralnik, the fact that $\mathrm{rk}(f) = \mathrm{rk}(f^{\top})$ can be proven using Proposition 10.15.

Proof. We know from Proposition 10.8 applied to $f^\top : F^* \to E^*$ that
$$\mathrm{Ker}\,(f^{\top\top}) = (\mathrm{Im}\, f^\top)^0,$$
and we showed in Proposition 10.15 that
$$\mathrm{Ker}\,(f^{\top\top}) = \mathrm{eval}_E(\mathrm{Ker}\,(f)).$$
It follows (since eval_E is an isomorphism) that
$$\dim((\mathrm{Im}\, f^\top)^0) = \dim(\mathrm{Ker}\,(f^{\top\top})) = \dim(\mathrm{Ker}\,(f)) = \dim(E) - \dim(\mathrm{Im}\, f),$$
and since
$$\dim(\mathrm{Im}\, f^\top) + \dim((\mathrm{Im}\, f^\top)^0) = \dim(E),$$
we get
$$\dim(\mathrm{Im}\, f^\top) = \dim(\mathrm{Im}\, f). \qquad \square$$

As indicated by Dan Guralnik, if $\dim(E)$ is finite, the above result can be used to prove the following result.

Proposition 10.16. *If* $\dim(E)$ *is finite, then for any linear map* $f : E \to F$, *we have*
$$\mathrm{Im}\, f^\top = (\mathrm{Ker}\,(f))^0.$$

Proof. From
$$\langle f^\top(\varphi), u \rangle = \langle \varphi, f(u) \rangle$$
for all $\varphi \in F^*$ and all $u \in E$, we see that if $u \in \mathrm{Ker}\,(f)$, then $\langle f^\top(\varphi), u \rangle = \langle \varphi, 0 \rangle = 0$, which means that $f^\top(\varphi) \in (\mathrm{Ker}\,(f))^0$, and thus, $\mathrm{Im}\, f^\top \subseteq (\mathrm{Ker}\,(f))^0$. For the converse, since $\dim(E)$ is finite, we have
$$\dim((\mathrm{Ker}\,(f))^0) = \dim(E) - \dim(\mathrm{Ker}\,(f)) = \dim(\mathrm{Im}\, f),$$
but we just proved that $\dim(\mathrm{Im}\, f^\top) = \dim(\mathrm{Im}\, f)$, so we get
$$\dim((\mathrm{Ker}\,(f))^0) = \dim(\mathrm{Im}\, f^\top),$$
and since $\mathrm{Im}\, f^\top \subseteq (\mathrm{Ker}\,(f))^0$, we obtain
$$\mathrm{Im}\, f^\top = (\mathrm{Ker}\,(f))^0,$$
as claimed. $\qquad \square$

Remarks:

(1) By the duality theorem, since $(\mathrm{Ker}\,(f))^{00} = \mathrm{Ker}\,(f)$, the above equation yields another proof of the fact that
$$\mathrm{Ker}\,(f) = (\mathrm{Im}\, f^\top)^0,$$
when E is finite-dimensional.

(2) The equation
$$\mathrm{Im}\, f^\top = (\mathrm{Ker}\,(f))^0$$
is actually valid even if when E if infinite-dimensional, but we will not prove this here.

10.8 The Four Fundamental Subspaces

Given a linear map $f\colon E \to F$ (where E and F are finite-dimensional), Proposition 10.8 revealed that the four spaces

$$\operatorname{Im} f, \ \operatorname{Im} f^\top, \ \operatorname{Ker} f, \ \operatorname{Ker} f^\top$$

play a special role. They are often called the *fundamental subspaces* associated with f. These spaces are related in an intimate manner, since Proposition 10.8 shows that

$$\operatorname{Ker} f = (\operatorname{Im} f^\top)^0$$
$$\operatorname{Ker} f^\top = (\operatorname{Im} f)^0,$$

and Theorem 10.2 shows that

$$\operatorname{rk}(f) = \operatorname{rk}(f^\top).$$

It is instructive to translate these relations in terms of matrices (actually, certain linear algebra books make a big deal about this!). If $\dim(E) = n$ and $\dim(F) = m$, given any basis $(u_1, \ldots, u_n)$ of E and a basis $(v_1, \ldots, v_m)$ of F, we know that f is represented by an $m \times n$ matrix $A = (a_{ij})$, where the jth column of A is equal to $f(u_j)$ over the basis $(v_1, \ldots, v_m)$. Furthermore, the transpose map $f^\top$ is represented by the $n \times m$ matrix $A^\top$ (with respect to the dual bases). Consequently, the four fundamental spaces

$$\operatorname{Im} f, \ \operatorname{Im} f^\top, \ \operatorname{Ker} f, \ \operatorname{Ker} f^\top$$

correspond to

(1) The *column space* of A, denoted by $\operatorname{Im} A$ or $\mathcal{R}(A)$; this is the subspace of $\mathbb{R}^m$ spanned by the columns of A, which corresponds to the image $\operatorname{Im} f$ of f.
(2) The *kernel* or *nullspace* of A, denoted by $\operatorname{Ker} A$ or $\mathcal{N}(A)$; this is the subspace of $\mathbb{R}^n$ consisting of all vectors $x \in \mathbb{R}^n$ such that $Ax = 0$.
(3) The *row space* of A, denoted by $\operatorname{Im} A^\top$ or $\mathcal{R}(A^\top)$; this is the subspace of $\mathbb{R}^n$ spanned by the rows of A, or equivalently, spanned by the columns of $A^\top$, which corresponds to the image $\operatorname{Im} f^\top$ of $f^\top$.
(4) The *left kernel* or *left nullspace* of A denoted by $\operatorname{Ker} A^\top$ or $\mathcal{N}(A^\top)$; this is the kernel (nullspace) of $A^\top$, the subspace of $\mathbb{R}^m$ consisting of all vectors $y \in \mathbb{R}^m$ such that $A^\top y = 0$, or equivalently, $y^\top A = 0$.

Recall that the dimension r of $\operatorname{Im} f$, which is also equal to the dimension of the column space $\operatorname{Im} A = \mathcal{R}(A)$, is the *rank* of A (and f). Then, some our previous results can be reformulated as follows:

(1) The column space $\mathcal{R}(A)$ of A has dimension r.
(2) The nullspace $\mathcal{N}(A)$ of A has dimension $n - r$.
(3) The row space $\mathcal{R}(A^\top)$ has dimension r.
(4) The left nullspace $\mathcal{N}(A^\top)$ of A has dimension $m - r$.

The above statements constitute what Strang calls the *Fundamental Theorem of Linear Algebra, Part I* (see Strang [Strang (1988)]).
The two statements

$$\mathrm{Ker}\, f = (\mathrm{Im}\, f^\top)^0$$
$$\mathrm{Ker}\, f^\top = (\mathrm{Im}\, f)^0$$

translate to

(1) The nullspace of A is the orthogonal of the row space of A.

(2) The left nullspace of A is the orthogonal of the column space of A.

The above statements constitute what Strang calls the *Fundamental Theorem of Linear Algebra, Part II* (see Strang [Strang (1988)]).
Since vectors are represented by column vectors and linear forms by row vectors (over a basis in E or F), a vector $x \in \mathbb{R}^n$ is orthogonal to a linear form y iff

$$yx = 0.$$

Then, a vector $x \in \mathbb{R}^n$ is orthogonal to the row space of A iff x is orthogonal to every row of A, namely $Ax = 0$, which is equivalent to the fact that x belong to the nullspace of A. Similarly, the column vector $y \in \mathbb{R}^m$ (representing a linear form over the dual basis of F^*) belongs to the nullspace of $A^\top$ iff $A^\top y = 0$, iff $y^\top A = 0$, which means that the linear form given by $y^\top$ (over the basis in F) is orthogonal to the column space of A.
Since (2) is equivalent to the fact that the column space of A is equal to the orthogonal of the left nullspace of A, we get the following criterion for the solvability of an equation of the form $Ax = b$:
  The equation $Ax = b$ has a solution iff for all $y \in \mathbb{R}^m$, if $A^\top y = 0$, then $y^\top b = 0$.
Indeed, the condition on the right-hand side says that b is orthogonal to the left nullspace of A; that is, b belongs to the column space of A.
This criterion can be cheaper to check that checking directly that b is spanned by the columns of A. For example, if we consider the system

$$x_1 - x_2 = b_1$$
$$x_2 - x_3 = b_2$$
$$x_3 - x_1 = b_3$$

which, in matrix form, is written $Ax = b$ as below:

$$\begin{pmatrix} 1 & -1 & 0 \\ 0 & 1 & -1 \\ -1 & 0 & 1 \end{pmatrix} \begin{pmatrix} x_1 \\ x_2 \\ x_3 \end{pmatrix} = \begin{pmatrix} b_1 \\ b_2 \\ b_3 \end{pmatrix},$$

we see that the rows of the matrix A add up to 0. In fact, it is easy to convince ourselves that the left nullspace of A is spanned by $y = (1, 1, 1)$, and so the system is solvable iff $y^\top b = 0$, namely

$$b_1 + b_2 + b_3 = 0.$$

Note that the above criterion can also be stated negatively as follows:

The equation $Ax = b$ has no solution iff there is some $y \in \mathbb{R}^m$ such that $A^\top y = 0$ and $y^\top b \neq 0$.

Since $A^\top y = 0$ iff $y^\top A = 0$, we can view $y^\top$ as a row vector representing a linear form, and $y^\top A = 0$ asserts that the linear form $y^\top$ vanishes on the columns $A^1, \ldots, A^n$ of A but does not vanish on b. Since the linear form $y^\top$ defines the hyperplane H of equation $y^\top z = 0$ (with $z \in \mathbb{R}^m$), geometrically the equation $Ax = b$ has no solution iff there is a hyperplane H containing $A^1, \ldots, A^n$ and not containing b.

10.9 Summary

The main concepts and results of this chapter are listed below:

- The *dual space* E^* and *linear forms* (*covector*). The *bidual* E^{**}.
- The *bilinear pairing* $\langle -, - \rangle \colon E^* \times E \to K$ (the *canonical pairing*).
- *Evaluation at v*: $\mathrm{eval}_v \colon E^* \to K$.
- The map $\mathrm{eval}_E \colon E \to E^{**}$.
- *Othogonality* between a subspace V of E and a subspace U of E^*; the *orthogonal* V^0 and the *orthogonal* U^0.
- *Coordinate forms*.
- The *Duality theorem* (Theorem 10.1).
- The *dual basis* of a basis.
- The isomorphism $\mathrm{eval}_E \colon E \to E^{**}$ when $\dim(E)$ is finite.
- *Pairing* between two vector spaces; *nondegenerate pairing*; Proposition 10.5.
- Hyperplanes and linear forms.
- The *transpose* $f^\top \colon F^* \to E^*$ of a linear map $f \colon E \to F$.
- The fundamental identities:

$$\operatorname{Ker} f^\top = (\operatorname{Im} f)^0 \quad \text{and} \quad \operatorname{Ker} f = (\operatorname{Im} f^\top)^0$$

(Proposition 10.8).

- If F is finite-dimensional, then

$$\text{rk}(f) = \text{rk}(f^\top).$$

  (Theorem 10.2).
- The matrix of the transpose map $f^\top$ is equal to the transpose of the matrix of the map f (Proposition 10.10).
- For any $m \times n$ matrix A,

$$\text{rk}(A) = \text{rk}(A^\top).$$

- Characterization of the rank of a matrix in terms of a maximal invertible submatrix (Proposition 10.12).
- The *four fundamental subspaces*:

$$\text{Im}\, f,\ \text{Im}\, f^\top,\ \text{Ker}\, f,\ \text{Ker}\, f^\top.$$

- The *column space*, the *nullspace*, the *row space*, and the *left nullspace* (of a matrix).
- Criterion for the solvability of an equation of the form $Ax = b$ in terms of the left nullspace.

10.10 Problems

Problem 10.1. Prove the following properties of transposition:

$$(f + g)^\top = f^\top + g^\top$$
$$(g \circ f)^\top = f^\top \circ g^\top$$
$$\text{id}_E^\top = \text{id}_{E^*}.$$

Problem 10.2. Let $(u_1, \ldots, u_{n-1})$ be $n - 1$ linearly independent vectors $u_i \in \mathbb{C}^n$. Prove that the hyperlane H spanned by $(u_1, \ldots, u_{n-1})$ is the nullspace of the linear form

$$x \mapsto \det(u_1, \ldots, u_{n-1}, x), \quad x \in \mathbb{C}^n.$$

Prove that if A is the $n \times n$ matrix whose columns are $(u_1, \ldots, u_{n-1}, x)$, and if $c_i = (-1)^{i+n} \det(A_{in})$ is the cofactor of $a_{in} = x_i$ for $i = 1, \ldots, n$, then H is defined by the equation

$$c_1 x_1 + \cdots + c_n x_n = 0.$$

Problem 10.3. (1) Let $\varphi \colon \mathbb{R}^n \times \mathbb{R}^n \to \mathbb{R}$ be the map defined by

$$\varphi((x_1, \ldots, x_n), (y_1, \ldots, y_n)) = x_1 y_1 + \cdots + x_n y_n.$$

Prove that φ is a bilinear nondegenerate pairing. Deduce that $(\mathbb{R}^n)^*$ is isomorphic to $\mathbb{R}^n$.

Prove that $\varphi(x, x) = 0$ iff $x = 0$.

(2) Let $\varphi_L \colon \mathbb{R}^4 \times \mathbb{R}^4 \to \mathbb{R}$ be the map defined by

$$\varphi_L((x_1, x_2, x_3, x_4), (y_1, y_2, y_3, , y_4)) = x_1 y_1 - x_2 y_2 - x_3 y_3 - x_4 y_4.$$

Prove that φ is a bilinear nondegenerate pairing.

Show that there exist nonzero vectors $x \in \mathbb{R}^4$ such that $\varphi_L(x, x) = 0$.

Remark: The vector space $\mathbb{R}^4$ equipped with the above bilinear form called the *Lorentz form* is called *Minkowski space*.

Problem 10.4. Given any two subspaces V_1, V_2 of a finite-dimensional vector space E, prove that

$$(V_1 + V_2)^0 = V_1^0 \cap V_2^0$$
$$(V_1 \cap V_2)^0 = V_1^0 + V_2^0.$$

Beware that in the second equation, V_1 and V_2 are subspaces of E, not E^*.

Hint. To prove the second equation, prove the inclusions $V_1^0 + V_2^0 \subseteq (V_1 \cap V_2)^0$ and $(V_1 \cap V_2)^0 \subseteq V_1^0 + V_2^0$. Proving the second inclusion is a little tricky. First, prove that we can pick a subspace W_1 of V_1 and a subspace W_2 of V_2 such that

(1) V_1 is the direct sum $V_1 = (V_1 \cap V_2) \oplus W_1$.
(2) V_2 is the direct sum $V_2 = (V_1 \cap V_2) \oplus W_2$.
(3) $V_1 + V_2$ is the direct sum $V_1 + V_2 = (V_1 \cap V_2) \oplus W_1 \oplus W_2$.

Problem 10.5. (1) Let A be any $n \times n$ matrix such that the sum of the entries of every row of A is the same (say c_1), and the sum of entries of every column of A is the same (say c_2). Prove that $c_1 = c_2$.

(2) Prove that for any $n \geq 2$, the $2n - 2$ equations asserting that the sum of the entries of every row of A is the same, and the sum of entries of every column of A is the same are linearly independent. For example,

when $n = 4$, we have the following 6 equations

$$a_{11} + a_{12} + a_{13} + a_{14} - a_{21} - a_{22} - a_{23} - a_{24} = 0$$
$$a_{21} + a_{22} + a_{23} + a_{24} - a_{31} - a_{32} - a_{33} - a_{34} = 0$$
$$a_{31} + a_{32} + a_{33} + a_{34} - a_{41} - a_{42} - a_{43} - a_{44} = 0$$
$$a_{11} + a_{21} + a_{31} + a_{41} - a_{12} - a_{22} - a_{32} - a_{42} = 0$$
$$a_{12} + a_{22} + a_{32} + a_{42} - a_{13} - a_{23} - a_{33} - a_{43} = 0$$
$$a_{13} + a_{23} + a_{33} + a_{43} - a_{14} - a_{24} - a_{34} - a_{44} = 0.$$

Hint. Group the equations as above; that is, first list the $n - 1$ equations relating the rows, and then list the $n - 1$ equations relating the columns. Prove that the first $n - 1$ equations are linearly independent, and that the last $n - 1$ equations are also linearly independent. Then, find a relationship between the two groups of equations that will allow you to prove that they span subspace V^r and V^c such that $V^r \cap V^c = (0)$.

(3) Now consider *magic squares*. Such matrices satisfy the two conditions about the sum of the entries in each row and in each column to be the same number, and also the additional two constraints that the main descending and the main ascending diagonals add up to this common number. Traditionally, it is also required that the entries in a magic square are positive integers, but we will consider generalized magic square with arbitrary real entries. For example, in the case $n = 4$, we have the following system of 8 equations:

$$a_{11} + a_{12} + a_{13} + a_{14} - a_{21} - a_{22} - a_{23} - a_{24} = 0$$
$$a_{21} + a_{22} + a_{23} + a_{24} - a_{31} - a_{32} - a_{33} - a_{34} = 0$$
$$a_{31} + a_{32} + a_{33} + a_{34} - a_{41} - a_{42} - a_{43} - a_{44} = 0$$
$$a_{11} + a_{21} + a_{31} + a_{41} - a_{12} - a_{22} - a_{32} - a_{42} = 0$$
$$a_{12} + a_{22} + a_{32} + a_{42} - a_{13} - a_{23} - a_{33} - a_{43} = 0$$
$$a_{13} + a_{23} + a_{33} + a_{43} - a_{14} - a_{24} - a_{34} - a_{44} = 0$$
$$a_{22} + a_{33} + a_{44} - a_{12} - a_{13} - a_{14} = 0$$
$$a_{41} + a_{32} + a_{23} - a_{11} - a_{12} - a_{13} = 0.$$

In general, the equation involving the descending diagonal is

$$a_{22} + a_{33} + \cdots + a_{nn} - a_{12} - a_{13} - \cdots - a_{1n} = 0 \qquad (10.3)$$

and the equation involving the ascending diagonal is

$$a_{n1} + a_{n-12} + \cdots + a_{2n-1} - a_{11} - a_{12} - \cdots - a_{1n-1} = 0. \qquad (10.4)$$

Prove that if $n \geq 3$, then the $2n$ equations asserting that a matrix is a generalized magic square are linearly independent.

Hint. Equations are really linear forms, so find some matrix annihilated by all equations except equation (10.3), and some matrix annihilated by all equations except equation (10.4).

Problem 10.6. Let $U_1, \ldots, U_p$ be some subspaces of a vector space E, and assume that they form a direct sum $U = U_1 \oplus \cdots \oplus U_p$. Let $j_i \colon U_i \to U_1 \oplus \cdots \oplus U_p$ be the canonical injections, and let $\pi_i \colon U_1^* \times \cdots \times U_p^* \to U_i^*$ be the canonical projections. Prove that there is an isomorphism f from $(U_1 \oplus \cdots \oplus U_p)^*$ to $U_1^* \times \cdots \times U_p^*$ such that

$$\pi_i \circ f = j_i^\top, \quad 1 \leq i \leq p.$$

Problem 10.7. Let U and V be two subspaces of a vector space E such that $E = U \oplus V$. Prove that

$$E^* = U^0 \oplus V^0.$$

Chapter 11

Euclidean Spaces

Rien n'est beau que le vrai.
—Hermann Minkowski

11.1 Inner Products, Euclidean Spaces

So far the framework of vector spaces allows us to deal with ratios of vectors and linear combinations, but there is no way to express the notion of angle or to talk about orthogonality of vectors. A Euclidean structure allows us to deal with *metric notions* such as angles, orthogonality, and length (or distance).

This chapter covers the bare bones of Euclidean geometry. Deeper aspects of Euclidean geometry are investigated in Chapter 12. One of our main goals is to give the basic properties of the transformations that preserve the Euclidean structure, rotations and reflections, since they play an important role in practice. Euclidean geometry is the study of properties invariant under certain affine maps called *rigid motions*. Rigid motions are the maps that preserve the distance between points.

We begin by defining inner products and Euclidean spaces. The Cauchy–Schwarz inequality and the Minkowski inequality are shown. We define orthogonality of vectors and of subspaces, orthogonal bases, and orthonormal bases. We prove that every finite-dimensional Euclidean space has orthonormal bases. The first proof uses duality and the second one the Gram–Schmidt orthogonalization procedure. The QR-decomposition for invertible matrices is shown as an application of the Gram–Schmidt procedure. Linear isometries (also called orthogonal transformations) are defined and studied briefly. We conclude with a short section in which some applications of Euclidean geometry are sketched. One of the most important applications, the method of least squares, is discussed in Chapter 21.

For a more detailed treatment of Euclidean geometry see Berger [Berger (1990a,b)], Snapper and Troyer [Snapper and Troyer (1989)], or any other book on geometry, such as Pedoe [Pedoe (1988)], Coxeter [Coxeter (1989)], Fresnel [Fresnel (1998)], Tisseron [Tisseron (1994)], or Cagnac, Ramis, and Commeau [Cagnac et al. (1965)]. Serious readers should consult Emil Artin's famous book [Artin (1957)], which contains an in-depth study of the orthogonal group, as well as other groups arising in geometry. It is still worth consulting some of the older classics, such as Hadamard [Hadamard (1947, 1949)] and Rouché and de Comberousse [Rouché and de Comberousse (1900)]. The first edition of [Hadamard (1947)] was published in 1898 and finally reached its thirteenth edition in 1947! In this chapter it is assumed that all vector spaces are defined over the field $\mathbb{R}$ of real numbers unless specified otherwise (in a few cases, over the complex numbers $\mathbb{C}$).

First we define a Euclidean structure on a vector space. Technically, a Euclidean structure over a vector space E is provided by a symmetric bilinear form on the vector space satisfying some extra properties. Recall that a bilinear form $\varphi \colon E \times E \to \mathbb{R}$ is *definite* if for every $u \in E$, $u \neq 0$ implies that $\varphi(u, u) \neq 0$, and *positive* if for every $u \in E$, $\varphi(u, u) \geq 0$.

Definition 11.1. A *Euclidean space* is a real vector space E equipped with a symmetric bilinear form $\varphi \colon E \times E \to \mathbb{R}$ that is *positive definite*. More explicitly, $\varphi \colon E \times E \to \mathbb{R}$ satisfies the following axioms:

$$\varphi(u_1 + u_2, v) = \varphi(u_1, v) + \varphi(u_2, v),$$
$$\varphi(u, v_1 + v_2) = \varphi(u, v_1) + \varphi(u, v_2),$$
$$\varphi(\lambda u, v) = \lambda \varphi(u, v),$$
$$\varphi(u, \lambda v) = \lambda \varphi(u, v),$$
$$\varphi(u, v) = \varphi(v, u),$$
$$u \neq 0 \text{ implies that } \varphi(u, u) > 0.$$

The real number $\varphi(u, v)$ is also called the *inner product (or scalar product) of u and v*. We also define the *quadratic form associated with φ* as the function $\Phi \colon E \to \mathbb{R}_+$ such that

$$\Phi(u) = \varphi(u, u),$$

for all $u \in E$.

Since φ is bilinear, we have $\varphi(0, 0) = 0$, and since it is positive definite, we have the stronger fact that

$$\varphi(u, u) = 0 \quad \text{iff} \quad u = 0,$$

that is, $\Phi(u) = 0$ iff $u = 0$.

Given an inner product $\varphi \colon E \times E \to \mathbb{R}$ on a vector space E, we also denote $\varphi(u, v)$ by

$$u \cdot v \quad \text{or} \quad \langle u, v \rangle \quad \text{or} \quad (u|v),$$

and $\sqrt{\Phi(u)}$ by $\|u\|$.

Example 11.1. The standard example of a Euclidean space is $\mathbb{R}^n$, under the inner product $\cdot$ defined such that

$$(x_1, \ldots, x_n) \cdot (y_1, \ldots, y_n) = x_1 y_1 + x_2 y_2 + \cdots + x_n y_n.$$

This Euclidean space is denoted by $\mathbb{E}^n$.

There are other examples.

Example 11.2. For instance, let E be a vector space of dimension 2, and let (e_1, e_2) be a basis of E. If $a > 0$ and $b^2 - ac < 0$, the bilinear form defined such that

$$\varphi(x_1 e_1 + y_1 e_2, x_2 e_1 + y_2 e_2) = a x_1 x_2 + b(x_1 y_2 + x_2 y_1) + c y_1 y_2$$

yields a Euclidean structure on E. In this case,

$$\Phi(x e_1 + y e_2) = a x^2 + 2 b x y + c y^2.$$

Example 11.3. Let $\mathcal{C}[a, b]$ denote the set of continuous functions $f \colon [a, b] \to \mathbb{R}$. It is easily checked that $\mathcal{C}[a, b]$ is a vector space of infinite dimension. Given any two functions $f, g \in \mathcal{C}[a, b]$, let

$$\langle f, g \rangle = \int_a^b f(t) g(t) dt.$$

We leave it as an easy exercise that $\langle -, - \rangle$ is indeed an inner product on $\mathcal{C}[a, b]$. In the case where $a = -\pi$ and $b = \pi$ (or $a = 0$ and $b = 2\pi$, this makes basically no difference), one should compute

$$\langle \sin px, \sin qx \rangle, \quad \langle \sin px, \cos qx \rangle, \quad \text{and} \quad \langle \cos px, \cos qx \rangle,$$

for all natural numbers $p, q \geq 1$. The outcome of these calculations is what makes Fourier analysis possible!

Example 11.4. Let $E = \mathrm{M}_n(\mathbb{R})$ be the vector space of real $n \times n$ matrices. If we view a matrix $A \in \mathrm{M}_n(\mathbb{R})$ as a "long" column vector obtained by concatenating together its columns, we can define the inner product of two matrices $A, B \in \mathrm{M}_n(\mathbb{R})$ as

$$\langle A, B \rangle = \sum_{i,j=1}^n a_{ij} b_{ij},$$

which can be conveniently written as

$$\langle A, B \rangle = \mathrm{tr}(A^\top B) = \mathrm{tr}(B^\top A).$$

Since this can be viewed as the Euclidean product on $\mathbb{R}^{n^2}$, it is an inner product on $\mathrm{M}_n(\mathbb{R})$. The corresponding norm

$$\|A\|_F = \sqrt{\mathrm{tr}(A^\top A)}$$

is the Frobenius norm (see Section 8.2).

Let us observe that φ can be recovered from Φ.

Proposition 11.1. *We have*

$$\varphi(u, v) = \frac{1}{2}[\Phi(u + v) - \Phi(u) - \Phi(v)]$$

for all $u, v \in E$. We say that φ is the **polar form of Φ**.

Proof. By bilinearity and symmetry, we have

$$
\begin{aligned}
\Phi(u + v) &= \varphi(u + v,\, u + v) \\
&= \varphi(u,\, u + v) + \varphi(v,\, u + v) \\
&= \varphi(u,\, u) + 2\varphi(u,\, v) + \varphi(v,\, v) \\
&= \Phi(u) + 2\varphi(u,\, v) + \Phi(v). \qquad \square
\end{aligned}
$$

If E is finite-dimensional and if $\varphi\colon E \times E \to \mathbb{R}$ is a bilinear form on E, given any basis $(e_1, \ldots, e_n)$ of E, we can write $x = \sum_{i=1}^{n} x_i e_i$ and $y = \sum_{j=1}^{n} y_j e_j$, and we have

$$\varphi(x, y) = \varphi\left(\sum_{i=1}^{n} x_i e_i,\, \sum_{j=1}^{n} y_j e_j\right) = \sum_{i,j=1}^{n} x_i y_j \varphi(e_i, e_j).$$

If we let G be the matrix $G = (\varphi(e_i, e_j))$, and if x and y are the column vectors associated with $(x_1, \ldots, x_n)$ and $(y_1, \ldots, y_n)$, then we can write

$$\varphi(x, y) = x^\top G y = y^\top G^\top x.$$

Note that we are committing an abuse of notation since $x = \sum_{i=1}^{n} x_i e_i$ is a vector in E, but the column vector associated with $(x_1, \ldots, x_n)$ belongs to $\mathbb{R}^n$. To avoid this minor abuse, we could denote the column vector associated with $(x_1, \ldots, x_n)$ by $\mathbf{x}$ (and similarly $\mathbf{y}$ for the column vector associated with $(y_1, \ldots, y_n)$), in which case the "correct" expression for $\varphi(x, y)$ is

$$\varphi(x, y) = \mathbf{x}^\top G \mathbf{y}.$$

However, in view of the isomorphism between E and $\mathbb{R}^n$, to keep notation as simple as possible, we will use x and y instead of $\mathbf{x}$ and $\mathbf{y}$.

Also observe that φ is symmetric iff $G = G^\top$, and φ is positive definite iff the matrix G is positive definite, that is,

$$x^\top G x > 0 \quad \text{for all } x \in \mathbb{R}^n, \ x \neq 0.$$

The matrix G associated with an inner product is called the *Gram matrix* of the inner product with respect to the basis $(e_1, \ldots, e_n)$.

Conversely, if A is a symmetric positive definite $n \times n$ matrix, it is easy to check that the bilinear form

$$\langle x, y \rangle = x^\top A y$$

is an inner product. If we make a change of basis from the basis $(e_1, \ldots, e_n)$ to the basis $(f_1, \ldots, f_n)$, and if the change of basis matrix is P (where the jth column of P consists of the coordinates of f_j over the basis $(e_1, \ldots, e_n)$), then with respect to coordinates x' and y' over the basis $(f_1, \ldots, f_n)$, we have

$$x^\top G y = x'^\top P^\top G P y',$$

so the matrix of our inner product over the basis $(f_1, \ldots, f_n)$ is $P^\top G P$. We summarize these facts in the following proposition.

Proposition 11.2. *Let E be a finite-dimensional vector space, and let $(e_1, \ldots, e_n)$ be a basis of E.*

(1) For any inner product $\langle -, - \rangle$ on E, if $G = (\langle e_i, e_j \rangle)$ is the Gram matrix of the inner product $\langle -, - \rangle$ w.r.t. the basis $(e_1, \ldots, e_n)$, then G is symmetric positive definite.

(2) For any change of basis matrix P, the Gram matrix of $\langle -, - \rangle$ with respect to the new basis is $P^\top G P$.

(3) If A is any $n \times n$ symmetric positive definite matrix, then

$$\langle x, y \rangle = x^\top A y$$

is an inner product on E.

We will see later that a symmetric matrix is positive definite iff its eigenvalues are all positive.

One of the very important properties of an inner product φ is that the map $u \mapsto \sqrt{\Phi(u)}$ is a norm.

Proposition 11.3. *Let E be a Euclidean space with inner product φ, and let Φ be the corresponding quadratic form. For all $u, v \in E$, we have the Cauchy–Schwarz inequality*

$$\varphi(u, v)^2 \leq \Phi(u) \Phi(v),$$

the equality holding iff u and v are linearly dependent.

We also have the Minkowski inequality

$$\sqrt{\Phi(u+v)} \leq \sqrt{\Phi(u)} + \sqrt{\Phi(v)},$$

the equality holding iff u and v are linearly dependent, where in addition if $u \neq 0$ and $v \neq 0$, then $u = \lambda v$ for some $\lambda > 0$.

Proof. For any vectors $u, v \in E$, we define the function $T \colon \mathbb{R} \to \mathbb{R}$ such that

$$T(\lambda) = \Phi(u + \lambda v),$$

for all $\lambda \in \mathbb{R}$. Using bilinearity and symmetry, we have

$$
\begin{aligned}
\Phi(u + \lambda v) &= \varphi(u + \lambda v, \, u + \lambda v) \\
&= \varphi(u, \, u + \lambda v) + \lambda \varphi(v, \, u + \lambda v) \\
&= \varphi(u, \, u) + 2\lambda \varphi(u, \, v) + \lambda^2 \varphi(v, \, v) \\
&= \Phi(u) + 2\lambda \varphi(u, \, v) + \lambda^2 \Phi(v).
\end{aligned}
$$

Since φ is positive definite, Φ is nonnegative, and thus $T(\lambda) \geq 0$ for all $\lambda \in \mathbb{R}$. If $\Phi(v) = 0$, then $v = 0$, and we also have $\varphi(u, v) = 0$. In this case, the Cauchy–Schwarz inequality is trivial, and $v = 0$ and u are linearly dependent.

Now assume $\Phi(v) > 0$. Since $T(\lambda) \geq 0$, the quadratic equation

$$\lambda^2 \Phi(v) + 2\lambda \varphi(u, \, v) + \Phi(u) = 0$$

cannot have distinct real roots, which means that its discriminant

$$\Delta = 4(\varphi(u, \, v)^2 - \Phi(u)\Phi(v))$$

is null or negative, which is precisely the Cauchy–Schwarz inequality

$$\varphi(u, v)^2 \leq \Phi(u)\Phi(v).$$

Let us now consider the case where we have the equality

$$\varphi(u, v)^2 = \Phi(u)\Phi(v).$$

There are two cases. If $\Phi(v) = 0$, then $v = 0$ and u and v are linearly dependent. If $\Phi(v) \neq 0$, then the above quadratic equation has a double root λ_0, and we have $\Phi(u + \lambda_0 v) = 0$. Since φ is positive definite, $\Phi(u + \lambda_0 v) = 0$ implies that $u + \lambda_0 v = 0$, which shows that u and v are linearly dependent. Conversely, it is easy to check that we have equality when u and v are linearly dependent.

The Minkowski inequality

$$\sqrt{\Phi(u+v)} \le \sqrt{\Phi(u)} + \sqrt{\Phi(v)}$$

is equivalent to

$$\Phi(u+v) \le \Phi(u) + \Phi(v) + 2\sqrt{\Phi(u)\Phi(v)}.$$

However, we have shown that

$$2\varphi(u,\, v) = \Phi(u+v) - \Phi(u) - \Phi(v),$$

and so the above inequality is equivalent to

$$\varphi(u,\, v) \le \sqrt{\Phi(u)\Phi(v)},$$

which is trivial when $\varphi(u,\, v) \le 0$, and follows from the Cauchy–Schwarz inequality when $\varphi(u,\, v) \ge 0$. Thus, the Minkowski inequality holds. Finally assume that $u \ne 0$ and $v \ne 0$, and that

$$\sqrt{\Phi(u+v)} = \sqrt{\Phi(u)} + \sqrt{\Phi(v)}.$$

When this is the case, we have

$$\varphi(u,\, v) = \sqrt{\Phi(u)\Phi(v)},$$

and we know from the discussion of the Cauchy–Schwarz inequality that the equality holds iff u and v are linearly dependent. The Minkowski inequality is an equality when u or v is null. Otherwise, if $u \ne 0$ and $v \ne 0$, then $u = \lambda v$ for some $\lambda \ne 0$, and since

$$\varphi(u,\, v) = \lambda\varphi(v,\, v) = \sqrt{\Phi(u)\Phi(v)},$$

by positivity, we must have $\lambda > 0$. $\qquad\square$

Note that the Cauchy–Schwarz inequality can also be written as

$$|\varphi(u,v)| \le \sqrt{\Phi(u)}\sqrt{\Phi(v)}.$$

Remark: It is easy to prove that the Cauchy–Schwarz and the Minkowski inequalities still hold for a symmetric bilinear form that is positive, but not necessarily definite (i.e., $\varphi(u,v) \ge 0$ for all $u, v \in E$). However, u and v need not be linearly dependent when the equality holds.

The Minkowski inequality

$$\sqrt{\Phi(u+v)} \le \sqrt{\Phi(u)} + \sqrt{\Phi(v)}$$

shows that the map $u \mapsto \sqrt{\Phi(u)}$ satisfies the convexity inequality (also known as triangle inequality), condition (N3) of Definition 8.1, and since φ is bilinear and positive definite, it also satisfies conditions (N1) and (N2) of Definition 8.1, and thus it is a *norm* on E. The norm induced by φ is called the *Euclidean norm induced by φ*.

The Cauchy–Schwarz inequality can be written as

$$|u \cdot v| \leq \|u\|\|v\|,$$

and the Minkowski inequality as

$$\|u + v\| \leq \|u\| + \|v\|.$$

If u and v are nonzero vectors then the Cauchy–Schwarz inequality implies that

$$-1 \leq \frac{u \cdot v}{\|u\|\,\|v\|} \leq +1.$$

Then there is a unique $\theta \in [0, \pi]$ such that

$$\cos\theta = \frac{u \cdot v}{\|u\|\,\|v\|}.$$

We have $u = v$ iff $\theta = 0$ and $u = -v$ iff $\theta = \pi$. For $0 < \theta < \pi$, the vectors u and v are linearly independent and there is an orientation of the plane spanned by u and v such that θ is the angle between u and v. See Problem 11.8 for the precise notion of orientation. If u is a unit vector (which means that $\|u\| = 1$), then the vector

$$(\|v\| \cos\theta)u = (u \cdot v)u = (v \cdot u)u$$

is called the *orthogonal projection* of v onto the space spanned by u.

Remark: One might wonder if every norm on a vector space is induced by some Euclidean inner product. In general this is false, but remarkably, there is a simple necessary and sufficient condition, which is that the norm must satisfy the *parallelogram law*:

$$\|u + v\|^2 + \|u - v\|^2 = 2(\|u\|^2 + \|v\|^2).$$

See Figure 11.1.

If $\langle -, - \rangle$ is an inner product, then we have

$$\|u + v\|^2 = \|u\|^2 + \|v\|^2 + 2\langle u, v \rangle$$
$$\|u - v\|^2 = \|u\|^2 + \|v\|^2 - 2\langle u, v \rangle,$$

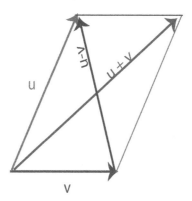

Fig. 11.1   The parallelogram law states that the sum of the lengths of the diagonals of the parallelogram determined by vectors u and v equals the sum of all the sides.

and by adding and subtracting these identities, we get the parallelogram law and the equation

$$\langle u, v \rangle = \frac{1}{4}(\|u + v\|^2 - \|u - v\|^2),$$

which allows us to recover $\langle -, - \rangle$ from the norm.

Conversely, if $\| \ \|$ is a norm satisfying the parallelogram law, and if it comes from an inner product, then this inner product must be given by

$$\langle u, v \rangle = \frac{1}{4}(\|u + v\|^2 - \|u - v\|^2).$$

We need to prove that the above form is indeed symmetric and bilinear.

Symmetry holds because $\|u - v\| = \|-(u - v)\| = \|v - u\|$. Let us prove additivity in the variable u. By the parallelogram law, we have

$$2(\|x + z\|^2 + \|y\|^2) = \|x + y + z\|^2 + \|x - y + z\|^2$$

which yields

$$\|x + y + z\|^2 = 2(\|x + z\|^2 + \|y\|^2) - \|x - y + z\|^2$$
$$\|x + y + z\|^2 = 2(\|y + z\|^2 + \|x\|^2) - \|y - x + z\|^2,$$

where the second formula is obtained by swapping x and y. Then by adding up these equations, we get

$$\|x + y + z\|^2 = \|x\|^2 + \|y\|^2 + \|x + z\|^2 + \|y + z\|^2$$
$$- \frac{1}{2}\|x - y + z\|^2 - \frac{1}{2}\|y - x + z\|^2.$$

Replacing z by $-z$ in the above equation, we get

$$\|x + y - z\|^2 = \|x\|^2 + \|y\|^2 + \|x - z\|^2 + \|y - z\|^2$$
$$- \frac{1}{2}\|x - y - z\|^2 - \frac{1}{2}\|y - x - z\|^2,$$

Since $\|x - y + z\| = \|-(x - y + z)\| = \|y - x - z\|$ and $\|y - x + z\| = \|-(y - x + z)\| = \|x - y - z\|$, by subtracting the last two equations, we get

$$\langle x + y, z \rangle = \frac{1}{4}(\|x + y + z\|^2 - \|x + y - z\|^2)$$
$$= \frac{1}{4}(\|x + z\|^2 - \|x - z\|^2) + \frac{1}{4}(\|y + z\|^2 - \|y - z\|^2)$$
$$= \langle x, z \rangle + \langle y, z \rangle,$$

as desired.

Proving that

$$\langle \lambda x, y \rangle = \lambda \langle x, y \rangle \quad \text{for all } \lambda \in \mathbb{R}$$

is a little tricky. The strategy is to prove the identity for $\lambda \in \mathbb{Z}$, then to promote it to $\mathbb{Q}$, and then to $\mathbb{R}$ by continuity.

Since

$$\langle -u, v \rangle = \frac{1}{4}(\|-u + v\|^2 - \|-u - v\|^2)$$
$$= \frac{1}{4}(\|u - v\|^2 - \|u + v\|^2)$$
$$= -\langle u, v \rangle,$$

the property holds for $\lambda = -1$. By linearity and by induction, for any $n \in \mathbb{N}$ with $n \geq 1$, writing $n = n - 1 + 1$, we get

$$\langle \lambda x, y \rangle = \lambda \langle x, y \rangle \quad \text{for all } \lambda \in \mathbb{N},$$

and since the above also holds for $\lambda = -1$, it holds for all $\lambda \in \mathbb{Z}$. For $\lambda = p/q$ with $p, q \in \mathbb{Z}$ and $q \neq 0$, we have

$$q\langle (p/q)u, v \rangle = \langle pu, v \rangle = p\langle u, v \rangle,$$

which shows that

$$\langle (p/q)u, v \rangle = (p/q)\langle u, v \rangle,$$

and thus

$$\langle \lambda x, y \rangle = \lambda \langle x, y \rangle \quad \text{for all } \lambda \in \mathbb{Q}.$$

To finish the proof, we use the fact that a norm is a continuous map $x \mapsto \|x\|$. Then, the continuous function $t \mapsto \frac{1}{t}\langle tu, v \rangle$ defined on $\mathbb{R} - \{0\}$ agrees with $\langle u, v \rangle$ on $\mathbb{Q} - \{0\}$, so it is equal to $\langle u, v \rangle$ on $\mathbb{R} - \{0\}$. The case $\lambda = 0$ is trivial, so we are done.

We now define orthogonality.

11.2 Orthogonality and Duality in Euclidean Spaces

An inner product on a vector space gives the ability to define the notion of orthogonality. Families of nonnull pairwise orthogonal vectors must be linearly independent. They are called orthogonal families. In a vector space of finite dimension it is always possible to find orthogonal bases. This is very useful theoretically and practically. Indeed, in an orthogonal basis, finding the coordinates of a vector is very cheap: It takes an inner product. Fourier series make crucial use of this fact. When E has finite dimension, we prove that the inner product on E induces a natural isomorphism between E and its dual space E^*. This allows us to define the adjoint of a linear map in an intrinsic fashion (i.e., independently of bases). It is also possible to orthonormalize any basis (certainly when the dimension is finite). We give two proofs, one using duality, the other more constructive using the Gram–Schmidt orthonormalization procedure.

Definition 11.2. Given a Euclidean space E, any two vectors $u, v \in E$ are *orthogonal, or perpendicular*, if $u \cdot v = 0$. Given a family $(u_i)_{i \in I}$ of vectors in E, we say that $(u_i)_{i \in I}$ is *orthogonal* if $u_i \cdot u_j = 0$ for all $i, j \in I$, where $i \neq j$. We say that the family $(u_i)_{i \in I}$ is *orthonormal* if $u_i \cdot u_j = 0$ for all $i, j \in I$, where $i \neq j$, and $\|u_i\| = u_i \cdot u_i = 1$, for all $i \in I$. For any subset F of E, the set

$$F^\perp = \{v \in E \mid u \cdot v = 0, \text{ for all } u \in F\},$$

of all vectors orthogonal to all vectors in F, is called the *orthogonal complement of F*.

Since inner products are positive definite, observe that for any vector $u \in E$, we have

$$u \cdot v = 0 \quad \text{for all } v \in E \quad \text{iff} \quad u = 0.$$

It is immediately verified that the orthogonal complement $F^\perp$ of F is a subspace of E.

Example 11.5. Going back to Example 11.3 and to the inner product

$$\langle f, g \rangle = \int_{-\pi}^{\pi} f(t)g(t)dt$$

on the vector space $\mathcal{C}[-\pi, \pi]$, it is easily checked that

$$\langle \sin px, \sin qx \rangle = \begin{cases} \pi & \text{if } p = q, \, p, q \geq 1, \\ 0 & \text{if } p \neq q, \, p, q \geq 1, \end{cases}$$

$$\langle \cos px, \cos qx \rangle = \begin{cases} \pi & \text{if } p = q, \ p, q \geq 1, \\ 0 & \text{if } p \neq q, \ p, q \geq 0, \end{cases}$$

and

$$\langle \sin px, \cos qx \rangle = 0,$$

for all $p \geq 1$ and $q \geq 0$, and of course, $\langle 1, 1 \rangle = \int_{-\pi}^{\pi} dx = 2\pi$.

As a consequence, the family $(\sin px)_{p \geq 1} \cup (\cos qx)_{q \geq 0}$ is orthogonal. It is not orthonormal, but becomes so if we divide every trigonometric function by $\sqrt{\pi}$, and 1 by $\sqrt{2\pi}$.

Proposition 11.4. *Given a Euclidean space E, for any family $(u_i)_{i \in I}$ of nonnull vectors in E, if $(u_i)_{i \in I}$ is orthogonal, then it is linearly independent.*

Proof. Assume there is a linear dependence

$$\sum_{j \in J} \lambda_j u_j = 0$$

for some $\lambda_j \in \mathbb{R}$ and some finite subset J of I. By taking the inner product with u_i for any $i \in J$, and using the the bilinearity of the inner product and the fact that $u_i \cdot u_j = 0$ whenever $i \neq j$, we get

$$0 = u_i \cdot 0 = u_i \cdot \left(\sum_{j \in J} \lambda_j u_j \right)$$

$$= \sum_{j \in J} \lambda_j (u_i \cdot u_j) = \lambda_i (u_i \cdot u_i),$$

so

$$\lambda_i (u_i \cdot u_i) = 0, \qquad \text{for all } i \in J,$$

and since $u_i \neq 0$ and an inner product is positive definite, $u_i \cdot u_i \neq 0$, so we obtain

$$\lambda_i = 0, \qquad \text{for all } i \in J,$$

which shows that the family $(u_i)_{i \in I}$ is linearly independent. $\square$

We leave the following simple result as an exercise.

Proposition 11.5. *Given a Euclidean space E, any two vectors $u, v \in E$ are orthogonal iff*

$$\|u + v\|^2 = \|u\|^2 + \|v\|^2.$$

See Figure 11.2 for a geometrical interpretation.

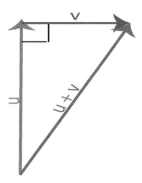

Fig. 11.2 The sum of the lengths of the two sides of a right triangle is equal to the length of the hypotenuse; i.e. the Pythagorean theorem.

One of the most useful features of orthonormal bases is that they afford a very simple method for computing the coordinates of a vector over any basis vector. Indeed, assume that $(e_1, \ldots, e_m)$ is an orthonormal basis. For any vector

$$x = x_1 e_1 + \cdots + x_m e_m,$$

if we compute the inner product $x \cdot e_i$, we get

$$x \cdot e_i = x_1 e_1 \cdot e_i + \cdots + x_i e_i \cdot e_i + \cdots + x_m e_m \cdot e_i = x_i,$$

since

$$e_i \cdot e_j = \begin{cases} 1 & \text{if } i = j, \\ 0 & \text{if } i \neq j \end{cases}$$

is the property characterizing an orthonormal family. Thus,

$$x_i = x \cdot e_i,$$

which means that $x_i e_i = (x \cdot e_i) e_i$ is the orthogonal projection of x onto the subspace generated by the basis vector e_i. See Figure 11.3. If the basis is orthogonal but not necessarily orthonormal, then

$$x_i = \frac{x \cdot e_i}{e_i \cdot e_i} = \frac{x \cdot e_i}{\|e_i\|^2}.$$

All this is true even for an infinite orthonormal (or orthogonal) basis $(e_i)_{i \in I}$.

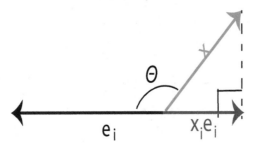

Fig. 11.3   The orthogonal projection of the red vector x onto the black basis vector e_i is the maroon vector $x_i e_i$. Observe that $x \cdot e_i = \|x\| \cos\theta$.

 However, remember that every vector x is expressed as a linear combination

$$x = \sum_{i \in I} x_i e_i$$

where the family of scalars $(x_i)_{i \in I}$ has **finite support**, which means that $x_i = 0$ for all $i \in I - J$, where J is a finite set. Thus, even though the family $(\sin px)_{p \geq 1} \cup (\cos qx)_{q \geq 0}$ is orthogonal (it is not orthonormal, but becomes so if we divide every trigonometric function by $\sqrt{\pi}$, and 1 by $\sqrt{2\pi}$; we won't because it looks messy!), the fact that a function $f \in \mathcal{C}^0[-\pi, \pi]$ can be written as a Fourier series as

$$f(x) = a_0 + \sum_{k=1}^{\infty} (a_k \cos kx + b_k \sin kx)$$

does not mean that $(\sin px)_{p \geq 1} \cup (\cos qx)_{q \geq 0}$ is a basis of this vector space of functions, because in general, the families (a_k) and (b_k) **do not** have finite support! In order for this infinite linear combination to make sense, it is necessary to prove that the partial sums

$$a_0 + \sum_{k=1}^{n} (a_k \cos kx + b_k \sin kx)$$

of the series converge to a limit when n goes to infinity. This requires a topology on the space.

A very important property of Euclidean spaces of finite dimension is that the inner product induces a canonical bijection (i.e., independent of the choice of bases) between the vector space E and its dual E^*. The reason is that an inner product $\cdot : E \times E \to \mathbb{R}$ defines a nondegenerate pairing, as defined in Definition 10.4. Indeed, if $u \cdot v = 0$ for all $v \in E$ then $u = 0$, and similarly if $u \cdot v = 0$ for all $u \in E$ then $v = 0$ (since an inner product is positive definite and symmetric). By Proposition 10.5, there is a canonical isomorphism between E and E^*. We feel that the reader will appreciate if we exhibit this mapping explicitly and reprove that it is an isomorphism.

The mapping from E to E^* is defined as follows.

Definition 11.3. For any vector $u \in E$, let $\varphi_u \colon E \to \mathbb{R}$ be the map defined such that

$$\varphi_u(v) = u \cdot v, \quad \text{for all } v \in E.$$

Since the inner product is bilinear, the map φ_u is a linear form in E^*. Thus, we have a map $\flat \colon E \to E^*$, defined such that

$$\flat(u) = \varphi_u.$$

Theorem 11.1. *Given a Euclidean space E, the map $\flat \colon E \to E^*$ defined such that*

$$\flat(u) = \varphi_u$$

is linear and injective. When E is also of finite dimension, the map $\flat \colon E \to E^*$ is a canonical isomorphism.

Proof. That $\flat \colon E \to E^*$ is a linear map follows immediately from the fact that the inner product is bilinear. If $\varphi_u = \varphi_v$, then $\varphi_u(w) = \varphi_v(w)$ for all $w \in E$, which by definition of φ_u means that $u \cdot w = v \cdot w$ for all $w \in E$, which by bilinearity is equivalent to

$$(v - u) \cdot w = 0$$

for all $w \in E$, which implies that $u = v$, since the inner product is positive definite. Thus, $\flat \colon E \to E^*$ is injective. Finally, when E is of finite dimension n, we know that E^* is also of dimension n, and then $\flat \colon E \to E^*$ is bijective. $\square$

The inverse of the isomorphism $\flat \colon E \to E^*$ is denoted by $\sharp \colon E^* \to E$.

As a consequence of Theorem 11.1 we have the following corollary.

Corollary 11.1. *If E is a Euclidean space of finite dimension, every linear form $f \in E^*$ corresponds to a unique $u \in E$ such that*

$$f(v) = u \cdot v, \quad \text{for every } v \in E.$$

In particular, if f is not the zero form, the kernel of f, which is a hyperplane H, is precisely the set of vectors that are orthogonal to u.

Remarks:

(1) The "musical map" $\flat \colon E \to E^*$ is not surjective when E has infinite dimension. The result can be salvaged by restricting our attention to continuous linear maps, and by assuming that the vector space E is a *Hilbert space* (i.e., E is a complete normed vector space w.r.t. the Euclidean norm). This is the famous "little" Riesz theorem (or Riesz representation theorem).

(2) Theorem 11.1 still holds if the inner product on E is replaced by a nondegenerate symmetric bilinear form φ. We say that a symmetric bilinear form $\varphi \colon E \times E \to \mathbb{R}$ is *nondegenerate* if for every $u \in E$,

$$\text{if} \quad \varphi(u, v) = 0 \quad \text{for all } v \in E, \quad \text{then} \quad u = 0.$$

For example, the symmetric bilinear form on $\mathbb{R}^4$ (the Lorentz form) defined such that

$$\varphi((x_1, x_2, x_3, x_4), (y_1, y_2, y_3, y_4)) = x_1 y_1 + x_2 y_2 + x_3 y_3 - x_4 y_4$$

is nondegenerate. However, there are nonnull vectors $u \in \mathbb{R}^4$ such that $\varphi(u, u) = 0$, which is impossible in a Euclidean space. Such vectors are called *isotropic*.

Example 11.6. Consider $\mathbb{R}^n$ with its usual Euclidean inner product. Given any differentiable function $f \colon U \to \mathbb{R}$, where U is some open subset of $\mathbb{R}^n$, by definition, for any $x \in U$, the *total derivative* df_x of f at x is the linear form defined so that for all $u = (u_1, \dots, u_n) \in \mathbb{R}^n$,

$$df_x(u) = \left(\frac{\partial f}{\partial x_1}(x) \; \cdots \; \frac{\partial f}{\partial x_n}(x) \right) \begin{pmatrix} u_1 \\ \vdots \\ u_n \end{pmatrix} = \sum_{i=1}^{n} \frac{\partial f}{\partial x_i}(x) \, u_i.$$

The unique vector $v \in \mathbb{R}^n$ such that

$$v \cdot u = df_x(u) \quad \text{for all } u \in \mathbb{R}^n$$

is the transpose of the *Jacobian matrix* of f at x, the $1 \times n$ matrix

$$\left(\frac{\partial f}{\partial x_1}(x) \cdots \frac{\partial f}{\partial x_n}(x) \right).$$

This is the *gradient* $\operatorname{grad}(f)_x$ of f at x, given by

$$\operatorname{grad}(f)_x = \begin{pmatrix} \dfrac{\partial f}{\partial x_1}(x) \\ \vdots \\ \dfrac{\partial f}{\partial x_n}(x) \end{pmatrix}.$$

Example 11.7. Given any two vectors $u, v \in \mathbb{R}^3$, let $c(u, v)$ be the linear form given by

$$c(u, v)(w) = \det(u, v, w) \quad \text{for all } w \in \mathbb{R}^3.$$

Since

$$\det(u, v, w) = \begin{vmatrix} u_1 & v_1 & w_1 \\ u_2 & v_2 & w_2 \\ u_3 & v_3 & w_3 \end{vmatrix} = w_1 \begin{vmatrix} u_2 & v_2 \\ u_3 & v_3 \end{vmatrix} - w_2 \begin{vmatrix} u_1 & v_1 \\ u_3 & v_3 \end{vmatrix} + w_3 \begin{vmatrix} u_1 & v_1 \\ u_2 & v_2 \end{vmatrix}$$

$$= w_1(u_2 v_3 - u_3 v_2) + w_2(u_3 v_1 - u_1 v_3) + w_3(u_1 v_2 - u_2 v_1),$$

we see that the unique vector $z \in \mathbb{R}^3$ such that

$$z \cdot w = c(u, v)(w) = \det(u, v, w) \quad \text{for all } w \in \mathbb{R}^3$$

is the vector

$$z = \begin{pmatrix} u_2 v_3 - u_3 v_2 \\ u_3 v_1 - u_1 v_3 \\ u_1 v_2 - u_2 v_1 \end{pmatrix}.$$

This is just the *cross-product* $u \times v$ of u and v. Since $\det(u, v, u) = \det(u, v, v) = 0$, we see that $u \times v$ is orthogonal to both u and v. The above allows us to generalize the cross-product to $\mathbb{R}^n$. Given any $n - 1$ vectors $u_1, \ldots, u_{n-1} \in \mathbb{R}^n$, the cross-product $u_1 \times \cdots \times u_{n-1}$ is the unique vector in $\mathbb{R}^n$ such that

$$(u_1 \times \cdots \times u_{n-1}) \cdot w = \det(u_1, \ldots, u_{n-1}, w) \quad \text{for all } w \in \mathbb{R}^n.$$

Example 11.8. Consider the vector space $M_n(\mathbb{R})$ of real $n \times n$ matrices with the inner product

$$\langle A, B \rangle = \operatorname{tr}(A^\top B).$$

Let $s\colon M_n(\mathbb{R}) \to \mathbb{R}$ be the function given by

$$s(A) = \sum_{i,j=1}^{n} a_{ij},$$

where $A = (a_{ij})$. It is immediately verified that s is a linear form. It is easy to check that the unique matrix Z such that

$$\langle Z, A \rangle = s(A) \quad \text{for all } A \in M_n(\mathbb{R})$$

is the matrix $Z = \mathbf{ones}(n, n)$ whose entries are all equal to 1.

11.3   Adjoint of a Linear Map

The existence of the isomorphism $\flat\colon E \to E^*$ is crucial to the existence of adjoint maps. The importance of adjoint maps stems from the fact that the linear maps arising in physical problems are often self-adjoint, which means that $f = f^*$. Moreover, self-adjoint maps can be diagonalized over orthonormal bases of eigenvectors. This is the key to the solution of many problems in mechanics and engineering in general (see Strang [Strang (1986)]).

Let E be a Euclidean space of finite dimension n, and let $f\colon E \to E$ be a linear map. For every $u \in E$, the map

$$v \mapsto u \cdot f(v)$$

is clearly a linear form in E^*, and by Theorem 11.1, there is a unique vector in E denoted by $f^*(u)$ such that

$$f^*(u) \cdot v = u \cdot f(v),$$

for every $v \in E$. The following simple proposition shows that the map f^* is linear.

Proposition 11.6. *Given a Euclidean space E of finite dimension, for every linear map $f\colon E \to E$, there is a unique linear map $f^*\colon E \to E$ such that*

$$f^*(u) \cdot v = u \cdot f(v), \quad \text{for all } u, v \in E.$$

Proof. Given $u_1, u_2 \in E$, since the inner product is bilinear, we have

$$(u_1 + u_2) \cdot f(v) = u_1 \cdot f(v) + u_2 \cdot f(v),$$

for all $v \in E$, and

$$(f^*(u_1) + f^*(u_2)) \cdot v = f^*(u_1) \cdot v + f^*(u_2) \cdot v,$$

for all $v \in E$, and since by assumption,

$$f^*(u_1) \cdot v = u_1 \cdot f(v) \quad \text{and} \quad f^*(u_2) \cdot v = u_2 \cdot f(v),$$

for all $v \in E$. Thus we get

$$(f^*(u_1) + f^*(u_2)) \cdot v = (u_1 + u_2) \cdot f(v) = f^*(u_1 + u_2) \cdot v,$$

for all $v \in E$. Since $\flat$ is bijective, this implies that

$$f^*(u_1 + u_2) = f^*(u_1) + f^*(u_2).$$

Similarly,

$$(\lambda u) \cdot f(v) = \lambda(u \cdot f(v)),$$

for all $v \in E$, and

$$(\lambda f^*(u)) \cdot v = \lambda(f^*(u) \cdot v),$$

for all $v \in E$, and since by assumption,

$$f^*(u) \cdot v = u \cdot f(v),$$

for all $v \in E$, we get

$$(\lambda f^*(u)) \cdot v = \lambda(u \cdot f(v)) = (\lambda u) \cdot f(v) = f^*(\lambda u) \cdot v$$

for all $v \in E$. Since $\flat$ is bijective, this implies that

$$f^*(\lambda u) = \lambda f^*(u).$$

Thus, f^* is indeed a linear map, and it is unique since $\flat$ is a bijection. $\quad\square$

Definition 11.4. Given a Euclidean space E of finite dimension, for every linear map $f \colon E \to E$, the unique linear map $f^* \colon E \to E$ such that

$$f^*(u) \cdot v = u \cdot f(v), \quad \text{for all } u, v \in E$$

given by Proposition 11.6 is called the *adjoint of* f *(w.r.t. to the inner product)*. Linear maps $f \colon E \to E$ such that $f = f^*$ are called *self-adjoint* maps.

Self-adjoint linear maps play a very important role because they have real eigenvalues, and because orthonormal bases arise from their eigenvectors. Furthermore, many physical problems lead to self-adjoint linear maps (in the form of symmetric matrices).

Remark: Proposition 11.6 still holds if the inner product on E is replaced by a nondegenerate symmetric bilinear form φ.

Linear maps such that $f^{-1} = f^*$, or equivalently
$$f^* \circ f = f \circ f^* = \mathrm{id},$$
also play an important role. They are *linear isometries*, or *isometries*. Rotations are special kinds of isometries. Another important class of linear maps are the linear maps satisfying the property
$$f^* \circ f = f \circ f^*,$$
called *normal linear maps*. We will see later on that normal maps can always be diagonalized over orthonormal bases of eigenvectors, but this will require using a Hermitian inner product (over $\mathbb{C}$).

Given two Euclidean spaces E and F, where the inner product on E is denoted by $\langle -, - \rangle_1$ and the inner product on F is denoted by $\langle -, - \rangle_2$, given any linear map $f \colon E \to F$, it is immediately verified that the proof of Proposition 11.6 can be adapted to show that there is a unique linear map $f^* \colon F \to E$ such that
$$\langle f(u), v \rangle_2 = \langle u, f^*(v) \rangle_1$$
for all $u \in E$ and all $v \in F$. The linear map f^* is also called the *adjoint of* f.

The following properties immediately follow from the definition of the adjoint map:

(1) For any linear map $f \colon E \to F$, we have
$$f^{**} = f.$$
(2) For any two linear maps $f, g \colon E \to F$ and any scalar $\lambda \in \mathbb{R}$:
$$(f + g)^* = f^* + g^*$$
$$(\lambda f)^* = \lambda f^*.$$
(3) If E, F, G are Euclidean spaces with respective inner products $\langle -, - \rangle_1, \langle -, - \rangle_2$, and $\langle -, - \rangle_3$, and if $f \colon E \to F$ and $g \colon F \to G$ are two linear maps, then
$$(g \circ f)^* = f^* \circ g^*.$$

Remark: Given any basis for E and any basis for F, it is possible to characterize the matrix of the adjoint f^* of f in terms of the matrix of f and the Gram matrices defining the inner products; see Problem 11.5. We will do so with respect to orthonormal bases in Proposition 11.12(2). Also, since inner products are symmetric, the adjoint f^* of f is also characterized by
$$f(u) \cdot v = u \cdot f^*(v),$$
for all $u, v \in E$.

11.4 Existence and Construction of Orthonormal Bases

We can also use Theorem 11.1 to show that any Euclidean space of finite dimension has an orthonormal basis.

Proposition 11.7. *Given any nontrivial Euclidean space E of finite dimension $n \geq 1$, there is an orthonormal basis $(u_1, \ldots, u_n)$ for E.*

Proof. We proceed by induction on n. When $n = 1$, take any nonnull vector $v \in E$, which exists since we assumed E nontrivial, and let

$$u = \frac{v}{\|v\|}.$$

If $n \geq 2$, again take any nonnull vector $v \in E$, and let

$$u_1 = \frac{v}{\|v\|}.$$

Consider the linear form φ_{u_1} associated with u_1. Since $u_1 \neq 0$, by Theorem 11.1, the linear form φ_{u_1} is nonnull, and its kernel is a hyperplane H. Since $\varphi_{u_1}(w) = 0$ iff $u_1 \cdot w = 0$, the hyperplane H is the orthogonal complement of $\{u_1\}$. Furthermore, since $u_1 \neq 0$ and the inner product is positive definite, $u_1 \cdot u_1 \neq 0$, and thus, $u_1 \notin H$, which implies that $E = H \oplus \mathbb{R}u_1$. However, since E is of finite dimension n, the hyperplane H has dimension $n - 1$, and by the induction hypothesis, we can find an orthonormal basis $(u_2, \ldots, u_n)$ for H. Now because H and the one dimensional space $\mathbb{R}u_1$ are orthogonal and $E = H \oplus \mathbb{R}u_1$, it is clear that $(u_1, \ldots, u_n)$ is an orthonormal basis for E. $\square$

As a consequence of Proposition 11.7, given any Euclidean space of finite dimension n, if $(e_1, \ldots, e_n)$ is an orthonormal basis for E, then for any two vectors $u = u_1 e_1 + \cdots + u_n e_n$ and $v = v_1 e_1 + \cdots + v_n e_n$, the inner product $u \cdot v$ is expressed as

$$u \cdot v = (u_1 e_1 + \cdots + u_n e_n) \cdot (v_1 e_1 + \cdots + v_n e_n) = \sum_{i=1}^{n} u_i v_i,$$

and the norm $\|u\|$ as

$$\|u\| = \|u_1 e_1 + \cdots + u_n e_n\| = \left(\sum_{i=1}^{n} u_i^2 \right)^{1/2}.$$

The fact that a Euclidean space always has an orthonormal basis implies that any Gram matrix G can be written as

$$G = Q^\top Q,$$

for some invertible matrix Q. Indeed, we know that in a change of basis matrix, a Gram matrix G becomes $G' = P^\top GP$. If the basis corresponding to G' is orthonormal, then $G' = I$, so $G = (P^{-1})^\top P^{-1}$.

There is a more constructive way of proving Proposition 11.7, using a procedure known as the *Gram–Schmidt orthonormalization procedure*. Among other things, the Gram–Schmidt orthonormalization procedure yields the *QR-decomposition for matrices*, an important tool in numerical methods.

Proposition 11.8. *Given any nontrivial Euclidean space E of finite dimension $n \geq 1$, from any basis $(e_1, \ldots, e_n)$ for E we can construct an orthonormal basis $(u_1, \ldots, u_n)$ for E, with the property that for every k, $1 \leq k \leq n$, the families $(e_1, \ldots, e_k)$ and $(u_1, \ldots, u_k)$ generate the same subspace.*

Proof. We proceed by induction on n. For $n = 1$, let

$$u_1 = \frac{e_1}{\|e_1\|}.$$

For $n \geq 2$, we also let

$$u_1 = \frac{e_1}{\|e_1\|},$$

and assuming that $(u_1, \ldots, u_k)$ is an orthonormal system that generates the same subspace as $(e_1, \ldots, e_k)$, for every k with $1 \leq k < n$, we note that the vector

$$u'_{k+1} = e_{k+1} - \sum_{i=1}^{k} (e_{k+1} \cdot u_i)\, u_i$$

is nonnull, since otherwise, because $(u_1, \ldots, u_k)$ and $(e_1, \ldots, e_k)$ generate the same subspace, $(e_1, \ldots, e_{k+1})$ would be linearly dependent, which is absurd, since $(e_1, \ldots, e_n)$ is a basis. Thus, the norm of the vector u'_{k+1} being nonzero, we use the following construction of the vectors u_k and u'_k:

$$u'_1 = e_1, \qquad u_1 = \frac{u'_1}{\|u'_1\|},$$

and for the inductive step

$$u'_{k+1} = e_{k+1} - \sum_{i=1}^{k} (e_{k+1} \cdot u_i)\, u_i, \qquad u_{k+1} = \frac{u'_{k+1}}{\|u'_{k+1}\|},$$

where $1 \leq k \leq n - 1$. It is clear that $\|u_{k+1}\| = 1$, and since $(u_1, \ldots, u_k)$ is an orthonormal system, we have

$$u'_{k+1} \cdot u_i = e_{k+1} \cdot u_i - (e_{k+1} \cdot u_i)u_i \cdot u_i = e_{k+1} \cdot u_i - e_{k+1} \cdot u_i = 0,$$

for all i with $1 \leq i \leq k$. This shows that the family $(u_1, \ldots, u_{k+1})$ is orthonormal, and since $(u_1, \ldots, u_k)$ and $(e_1, \ldots, e_k)$ generates the same subspace, it is clear from the definition of u_{k+1} that $(u_1, \ldots, u_{k+1})$ and $(e_1, \ldots, e_{k+1})$ generate the same subspace. This completes the induction step and the proof of the proposition. □

Note that u'_{k+1} is obtained by subtracting from e_{k+1} the projection of e_{k+1} itself onto the orthonormal vectors $u_1, \ldots, u_k$ that have already been computed. Then u'_{k+1} is normalized.

Example 11.9. For a specific example of this procedure, let $E = \mathbb{R}^3$ with the standard Euclidean norm. Take the basis

$$e_1 = \begin{pmatrix} 1 \\ 1 \\ 1 \end{pmatrix} \qquad e_2 = \begin{pmatrix} 1 \\ 0 \\ 1 \end{pmatrix} \qquad e_3 = \begin{pmatrix} 1 \\ 1 \\ 0 \end{pmatrix}.$$

Then

$$u_1 = \frac{1}{\sqrt{3}} \begin{pmatrix} 1 \\ 1 \\ 1 \end{pmatrix},$$

and

$$u'_2 = e_2 - (e_2 \cdot u_1)u_1 = \begin{pmatrix} 1 \\ 0 \\ 1 \end{pmatrix} - \frac{2}{3} \begin{pmatrix} 1 \\ 1 \\ 1 \end{pmatrix} = \frac{1}{3} \begin{pmatrix} 1 \\ -2 \\ 1 \end{pmatrix}.$$

This implies that

$$u_2 = \frac{1}{\sqrt{6}} \begin{pmatrix} 1 \\ -2 \\ 1 \end{pmatrix},$$

and that

$$u'_3 = e_3 - (e_3 \cdot u_1)u_1 - (e_3 \cdot u_2)u_2 = \begin{pmatrix} 1 \\ 1 \\ 0 \end{pmatrix} - \frac{2}{3} \begin{pmatrix} 1 \\ 1 \\ 1 \end{pmatrix} + \frac{1}{6} \begin{pmatrix} 1 \\ -2 \\ 1 \end{pmatrix} = \frac{1}{2} \begin{pmatrix} 1 \\ 0 \\ -1 \end{pmatrix}.$$

To complete the orthonormal basis, normalize u'_3 to obtain

$$u_3 = \frac{1}{\sqrt{2}} \begin{pmatrix} 1 \\ 0 \\ -1 \end{pmatrix}.$$

An illustration of this example is provided by Figure 11.4.

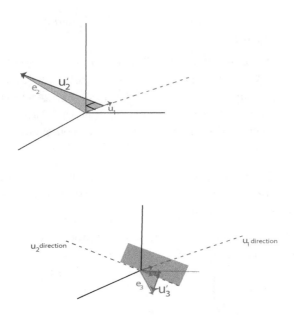

Fig. 11.4   The top figure shows the construction of the blue u'_2 as perpendicular to the orthogonal projection of e_2 onto u_1, while the bottom figure shows the construction of the green u'_3 as normal to the plane determined by u_1 and u_2.

Remarks:

(1) The QR-decomposition can now be obtained very easily, but we postpone this until Section 11.6.
(2) The proof of Proposition 11.8 also works for a countably infinite basis for E, producing a countably infinite orthonormal basis.

It should also be said that the Gram–Schmidt orthonormalization procedure that we have presented is not very stable numerically, and instead, one should use the *modified Gram–Schmidt method*. To compute u'_{k+1}, instead of projecting e_{k+1} onto $u_1, \ldots, u_k$ in a single step, it is better to perform k projections. We compute $u_1^{k+1}, u_2^{k+1}, \ldots, u_k^{k+1}$ as follows:

$$u_1^{k+1} = e_{k+1} - (e_{k+1} \cdot u_1)\, u_1,$$

$$u_{i+1}^{k+1} = u_i^{k+1} - (u_i^{k+1} \cdot u_{i+1})\, u_{i+1},$$

where $1 \leq i \leq k - 1$. It is easily shown that $u'_{k+1} = u_k^{k+1}$.

Example 11.10. Let us apply the modified Gram–Schmidt method to the

(e_1, e_2, e_3) basis of Example 11.9. The only change is the computation of u_3'. For the modified Gram–Schmidt procedure, we first calculate

$$u_1^3 = e_3 - (e_3 \cdot u_1)u_1 = \begin{pmatrix} 1 \\ 1 \\ 0 \end{pmatrix} - \frac{2}{3}\begin{pmatrix} 1 \\ 1 \\ 1 \end{pmatrix} = \frac{1}{3}\begin{pmatrix} 1 \\ 1 \\ -2 \end{pmatrix}.$$

Then

$$u_2^3 = u_1^3 - (u_1^3 \cdot u_2)u_2 = \frac{1}{3}\begin{pmatrix} 1 \\ 1 \\ -2 \end{pmatrix} + \frac{1}{6}\begin{pmatrix} 1 \\ -2 \\ 1 \end{pmatrix} = \frac{1}{2}\begin{pmatrix} 1 \\ 0 \\ -1 \end{pmatrix},$$

and observe that $u_2^3 = u_3'$. See Figure 11.5.

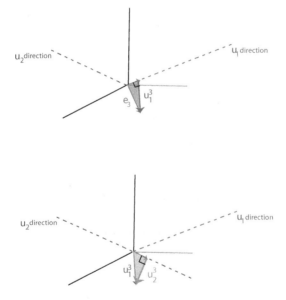

Fig. 11.5 The top figure shows the construction of the blue u_1^3 as perpendicular to the orthogonal projection of e_3 onto u_1, while the bottom figure shows the construction of the sky blue u_2^3 as perpendicular to the orthogonal projection of u_1^3 onto u_2.

The following `Matlab` program implements the modified Gram–Schmidt procedure.

```
function q = gramschmidt4(e)
n = size(e,1);
```

```
for i = 1:n
    q(:,i) = e(:,i);
    for j = 1:i-1
        r = q(:,j)'*q(:,i);
        q(:,i) = q(:,i) - r*q(:,j);
    end
    r = sqrt(q(:,i)'*q(:,i));
    q(:,i) = q(:,i)/r;
end
end
```

If we apply the above function to the matrix

$$\begin{pmatrix} 1\ 1\ 1 \\ 1\ 0\ 1 \\ 1\ 1\ 0 \end{pmatrix},$$

the output is the matrix

$$\begin{pmatrix} 0.5774 & 0.4082 & 0.7071 \\ 0.5774 & -0.8165 & -0.0000 \\ 0.5774 & 0.4082 & -0.7071 \end{pmatrix},$$

which matches the result of Example 11.9.

Example 11.11. If we consider polynomials and the inner product

$$\langle f, g \rangle = \int_{-1}^{1} f(t)g(t)dt,$$

applying the Gram–Schmidt orthonormalization procedure to the polynomials

$$1, x, x^2, \ldots, x^n, \ldots,$$

which form a basis of the polynomials in one variable with real coefficients, we get a family of orthonormal polynomials $Q_n(x)$ related to the *Legendre polynomials*.

The Legendre polynomials $P_n(x)$ have many nice properties. They are orthogonal, but their norm is not always 1. The Legendre polynomials $P_n(x)$ can be defined as follows. Letting f_n be the function

$$f_n(x) = (x^2 - 1)^n,$$

we define $P_n(x)$ as follows:

$$P_0(x) = 1, \quad \text{and} \quad P_n(x) = \frac{1}{2^n n!} f_n^{(n)}(x),$$

where $f_n^{(n)}$ is the nth derivative of f_n.

They can also be defined inductively as follows:

$$P_0(x) = 1,$$
$$P_1(x) = x,$$
$$P_{n+1}(x) = \frac{2n+1}{n+1} x P_n(x) - \frac{n}{n+1} P_{n-1}(x).$$

Here is an explicit summation for $P_n(x)$:

$$P_n(x) = \frac{1}{2^n} \sum_{k=0}^{\lfloor n/2 \rfloor} (-1)^k \binom{n}{k} \binom{2n-2k}{n} x^{n-2k}.$$

The polynomials Q_n are related to the Legendre polynomials P_n as follows:

$$Q_n(x) = \sqrt{\frac{2n+1}{2}} P_n(x).$$

Example 11.12. Consider polynomials over $[-1, 1]$, with the symmetric bilinear form

$$\langle f, g \rangle = \int_{-1}^{1} \frac{1}{\sqrt{1-t^2}} f(t) g(t) dt.$$

We leave it as an exercise to prove that the above defines an inner product. It can be shown that the polynomials $T_n(x)$ given by

$$T_n(x) = \cos(n \arccos x), \quad n \geq 0,$$

(equivalently, with $x = \cos\theta$, we have $T_n(\cos\theta) = \cos(n\theta)$) are orthogonal with respect to the above inner product. These polynomials are the *Chebyshev polynomials*. Their norm is not equal to 1. Instead, we have

$$\langle T_n, T_n \rangle = \begin{cases} \frac{\pi}{2} & \text{if } n > 0, \\ \pi & \text{if } n = 0. \end{cases}$$

Using the identity $(\cos\theta + i\sin\theta)^n = \cos n\theta + i\sin n\theta$ and the binomial formula, we obtain the following expression for $T_n(x)$:

$$T_n(x) = \sum_{k=0}^{\lfloor n/2 \rfloor} \binom{n}{2k} (x^2 - 1)^k x^{n-2k}.$$

The Chebyshev polynomials are defined inductively as follows:

$$T_0(x) = 1$$
$$T_1(x) = x$$
$$T_{n+1}(x) = 2x T_n(x) - T_{n-1}(x), \quad n \geq 1.$$

Using these recurrence equations, we can show that

$$T_n(x) = \frac{(x - \sqrt{x^2 - 1})^n + (x + \sqrt{x^2 - 1})^n}{2}.$$

The polynomial T_n has n distinct roots in the interval $[-1, 1]$. The Chebyshev polynomials play an important role in approximation theory. They are used as an approximation to a best polynomial approximation of a continuous function under the sup-norm (∞-norm).

The inner products of the last two examples are special cases of an inner product of the form

$$\langle f, g \rangle = \int_{-1}^{1} W(t) f(t) g(t) dt,$$

where $W(t)$ is a *weight function*. If W is a nonzero continuous function such that $W(x) \geq 0$ on $(-1, 1)$, then the above bilinear form is indeed positive definite. Families of orthogonal polynomials used in approximation theory and in physics arise by a suitable choice of the weight function W. Besides the previous two examples, the *Hermite polynomials* correspond to $W(x) = e^{-x^2}$, the *Laguerre polynomials* to $W(x) = e^{-x}$, and the *Jacobi polynomials* to $W(x) = (1 - x)^\alpha (1 + x)^\beta$, with $\alpha, \beta > -1$. Comprehensive treatments of orthogonal polynomials can be found in Lebedev [Lebedev (1972)], Sansone [Sansone (1991)], and Andrews, Askey and Roy [Andrews *et al.* (2000)].

We can also prove the following proposition regarding orthogonal spaces.

Proposition 11.9. *Given any nontrivial Euclidean space E of finite dimension $n \geq 1$, for any subspace F of dimension k, the orthogonal complement $F^\perp$ of F has dimension $n - k$, and $E = F \oplus F^\perp$. Furthermore, we have $F^{\perp\perp} = F$.*

Proof. From Proposition 11.7, the subspace F has some orthonormal basis $(u_1, \ldots, u_k)$. This linearly independent family $(u_1, \ldots, u_k)$ can be extended to a basis $(u_1, \ldots, u_k, v_{k+1}, \ldots, v_n)$, and by Proposition 11.8, it can be converted to an orthonormal basis $(u_1, \ldots, u_n)$, which contains $(u_1, \ldots, u_k)$ as an orthonormal basis of F. Now any vector $w = w_1 u_1 + \cdots + w_n u_n \in E$ is orthogonal to F iff $w \cdot u_i = 0$, for every i, where $1 \leq i \leq k$, iff $w_i = 0$ for every i, where $1 \leq i \leq k$. Clearly, this shows that $(u_{k+1}, \ldots, u_n)$ is a basis of $F^\perp$, and thus $E = F \oplus F^\perp$, and $F^\perp$ has dimension $n - k$. Similarly, any vector $w = w_1 u_1 + \cdots + w_n u_n \in E$ is orthogonal to $F^\perp$ iff $w \cdot u_i = 0$, for every i, where $k + 1 \leq i \leq n$, iff $w_i = 0$ for every i, where $k + 1 \leq i \leq n$. Thus, $(u_1, \ldots, u_k)$ is a basis of $F^{\perp\perp}$, and $F^{\perp\perp} = F$. $\qquad\square$

11.5   Linear Isometries (Orthogonal Transformations)

In this section we consider linear maps between Euclidean spaces that preserve the Euclidean norm. These transformations, sometimes called *rigid motions*, play an important role in geometry.

Definition 11.5. Given any two nontrivial Euclidean spaces E and F of the same finite dimension n, a function $f \colon E \to F$ is *an orthogonal transformation, or a linear isometry*, if it is linear and

$$\|f(u)\| = \|u\|, \quad \text{for all } u \in E.$$

Remarks:

(1) A linear isometry is often defined as a linear map such that

$$\|f(v) - f(u)\| = \|v - u\|,$$

for all $u, v \in E$. Since the map f is linear, the two definitions are equivalent. The second definition just focuses on preserving the distance between vectors.

(2) Sometimes, a linear map satisfying the condition of Definition 11.5 is called a *metric map*, and a linear isometry is defined as a *bijective* metric map.

An isometry (without the word linear) is sometimes defined as a function $f \colon E \to F$ (not necessarily linear) such that

$$\|f(v) - f(u)\| = \|v - u\|,$$

for all $u, v \in E$, i.e., as a function that preserves the distance. This requirement turns out to be very strong. Indeed, the next proposition shows that all these definitions are equivalent when E and F are of finite dimension, and for functions such that $f(0) = 0$.

Proposition 11.10. *Given any two nontrivial Euclidean spaces E and F of the same finite dimension n, for every function $f \colon E \to F$, the following properties are equivalent:*

(1) f is a linear map and $\|f(u)\| = \|u\|$, for all $u \in E$;
(2) $\|f(v) - f(u)\| = \|v - u\|$, for all $u, v \in E$, and $f(0) = 0$;
(3) $f(u) \cdot f(v) = u \cdot v$, for all $u, v \in E$.

Furthermore, such a map is bijective.

Proof. Clearly, (1) implies (2), since in (1) it is assumed that f is linear.

Assume that (2) holds. In fact, we shall prove a slightly stronger result. We prove that if

$$\|f(v) - f(u)\| = \|v - u\|$$

for all $u, v \in E$, then for any vector $\tau \in E$, the function $g \colon E \to F$ defined such that

$$g(u) = f(\tau + u) - f(\tau)$$

for all $u \in E$ is a linear map such that $g(0) = 0$ and (3) holds. Clearly, $g(0) = f(\tau) - f(\tau) = 0$.

Note that from the hypothesis

$$\|f(v) - f(u)\| = \|v - u\|$$

for all $u, v \in E$, we conclude that

$$
\begin{aligned}
\|g(v) - g(u)\| &= \|f(\tau + v) - f(\tau) - (f(\tau + u) - f(\tau))\|, \\
&= \|f(\tau + v) - f(\tau + u)\|, \\
&= \|\tau + v - (\tau + u)\|, \\
&= \|v - u\|,
\end{aligned}
$$

for all $u, v \in E$. Since $g(0) = 0$, by setting $u = 0$ in

$$\|g(v) - g(u)\| = \|v - u\|,$$

we get

$$\|g(v)\| = \|v\|$$

for all $v \in E$. In other words, g preserves both the distance and the norm.

To prove that g preserves the inner product, we use the simple fact that

$$2u \cdot v = \|u\|^2 + \|v\|^2 - \|u - v\|^2$$

for all $u, v \in E$. Then since g preserves distance and norm, we have

$$
\begin{aligned}
2g(u) \cdot g(v) &= \|g(u)\|^2 + \|g(v)\|^2 - \|g(u) - g(v)\|^2 \\
&= \|u\|^2 + \|v\|^2 - \|u - v\|^2 \\
&= 2u \cdot v,
\end{aligned}
$$

and thus $g(u) \cdot g(v) = u \cdot v$, for all $u, v \in E$, which is (3). In particular, if $f(0) = 0$, by letting $\tau = 0$, we have $g = f$, and f preserves the scalar product, i.e., (3) holds.

Now assume that (3) holds. Since E is of finite dimension, we can pick an orthonormal basis $(e_1, \ldots, e_n)$ for E. Since f preserves inner products, $(f(e_1), \ldots, f(e_n))$ is also orthonormal, and since F also has dimension n, it is a basis of F. Then note that since $(e_1, \ldots, e_n)$ and $(f(e_1), \ldots, f(e_n))$ are orthonormal bases, for any $u \in E$ we have

$$u = \sum_{i=1}^{n} (u \cdot e_i) e_i = \sum_{i=1}^{n} u_i e_i$$

and

$$f(u) = \sum_{i=1}^{n} (f(u) \cdot f(e_i)) f(e_i),$$

and since f preserves inner products, this shows that

$$f(u) = \sum_{i=1}^{n} (f(u) \cdot f(e_i)) f(e_i) = \sum_{i=1}^{n} (u \cdot e_i) f(e_i) = \sum_{i=1}^{n} u_i f(e_i),$$

which proves that f is linear. Obviously, f preserves the Euclidean norm, and (3) implies (1).

Finally, if $f(u) = f(v)$, then by linearity $f(v - u) = 0$, so that $\|f(v - u)\| = 0$, and since f preserves norms, we must have $\|v - u\| = 0$, and thus $u = v$. Thus, f is injective, and since E and F have the same finite dimension, f is bijective. □

Remarks:

(i) The dimension assumption is needed only to prove that (3) implies (1) when f is not known to be linear, and to prove that f is surjective, but the proof shows that (1) implies that f is injective.

(ii) The implication that (3) implies (1) holds if we also assume that f is surjective, even if E has infinite dimension.

In (2), when f does not satisfy the condition $f(0) = 0$, the proof shows that f is an affine map. Indeed, taking any vector τ as an origin, the map g is linear, and

$$f(\tau + u) = f(\tau) + g(u) \quad \text{for all } u \in E.$$

By Proposition 5.14, this shows that f is affine with associated linear map g.

This fact is worth recording as the following proposition.

Proposition 11.11. *Given any two nontrivial Euclidean spaces E and F of the same finite dimension n, for every function $f\colon E \to F$, if*

$$\|f(v) - f(u)\| = \|v - u\| \quad \text{for all } u, v \in E,$$

then f is an affine map, and its associated linear map g is an isometry.

In view of Proposition 11.10, we usually abbreviate "linear isometry" as "isometry," unless we wish to emphasize that we are dealing with a map between vector spaces.

We are now going to take a closer look at the isometries $f\colon E \to E$ of a Euclidean space of finite dimension.

11.6   The Orthogonal Group, Orthogonal Matrices

In this section we explore some of the basic properties of the orthogonal group and of orthogonal matrices.

Proposition 11.12. *Let E be any Euclidean space of finite dimension n, and let $f\colon E \to E$ be any linear map. The following properties hold:*

(1) The linear map $f\colon E \to E$ is an isometry iff

$$f \circ f^* = f^* \circ f = \text{id}.$$

(2) For every orthonormal basis $(e_1, \ldots, e_n)$ of E, if the matrix of f is A, then the matrix of f^* is the transpose $A^\top$ of A, and f is an isometry iff A satisfies the identities

$$A\,A^\top = A^\top A = I_n,$$

where I_n denotes the identity matrix of order n, iff the columns of A form an orthonormal basis of $\mathbb{R}^n$, iff the rows of A form an orthonormal basis of $\mathbb{R}^n$.

Proof. (1) The linear map $f\colon E \to E$ is an isometry iff

$$f(u) \cdot f(v) = u \cdot v,$$

for all $u, v \in E$, iff

$$f^*(f(u)) \cdot v = f(u) \cdot f(v) = u \cdot v$$

for all $u, v \in E$, which implies

$$(f^*(f(u)) - u) \cdot v = 0$$

for all $u, v \in E$. Since the inner product is positive definite, we must have

$$f^*(f(u)) - u = 0$$

for all $u \in E$, that is,

$$f^* \circ f = \mathrm{id}.$$

But an endomorphism f of a finite-dimensional vector space that has a left inverse is an isomorphism, so $f \circ f^* = \mathrm{id}$. The converse is established by doing the above steps backward.

(2) If $(e_1, \ldots, e_n)$ is an orthonormal basis for E, let $A = (a_{i\,j})$ be the matrix of f, and let $B = (b_{i\,j})$ be the matrix of f^*. Since f^* is characterized by

$$f^*(u) \cdot v = u \cdot f(v)$$

for all $u, v \in E$, using the fact that if $w = w_1 e_1 + \cdots + w_n e_n$ we have $w_k = w \cdot e_k$ for all k, $1 \leq k \leq n$, letting $u = e_i$ and $v = e_j$, we get

$$b_{j\,i} = f^*(e_i) \cdot e_j = e_i \cdot f(e_j) = a_{i\,j},$$

for all i, j, $1 \leq i, j \leq n$. Thus, $B = A^\top$. Now if X and Y are arbitrary matrices over the basis $(e_1, \ldots, e_n)$, denoting as usual the jth column of X by X^j, and similarly for Y, a simple calculation shows that

$$X^\top Y = (X^i \cdot Y^j)_{1 \leq i, j \leq n}.$$

Then it is immediately verified that if $X = Y = A$, then

$$A^\top A = A A^\top = I_n$$

iff the column vectors $(A^1, \ldots, A^n)$ form an orthonormal basis. Thus, from (1), we see that (2) is clear (also because the rows of A are the columns of $A^\top$). □

Proposition 11.12 shows that the inverse of an isometry f is its adjoint f^*. Recall that the set of all real $n \times n$ matrices is denoted by $\mathrm{M}_n(\mathbb{R})$. Proposition 11.12 also motivates the following definition.

Definition 11.6. A real $n \times n$ matrix is an *orthogonal matrix* if

$$A A^\top = A^\top A = I_n.$$

Remark: It is easy to show that the conditions $A A^\top = I_n$, $A^\top A = I_n$, and $A^{-1} = A^\top$, are equivalent. Given any two orthonormal bases $(u_1, \ldots, u_n)$ and $(v_1, \ldots, v_n)$, if P is the change of basis matrix from $(u_1, \ldots, u_n)$ to

$(v_1, \ldots, v_n)$, since the columns of P are the coordinates of the vectors v_j with respect to the basis $(u_1, \ldots, u_n)$, and since $(v_1, \ldots, v_n)$ is orthonormal, the columns of P are orthonormal, and by Proposition 11.12 (2), the matrix P is orthogonal.

The proof of Proposition 11.10 (3) also shows that if f is an isometry, then the image of an orthonormal basis $(u_1, \ldots, u_n)$ is an orthonormal basis. Students often ask why ortho*gon*al matrices are not called ortho*norm*al matrices, since their columns (and rows) are orthonormal bases! I have no good answer, but isometries do preserve orthogonality, and orthogonal matrices correspond to isometries.

Recall that the determinant $\det(f)$ of a linear map $f \colon E \to E$ is independent of the choice of a basis in E. Also, for every matrix $A \in M_n(\mathbb{R})$, we have $\det(A) = \det(A^\top)$, and for any two $n \times n$ matrices A and B, we have $\det(AB) = \det(A)\det(B)$. Then if f is an isometry, and A is its matrix with respect to any orthonormal basis, $A A^\top = A^\top A = I_n$ implies that $\det(A)^2 = 1$, that is, either $\det(A) = 1$, or $\det(A) = -1$. It is also clear that the isometries of a Euclidean space of dimension n form a group, and that the isometries of determinant $+1$ form a subgroup. This leads to the following definition.

Definition 11.7. Given a Euclidean space E of dimension n, the set of isometries $f \colon E \to E$ forms a subgroup of $\mathbf{GL}(E)$ denoted by $\mathbf{O}(E)$, or $\mathbf{O}(n)$ when $E = \mathbb{R}^n$, called the *orthogonal group (of E)*. For every isometry f, we have $\det(f) = \pm 1$, where $\det(f)$ denotes the determinant of f. The isometries such that $\det(f) = 1$ are called *rotations, or proper isometries, or proper orthogonal transformations*, and they form a subgroup of the special linear group $\mathbf{SL}(E)$ (and of $\mathbf{O}(E)$), denoted by $\mathbf{SO}(E)$, or $\mathbf{SO}(n)$ when $E = \mathbb{R}^n$, called the *special orthogonal group (of E)*. The isometries such that $\det(f) = -1$ are called *improper isometries, or improper orthogonal transformations, or flip transformations*.

11.7 The Rodrigues Formula

When $n = 3$ and A is a skew symmetric matrix, it is possible to work out an explicit formula for e^A. For any 3×3 real skew symmetric matrix

$$A = \begin{pmatrix} 0 & -c & b \\ c & 0 & -a \\ -b & a & 0 \end{pmatrix},$$

if we let $\theta = \sqrt{a^2 + b^2 + c^2}$ and

$$B = \begin{pmatrix} a^2 & ab & ac \\ ab & b^2 & bc \\ ac & bc & c^2 \end{pmatrix},$$

then we have the following result known as *Rodrigues' formula* (1840). The (real) vector space of $n \times n$ skew symmetric matrices is denoted by $\mathfrak{so}(n)$.

Proposition 11.13. *The exponential map* $\exp \colon \mathfrak{so}(3) \to \mathbf{SO}(3)$ *is given by*

$$e^A = \cos\theta\, I_3 + \frac{\sin\theta}{\theta} A + \frac{(1 - \cos\theta)}{\theta^2} B,$$

or, equivalently, by

$$e^A = I_3 + \frac{\sin\theta}{\theta} A + \frac{(1 - \cos\theta)}{\theta^2} A^2$$

if $\theta \neq 0$, with $e^{0_3} = I_3$.

Proof sketch. First observe that

$$A^2 = -\theta^2 I_3 + B,$$

since

$$A^2 = \begin{pmatrix} 0 & -c & b \\ c & 0 & -a \\ -b & a & 0 \end{pmatrix} \begin{pmatrix} 0 & -c & b \\ c & 0 & -a \\ -b & a & 0 \end{pmatrix} = \begin{pmatrix} -c^2 - b^2 & ba & ca \\ ab & -c^2 - a^2 & cb \\ ac & cb & -b^2 - a^2 \end{pmatrix}$$

$$= \begin{pmatrix} -a^2 - b^2 - c^2 & 0 & 0 \\ 0 & -a^2 - b^2 - c^2 & 0 \\ 0 & 0 & -a^2 - b^2 - c^2 \end{pmatrix} + \begin{pmatrix} a^2 & ba & ca \\ ab & b^2 & cb \\ ac & cb & c^2 \end{pmatrix}$$

$$= -\theta^2 I_3 + B,$$

and that

$$AB = BA = 0.$$

From the above, deduce that

$$A^3 = -\theta^2 A,$$

and for any $k \geq 0$,

$$A^{4k+1} = \theta^{4k} A,$$
$$A^{4k+2} = \theta^{4k} A^2,$$
$$A^{4k+3} = -\theta^{4k+2} A,$$
$$A^{4k+4} = -\theta^{4k+2} A^2.$$

Then prove the desired result by writing the power series for e^A and re-grouping terms so that the power series for $\cos\theta$ and $\sin\theta$ show up. In particular

$$e^A = I_3 + \sum_{p \geq 1} \frac{A^p}{p!} = I_3 + \sum_{p \geq 0} \frac{A^{2p+1}}{(2p+1)!} + \sum_{p \geq 1} \frac{A^{2p}}{(2p)!}$$

$$= I_3 + \sum_{p \geq 0} \frac{(-1)^p \theta^{2p}}{(2p+1)!} A + \sum_{p \geq 1} \frac{(-1)^{p-1} \theta^{2(p-1)}}{(2p)!} A^2$$

$$= I_3 + \frac{A}{\theta} \sum_{p \geq 0} \frac{(-1)^p \theta^{2p+1}}{(2p+1)!} - \frac{A^2}{\theta^2} \sum_{p \geq 1} \frac{(-1)^p \theta^{2p}}{(2p)!}$$

$$= I_3 + \frac{\sin\theta}{\theta} A - \frac{A^2}{\theta^2} \sum_{p \geq 0} \frac{(-1)^p \theta^{2p}}{(2p)!} + \frac{A^2}{\theta^2}$$

$$= I_3 + \frac{\sin\theta}{\theta} A + \frac{(1 - \cos\theta)}{\theta^2} A^2,$$

as claimed.   □

The above formulae are the well-known formulae expressing a rotation of axis specified by the vector (a, b, c) and angle θ.

The Rodrigues formula can used to show that the exponential map $\exp\colon \mathfrak{so}(3) \to \mathbf{SO}(3)$ is surjective.

Given any rotation matrix $R \in \mathbf{SO}(3)$, we have the following cases:

(1) The case $R = I$ is trivial.
(2) If $R \neq I$ and $\operatorname{tr}(R) \neq -1$, then

$$\exp^{-1}(R) = \left\{ \frac{\theta}{2\sin\theta}(R - R^T) \,\middle|\, 1 + 2\cos\theta = \operatorname{tr}(R) \right\}.$$

(Recall that $\operatorname{tr}(R) = r_{11} + r_{22} + r_{33}$, the *trace* of the matrix R.)
Then there is a unique skew-symmetric B with corresponding θ satisfying $0 < \theta < \pi$ such that $e^B = R$.
(3) If $R \neq I$ and $\operatorname{tr}(R) = -1$, then R is a rotation by the angle π and things are more complicated, but a matrix B can be found. We leave this part as a good exercise: see Problem 16.8.

The computation of a logarithm of a rotation in $\mathbf{SO}(3)$ as sketched above has applications in kinematics, robotics, and motion interpolation.

As an immediate corollary of the Gram–Schmidt orthonormalization procedure, we obtain the QR-decomposition for invertible matrices.

11.8 QR-Decomposition for Invertible Matrices

Now that we have the definition of an orthogonal matrix, we can explain how the Gram–Schmidt orthonormalization procedure immediately yields the QR-decomposition for matrices.

Definition 11.8. Given any real $n \times n$ matrix A, a QR-decomposition of A is any pair of $n \times n$ matrices (Q, R), where Q is an orthogonal matrix and R is an upper triangular matrix such that $A = QR$.

Note that if A is not invertible, then some diagonal entry in R must be zero.

Proposition 11.14. *Given any real $n \times n$ matrix A, if A is invertible, then there is an orthogonal matrix Q and an upper triangular matrix R with positive diagonal entries such that $A = QR$.*

Proof. We can view the columns of A as vectors $A^1, \ldots, A^n$ in $\mathbb{E}^n$. If A is invertible, then they are linearly independent, and we can apply Proposition 11.8 to produce an orthonormal basis using the Gram–Schmidt orthonormalization procedure. Recall that we construct vectors Q^k and Q'^k as follows:

$$Q'^1 = A^1, \qquad Q^1 = \frac{Q'^1}{\|Q'^1\|},$$

and for the inductive step

$$Q'^{k+1} = A^{k+1} - \sum_{i=1}^{k}(A^{k+1} \cdot Q^i)\, Q^i, \qquad Q^{k+1} = \frac{Q'^{k+1}}{\|Q'^{k+1}\|},$$

where $1 \leq k \leq n - 1$. If we express the vectors A^k in terms of the Q^i and Q'^i, we get the triangular system

$$A^1 = \|Q'^1\|Q^1,$$

$$\vdots$$

$$A^j = (A^j \cdot Q^1)\, Q^1 + \cdots + (A^j \cdot Q^i)\, Q^i + \cdots + (A^j \cdot Q^{j-1})\, Q^{j-1} + \|Q'^j\|Q^j,$$

$$\vdots$$

$$A^n = (A^n \cdot Q^1)\, Q^1 + \cdots + (A^n \cdot Q^{n-1})\, Q^{n-1} + \|Q'^n\|Q^n.$$

Letting $r_{k\,k} = \|Q'^k\|$, and $r_{i\,j} = A^j \cdot Q^i$ (the reversal of i and j on the right-hand side *is* intentional!), where $1 \leq k \leq n$, $2 \leq j \leq n$, and

$1 \leq i \leq j - 1$, and letting q_{ij} be the ith component of Q^j, we note that a_{ij}, the ith component of A^j, is given by

$$a_{ij} = r_{1j}q_{i1} + \cdots + r_{ij}q_{ii} + \cdots + r_{jj}q_{ij} = q_{i1}r_{1j} + \cdots + q_{ii}r_{ij} + \cdots + q_{ij}r_{jj}.$$

If we let $Q = (q_{ij})$, the matrix whose columns are the components of the Q^j, and $R = (r_{ij})$, the above equations show that $A = QR$, where R is upper triangular. The diagonal entries $r_{kk} = \|Q'^k\| = A^k \cdot Q^k$ are indeed positive.   □

The reader should try the above procedure on some concrete examples for 2×2 and 3×3 matrices.

Remarks:

(1) Because the diagonal entries of R are positive, it can be shown that Q and R are unique. More generally, if A is invertible and if $A = Q_1 R_1 = Q_2 R_2$ are two QR-decompositions for A, then

$$R_1 R_2^{-1} = Q_1^\top Q_2.$$

The matrix $Q_1^\top Q_2$ is orthogonal and it is easy to see that $R_1 R_2^{-1}$ is upper triangular. But an upper triangular matrix which is orthogonal must be a diagonal matrix D with diagonal entries ± 1, so $Q_2 = Q_1 D$ and $R_2 = D R_1$.

(2) The QR-decomposition holds even when A is not invertible. In this case, R has some zero on the diagonal. However, a different proof is needed. We will give a nice proof using Householder matrices (see Proposition 12.1, and also Strang [Strang (1986, 1988)], Golub and Van Loan [Golub and Van Loan (1996)], Trefethen and Bau [Trefethen and Bau III (1997)], Demmel [Demmel (1997)], Kincaid and Cheney [Kincaid and Cheney (1996)], or Ciarlet [Ciarlet (1989)]).

For better numerical stability, it is preferable to use the modified Gram–Schmidt method to implement the QR-factorization method. Here is a Matlab program implementing QR-factorization using modified Gram–Schmidt.

```
function [Q,R] = qrv4(A)
n = size(A,1);
for i = 1:n
    Q(:,i) = A(:,i);
    for j = 1:i-1
```

```
    R(j,i) = Q(:,j)'*Q(:,i);
    Q(:,i) = Q(:,i) - R(j,i)*Q(:,j);
  end
  R(i,i) = sqrt(Q(:,i)'*Q(:,i));
  Q(:,i) = Q(:,i)/R(i,i);
end
end
```

Example 11.13. Consider the matrix

$$A = \begin{pmatrix} 0 & 0 & 5 \\ 0 & 4 & 1 \\ 1 & 1 & 1 \end{pmatrix}.$$

To determine the QR-decomposition of A, we first use the Gram-Schmidt orthonormalization procedure to calculate $Q = (Q^1 Q^2 Q^3)$. By definition

$$A^1 = Q'^1 = Q^1 = \begin{pmatrix} 0 \\ 0 \\ 1 \end{pmatrix},$$

and since $A^2 = \begin{pmatrix} 0 \\ 4 \\ 1 \end{pmatrix}$, we discover that

$$Q'^2 = A^2 - (A^2 \cdot Q^1)Q^1 = \begin{pmatrix} 0 \\ 4 \\ 1 \end{pmatrix} - \begin{pmatrix} 0 \\ 0 \\ 1 \end{pmatrix} = \begin{pmatrix} 0 \\ 4 \\ 0 \end{pmatrix}.$$

Hence, $Q^2 = \begin{pmatrix} 0 \\ 1 \\ 0 \end{pmatrix}$. Finally,

$$Q'^3 = A_3 - (A^3 \cdot Q^1)Q^1 - (A^3 \cdot Q^2)Q^2 = \begin{pmatrix} 5 \\ 1 \\ 1 \end{pmatrix} - \begin{pmatrix} 0 \\ 0 \\ 1 \end{pmatrix} - \begin{pmatrix} 0 \\ 1 \\ 0 \end{pmatrix} = \begin{pmatrix} 5 \\ 0 \\ 0 \end{pmatrix},$$

which implies that $Q^3 = \begin{pmatrix} 1 \\ 0 \\ 0 \end{pmatrix}$. According to Proposition 11.14, in order to determine R we need to calculate

$$r_{11} = \left\| Q'^1 \right\| = 1 \qquad r_{12} = A^2 \cdot Q^1 = 1 \qquad r_{13} = A^3 \cdot Q^1 = 1$$
$$r_{22} = \left\| Q'^2 \right\| = 4 \qquad r_{23} = A_3 \cdot Q^2 = 1$$
$$r_{33} = \left\| Q'^3 \right\| = 5.$$

In summary, we have found that the QR-decomposition of $A = \begin{pmatrix} 0 & 0 & 5 \\ 0 & 4 & 1 \\ 1 & 1 & 1 \end{pmatrix}$ is

$$Q = \begin{pmatrix} 0 & 0 & 1 \\ 0 & 1 & 0 \\ 1 & 0 & 0 \end{pmatrix} \quad \text{and} \quad R = \begin{pmatrix} 1 & 1 & 1 \\ 0 & 4 & 1 \\ 0 & 0 & 5 \end{pmatrix}.$$

Example 11.14. Another example of QR-decomposition is

$$A = \begin{pmatrix} 1 & 1 & 2 \\ 0 & 0 & 1 \\ 1 & 0 & 0 \end{pmatrix} = \begin{pmatrix} 1/\sqrt{2} & 1/\sqrt{2} & 0 \\ 0 & 0 & 1 \\ 1/\sqrt{2} & -1/\sqrt{2} & 0 \end{pmatrix} \begin{pmatrix} \sqrt{2} & 1/\sqrt{2} & \sqrt{2} \\ 0 & 1/\sqrt{2} & \sqrt{2} \\ 0 & 0 & 1 \end{pmatrix}.$$

Example 11.15. If we apply the above `Matlab` function to the matrix

$$A = \begin{pmatrix} 4 & 1 & 0 & 0 & 0 \\ 1 & 4 & 1 & 0 & 0 \\ 0 & 1 & 4 & 1 & 0 \\ 0 & 0 & 1 & 4 & 1 \\ 0 & 0 & 0 & 1 & 4 \end{pmatrix},$$

we obtain

$$Q = \begin{pmatrix} 0.9701 & -0.2339 & 0.0619 & -0.0166 & 0.0046 \\ 0.2425 & 0.9354 & -0.2477 & 0.0663 & -0.0184 \\ 0 & 0.2650 & 0.9291 & -0.2486 & 0.0691 \\ 0 & 0 & 0.2677 & 0.9283 & -0.2581 \\ 0 & 0 & 0 & 0.2679 & 0.9634 \end{pmatrix}$$

and

$$R = \begin{pmatrix} 4.1231 & 1.9403 & 0.2425 & 0 & 0 \\ 0 & 3.7730 & 1.9956 & 0.2650 & 0 \\ 0 & 0 & 3.7361 & 1.9997 & 0.2677 \\ 0 & 0 & & 073.7324 & 2.0000 \\ 0 & 0 & 0 & 0 & 3.5956 \end{pmatrix}.$$

Remark: The `Matlab` function `qr`, called by `[Q, R] = qr(A)`, does not necessarily return an upper-triangular matrix whose diagonal entries are positive.

The QR-decomposition yields a rather efficient and numerically stable method for solving systems of linear equations. Indeed, given a system

$Ax = b$, where A is an $n \times n$ invertible matrix, writing $A = QR$, since Q is orthogonal, we get

$$Rx = Q^\top b,$$

and since R is upper triangular, we can solve it by Gaussian elimination, by solving for the last variable x_n first, substituting its value into the system, then solving for x_{n-1}, etc. The QR-decomposition is also very useful in solving least squares problems (we will come back to this in Chapter 21), and for finding eigenvalues; see Chapter 17. It can be easily adapted to the case where A is a rectangular $m \times n$ matrix with independent columns (thus, $n \leq m$). In this case, Q is not quite orthogonal. It is an $m \times n$ matrix whose columns are orthogonal, and R is an invertible $n \times n$ upper triangular matrix with positive diagonal entries. For more on QR, see Strang [Strang (1986, 1988)], Golub and Van Loan [Golub and Van Loan (1996)], Demmel [Demmel (1997)], Trefethen and Bau [Trefethen and Bau III (1997)], or Serre [Serre (2010)].

A somewhat surprising consequence of the QR-decomposition is a famous determinantal inequality due to Hadamard.

Proposition 11.15. *(Hadamard) For any real $n \times n$ matrix $A = (a_{ij})$, we have*

$$|\det(A)| \leq \prod_{i=1}^{n} \left(\sum_{j=1}^{n} a_{ij}^2 \right)^{1/2} \quad and \quad |\det(A)| \leq \prod_{j=1}^{n} \left(\sum_{i=1}^{n} a_{ij}^2 \right)^{1/2}.$$

Moreover, equality holds iff either A has a zero row in the left inequality or a zero column in the right inequality, or A is orthogonal.

Proof. If $\det(A) = 0$, then the inequality is trivial. In addition, if the right-hand side is also 0, then either some column or some row is zero. If $\det(A) \neq 0$, then we can factor A as $A = QR$, with Q is orthogonal and $R = (r_{ij})$ upper triangular with positive diagonal entries. Then since Q is orthogonal $\det(Q) = \pm 1$, so

$$|\det(A)| = |\det(Q)| \, |\det(R)| = \prod_{j=1}^{n} r_{jj}.$$

Now as Q is orthogonal, it preserves the Euclidean norm, so

$$\sum_{i=1}^{n} a_{ij}^2 = \left\| A^j \right\|_2^2 = \left\| QR^j \right\|_2^2 = \left\| R^j \right\|_2^2 = \sum_{i=1}^{n} r_{ij}^2 \geq r_{jj}^2,$$

which implies that

$$|\det(A)| = \prod_{j=1}^{n} r_{jj} \leq \prod_{j=1}^{n} \left\| R^j \right\|_2 = \prod_{j=1}^{n} \left(\sum_{i=1}^{n} a_{ij}^2 \right)^{1/2}.$$

The other inequality is obtained by replacing A by $A^\top$. Finally, if $\det(A) \neq 0$ and equality holds, then we must have

$$r_{jj} = \left\| A^j \right\|_2, \quad 1 \leq j \leq n,$$

which can only occur if A is orthogonal. $\square$

Another version of Hadamard's inequality applies to symmetric positive semidefinite matrices.

Proposition 11.16. *(Hadamard) For any real $n \times n$ matrix $A = (a_{ij})$, if A is symmetric positive semidefinite, then we have*

$$\det(A) \leq \prod_{i=1}^{n} a_{ii}.$$

Moreover, if A is positive definite, then equality holds iff A is a diagonal matrix.

Proof. If $\det(A) = 0$, the inequality is trivial. Otherwise, A is positive definite, and by Theorem 7.4 (the Cholesky Factorization), there is a unique upper triangular matrix B with positive diagonal entries such that

$$A = B^\top B.$$

Thus, $\det(A) = \det(B^\top B) = \det(B^\top)\det(B) = \det(B)^2$. If we apply the Hadamard inequality (Proposition 11.15) to B, we obtain

$$\det(B) \leq \prod_{j=1}^{n} \left(\sum_{i=1}^{n} b_{ij}^2 \right)^{1/2}. \tag{11.1}$$

However, the diagonal entries a_{jj} of $A = B^\top B$ are precisely the square norms $\left\| B^j \right\|_2^2 = \sum_{i=1}^{n} b_{ij}^2$, so by squaring (11.1), we obtain

$$\det(A) = \det(B)^2 \leq \prod_{j=1}^{n} \left(\sum_{i=1}^{n} b_{ij}^2 \right) = \prod_{j=1}^{n} a_{jj}.$$

If $\det(A) \neq 0$ and equality holds, then B must be orthogonal, which implies that B is a diagonal matrix, and so is A. $\square$

We derived the second Hadamard inequality (Proposition 11.16) from the first (Proposition 11.15). We leave it as an exercise to prove that the first Hadamard inequality can be deduced from the second Hadamard inequality.

11.9   Some Applications of Euclidean Geometry

Euclidean geometry has applications in computational geometry, in particular Voronoi diagrams and Delaunay triangulations. In turn, Voronoi diagrams have applications in motion planning (see O'Rourke [O'Rourke (1998)]).

Euclidean geometry also has applications to matrix analysis. Recall that a real $n \times n$ matrix A is *symmetric* if it is equal to its transpose $A^\top$. One of the most important properties of symmetric matrices is that they have real eigenvalues and that they can be diagonalized by an orthogonal matrix (see Chapter 16). This means that for every symmetric matrix A, there is a diagonal matrix D and an orthogonal matrix P such that

$$A = PDP^\top.$$

Even though it is not always possible to diagonalize an arbitrary matrix, there are various decompositions involving orthogonal matrices that are of great practical interest. For example, for every real matrix A, there is the *QR-decomposition*, which says that a real matrix A can be expressed as

$$A = QR,$$

where Q is orthogonal and R is an upper triangular matrix. This can be obtained from the Gram–Schmidt orthonormalization procedure, as we saw in Section 11.8, or better, using Householder matrices, as shown in Section 12.2. There is also the *polar decomposition*, which says that a real matrix A can be expressed as

$$A = QS,$$

where Q is orthogonal and S is symmetric positive semidefinite (which means that the eigenvalues of S are nonnegative). Such a decomposition is important in continuum mechanics and in robotics, since it separates stretching from rotation. Finally, there is the wonderful *singular value decomposition*, abbreviated as SVD, which says that a real matrix A can be expressed as

$$A = VDU^\top,$$

where U and V are orthogonal and D is a diagonal matrix with nonnegative entries (see Chapter 20). This decomposition leads to the notion of *pseudo-inverse*, which has many applications in engineering (least squares solutions, etc.). For an excellent presentation of all these notions, we highly recommend Strang [Strang (1988, 1986)], Golub and Van Loan [Golub and

Van Loan (1996)], Demmel [Demmel (1997)], Serre [Serre (2010)], and Trefethen and Bau [Trefethen and Bau III (1997)].

The method of least squares, invented by Gauss and Legendre around 1800, is another great application of Euclidean geometry. Roughly speaking, the method is used to solve inconsistent linear systems $Ax = b$, where the number of equations is greater than the number of variables. Since this is generally impossible, the method of least squares consists in finding a solution x minimizing the Euclidean norm $\|Ax - b\|^2$, that is, the sum of the squares of the "errors." It turns out that there is always a unique solution x^+ of smallest norm minimizing $\|Ax - b\|^2$, and that it is a solution of the square system

$$A^\top Ax = A^\top b,$$

called the system of *normal equations*. The solution x^+ can be found either by using the QR-decomposition in terms of Householder transformations, or by using the notion of pseudo-inverse of a matrix. The pseudo-inverse can be computed using the SVD decomposition. Least squares methods are used extensively in computer vision. More details on the method of least squares and pseudo-inverses can be found in Chapter 21.

11.10   Summary

The main concepts and results of this chapter are listed below:

- Bilinear forms; *positive definite* bilinear forms.
- *Inner products, scalar products, Euclidean spaces.*
- *Quadratic form* associated with a bilinear form.
- The Euclidean space $\mathbb{E}^n$.
- The *polar form* of a quadratic form.
- *Gram matrix* associated with an inner product.
- The *Cauchy–Schwarz inequality*; the *Minkowski inequality*.
- The *parallelogram law*.
- *Orthogonality, orthogonal complement $F^\perp$; orthonormal family.*
- The *musical isomorphisms* $\flat\colon E \to E^*$ and $\sharp\colon E^* \to E$ (when E is finite-dimensional); Theorem 11.1.
- The *adjoint* of a linear map (with respect to an inner product).
- Existence of an orthonormal basis in a finite-dimensional Euclidean space (Proposition 11.7).
- The *Gram–Schmidt orthonormalization procedure* (Proposition 11.8).

- The *Legendre* and the *Chebyshev* polynomials.
- *Linear isometries* (*orthogonal transformations, rigid motions*).
- The *orthogonal group, orthogonal matrices.*
- The matrix representing the adjoint f^* of a linear map f is the transpose of the matrix representing f.
- The *orthogonal group* $\mathbf{O}(n)$ and the *special orthogonal group* $\mathbf{SO}(n)$.
- *QR-decomposition* for invertible matrices.
- The *Hadamard inequality* for arbitrary real matrices.
- The *Hadamard inequality* for symmetric positive semidefinite matrices.
- The *Rodrigues formula* for rotations in $\mathbf{SO}(3)$.

11.11 Problems

Problem 11.1. E be a vector space of dimension 2, and let (e_1, e_2) be a basis of E. Prove that if $a > 0$ and $b^2 - ac < 0$, then the bilinear form defined such that

$$\varphi(x_1e_1 + y_1e_2, x_2e_1 + y_2e_2) = ax_1x_2 + b(x_1y_2 + x_2y_1) + cy_1y_2$$

is a Euclidean inner product.

Problem 11.2. Let $\mathcal{C}[a, b]$ denote the set of continuous functions $f\colon [a, b] \to \mathbb{R}$. Given any two functions $f, g \in \mathcal{C}[a, b]$, let

$$\langle f, g \rangle = \int_a^b f(t)g(t)dt.$$

Prove that the above bilinear form is indeed a Euclidean inner product.

Problem 11.3. Consider the inner product

$$\langle f, g \rangle = \int_{-\pi}^{\pi} f(t)g(t)dt$$

of Problem 11.2 on the vector space $\mathcal{C}[-\pi, \pi]$. Prove that

$$\langle \sin px, \sin qx \rangle = \begin{cases} \pi & \text{if } p = q, \; p, q \geq 1, \\ 0 & \text{if } p \neq q, \; p, q \geq 1, \end{cases}$$

$$\langle \cos px, \cos qx \rangle = \begin{cases} \pi & \text{if } p = q, \; p, q \geq 1, \\ 0 & \text{if } p \neq q, \; p, q \geq 0, \end{cases}$$

$$\langle \sin px, \cos qx \rangle = 0,$$

for all $p \geq 1$ and $q \geq 0$, and $\langle 1, 1 \rangle = \int_{-\pi}^{\pi} dx = 2\pi$.

Problem 11.4. Prove that the following matrix is orthogonal and skew-symmetric:

$$M = \frac{1}{\sqrt{3}} \begin{pmatrix} 0 & 1 & 1 & 1 \\ -1 & 0 & -1 & 1 \\ -1 & 1 & 0 & -1 \\ -1 & -1 & 1 & 0 \end{pmatrix}.$$

Problem 11.5. Let E and F be two finite Euclidean spaces, let $(u_1, \ldots, u_n)$ be a basis of E, and let $(v_1, \ldots, v_m)$ be a basis of F. For any linear map $f \colon E \to F$, if A is the matrix of f w.r.t. the basis $(u_1, \ldots, u_n)$ and B is the matrix of f^* w.r.t. the basis $(v_1, \ldots, v_m)$, if G_1 is the Gram matrix of the inner product on E (w.r.t. $(u_1, \ldots, u_n)$) and if G_2 is the Gram matrix of the inner product on F (w.r.t. $(v_1, \ldots, v_m)$), then

$$B = G_1^{-1} A^\top G_2.$$

Problem 11.6. Let A be an invertible matrix. Prove that if $A = Q_1 R_1 = Q_2 R_2$ are two QR-decompositions of A and if the diagonal entries of R_1 and R_2 are positive, then $Q_1 = Q_2$ and $R_1 = R_2$.

Problem 11.7. Prove that the first Hadamard inequality can be deduced from the second Hadamard inequality.

Problem 11.8. Let E be a real vector space of finite dimension, $n \geq 1$. Say that two bases, $(u_1, \ldots, u_n)$ and $(v_1, \ldots, v_n)$, of E have the *same orientation* iff $\det(P) > 0$, where P the change of basis matrix from $(u_1, \ldots, u_n)$ and $(v_1, \ldots, v_n)$, namely, the matrix whose jth columns consist of the coordinates of v_j over the basis $(u_1, \ldots, u_n)$.

(1) Prove that having the same orientation is an equivalence relation with two equivalence classes.

An *orientation* of a vector space, E, is the choice of any fixed basis, say $(e_1, \ldots, e_n)$, of E. Any other basis, $(v_1, \ldots, v_n)$, has the *same orientation* as $(e_1, \ldots, e_n)$ (and is said to be *positive* or *direct*) iff $\det(P) > 0$, else it is said to have the *opposite orientation* of $(e_1, \ldots, e_n)$ (or to be *negative* or *indirect*), where P is the change of basis matrix from $(e_1, \ldots, e_n)$ to $(v_1, \ldots, v_n)$. An *oriented* vector space is a vector space with some chosen orientation (a positive basis).

(2) Let $B_1 = (u_1, \ldots, u_n)$ and $B_2 = (v_1, \ldots, v_n)$ be two orthonormal bases. For any sequence of vectors, $(w_1, \ldots, w_n)$, in E, let $\det_{B_1}(w_1, \ldots, w_n)$ be the determinant of the matrix whose columns

are the coordinates of the w_j's over the basis B_1 and similarly for $\det_{B_2}(w_1, \ldots, w_n)$.

Prove that if B_1 and B_2 have the same orientation, then

$$\det_{B_1}(w_1, \ldots, w_n) = \det_{B_2}(w_1, \ldots, w_n).$$

Given any oriented vector space, E, for any sequence of vectors, $(w_1, \ldots, w_n)$, in E, the common value, $\det_B(w_1, \ldots, w_n)$, for all positive orthonormal bases, B, of E is denoted

$$\lambda_E(w_1, \ldots, w_n)$$

and called a *volume form* of $(w_1, \ldots, w_n)$.

(3) Given any Euclidean oriented vector space, E, of dimension n for any $n - 1$ vectors, $w_1, \ldots, w_{n-1}$, in E, check that the map

$$x \mapsto \lambda_E(w_1, \ldots, w_{n-1}, x)$$

is a linear form. Then prove that there is a unique vector, denoted $w_1 \times \cdots \times w_{n-1}$, such that

$$\lambda_E(w_1, \ldots, w_{n-1}, x) = (w_1 \times \cdots \times w_{n-1}) \cdot x,$$

for all $x \in E$. The vector $w_1 \times \cdots \times w_{n-1}$ is called the *cross-product* of $(w_1, \ldots, w_{n-1})$. It is a generalization of the cross-product in $\mathbb{R}^3$ (when $n = 3$).

Problem 11.9. Given p vectors $(u_1, \ldots, u_p)$ in a Euclidean space E of dimension $n \geq p$, the *Gram determinant (or Gramian)* of the vectors $(u_1, \ldots, u_p)$ is the determinant

$$\mathrm{Gram}(u_1, \ldots, u_p) = \begin{vmatrix} \|u_1\|^2 & \langle u_1, u_2 \rangle & \ldots & \langle u_1, u_p \rangle \\ \langle u_2, u_1 \rangle & \|u_2\|^2 & \ldots & \langle u_2, u_p \rangle \\ \vdots & \vdots & \ddots & \vdots \\ \langle u_p, u_1 \rangle & \langle u_p, u_2 \rangle & \ldots & \|u_p\|^2 \end{vmatrix}.$$

(1) Prove that

$$\mathrm{Gram}(u_1, \ldots, u_n) = \lambda_E(u_1, \ldots, u_n)^2.$$

Hint. If $(e_1, \ldots, e_n)$ is an orthonormal basis and A is the matrix of the vectors $(u_1, \ldots, u_n)$ over this basis,

$$\det(A)^2 = \det(A^\top A) = \det(A^i \cdot A^j),$$

where A^i denotes the ith column of the matrix A, and $(A^i \cdot A^j)$ denotes the $n \times n$ matrix with entries $A^i \cdot A^j$.

(2) Prove that

$$\|u_1 \times \cdots \times u_{n-1}\|^2 = \mathrm{Gram}(u_1, \ldots, u_{n-1}).$$

Hint. Letting $w = u_1 \times \cdots \times u_{n-1}$, observe that

$$\lambda_E(u_1, \ldots, u_{n-1}, w) = \langle w, w \rangle = \|w\|^2,$$

and show that

$$\|w\|^4 = \lambda_E(u_1, \ldots, u_{n-1}, w)^2 = \mathrm{Gram}(u_1, \ldots, u_{n-1}, w)$$
$$= \mathrm{Gram}(u_1, \ldots, u_{n-1})\|w\|^2.$$

Problem 11.10. Let $\varphi \colon E \times E \to \mathbb{R}$ be a bilinear form on a real vector space E of finite dimension n. Given any basis $(e_1, \ldots, e_n)$ of E, let $A = (a_{ij})$ be the matrix defined such that

$$a_{ij} = \varphi(e_i, e_j),$$

$1 \le i, j \le n$. We call A *the matrix of φ w.r.t. the basis* $(e_1, \ldots, e_n)$.

(1) For any two vectors x and y, if X and Y denote the column vectors of coordinates of x and y w.r.t. the basis $(e_1, \ldots, e_n)$, prove that

$$\varphi(x, y) = X^\top A Y.$$

(2) Recall that A is a *symmetric* matrix if $A = A^\top$. Prove that φ is symmetric if A is a symmetric matrix.

(3) If $(f_1, \ldots, f_n)$ is another basis of E and P is the change of basis matrix from $(e_1, \ldots, e_n)$ to $(f_1, \ldots, f_n)$, prove that the matrix of φ w.r.t. the basis $(f_1, \ldots, f_n)$ is

$$P^\top A P.$$

The common rank of all matrices representing φ is called the *rank* of φ.

Problem 11.11. Let $\varphi \colon E \times E \to \mathbb{R}$ be a symmetric bilinear form on a real vector space E of finite dimension n. Two vectors x and y are said to be *conjugate or orthogonal w.r.t.* φ if $\varphi(x, y) = 0$. The main purpose of this problem is to prove that there is a basis of vectors that are pairwise conjugate w.r.t. φ.

(1) Prove that if $\varphi(x, x) = 0$ for all $x \in E$, then φ is identically null on E.

Otherwise, we can assume that there is some vector $x \in E$ such that $\varphi(x, x) \ne 0$.

Use induction to prove that there is a basis of vectors $(u_1, \ldots, u_n)$ that are pairwise conjugate w.r.t. φ.

Hint. For the induction step, proceed as follows. Let $(u_1, e_2, \ldots, e_n)$ be a basis of E, with $\varphi(u_1, u_1) \neq 0$. Prove that there are scalars $\lambda_2, \ldots, \lambda_n$ such that each of the vectors

$$v_i = e_i + \lambda_i u_1$$

is conjugate to u_1 w.r.t. φ, where $2 \leq i \leq n$, and that $(u_1, v_2, \ldots, v_n)$ is a basis.

(2) Let $(e_1, \ldots, e_n)$ be a basis of vectors that are pairwise conjugate w.r.t. φ and assume that they are ordered such that

$$\varphi(e_i, e_i) = \begin{cases} \theta_i \neq 0 & \text{if } 1 \leq i \leq r, \\ 0 & \text{if } r+1 \leq i \leq n, \end{cases}$$

where r is the rank of φ. Show that the matrix of φ w.r.t. $(e_1, \ldots, e_n)$ is a diagonal matrix, and that

$$\varphi(x, y) = \sum_{i=1}^{r} \theta_i x_i y_i,$$

where $x = \sum_{i=1}^{n} x_i e_i$ and $y = \sum_{i=1}^{n} y_i e_i$.

Prove that for every symmetric matrix A, there is an invertible matrix P such that

$$P^\top A P = D,$$

where D is a diagonal matrix.

(3) Prove that there is an integer p, $0 \leq p \leq r$ (where r is the rank of φ), such that $\varphi(u_i, u_i) > 0$ for exactly p vectors of every basis $(u_1, \ldots, u_n)$ of vectors that are pairwise conjugate w.r.t. φ (*Sylvester's inertia theorem*).

Proceed as follows. Assume that in the basis $(u_1, \ldots, u_n)$, for any $x \in E$, we have

$$\varphi(x, x) = \alpha_1 x_1^2 + \cdots + \alpha_p x_p^2 - \alpha_{p+1} x_{p+1}^2 - \cdots - \alpha_r x_r^2,$$

where $x = \sum_{i=1}^{n} x_i u_i$, and that in the basis $(v_1, \ldots, v_n)$, for any $x \in E$, we have

$$\varphi(x, x) = \beta_1 y_1^2 + \cdots + \beta_q y_q^2 - \beta_{q+1} y_{q+1}^2 - \cdots - \beta_r y_r^2,$$

where $x = \sum_{i=1}^{n} y_i v_i$, with $\alpha_i > 0$, $\beta_i > 0$, $1 \leq i \leq r$.

Assume that $p > q$ and derive a contradiction. First consider x in the subspace F spanned by

$$(u_1, \ldots, u_p, u_{r+1}, \ldots, u_n),$$

and observe that $\varphi(x, x) \geq 0$ if $x \neq 0$. Next consider x in the subspace G spanned by

$$(v_{q+1}, \ldots, v_r),$$

and observe that $\varphi(x, x) < 0$ if $x \neq 0$. Prove that $F \cap G$ is nontrivial (i.e., contains some nonnull vector), and derive a contradiction. This implies that $p \leq q$. Finish the proof.

The pair $(p, r - p)$ is called the *signature* of φ.

(4) A symmetric bilinear form φ is *definite* if for every $x \in E$, if $\varphi(x, x) = 0$, then $x = 0$.

Prove that a symmetric bilinear form is definite iff its signature is either $(n, 0)$ or $(0, n)$. In other words, a symmetric definite bilinear form has rank n and is either positive or negative.

Problem 11.12. Consider the $n \times n$ matrices $R^{i,j}$ defined for all i, j with $1 \leq i < j \leq n$ and $n \geq 3$, such that the only nonzero entries are

$$R^{i,j}(i, j) = -1$$
$$R^{i,j}(i, i) = 0$$
$$R^{i,j}(j, i) = 1$$
$$R^{i,j}(j, j) = 0$$
$$R^{i,j}(k, k) = 1, \quad 1 \leq k \leq n, k \neq i, j.$$

For example,

$$R^{i,j} = \begin{pmatrix} 1 & & & & & & & & & \\ & \ddots & & & & & & & & \\ & & 1 & & & & & & & \\ & & & 0\ 0 \cdots 0 -1 & & & & \\ & & & 0\ 1 \cdots 0\ \ 0 & & & & \\ & & & \vdots\ \vdots\ \ddots\ \vdots\ \ \vdots & & & \\ & & & 0\ 0 \cdots 1\ \ 0 & & & & \\ & & & 1\ 0 \cdots 0\ \ 0 & & & & \\ & & & & & 1 & & & \\ & & & & & & \ddots & \\ & & & & & & & 1 \end{pmatrix}.$$

(1) Prove that the $R^{i,j}$ are rotation matrices. Use the matrices R^{ij} to form a basis of the $n \times n$ skew-symmetric matrices.

(2) Consider the $n \times n$ symmetric matrices $S^{i,j}$ defined for all i, j with $1 \leq i < j \leq n$ and $n \geq 3$, such that the only nonzero entries are

$$S^{i,j}(i,j) = 1$$
$$S^{i,j}(i,i) = 0$$
$$S^{i,j}(j,i) = 1$$
$$S^{i,j}(j,j) = 0$$
$$S^{i,j}(k,k) = 1, \quad 1 \leq k \leq n, k \neq i, j,$$

and if $i + 2 \leq j$ then $S^{i,j}(i+1, i+1) = -1$, else if $i > 1$ and $j = i+1$ then $S^{i,j}(1,1) = -1$, and if $i = 1$ and $j = 2$, then $S^{i,j}(3,3) = -1$.

For example,

$$S^{i,j} = \begin{pmatrix} 1 & & & & & & & & \\ & \ddots & & & & & & & \\ & & 1 & & & & & & \\ & & & 0 & 0 & \cdots & 0 & 1 & \\ & & & 0 & -1 & \cdots & 0 & 0 & \\ & & & \vdots & \vdots & \ddots & \vdots & \vdots & \\ & & & 0 & 0 & \cdots & 1 & 0 & \\ & & & 1 & 0 & \cdots & 0 & 0 & \\ & & & & & & & & 1 \\ & & & & & & & & & \ddots \\ & & & & & & & & & & 1 \end{pmatrix}.$$

Note that $S^{i,j}$ has a single diagonal entry equal to -1. Prove that the $S^{i,j}$ are rotations matrices.

Use Problem 2.15 together with the $S^{i,j}$ to form a basis of the $n \times n$ symmetric matrices.

(3) Prove that if $n \geq 3$, the set of all linear combinations of matrices in $\mathbf{SO}(n)$ is the space $\mathrm{M}_n(\mathbb{R})$ of all $n \times n$ matrices.

Prove that if $n \geq 3$ and if a matrix $A \in \mathrm{M}_n(\mathbb{R})$ commutes with all rotations matrices, then A commutes with all matrices in $\mathrm{M}_n(\mathbb{R})$.

What happens for $n = 2$?

Problem 11.13. Let A be an $n \times n$ real invertible matrix. Prove that if $A = Q_1 R_1$ and $A = Q_2 R_2$ are two QR-decompositions of A where R_1 and R_2 are upper-triangular with positive diagonal entries, then $Q_1 = Q_2$ and $R_1 = R_2$.

Problem 11.14. (1) Let H be the affine hyperplane in $\mathbb{R}^n$ given by the equation

$$a_1 x_1 + \cdots + a_n x_n = c,$$

with $a_i \neq 0$ for some $i, 1 \leq i \leq n$. The linear hyperplane H_0 parallel to H is given by the equation

$$a_1 x_1 + \cdots + a_n x_n = 0,$$

and we say that a vector $y \in \mathbb{R}^n$ is *orthogonal* (or *perpendicular*) to H iff y is orthogonal to H_0. Let h be the intersection of H with the line through the origin and perpendicular to H. Prove that the coordinates of h are given by

$$\frac{c}{a_1^2 + \cdots + a_n^2}(a_1, \ldots, a_n).$$

(2) For any point $p \in H$, prove that $\|h\| \leq \|p\|$. Thus, it is natural to define the *distance* $d(O, H)$ from the origin O to the hyperplane H as $d(O, H) = \|h\|$. Prove that

$$d(O, H) = \frac{|c|}{(a_1^2 + \cdots + a_n^2)^{\frac{1}{2}}}.$$

(3) Let S be a finite set of $n \geq 3$ points in the plane ($\mathbb{R}^2$). Prove that if for every pair of distinct points $p_i, p_j \in S$, there is a third point $p_k \in S$ (distinct from p_i and p_j) such that p_i, p_j, p_k belong to the same (affine) line, then all points in S belong to a common (affine) line.

Hint. Proceed by contradiction and use a minimality argument. This is either ∞-hard or relatively easy, depending how you proceed!

Problem 11.15. (The space of closed polygons in $\mathbb{R}^2$, after Hausmann and Knutson.)

An *open polygon* P in the plane is a sequence $P = (v_1, \ldots, v_{n+1})$ of points $v_i \in \mathbb{R}^2$ called *vertices* (with $n \geq 1$). A *closed polygon*, for short a *polygon*, is an open polygon $P = (v_1, \ldots, v_{n+1})$ such that $v_{n+1} = v_1$. The sequence of *edge vectors* $(e_1, \ldots, e_n)$ associated with the open (or closed) polygon $P = (v_1, \ldots, v_{n+1})$ is defined by

$$e_i = v_{i+1} - v_i, \quad i = 1, \ldots, n.$$

Thus, a closed or open polygon is also defined by a pair $(v_1, (e_1, \ldots, e_n))$, with the vertices given by

$$v_{i+1} = v_i + e_i, \quad i = 1, \ldots, n.$$

Observe that a polygon $(v_1, (e_1, \ldots, e_n))$ is closed iff

$$e_1 + \cdots + e_n = 0.$$

Since every polygon $(v_1, (e_1, \ldots, e_n))$ can be translated by $-v_1$, so that $v_1 = (0, 0)$, we may assume that our polygons are specified by a sequence of edge vectors.

Recall that the plane $\mathbb{R}^2$ is isomorphic to $\mathbb{C}$, via the isomorphism

$$(x, y) \mapsto x + iy.$$

We will represent each edge vector e_k by the square of a complex number $w_k = a_k + ib_k$. Thus, every sequence of complex numbers $(w_1, \ldots, w_n)$ defines a polygon (namely, $(w_1^2, \ldots, w_n^2)$). This representation is many-to-one: the sequences $(\pm w_1, \ldots, \pm w_n)$ describe the same polygon. To every sequence of complex numbers $(w_1, \ldots, w_n)$, we associate the pair of vectors (a, b), with $a, b \in \mathbb{R}^n$, such that if $w_k = a_k + ib_k$, then

$$a = (a_1, \ldots, a_n), \quad b = (b_1, \ldots, b_n).$$

The mapping

$$(w_1, \ldots, w_n) \mapsto (a, b)$$

is clearly a bijection, so we can also represent polygons by pairs of vectors $(a, b) \in \mathbb{R}^n \times \mathbb{R}^n$.

(1) Prove that a polygon P represented by a pair of vectors $(a, b) \in \mathbb{R}^n \times \mathbb{R}^n$ is closed iff $a \cdot b = 0$ and $\|a\|_2 = \|b\|_2$.

(2) Given a polygon P represented by a pair of vectors $(a, b) \in \mathbb{R}^n \times \mathbb{R}^n$, the length $l(P)$ of the polygon P is defined by $l(P) = |w_1|^2 + \cdots + |w_n|^2$, with $w_k = a_k + ib_k$. Prove that

$$l(P) = \|a\|_2^2 + \|b\|_2^2.$$

Deduce from (a) and (b) that every closed polygon of length 2 with n edges is represented by a $n \times 2$ matrix A such that $A^\top A = I$.

Remark: The space of all a $n \times 2$ real matrices A such that $A^\top A = I$ is a space known as the *Stiefel manifold* $S(2, n)$.

(3) Recall that in $\mathbb{R}^2$, the rotation of angle θ specified by the matrix

$$R_\theta = \begin{pmatrix} \cos\theta & -\sin\theta \\ \sin\theta & \cos\theta \end{pmatrix}$$

is expressed in terms of complex numbers by the map

$$z \mapsto ze^{i\theta}.$$

Let P be a polygon represented by a pair of vectors $(a, b) \in \mathbb{R}^n \times \mathbb{R}^n$. Prove that the polygon $R_\theta(P)$ obtained by applying the rotation R_θ to every vertex $w_k^2 = (a_k + ib_k)^2$ of P is specified by the pair of vectors

$$(\cos(\theta/2)a - \sin(\theta/2)b, \; \sin(\theta/2)a + \cos(\theta/2)b)$$

$$= \begin{pmatrix} a_1 & b_1 \\ a_2 & b_2 \\ \vdots & \vdots \\ a_n & b_n \end{pmatrix} \begin{pmatrix} \cos(\theta/2) & \sin(\theta/2) \\ -\sin(\theta/2) & \cos(\theta/2) \end{pmatrix}.$$

(4) The reflection ρ_x about the x-axis corresponds to the map

$$z \mapsto \bar{z},$$

whose matrix is,

$$\begin{pmatrix} 1 & 0 \\ 0 & -1 \end{pmatrix}.$$

Prove that the polygon $\rho_x(P)$ obtained by applying the reflection ρ_x to every vertex $w_k^2 = (a_k + ib_k)^2$ of P is specified by the pair of vectors

$$(a, -b) = \begin{pmatrix} a_1 & b_1 \\ a_2 & b_2 \\ \vdots & \vdots \\ a_n & b_n \end{pmatrix} \begin{pmatrix} 1 & 0 \\ 0 & -1 \end{pmatrix}.$$

(5) Let $Q \in \mathbf{O}(2)$ be any isometry such that $\det(Q) = -1$ (a reflection). Prove that there is a rotation $R_{-\theta} \in \mathbf{SO}(2)$ such that

$$Q = \rho_x \circ R_{-\theta}.$$

Prove that the isometry Q, which is given by the matrix

$$Q = \begin{pmatrix} \cos\theta & \sin\theta \\ \sin\theta & -\cos\theta \end{pmatrix},$$

is the reflection about the line corresponding to the angle $\theta/2$ (the line of equation $y = \tan(\theta/2)x$).

Prove that the polygon $Q(P)$ obtained by applying the reflection $Q = \rho_x \circ R_{-\theta}$ to every vertex $w_k^2 = (a_k + ib_k)^2$ of P, is specified by the pair of vectors

$$(\cos(\theta/2)a + \sin(\theta/2)b, \; \sin(\theta/2)a - \cos(\theta/2)b)$$

$$= \begin{pmatrix} a_1 & b_1 \\ a_2 & b_2 \\ \vdots & \vdots \\ a_n & b_n \end{pmatrix} \begin{pmatrix} \cos(\theta/2) & \sin(\theta/2) \\ \sin(\theta/2) & -\cos(\theta/2) \end{pmatrix}.$$

(6) Define an equivalence relation $\sim$ on $S(2, n)$ such that if $A_1, A_2 \in S(2, n)$ are any $n \times 2$ matrices such that $A_1^\top A_1 = A_2^\top A_2 = I$, then

$$A_1 \sim A_2 \quad \text{iff} \quad A_2 = A_1 Q \quad \text{for some } Q \in \mathbf{O}(2).$$

Prove that the quotient $G(2, n) = S(2, n)/\sim$ is in bijection with the set of all 2-dimensional subspaces (the planes) of $\mathbb{R}^n$. The space $G(2, n)$ is called a *Grassmannian manifold*.

Prove that up to translations and isometries in $\mathbf{O}(2)$ (rotations and reflections), the n-sided closed polygons of length 2 are represented by planes in $G(2, n)$.

Problem 11.16. (1) Find two symmetric matrices, A and B, such that AB is not symmetric.

(2) Find two matrices A and B such that

$$e^A e^B \neq e^{A+B}.$$

Hint. Try

$$A = \pi \begin{pmatrix} 0 & 0 & 0 \\ 0 & 0 & -1 \\ 0 & 1 & 0 \end{pmatrix} \quad \text{and} \quad B = \pi \begin{pmatrix} 0 & 0 & 1 \\ 0 & 0 & 0 \\ -1 & 0 & 0 \end{pmatrix},$$

and use the Rodrigues formula.

(3) Find some square matrices A, B such that $AB \neq BA$, yet

$$e^A e^B = e^{A+B}.$$

Hint. Look for 2×2 matrices with zero trace and use Problem 8.15.

Problem 11.17. Given a field K and any nonempty set I, let $K^{(I)}$ be the subset of the cartesian product K^I consisting of all functions $\lambda \colon I \to K$ with *finite support*, which means that $\lambda(i) = 0$ for all but finitely many $i \in I$. We usually denote the function defined by λ as $(\lambda_i)_{i \in I}$, and call is a *family indexed by I*. We define addition and multiplication by a scalar as follows:

$$(\lambda_i)_{i \in I} + (\mu_i)_{i \in I} = (\lambda_i + \mu_i)_{i \in I},$$

and

$$\alpha \cdot (\mu_i)_{i \in I} = (\alpha \mu_i)_{i \in I}.$$

(1) Check that $K^{(I)}$ is a vector space.

(2) If I is any nonempty subset, for any $i \in I$, we denote by e_i the family $(e_j)_{j \in I}$ defined so that

$$e_j = \begin{cases} 1 & \text{if } j = i \\ 0 & \text{if } j \neq i. \end{cases}$$

Prove that the family $(e_i)_{i \in I}$ is linearly independent and spans $K^{(I)}$, so that it is a basis of $K^{(I)}$ called the *canonical basis* of $K^{(I)}$. When I is finite, say of cardinality n, then prove that $K^{(I)}$ is isomorphic to K^n.

(3) The function $\iota \colon I \to K^{(I)}$, such that $\iota(i) = e_i$ for every $i \in I$, is clearly an injection.

For any other vector space F, for any function $f \colon I \to F$, prove that there is a *unique linear map* $\overline{f} \colon K^{(I)} \to F$, such that

$$f = \overline{f} \circ \iota,$$

as in the following commutative diagram:

We call the vector space $K^{(I)}$ the vector space *freely generated* by the set I.

Problem 11.18. (Some pitfalls of infinite dimension) Let E be the vector space freely generated by the set of natural numbers, $\mathbb{N} = \{0, 1, 2, \ldots\}$, and let $(e_0, e_1, e_2, \ldots, e_n, \ldots)$ be its canonical basis. We define the function φ such that

$$\varphi(e_i, e_j) = \begin{cases} \delta_{ij} & \text{if } i, j \geq 1, \\ 1 & \text{if } i = j = 0, \\ 1/2^j & \text{if } i = 0, j \geq 1, \\ 1/2^i & \text{if } i \geq 1, j = 0, \end{cases}$$

and we extend φ by bilinearity to a function $\varphi \colon E \times E \to K$. This means that if $u = \sum_{i \in \mathbb{N}} \lambda_i e_i$ and $v = \sum_{j \in \mathbb{N}} \mu_j e_j$, then

$$\varphi\left(\sum_{i \in \mathbb{N}} \lambda_i e_i, \sum_{j \in \mathbb{N}} \mu_j e_j\right) = \sum_{i,j \in \mathbb{N}} \lambda_i \mu_j \varphi(e_i, e_j),$$

but remember that $\lambda_i \neq 0$ and $\mu_j \neq 0$ *only for finitely many indices* i, j.

(1) Prove that φ is positive definite, so that it is an inner product on E.

What would happen if we changed $1/2^j$ to 1 (or any constant)?

(2) Let H be the subspace of E spanned by the family $(e_i)_{i \geq 1}$, a hyperplane in E. Find $H^\perp$ and $H^{\perp\perp}$, and prove that

$$H \neq H^{\perp\perp}.$$

(3) Let U be the subspace of E spanned by the family $(e_{2i})_{i \geq 1}$, and let V be the subspace of E spanned by the family $(e_{2i-1})_{i \geq 1}$. Prove that

$$U^\perp = V$$
$$V^\perp = U$$
$$U^{\perp\perp} = U$$
$$V^{\perp\perp} = V,$$

yet

$$(U \cap V)^\perp \neq U^\perp + V^\perp$$

and

$$(U + V)^{\perp\perp} \neq U + V.$$

If W is the subspace spanned by e_0 and e_1, prove that

$$(W \cap H)^\perp \neq W^\perp + H^\perp.$$

(4) Consider the dual space E^* of E, and let $(e_i^*)_{i \in \mathbb{N}}$ be the family of dual forms of the basis $(e_i)_{i \in N}$. Check that the family $(e_i^*)_{i \in \mathbb{N}}$ is linearly independent.

(5) Let $f \in E^*$ be the linear form defined by

$$f(e_i) = 1 \quad \text{for all } i \in \mathbb{N}.$$

Prove that f is not in the subspace spanned by the e_i^*. If F is the subspace of E^* spanned by the e_i^* and f, find F^0 and F^{00}, and prove that

$$F \neq F^{00}.$$

Chapter 12

QR-Decomposition for Arbitrary Matrices

12.1 Orthogonal Reflections

Hyperplane reflections are represented by matrices called Householder matrices. These matrices play an important role in numerical methods, for instance for solving systems of linear equations, solving least squares problems, for computing eigenvalues, and for transforming a symmetric matrix into a tridiagonal matrix. We prove a simple geometric lemma that immediately yields the QR-decomposition of arbitrary matrices in terms of Householder matrices.

Orthogonal symmetries are a very important example of isometries. First let us review the definition of projections, introduced in Section 5.2, just after Proposition 5.5. Given a vector space E, let F and G be subspaces of E that form a direct sum $E = F \oplus G$. Since every $u \in E$ can be written uniquely as $u = v + w$, where $v \in F$ and $w \in G$, we can define the two *projections* $p_F \colon E \to F$ and $p_G \colon E \to G$ such that $p_F(u) = v$ and $p_G(u) = w$. In Section 5.2 we used the notation π_1 and π_2, but in this section it is more convenient to use p_F and p_G.

It is immediately verified that p_G and p_F are linear maps, and that

$$p_F^2 = p_F,\ p_G^2 = p_G,\ p_F \circ p_G = p_G \circ p_F = 0, \quad \text{and} \quad p_F + p_G = \mathrm{id}.$$

Definition 12.1. Given a vector space E, for any two subspaces F and G that form a direct sum $E = F \oplus G$, the *symmetry (or reflection) with respect to F and parallel to G* is the linear map $s \colon E \to E$ defined such that

$$s(u) = 2p_F(u) - u,$$

for every $u \in E$.

463

Because $p_F + p_G = \mathrm{id}$, note that we also have

$$s(u) = p_F(u) - p_G(u)$$

and

$$s(u) = u - 2p_G(u),$$

$s^2 = \mathrm{id}$, s is the identity on F, and $s = -\mathrm{id}$ on G.

We now assume that E is a Euclidean space of *finite* dimension.

Definition 12.2. Let E be a Euclidean space of finite dimension n. For any two subspaces F and G, if F and G form a direct sum $E = F \oplus G$ and F and G are orthogonal, i.e., $F = G^{\perp}$, the *orthogonal symmetry (or reflection) with respect to F and parallel to G* is the linear map $s \colon E \to E$ defined such that

$$s(u) = 2p_F(u) - u = p_F(u) - p_G(u),$$

for every $u \in E$. When F is a hyperplane, we call s a *hyperplane symmetry with respect to F (or reflection about F)*, and when G is a plane (and thus $\dim(F) = n - 2$), we call s a *flip about F*.

A reflection about a hyperplane F is shown in Figure 12.1.

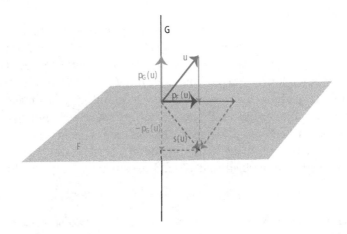

Fig. 12.1 A reflection about the peach hyperplane F. Note that u is purple, $p_F(u)$ is blue and $p_G(u)$ is red.

For any two vectors $u, v \in E$, it is easily verified using the bilinearity of the inner product that

$$\|u + v\|^2 - \|u - v\|^2 = 4(u \cdot v). \tag{12.1}$$

In particular, if $u \cdot v = 0$, then $\|u + v\| = \|u - v\|$. Then since

$$u = p_F(u) + p_G(u)$$

and

$$s(u) = p_F(u) - p_G(u),$$

and since F and G are orthogonal, it follows that

$$p_F(u) \cdot p_G(v) = 0,$$

and thus by (12.1)

$$\|s(u)\| = \|p_F(u) - p_G(u)\| = \|p_F(u) + p_G(u)\| = \|u\|,$$

so that s *is an isometry*.

Using Proposition 11.8, it is possible to find an orthonormal basis $(e_1, \ldots, e_n)$ of E consisting of an orthonormal basis of F and an orthonormal basis of G. Assume that F has dimension p, so that G has dimension $n - p$. With respect to the orthonormal basis $(e_1, \ldots, e_n)$, the symmetry s has a matrix of the form

$$\begin{pmatrix} I_p & 0 \\ 0 & -I_{n-p} \end{pmatrix}.$$

Thus, $\det(s) = (-1)^{n-p}$, and s is a rotation iff $n - p$ is even. In particular, when F is a hyperplane H, we have $p = n - 1$ and $n - p = 1$, so that s is an improper orthogonal transformation. When $F = \{0\}$, we have $s = -\mathrm{id}$, which is called the *symmetry with respect to the origin*. The symmetry with respect to the origin is a rotation iff n is even, and an improper orthogonal transformation iff n is odd. When n is odd, since $s \circ s = \mathrm{id}$ and $\det(s) = (-1)^n = -1$, we observe that every improper orthogonal transformation f is the composition $f = (f \circ s) \circ s$ of the rotation $f \circ s$ with s, the symmetry with respect to the origin. When G is a plane, $p = n - 2$, and $\det(s) = (-1)^2 = 1$, so that a flip about F is a rotation. In particular, when $n = 3$, F is a line, and a flip about the line F is indeed a rotation of measure π as illustrated by Figure 12.2.

Remark: Given any two orthogonal subspaces F, G forming a direct sum $E = F \oplus G$, let f be the symmetry with respect to F and parallel to G, and let g be the symmetry with respect to G and parallel to F. We leave as an exercise to show that

$$f \circ g = g \circ f = -\mathrm{id}.$$

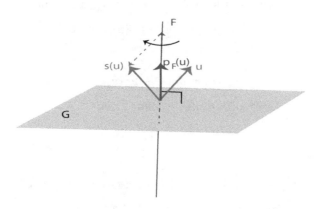

Fig. 12.2   A flip in $\mathbb{R}^3$ is a rotation of π about the F axis.

When $F = H$ is a hyperplane, we can give an explicit formula for $s(u)$ in terms of any nonnull vector w orthogonal to H. Indeed, from

$$u = p_H(u) + p_G(u),$$

since $p_G(u) \in G$ and G is spanned by w, which is orthogonal to H, we have

$$p_G(u) = \lambda w$$

for some $\lambda \in \mathbb{R}$, and we get

$$u \cdot w = \lambda \|w\|^2,$$

and thus

$$p_G(u) = \frac{(u \cdot w)}{\|w\|^2}\, w.$$

Since

$$s(u) = u - 2p_G(u),$$

we get

$$s(u) = u - 2\,\frac{(u \cdot w)}{\|w\|^2}\, w.$$

Since the above formula is important, we record it in the following proposition.

Proposition 12.1. *Let E be a finite-dimensional Euclidean space and let H be a hyperplane in E. For any nonzero vector w orthogonal to H, the hyperplane reflection s about H is given by*

$$s(u) = u - 2\,\frac{(u \cdot w)}{\|w\|^2}\, w, \quad u \in E.$$

Such reflections are represented by matrices called *Householder matrices*, which play an important role in numerical matrix analysis (see Kincaid and Cheney [Kincaid and Cheney (1996)] or Ciarlet [Ciarlet (1989)]).

Definition 12.3. A *Householder matrix* if a matrix of the form

$$H = I_n - 2\frac{WW^\top}{\|W\|^2} = I_n - 2\frac{WW^\top}{W^\top W},$$

where $W \in \mathbb{R}^n$ is a nonzero vector.

Householder matrices are symmetric and orthogonal. It is easily checked that over an orthonormal basis $(e_1, \ldots, e_n)$, a hyperplane reflection about a hyperplane H orthogonal to a nonzero vector w is represented by the matrix

$$H = I_n - 2\frac{WW^\top}{\|W\|^2},$$

where W is the column vector of the coordinates of w over the basis $(e_1, \ldots, e_n)$. Since

$$p_G(u) = \frac{(u \cdot w)}{\|w\|^2}\, w,$$

the matrix representing p_G is

$$\frac{WW^\top}{W^\top W},$$

and since $p_H + p_G = \mathrm{id}$, the matrix representing p_H is

$$I_n - \frac{WW^\top}{W^\top W}.$$

These formulae can be used to derive a formula for a rotation of $\mathbb{R}^3$, given the direction w of its axis of rotation and given the angle θ of rotation.

The following fact is the key to the proof that every isometry can be decomposed as a product of reflections.

Proposition 12.2. *Let E be any nontrivial Euclidean space. For any two vectors $u, v \in E$, if $\|u\| = \|v\|$, then there is a hyperplane H such that the reflection s about H maps u to v, and if $u \neq v$, then this reflection is unique. See Figure 12.3.*

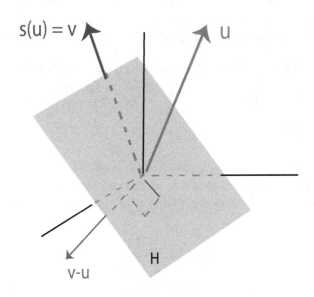

Fig. 12.3   In $\mathbb{R}^3$, the (hyper)plane perpendicular to $v - u$ reflects u onto v.

Proof. If $u = v$, then any hyperplane containing u does the job. Otherwise, we must have $H = \{v - u\}^\perp$, and by the above formula,

$$s(u) = u - 2\,\frac{(u \cdot (v - u))}{\|(v - u)\|^2}\,(v - u) = u + \frac{2\|u\|^2 - 2u \cdot v}{\|(v - u)\|^2}\,(v - u),$$

and since

$$\|(v - u)\|^2 = \|u\|^2 + \|v\|^2 - 2u \cdot v$$

and $\|u\| = \|v\|$, we have

$$\|(v - u)\|^2 = 2\|u\|^2 - 2u \cdot v,$$

and thus, $s(u) = v$. $\square$

If E is a complex vector space and the inner product is Hermitian, Proposition 12.2 is false. The problem is that the vector $v - u$ does not work unless the inner product $u \cdot v$ is real! The proposition can be salvaged enough to yield the QR-decomposition in terms of Householder transformations; see Section 13.5.

We now show that hyperplane reflections can be used to obtain another proof of the QR-decomposition.

12.2 QR-Decomposition Using Householder Matrices

First we state the result geometrically. When translated in terms of Householder matrices, we obtain the fact advertised earlier that every matrix (not necessarily invertible) has a QR-decomposition.

Proposition 12.3. *Let E be a nontrivial Euclidean space of dimension n. For any orthonormal basis $(e_1, \ldots, e_n)$ and for any n-tuple of vectors $(v_1, \ldots, v_n)$, there is a sequence of n isometries $h_1, \ldots, h_n$ such that h_i is a hyperplane reflection or the identity, and if $(r_1, \ldots, r_n)$ are the vectors given by*

$$r_j = h_n \circ \cdots \circ h_2 \circ h_1(v_j),$$

then every r_j is a linear combination of the vectors $(e_1, \ldots, e_j)$, $1 \leq j \leq n$. Equivalently, the matrix R whose columns are the components of the r_j over the basis $(e_1, \ldots, e_n)$ is an upper triangular matrix. Furthermore, the h_i can be chosen so that the diagonal entries of R are nonnegative.

Proof. We proceed by induction on n. For $n = 1$, we have $v_1 = \lambda e_1$ for some $\lambda \in \mathbb{R}$. If $\lambda \geq 0$, we let $h_1 = \mathrm{id}$, else if $\lambda < 0$, we let $h_1 = -\mathrm{id}$, the reflection about the origin.

For $n \geq 2$, we first have to find h_1. Let

$$r_{1,1} = \|v_1\|.$$

If $v_1 = r_{1,1} e_1$, we let $h_1 = \mathrm{id}$. Otherwise, there is a unique hyperplane reflection h_1 such that

$$h_1(v_1) = r_{1,1} e_1,$$

defined such that

$$h_1(u) = u - 2 \frac{(u \cdot w_1)}{\|w_1\|^2} w_1$$

for all $u \in E$, where

$$w_1 = r_{1,1} e_1 - v_1.$$

The map h_1 is the reflection about the hyperplane H_1 orthogonal to the vector $w_1 = r_{1,1} e_1 - v_1$. See Figure 12.4. Letting

$$r_1 = h_1(v_1) = r_{1,1} e_1,$$

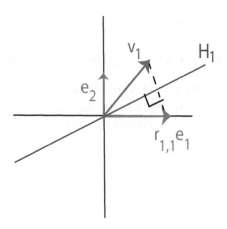

Fig. 12.4   The construction of h_1 in Proposition 12.3.

it is obvious that r_1 belongs to the subspace spanned by e_1, and $r_{1,1} = \|v_1\|$ is nonnegative.

Next assume that we have found k linear maps $h_1, \ldots, h_k$, hyperplane reflections or the identity, where $1 \le k \le n - 1$, such that if $(r_1, \ldots, r_k)$ are the vectors given by

$$r_j = h_k \circ \cdots \circ h_2 \circ h_1(v_j),$$

then every r_j is a linear combination of the vectors $(e_1, \ldots, e_j)$, $1 \le j \le k$. See Figure 12.5. The vectors $(e_1, \ldots, e_k)$ form a basis for the subspace denoted by U_k', the vectors $(e_{k+1}, \ldots, e_n)$ form a basis for the subspace denoted by U_k'', the subspaces U_k' and U_k'' are orthogonal, and $E = U_k' \oplus U_k''$. Let

$$u_{k+1} = h_k \circ \cdots \circ h_2 \circ h_1(v_{k+1}).$$

We can write

$$u_{k+1} = u_{k+1}' + u_{k+1}'',$$

where $u_{k+1}' \in U_k'$ and $u_{k+1}'' \in U_k''$. See Figure 12.6. Let

$$r_{k+1,k+1} = \|u_{k+1}''\|.$$

If $u_{k+1}'' = r_{k+1,k+1} \, e_{k+1}$, we let $h_{k+1} = \text{id}$. Otherwise, there is a unique hyperplane reflection h_{k+1} such that

$$h_{k+1}(u_{k+1}'') = r_{k+1,k+1} \, e_{k+1},$$

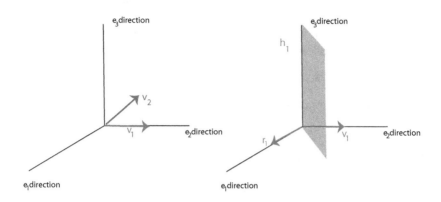

Fig. 12.5   The construction of $r_1 = h_1(v_1)$ in Proposition 12.3.

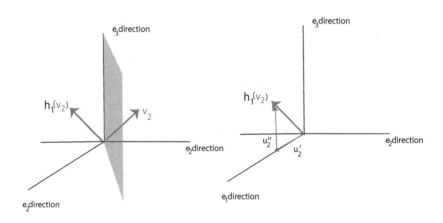

Fig. 12.6   The construction of $u_2 = h_1(v_2)$ and its decomposition as $u_2 = u'_2 + u''_2$.

defined such that

$$h_{k+1}(u) = u - 2\,\frac{(u \cdot w_{k+1})}{\|w_{k+1}\|^2}\,w_{k+1}$$

for all $u \in E$, where

$$w_{k+1} = r_{k+1,k+1}\,e_{k+1} - u''_{k+1}.$$

The map h_{k+1} is the reflection about the hyperplane H_{k+1} orthogonal to the vector $w_{k+1} = r_{k+1,k+1}\, e_{k+1} - u''_{k+1}$. However, since $u''_{k+1}, e_{k+1} \in U''_k$ and U'_k is orthogonal to U''_k, the subspace U'_k is contained in H_{k+1}, and thus, the vectors $(r_1, \ldots, r_k)$ and u'_{k+1}, which belong to U'_k, are invariant under h_{k+1}. This proves that

$$h_{k+1}(u_{k+1}) = h_{k+1}(u'_{k+1}) + h_{k+1}(u''_{k+1}) = u'_{k+1} + r_{k+1,k+1}\, e_{k+1}$$

is a linear combination of $(e_1, \ldots, e_{k+1})$. Letting

$$r_{k+1} = h_{k+1}(u_{k+1}) = u'_{k+1} + r_{k+1,k+1}\, e_{k+1},$$

since $u_{k+1} = h_k \circ \cdots \circ h_2 \circ h_1(v_{k+1})$, the vector

$$r_{k+1} = h_{k+1} \circ \cdots \circ h_2 \circ h_1(v_{k+1})$$

is a linear combination of $(e_1, \ldots, e_{k+1})$. See Figure 12.7. The coefficient of r_{k+1} over e_{k+1} is $r_{k+1,k+1} = \|u''_{k+1}\|$, which is nonnegative. This concludes the induction step, and thus the proof. $\square$

Remarks:

(1) Since every h_i is a hyperplane reflection or the identity,

$$\rho = h_n \circ \cdots \circ h_2 \circ h_1$$

is an isometry.

(2) If we allow negative diagonal entries in R, the last isometry h_n may be omitted.

(3) Instead of picking $r_{k,k} = \|u''_k\|$, which means that

$$w_k = r_{k,k}\, e_k - u''_k,$$

where $1 \leq k \leq n$, it might be preferable to pick $r_{k,k} = -\|u''_k\|$ if this makes $\|w_k\|^2$ larger, in which case

$$w_k = r_{k,k}\, e_k + u''_k.$$

Indeed, since the definition of h_k involves division by $\|w_k\|^2$, it is desirable to avoid division by very small numbers.

(4) The method also applies to any m-tuple of vectors $(v_1, \ldots, v_m)$, with $m \leq n$. Then R is an upper triangular $m \times m$ matrix and Q is an $n \times m$ matrix with orthogonal columns ($Q^\top Q = I_m$). We leave the minor adjustments to the method as an exercise to the reader

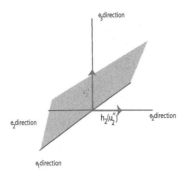

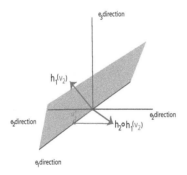

Fig. 12.7 The construction of h_2 and $r_2 = h_2 \circ h_1(v_2)$ in Proposition 12.3.

Proposition 12.3 directly yields the QR-decomposition in terms of Householder transformations (see Strang [Strang (1986, 1988)], Golub and Van Loan [Golub and Van Loan (1996)], Trefethen and Bau [Trefethen and Bau III (1997)], Kincaid and Cheney [Kincaid and Cheney (1996)], or Ciarlet [Ciarlet (1989)]).

Theorem 12.1. *For every real $n \times n$ matrix A, there is a sequence $H_1, \ldots,$ H_n of matrices, where each H_i is either a Householder matrix or the identity, and an upper triangular matrix R such that*

$$R = H_n \cdots H_2 H_1 A.$$

As a corollary, there is a pair of matrices Q, R, where Q is orthogonal and R is upper triangular, such that $A = QR$ (a QR-decomposition of A).

Furthermore, R can be chosen so that its diagonal entries are nonnegative.

Proof. The jth column of A can be viewed as a vector v_j over the canonical basis $(e_1, \ldots, e_n)$ of $\mathbb{E}^n$ (where $(e_j)_i = 1$ if $i = j$, and 0 otherwise, $1 \leq i, j \leq n$). Applying Proposition 12.3 to $(v_1, \ldots, v_n)$, there is a sequence of n isometries $h_1, \ldots, h_n$ such that h_i is a hyperplane reflection or the identity, and if $(r_1, \ldots, r_n)$ are the vectors given by

$$r_j = h_n \circ \cdots \circ h_2 \circ h_1(v_j),$$

then every r_j is a linear combination of the vectors $(e_1, \ldots, e_j)$, $1 \leq j \leq n$. Letting R be the matrix whose columns are the vectors r_j, and H_i the matrix associated with h_i, it is clear that

$$R = H_n \cdots H_2 H_1 A,$$

where R is upper triangular and every H_i is either a Householder matrix or the identity. However, $h_i \circ h_i = \mathrm{id}$ for all i, $1 \leq i \leq n$, and so

$$v_j = h_1 \circ h_2 \circ \cdots \circ h_n(r_j)$$

for all j, $1 \leq j \leq n$. But $\rho = h_1 \circ h_2 \circ \cdots \circ h_n$ is an isometry represented by the orthogonal matrix $Q = H_1 H_2 \cdots H_n$. It is clear that $A = QR$, where R is upper triangular. As we noted in Proposition 12.3, the diagonal entries of R can be chosen to be nonnegative. $\qquad\square$

Remarks:

(1) Letting

$$A_{k+1} = H_k \cdots H_2 H_1 A,$$

with $A_1 = A$, $1 \leq k \leq n$, the proof of Proposition 12.3 can be interpreted in terms of the computation of the sequence of matrices $A_1, \ldots, A_{n+1} = R$. The matrix A_{k+1} has the shape

$$A_{k+1} = \begin{pmatrix} \times & \times & \times & u_1^{k+1} & \times & \times & \times & \times \\ 0 & \times & \vdots & \vdots & \vdots & \vdots & \vdots & \vdots \\ 0 & 0 & \times & u_k^{k+1} & \times & \times & \times & \times \\ 0 & 0 & 0 & u_{k+1}^{k+1} & \times & \times & \times & \times \\ 0 & 0 & 0 & u_{k+2}^{k+1} & \times & \times & \times & \times \\ \vdots & \vdots & \vdots & \vdots & \vdots & \vdots & \vdots & \vdots \\ 0 & 0 & 0 & u_{n-1}^{k+1} & \times & \times & \times & \times \\ 0 & 0 & 0 & u_n^{k+1} & \times & \times & \times & \times \end{pmatrix},$$

where the $(k+1)$th column of the matrix is the vector

$$u_{k+1} = h_k \circ \cdots \circ h_2 \circ h_1(v_{k+1}),$$

and thus

$$u'_{k+1} = \left(u_1^{k+1}, \ldots, u_k^{k+1} \right)$$

and

$$u''_{k+1} = \left(u_{k+1}^{k+1}, u_{k+2}^{k+1}, \ldots, u_n^{k+1} \right).$$

If the last $n - k - 1$ entries in column $k+1$ are all zero, there is nothing to do, and we let $H_{k+1} = I$. Otherwise, we kill these $n-k-1$ entries by multiplying A_{k+1} on the left by the Householder matrix H_{k+1} sending

$$\left(0, \ldots, 0, u_{k+1}^{k+1}, \ldots, u_n^{k+1} \right) \quad \text{to} \quad (0, \ldots, 0, r_{k+1,k+1}, 0, \ldots, 0),$$

where $r_{k+1,k+1} = \| (u_{k+1}^{k+1}, \ldots, u_n^{k+1}) \|$.

(2) If A is invertible and the diagonal entries of R are positive, it can be shown that Q and R are unique.

(3) If we allow negative diagonal entries in R, the matrix H_n may be omitted ($H_n = I$).

(4) The method allows the computation of the determinant of A. We have

$$\det(A) = (-1)^m r_{1,1} \cdots r_{n,n},$$

where m is the number of Householder matrices (not the identity) among the H_i.

(5) The "condition number" of the matrix A is preserved (see Strang [Strang (1988)], Golub and Van Loan [Golub and Van Loan (1996)], Trefethen and Bau [Trefethen and Bau III (1997)], Kincaid and Cheney [Kincaid and Cheney (1996)], or Ciarlet [Ciarlet (1989)]). This is very good for numerical stability.

(6) The method also applies to a rectangular $m \times n$ matrix. If $m \geq n$, then R is an $n \times n$ upper triangular matrix and Q is an $m \times n$ matrix such that $Q^\top Q = I_n$.

The following Matlab functions implement the QR-factorization method of a real square (possibly singular) matrix A using Householder reflections

The main function houseqr computes the upper triangular matrix R obtained by applying Householder reflections to A. It makes use of the function house, which computes a unit vector u such that given a vector $x \in \mathbb{R}^p$, the Householder transformation $P = I - 2uu^\top$ sets to zero all entries in x but the first entry x_1. It only applies if $\|x(2:p)\|_1 = |x_2| + \cdots + |x_p| > 0$.

Since computations are done in floating point, we use a tolerance factor
tol, and if $\|x(2:p)\|_1 \leq tol$, then we return $u = 0$, which indicates that the
corresponding Householder transformation is the identity. To make sure
that $\|Px\|$ is as large as possible, we pick $uu = x + \text{sign}(x_1)\|x\|_2\, e_1$, where
$\text{sign}(z) = 1$ if $z \geq 0$ and $\text{sign}(z) = -1$ if $z < 0$. Note that as a result,
diagonal entries in R may be negative. We will take care of this issue later.

```
function s = signe(x)
%  if x >= 0, then signe(x) = 1
%  else if x < 0 then signe(x) = -1
%

if x < 0
   s = -1;
else
    s = 1;
end
end


function [uu, u] = house(x)
% This constructs the unnormalized  vector uu
% defining the Householder reflection that
% zeros all but the first entries in x.
% u is the normalized vector uu/||uu||
%

tol = 2*10^(-15);   % tolerance
uu = x;
p = size(x,1);
% computes l^1-norm of x(2:p,1)
n1 = sum(abs(x(2:p,1)));
if n1 <= tol
   u = zeros(p,1);   uu = u;
else
   l = sqrt(x'*x);  % l^2 norm of x
   uu(1) = x(1) + signe(x(1))*l;
   u = uu/sqrt(uu'*uu);
end
end
```

The Householder transformations are recorded in an array u of $n - 1$ vectors. There are more efficient implementations, but for the sake of clarity we present the following version.

```
function [R,  u] = houseqr(A)
%  This function computes the upper triangular R in the QR
%  factorization of A using Householder reflections, and an
%  implicit representation of Q as  a sequence of n - 1
%  vectors u_i representing Householder reflections
n = size(A, 1);
R = A;
u = zeros(n,n-1);
for i = 1:n-1
    [~, u(i:n,i)] = house(R(i:n,i));
    if u(i:n,i) == zeros(n - i + 1,1)
       R(i+1:n,i) = zeros(n - i,1);
    else
       R(i:n,i:n) = R(i:n,i:n)
                   - 2*u(i:n,i)*(u(i:n,i)'*R(i:n,i:n));
    end
end
end
```

If only R is desired, then **houseqr** does the job. In order to obtain R, we need to compose the Householder transformations. We present a simple method which is not the most efficient (there is a way to avoid multiplying explicity the Householder matrices).

The function **buildhouse** creates a Householder reflection from a vector v.

```
function P = buildhouse(v,i)
% This function builds a Householder reflection
%    [I 0 ]
%    [0 PP]
%    from a Householder reflection
%    PP = I - 2uu*uu'
%    where uu = v(i:n)
%    If uu = 0 then P - I
%
```

```
n = size(v,1);
if v(i:n) == zeros(n - i + 1,1)
   P = eye(n);
else
   PP = eye(n - i + 1) - 2*v(i:n)*v(i:n)';
   P = [eye(i-1) zeros(i-1, n - i + 1);
   zeros(n - i + 1, i - 1) PP];
end
end
```

The function `buildQ` builds the matrix Q in the QR-decomposition of A.

```
function Q = buildQ(u)
% Builds the matrix Q in the QR decomposition
% of an nxn  matrix A using Householder matrices,
% where u is a representation of the n - 1
% Householder reflection by a list u of vectors produced by
% houseqr

n = size(u,1);
Q = buildhouse(u(:,1),1);
for i = 2:n-1
  Q = Q*buildhouse(u(:,i),i);
end
end
```

The function `buildhouseQR` computes a QR-factorization of A. At the end, if some entries on the diagonal of R are negative, it creates a diagonal orthogonal matrix P such that PR has nonnegative diagonal entries, so that $A = (QP)(PR)$ is the desired QR-factorization of A.

```
function [Q,R] = buildhouseQR(A)
%
%    Computes the QR decomposition of a square
%    matrix A (possibly singular) using Householder reflections

n = size(A,1);
[R,u] = houseqr(A);
Q = buildQ(u);
% Produces a matrix R whose diagonal entries are
```

```
% nonnegative
P = eye(n);
for i = 1:n
   if R(i,i) < 0
      P(i,i) = -1;
   end
end
Q = Q*P; R = P*R;
end
```

Example 12.1. Consider the matrix

$$A = \begin{pmatrix} 1\ 2\ 3\ 4 \\ 2\ 3\ 4\ 5 \\ 3\ 4\ 5\ 6 \\ 4\ 5\ 6\ 7 \end{pmatrix}.$$

Running the function buildhouseQR, we get

$$Q = \begin{pmatrix} 0.1826 & 0.8165 & 0.4001 & 0.3741 \\ 0.3651 & 0.4082 & -0.2546 & -0.7970 \\ 0.5477 & -0.0000 & -0.6910 & 0.4717 \\ 0.7303 & -0.4082 & 0.5455 & -0.0488 \end{pmatrix}$$

and

$$R = \begin{pmatrix} 5.4772 & 7.3030 & 9.1287 & 10.9545 \\ 0 & 0.8165 & 1.6330 & 2.4495 \\ 0 & -0.0000 & 0.0000 & 0.0000 \\ 0 & -0.0000 & 0 & 0.0000 \end{pmatrix}.$$

Observe that A has rank 2. The reader should check that $A = QR$.

Remark: Curiously, running Matlab built-in function qr, the same R is obtained (up to column signs) but a different Q is obtained (the last two columns are different).

12.3   Summary

The main concepts and results of this chapter are listed below:

- *Symmetry (or reflection) with respect to F and parallel to G.*
- *Orthogonal symmetry (or reflection) with respect to F and parallel to G; reflections, flips.*

- Hyperplane reflections and *Householder matrices*.
- A key fact about reflections (Proposition 12.2).
- *QR-decomposition in terms of Householder transformations* (Theorem 12.1).

12.4   Problems

Problem 12.1. (1) Given a unit vector $(-\sin\theta, \cos\theta)$, prove that the Householder matrix determined by the vector $(-\sin\theta, \cos\theta)$ is

$$\begin{pmatrix} \cos 2\theta & \sin 2\theta \\ \sin 2\theta & -\cos 2\theta \end{pmatrix}.$$

Give a geometric interpretation (i.e., why the choice $(-\sin\theta, \cos\theta)$?).

(2) Given any matrix

$$A = \begin{pmatrix} a & b \\ c & d \end{pmatrix},$$

Prove that there is a Householder matrix H such that AH is lower triangular, i.e.,

$$AH = \begin{pmatrix} a' & 0 \\ c' & d' \end{pmatrix}$$

for some $a', c', d' \in \mathbb{R}$.

Problem 12.2. Given a Euclidean space E of dimension n, if h is a reflection about some hyperplane orthogonal to a nonzero vector u and f is any isometry, prove that $f \circ h \circ f^{-1}$ is the reflection about the hyperplane orthogonal to $f(u)$.

Problem 12.3. (1) Given a matrix

$$A = \begin{pmatrix} a & b \\ c & d \end{pmatrix},$$

prove that there are Householder matrices G, H such that

$$GAH = \begin{pmatrix} \cos\theta & \sin\theta \\ \sin\theta & -\cos\theta \end{pmatrix} \begin{pmatrix} a & b \\ c & d \end{pmatrix} \begin{pmatrix} \cos\varphi & \sin\varphi \\ \sin\varphi & -\cos\varphi \end{pmatrix} = D,$$

where D is a diagonal matrix, iff the following equations hold:

$$(b + c)\cos(\theta + \varphi) = (a - d)\sin(\theta + \varphi),$$
$$(c - b)\cos(\theta - \varphi) = (a + d)\sin(\theta - \varphi).$$

(2) Discuss the solvability of the system. Consider the following cases:

Case 1: $a - d = a + d = 0$.

Case 2a: $a - d = b + c = 0$, $a + d \neq 0$.

Case 2b: $a - d = 0$, $b + c \neq 0$, $a + d \neq 0$.

Case 3a: $a + d = c - b = 0$, $a - d \neq 0$.

Case 3b: $a + d = 0$, $c - b \neq 0$, $a - d \neq 0$.

Case 4: $a + d \neq 0$, $a - d \neq 0$. Show that the solution in this case is

$$\theta = \frac{1}{2}\left[\arctan\left(\frac{b+c}{a-d}\right) + \arctan\left(\frac{c-b}{a+d}\right)\right],$$

$$\varphi = \frac{1}{2}\left[\arctan\left(\frac{b+c}{a-d}\right) - \arctan\left(\frac{c-b}{a+d}\right)\right].$$

If $b = 0$, show that the discussion is simpler: basically, consider $c = 0$ or $c \neq 0$.

(3) Expressing everything in terms of $u = \cot\theta$ and $v = \cot\varphi$, show that the equations in (2) become

$$(b+c)(uv - 1) = (u+v)(a-d),$$

$$(c-b)(uv + 1) = (-u+v)(a+d).$$

Problem 12.4. Let A be an $n \times n$ real invertible matrix.

(1) Prove that $A^\top A$ is symmetric positive definite.

(2) Use the Cholesky factorization $A^\top A = R^\top R$ with R upper triangular with positive diagonal entries to prove that $Q = AR^{-1}$ is orthogonal, so that $A = QR$ is the QR-factorization of A.

Problem 12.5. Modify the function houseqr so that it applies to an $m \times n$ matrix with $m \geq n$, to produce an $m \times n$ upper-triangular matrix whose last $m - n$ rows are zeros.

Problem 12.6. The purpose of this problem is to prove that given any self-adjoint linear map $f \colon E \to E$ (i.e., such that $f^* = f$), where E is a Euclidean space of dimension $n \geq 3$, given an orthonormal basis $(e_1, \ldots, e_n)$, there are $n - 2$ isometries h_i, hyperplane reflections or the identity, such that the matrix of

$$h_{n-2} \circ \cdots \circ h_1 \circ f \circ h_1 \circ \cdots \circ h_{n-2}$$

is a symmetric tridiagonal matrix.

(1) Prove that for any isometry $f \colon E \to E$, we have $f = f^* = f^{-1}$ iff $f \circ f = \mathrm{id}$.

Prove that if f and h are self-adjoint linear maps ($f^* = f$ and $h^* = h$), then $h \circ f \circ h$ is a self-adjoint linear map.

(2) Let V_k be the subspace spanned by $(e_{k+1}, \ldots, e_n)$. Proceed by induction. For the base case, proceed as follows.

Let
$$f(e_1) = a_1^0 e_1 + \cdots + a_n^0 e_n,$$
and let
$$r_{1,2} = \left\| a_2^0 e_2 + \cdots + a_n^0 e_n \right\|.$$
Find an isometry h_1 (reflection or id) such that
$$h_1(f(e_1) - a_1^0 e_1) = r_{1,2}\, e_2.$$
Observe that
$$w_1 = r_{1,2}\, e_2 + a_1^0 e_1 - f(e_1) \in V_1,$$
and prove that $h_1(e_1) = e_1$, so that
$$h_1 \circ f \circ h_1(e_1) = a_1^0 e_1 + r_{1,2}\, e_2.$$
Let $f_1 = h_1 \circ f \circ h_1$.

Assuming by induction that
$$f_k = h_k \circ \cdots \circ h_1 \circ f \circ h_1 \circ \cdots \circ h_k$$
has a tridiagonal matrix up to the kth row and column, $1 \leq k \leq n - 3$, let
$$f_k(e_{k+1}) = a_k^k e_k + a_{k+1}^k e_{k+1} + \cdots + a_n^k e_n,$$
and let
$$r_{k+1,\,k+2} = \left\| a_{k+2}^k e_{k+2} + \cdots + a_n^k e_n \right\|.$$
Find an isometry h_{k+1} (reflection or id) such that
$$h_{k+1}(f_k(e_{k+1}) - a_k^k e_k - a_{k+1}^k e_{k+1}) = r_{k+1,\,k+2}\, e_{k+2}.$$
Observe that
$$w_{k+1} = r_{k+1,\,k+2}\, e_{k+2} + a_k^k e_k + a_{k+1}^k e_{k+1} - f_k(e_{k+1}) \in V_{k+1},$$
and prove that $h_{k+1}(e_k) = e_k$ and $h_{k+1}(e_{k+1}) = e_{k+1}$, so that
$$h_{k+1} \circ f_k \circ h_{k+1}(e_{k+1}) = a_k^k e_k + a_{k+1}^k e_{k+1} + r_{k+1,\,k+2}\, e_{k+2}.$$
Let $f_{k+1} = h_{k+1} \circ f_k \circ h_{k+1}$, and finish the proof.

(3) Prove that given any symmetric $n \times n$-matrix A, there are $n - 2$ matrices $H_1, \ldots, H_{n-2}$, Householder matrices or the identity, such that
$$B = H_{n-2} \cdots H_1 A H_1 \cdots H_{n-2}$$
is a symmetric tridiagonal matrix.

(4) Write a computer program implementing the above method.

Problem 12.7. Recall from Problem 5.6 that an $n \times n$ matrix H is *upper Hessenberg* if $h_{jk} = 0$ for all (j, k) such that $j - k \geq 0$. Adapt the proof of Problem 12.6 to prove that given any $n \times n$-matrix A, there are $n - 2 \geq 1$ matrices $H_1, \ldots, H_{n-2}$, Householder matrices or the identity, such that

$$B = H_{n-2} \cdots H_1 A H_1 \cdots H_{n-2}$$

is upper Hessenberg.

Problem 12.8. The purpose of this problem is to prove that given any linear map $f : E \to E$, where E is a Euclidean space of dimension $n \geq 2$, given an orthonormal basis $(e_1, \ldots, e_n)$, there are isometries g_i, h_i, hyperplane reflections or the identity, such that the matrix of

$$g_n \circ \cdots \circ g_1 \circ f \circ h_1 \circ \cdots \circ h_n$$

is a lower bidiagonal matrix, which means that the nonzero entries (if any) are on the main descending diagonal and on the diagonal below it.

(1) Let U_k' be the subspace spanned by $(e_1, \ldots, e_k)$ and U_k'' be the subspace spanned by $(e_{k+1}, \ldots, e_n)$, $1 \leq k \leq n - 1$. Proceed by induction. For the base case, proceed as follows.

Let $v_1 = f^*(e_1)$ and $r_{1,1} = \|v_1\|$. Find an isometry h_1 (reflection or id) such that

$$h_1(f^*(e_1)) = r_{1,1} e_1.$$

Observe that $h_1(f^*(e_1)) \in U_1'$, so that

$$\langle h_1(f^*(e_1)), e_j \rangle = 0$$

for all $j, 2 \leq j \leq n$, and conclude that

$$\langle e_1, f \circ h_1(e_j) \rangle = 0$$

for all $j, 2 \leq j \leq n$.

Next let

$$u_1 = f \circ h_1(e_1) = u_1' + u_1'',$$

where $u_1' \in U_1'$ and $u_1'' \in U_1''$, and let $r_{2,1} = \|u_1''\|$. Find an isometry g_1 (reflection or id) such that

$$g_1(u_1'') = r_{2,1} e_2.$$

Show that $g_1(e_1) = e_1$,

$$g_1 \circ f \circ h_1(e_1) = u_1' + r_{2,1} e_2,$$

and that

$$\langle e_1, g_1 \circ f \circ h_1(e_j) \rangle = 0$$

for all $j, 2 \leq j \leq n$. At the end of this stage, show that $g_1 \circ f \circ h_1$ has a matrix such that all entries on its first row except perhaps the first are zero, and that all entries on the first column, except perhaps the first two, are zero.

Assume by induction that some isometries $g_1, \ldots, g_k$ and $h_1, \ldots, h_k$ have been found, either reflections or the identity, and such that

$$f_k = g_k \circ \cdots \circ g_1 \circ f \circ h_1 \cdots \circ h_k$$

has a matrix which is lower bidiagonal up to and including row and column k, where $1 \leq k \leq n - 2$.

Let

$$v_{k+1} = f_k^*(e_{k+1}) = v'_{k+1} + v''_{k+1},$$

where $v'_{k+1} \in U'_k$ and $v''_{k+1} \in U''_k$, and let $r_{k+1, k+1} = \left\| v''_{k+1} \right\|$. Find an isometry h_{k+1} (reflection or id) such that

$$h_{k+1}(v''_{k+1}) = r_{k+1, k+1} e_{k+1}.$$

Show that if h_{k+1} is a reflection, then $U'_k \subseteq H_{k+1}$, where H_{k+1} is the hyperplane defining the reflection h_{k+1}. Deduce that $h_{k+1}(v'_{k+1}) = v'_{k+1}$, and that

$$h_{k+1}(f_k^*(e_{k+1})) = v'_{k+1} + r_{k+1, k+1} e_{k+1}.$$

Observe that $h_{k+1}(f_k^*(e_{k+1})) \in U'_{k+1}$, so that

$$\langle h_{k+1}(f_k^*(e_{k+1})), e_j \rangle = 0$$

for all $j, k + 2 \leq j \leq n$, and thus,

$$\langle e_{k+1}, f_k \circ h_{k+1}(e_j) \rangle = 0$$

for all $j, k + 2 \leq j \leq n$.

Next let

$$u_{k+1} = f_k \circ h_{k+1}(e_{k+1}) = u'_{k+1} + u''_{k+1},$$

where $u'_{k+1} \in U'_{k+1}$ and $u''_{k+1} \in U''_{k+1}$, and let $r_{k+2, k+1} = \left\| u''_{k+1} \right\|$. Find an isometry g_{k+1} (reflection or id) such that

$$g_{k+1}(u''_{k+1}) = r_{k+2, k+1} e_{k+2}.$$

Show that if g_{k+1} is a reflection, then $U'_{k+1} \subseteq G_{k+1}$, where G_{k+1} is the hyperplane defining the reflection g_{k+1}. Deduce that $g_{k+1}(e_i) = e_i$ for all i, $1 \leq i \leq k+1$, and that

$$g_{k+1} \circ f_k \circ h_{k+1}(e_{k+1}) = u'_{k+1} + r_{k+2,\,k+1}e_{k+2}.$$

Since by induction hypothesis,

$$\langle e_i, f_k \circ h_{k+1}(e_j) \rangle = 0$$

for all i, j, $1 \leq i \leq k+1$, $k+2 \leq j \leq n$, and since $g_{k+1}(e_i) = e_i$ for all i, $1 \leq i \leq k+1$, conclude that

$$\langle e_i, g_{k+1} \circ f_k \circ h_{k+1}(e_j) \rangle = 0$$

for all i, j, $1 \leq i \leq k+1$, $k+2 \leq j \leq n$. Finish the proof.

Chapter 13

Hermitian Spaces

13.1 Sesquilinear and Hermitian Forms, Pre-Hilbert Spaces and Hermitian Spaces

In this chapter we generalize the basic results of Euclidean geometry presented in Chapter 11 to vector spaces over the complex numbers. Such a generalization is inevitable and not simply a luxury. For example, linear maps may not have real eigenvalues, but they always have complex eigenvalues. Furthermore, some very important classes of linear maps can be diagonalized if they are extended to the complexification of a real vector space. This is the case for orthogonal matrices and, more generally, normal matrices. Also, complex vector spaces are often the natural framework in physics or engineering, and they are more convenient for dealing with Fourier series. However, some complications arise due to complex conjugation.

Recall that for any complex number $z \in \mathbb{C}$, if $z = x + iy$ where $x, y \in \mathbb{R}$, we let $\Re z = x$, the real part of z, and $\Im z = y$, the imaginary part of z. We also denote the conjugate of $z = x + iy$ by $\bar{z} = x - iy$, and the absolute value (or length, or modulus) of z by $|z|$. Recall that $|z|^2 = z\bar{z} = x^2 + y^2$.

There are many natural situations where a map $\varphi \colon E \times E \to \mathbb{C}$ is linear in its first argument and only semilinear in its second argument, which means that $\varphi(u, \mu v) = \bar{\mu}\varphi(u, v)$, as opposed to $\varphi(u, \mu v) = \mu\varphi(u, v)$. For example, the natural inner product to deal with functions $f \colon \mathbb{R} \to \mathbb{C}$, especially Fourier series, is

$$\langle f, g \rangle = \int_{-\pi}^{\pi} f(x)\overline{g(x)}dx,$$

which is semilinear (but not linear) in g. Thus, when generalizing a result from the real case of a Euclidean space to the complex case, we always

487

have to check very carefully that our proofs do not rely on linearity in the second argument. Otherwise, we need to revise our proofs, and sometimes the result is simply wrong!

Before defining the natural generalization of an inner product, it is convenient to define semilinear maps.

Definition 13.1. Given two vector spaces E and F over the complex field $\mathbb{C}$, a function $f: E \to F$ is *semilinear* if

$$f(u + v) = f(u) + f(v),$$
$$f(\lambda u) = \overline{\lambda} f(u),$$

for all $u, v \in E$ and all $\lambda \in \mathbb{C}$.

Remark: Instead of defining semilinear maps, we could have defined the vector space $\overline{E}$ as the vector space with the same carrier set E whose addition is the same as that of E, but whose multiplication by a complex number is given by

$$(\lambda, u) \mapsto \overline{\lambda} u.$$

Then it is easy to check that a function $f: E \to \mathbb{C}$ is semilinear iff $f: \overline{E} \to \mathbb{C}$ is linear.

We can now define sesquilinear forms and Hermitian forms.

Definition 13.2. Given a complex vector space E, a function $\varphi: E \times E \to \mathbb{C}$ is a *sesquilinear form* if it is linear in its first argument and semilinear in its second argument, which means that

$$\varphi(u_1 + u_2, v) = \varphi(u_1, v) + \varphi(u_2, v),$$
$$\varphi(u, v_1 + v_2) = \varphi(u, v_1) + \varphi(u, v_2),$$
$$\varphi(\lambda u, v) = \lambda \varphi(u, v),$$
$$\varphi(u, \mu v) = \overline{\mu} \varphi(u, v),$$

for all $u, v, u_1, u_2, v_1, v_2 \in E$, and all $\lambda, \mu \in \mathbb{C}$. A function $\varphi: E \times E \to \mathbb{C}$ is a *Hermitian form* if it is sesquilinear and if

$$\varphi(v, u) = \overline{\varphi(u, v)}$$

for all all $u, v \in E$.

Obviously, $\varphi(0, v) = \varphi(u, 0) = 0$. Also note that if $\varphi \colon E \times E \to \mathbb{C}$ is sesquilinear, we have

$$\varphi(\lambda u + \mu v, \lambda u + \mu v) = |\lambda|^2 \varphi(u, u) + \lambda \overline{\mu} \varphi(u, v) + \overline{\lambda} \mu \varphi(v, u) + |\mu|^2 \varphi(v, v),$$

and if $\varphi \colon E \times E \to \mathbb{C}$ is Hermitian, we have

$$\varphi(\lambda u + \mu v, \lambda u + \mu v) = |\lambda|^2 \varphi(u, u) + 2 \Re(\lambda \overline{\mu} \varphi(u, v)) + |\mu|^2 \varphi(v, v).$$

Note that restricted to real coefficients, a sesquilinear form is bilinear (we sometimes say $\mathbb{R}$-bilinear).

Definition 13.3. Given a sesquilinear form $\varphi \colon E \times E \to \mathbb{C}$, the function $\Phi \colon E \to \mathbb{C}$ defined such that $\Phi(u) = \varphi(u, u)$ for all $u \in E$ is called the *quadratic form* associated with φ.

The standard example of a Hermitian form on $\mathbb{C}^n$ is the map φ defined such that

$$\varphi((x_1, \ldots, x_n), (y_1, \ldots, y_n)) = x_1 \overline{y_1} + x_2 \overline{y_2} + \cdots + x_n \overline{y_n}.$$

This map is also positive definite, but before dealing with these issues, we show the following useful proposition.

Proposition 13.1. *Given a complex vector space E, the following properties hold:*

(1) A sesquilinear form $\varphi \colon E \times E \to \mathbb{C}$ is a Hermitian form iff $\varphi(u, u) \in \mathbb{R}$ for all $u \in E$.

(2) If $\varphi \colon E \times E \to \mathbb{C}$ is a sesquilinear form, then

$$4 \varphi(u, v) = \varphi(u + v, u + v) - \varphi(u - v, u - v)$$
$$+ i \varphi(u + iv, u + iv) - i \varphi(u - iv, u - iv),$$

and

$$2 \varphi(u, v) = (1 + i)(\varphi(u, u) + \varphi(v, v)) - \varphi(u - v, u - v) - i \varphi(u - iv, u - iv).$$

*These are called **polarization identities**.*

Proof. (1) If φ is a Hermitian form, then

$$\varphi(v, u) = \overline{\varphi(u, v)}$$

implies that

$$\varphi(u, u) = \overline{\varphi(u, u)},$$

and thus $\varphi(u, u) \in \mathbb{R}$. If φ is sesquilinear and $\varphi(u, u) \in \mathbb{R}$ for all $u \in E$, then

$$\varphi(u + v, u + v) = \varphi(u, u) + \varphi(u, v) + \varphi(v, u) + \varphi(v, v),$$

which proves that

$$\varphi(u, v) + \varphi(v, u) = \alpha,$$

where α is real, and changing u to iu, we have

$$i(\varphi(u, v) - \varphi(v, u)) = \beta,$$

where β is real, and thus

$$\varphi(u, v) = \frac{\alpha - i\beta}{2} \quad \text{and} \quad \varphi(v, u) = \frac{\alpha + i\beta}{2},$$

proving that φ is Hermitian.

(2) These identities are verified by expanding the right-hand side, and we leave them as an exercise. $\qquad \square$

Proposition 13.1 shows that a sesquilinear form is completely determined by the quadratic form $\Phi(u) = \varphi(u, u)$, even if φ is not Hermitian. This is false for a real bilinear form, unless it is symmetric. For example, the bilinear form $\varphi \colon \mathbb{R}^2 \times \mathbb{R}^2 \to \mathbb{R}$ defined such that

$$\varphi((x_1, y_1), (x_2, y_2)) = x_1 y_2 - x_2 y_1$$

is not identically zero, and yet it is null on the diagonal. However, a real symmetric bilinear form is indeed determined by its values on the diagonal, as we saw in Chapter 11.

As in the Euclidean case, Hermitian forms for which $\varphi(u, u) \geq 0$ play an important role.

Definition 13.4. Given a complex vector space E, a Hermitian form $\varphi \colon E \times E \to \mathbb{C}$ is *positive* if $\varphi(u, u) \geq 0$ for all $u \in E$, and *positive definite* if $\varphi(u, u) > 0$ for all $u \neq 0$. A pair $\langle E, \varphi \rangle$ where E is a complex vector space and φ is a Hermitian form on E is called a *pre-Hilbert space* if φ is positive, and a *Hermitian (or unitary) space* if φ is positive definite.

We warn our readers that some authors, such as Lang [Lang (1996)], define a pre-Hilbert space as what we define as a Hermitian space. We prefer following the terminology used in Schwartz [Schwartz (1991)] and Bourbaki [Bourbaki (1981b)]. The quantity $\varphi(u, v)$ is usually called the *Hermitian product* of u and v. We will occasionally call it the *inner product* of u and v.

Given a pre-Hilbert space $\langle E, \varphi \rangle$, as in the case of a Euclidean space, we also denote $\varphi(u, v)$ by

$$u \cdot v \quad \text{or} \quad \langle u, v \rangle \quad \text{or} \quad (u|v),$$

and $\sqrt{\Phi(u)}$ by $\|u\|$.

Example 13.1. The complex vector space $\mathbb{C}^n$ under the Hermitian form

$$\varphi((x_1, \ldots, x_n), (y_1, \ldots, y_n)) = x_1\overline{y_1} + x_2\overline{y_2} + \cdots + x_n\overline{y_n}$$

is a Hermitian space.

Example 13.2. Let ℓ^2 denote the set of all countably infinite sequences $x = (x_i)_{i \in \mathbb{N}}$ of complex numbers such that $\sum_{i=0}^{\infty} |x_i|^2$ is defined (i.e., the sequence $\sum_{i=0}^{n} |x_i|^2$ converges as $n \to \infty$). It can be shown that the map $\varphi \colon \ell^2 \times \ell^2 \to \mathbb{C}$ defined such that

$$\varphi((x_i)_{i \in \mathbb{N}}, (y_i)_{i \in \mathbb{N}}) = \sum_{i=0}^{\infty} x_i\overline{y_i}$$

is well defined, and ℓ^2 is a Hermitian space under φ. Actually, ℓ^2 is even a Hilbert space.

Example 13.3. Let $\mathcal{C}_{\text{piece}}[a, b]$ be the set of bounded piecewise continuous functions $f \colon [a, b] \to \mathbb{C}$ under the Hermitian form

$$\langle f, g \rangle = \int_a^b f(x)\overline{g(x)}dx.$$

It is easy to check that this Hermitian form is positive, but it is not definite. Thus, under this Hermitian form, $\mathcal{C}_{\text{piece}}[a, b]$ is only a pre-Hilbert space.

Example 13.4. Let $\mathcal{C}[a, b]$ be the set of complex-valued continuous functions $f \colon [a, b] \to \mathbb{C}$ under the Hermitian form

$$\langle f, g \rangle = \int_a^b f(x)\overline{g(x)}dx.$$

It is easy to check that this Hermitian form is positive definite. Thus, $\mathcal{C}[a, b]$ is a Hermitian space.

Example 13.5. Let $E = \mathrm{M}_n(\mathbb{C})$ be the vector space of complex $n \times n$ matrices. If we view a matrix $A \in \mathrm{M}_n(\mathbb{C})$ as a "long" column vector obtained by concatenating together its columns, we can define the Hermitian product of two matrices $A, B \in \mathrm{M}_n(\mathbb{C})$ as

$$\langle A, B \rangle = \sum_{i,j=1}^{n} a_{ij}\overline{b}_{ij},$$

which can be conveniently written as

$$\langle A, B \rangle = \operatorname{tr}(A^\top \overline{B}) = \operatorname{tr}(B^* A).$$

Since this can be viewed as the standard Hermitian product on $\mathbb{C}^{n^2}$, it is a Hermitian product on $\operatorname{M}_n(\mathbb{C})$. The corresponding norm

$$\|A\|_F = \sqrt{\operatorname{tr}(A^* A)}$$

is the Frobenius norm (see Section 8.2).

If E is finite-dimensional and if $\varphi \colon E \times E \to \mathbb{R}$ is a sequilinear form on E, given any basis $(e_1, \ldots, e_n)$ of E, we can write $x = \sum_{i=1}^n x_i e_i$ and $y = \sum_{j=1}^n y_j e_j$, and we have

$$\varphi(x, y) = \varphi\left(\sum_{i=1}^n x_i e_i, \sum_{j=1}^n y_j e_j \right) = \sum_{i,j=1}^n x_i \overline{y}_j \varphi(e_i, e_j).$$

If we let $G = (g_{ij})$ be the matrix given by $g_{ij} = \varphi(e_j, e_i)$, and if x and y are the column vectors associated with $(x_1, \ldots, x_n)$ and $(y_1, \ldots, y_n)$, then we can write

$$\varphi(x, y) = x^\top G^\top \overline{y} = y^* G x,$$

where $\overline{y}$ corresponds to $(\overline{y}_1, \ldots, \overline{y}_n)$. As in Section 11.1, we are committing the slight abuse of notation of letting x denote both the vector $x = \sum_{i=1}^n x_i e_i$ and the column vector associated with $(x_1, \ldots, x_n)$ (and similarly for y). The "correct" expression for $\varphi(x, y)$ is

$$\varphi(x, y) = \mathbf{y}^* G \mathbf{x} = \mathbf{x}^\top G^\top \overline{\mathbf{y}}.$$

Observe that in $\varphi(x, y) = y^* G x$, the matrix involved is the transpose of the matrix $(\varphi(e_i, e_j))$. The reason for this is that we want G to be positive definite when φ is positive definite, not $G^\top$.

Furthermore, observe that φ is Hermitian iff $G = G^*$, and φ is positive definite iff the matrix G is positive definite, that is,

$$(Gx)^\top \overline{x} = x^* G x > 0 \quad \text{for all } x \in \mathbb{C}^n, \ x \neq 0.$$

Definition 13.5. The matrix G associated with a Hermitian product is called the *Gram matrix* of the Hermitian product with respect to the basis $(e_1, \ldots, e_n)$.

Conversely, if A is a Hermitian positive definite $n \times n$ matrix, it is easy to check that the Hermitian form

$$\langle x, y \rangle = y^* A x$$

is positive definite. If we make a change of basis from the basis $(e_1, \ldots, e_n)$ to the basis $(f_1, \ldots, f_n)$, and if the change of basis matrix is P (where the jth column of P consists of the coordinates of f_j over the basis $(e_1, \ldots, e_n)$), then with respect to coordinates x' and y' over the basis $(f_1, \ldots, f_n)$, we have

$$y^* G x = (y')^* P^* G P x',$$

so the matrix of our inner product over the basis $(f_1, \ldots, f_n)$ is $P^* G P$. We summarize these facts in the following proposition.

Proposition 13.2. *Let E be a finite-dimensional vector space, and let $(e_1, \ldots, e_n)$ be a basis of E.*

(1) For any Hermitian inner product $\langle -, - \rangle$ on E, if $G = (g_{ij})$ with $g_{ij} = \langle e_j, e_i \rangle$ is the Gram matrix of the Hermitian product $\langle -, - \rangle$ w.r.t. the basis $(e_1, \ldots, e_n)$, then G is Hermitian positive definite.

(2) For any change of basis matrix P, the Gram matrix of $\langle -, - \rangle$ with respect to the new basis is $P^* G P$.

(3) If A is any $n \times n$ Hermitian positive definite matrix, then

$$\langle x, y \rangle = y^* A x$$

is a Hermitian product on E.

We will see later that a Hermitian matrix is positive definite iff its eigenvalues are all positive.

The following result reminiscent of the first polarization identity of Proposition 13.1 can be used to prove that two linear maps are identical.

Proposition 13.3. *Given any Hermitian space E with Hermitian product $\langle -, - \rangle$, for any linear map $f \colon E \to E$, if $\langle f(x), x \rangle = 0$ for all $x \in E$, then $f = 0$.*

Proof. Compute $\langle f(x + y), x + y \rangle$ and $\langle f(x - y), x - y \rangle$:

$$\langle f(x+y), x+y \rangle = \langle f(x), x \rangle + \langle f(x), y \rangle + \langle f(y), x \rangle + \langle y, y \rangle$$
$$\langle f(x-y), x-y \rangle = \langle f(x), x \rangle - \langle f(x), y \rangle - \langle f(y), x \rangle + \langle y, y \rangle;$$

then subtract the second equation from the first to obtain

$$\langle f(x+y), x+y \rangle - \langle f(x-y), x-y \rangle = 2(\langle f(x), y \rangle + \langle f(y), x \rangle).$$

If $\langle f(u), u \rangle = 0$ for all $u \in E$, we get

$$\langle f(x), y \rangle + \langle f(y), x \rangle = 0 \quad \text{for all } x, y \in E.$$

Then the above equation also holds if we replace x by ix, and we obtain

$$i\langle f(x), y \rangle - i\langle f(y), x \rangle = 0, \quad \text{for all } x, y \in E,$$

so we have

$$\langle f(x), y \rangle + \langle f(y), x \rangle = 0$$
$$\langle f(x), y \rangle - \langle f(y), x \rangle = 0,$$

which implies that $\langle f(x), y \rangle = 0$ for all $x, y \in E$. Since $\langle -, - \rangle$ is positive definite, we have $f(x) = 0$ for all $x \in E$; that is, $f = 0$. $\qquad \square$

One should be careful not to apply Proposition 13.3 to a linear map on a real Euclidean space because it is false! The reader should find a counterexample.

The Cauchy–Schwarz inequality and the Minkowski inequalities extend to pre-Hilbert spaces and to Hermitian spaces.

Proposition 13.4. *Let $\langle E, \varphi \rangle$ be a pre-Hilbert space with associated quadratic form Φ. For all $u, v \in E$, we have the* Cauchy–Schwarz *inequality*

$$|\varphi(u, v)| \leq \sqrt{\Phi(u)}\sqrt{\Phi(v)}.$$

Furthermore, if $\langle E, \varphi \rangle$ is a Hermitian space, the equality holds iff u and v are linearly dependent.

We also have the Minkowski *inequality*

$$\sqrt{\Phi(u + v)} \leq \sqrt{\Phi(u)} + \sqrt{\Phi(v)}.$$

Furthermore, if $\langle E, \varphi \rangle$ is a Hermitian space, the equality holds iff u and v are linearly dependent, where in addition, if $u \neq 0$ and $v \neq 0$, then $u = \lambda v$ for some real λ such that $\lambda > 0$.

Proof. For all $u, v \in E$ and all $\mu \in \mathbb{C}$, we have observed that

$$\varphi(u + \mu v, u + \mu v) = \varphi(u, u) + 2\Re(\overline{\mu}\varphi(u, v)) + |\mu|^2 \varphi(v, v).$$

Let $\varphi(u, v) = \rho e^{i\theta}$, where $|\varphi(u, v)| = \rho$ $(\rho \geq 0)$. Let $F \colon \mathbb{R} \to \mathbb{R}$ be the function defined such that

$$F(t) = \Phi(u + te^{i\theta}v),$$

for all $t \in \mathbb{R}$. The above shows that

$$F(t) = \varphi(u, u) + 2t|\varphi(u, v)| + t^2\varphi(v, v) = \Phi(u) + 2t|\varphi(u, v)| + t^2\Phi(v).$$

Since φ is assumed to be positive, we have $F(t) \geq 0$ for all $t \in \mathbb{R}$. If $\Phi(v) = 0$, we must have $\varphi(u, v) = 0$, since otherwise, $F(t)$ could be made negative by choosing t negative and small enough. If $\Phi(v) > 0$, in order for $F(t)$ to be nonnegative, the equation

$$\Phi(u) + 2t|\varphi(u, v)| + t^2\Phi(v) = 0$$

must not have distinct real roots, which is equivalent to

$$|\varphi(u, v)|^2 \leq \Phi(u)\Phi(v).$$

Taking the square root on both sides yields the Cauchy–Schwarz inequality.

For the second part of the claim, if φ is positive definite, we argue as follows. If u and v are linearly dependent, it is immediately verified that we get an equality. Conversely, if

$$|\varphi(u, v)|^2 = \Phi(u)\Phi(v),$$

then there are two cases. If $\Phi(v) = 0$, since φ is positive definite, we must have $v = 0$, so u and v are linearly dependent. Otherwise, the equation

$$\Phi(u) + 2t|\varphi(u, v)| + t^2\Phi(v) = 0$$

has a double root t_0, and thus

$$\Phi(u + t_0 e^{i\theta}v) = 0.$$

Since φ is positive definite, we must have

$$u + t_0 e^{i\theta}v = 0,$$

which shows that u and v are linearly dependent.

If we square the Minkowski inequality, we get

$$\Phi(u + v) \leq \Phi(u) + \Phi(v) + 2\sqrt{\Phi(u)}\sqrt{\Phi(v)}.$$

However, we observed earlier that

$$\Phi(u + v) = \Phi(u) + \Phi(v) + 2\Re(\varphi(u, v)).$$

Thus, it is enough to prove that

$$\Re(\varphi(u, v)) \leq \sqrt{\Phi(u)}\sqrt{\Phi(v)},$$

but this follows from the Cauchy–Schwarz inequality

$$|\varphi(u, v)| \leq \sqrt{\Phi(u)}\sqrt{\Phi(v)}$$

and the fact that $\Re z \leq |z|$.

If φ is positive definite and u and v are linearly dependent, it is immediately verified that we get an equality. Conversely, if equality holds in the Minkowski inequality, we must have

$$\Re(\varphi(u,v)) = \sqrt{\Phi(u)}\sqrt{\Phi(v)},$$

which implies that

$$|\varphi(u,v)| = \sqrt{\Phi(u)}\sqrt{\Phi(v)},$$

since otherwise, by the Cauchy–Schwarz inequality, we would have

$$\Re(\varphi(u,v)) \leq |\varphi(u,v)| < \sqrt{\Phi(u)}\sqrt{\Phi(v)}.$$

Thus, equality holds in the Cauchy–Schwarz inequality, and

$$\Re(\varphi(u,v)) = |\varphi(u,v)|.$$

But then we proved in the Cauchy–Schwarz case that u and v are linearly dependent. Since we also just proved that $\varphi(u,v)$ is real and nonnegative, the coefficient of proportionality between u and v is indeed nonnegative. $\square$

As in the Euclidean case, if $\langle E, \varphi \rangle$ is a Hermitian space, the Minkowski inequality

$$\sqrt{\Phi(u+v)} \leq \sqrt{\Phi(u)} + \sqrt{\Phi(v)}$$

shows that the map $u \mapsto \sqrt{\Phi(u)}$ is a *norm* on E. The norm induced by φ is called the *Hermitian norm induced by* φ. We usually denote $\sqrt{\Phi(u)}$ by $\|u\|$, and the Cauchy–Schwarz inequality is written as

$$|u \cdot v| \leq \|u\|\|v\|.$$

Since a Hermitian space is a normed vector space, it is a topological space under the topology induced by the norm (a basis for this topology is given by the open balls $B_0(u,\rho)$ of center u and radius $\rho > 0$, where

$$B_0(u,\rho) = \{v \in E \mid \|v - u\| < \rho\}.$$

If E has finite dimension, every linear map is continuous; see Chapter 8 (or Lang [Lang (1996, 1997)], Dixmier [Dixmier (1984)], or Schwartz [Schwartz (1991, 1992)]). The Cauchy–Schwarz inequality

$$|u \cdot v| \leq \|u\|\|v\|$$

shows that $\varphi \colon E \times E \to \mathbb{C}$ is continuous, and thus, that $\|\ \|$ is continuous.

If $\langle E, \varphi \rangle$ is only pre-Hilbertian, $\|u\|$ is called a *seminorm*. In this case, the condition

$$\|u\| = 0 \quad \text{implies} \quad u = 0$$

is not necessarily true. However, the Cauchy–Schwarz inequality shows that if $\|u\| = 0$, then $u \cdot v = 0$ for all $v \in E$.

Remark: As in the case of real vector spaces, a norm on a complex vector space is induced by some positive definite Hermitian product $\langle -, - \rangle$ iff it satisfies the *parallelogram law*:

$$\|u + v\|^2 + \|u - v\|^2 = 2(\|u\|^2 + \|v\|^2).$$

This time the Hermitian product is recovered using the polarization identity from Proposition 13.1:

$$4\langle u, v \rangle = \|u + v\|^2 - \|u - v\|^2 + i \|u + iv\|^2 - i \|u - iv\|^2.$$

It is easy to check that $\langle u, u \rangle = \|u\|^2$, and

$$\langle v, u \rangle = \overline{\langle u, v \rangle}$$
$$\langle iu, v \rangle = i\langle u, v \rangle,$$

so it is enough to check linearity in the variable u, and only for real scalars. This is easily done by applying the proof from Section 11.1 to the real and imaginary part of $\langle u, v \rangle$; the details are left as an exercise.

We will now basically mirror the presentation of Euclidean geometry given in Chapter 11 rather quickly, leaving out most proofs, except when they need to be seriously amended.

13.2 Orthogonality, Duality, Adjoint of a Linear Map

In this section we assume that we are dealing with Hermitian spaces. We denote the Hermitian inner product by $u \cdot v$ or $\langle u, v \rangle$. The concepts of orthogonality, orthogonal family of vectors, orthonormal family of vectors, and orthogonal complement of a set of vectors are unchanged from the Euclidean case (Definition 11.2).

For example, the set $\mathcal{C}[-\pi, \pi]$ of continuous functions $f \colon [-\pi, \pi] \to \mathbb{C}$ is a Hermitian space under the product

$$\langle f, g \rangle = \int_{-\pi}^{\pi} f(x)\overline{g(x)}dx,$$

and the family $(e^{ikx})_{k \in \mathbb{Z}}$ is orthogonal.

Propositions 11.4 and 11.5 hold without any changes. It is easy to show that

$$\left\| \sum_{i=1}^{n} u_i \right\|^2 = \sum_{i=1}^{n} \|u_i\|^2 + \sum_{1 \leq i < j \leq n} 2\Re(u_i \cdot u_j).$$

Analogously to the case of Euclidean spaces of finite dimension, the Hermitian product induces a canonical bijection (i.e., independent of the choice of bases) between the vector space E and the space E^*. This is one of the places where conjugation shows up, but in this case, troubles are minor.

Given a Hermitian space E, for any vector $u \in E$, let $\varphi_u^l \colon E \to \mathbb{C}$ be the map defined such that

$$\varphi_u^l(v) = \overline{u \cdot v}, \quad \text{for all } v \in E.$$

Similarly, for any vector $v \in E$, let $\varphi_v^r \colon E \to \mathbb{C}$ be the map defined such that

$$\varphi_v^r(u) = u \cdot v, \quad \text{for all } u \in E.$$

Since the Hermitian product is linear in its first argument u, the map φ_v^r is a linear form in E^*, and since it is semilinear in its second argument v, the map φ_u^l is also a linear form in E^*. Thus, we have two maps $\flat^l \colon E \to E^*$ and $\flat^r \colon E \to E^*$, defined such that

$$\flat^l(u) = \varphi_u^l, \quad \text{and} \quad \flat^r(v) = \varphi_v^r.$$

Proposition 13.5. *The equations $\varphi_u^l = \varphi_u^r$ and $\flat^l = \flat^r$ hold.*

Proof. Indeed, for all $u, v \in E$, we have

$$\begin{aligned}
\flat^l(u)(v) &= \varphi_u^l(v) \\
&= \overline{u \cdot v} \\
&= v \cdot u \\
&= \varphi_u^r(v) \\
&= \flat^r(u)(v).
\end{aligned}$$
$\square$

Therefore, we use the notation φ_u for both φ_u^l and φ_u^r, and $\flat$ for both $\flat^l$ and $\flat^r$.

Theorem 13.1. *Let E be a Hermitian space E. The map $\flat \colon E \to E^*$ defined such that*

$$\flat(u) = \varphi_u^l = \varphi_u^r \quad \text{for all } u \in E$$

is semilinear and injective. When E is also of finite dimension, the map $\flat \colon \overline{E} \to E^*$ is a canonical isomorphism.

Proof. That $\flat\colon E \to E^*$ is a semilinear map follows immediately from the fact that $\flat = \flat^r$, and that the Hermitian product is semilinear in its second argument. If $\varphi_u = \varphi_v$, then $\varphi_u(w) = \varphi_v(w)$ for all $w \in E$, which by definition of φ_u and φ_v means that

$$w \cdot u = w \cdot v$$

for all $w \in E$, which by semilinearity on the right is equivalent to

$$w \cdot (v - u) = 0 \quad \text{for all } w \in E,$$

which implies that $u = v$, since the Hermitian product is positive definite. Thus, $\flat\colon E \to E^*$ is injective. Finally, when E is of finite dimension n, E^* is also of dimension n, and then $\flat\colon E \to E^*$ is bijective. Since $\flat$ is semilinar, the map $\flat\colon \overline{E} \to E^*$ is an isomorphism. $\qquad\square$

The inverse of the isomorphism $\flat\colon \overline{E} \to E^*$ is denoted by $\sharp\colon E^* \to \overline{E}$.

As a corollary of the isomorphism $\flat\colon \overline{E} \to E^*$ we have the following result.

Proposition 13.6. *If E is a Hermitian space of finite dimension, then every linear form $f \in E^*$ corresponds to a unique $v \in E$, such that*

$$f(u) = u \cdot v, \quad \text{for every } u \in E.$$

In particular, if f is not the zero form, the kernel of f, which is a hyperplane H, is precisely the set of vectors that are orthogonal to v.

Remarks:

(1) The "musical map" $\flat\colon \overline{E} \to E^*$ is not surjective when E has infinite dimension. This result can be salvaged by restricting our attention to continuous linear maps and by assuming that the vector space E is a *Hilbert space*.

(2) *Dirac's "bra-ket" notation.* Dirac invented a notation widely used in quantum mechanics for denoting the linear form $\varphi_u = \flat(u)$ associated to the vector $u \in E$ *via* the duality induced by a Hermitian inner product. Dirac's proposal is to denote the vectors u in E by $|u\rangle$, and call them *kets*; the notation $|u\rangle$ is pronounced "ket u." Given two kets (vectors) $|u\rangle$ and $|v\rangle$, their inner product is denoted by

$$\langle u|v \rangle$$

(instead of $|u\rangle \cdot |v\rangle$). The notation $\langle u|v \rangle$ for the inner product of $|u\rangle$ and $|v\rangle$ anticipates duality. Indeed, we define the dual (usually called adjoint) *bra u* of ket u, denoted by $\langle u|$, as the linear form whose value on any ket v is given by the inner product, so

$$\langle u|(|v\rangle) = \langle u|v\rangle.$$

Thus, bra $u = \langle u|$ is Dirac's notation for our $\flat(u)$. Since the map $\flat$ is semi-linear, we have

$$\langle \lambda u| = \overline{\lambda}\langle u|.$$

Using the bra-ket notation, given an orthonormal basis $(|u_1\rangle, \ldots, |u_n\rangle)$, ket v (a vector) is written as

$$|v\rangle = \sum_{i=1}^{n} \langle v|u_i\rangle |u_i\rangle,$$

and the corresponding linear form bra v is written as

$$\langle v| = \sum_{i=1}^{n} \overline{\langle v|u_i\rangle}\langle u_i| = \sum_{i=1}^{n} \langle u_i|v\rangle \langle u_i|$$

over the dual basis $(\langle u_1|, \ldots, \langle u_n|)$. As cute as it looks, we do not recommend using the Dirac notation.

The existence of the isomorphism $\flat\colon \overline{E} \to E^*$ is crucial to the existence of adjoint maps. Indeed, Theorem 13.1 allows us to define the adjoint of a linear map on a Hermitian space. Let E be a Hermitian space of finite dimension n, and let $f\colon E \to E$ be a linear map. For every $u \in E$, the map

$$v \mapsto \overline{u \cdot f(v)}$$

is clearly a linear form in E^*, and by Theorem 13.1, there is a unique vector in E denoted by $f^*(u)$, such that

$$\overline{f^*(u) \cdot v} = \overline{u \cdot f(v)},$$

that is,

$$f^*(u) \cdot v = u \cdot f(v), \quad \text{for every } v \in E.$$

The following proposition shows that the map f^* is linear.

Proposition 13.7. *Given a Hermitian space E of finite dimension, for every linear map $f\colon E \to E$ there is a unique linear map $f^*\colon E \to E$ such that*

$$f^*(u) \cdot v = u \cdot f(v), \quad \text{for all } u, v \in E.$$

Proof. Careful inspection of the proof of Proposition 11.6 reveals that it applies unchanged. The only potential problem is in proving that $f^*(\lambda u) = \bar\lambda f^*(u)$, but everything takes place in the first argument of the Hermitian product, and there, we have linearity. □

Definition 13.6. Given a Hermitian space E of finite dimension, for every linear map $f \colon E \to E$, the unique linear map $f^* \colon E \to E$ such that

$$f^*(u) \cdot v = u \cdot f(v), \quad \text{for all } u, v \in E$$

given by Proposition 13.7 is called the *adjoint of f (w.r.t. to the Hermitian product)*.

The fact that

$$v \cdot u = \overline{u \cdot v}$$

implies that the adjoint f^* of f is also characterized by

$$f(u) \cdot v = u \cdot f^*(v),$$

for all $u, v \in E$.

Given two Hermitian spaces E and F, where the Hermitian product on E is denoted by $\langle -, - \rangle_1$ and the Hermitian product on F is denoted by $\langle -, - \rangle_2$, given any linear map $f \colon E \to F$, it is immediately verified that the proof of Proposition 13.7 can be adapted to show that there is a unique linear map $f^* \colon F \to E$ such that

$$\langle f(u), v \rangle_2 = \langle u, f^*(v) \rangle_1$$

for all $u \in E$ and all $v \in F$. The linear map f^* is also called the *adjoint* of f.

As in the Euclidean case, the following properties immediately follow from the definition of the adjoint map.

Proposition 13.8.

(1) For any linear map $f \colon E \to F$, we have

$$f^{**} = f.$$

(2) For any two linear maps $f, g \colon E \to F$ and any scalar $\lambda \in \mathbb{R}$:

$$(f + g)^* = f^* + g^*$$
$$(\lambda f)^* = \bar\lambda f^*.$$

(3) If E, F, G are Hermitian spaces with respective inner products $\langle -, - \rangle_1, \langle -, - \rangle_2$, and $\langle -, - \rangle_3$, and if $f \colon E \to F$ and $g \colon F \to G$ are two linear maps, then

$$(g \circ f)^* = f^* \circ g^*.$$

As in the Euclidean case, a linear map $f \colon E \to E$ (where E is a finite-dimensional Hermitian space) is *self-adjoint* if $f = f^*$. The map f is *positive semidefinite* iff

$$\langle f(x), x \rangle \geq 0 \quad \text{all } x \in E;$$

positive definite iff

$$\langle f(x), x \rangle > 0 \quad \text{all } x \in E, \, x \neq 0.$$

An interesting corollary of Proposition 13.3 is that a positive semidefinite linear map must be self-adjoint. In fact, we can prove a slightly more general result.

Proposition 13.9. *Given any finite-dimensional Hermitian space E with Hermitian product $\langle -, - \rangle$, for any linear map $f \colon E \to E$, if $\langle f(x), x \rangle \in \mathbb{R}$ for all $x \in E$, then f is self-adjoint. In particular, any positive semidefinite linear map $f \colon E \to E$ is self-adjoint.*

Proof. Since $\langle f(x), x \rangle \in \mathbb{R}$ for all $x \in E$, we have

$$\begin{aligned}
\langle f(x), x \rangle &= \overline{\langle f(x), x \rangle} \\
&= \langle x, f(x) \rangle \\
&= \langle f^*(x), x \rangle,
\end{aligned}$$

so we have

$$\langle (f - f^*)(x), x \rangle = 0 \quad \text{all } x \in E,$$

and Proposition 13.3 implies that $f - f^* = 0$.   □

Beware that Proposition 13.9 is false if E is a real Euclidean space.

As in the Euclidean case, Theorem 13.1 can be used to show that any Hermitian space of finite dimension has an orthonormal basis. The proof is unchanged.

Proposition 13.10. *Given any nontrivial Hermitian space E of finite dimension $n \geq 1$, there is an orthonormal basis $(u_1, \ldots, u_n)$ for E.*

The *Gram–Schmidt orthonormalization procedure* also applies to Hermitian spaces of finite dimension, without any changes from the Euclidean case!

Proposition 13.11. *Given a nontrivial Hermitian space E of finite dimension $n \geq 1$, from any basis $(e_1, \ldots, e_n)$ for E we can construct an orthonormal basis $(u_1, \ldots, u_n)$ for E with the property that for every k, $1 \leq k \leq n$, the families $(e_1, \ldots, e_k)$ and $(u_1, \ldots, u_k)$ generate the same subspace.*

Remark: The remarks made after Proposition 11.8 also apply here, except that in the QR-decomposition, Q is a unitary matrix.

As a consequence of Proposition 11.7 (or Proposition 13.11), given any Hermitian space of finite dimension n, if $(e_1, \ldots, e_n)$ is an orthonormal basis for E, then for any two vectors $u = u_1 e_1 + \cdots + u_n e_n$ and $v = v_1 e_1 + \cdots + v_n e_n$, the Hermitian product $u \cdot v$ is expressed as

$$u \cdot v = (u_1 e_1 + \cdots + u_n e_n) \cdot (v_1 e_1 + \cdots + v_n e_n) = \sum_{i=1}^{n} u_i \overline{v_i},$$

and the norm $\|u\|$ as

$$\|u\| = \|u_1 e_1 + \cdots + u_n e_n\| = \left(\sum_{i=1}^{n} |u_i|^2 \right)^{1/2}.$$

The fact that a Hermitian space always has an orthonormal basis implies that any Gram matrix G can be written as

$$G = Q^* Q,$$

for some invertible matrix Q. Indeed, we know that in a change of basis matrix, a Gram matrix G becomes $G' = P^* G P$. If the basis corresponding to G' is orthonormal, then $G' = I$, so $G = (P^{-1})^* P^{-1}$.

Proposition 11.9 also holds unchanged.

Proposition 13.12. *Given any nontrivial Hermitian space E of finite dimension $n \geq 1$, for any subspace F of dimension k, the orthogonal complement $F^\perp$ of F has dimension $n - k$, and $E = F \oplus F^\perp$. Furthermore, we have $F^{\perp\perp} = F$.*

13.3 Linear Isometries (Also Called Unitary Transformations)

In this section we consider linear maps between Hermitian spaces that preserve the Hermitian norm. All definitions given for Euclidean spaces in Section 11.5 extend to Hermitian spaces, except that orthogonal transformations are called unitary transformation, but Proposition 11.10 extends only with a modified Condition (2). Indeed, the old proof that (2) implies (3) does not work, and the implication is in fact false! It can be repaired by strengthening Condition (2). For the sake of completeness, we state the Hermitian version of Definition 11.5.

Definition 13.7. Given any two nontrivial Hermitian spaces E and F of the same finite dimension n, a function $f\colon E \to F$ is *a unitary transformation, or a linear isometry*, if it is linear and
$$\|f(u)\| = \|u\|, \quad \text{for all } u \in E.$$

Proposition 11.10 can be salvaged by strengthening Condition (2).

Proposition 13.13. *Given any two nontrivial Hermitian spaces E and F of the same finite dimension n, for every function $f\colon E \to F$, the following properties are equivalent:*

(1) f is a linear map and $\|f(u)\| = \|u\|$, for all $u \in E$;
(2) $\|f(v) - f(u)\| = \|v - u\|$ and $f(iu) = if(u)$, for all $u, v \in E$.
(3) $f(u) \cdot f(v) = u \cdot v$, for all $u, v \in E$.

Furthermore, such a map is bijective.

Proof. The proof that (2) implies (3) given in Proposition 11.10 needs to be revised as follows. We use the polarization identity
$$2\varphi(u, v) = (1 + i)(\|u\|^2 + \|v\|^2) - \|u - v\|^2 - i\|u - iv\|^2.$$
Since $f(iv) = if(v)$, we get $f(0) = 0$ by setting $v = 0$, so the function f preserves distance and norm, and we get
$$\begin{aligned}
2\varphi(f(u), f(v)) &= (1 + i)(\|f(u)\|^2 + \|f(v)\|^2) - \|f(u) - f(v)\|^2 \\
&\quad - i\|f(u) - if(v)\|^2 \\
&= (1 + i)(\|f(u)\|^2 + \|f(v)\|^2) - \|f(u) - f(v)\|^2 \\
&\quad - i\|f(u) - f(iv)\|^2 \\
&= (1 + i)(\|u\|^2 + \|v\|^2) - \|u - v\|^2 - i\|u - iv\|^2 \\
&= 2\varphi(u, v),
\end{aligned}$$

which shows that f preserves the Hermitian inner product as desired. The rest of the proof is unchanged. $\square$

Remarks:

(i) In the Euclidean case, we proved that the assumption

$$\|f(v) - f(u)\| = \|v - u\| \quad \text{for all } u, v \in E \text{ and } f(0) = 0 \qquad (13.1)$$

implies (3). For this we used the polarization identity

$$2u \cdot v = \|u\|^2 + \|v\|^2 - \|u - v\|^2.$$

In the Hermitian case the polarization identity involves the complex number i. In fact, the implication (13.1) implies (3) is false in the Hermitian case! Conjugation $z \mapsto \overline{z}$ satisfies (2′) since

$$|\overline{z_2} - \overline{z_1}| = |\overline{z_2 - z_1}| = |z_2 - z_1|,$$

and yet, it is not linear!

(ii) If we modify (2) by changing the second condition by now requiring that there be some $\tau \in E$ such that

$$f(\tau + iu) = f(\tau) + i(f(\tau + u) - f(\tau))$$

for all $u \in E$, then the function $g \colon E \to E$ defined such that

$$g(u) = f(\tau + u) - f(\tau)$$

satisfies the old conditions of (2), and the implications (2) $\to$ (3) and (3) $\to$ (1) prove that g is linear, and thus that f is affine. In view of the first remark, some condition involving i is needed on f, in addition to the fact that f is distance-preserving.

13.4   The Unitary Group, Unitary Matrices

In this section, as a mirror image of our treatment of the isometries of a Euclidean space, we explore some of the fundamental properties of the unitary group and of unitary matrices. As an immediate corollary of the Gram–Schmidt orthonormalization procedure, we obtain the QR-decomposition for invertible matrices.

In the Hermitian framework, the matrix of the adjoint of a linear map is not given by the transpose of the original matrix, but by its conjugate.

Definition 13.8. Given a complex $m \times n$ matrix A, the *transpose* $A^\top$ of A is the $n \times m$ matrix $A^\top = \left(a_{ij}^\top\right)$ defined such that

$$a_{ij}^\top = a_{ji},$$

and the *conjugate* $\overline{A}$ of A is the $m \times n$ matrix $\overline{A} = (b_{ij})$ defined such that

$$b_{ij} = \overline{a}_{ij}$$

for all i, j, $1 \leq i \leq m$, $1 \leq j \leq n$. The *adjoint* A^* *of* A is the matrix defined such that

$$A^* = \overline{(A^\top)} = \left(\overline{A}\right)^\top.$$

Proposition 13.14. *Let E be any Hermitian space of finite dimension n, and let $f \colon E \to E$ be any linear map. The following properties hold:*

(1) The linear map $f \colon E \to E$ is an isometry iff

$$f \circ f^* = f^* \circ f = \mathrm{id}.$$

(2) For every orthonormal basis $(e_1, \ldots, e_n)$ of E, if the matrix of f is A, then the matrix of f^* is the adjoint A^* of A, and f is an isometry iff A satisfies the identities

$$A A^* = A^* A = I_n,$$

where I_n denotes the identity matrix of order n, iff the columns of A form an orthonormal basis of $\mathbb{C}^n$, iff the rows of A form an orthonormal basis of $\mathbb{C}^n$.

Proof. (1) The proof is identical to that of Proposition 11.12 (1).

(2) If $(e_1, \ldots, e_n)$ is an orthonormal basis for E, let $A = (a_{ij})$ be the matrix of f, and let $B = (b_{ij})$ be the matrix of f^*. Since f^* is characterized by

$$f^*(u) \cdot v = u \cdot f(v)$$

for all $u, v \in E$, using the fact that if $w = w_1 e_1 + \cdots + w_n e_n$, we have $w_k = w \cdot e_k$, for all k, $1 \leq k \leq n$; letting $u = e_i$ and $v = e_j$, we get

$$b_{ji} = f^*(e_i) \cdot e_j = e_i \cdot f(e_j) = \overline{f(e_j) \cdot e_i} = \overline{a_{ij}},$$

for all i, j, $1 \leq i, j \leq n$. Thus, $B = A^*$. Now if X and Y are arbitrary matrices over the basis $(e_1, \ldots, e_n)$, denoting as usual the jth column of X by X^j, and similarly for Y, a simple calculation shows that

$$Y^* X = (X^j \cdot Y^i)_{1 \leq i,j \leq n}.$$

Then it is immediately verified that if $X = Y = A$, then $A^* A = A A^* = I_n$ iff the column vectors $(A^1, \ldots, A^n)$ form an orthonormal basis. Thus, from (1), we see that (2) is clear. $\qquad\square$

Proposition 11.12 shows that the inverse of an isometry f is its adjoint
f^*. Proposition 11.12 also motivates the following definition.

Definition 13.9. A complex $n \times n$ matrix is a *unitary matrix* if

$$A A^* = A^* A = I_n.$$

Remarks:

(1) The conditions $A A^* = I_n$, $A^* A = I_n$, and $A^{-1} = A^*$ are equivalent. Given any two orthonormal bases $(u_1, \ldots, u_n)$ and $(v_1, \ldots, v_n)$, if P is the change of basis matrix from $(u_1, \ldots, u_n)$ to $(v_1, \ldots, v_n)$, it is easy to show that the matrix P is unitary. The proof of Proposition 13.13 (3) also shows that if f is an isometry, then the image of an orthonormal basis $(u_1, \ldots, u_n)$ is an orthonormal basis.

(2) Using the explicit formula for the determinant, we see immediately that

$$\det(\overline{A}) = \overline{\det(A)}.$$

If f is a unitary transformation and A is its matrix with respect to any orthonormal basis, from $AA^* = I$, we get

$$\det(AA^*) = \det(A)\det(A^*) = \det(A)\overline{\det(A^\top)}$$
$$= \det(A)\overline{\det(A)} = |\det(A)|^2,$$

and so $|\det(A)| = 1$. It is clear that the isometries of a Hermitian space of dimension n form a group, and that the isometries of determinant $+1$ form a subgroup.

This leads to the following definition.

Definition 13.10. Given a Hermitian space E of dimension n, the set of isometries $f \colon E \to E$ forms a subgroup of $\mathbf{GL}(E, \mathbb{C})$ denoted by $\mathbf{U}(E)$, or $\mathbf{U}(n)$ when $E = \mathbb{C}^n$, called the *unitary group (of E)*. For every isometry f we have $|\det(f)| = 1$, where $\det(f)$ denotes the determinant of f. The isometries such that $\det(f) = 1$ are called *rotations, or proper isometries, or proper unitary transformations*, and they form a subgroup of the special linear group $\mathbf{SL}(E, \mathbb{C})$ (and of $\mathbf{U}(E)$), denoted by $\mathbf{SU}(E)$, or $\mathbf{SU}(n)$ when $E = \mathbb{C}^n$, called the *special unitary group (of E)*. The isometries such that $\det(f) \neq 1$ are called *improper isometries, or improper unitary transformations, or flip transformations*.

A very important example of unitary matrices is provided by Fourier matrices (up to a factor of $\sqrt{n}$), matrices that arise in the various versions of the discrete Fourier transform. For more on this topic, see the problems, and Strang [Strang (1986); Strang and Truong (1997)].

The group $\mathbf{SU}(2)$ turns out to be the group of *unit quaternions*, invented by Hamilton. This group plays an important role in the representation of rotations in $\mathbf{SO}(3)$ used in computer graphics and robotics; see Chapter 15.

Now that we have the definition of a unitary matrix, we can explain how the Gram–Schmidt orthonormalization procedure immediately yields the QR-decomposition for matrices.

Definition 13.11. Given any complex $n \times n$ matrix A, a *QR-decomposition* of A is any pair of $n \times n$ matrices (U, R), where U is a unitary matrix and R is an upper triangular matrix such that $A = UR$.

Proposition 13.15. *Given any $n \times n$ complex matrix A, if A is invertible, then there is a unitary matrix U and an upper triangular matrix R with positive diagonal entries such that $A = UR$.*

The proof is absolutely the same as in the real case!

Remark: If A is invertible and if $A = U_1 R_1 = U_2 R_2$ are two QR-decompositions for A, then

$$R_1 R_2^{-1} = U_1^* U_2.$$

Then it is easy to show that there is a diagonal matrix D with diagonal entries such that $|d_{ii}| = 1$ for $i = 1, \ldots, n$, and $U_2 = U_1 D$, $R_2 = D^* R_1$.

We have the following version of the Hadamard inequality for complex matrices. The proof is essentially the same as in the Euclidean case but it uses Proposition 13.15 instead of Proposition 11.14.

Proposition 13.16. *(Hadamard) For any complex $n \times n$ matrix $A = (a_{ij})$, we have*

$$|\det(A)| \leq \prod_{i=1}^{n} \left(\sum_{j=1}^{n} |a_{ij}|^2 \right)^{1/2} \quad and \quad |\det(A)| \leq \prod_{j=1}^{n} \left(\sum_{i=1}^{n} |a_{ij}|^2 \right)^{1/2}.$$

Moreover, equality holds iff either A has a zero row in the left inequality or a zero column in the right inequality, or A is unitary.

We also have the following version of Proposition 11.16 for Hermitian matrices. The proof of Proposition 11.16 goes through because the Cholesky decomposition for a Hermitian positive definite A matrix holds in the form $A = B^*B$, where B is upper triangular with positive diagonal entries. The details are left to the reader.

Proposition 13.17. *(Hadamard) For any complex $n \times n$ matrix $A = (a_{ij})$, if A is Hermitian positive semidefinite, then we have*

$$\det(A) \leq \prod_{i=1}^{n} a_{ii}.$$

Moreover, if A is positive definite, then equality holds iff A is a diagonal matrix.

13.5 Hermitian Reflections and QR-Decomposition

If A is an $n \times n$ complex singular matrix, there is some (not necessarily unique) QR-decomposition $A = QR$ with Q a unitary matrix which is a product of Householder reflections and R an upper triangular matrix, but the proof is more involved. One way to proceed is to generalize the notion of hyperplane reflection. This is not really surprising since in the Hermitian case there are improper isometries whose determinant can be any unit complex number. Hyperplane reflections are generalized as follows.

Definition 13.12. Let E be a Hermitian space of finite dimension. For any hyperplane H, for any nonnull vector w orthogonal to H, so that $E = H \oplus G$, where $G = \mathbb{C}w$, a *Hermitian reflection about H of angle θ* is a linear map of the form $\rho_{H,\theta} \colon E \to E$, defined such that

$$\rho_{H,\theta}(u) = p_H(u) + e^{i\theta}p_G(u),$$

for any unit complex number $e^{i\theta} \neq 1$ (i.e. $\theta \neq k2\pi$). For any nonzero vector $w \in E$, we denote by $\rho_{w,\theta}$ the Hermitian reflection given by $\rho_{H,\theta}$, where H is the hyperplane orthogonal to w.

Since $u = p_H(u) + p_G(u)$, the Hermitian reflection $\rho_{w,\theta}$ is also expressed as

$$\rho_{w,\theta}(u) = u + (e^{i\theta} - 1)p_G(u),$$

or as

$$\rho_{w,\theta}(u) = u + (e^{i\theta} - 1)\frac{(u \cdot w)}{\|w\|^2}w.$$

Note that the case of a standard hyperplane reflection is obtained when $e^{i\theta} = -1$, i.e., $\theta = \pi$. In this case,

$$\rho_{w,\pi}(u) = u - 2\frac{(u \cdot w)}{\|w\|^2}\,w,$$

and the matrix of such a reflection is a Householder matrix, as in Section 12.1, except that w may be a complex vector.

We leave as an easy exercise to check that $\rho_{w,\theta}$ is indeed an isometry, and that the inverse of $\rho_{w,\theta}$ is $\rho_{w,-\theta}$. If we pick an orthonormal basis $(e_1, \ldots, e_n)$ such that $(e_1, \ldots, e_{n-1})$ is an orthonormal basis of H, the matrix of $\rho_{w,\theta}$ is

$$\begin{pmatrix} I_{n-1} & 0 \\ 0 & e^{i\theta} \end{pmatrix}$$

We now come to the main surprise. Given any two distinct vectors u and v such that $\|u\| = \|v\|$, there isn't always a hyperplane reflection mapping u to v, but this can be done using two Hermitian reflections!

Proposition 13.18. *Let E be any nontrivial Hermitian space.*

(1) For any two vectors $u, v \in E$ such that $u \neq v$ and $\|u\| = \|v\|$, if $u \cdot v = e^{i\theta}|u \cdot v|$, then the (usual) reflection s about the hyperplane orthogonal to the vector $v - e^{-i\theta}u$ is such that $s(u) = e^{i\theta}v$.

(2) For any nonnull vector $v \in E$, for any unit complex number $e^{i\theta} \neq 1$, there is a Hermitian reflection $\rho_{v,\theta}$ such that

$$\rho_{v,\theta}(v) = e^{i\theta}v.$$

As a consequence, for u and v as in (1), we have $\rho_{v,-\theta} \circ s(u) = v$.

Proof. (1) Consider the (usual) reflection about the hyperplane orthogonal to $w = v - e^{-i\theta}u$. We have

$$s(u) = u - 2\frac{(u \cdot (v - e^{-i\theta}u))}{\|v - e^{-i\theta}u\|^2}\,(v - e^{-i\theta}u).$$

We need to compute

$$-2u \cdot (v - e^{-i\theta}u) \quad \text{and} \quad (v - e^{-i\theta}u) \cdot (v - e^{-i\theta}u).$$

Since $u \cdot v = e^{i\theta}|u \cdot v|$, we have

$$e^{-i\theta}u \cdot v = |u \cdot v| \quad \text{and} \quad e^{i\theta}v \cdot u = |u \cdot v|.$$

Using the above and the fact that $\|u\| = \|v\|$, we get

$$-2u \cdot (v - e^{-i\theta}u) = 2e^{i\theta}\|u\|^2 - 2u \cdot v,$$
$$= 2e^{i\theta}(\|u\|^2 - |u \cdot v|),$$

and

$$(v - e^{-i\theta}u) \cdot (v - e^{-i\theta}u) = \|v\|^2 + \|u\|^2 - e^{-i\theta}u \cdot v - e^{i\theta}v \cdot u,$$
$$= 2(\|u\|^2 - |u \cdot v|),$$

and thus,

$$-2\frac{(u \cdot (v - e^{-i\theta}u))}{\|(v - e^{-i\theta}u)\|^2}(v - e^{-i\theta}u) = e^{i\theta}(v - e^{-i\theta}u).$$

But then,

$$s(u) = u + e^{i\theta}(v - e^{-i\theta}u) = u + e^{i\theta}v - u = e^{i\theta}v,$$

and $s(u) = e^{i\theta}v$, as claimed.

(2) This part is easier. Consider the Hermitian reflection

$$\rho_{v,\theta}(u) = u + (e^{i\theta} - 1)\frac{(u \cdot v)}{\|v\|^2}v.$$

We have

$$\rho_{v,\theta}(v) = v + (e^{i\theta} - 1)\frac{(v \cdot v)}{\|v\|^2}v,$$
$$= v + (e^{i\theta} - 1)v,$$
$$= e^{i\theta}v.$$

Thus, $\rho_{v,\theta}(v) = e^{i\theta}v$. Since $\rho_{v,\theta}$ is linear, changing the argument v to $e^{i\theta}v$, we get

$$\rho_{v,-\theta}(e^{i\theta}v) = v,$$

and thus, $\rho_{v,-\theta} \circ s(u) = v$. $\qquad\square$

Remarks:

(1) If we use the vector $v + e^{-i\theta}u$ instead of $v - e^{-i\theta}u$, we get $s(u) = -e^{i\theta}v$.

(2) Certain authors, such as Kincaid and Cheney [Kincaid and Cheney (1996)] and Ciarlet [Ciarlet (1989)], use the vector $u + e^{i\theta}v$ instead of the vector $v + e^{-i\theta}u$. The effect of this choice is that they also get $s(u) = -e^{i\theta}v$.

(3) If $v = \|u\|e_1$, where e_1 is a basis vector, $u \cdot e_1 = a_1$, where a_1 is just the coefficient of u over the basis vector e_1. Then, since $u \cdot e_1 = e^{i\theta}|a_1|$, the choice of the plus sign in the vector $\|u\|e_1 + e^{-i\theta}u$ has the effect that the coefficient of this vector over e_1 is $\|u\| + |a_1|$, and no cancellations takes place, which is preferable for numerical stability (we need to divide by the square norm of this vector).

We now show that the QR-decomposition in terms of (complex) House-holder matrices holds for complex matrices. We need the version of Proposition 13.18 and a trick at the end of the argument, but the proof is basically unchanged.

Proposition 13.19. *Let E be a nontrivial Hermitian space of dimension n. Given any orthonormal basis $(e_1, \ldots, e_n)$, for any n-tuple of vectors $(v_1, \ldots, v_n)$, there is a sequence of $n-1$ isometries $h_1, \ldots, h_{n-1}$, such that h_i is a (standard) hyperplane reflection or the identity, and if $(r_1, \ldots, r_n)$ are the vectors given by*

$$r_j = h_{n-1} \circ \cdots \circ h_2 \circ h_1(v_j), \quad 1 \leq j \leq n,$$

then every r_j is a linear combination of the vectors $(e_1, \ldots, e_j)$, $(1 \leq j \leq n)$. Equivalently, the matrix R whose columns are the components of the r_j over the basis $(e_1, \ldots, e_n)$ is an upper triangular matrix. Furthermore, if we allow one more isometry h_n of the form

$$h_n = \rho_{e_n, \varphi_n} \circ \cdots \circ \rho_{e_1, \varphi_1}$$

after $h_1, \ldots, h_{n-1}$, we can ensure that the diagonal entries of R are non-negative.

Proof. The proof is very similar to the proof of Proposition 12.3, but it needs to be modified a little bit since Proposition 13.18 is weaker than Proposition 12.2. We explain how to modify the induction step, leaving the base case and the rest of the proof as an exercise.

As in the proof of Proposition 12.3, the vectors $(e_1, \ldots, e_k)$ form a basis for the subspace denoted as U_k', the vectors $(e_{k+1}, \ldots, e_n)$ form a basis for the subspace denoted as U_k'', the subspaces U_k' and U_k'' are orthogonal, and $E = U_k' \oplus U_k''$. Let

$$u_{k+1} = h_k \circ \cdots \circ h_2 \circ h_1(v_{k+1}).$$

We can write

$$u_{k+1} = u_{k+1}' + u_{k+1}'',$$

where $u_{k+1}' \in U_k'$ and $u_{k+1}'' \in U_k''$. Let

$$r_{k+1, k+1} = \left\| u_{k+1}'' \right\|, \quad \text{and} \quad e^{i\theta_{k+1}} \left| u_{k+1}'' \cdot e_{k+1} \right| = u_{k+1}'' \cdot e_{k+1}.$$

If $u_{k+1}'' = e^{i\theta_{k+1}} r_{k+1, k+1} e_{k+1}$, we let $h_{k+1} = \mathrm{id}$. Otherwise, by Proposition 13.18(1) (with $u = u_{k+1}''$ and $v = r_{k+1, k+1} e_{k+1}$), there is a unique hyperplane reflection h_{k+1} such that

$$h_{k+1}(u_{k+1}'') = e^{i\theta_{k+1}} r_{k+1, k+1} e_{k+1},$$

where h_{k+1} is the reflection about the hyperplane H_{k+1} orthogonal to the vector

$$w_{k+1} = r_{k+1,k+1}\, e_{k+1} - e^{-i\theta_{k+1}} u''_{k+1}.$$

At the end of the induction, we have a triangular matrix R, but the diagonal entries $e^{i\theta_j} r_{j,j}$ of R may be complex. Letting

$$h_n = \rho_{e_n,\, -\theta_n} \circ \cdots \circ \rho_{e_1,\, -\theta_1},$$

we observe that the diagonal entries of the matrix of vectors

$$r'_j = h_n \circ h_{n-1} \circ \cdots \circ h_2 \circ h_1(v_j)$$

is triangular with nonnegative entries. $\qquad\square$

Remark: For numerical stability, it is preferable to use $w_{k+1} = r_{k+1,k+1}\, e_{k+1} + e^{-i\theta_{k+1}} u''_{k+1}$ instead of $w_{k+1} = r_{k+1,k+1}\, e_{k+1} - e^{-i\theta_{k+1}} u''_{k+1}$. The effect of that choice is that the diagonal entries in R will be of the form $-e^{i\theta_j} r_{j,\,j} = e^{i(\theta_j + \pi)} r_{j,\,j}$. Of course, we can make these entries nonnegative by applying

$$h_n = \rho_{e_n,\, \pi - \theta_n} \circ \cdots \circ \rho_{e_1,\, \pi - \theta_1}$$

after h_{n-1}.

As in the Euclidean case, Proposition 13.19 immediately implies the QR-decomposition for arbitrary complex $n \times n$-matrices, where Q is now unitary (see Kincaid and Cheney [Kincaid and Cheney (1996)] and Ciarlet [Ciarlet (1989)]).

Proposition 13.20. *For every complex $n \times n$-matrix A, there is a sequence $H_1, \dots, H_{n-1}$ of matrices, where each H_i is either a Householder matrix or the identity, and an upper triangular matrix R, such that*

$$R = H_{n-1} \cdots H_2 H_1 A.$$

As a corollary, there is a pair of matrices Q, R, where Q is unitary and R is upper triangular, such that $A = QR$ (a QR-decomposition of A). Furthermore, R can be chosen so that its diagonal entries are nonnegative. This can be achieved by a diagonal matrix D with entries such that $|d_{ii}| = 1$ for $i = 1, \dots, n$, and we have $A = \widetilde{Q}\widetilde{R}$ with

$$\widetilde{Q} = H_1 \cdots H_{n-1} D, \quad \widetilde{R} = D^* R,$$

where $\widetilde{R}$ is upper triangular and has nonnegative diagonal entries.

Proof. It is essentially identical to the proof of Proposition 12.1, and we leave the details as an exercise. For the last statement, observe that $h_n \circ \cdots \circ h_1$ is also an isometry. $\qquad\square$

13.6   Orthogonal Projections and Involutions

In this section we begin by assuming that the field K is not a field of characteristic 2. Recall that a linear map $f \colon E \to E$ is an *involution* iff $f^2 = \mathrm{id}$, and is *idempotent* iff $f^2 = f$. We know from Proposition 5.7 that if f is idempotent, then

$$E = \mathrm{Im}(f) \oplus \mathrm{Ker}\,(f),$$

and that the restriction of f to its image is the identity. For this reason, a linear involution is called a *projection*. The connection between involutions and projections is given by the following simple proposition.

Proposition 13.21. *For any linear map $f \colon E \to E$, we have $f^2 = \mathrm{id}$ iff $\frac{1}{2}(\mathrm{id} - f)$ is a projection iff $\frac{1}{2}(\mathrm{id} + f)$ is a projection; in this case, f is equal to the difference of the two projections $\frac{1}{2}(\mathrm{id} + f)$ and $\frac{1}{2}(\mathrm{id} - f)$.*

Proof. We have

$$\left(\frac{1}{2}(\mathrm{id} - f) \right)^2 = \frac{1}{4}(\mathrm{id} - 2f + f^2)$$

so

$$\left(\frac{1}{2}(\mathrm{id} - f) \right)^2 = \frac{1}{2}(\mathrm{id} - f) \quad \text{iff} \quad f^2 = \mathrm{id}.$$

We also have

$$\left(\frac{1}{2}(\mathrm{id} + f) \right)^2 = \frac{1}{4}(\mathrm{id} + 2f + f^2),$$

so

$$\left(\frac{1}{2}(\mathrm{id} + f) \right)^2 = \frac{1}{2}(\mathrm{id} + f) \quad \text{iff} \quad f^2 = \mathrm{id}.$$

Obviously, $f = \frac{1}{2}(\mathrm{id} + f) - \frac{1}{2}(\mathrm{id} - f)$. $\square$

Proposition 13.22. *For any linear map $f \colon E \to E$, let $U^+ = \mathrm{Ker}\,(\frac{1}{2}(\mathrm{id} - f))$ and let $U^- = \mathrm{Im}(\frac{1}{2}(\mathrm{id} - f))$. If $f^2 = \mathrm{id}$, then*

$$U^+ = \mathrm{Ker}\left(\frac{1}{2}(\mathrm{id} - f) \right) = \mathrm{Im}\left(\frac{1}{2}(\mathrm{id} + f) \right),$$

and so, $f(u) = u$ on U^+ and $f(u) = -u$ on U^-.

Proof. If $f^2 = \text{id}$, then
$$(\text{id} + f) \circ (\text{id} - f) = \text{id} - f^2 = \text{id} - \text{id} = 0,$$
which implies that
$$\text{Im}\left(\frac{1}{2}(\text{id} + f)\right) \subseteq \text{Ker}\left(\frac{1}{2}(\text{id} - f)\right).$$
Conversely, if $u \in \text{Ker}\left(\frac{1}{2}(\text{id} - f)\right)$, then $f(u) = u$, so
$$\frac{1}{2}(\text{id} + f)(u) = \frac{1}{2}(u + u) = u,$$
and thus
$$\text{Ker}\left(\frac{1}{2}(\text{id} - f)\right) \subseteq \text{Im}\left(\frac{1}{2}(\text{id} + f)\right).$$
Therefore,
$$U^+ = \text{Ker}\left(\frac{1}{2}(\text{id} - f)\right) = \text{Im}\left(\frac{1}{2}(\text{id} + f)\right),$$
and so, $f(u) = u$ on U^+ and $f(u) = -u$ on U^-. $\square$

We now assume that $K = \mathbb{C}$. The involutions of E that are unitary transformations are characterized as follows.

Proposition 13.23. *Let $f \in \mathbf{GL}(E)$ be an involution. The following properties are equivalent:*

(a) The map f is unitary; that is, $f \in \mathbf{U}(E)$.

(b) The subspaces $U^- = \text{Im}(\frac{1}{2}(\text{id} - f))$ and $U^+ = \text{Im}(\frac{1}{2}(\text{id} + f))$ are orthogonal.

Furthermore, if E is finite-dimensional, then (a) and (b) are equivalent to (c) below:

(c) The map is self-adjoint; that is, $f = f^*$.

Proof. If f is unitary, then from $\langle f(u), f(v) \rangle = \langle u, v \rangle$ for all $u, v \in E$, we see that if $u \in U^+$ and $v \in U^-$, we get
$$\langle u, v \rangle = \langle f(u), f(v) \rangle = \langle u, -v \rangle = -\langle u, v \rangle,$$
so $2\langle u, v \rangle = 0$, which implies $\langle u, v \rangle = 0$, that is, U^+ and U^- are orthogonal. Thus, (a) implies (b).

Conversely, if (b) holds, since $f(u) = u$ on U^+ and $f(u) = -u$ on U^-, we see that $\langle f(u), f(v) \rangle = \langle u, v \rangle$ if $u, v \in U^+$ or if $u, v \in U^-$. Since $E = U^+ \oplus U^-$ and since U^+ and U^- are orthogonal, we also have $\langle f(u), f(v) \rangle = \langle u, v \rangle$ for all $u, v \in E$, and (b) implies (a).

If E is finite-dimensional, the adjoint f^* of f exists, and we know that $f^{-1} = f^*$. Since f is an involution, $f^2 = \text{id}$, which implies that $f^* = f^{-1} = f$. $\square$

A unitary involution is the identity on $U^+ = \text{Im}(\frac{1}{2}(\text{id} + f))$, and $f(v) = -v$ for all $v \in U^- = \text{Im}(\frac{1}{2}(\text{id} - f))$. Furthermore, E is an orthogonal direct sum $E = U^+ \oplus U^-$. We say that f is an *orthogonal reflection* about U^+. In the special case where U^+ is a hyperplane, we say that f is a *hyperplane reflection*. We already studied hyperplane reflections in the Euclidean case; see Chapter 12.

If $f: E \to E$ is a projection ($f^2 = f$), then

$$(\text{id} - 2f)^2 = \text{id} - 4f + 4f^2 = \text{id} - 4f + 4f = \text{id},$$

so $\text{id} - 2f$ is an involution. As a consequence, we get the following result.

Proposition 13.24. *If $f: E \to E$ is a projection ($f^2 = f$), then $\text{Ker}(f)$ and $\text{Im}(f)$ are orthogonal iff $f^* = f$.*

Proof. Apply Proposition 13.23 to $g = \text{id} - 2f$. Since $\text{id} - g = 2f$ we have

$$U^+ = \text{Ker}\left(\frac{1}{2}(\text{id} - g)\right) = \text{Ker}(f)$$

and

$$U^- = \text{Im}\left(\frac{1}{2}(\text{id} - g)\right) = \text{Im}(f),$$

which proves the proposition. $\qquad\square$

A projection such that $f = f^*$ is called an *orthogonal projection*.

If $(a_1 \dots, a_k)$ are k linearly independent vectors in $\mathbb{R}^n$, let us determine the matrix P of the orthogonal projection onto the subspace of $\mathbb{R}^n$ spanned by $(a_1, \dots, a_k)$. Let A be the $n \times k$ matrix whose jth column consists of the coordinates of the vector a_j over the canonical basis $(e_1, \dots, e_n)$.

Any vector in the subspace $(a_1, \dots, a_k)$ is a linear combination of the form Ax, for some $x \in \mathbb{R}^k$. Given any $y \in \mathbb{R}^n$, the orthogonal projection $Py = Ax$ of y onto the subspace spanned by $(a_1, \dots, a_k)$ is the vector Ax such that $y - Ax$ is orthogonal to the subspace spanned by $(a_1, \dots, a_k)$ (prove it). This means that $y - Ax$ is orthogonal to every a_j, which is expressed by

$$A^\top(y - Ax) = 0;$$

that is,

$$A^\top Ax = A^\top y.$$

The matrix $A^\top A$ is invertible because A has full rank k, thus we get

$$x = (A^\top A)^{-1} A^\top y,$$

and so

$$Py = Ax = A(A^\top A)^{-1} A^\top y.$$

Therefore, the matrix P of the projection onto the subspace spanned by $(a_1 \ldots, a_k)$ is given by

$$P = A(A^\top A)^{-1} A^\top.$$

The reader should check that $P^2 = P$ and $P^\top = P$.

13.7 Dual Norms

In the remark following the proof of Proposition 8.7, we explained that if $(E, \| \ \|)$ and $(F, \| \ \|)$ are two normed vector spaces and if we let $\mathcal{L}(E; F)$ denote the set of all continuous (equivalently, bounded) linear maps from E to F, then, we can define the *operator norm* (or *subordinate norm*) $\| \ \|$ on $\mathcal{L}(E; F)$ as follows: for every $f \in \mathcal{L}(E; F)$,

$$\|f\| = \sup_{\substack{x \in E \\ x \neq 0}} \frac{\|f(x)\|}{\|x\|} = \sup_{\substack{x \in E \\ \|x\|=1}} \|f(x)\|.$$

In particular, if $F = \mathbb{C}$, then $\mathcal{L}(E; F) = E'$ is the *dual space* of E, and we get the operator norm denoted by $\| \ \|_*$ given by

$$\|f\|_* = \sup_{\substack{x \in E \\ \|x\|=1}} |f(x)|.$$

The norm $\| \ \|_*$ is called the *dual norm* of $\| \ \|$ on E'.

Let us now assume that E is a finite-dimensional Hermitian space, in which case $E' = E^*$. Theorem 13.1 implies that for every linear form $f \in E^*$, there is a unique vector $y \in E$ so that

$$f(x) = \langle x, y \rangle,$$

for all $x \in E$, and so we can write

$$\|f\|_* = \sup_{\substack{x \in E \\ \|x\|=1}} |\langle x, y \rangle|.$$

The above suggests defining a norm $\| \ \|^D$ on E.

Definition 13.13. If E is a finite-dimensional Hermitian space and $\| \ \|$ is any norm on E, for any $y \in E$ we let

$$\|y\|^D = \sup_{\substack{x \in E \\ \|x\|=1}} |\langle x, y \rangle|,$$

be the *dual norm* of $\| \; \|$ (on E). If E is a real Euclidean space, then the dual norm is defined by

$$\|y\|^D = \sup_{\substack{x \in E \\ \|x\|=1}} \langle x, y \rangle$$

for all $y \in E$.

Beware that $\| \; \|$ is generally *not* the Hermitian norm associated with the Hermitian inner product. The dual norm shows up in convex programming; see Boyd and Vandenberghe [Boyd and Vandenberghe (2004)], Chapters 2, 3, 6, 9.

The fact that $\| \; \|^D$ is a norm follows from the fact that $\| \; \|_*$ is a norm and can also be checked directly. It is worth noting that the triangle inequality for $\| \; \|^D$ comes "for free," in the sense that it holds for any function $p \colon E \to \mathbb{R}$.

Proposition 13.25. *For any function $p \colon E \to \mathbb{R}$, if we define p^D by*

$$p^D(x) = \sup_{p(z)=1} |\langle z, x \rangle|,$$

then we have

$$p^D(x + y) \leq p^D(x) + p^D(y).$$

Proof. We have

$$\begin{aligned}
p^D(x + y) &= \sup_{p(z)=1} |\langle z, x + y \rangle| \\
&= \sup_{p(z)=1} \left(|\langle z, x \rangle + \langle z, y \rangle| \right) \\
&\leq \sup_{p(z)=1} \left(|\langle z, x \rangle| + |\langle z, y \rangle| \right) \\
&\leq \sup_{p(z)=1} |\langle z, x \rangle| + \sup_{p(z)=1} |\langle z, y \rangle| \\
&= p^D(x) + p^D(y). \qquad \square
\end{aligned}$$

Definition 13.14. If $p \colon E \to \mathbb{R}$ is a function such that

(1) $p(x) \geq 0$ for all $x \in E$, and $p(x) = 0$ iff $x = 0$;
(2) $p(\lambda x) = |\lambda| p(x)$, for all $x \in E$ and all $\lambda \in \mathbb{C}$;
(3) p is continuous, in the sense that for some basis $(e_1, \ldots, e_n)$ of E, the function

$$(x_1, \ldots, x_n) \mapsto p(x_1 e_1 + \cdots + x_n e_n)$$

from $\mathbb{C}^n$ to $\mathbb{R}$ is continuous,

then we say that p is a *pre-norm*.

Obviously, every norm is a pre-norm, but a pre-norm may not satisfy the triangle inequality.

Corollary 13.1. *The dual norm of any pre-norm is actually a norm.*

Proposition 13.26. *For all $y \in E$, we have*

$$\|y\|^D = \sup_{\substack{x \in E \\ \|x\|=1}} |\langle x, y \rangle| = \sup_{\substack{x \in E \\ \|x\|=1}} \Re\langle x, y \rangle.$$

Proof. Since E is finite dimensional, the unit sphere $S^{n-1} = \{x \in E \mid \|x\| = 1\}$ is compact, so there is some $x_0 \in S^{n-1}$ such that

$$\|y\|^D = |\langle x_0, y \rangle|.$$

If $\langle x_0, y \rangle = \rho e^{i\theta}$, with $\rho \geq 0$, then

$$|\langle e^{-i\theta} x_0, y \rangle| = |e^{-i\theta} \langle x_0, y \rangle| = |e^{-i\theta} \rho e^{i\theta}| = \rho,$$

so

$$\|y\|^D = \rho = \langle e^{-i\theta} x_0, y \rangle, \tag{13.2}$$

with $\|e^{-i\theta} x_0\| = \|x_0\| = 1$. On the other hand,

$$\Re\langle x, y \rangle \leq |\langle x, y \rangle|,$$

so by (13.2) we get

$$\|y\|^D = \sup_{\substack{x \in E \\ \|x\|=1}} |\langle x, y \rangle| = \sup_{\substack{x \in E \\ \|x\|=1}} \Re\langle x, y \rangle,$$

as claimed. $\qquad \square$

Proposition 13.27. *For all $x, y \in E$, we have*

$$|\langle x, y \rangle| \leq \|x\| \, \|y\|^D$$
$$|\langle x, y \rangle| \leq \|x\|^D \, \|y\|.$$

Proof. If $x = 0$, then $\langle x, y \rangle = 0$ and these inequalities are trivial. If $x \neq 0$, since $\|x/\|x\|\| = 1$, by definition of $\|y\|^D$, we have

$$|\langle x/\|x\|, y \rangle| \leq \sup_{\|z\|=1} |\langle z, y \rangle| = \|y\|^D,$$

which yields

$$|\langle x, y \rangle| \leq \|x\| \, \|y\|^D.$$

The second inequality holds because $|\langle x, y \rangle| = |\langle y, x \rangle|$. $\qquad \square$

It is not hard to show that for all $y \in \mathbb{C}^n$,

$$\|y\|_1^D = \|y\|_\infty$$
$$\|y\|_\infty^D = \|y\|_1$$
$$\|y\|_2^D = \|y\|_2 \,.$$

Thus, the Euclidean norm is autodual. More generally, the following proposition holds.

Proposition 13.28. *If $p, q \geq 1$ and $1/p + 1/q = 1$, then for all $y \in \mathbb{C}^n$, we have*

$$\|y\|_p^D = \|y\|_q \,.$$

Proof. By Hölder's inequality (Corollary 8.1), for all $x, y \in \mathbb{C}^n$, we have

$$|\langle x, y \rangle| \leq \|x\|_p \|y\|_q \,,$$

so

$$\|y\|_p^D = \sup_{\substack{x \in \mathbb{C}^n \\ \|x\|_p = 1}} |\langle x, y \rangle| \leq \|y\|_q \,.$$

For the converse, we consider the cases $p = 1$, $1 < p < +\infty$, and $p = +\infty$. First assume $p = 1$. The result is obvious for $y = 0$, so assume $y \neq 0$. Given y, if we pick $x_j = 1$ for some index j such that $\|y\|_\infty = \max_{1 \leq i \leq n} |y_i| = |y_j|$, and $x_k = 0$ for $k \neq j$, then $|\langle x, y \rangle| = |y_j| = \|y\|_\infty$, so $\|y\|_1^D = \|y\|_\infty$.

Now we turn to the case $1 < p < +\infty$. Then we also have $1 < q < +\infty$, and the equation $1/p + 1/q = 1$ is equivalent to $pq = p + q$, that is, $p(q - 1) = q$. Pick $z_j = y_j|y_j|^{q-2}$ for $j = 1, \ldots, n$, so that

$$\|z\|_p = \left(\sum_{j=1}^n |z_j|^p \right)^{1/p} = \left(\sum_{j=1}^n |y_j|^{(q-1)p} \right)^{1/p} = \left(\sum_{j=1}^n |y_j|^q \right)^{1/p} \,.$$

Then if $x = z/\|z\|_p$, we have

$$|\langle x, y \rangle| = \frac{\left| \sum_{j=1}^n z_j \overline{y_j} \right|}{\|z\|_p} = \frac{\left| \sum_{j=1}^n y_j \overline{y_j} |y_j|^{q-2} \right|}{\|z\|_p}$$

$$= \frac{\sum_{j=1}^n |y_j|^q}{\left(\sum_{j=1}^n |y_j|^q \right)^{1/p}} = \left(\sum_{j=1}^n |y_j|^q \right)^{1/q} = \|y\|_q \,.$$

Thus $\|y\|_p^D = \|y\|_q$.

Finally, if $p = \infty$, then pick $x_j = y_j/|y_j|$ if $y_j \neq 0$, and $x_j = 0$ if $y_j = 0$. Then

$$|\langle x, y \rangle| = \left| \sum_{y_j \neq 0}^{n} y_j \overline{y_j}/|y_j| \right| = \sum_{y_j \neq 0} |y_j| = \|y\|_1.$$

Thus $\|y\|_\infty^D = \|y\|_1$. $\qquad\qquad\qquad\qquad\qquad\qquad\qquad\qquad\qquad\square$

We can show that the dual of the spectral norm is the *trace norm* (or *nuclear norm*) also discussed in Section 20.5. Recall from Proposition 8.7 that the spectral norm $\|A\|_2$ of a matrix A is the square root of the largest eigenvalue of A^*A, that is, the largest singular value of A.

Proposition 13.29. *The dual of the spectral norm is given by*

$$\|A\|_2^D = \sigma_1 + \cdots + \sigma_r,$$

where $\sigma_1 > \cdots > \sigma_r > 0$ are the singular values of $A \in M_n(\mathbb{C})$ (which has rank r).

Proof. In this case the inner product on $M_n(\mathbb{C})$ is the Frobenius inner product $\langle A, B \rangle = \mathrm{tr}(B^*A)$, and the dual norm of the spectral norm is given by

$$\|A\|_2^D = \sup\{|\mathrm{tr}(A^*B)| \mid \|B\|_2 = 1\}.$$

If we factor A using an SVD as $A = V\Sigma U^*$, where U and V are unitary and Σ is a diagonal matrix whose r nonzero entries are the singular values $\sigma_1 > \cdots > \sigma_r > 0$, where r is the rank of A, then

$$|\mathrm{tr}(A^*B)| = |\mathrm{tr}(U\Sigma V^*B)| = |\mathrm{tr}(\Sigma V^*BU)|,$$

so if we pick $B = VU^*$, a unitary matrix such that $\|B\|_2 = 1$, we get

$$|\mathrm{tr}(A^*B)| = \mathrm{tr}(\Sigma) = \sigma_1 + \cdots + \sigma_r,$$

and thus

$$\|A\|_2^D \geq \sigma_1 + \cdots + \sigma_r.$$

Since $\|B\|_2 = 1$ and U and V are unitary, by Proposition 8.7 we have $\|V^*BU\|_2 = \|B\|_2 = 1$. If $Z = V^*BU$, by definition of the operator norm

$$1 = \|Z\|_2 = \sup\{\|Zx\|_2 \mid \|x\|_2 = 1\},$$

so by picking x to be the canonical vector e_j, we see that $\|Z^j\|_2 \leq 1$ where Z^j is the jth column of Z, so $|z_{jj}| \leq 1$, and since

$$|\mathrm{tr}(\Sigma V^*BU)| = |\mathrm{tr}(\Sigma Z)| = \left| \sum_{j=1}^{r} \sigma_j z_{jj} \right| \leq \sum_{j=1}^{r} \sigma_j |z_{jj}| \leq \sum_{j=1}^{r} \sigma_j,$$

and we conclude that

$$|\text{tr}(\Sigma V^* B U)| \leq \sum_{j=1}^{r} \sigma_j.$$

The above implies that

$$\|A\|_2^D \leq \sigma_1 + \cdots + \sigma_r,$$

and since we also have $\|A\|_2^D \geq \sigma_1 + \cdots + \sigma_r$, we conclude that

$$\|A\|_2^D = \sigma_1 + \cdots + \sigma_r,$$

proving our proposition.   □

Definition 13.15. Given any complex matrix $n \times n$ matrix A of rank r, its *nuclear norm* (or *trace norm*) is given by

$$\|A\|_N = \sigma_1 + \cdots + \sigma_r.$$

The nuclear norm can be generalized to $m \times n$ matrices (see Section 20.5). The nuclear norm $\sigma_1 + \cdots + \sigma_r$ of an $m \times n$ matrix A (where r is the rank of A) is denoted by $\|A\|_N$. The nuclear norm plays an important role in *matrix completion*. The problem is this. Given a matrix A_0 with missing entries (missing data), one would like to fill in the missing entries in A_0 to obtain a matrix A of minimal rank. For example, consider the matrices

$$A_0 = \begin{pmatrix} 1 & 2 \\ * & * \end{pmatrix}, \qquad B_0 = \begin{pmatrix} 1 & * \\ * & 4 \end{pmatrix}, \qquad C_0 = \begin{pmatrix} 1 & 2 \\ 3 & * \end{pmatrix}.$$

All can be completed with rank 1. For A_0, use any multiple of $(1, 2)$ for the second row. For B_0, use any numbers b and c such that $bc = 4$. For C_0, the only possibility is $d = 6$.

A famous example of this problem is the *Netflix competition*. The ratings of m films by n viewers goes into A_0. But the customers didn't see all the movies. Many ratings were missing. Those had to be predicted by a recommender system. The nuclear norm gave a good solution that needed to be adjusted for human psychology.

Since the rank of a matrix is not a norm, in order to solve the matrix completion problem we can use the following "convex relaxation." Let A_0 be an incomplete $m \times n$ matrix:

Minimize $\|A\|_N$ subject to $A = A_0$ in the known entries.

The above problem has been extensively studied, in particular by Candès and Recht. Roughly, they showed that if A is an $n \times n$ matrix of rank r and K entries are known in A, then if K is large enough

$(K > Cn^{5/4}r \log n)$, with high probability, the recovery of A is perfect. See Strang [Strang (2019)] for details (Section III.5).

We close this section by stating the following duality theorem.

Theorem 13.2. *If E is a finite-dimensional Hermitian space, then for any norm $\| \; \|$ on E, we have*

$$\|y\|^{DD} = \|y\|$$

for all $y \in E$.

Proof. By Proposition 13.27, we have

$$|\langle x, y \rangle| \leq \|x\|^D \|y\|,$$

so we get

$$\|y\|^{DD} = \sup_{\|x\|^D = 1} |\langle x, y \rangle| \leq \|y\|, \quad \text{for all } y \in E.$$

It remains to prove that

$$\|y\| \leq \|y\|^{DD}, \quad \text{for all } y \in E.$$

Proofs of this fact can be found in Horn and Johnson [Horn and Johnson (1990)] (Section 5.5), and in Serre [Serre (2010)] (Chapter 7). The proof makes use of the fact that a nonempty, closed, convex set has a supporting hyperplane through each of its boundary points, a result known as *Minkowski's lemma*. For a geometric interpretation of supporting hyperplane see Figure 13.1. This result is a consequence of the *Hahn–Banach theorem*; see Gallier [Gallier (2011b)]. We give the proof in the case where E is a real Euclidean space. Some minor modifications have to be made when dealing with complex vector spaces and are left as an exercise.

Since the unit ball $B = \{z \in E \mid \|z\| \leq 1\}$ is closed and convex, the Minkowski lemma says for every x such that $\|x\| = 1$, there is an affine map g of the form

$$g(z) = \langle z, w \rangle - \langle x, w \rangle$$

with $\|w\| = 1$, such that $g(x) = 0$ and $g(z) \leq 0$ for all z such that $\|z\| \leq 1$. Then it is clear that

$$\sup_{\|z\|=1} \langle z, w \rangle = \langle x, w \rangle,$$

and so

$$\|w\|^D = \langle x, w \rangle.$$

Fig. 13.1   The orange tangent plane is a supporting hyperplane to the unit ball in $\mathbb{R}^3$ since this ball is entirely contained in "one side" of the tangent plane.

It follows that

$$\|x\|^{DD} \geq \langle w/\|w\|^D, x\rangle = \frac{\langle x, w\rangle}{\|w\|^D} = 1 = \|x\|$$

for all x such that $\|x\| = 1$. By homogeneity, this is true for all $y \in E$, which completes the proof in the real case. When E is a complex vector space, we have to view the unit ball B as a closed convex set in $\mathbb{R}^{2n}$ and we use the fact that there is real affine map of the form

$$g(z) = \Re\langle z, w\rangle - \Re\langle x, w\rangle$$

such that $g(x) = 0$ and $g(z) \leq 0$ for all z with $\|z\| = 1$, so that $\|w\|^D = \Re\langle x, w\rangle$. $\square$

More details on dual norms and unitarily invariant norms can be found in Horn and Johnson [Horn and Johnson (1990)] (Chapters 5 and 7).

13.8   Summary

The main concepts and results of this chapter are listed below:

- *Semilinear maps.*
- *Sesquilinear forms*; *Hermitian forms.*
- *Quadratic form* associated with a sesquilinear form.

- *Polarization identities.*
- *Positive* and *positive definite* Hermitian forms; *pre-Hilbert spaces, Hermitian spaces.*
- *Gram matrix* associated with a Hermitian product.
- The *Cauchy–Schwarz inequality* and the *Minkowski inequality.*
- *Hermitian inner product, Hermitian norm.*
- The *parallelogram law.*
- The musical isomorphisms $\flat\colon \overline{E} \to E^*$ and $\sharp\colon E^* \to \overline{E}$; Theorem 13.1 ($E$ is finite-dimensional).
- The *adjoint* of a linear map (with respect to a Hermitian inner product).
- Existence of orthonormal bases in a Hermitian space (Proposition 13.10).
- *Gram–Schmidt orthonormalization procedure.*
- *Linear isometries (unitary transformations).*
- The *unitary group, unitary matrices.*
- The *unitary group* $\mathbf{U}(n)$.
- The *special unitary group* $\mathbf{SU}(n)$.
- QR-Decomposition for arbitrary complex matrices.
- The *Hadamard inequality* for complex matrices.
- The *Hadamard inequality* for Hermitian positive semidefinite matrices.
- Orthogonal projections and involutions; orthogonal reflections.
- Dual norms.
- Nuclear norm (also called trace norm).
- Matrix completion.

13.9 Problems

Problem 13.1. Let $(E, \langle -, - \rangle)$ be a Hermitian space of finite dimension. Prove that if $f\colon E \to E$ is a self-adjoint linear map (that is, $f^* = f$), then $\langle f(x), x \rangle \in \mathbb{R}$ for all $x \in E$.

Problem 13.2. Prove the polarization identities of Proposition 13.1.

Problem 13.3. Let E be a real Euclidean space. Give an example of a nonzero linear map $f\colon E \to E$ such that $\langle f(u), u \rangle = 0$ for all $u \in E$.

Problem 13.4. Prove Proposition 13.8.

Problem 13.5. (1) Prove that every matrix in $\mathbf{SU}(2)$ is of the form

$$A = \begin{pmatrix} a + ib & c + id \\ -c + id & a - ib \end{pmatrix}, \quad a^2 + b^2 + c^2 + d^2 = 1, \ a, b, c, d \in \mathbb{R},$$

(2) Prove that the matrices

$$\begin{pmatrix} 1 & 0 \\ 0 & 1 \end{pmatrix}, \quad \begin{pmatrix} i & 0 \\ 0 & -i \end{pmatrix}, \quad \begin{pmatrix} 0 & 1 \\ -1 & 0 \end{pmatrix}, \quad \begin{pmatrix} 0 & i \\ i & 0 \end{pmatrix}$$

all belong to $\mathbf{SU}(2)$ and are linearly independent over $\mathbb{C}$.

(3) Prove that the linear span of $\mathbf{SU}(2)$ over $\mathbb{C}$ is the complex vector space $\mathrm{M}_2(\mathbb{C})$ of all complex 2×2 matrices.

Problem 13.6. The purpose of this problem is to prove that the linear span of $\mathbf{SU}(n)$ over $\mathbb{C}$ is $\mathrm{M}_n(\mathbb{C})$ for all $n \geq 3$. One way to prove this result is to adapt the method of Problem 11.12, so please review this problem.

Every complex matrix $A \in \mathrm{M}_n(\mathbb{C})$ can be written as

$$A = \frac{A + A^*}{2} + \frac{A - A^*}{2}$$

where the first matrix is Hermitian and the second matrix is skew-Hermitian. Observe that if $A = (z_{ij})$ is a Hermitian matrix, that is $A^* = A$, then $z_{ji} = \bar{z}_{ij}$, so if $z_{ij} = a_{ij} + ib_{ij}$ with $a_{ij}, b_{ij} \in \mathbb{R}$, then $a_{ij} = a_{ji}$ and $b_{ij} = -b_{ji}$. On the other hand, if $A = (z_{ij})$ is a skew-Hermitian matrix, that is $A^* = -A$, then $z_{ji} = -\bar{z}_{ij}$, so $a_{ij} = -a_{ji}$ and $b_{ij} = b_{ji}$.

The Hermitian and the skew-Hermitian matrices do not form complex vector spaces because they are not closed under multiplication by a complex number, but we can get around this problem by treating the real part and the complex part of these matrices separately and using multiplication by reals.

(1) Consider the matrices of the form

$$R_c^{i,j} = \begin{pmatrix} 1 & & & & & & & & \\ & \ddots & & & & & & & \\ & & 1 & & & & & & \\ & & & 0 & 0 & \cdots & 0 & i & \\ & & & 0 & 1 & \cdots & 0 & 0 & \\ & & & \vdots & \vdots & \ddots & \vdots & \vdots & \\ & & & 0 & 0 & \cdots & 1 & 0 & \\ & & & i & 0 & \cdots & 0 & 0 & \\ & & & & & & & & 1 & \\ & & & & & & & & & \ddots & \\ & & & & & & & & & & 1 \end{pmatrix}.$$

Prove that $(R_c^{i,j})^* R_c^{i,j} = I$ and $\det(R_c^{i,j}) = +1$. Use the matrices $R^{i,j}, R_c^{i,j} \in \mathbf{SU}(n)$ and the matrices $(R^{i,j} - (R^{i,j})^*)/2$ (from Problem 11.12) to form the real part of a skew-Hermitian matrix and the matrices $(R_c^{i,j} - (R_c^{i,j})^*)/2$ to form the imaginary part of a skew-Hermitian matrix. Deduce that the matrices in $\mathbf{SU}(n)$ span all skew-Hermitian matrices.

(2) Consider matrices of the form

Type 1

$$
S_c^{1,2} = \begin{pmatrix}
0 & -i & 0 & 0 & \dots & 0 \\
i & 0 & 0 & 0 & \dots & 0 \\
0 & 0 & -1 & 0 & \dots & 0 \\
0 & 0 & 0 & 1 & \dots & 0 \\
\vdots & \vdots & \vdots & \vdots & \ddots & \vdots \\
0 & 0 & 0 & 0 & \dots & 1
\end{pmatrix}.
$$

Type 2

$$
S_c^{i,i+1} = \begin{pmatrix}
-1 & & & & & & & & \\
& 1 & & & & & & & \\
& & \ddots & & & & & & \\
& & & 1 & & & & & \\
& & & & 0 & -i & & & \\
& & & & i & 0 & & & \\
& & & & & & 1 & & \\
& & & & & & & \ddots & \\
& & & & & & & & 1
\end{pmatrix}.
$$

Type 3

$$
S_c^{i,j} = \begin{pmatrix}
1 & & & & & & & & & \\
& \ddots & & & & & & & & \\
& & 1 & & & & & & & \\
& & & 0 & 0 & \cdots & 0 & -i & & \\
& & & 0 & -1 & \cdots & 0 & 0 & & \\
& & & \vdots & \vdots & \ddots & \vdots & \vdots & & \\
& & & 0 & 0 & \cdots & 1 & 0 & & \\
& & & i & 0 & \cdots & 0 & 0 & & \\
& & & & & & & & 1 & \\
& & & & & & & & & \ddots \\
& & & & & & & & & & 1
\end{pmatrix}.
$$

Prove that $S^{i,j}, S_c^{i,j} \in \mathbf{SU}(n)$, and using diagonal matrices as in Problem 11.12, prove that the matrices $S^{i,j}$ can be used to form the real part of a Hermitian matrix and the matrices $S_c^{i,j}$ can be used to form the imaginary part of a Hermitian matrix.

(3) Use (1) and (2) to prove that the matrices in $\mathbf{SU}(n)$ span all Hermitian matrices. It follows that $\mathbf{SU}(n)$ spans $M_n(\mathbb{C})$ for $n \geq 3$.

Problem 13.7. Consider the complex matrix

$$A = \begin{pmatrix} i & 1 \\ 1 & -i \end{pmatrix}.$$

Check that this matrix is symmetric but not Hermitian. Prove that

$$\det(\lambda I - A) = \lambda^2,$$

and so the eigenvalues of A are $0, 0$.

Problem 13.8. Let $(E, \langle -, - \rangle)$ be a Hermitian space of finite dimension and let $f \colon E \to E$ be a linear map. Prove that the following conditions are equivalent.

(1) $f \circ f^* = f^* \circ f$ (f is normal).
(2) $\langle f(x), f(y) \rangle = \langle f^*(x), f^*(y) \rangle$ for all $x, y \in E$.
(3) $\|f(x)\| = \|f^*(x)\|$ for all $x \in E$.
(4) The map f can be diagonalized with respect to an orthonormal basis of eigenvectors.
(5) There exist some linear maps $g, h \colon E \to E$ such that, $g = g^*$, $\langle x, g(x) \rangle \geq 0$ for all $x \in E$, $h^{-1} = h^*$, and $f = g \circ h = h \circ g$.
(6) There exist some linear map $h \colon E \to E$ such that $h^{-1} = h^*$ and $f^* = h \circ f$.
(7) There is a polynomial P (with complex coefficients) such that $f^* = P(f)$.

Problem 13.9. Recall from Problem 12.7 that a complex $n \times n$ matrix H is *upper Hessenberg* if $h_{jk} = 0$ for all (j, k) such that $j - k \geq 0$. Adapt the proof of Problem 12.7 to prove that given any complex $n \times n$-matrix A, there are $n - 2 \geq 1$ complex matrices $H_1, \ldots, H_{n-2}$, Householder matrices or the identity, such that

$$B = H_{n-2} \cdots H_1 A H_1 \cdots H_{n-2}$$

is upper Hessenberg.

Problem 13.10. Prove that all $y \in \mathbb{C}^n$,

$$\|y\|_1^D = \|y\|_\infty$$
$$\|y\|_\infty^D = \|y\|_1$$
$$\|y\|_2^D = \|y\|_2 .$$

Problem 13.11. The purpose of this problem is to complete each of the matrices A_0, B_0, C_0 of Section 13.7 to a matrix A in such way that the nuclear norm $\|A\|_N$ is minimized.

(1) Prove that the squares σ_1^2 and σ_2^2 of the singular values of

$$A = \begin{pmatrix} 1 & 2 \\ c & d \end{pmatrix}$$

are the zeros of the equation

$$\lambda^2 - (5 + c^2 + d^2)\lambda + (2c - d)^2 = 0.$$

(2) Using the fact that

$$\|A\|_N = \sigma_1 + \sigma_2 = \sqrt{\sigma_1^2 + \sigma_2^2 + 2\sigma_1\sigma_2},$$

prove that

$$\|A\|_N^2 = 5 + c^2 + d^2 + 2|2c - d|.$$

Consider the cases where $2c - d \geq 0$ and $2c - d \leq 0$, and show that in both cases we must have $c = -2d$, and that the minimum of $f(c, d) = 5 + c^2 + d^2 + 2|2c - d|$ is achieved by $c = d = 0$. Conclude that the matrix A completing A_0 that minimizes $\|A\|_N$ is

$$A = \begin{pmatrix} 1 & 2 \\ 0 & 0 \end{pmatrix} .$$

(3) Prove that the squares σ_1^2 and σ_2^2 of the singular values of

$$A = \begin{pmatrix} 1 & b \\ c & 4 \end{pmatrix}$$

are the zeros of the equation

$$\lambda^2 - (17 + b^2 + c^2)\lambda + (4 - bc)^2 = 0.$$

(4) Prove that

$$\|A\|_N^2 = 17 + b^2 + c^2 + 2|4 - bc|.$$

Consider the cases where $4 - bc \geq 0$ and $4 - bc \leq 0$, and show that in both cases we must have $b^2 = c^2$. Then show that the minimum of $f(c, d) =$

$17 + b^2 + c^2 + 2|4 - bc|$ is achieved by $b = c$ with $-2 \leq b \leq 2$. Conclude that the matrices A completing B_0 that minimize $\|A\|_N$ are given by

$$A = \begin{pmatrix} 1 & b \\ b & 4 \end{pmatrix}, \qquad -2 \leq b \leq 2.$$

(5) Prove that the squares σ_1^2 and σ_2^2 of the singular values of

$$A = \begin{pmatrix} 1 & 2 \\ 3 & d \end{pmatrix}$$

are the zeros of the equation

$$\lambda^2 - (14 + d^2)\lambda + (6 - d)^2 = 0$$

(6) Prove that

$$\|A\|_N^2 = 14 + d^2 + 2|6 - d|.$$

Consider the cases where $6 - d \geq 0$ and $6 - d \leq 0$, and show that the minimum of $f(c, d) = 14 + d^2 + 2|6 - d|$ is achieved by $d = 1$. Conclude that the the matrix A completing C_0 that minimizes $\|A\|_N$ is given by

$$A = \begin{pmatrix} 1 & 2 \\ 3 & 1 \end{pmatrix}.$$

Problem 13.12. Prove Theorem 13.2 when E is a finite-dimensional Hermitian space.

Chapter 14

Eigenvectors and Eigenvalues

In this chapter all vector spaces are defined over an arbitrary field K. For the sake of concreteness, the reader may safely assume that $K = \mathbb{R}$ or $K = \mathbb{C}$.

14.1 Eigenvectors and Eigenvalues of a Linear Map

Given a finite-dimensional vector space E, let $f \colon E \to E$ be any linear map. If by luck there is a basis $(e_1, \ldots, e_n)$ of E with respect to which f is represented by a *diagonal matrix*

$$
D = \begin{pmatrix} \lambda_1 & 0 & \ldots & 0 \\ 0 & \lambda_2 & \ddots & \vdots \\ \vdots & \ddots & \ddots & 0 \\ 0 & \ldots & 0 & \lambda_n \end{pmatrix},
$$

then the action of f on E is very simple; in every "direction" e_i, we have

$$
f(e_i) = \lambda_i e_i.
$$

We can think of f as a transformation that stretches or shrinks space along the direction $e_1, \ldots, e_n$ (at least if E is a real vector space). In terms of matrices, the above property translates into the fact that there is an invertible matrix P and a diagonal matrix D such that a matrix A can be factored as

$$
A = PDP^{-1}.
$$

When this happens, we say that f (or A) is *diagonalizable*, the λ_i's are called the *eigenvalues* of f, and the e_i's are *eigenvectors* of f. For example, we will see that every symmetric matrix can be diagonalized. Unfortunately, not every matrix can be diagonalized. For example, the matrix

$$A_1 = \begin{pmatrix} 1 & 1 \\ 0 & 1 \end{pmatrix}$$

can't be diagonalized. Sometimes a matrix fails to be diagonalizable because its eigenvalues do not belong to the field of coefficients, such as

$$A_2 = \begin{pmatrix} 0 & -1 \\ 1 & 0 \end{pmatrix},$$

whose eigenvalues are $\pm i$. This is not a serious problem because A_2 can be diagonalized over the complex numbers. However, A_1 is a "fatal" case! Indeed, its eigenvalues are both 1 and the problem is that A_1 does not have enough eigenvectors to span E.

The next best thing is that there is a basis with respect to which f is represented by an *upper triangular* matrix. In this case we say that f can be *triangularized*, or that f is *triangulable*. As we will see in Section 14.2, if all the eigenvalues of f belong to the field of coefficients K, then f can be triangularized. In particular, this is the case if $K = \mathbb{C}$.

Now an alternative to triangularization is to consider the representation of f with respect to *two* bases $(e_1, \ldots, e_n)$ and $(f_1, \ldots, f_n)$, rather than a single basis. In this case, if $K = \mathbb{R}$ or $K = \mathbb{C}$, it turns out that we can even pick these bases to be *orthonormal*, and we get a diagonal matrix Σ with *nonnegative entries*, such that

$$f(e_i) = \sigma_i f_i, \quad 1 \leq i \leq n.$$

The nonzero σ_i's are the *singular values* of f, and the corresponding representation is the *singular value decomposition*, or *SVD*. The SVD plays a very important role in applications, and will be considered in detail in Chapter 20.

In this section we focus on the possibility of diagonalizing a linear map, and we introduce the relevant concepts to do so. Given a vector space E over a field K, let id denote the identity map on E.

The notion of eigenvalue of a linear map $f \colon E \to E$ defined on an infinite-dimensional space E is quite subtle because it cannot be defined in terms of eigenvectors as in the finite-dimensional case. The problem is that the map $\lambda \, \mathrm{id} - f$ (with $\lambda \in \mathbb{C}$) could be noninvertible (because it is not surjective) and yet injective. In finite dimension this cannot happen, so until further notice we *assume that E is of finite dimension n.*

Definition 14.1. Given any vector space E of finite dimension n and any linear map $f \colon E \to E$, a scalar $\lambda \in K$ is called an *eigenvalue, or proper* *value, or characteristic value of* f if there is some *nonzero* vector $u \in E$ such that

$$f(u) = \lambda u.$$

Equivalently, λ is an eigenvalue of f if $\mathrm{Ker}\,(\lambda\,\mathrm{id} - f)$ is nontrivial (i.e., $\mathrm{Ker}\,(\lambda\,\mathrm{id} - f) \neq \{0\}$) iff $\lambda\,\mathrm{id} - f$ is *not* invertible (this is where the fact that E is finite-dimensional is used; a linear map from E to itself is injective iff it is invertible). A vector $u \in E$ is called an *eigenvector, or proper vector, or characteristic vector of* f if $u \neq 0$ and if there is some $\lambda \in K$ such that

$$f(u) = \lambda u;$$

the scalar λ is then an eigenvalue, and we say that u is an *eigenvector associated with* λ. Given any eigenvalue $\lambda \in K$, the nontrivial subspace $\mathrm{Ker}\,(\lambda\,\mathrm{id} - f)$ consists of all the eigenvectors associated with λ together with the zero vector; this subspace is denoted by $E_\lambda(f)$, or $E(\lambda, f)$, or even by E_λ, and is called the *eigenspace associated with* λ, *or proper subspace associated with* λ.

Note that distinct eigenvectors may correspond to the same eigenvalue, but distinct eigenvalues correspond to disjoint sets of eigenvectors.

Remark: As we emphasized in the remark following Definition 8.4, we *require an eigenvector to be nonzero*. This requirement seems to have more benefits than inconveniences, even though it may considered somewhat inelegant because the set of all eigenvectors associated with an eigenvalue is not a subspace since the zero vector is excluded.

The next proposition shows that the eigenvalues of a linear map $f : E \to E$ are the roots of a polynomial associated with f.

Proposition 14.1. *Let* E *be any vector space of finite dimension* n *and let* f *be any linear map* $f : E \to E$. *The eigenvalues of* f *are the roots (in* K) *of the polynomial*

$$\det(\lambda\,\mathrm{id} - f).$$

Proof. A scalar $\lambda \in K$ is an eigenvalue of f iff there is some vector $u \neq 0$ in E such that

$$f(u) = \lambda u$$

iff

$$(\lambda\,\mathrm{id} - f)(u) = 0$$

iff $(\lambda\,\mathrm{id} - f)$ is not invertible iff, by Proposition 6.10,

$$\det(\lambda\,\mathrm{id} - f) = 0. \qquad \square$$

In view of the importance of the polynomial $\det(\lambda \, \mathrm{id} - f)$, we have the following definition.

Definition 14.2. Given any vector space E of dimension n, for any linear map $f \colon E \to E$, the polynomial $P_f(X) = \chi_f(X) = \det(X \, \mathrm{id} - f)$ is called the *characteristic polynomial of f*. For any square matrix A, the polynomial $P_A(X) = \chi_A(X) = \det(XI - A)$ is called the *characteristic polynomial of A*.

Note that we already encountered the characteristic polynomial in Section 6.7; see Definition 6.11.

Given any basis $(e_1, \ldots, e_n)$, if $A = M(f)$ is the matrix of f w.r.t. $(e_1, \ldots, e_n)$, we can compute the characteristic polynomial $\chi_f(X) = \det(X \, \mathrm{id} - f)$ of f by expanding the following determinant:

$$\det(XI - A) = \begin{vmatrix} X - a_{1\,1} & -a_{1\,2} & \ldots & -a_{1\,n} \\ -a_{2\,1} & X - a_{2\,2} & \ldots & -a_{2\,n} \\ \vdots & \vdots & \ddots & \vdots \\ -a_{n\,1} & -a_{n\,2} & \ldots & X - a_{n\,n} \end{vmatrix}.$$

If we expand this determinant, we find that

$$\chi_A(X) = \det(XI - A) = X^n - (a_{1\,1} + \cdots + a_{n\,n})X^{n-1} + \cdots + (-1)^n \det(A).$$

The sum $\operatorname{tr}(A) = a_{1\,1} + \cdots + a_{n\,n}$ of the diagonal elements of A is called the *trace of A*. Since we proved in Section 6.7 that the characteristic polynomial only depends on the linear map f, the above shows that $\operatorname{tr}(A)$ has the same value for all matrices A representing f. Thus, the *trace of a linear map* is well-defined; we have $\operatorname{tr}(f) = \operatorname{tr}(A)$ for any matrix A representing f.

Remark: The characteristic polynomial of a linear map is sometimes defined as $\det(f - X \, \mathrm{id})$. Since

$$\det(f - X \, \mathrm{id}) = (-1)^n \det(X \, \mathrm{id} - f),$$

this makes essentially no difference but the version $\det(X \, \mathrm{id} - f)$ has the small advantage that the coefficient of X^n is $+1$.

If we write

$$\begin{aligned} \chi_A(X) &= \det(XI - A) \\ &= X^n - \tau_1(A)X^{n-1} + \cdots + (-1)^k \tau_k(A)X^{n-k} + \cdots + (-1)^n \tau_n(A), \end{aligned}$$

then we just proved that

$$\tau_1(A) = \text{tr}(A) \quad \text{and} \quad \tau_n(A) = \det(A).$$

It is also possible to express $\tau_k(A)$ in terms of determinants of certain submatrices of A. For any nonempty subset, $I \subseteq \{1, \ldots, n\}$, say $I = \{i_1 < \ldots < i_k\}$, let $A_{I,I}$ be the $k \times k$ submatrix of A whose jth column consists of the elements $a_{i_h\, i_j}$, where $h = 1, \ldots, k$. Equivalently, $A_{I,I}$ is the matrix obtained from A by first selecting the columns whose indices belong to I, and then the rows whose indices also belong to I. Then it can be shown that

$$\tau_k(A) = \sum_{\substack{I \subseteq \{1,\ldots,n\} \\ |I|=k}} \det(A_{I,I}).$$

If all the roots, $\lambda_1, \ldots, \lambda_n$, of the polynomial $\det(XI - A)$ belong to the field K, then we can write

$$\chi_A(X) = \det(XI - A) = (X - \lambda_1) \cdots (X - \lambda_n),$$

where some of the λ_i's may appear more than once. Consequently,

$$\chi_A(X) = \det(XI - A)$$
$$= X^n - \sigma_1(\lambda)X^{n-1} + \cdots + (-1)^k \sigma_k(\lambda)X^{n-k} + \cdots + (-1)^n \sigma_n(\lambda),$$

where

$$\sigma_k(\lambda) = \sum_{\substack{I \subseteq \{1,\ldots,n\} \\ |I|=k}} \prod_{i \in I} \lambda_i,$$

the *kth elementary symmetric polynomial (or function)* of the λ_i's, where $\lambda = (\lambda_1, \ldots, \lambda_n)$. The elementary symmetric polynomial $\sigma_k(\lambda)$ is often denoted $E_k(\lambda)$, but this notation may be confusing in the context of linear algebra. For $n = 5$, the elementary symmetric polynomials are listed below:

$$\sigma_0(\lambda) = 1$$
$$\sigma_1(\lambda) = \lambda_1 + \lambda_2 + \lambda_3 + \lambda_4 + \lambda_5$$
$$\sigma_2(\lambda) = \lambda_1\lambda_2 + \lambda_1\lambda_3 + \lambda_1\lambda_4 + \lambda_1\lambda_5 + \lambda_2\lambda_3 + \lambda_2\lambda_4 + \lambda_2\lambda_5$$
$$\qquad + \lambda_3\lambda_4 + \lambda_3\lambda_5 + \lambda_4\lambda_5$$
$$\sigma_3(\lambda) = \lambda_3\lambda_4\lambda_5 + \lambda_2\lambda_4\lambda_5 + \lambda_2\lambda_3\lambda_5 + \lambda_2\lambda_3\lambda_4 + \lambda_1\lambda_4\lambda_5$$
$$\qquad + \lambda_1\lambda_3\lambda_5 + \lambda_1\lambda_3\lambda_4 + \lambda_1\lambda_2\lambda_5 + \lambda_1\lambda_2\lambda_4 + \lambda_1\lambda_2\lambda_3$$
$$\sigma_4(\lambda) = \lambda_1\lambda_2\lambda_3\lambda_4 + \lambda_1\lambda_2\lambda_3\lambda_5 + \lambda_1\lambda_2\lambda_4\lambda_5 + \lambda_1\lambda_3\lambda_4\lambda_5 + \lambda_2\lambda_3\lambda_4\lambda_5$$
$$\sigma_5(\lambda) = \lambda_1\lambda_2\lambda_3\lambda_4\lambda_5.$$

Since

$$\chi_A(X) = X^n - \tau_1(A)X^{n-1} + \cdots + (-1)^k \tau_k(A)X^{n-k} + \cdots + (-1)^n \tau_n(A)$$
$$= X^n - \sigma_1(\lambda)X^{n-1} + \cdots + (-1)^k \sigma_k(\lambda)X^{n-k} + \cdots + (-1)^n \sigma_n(\lambda),$$

we have

$$\sigma_k(\lambda) = \tau_k(A), \quad k = 1, \ldots, n,$$

and in particular, the product of the eigenvalues of f is equal to $\det(A) = \det(f)$, and the sum of the eigenvalues of f is equal to the trace $\operatorname{tr}(A) = \operatorname{tr}(f)$, of f; for the record,

$$\operatorname{tr}(f) = \lambda_1 + \cdots + \lambda_n$$
$$\det(f) = \lambda_1 \cdots \lambda_n,$$

where $\lambda_1, \ldots, \lambda_n$ are the eigenvalues of f (and A), where some of the λ_i's may appear more than once. In particular, f is not invertible iff it admits 0 has an eigenvalue (since f is singular iff $\lambda_1 \cdots \lambda_n = \det(f) = 0$).

Remark: Depending on the field K, the characteristic polynomial $\chi_A(X) = \det(XI - A)$ may or may not have roots in K. This motivates considering *algebraically closed fields*, which are fields K such that every polynomial with coefficients in K has all its root in K. For example, over $K = \mathbb{R}$, not every polynomial has real roots. If we consider the matrix

$$A = \begin{pmatrix} \cos\theta & -\sin\theta \\ \sin\theta & \cos\theta \end{pmatrix},$$

then the characteristic polynomial $\det(XI - A)$ has no real roots unless $\theta = k\pi$. However, over the field $\mathbb{C}$ of complex numbers, every polynomial has roots. For example, the matrix above has the roots $\cos\theta \pm i\sin\theta = e^{\pm i\theta}$.

Remark: It is possible to show that every linear map f over a complex vector space E must have some (complex) eigenvalue without having recourse to determinants (and the characteristic polynomial). Let $n = \dim(E)$, pick any nonzero vector $u \in E$, and consider the sequence

$$u, f(u), f^2(u), \ldots, f^n(u).$$

Since the above sequence has $n + 1$ vectors and E has dimension n, these vectors must be linearly dependent, so there are some complex numbers $c_0, \ldots, c_m$, not all zero, such that

$$c_0 f^m(u) + c_1 f^{m-1}(u) + \cdots + c_m u = 0,$$

where $m \leq n$ is the largest integer such that the coefficient of $f^m(u)$ is nonzero (m must exits since we have a nontrivial linear dependency). Now because the field $\mathbb{C}$ is algebraically closed, the polynomial

$$c_0 X^m + c_1 X^{m-1} + \cdots + c_m$$

can be written as a product of linear factors as

$$c_0 X^m + c_1 X^{m-1} + \cdots + c_m = c_0 (X - \lambda_1) \cdots (X - \lambda_m)$$

for some complex numbers $\lambda_1, \ldots, \lambda_m \in \mathbb{C}$, not necessarily distinct. But then since $c_0 \neq 0$,

$$c_0 f^m(u) + c_1 f^{m-1}(u) + \cdots + c_m u = 0$$

is equivalent to

$$(f - \lambda_1 \, \mathrm{id}) \circ \cdots \circ (f - \lambda_m \, \mathrm{id})(u) = 0.$$

If all the linear maps $f - \lambda_i \, \mathrm{id}$ were injective, then $(f - \lambda_1 \, \mathrm{id}) \circ \cdots \circ (f - \lambda_m \, \mathrm{id})$ would be injective, contradicting the fact that $u \neq 0$. Therefore, some linear map $f - \lambda_i \, \mathrm{id}$ must have a nontrivial kernel, which means that there is some $v \neq 0$ so that

$$f(v) = \lambda_i v;$$

that is, λ_i is some eigenvalue of f and v is some eigenvector of f.

As nice as the above argument is, it does not provide a method for *finding* the eigenvalues of f, and even if we prefer avoiding determinants as a much as possible, we are forced to deal with the characteristic polynomial $\det(X \, \mathrm{id} - f)$.

Definition 14.3. Let A be an $n \times n$ matrix over a field K. Assume that all the roots of the characteristic polynomial $\chi_A(X) = \det(XI - A)$ of A belong to K, which means that we can write

$$\det(XI - A) = (X - \lambda_1)^{k_1} \cdots (X - \lambda_m)^{k_m},$$

where $\lambda_1, \ldots, \lambda_m \in K$ are the distinct roots of $\det(XI - A)$ and $k_1 + \cdots + k_m = n$. The integer k_i is called the *algebraic multiplicity* of the eigenvalue λ_i, and the dimension of the eigenspace $E_{\lambda_i} = \mathrm{Ker}(\lambda_i I - A)$ is called the *geometric multiplicity* of λ_i. We denote the algebraic multiplicity of λ_i by $\mathrm{alg}(\lambda_i)$, and its geometric multiplicity by $\mathrm{geo}(\lambda_i)$.

By definition, the sum of the algebraic multiplicities is equal to n, but the sum of the geometric multiplicities can be strictly smaller.

Proposition 14.2. *Let A be an $n \times n$ matrix over a field K and assume that all the roots of the characteristic polynomial $\chi_A(X) = \det(XI - A)$ of A belong to K. For every eigenvalue λ_i of A, the geometric multiplicity of λ_i is always less than or equal to its algebraic multiplicity, that is,*

$$\mathrm{geo}(\lambda_i) \leq \mathrm{alg}(\lambda_i).$$

Proof. To see this, if n_i is the dimension of the eigenspace E_{λ_i} associated with the eigenvalue λ_i, we can form a basis of K^n obtained by picking a basis of E_{λ_i} and completing this linearly independent family to a basis of K^n. With respect to this new basis, our matrix is of the form

$$A' = \begin{pmatrix} \lambda_i I_{n_i} & B \\ 0 & D \end{pmatrix},$$

and a simple determinant calculation shows that

$$\det(XI - A) = \det(XI - A') = (X - \lambda_i)^{n_i} \det(XI_{n-n_i} - D).$$

Therefore, $(X - \lambda_i)^{n_i}$ divides the characteristic polynomial of A', and thus, the characteristic polynomial of A. It follows that n_i is less than or equal to the algebraic multiplicity of λ_i. $\qquad\square$

The following proposition shows an interesting property of eigenspaces.

Proposition 14.3. *Let E be any vector space of finite dimension n and let f be any linear map. If $u_1, \ldots, u_m$ are eigenvectors associated with pairwise distinct eigenvalues $\lambda_1, \ldots, \lambda_m$, then the family $(u_1, \ldots, u_m)$ is linearly independent.*

Proof. Assume that $(u_1, \ldots, u_m)$ is linearly dependent. Then there exists $\mu_1, \ldots, \mu_k \in K$ such that

$$\mu_1 u_{i_1} + \cdots + \mu_k u_{i_k} = 0,$$

where $1 \leq k \leq m$, $\mu_i \neq 0$ for all i, $1 \leq i \leq k$, $\{i_1, \ldots, i_k\} \subseteq \{1, \ldots, m\}$, and no proper subfamily of $(u_{i_1}, \ldots, u_{i_k})$ is linearly dependent (in other words, we consider a dependency relation with k minimal). Applying f to this dependency relation, we get

$$\mu_1 \lambda_{i_1} u_{i_1} + \cdots + \mu_k \lambda_{i_k} u_{i_k} = 0,$$

and if we multiply the original dependency relation by λ_{i_1} and subtract it from the above, we get

$$\mu_2(\lambda_{i_2} - \lambda_{i_1})u_{i_2} + \cdots + \mu_k(\lambda_{i_k} - \lambda_{i_1})u_{i_k} = 0,$$

which is a nontrivial linear dependency among a proper subfamily of $(u_{i_1}, \ldots, u_{i_k})$ since the λ_j are all distinct and the μ_i are nonzero, a contradiction. $\square$

As a corollary of Proposition 14.3 we have the following result.

Corollary 14.1. *If* $\lambda_1, \ldots, \lambda_m$ *are all the pairwise distinct eigenvalues of* f *(where* $m \leq n$*), we have a direct sum*

$$E_{\lambda_1} \oplus \cdots \oplus E_{\lambda_m}$$

of the eigenspaces E_{λ_i}.

Unfortunately, it is not always the case that

$$E = E_{\lambda_1} \oplus \cdots \oplus E_{\lambda_m}.$$

Definition 14.4. When

$$E = E_{\lambda_1} \oplus \cdots \oplus E_{\lambda_m},$$

we say that f is *diagonalizable* (and similarly for any matrix associated with f).

Indeed, picking a basis in each E_{λ_i}, we obtain a matrix which is a diagonal matrix consisting of the eigenvalues, each λ_i occurring a number of times equal to the dimension of E_{λ_i}. This happens if the algebraic multiplicity and the geometric multiplicity of every eigenvalue are equal. *In particular, when the characteristic polynomial has* n *distinct roots, then* f *is diagonalizable.* It can also be shown that symmetric matrices have real eigenvalues and can be diagonalized.

For a negative example, we leave it as exercise to show that the matrix

$$M = \begin{pmatrix} 1 & 1 \\ 0 & 1 \end{pmatrix}$$

cannot be diagonalized, even though 1 is an eigenvalue. The problem is that the eigenspace of 1 only has dimension 1. The matrix

$$A = \begin{pmatrix} \cos\theta & -\sin\theta \\ \sin\theta & \cos\theta \end{pmatrix}$$

cannot be diagonalized either, because it has no real eigenvalues, unless $\theta = k\pi$. However, over the field of complex numbers, it can be diagonalized.

14.2 Reduction to Upper Triangular Form

Unfortunately, not every linear map on a complex vector space can be diagonalized. The next best thing is to "triangularize," which means to find a basis over which the matrix has zero entries below the main diagonal. Fortunately, such a basis always exist.

We say that a square matrix A is an *upper triangular matrix* if it has the following shape,

$$\begin{pmatrix} a_{1\,1} & a_{1\,2} & a_{1\,3} & \cdots & a_{1\,n-1} & a_{1\,n} \\ 0 & a_{2\,2} & a_{2\,3} & \cdots & a_{2\,n-1} & a_{2\,n} \\ 0 & 0 & a_{3\,3} & \cdots & a_{3\,n-1} & a_{3\,n} \\ \vdots & \vdots & \vdots & \ddots & \vdots & \vdots \\ 0 & 0 & 0 & \cdots & a_{n-1\,n-1} & a_{n-1\,n} \\ 0 & 0 & 0 & \cdots & 0 & a_{n\,n} \end{pmatrix},$$

i.e., $a_{i\,j} = 0$ whenever $j < i$, $1 \le i, j \le n$.

Theorem 14.1. *Given any finite dimensional vector space over a field K, for any linear map $f \colon E \to E$, there is a basis $(u_1, \ldots, u_n)$ with respect to which f is represented by an upper triangular matrix (in $\mathrm{M}_n(K)$) iff all the eigenvalues of f belong to K. Equivalently, for every $n \times n$ matrix $A \in \mathrm{M}_n(K)$, there is an invertible matrix P and an upper triangular matrix T (both in $\mathrm{M}_n(K)$) such that*

$$A = PTP^{-1}$$

iff all the eigenvalues of A belong to K.

Proof. If there is a basis $(u_1, \ldots, u_n)$ with respect to which f is represented by an upper triangular matrix T in $\mathrm{M}_n(K)$, then since the eigenvalues of f are the diagonal entries of T, all the eigenvalues of f belong to K.

For the converse, we proceed by induction on the dimension n of E. For $n = 1$ the result is obvious. If $n > 1$, since by assumption f has all its eigenvalue in K, pick some eigenvalue $\lambda_1 \in K$ of f, and let u_1 be some corresponding (nonzero) eigenvector. We can find $n-1$ vectors $(v_2, \ldots, v_n)$ such that $(u_1, v_2, \ldots, v_n)$ is a basis of E, and let F be the subspace of dimension $n - 1$ spanned by $(v_2, \ldots, v_n)$. In the basis $(u_1, v_2 \ldots, v_n)$, the matrix of f is of the form

$$U = \begin{pmatrix} \lambda_1 & a_{1\,2} & \cdots & a_{1\,n} \\ 0 & a_{2\,2} & \cdots & a_{2\,n} \\ \vdots & \vdots & \ddots & \vdots \\ 0 & a_{n\,2} & \cdots & a_{n\,n} \end{pmatrix},$$

since its first column contains the coordinates of $\lambda_1 u_1$ over the basis $(u_1, v_2, \ldots, v_n)$. If we let $p \colon E \to F$ be the projection defined such that $p(u_1) = 0$ and $p(v_i) = v_i$ when $2 \le i \le n$, the linear map $g \colon F \to F$ defined as the restriction of $p \circ f$ to F is represented by the $(n-1) \times (n-1)$ matrix $V = (a_{ij})_{2 \le i,j \le n}$ over the basis $(v_2, \ldots, v_n)$. We need to prove that all the eigenvalues of g belong to K. However, since the first column of U has a single nonzero entry, we get

$$\chi_U(X) = \det(XI - U) = (X - \lambda_1)\det(XI - V) = (X - \lambda_1)\chi_V(X),$$

where $\chi_U(X)$ is the characteristic polynomial of U and $\chi_V(X)$ is the characteristic polynomial of V. It follows that $\chi_V(X)$ divides $\chi_U(X)$, and since all the roots of $\chi_U(X)$ are in K, all the roots of $\chi_V(X)$ are also in K. Consequently, we can apply the induction hypothesis, and there is a basis $(u_2, \ldots, u_n)$ of F such that g is represented by an upper triangular matrix $(b_{ij})_{1 \le i,j \le n-1}$. However,

$$E = Ku_1 \oplus F,$$

and thus $(u_1, \ldots, u_n)$ is a basis for E. Since p is the projection from $E = Ku_1 \oplus F$ onto F and $g \colon F \to F$ is the restriction of $p \circ f$ to F, we have

$$f(u_1) = \lambda_1 u_1$$

and

$$f(u_{i+1}) = a_{1\,i}u_1 + \sum_{j=1}^{i} b_{ij}u_{j+1}$$

for some $a_{1\,i} \in K$, when $1 \le i \le n-1$. But then the matrix of f with respect to $(u_1, \ldots, u_n)$ is upper triangular.

For the matrix version, we assume that A is the matrix of f with respect to some basis. Then we just proved that there is a change of basis matrix P such that $A = PTP^{-1}$ where T is upper triangular. $\square$

If $A = PTP^{-1}$ where T is upper triangular, note that the diagonal entries of T are the eigenvalues $\lambda_1, \ldots, \lambda_n$ of A. Indeed, A and T have the same characteristic polynomial. Also, if A is a real matrix whose eigenvalues are all real, then P can be chosen to real, and if A is a rational matrix whose eigenvalues are all rational, then P can be chosen rational. *Since any polynomial over $\mathbb{C}$ has all its roots in $\mathbb{C}$, Theorem 14.1 implies that every complex $n \times n$ matrix can be triangularized.*

If E is a Hermitian space (see Chapter 13), the proof of Theorem 14.1 can be easily adapted to prove that there is an *orthonormal* basis

$(u_1, \ldots, u_n)$ with respect to which the matrix of f is upper triangular. This is usually known as *Schur's lemma*.

Theorem 14.2. *(Schur decomposition) Given any linear map $f \colon E \to E$ over a complex Hermitian space E, there is an orthonormal basis $(u_1, \ldots, u_n)$ with respect to which f is represented by an upper triangular matrix. Equivalently, for every $n \times n$ matrix $A \in \mathrm{M}_n(\mathbb{C})$, there is a unitary matrix U and an upper triangular matrix T such that*

$$A = UTU^*.$$

If A is real and if all its eigenvalues are real, then there is an orthogonal matrix Q and a real upper triangular matrix T such that

$$A = QTQ^\top.$$

Proof. During the induction, we choose F to be the orthogonal complement of $\mathbb{C}u_1$ and we pick orthonormal bases (use Propositions 13.12 and 13.11). If E is a real Euclidean space and if the eigenvalues of f are all real, the proof also goes through with real matrices (use Propositions 11.9 and 11.8). $\qquad\square$

If λ is an eigenvalue of the matrix A and if u is an eigenvector associated with λ, from

$$Au = \lambda u,$$

we obtain

$$A^2 u = A(Au) = A(\lambda u) = \lambda Au = \lambda^2 u,$$

which shows that λ^2 is an eigenvalue of A^2 for the eigenvector u. An obvious induction shows that λ^k is an eigenvalue of A^k for the eigenvector u, for all $k \geq 1$. Now, if all eigenvalues $\lambda_1, \ldots, \lambda_n$ of A are in K, it follows that $\lambda_1^k, \ldots, \lambda_n^k$ are eigenvalues of A^k. However, it is not obvious that A^k does not have other eigenvalues. In fact, this can't happen, and this can be proven using Theorem 14.1.

Proposition 14.4. *Given any $n \times n$ matrix $A \in \mathrm{M}_n(K)$ with coefficients in a field K, if all eigenvalues $\lambda_1, \ldots, \lambda_n$ of A are in K, then for every polynomial $q(X) \in K[X]$, the eigenvalues of $q(A)$ are exactly $(q(\lambda_1), \ldots, q(\lambda_n))$.*

Proof. By Theorem 14.1, there is an upper triangular matrix T and an invertible matrix P (both in $\mathrm{M}_n(K)$) such that

$$A = PTP^{-1}.$$

Since A and T are similar, they have the same eigenvalues (with the same multiplicities), so the diagonal entries of T are the eigenvalues of A. Since

$$A^k = PT^k P^{-1}, \quad k \geq 1,$$

for any polynomial $q(X) = c_0 X^m + \cdots + c_{m-1} X + c_m$, we have

$$
\begin{aligned}
q(A) &= c_0 A^m + \cdots + c_{m-1} A + c_m I \\
&= c_0 PT^m P^{-1} + \cdots + c_{m-1} PTP^{-1} + c_m PIP^{-1} \\
&= P(c_0 T^m + \cdots + c_{m-1} T + c_m I) P^{-1} \\
&= Pq(T)P^{-1}.
\end{aligned}
$$

Furthermore, it is easy to check that $q(T)$ is upper triangular and that its diagonal entries are $q(\lambda_1), \ldots, q(\lambda_n)$, where $\lambda_1, \ldots, \lambda_n$ are the diagonal entries of T, namely the eigenvalues of A. It follows that $q(\lambda_1), \ldots, q(\lambda_n)$ are the eigenvalues of $q(A)$. $\qquad\square$

Remark: There is another way to prove Proposition 14.4 that does not use Theorem 14.1, but instead uses the fact that given any field K, there is field extension $\overline{K}$ of K ($K \subseteq \overline{K}$) such that every polynomial $q(X) = c_0 X^m + \cdots + c_{m-1} X + c_m$ (of degree $m \geq 1$) with coefficients $c_i \in K$ factors as

$$q(X) = c_0(X - \alpha_1) \cdots (X - \alpha_n), \quad \alpha_i \in \overline{K}, i = 1, \ldots, n.$$

The field $\overline{K}$ is called an *algebraically closed field* (and an algebraic closure of K).

Assume that all eigenvalues $\lambda_1, \ldots, \lambda_n$ of A belong to K. Let $q(X)$ be any polynomial (in $K[X]$) and let $\mu \in \overline{K}$ be any eigenvalue of $q(A)$ (this means that μ is a zero of the characteristic polynomial $\chi_{q(A)}(X) \in K[X]$ of $q(A)$. Since $\overline{K}$ is algebraically closed, $\chi_{q(A)}(X)$ has all its roots in $\overline{K}$). We claim that $\mu = q(\lambda_i)$ for some eigenvalue λ_i of A.

Proof. (After Lax [Lax (2007)], Chapter 6). Since $\overline{K}$ is algebraically closed, the polynomial $\mu - q(X)$ factors as

$$\mu - q(X) = c_0(X - \alpha_1) \cdots (X - \alpha_n),$$

for some $\alpha_i \in \overline{K}$. Now $\mu I - q(A)$ is a matrix in $M_n(\overline{K})$, and since μ is an eigenvalue of $q(A)$, it must be singular. We have

$$\mu I - q(A) = c_0(A - \alpha_1 I) \cdots (A - \alpha_n I),$$

and since the left-hand side is singular, so is the right-hand side, which implies that some factor $A - \alpha_i I$ is singular. This means that α_i is an eigenvalue of A, say $\alpha_i = \lambda_i$. As $\alpha_i = \lambda_i$ is a zero of $\mu - q(X)$, we get

$$\mu = q(\lambda_i),$$

which proves that μ is indeed of the form $q(\lambda_i)$ for some eigenvalue λ_i of A. $\qquad\square$

Using Theorem 14.2, we can derive two very important results.

Proposition 14.5. *If A is a Hermitian matrix (i.e. $A^* = A$), then its eigenvalues are real and A can be diagonalized with respect to an orthonormal basis of eigenvectors. In matrix terms, there is a unitary matrix U and a real diagonal matrix D such that $A = UDU^*$. If A is a real symmetric matrix (i.e. $A^\top = A$), then its eigenvalues are real and A can be diagonalized with respect to an orthonormal basis of eigenvectors. In matrix terms, there is an orthogonal matrix Q and a real diagonal matrix D such that $A = QDQ^\top$.*

Proof. By Theorem 14.2, we can write $A = UTU^*$ where $T = (t_{ij})$ is upper triangular and U is a unitary matrix. If $A^* = A$, we get

$$UTU^* = UT^*U^*,$$

and this implies that $T = T^*$. Since T is an upper triangular matrix, T^* is a lower triangular matrix, which implies that T is a diagonal matrix. Furthermore, since $T = T^*$, we have $t_{ii} = \overline{t_{ii}}$ for $i = 1, \ldots, n$, which means that the t_{ii} are real, so T is indeed a real diagonal matrix, say D.

If we apply this result to a (real) symmetric matrix A, we obtain the fact that all the eigenvalues of a symmetric matrix are real, and by applying Theorem 14.2 again, we conclude that $A = QDQ^\top$, where Q is orthogonal and D is a real diagonal matrix. $\qquad\square$

More general versions of Proposition 14.5 are proven in Chapter 16.

When a real matrix A has complex eigenvalues, there is a version of Theorem 14.2 involving only real matrices provided that we allow T to be block upper-triangular (the diagonal entries may be 2×2 matrices or real entries).

Theorem 14.2 is not a very practical result but it is a useful theoretical result to cope with matrices that cannot be diagonalized. For example, it can be used to prove that *every* complex matrix is the limit of a sequence of diagonalizable matrices that have distinct eigenvalues!

14.3   Location of Eigenvalues

If A is an $n \times n$ complex (or real) matrix A, it would be useful to know, even roughly, where the eigenvalues of A are located in the complex plane $\mathbb{C}$. The Gershgorin discs provide some precise information about this.

Definition 14.5. For any complex $n \times n$ matrix A, for $i = 1, \ldots, n$, let

$$R_i'(A) = \sum_{\substack{j=1 \\ j \neq i}}^{n} |a_{ij}|$$

and let

$$G(A) = \bigcup_{i=1}^{n} \{z \in \mathbb{C} \mid |z - a_{ii}| \leq R_i'(A)\}.$$

Each disc $\{z \in \mathbb{C} \mid |z - a_{ii}| \leq R_i'(A)\}$ is called a *Gershgorin disc* and their union $G(A)$ is called the *Gershgorin domain*. An example of Gershgorin domain for $A = \begin{pmatrix} 1 & 2 & 3 \\ 4 & i & 6 \\ 7 & 8 & 1+i \end{pmatrix}$ is illustrated in Figure 14.1.

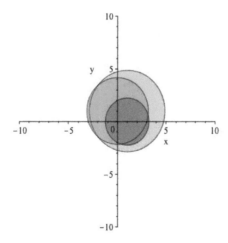

Fig. 14.1 Let A be the 3×3 matrix specified at the end of Definition 14.5. For this particular A, we find that $R_1'(A) = 5$, $R_2'(A) = 10$, and $R_3'(A) = 15$. The blue/purple disk is $|z - 1| \leq 5$, the pink disk is $|z - i| \leq 10$, the peach disk is $|z - 1 - i| \leq 15$, and $G(A)$ is the union of these three disks.

Although easy to prove, the following theorem is very useful:

Theorem 14.3. *(Gershgorin's disc theorem) For any complex $n \times n$ matrix A, all the eigenvalues of A belong to the Gershgorin domain $G(A)$. Furthermore the following properties hold:*

(1) If A is strictly row diagonally dominant, that is

$$|a_{i\,i}| > \sum_{j=1,\, j \neq i}^{n} |a_{i\,j}|, \quad for\ i = 1, \ldots, n,$$

then A is invertible.

(2) If A is strictly row diagonally dominant, and if $a_{i\,i} > 0$ for $i = 1, \ldots, n$, then every eigenvalue of A has a strictly positive real part.

Proof. Let λ be any eigenvalue of A and let u be a corresponding eigenvector (recall that we must have $u \neq 0$). Let k be an index such that

$$|u_k| = \max_{1 \leq i \leq n} |u_i|.$$

Since $Au = \lambda u$, we have

$$(\lambda - a_{k\,k})u_k = \sum_{\substack{j=1 \\ j \neq k}}^{n} a_{k\,j} u_j,$$

which implies that

$$|\lambda - a_{k\,k}||u_k| \leq \sum_{\substack{j=1 \\ j \neq k}}^{n} |a_{k\,j}||u_j| \leq |u_k| \sum_{\substack{j=1 \\ j \neq k}}^{n} |a_{k\,j}|.$$

Since $u \neq 0$ and $|u_k| = \max_{1 \leq i \leq n} |u_i|$, we must have $|u_k| \neq 0$, and it follows that

$$|\lambda - a_{k\,k}| \leq \sum_{\substack{j=1 \\ j \neq k}}^{n} |a_{k\,j}| = R'_k(A),$$

and thus

$$\lambda \in \{z \in \mathbb{C} \mid |z - a_{k\,k}| \leq R'_k(A)\} \subseteq G(A),$$

as claimed.

(1) Strict row diagonal dominance implies that 0 does not belong to any of the Gershgorin discs, so all eigenvalues of A are nonzero, and A is invertible.

(2) If A is strictly row diagonally dominant and $a_{i\,i} > 0$ for $i = 1, \ldots, n$, then each of the Gershgorin discs lies strictly in the right half-plane, so every eigenvalue of A has a strictly positive real part. $\qquad\square$

In particular, Theorem 14.3 implies that if a symmetric matrix is strictly row diagonally dominant and has strictly positive diagonal entries, then it is positive definite. Theorem 14.3 is sometimes called the *Gershgorin–Hadamard theorem*.

Since A and $A^\top$ have the same eigenvalues (even for complex matrices) we also have a version of Theorem 14.3 for the discs of radius

$$C'_j(A) = \sum_{\substack{i=1 \\ i \neq j}}^{n} |a_{i\,j}|,$$

whose domain $G(A^\top)$ is given by

$$G(A^\top) = \bigcup_{i=1}^{n} \{ z \in \mathbb{C} \mid |z - a_{i\,i}| \leq C'_i(A) \}.$$

Figure 14.2 shows $G(A^\top)$ for $A = \begin{pmatrix} 1 & 2 & 3 \\ 4 & i & 6 \\ 7 & 8 & 1+i \end{pmatrix}$.

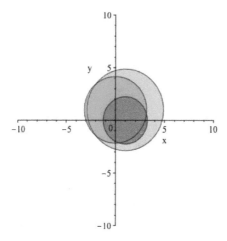

Fig. 14.2   Let A be the 3×3 matrix specified at the end of Definition 14.5. For this particular A, we find that $C'_1(A) = 11$, $C'_2(A) = 10$, and $C'_3(A) = 9$. The pale blue disk is $|z - 1| \leq 1$, the pink disk is $|z - i| \leq 10$, the ocher disk is $|z - 1 - i| \leq 9$, and $G(A^\top)$ is the union of these three disks.

Thus we get the following:

Theorem 14.4. *For any complex $n \times n$ matrix A, all the eigenvalues of A belong to the intersection of the Gershgorin domains $G(A) \cap G(A^\top)$. See Figure 14.3. Furthermore the following properties hold:*

(1) If A is strictly column diagonally dominant, that is

$$|a_{ii}| > \sum_{i=1,\, i\neq j}^{n} |a_{ij}|, \quad for \; j = 1,\dots,n,$$

then A is invertible.

(2) If A is strictly column diagonally dominant, and if $a_{ii} > 0$ for $i = 1,\dots,n$, then every eigenvalue of A has a strictly positive real part.

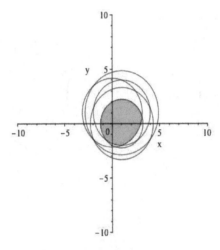

Fig. 14.3   Let A be the 3×3 matrix specified at the end of Definition 14.5. The dusty rose region is $G(A) \cap G(A^{\top})$.

There are refinements of Gershgorin's theorem and eigenvalue location results involving other domains besides discs; for more on this subject, see Horn and Johnson [Horn and Johnson (1990)], Sections 6.1 and 6.2.

Remark: Neither strict row diagonal dominance nor strict column diagonal dominance are necessary for invertibility. Also, if we relax all strict inequalities to inequalities, then row diagonal dominance (or column diagonal dominance) is not a sufficient condition for invertibility.

14.4 Conditioning of Eigenvalue Problems

The following $n \times n$ matrix

$$A = \begin{pmatrix} 0 & & & & & \\ 1 & 0 & & & & \\ & 1 & 0 & & & \\ & & \ddots & \ddots & & \\ & & & 1 & 0 & \\ & & & & 1 & 0 \end{pmatrix}$$

has the eigenvalue 0 with multiplicity n. However, if we perturb the top rightmost entry of A by ϵ, it is easy to see that the characteristic polynomial of the matrix

$$A(\epsilon) = \begin{pmatrix} 0 & & & & & \epsilon \\ 1 & 0 & & & & \\ & 1 & 0 & & & \\ & & \ddots & \ddots & & \\ & & & 1 & 0 & \\ & & & & 1 & 0 \end{pmatrix}$$

is $X^n - \epsilon$. It follows that if $n = 40$ and $\epsilon = 10^{-40}$, $A(10^{-40})$ has the eigenvalues $10^{-1}e^{k2\pi i/40}$ with $k = 1,\ldots,40$. Thus, we see that a very small change ($\epsilon = 10^{-40}$) to the matrix A causes a significant change to the eigenvalues of A (from 0 to $10^{-1}e^{k2\pi i/40}$). Indeed, the relative error is 10^{-39}. Worse, due to machine precision, since very small numbers are treated as 0, the error on the computation of eigenvalues (for example, of the matrix $A(10^{-40})$) can be very large.

This phenomenon is similar to the phenomenon discussed in Section 8.5 where we studied the effect of a small perturbation of the coefficients of a linear system $Ax = b$ on its solution. In Section 8.5, we saw that the behavior of a linear system under small perturbations is governed by the condition number cond(A) of the matrix A. In the case of the eigenvalue problem (finding the eigenvalues of a matrix), we will see that the conditioning of the problem depends on the condition number of the change of basis matrix P used in reducing the matrix A to its diagonal form $D = P^{-1}AP$, rather than on the condition number of A itself. The following proposition in which we assume that A is diagonalizable and that the matrix norm $\| \ \|$ satisfies a special condition (satisfied by the operator norms $\| \ \|_p$ for

$p = 1, 2, \infty$), is due to Bauer and Fike (1960).

Proposition 14.6. *Let* $A \in M_n(\mathbb{C})$ *be a diagonalizable matrix, P be an invertible matrix, and D be a diagonal matrix* $D = \mathrm{diag}(\lambda_1, \ldots, \lambda_n)$ *such that*

$$A = PDP^{-1},$$

and let $\| \ \|$ *be a matrix norm such that*

$$\|\mathrm{diag}(\alpha_1, \ldots, \alpha_n)\| = \max_{1 \le i \le n} |\alpha_i|,$$

for every diagonal matrix. Then for every perturbation matrix ΔA, if we write

$$B_i = \{z \in \mathbb{C} \mid |z - \lambda_i| \le \mathrm{cond}(P)\, \|\Delta A\|\},$$

for every eigenvalue λ of $A + \Delta A$, we have

$$\lambda \in \bigcup_{k=1}^{n} B_k.$$

Proof. Let λ be any eigenvalue of the matrix $A + \Delta A$. If $\lambda = \lambda_j$ for some j, then the result is trivial. Thus assume that $\lambda \neq \lambda_j$ for $j = 1, \ldots, n$. In this case the matrix $D - \lambda I$ is invertible (since its eigenvalues are $\lambda - \lambda_j$ for $j = 1, \ldots, n$), and we have

$$P^{-1}(A + \Delta A - \lambda I)P = D - \lambda I + P^{-1}(\Delta A)P$$
$$= (D - \lambda I)(I + (D - \lambda I)^{-1}P^{-1}(\Delta A)P).$$

Since λ is an eigenvalue of $A + \Delta A$, the matrix $A + \Delta A - \lambda I$ is singular, so the matrix

$$I + (D - \lambda I)^{-1}P^{-1}(\Delta A)P$$

must also be singular. By Proposition 8.8(2), we have

$$1 \le \left\| (D - \lambda I)^{-1}P^{-1}(\Delta A)P \right\|,$$

and since $\| \ \|$ is a matrix norm,

$$\left\| (D - \lambda I)^{-1}P^{-1}(\Delta A)P \right\| \le \left\| (D - \lambda I)^{-1} \right\| \left\| P^{-1} \right\| \|\Delta A\| \|P\|,$$

so we have

$$1 \le \left\| (D - \lambda I)^{-1} \right\| \left\| P^{-1} \right\| \|\Delta A\| \|P\|.$$

Now $(D - \lambda I)^{-1}$ is a diagonal matrix with entries $1/(\lambda_i - \lambda)$, so by our assumption on the norm,

$$\left\| (D - \lambda I)^{-1} \right\| = \frac{1}{\min_i(|\lambda_i - \lambda|)}.$$

As a consequence, since there is some index k for which $\min_i(|\lambda_i - \lambda|) = |\lambda_k - \lambda|$, we have

$$\left\|(D - \lambda I)^{-1}\right\| = \frac{1}{|\lambda_k - \lambda|},$$

and we obtain

$$|\lambda - \lambda_k| \leq \left\|P^{-1}\right\| \|\Delta A\| \|P\| = \operatorname{cond}(P) \|\Delta A\|,$$

which proves our result. ☐

Proposition 14.6 implies that for any diagonalizable matrix A, if we define $\Gamma(A)$ by

$$\Gamma(A) = \inf\{\operatorname{cond}(P) \mid P^{-1}AP = D\},$$

then for every eigenvalue λ of $A + \Delta A$, we have

$$\lambda \in \bigcup_{k=1}^{n}\{z \in \mathbb{C}^n \mid |z - \lambda_k| \leq \Gamma(A) \|\Delta A\|\}.$$

Definition 14.6. The number $\Gamma(A) = \inf\{\operatorname{cond}(P) \mid P^{-1}AP = D\}$ is called the *conditioning of A relative to the eigenvalue problem.*

If A is a normal matrix, since by Theorem 16.12, A can be diagonalized with respect to a unitary matrix U, and since for the spectral norm $\|U\|_2 = 1$, we see that $\Gamma(A) = 1$. Therefore, normal matrices are very well conditioned w.r.t. the eigenvalue problem. In fact, for every eigenvalue λ of $A + \Delta A$ (with A normal), we have

$$\lambda \in \bigcup_{k=1}^{n}\{z \in \mathbb{C}^n \mid |z - \lambda_k| \leq \|\Delta A\|_2\}.$$

If A and $A + \Delta A$ are both symmetric (or Hermitian), there are sharper results; see Proposition 16.15.

Note that the matrix $A(\epsilon)$ from the beginning of the section is not normal.

14.5 Eigenvalues of the Matrix Exponential

The Schur decomposition yields a characterization of the eigenvalues of the matrix exponential e^A in terms of the eigenvalues of the matrix A. First we have the following proposition.

Proposition 14.7. *Let A and U be (real or complex) matrices and assume that U is invertible. Then*

$$e^{UAU^{-1}} = Ue^AU^{-1}.$$

Proof. A trivial induction shows that

$$UA^pU^{-1} = (UAU^{-1})^p,$$

and thus

$$e^{UAU^{-1}} = \sum_{p\geq 0} \frac{(UAU^{-1})^p}{p!} = \sum_{p\geq 0} \frac{UA^pU^{-1}}{p!}$$

$$= U\left(\sum_{p\geq 0} \frac{A^p}{p!}\right) U^{-1} = Ue^AU^{-1},$$

as claimed.   □

Proposition 14.8. *Given any complex $n \times n$ matrix A, if $\lambda_1,\ldots,\lambda_n$ are the eigenvalues of A, then $e^{\lambda_1},\ldots,e^{\lambda_n}$ are the eigenvalues of e^A. Furthermore, if u is an eigenvector of A for λ_i, then u is an eigenvector of e^A for e^{λ_i}.*

Proof. By Theorem 14.1, there is an invertible matrix P and an upper triangular matrix T such that

$$A = PTP^{-1}.$$

By Proposition 14.7,

$$e^{PTP^{-1}} = Pe^TP^{-1}.$$

Note that $e^T = \sum_{p\geq 0} \frac{T^p}{p!}$ is upper triangular since T^p is upper triangular for all $p \geq 0$. If $\lambda_1,\lambda_2,\ldots,\lambda_n$ are the diagonal entries of T, the properties of matrix multiplication, when combined with an induction on p, imply that the diagonal entries of T^p are $\lambda_1^p,\lambda_2^p,\ldots,\lambda_n^p$. This in turn implies that the diagonal entries of e^T are $\sum_{p\geq 0} \frac{\lambda_i^p}{p!} = e^{\lambda_i}$ for $1 \leq i \leq n$. Since A and T are similar matrices, we know that they have the same eigenvalues, namely the diagonal entries $\lambda_1,\ldots,\lambda_n$ of T. Since $e^A = e^{PTP^{-1}} = Pe^TP^{-1}$, and e^T is upper triangular, we use the same argument to conclude that both e^A and e^T have the same eigenvalues, which are the diagonal entries of e^T, where the diagonal entries of e^T are of the form $e^{\lambda_1},\ldots,e^{\lambda_n}$. Now, if u is an eigenvector of A for the eigenvalue λ, a simple induction shows that u is an eigenvector of A^n for the eigenvalue λ^n, from which is follows that

$$e^Au = \left[I + \frac{A}{1!} + \frac{A^2}{2!} + \frac{A^3}{3!} + \ldots\right] u = u + Au + \frac{A^2}{2!}u + \frac{A^3}{3!}u + \ldots$$

$$= u + \lambda u + \frac{\lambda^2}{2!}u + \frac{\lambda^3}{3!}u + \cdots = \left[1 + \lambda + \frac{\lambda^2}{2!} + \frac{\lambda^3}{3!} + \ldots\right] u = e^\lambda u,$$

which shows that u is an eigenvector of e^A for e^λ.   □

As a consequence, we obtain the following result.

Proposition 14.9. *For every complex (or real) square matrix A, we have*

$$\det(e^A) = e^{\operatorname{tr}(A)},$$

where $\operatorname{tr}(A)$ is the trace of A, i.e., the sum $a_{11} + \cdots + a_{nn}$ of its diagonal entries.

Proof. The trace of a matrix A is equal to the sum of the eigenvalues of A. The determinant of a matrix is equal to the product of its eigenvalues, and if $\lambda_1, \ldots, \lambda_n$ are the eigenvalues of A, then by Proposition 14.8, $e^{\lambda_1}, \ldots, e^{\lambda_n}$ are the eigenvalues of e^A, and thus

$$\det\left(e^A\right) = e^{\lambda_1} \cdots e^{\lambda_n} = e^{\lambda_1 + \cdots + \lambda_n} = e^{\operatorname{tr}(A)},$$

as desired. $\qquad\square$

If B is a skew symmetric matrix, since $\operatorname{tr}(B) = 0$, we deduce that $\det(e^B) = e^0 = 1$. This allows us to obtain the following result. Recall that the (real) vector space of skew symmetric matrices is denoted by $\mathfrak{so}(n)$.

Proposition 14.10. *For every skew symmetric matrix $B \in \mathfrak{so}(n)$, we have $e^B \in \mathbf{SO}(n)$, that is, e^B is a rotation.*

Proof. By Proposition 8.18, e^B is an orthogonal matrix. Since $\operatorname{tr}(B) = 0$, we deduce that $\det(e^B) = e^0 = 1$. Therefore, $e^B \in \mathbf{SO}(n)$. $\qquad\square$

Proposition 14.10 shows that the map $B \mapsto e^B$ is a map $\exp \colon \mathfrak{so}(n) \to \mathbf{SO}(n)$. It is not injective, but it can be shown (using one of the spectral theorems) that it is surjective.

If B is a (real) symmetric matrix, then

$$(e^B)^\top = e^{B^\top} = e^B,$$

so e^B is also symmetric. Since the eigenvalues $\lambda_1, \ldots, \lambda_n$ of B are real, by Proposition 14.8, since the eigenvalues of e^B are $e^{\lambda_1}, \ldots, e^{\lambda_n}$ and the λ_i are real, we have $e^{\lambda_i} > 0$ for $i = 1, \ldots, n$, which implies that e^B is symmetric positive definite. In fact, it can be shown that for every symmetric positive definite matrix A, there is a *unique* symmetric matrix B such that $A = e^B$; see Gallier [Gallier (2011b)].

14.6   Summary

The main concepts and results of this chapter are listed below:

- *Diagonal matrix.*
- *Eigenvalues, eigenvectors*; the *eigenspace associated* with an eigenvalue.
- *Characteristic polynomial.*
- *Trace.*
- *Algebraic and geometric multiplicity.*
- Eigenspaces associated with distinct eigenvalues form a direct sum (Proposition 14.3).
- Reduction of a matrix to an upper-triangular matrix.
- *Schur decomposition.*
- The *Gershgorin's discs* can be used to locate the eigenvalues of a complex matrix; see Theorems 14.3 and 14.4.
- The conditioning of eigenvalue problems.
- Eigenvalues of the matrix exponential. The formula $\det(e^A) = e^{\text{tr}(A)}$.

14.7   Problems

Problem 14.1. Let A be the following 2×2 matrix

$$A = \begin{pmatrix} 1 & -1 \\ 1 & -1 \end{pmatrix}.$$

(1) Prove that A has the eigenvalue 0 with multiplicity 2 and that $A^2 = 0$.

(2) Let A be any real 2×2 matrix

$$A = \begin{pmatrix} a & b \\ c & d \end{pmatrix}.$$

Prove that if $bc > 0$, then A has two distinct real eigenvalues. Prove that if $a, b, c, d > 0$, then there is a positive eigenvector u associated with the largest of the two eigenvalues of A, which means that if $u = (u_1, u_2)$, then $u_1 > 0$ and $u_2 > 0$.

(3) Suppose now that A is any complex 2×2 matrix as in (2). Prove that if A has the eigenvalue 0 with multiplicity 2, then $A^2 = 0$. Prove that if A is real symmetric, then $A = 0$.

Problem 14.2. Let A be any complex $n \times n$ matrix. Prove that if A has the eigenvalue 0 with multiplicity n, then $A^n = 0$. Give an example of a matrix A such that $A^n = 0$ but $A \neq 0$.

Problem 14.3. Let A be a complex 2×2 matrix, and let λ_1 and λ_2 be the eigenvalues of A. Prove that if $\lambda_1 \neq \lambda_2$, then

$$e^A = \frac{\lambda_1 e^{\lambda_2} - \lambda_2 e^{\lambda_1}}{\lambda_1 - \lambda_2} I + \frac{e^{\lambda_1} - e^{\lambda_2}}{\lambda_1 - \lambda_2} A.$$

Problem 14.4. Let A be the real symmetric 2×2 matrix

$$A = \begin{pmatrix} a & b \\ b & c \end{pmatrix}.$$

(1) Prove that the eigenvalues of A are real and given by

$$\lambda_1 = \frac{a + c + \sqrt{4b^2 + (a-c)^2}}{2}, \quad \lambda_2 = \frac{a + c - \sqrt{4b^2 + (a-c)^2}}{2}.$$

(2) Prove that A has a double eigenvalue ($\lambda_1 = \lambda_2 = a$) if and only if $b = 0$ and $a = c$; that is, A is a diagonal matrix.

(3) Prove that the eigenvalues of A are nonnegative iff $b^2 \leq ac$ and $a + c \geq 0$.

(4) Prove that the eigenvalues of A are positive iff $b^2 < ac$, $a > 0$ and $c > 0$.

Problem 14.5. Find the eigenvalues of the matrices

$$A = \begin{pmatrix} 3 & 0 \\ 1 & 1 \end{pmatrix}, \quad B = \begin{pmatrix} 1 & 1 \\ 0 & 3 \end{pmatrix}, \quad C = A + B = \begin{pmatrix} 4 & 1 \\ 1 & 4 \end{pmatrix}.$$

Check that the eigenvalues of $A + B$ are not equal to the sums of eigenvalues of A plus eigenvalues of B.

Problem 14.6. Let A be a real symmetric $n \times n$ matrix and B be a real symmetric positive definite $n \times n$ matrix. We would like to solve the *generalized eigenvalue problem*: find $\lambda \in \mathbb{R}$ and $u \neq 0$ such that

$$Au = \lambda Bu. \tag{14.1}$$

(1) Use the Choleseky decomposition $B = CC^\top$ to show that λ and u are solutions of the generalized eigenvalue problem (14.1) iff λ and v are solutions the (ordinary) eigenvalue problem

$$C^{-1}A(C^\top)^{-1}v = \lambda v, \quad \text{with } v = C^\top u.$$

Check that $C^{-1}A(C^\top)^{-1}$ is symmetric.

(2) Prove that if $Au_1 = \lambda_1 Bu_1$, $Au_2 = \lambda_2 Bu_2$, with $u_1 \neq 0$, $u_2 \neq 0$ and $\lambda_1 \neq \lambda_2$, then $u_1^\top Bu_2 = 0$.

(3) Prove that $B^{-1}A$ and $C^{-1}A(C^\top)^{-1}$ have the same eigenvalues.

Problem 14.7. The sequence of *Fibonacci numbers*, $0, 1, 1, 2, 3, 5, 8, 13,$ $21, 34, 55, \ldots,$ is given by the recurrence

$$F_{n+2} = F_{n+1} + F_n,$$

with $F_0 = 0$ and $F_1 = 1$. In matrix form, we can write

$$\begin{pmatrix} F_{n+1} \\ F_n \end{pmatrix} = \begin{pmatrix} 1 & 1 \\ 1 & 0 \end{pmatrix} \begin{pmatrix} F_n \\ F_{n-1} \end{pmatrix}, \quad n \geq 1, \quad \begin{pmatrix} F_1 \\ F_0 \end{pmatrix} = \begin{pmatrix} 1 \\ 0 \end{pmatrix}.$$

(1) Show that

$$\begin{pmatrix} F_{n+1} \\ F_n \end{pmatrix} = \begin{pmatrix} 1 & 1 \\ 1 & 0 \end{pmatrix}^n \begin{pmatrix} 1 \\ 0 \end{pmatrix}.$$

(2) Prove that the eigenvalues of the matrix

$$A = \begin{pmatrix} 1 & 1 \\ 1 & 0 \end{pmatrix}$$

are

$$\lambda = \frac{1 \pm \sqrt{5}}{2}.$$

The number

$$\varphi = \frac{1 + \sqrt{5}}{2}$$

is called the *golden ratio*. Show that the eigenvalues of A are φ and $-\varphi^{-1}$.

(3) Prove that A is diagonalized as

$$A = \begin{pmatrix} 1 & 1 \\ 1 & 0 \end{pmatrix} = \frac{1}{\sqrt{5}} \begin{pmatrix} \varphi & -\varphi^{-1} \\ 1 & 1 \end{pmatrix} \begin{pmatrix} \varphi & 0 \\ 0 & -\varphi^{-1} \end{pmatrix} \begin{pmatrix} 1 & \varphi^{-1} \\ -1 & \varphi \end{pmatrix}.$$

Prove that

$$\begin{pmatrix} F_{n+1} \\ F_n \end{pmatrix} = \frac{1}{\sqrt{5}} \begin{pmatrix} \varphi & -\varphi^{-1} \\ 1 & 1 \end{pmatrix} \begin{pmatrix} \varphi^n \\ -(-\varphi^{-1})^n \end{pmatrix},$$

and thus

$$F_n = \frac{1}{\sqrt{5}}(\varphi^n - (-\varphi^{-1})^n) = \frac{1}{\sqrt{5}}\left[\left(\frac{1+\sqrt{5}}{2}\right)^n - \left(\frac{1-\sqrt{5}}{2}\right)^n\right], \quad n \geq 0.$$

Problem 14.8. Let A be an $n \times n$ matrix. For any subset I of $\{1, \ldots, n\}$, let $A_{I,I}$ be the matrix obtained from A by first selecting the columns whose indices belong to I, and then the rows whose indices also belong to I. Prove that

$$\tau_k(A) = \sum_{\substack{I \subseteq \{1, \ldots, n\} \\ |I| = k}} \det(A_{I,I}).$$

Problem 14.9. (1) Consider the matrix

$$A = \begin{pmatrix} 0 & 0 & -a_3 \\ 1 & 0 & -a_2 \\ 0 & 1 & -a_1 \end{pmatrix}.$$

Prove that the characteristic polynomial $\chi_A(z) = \det(zI - A)$ of A is given by

$$\chi_A(z) = z^3 + a_1 z^2 + a_2 z + a_3.$$

(2) Consider the matrix

$$A = \begin{pmatrix} 0 & 0 & 0 & -a_4 \\ 1 & 0 & 0 & -a_3 \\ 0 & 1 & 0 & -a_2 \\ 0 & 0 & 1 & -a_1 \end{pmatrix}.$$

Prove that the characteristic polynomial $\chi_A(z) = \det(zI - A)$ of A is given by

$$\chi_A(z) = z^4 + a_1 z^3 + a_2 z^2 + a_3 z + a_4.$$

(3) Consider the $n \times n$ matrix (called a *companion matrix*)

$$A = \begin{pmatrix} 0 & 0 & 0 & \cdots & 0 & -a_n \\ 1 & 0 & 0 & \cdots & 0 & -a_{n-1} \\ 0 & 1 & 0 & \cdots & 0 & -a_{n-2} \\ \vdots & \ddots & \ddots & \ddots & \vdots & \vdots \\ 0 & 0 & 0 & \ddots & 0 & -a_2 \\ 0 & 0 & 0 & \cdots & 1 & -a_1 \end{pmatrix}.$$

Prove that the characteristic polynomial $\chi_A(z) = \det(zI - A)$ of A is given by

$$\chi_A(z) = z^n + a_1 z^{n-1} + a_2 z^{n-2} + \cdots + a_{n-1} z + a_n.$$

Hint. Use induction.

Explain why finding the roots of a polynomial (with real or complex coefficients) and finding the eigenvalues of a (real or complex) matrix are equivalent problems, in the sense that if we have a method for solving one of these problems, then we have a method to solve the other.

Problem 14.10. Let A be a complex $n \times n$ matrix. Prove that if A is invertible and if the eigenvalues of A are $(\lambda_1, \ldots, \lambda_n)$, then the eigenvalues of A^{-1} are $(\lambda_1^{-1}, \ldots, \lambda_n^{-1})$. Prove that if u is an eigenvector of A for λ_i, then u is an eigenvector of A^{-1} for λ_i^{-1}.

Problem 14.11. Prove that every complex matrix is the limit of a sequence of diagonalizable matrices that have distinct eigenvalues

Problem 14.12. Consider the following tridiagonal $n \times n$ matrices

$$A = \begin{pmatrix} 2 & -1 & 0 & & \\ -1 & 2 & -1 & & \\ & \ddots & \ddots & \ddots & \\ & & -1 & 2 & -1 \\ & & 0 & -1 & 2 \end{pmatrix}, \quad S = \begin{pmatrix} 0 & 1 & 0 & & \\ 1 & 0 & 1 & & \\ & \ddots & \ddots & \ddots & \\ & & 1 & 0 & 1 \\ & & 0 & 1 & 0 \end{pmatrix}.$$

Observe that $A = 2I - S$ and show that the eigenvalues of A are $\lambda_k = 2 - \mu_k$, where the μ_k are the eigenvalues of S.

(2) Using Problem 9.6, prove that the eigenvalues of the matrix A are given by

$$\lambda_k = 4\sin^2\left(\frac{k\pi}{2(n+1)}\right), \quad k = 1, \ldots, n.$$

Show that A is symmetric positive definite.

(3) Find the condition number of A with respect to the 2-norm.

(4) Show that an eigenvector $(y_1^{(k)}, \ldots, y_n^{(k)})$ associated with the eigenvalue λ_k is given by

$$y_j^{(k)} = \sin\left(\frac{kj\pi}{n+1}\right), \quad j = 1, \ldots, n.$$

Problem 14.13. Consider the following real tridiagonal symmetric $n \times n$ matrix

$$A = \begin{pmatrix} c & 1 & 0 & & \\ 1 & c & 1 & & \\ & \ddots & \ddots & \ddots & \\ & & 1 & c & 1 \\ & & 0 & 1 & c \end{pmatrix}.$$

(1) Using Problem 9.6, prove that the eigenvalues of the matrix A are given by

$$\lambda_k = c + 2\cos\left(\frac{k\pi}{n+1}\right), \quad k = 1, \ldots, n.$$

(2) Find a condition on c so that A is positive definite. It is satisfied by $c = 4$?

Problem 14.14. Let A be an $m \times n$ matrix and B be an $n \times m$ matrix (over $\mathbb{C}$).

(1) Prove that

$$\det(I_m - AB) = \det(I_n - BA).$$

Hint. Consider the matrices

$$X = \begin{pmatrix} I_m & A \\ B & I_n \end{pmatrix} \quad \text{and} \quad Y = \begin{pmatrix} I_m & 0 \\ -B & I_n \end{pmatrix}.$$

(2) Prove that

$$\lambda^n \det(\lambda I_m - AB) = \lambda^m \det(\lambda I_n - BA).$$

Hint. Consider the matrices

$$X = \begin{pmatrix} \lambda I_m & A \\ B & I_n \end{pmatrix} \quad \text{and} \quad Y = \begin{pmatrix} I_m & 0 \\ -B & \lambda I_n \end{pmatrix}.$$

Deduce that AB and BA have the same nonzero eigenvalues with the same multiplicity.

Problem 14.15. The purpose of this problem is to prove that the characteristic polynomial of the matrix

$$A = \begin{pmatrix} 1 & 2 & 3 & 4 & \cdots & n \\ 2 & 3 & 4 & 5 & \cdots & n+1 \\ 3 & 4 & 5 & 6 & \cdots & n+2 \\ \vdots & \vdots & \vdots & & \ddots & \vdots \\ n & n+1 & n+2 & n+3 & \cdots & 2n-1 \end{pmatrix}$$

is

$$P_A(\lambda) = \lambda^{n-2}\left(\lambda^2 - n^2\lambda - \frac{1}{12}n^2(n^2 - 1)\right).$$

(1) Prove that the characteristic polynomial $P_A(\lambda)$ is given by

$$P_A(\lambda) = \lambda^{n-2}P(\lambda),$$

with

$$P(\lambda) = \begin{vmatrix} \lambda-1 & -2 & -3 & -4 & \cdots & -n+3 & -n+2 & -n+1 & -n \\ -\lambda-1 & \lambda-1 & -1 & -1 & \cdots & -1 & -1 & -1 & -1 \\ 1 & -2 & 1 & 0 & \cdots & 0 & 0 & 0 & 0 \\ 0 & 1 & -2 & 1 & \cdots & 0 & 0 & 0 & 0 \\ \vdots & \vdots & \ddots & \ddots & \ddots & \vdots & \vdots & \vdots & \vdots \\ 0 & 0 & 0 & 0 & \ddots & 1 & 0 & 0 & 0 \\ 0 & 0 & 0 & 0 & \ddots & -2 & 1 & 0 & 0 \\ 0 & 0 & 0 & 0 & \cdots & 1 & -2 & 1 & 0 \\ 0 & 0 & 0 & 0 & \cdots & 0 & 1 & -2 & 1 \end{vmatrix}.$$

(2) Prove that the sum of the roots λ_1, λ_2 of the (degree two) polynomial $P(\lambda)$ is

$$\lambda_1 + \lambda_2 = n^2.$$

The problem is thus to compute the product $\lambda_1 \lambda_2$ of these roots. Prove that

$$\lambda_1 \lambda_2 = P(0).$$

(3) The problem is now to evaluate $d_n = P(0)$, where

$$d_n = \begin{vmatrix} -1 & -2 & -3 & -4 & \cdots & -n+3 & -n+2 & -n+1 & -n \\ -1 & -1 & -1 & -1 & \cdots & -1 & -1 & -1 & -1 \\ 1 & -2 & 1 & 0 & \cdots & 0 & 0 & 0 & 0 \\ 0 & 1 & -2 & 1 & \cdots & 0 & 0 & 0 & 0 \\ \vdots & \vdots & \ddots & \ddots & \ddots & \vdots & \vdots & \vdots & \vdots \\ 0 & 0 & 0 & 0 & \ddots & 1 & 0 & 0 & 0 \\ 0 & 0 & 0 & 0 & \ddots & -2 & 1 & 0 & 0 \\ 0 & 0 & 0 & 0 & \cdots & 1 & -2 & 1 & 0 \\ 0 & 0 & 0 & 0 & \cdots & 0 & 1 & -2 & 1 \end{vmatrix}$$

I suggest the following strategy: cancel out the first entry in row 1 and row 2 by adding a suitable multiple of row 3 to row 1 and row 2, and then subtract row 2 from row 1.

Do this twice.

You will notice that the first two entries on row 1 and the first two entries on row 2 change, but the rest of the matrix looks the same, except that the dimension is reduced.

This suggests setting up a recurrence involving the entries u_k, v_k, x_k, y_k in the determinant

$$D_k = \begin{vmatrix} u_k & x_k & -3 & -4 & \cdots & -n+k-3 & -n+k-2 & -n+k-1 & -n+k \\ v_k & y_k & -1 & -1 & \cdots & -1 & -1 & -1 & -1 \\ 1 & -2 & 1 & 0 & \cdots & 0 & 0 & 0 & 0 \\ 0 & 1 & -2 & 1 & \cdots & 0 & 0 & 0 & 0 \\ \vdots & \vdots & \ddots & \ddots & \ddots & \vdots & \vdots & \vdots & \vdots \\ 0 & 0 & 0 & 0 & \ddots & 1 & 0 & 0 & 0 \\ 0 & 0 & 0 & 0 & \ddots & -2 & 1 & 0 & 0 \\ 0 & 0 & 0 & 0 & \cdots & 1 & -2 & 1 & 0 \\ 0 & 0 & 0 & 0 & \cdots & 0 & 1 & -2 & 1 \end{vmatrix},$$

starting with $k = 0$, with

$$u_0 = -1, \quad v_0 = -1, \quad x_0 = -2, \quad y_0 = -1,$$

and ending with $k = n - 2$, so that

$$d_n = D_{n-2} = \begin{vmatrix} u_{n-3} & x_{n-3} & -3 \\ v_{n-3} & y_{n-3} & -1 \\ 1 & -2 & 1 \end{vmatrix} = \begin{vmatrix} u_{n-2} & x_{n-2} \\ v_{n-2} & y_{n-2} \end{vmatrix}.$$

Prove that we have the recurrence relations

$$\begin{pmatrix} u_{k+1} \\ v_{k+1} \\ x_{k+1} \\ y_{k+1} \end{pmatrix} = \begin{pmatrix} 2 & -2 & 1 & -1 \\ 0 & 2 & 0 & 1 \\ -1 & 1 & 0 & 0 \\ 0 & -1 & 0 & 0 \end{pmatrix} \begin{pmatrix} u_k \\ v_k \\ x_k \\ y_k \end{pmatrix} + \begin{pmatrix} 0 \\ 0 \\ -2 \\ -1 \end{pmatrix}.$$

These appear to be nasty affine recurrence relations, so we will use the trick to convert this affine map to a linear map.

(4) Consider the linear map given by

$$\begin{pmatrix} u_{k+1} \\ v_{k+1} \\ x_{k+1} \\ y_{k+1} \\ 1 \end{pmatrix} = \begin{pmatrix} 2 & -2 & 1 & -1 & 0 \\ 0 & 2 & 0 & 1 & 0 \\ -1 & 1 & 0 & 0 & -2 \\ 0 & -1 & 0 & 0 & -1 \\ 0 & 0 & 0 & 0 & 1 \end{pmatrix} \begin{pmatrix} u_k \\ v_k \\ x_k \\ y_k \\ 1 \end{pmatrix},$$

and show that its action on u_k, v_k, x_k, y_k is the same as the affine action of Part (3).

Use `Matlab` to find the eigenvalues of the matrix

$$T = \begin{pmatrix} 2 & -2 & 1 & -1 & 0 \\ 0 & 2 & 0 & 1 & 0 \\ -1 & 1 & 0 & 0 & -2 \\ 0 & -1 & 0 & 0 & -1 \\ 0 & 0 & 0 & 0 & 1 \end{pmatrix}.$$

You will be stunned!

Let N be the matrix given by

$$N = T - I.$$

Prove that

$$N^4 = 0.$$

Use this to prove that

$$T^k = I + kN + \frac{1}{2}k(k-1)N^2 + \frac{1}{6}k(k-1)(k-2)N^3,$$

for all $k \geq 0$.

(5) Prove that

$$\begin{pmatrix} u_k \\ v_k \\ x_k \\ y_k \\ 1 \end{pmatrix} = T^k \begin{pmatrix} -1 \\ -1 \\ -2 \\ -1 \\ 1 \end{pmatrix} = \begin{pmatrix} 2 & -2 & 1 & -1 & 0 \\ 0 & 2 & 0 & 1 & 0 \\ -1 & 1 & 0 & 0 & -2 \\ 0 & -1 & 0 & 0 & -1 \\ 0 & 0 & 0 & 0 & 1 \end{pmatrix}^k \begin{pmatrix} -1 \\ -1 \\ -2 \\ -1 \\ 1 \end{pmatrix},$$

for $k \geq 0$.

Prove that

$$T^k = \begin{pmatrix} k+1 & -k(k+1) & k & -k^2 & \frac{1}{6}(k-1)k(2k-7) \\ 0 & k+1 & 0 & k & -\frac{1}{2}(k-1)k \\ -k & k^2 & 1-k & (k-1)k & -\frac{1}{3}k((k-6)k+11) \\ 0 & -k & 0 & 1-k & \frac{1}{2}(k-3)k \\ 0 & 0 & 0 & 0 & 1 \end{pmatrix},$$

and thus that

$$
\begin{pmatrix} u_k \\ v_k \\ x_k \\ y_k \end{pmatrix} = \begin{pmatrix} \frac{1}{6}(2k^3 + 3k^2 - 5k - 6) \\ -\frac{1}{2}(k^2 + 3k + 2) \\ \frac{1}{3}(-k^3 + k - 6) \\ \frac{1}{2}(k^2 + k - 2) \end{pmatrix},
$$

and that

$$
\begin{vmatrix} u_k & x_k \\ v_k & y_k \end{vmatrix} = -1 - \frac{7}{3}k - \frac{23}{12}k^2 - \frac{2}{3}k^3 - \frac{1}{12}k^4.
$$

As a consequence, prove that amazingly

$$
d_n = D_{n-2} = -\frac{1}{12}n^2(n^2 - 1).
$$

(6) Prove that the characteristic polynomial of A is indeed

$$
P_A(\lambda) = \lambda^{n-2}\left(\lambda^2 - n^2\lambda - \frac{1}{12}n^2(n^2 - 1)\right).
$$

Use the above to show that the two nonzero eigenvalues of A are

$$
\lambda = \frac{n}{2}\left(n \pm \frac{\sqrt{3}}{3}\sqrt{4n^2 - 1}\right).
$$

The negative eigenvalue λ_1 can also be expressed as

$$
\lambda_1 = n^2\frac{(3 - 2\sqrt{3})}{6}\sqrt{1 - \frac{1}{4n^2}}.
$$

Use this expression to explain the following phenomenon: if we add any number greater than or equal to $(2/25)n^2$ to every diagonal entry of A we get an invertible matrix. What about $0.077351n^2$? Try it!

Problem 14.16. Let A be a symmetric tridiagonal $n \times n$-matrix

$$
A = \begin{pmatrix}
b_1 & c_1 & & & & \\
c_1 & b_2 & c_2 & & & \\
& c_2 & b_3 & c_3 & & \\
& & \ddots & \ddots & \ddots & \\
& & & c_{n-2} & b_{n-1} & c_{n-1} \\
& & & & c_{n-1} & b_n
\end{pmatrix},
$$

where it is assumed that $c_i \neq 0$ for all i, $1 \leq i \leq n - 1$, and let A_k be the $k \times k$-submatrix consisting of the first k rows and columns of A, $1 \leq k \leq n$. We define the polynomials $P_k(x)$ as follows: $(0 \leq k \leq n)$.

$$P_0(x) = 1,$$
$$P_1(x) = b_1 - x,$$
$$P_k(x) = (b_k - x)P_{k-1}(x) - c_{k-1}^2 P_{k-2}(x),$$

where $2 \leq k \leq n$.

(1) Prove the following properties:

(i) $P_k(x)$ is the characteristic polynomial of A_k, where $1 \leq k \leq n$.

(ii) $\lim_{x \to -\infty} P_k(x) = +\infty$, where $1 \leq k \leq n$.

(iii) If $P_k(x) = 0$, then $P_{k-1}(x)P_{k+1}(x) < 0$, where $1 \leq k \leq n - 1$.

(iv) $P_k(x)$ has k distinct real roots that separate the $k + 1$ roots of $P_{k+1}(x)$, where $1 \leq k \leq n - 1$.

(2) Given any real number $\mu > 0$, for every k, $1 \leq k \leq n$, define the function $sg_k(\mu)$ as follows:

$$sg_k(\mu) = \begin{cases} \text{sign of } P_k(\mu) & \text{if } P_k(\mu) \neq 0, \\ \text{sign of } P_{k-1}(\mu) & \text{if } P_k(\mu) = 0. \end{cases}$$

We encode the sign of a positive number as $+$, and the sign of a negative number as $-$. Then let $E(k, \mu)$ be the ordered list

$$E(k, \mu) = \langle +, \, sg_1(\mu), \, sg_2(\mu), \, \ldots, \, sg_k(\mu) \rangle,$$

and let $N(k, \mu)$ be the number changes of sign between consecutive signs in $E(k, \mu)$.

Prove that $sg_k(\mu)$ is well defined and that $N(k, \mu)$ is the number of roots λ of $P_k(x)$ such that $\lambda < \mu$.

Remark: The above can be used to compute the eigenvalues of a (tridiagonal) symmetric matrix (the method of Givens-Householder).

Chapter 15

Unit Quaternions and Rotations in SO(3)

This chapter is devoted to the representation of rotations in $\mathbf{SO}(3)$ in terms of unit quaternions. Since we already defined the unitary groups $\mathbf{SU}(n)$, the quickest way to introduce the *unit quaternions* is to define them as the elements of the group $\mathbf{SU}(2)$.

The skew field $\mathbb{H}$ of quaternions and the group $\mathbf{SU}(2)$ of unit quaternions are discussed in Section 15.1. In Section 15.2, we define a homomorphism $r\colon \mathbf{SU}(2) \to \mathbf{SO}(3)$ and prove that its kernel is $\{-I, I\}$. We compute the rotation matrix R_q associated with the rotation r_q induced by a unit quaternion q in Section 15.3. In Section 15.4, we prove that the homomorphism $r\colon \mathbf{SU}(2) \to \mathbf{SO}(3)$ is surjective by providing an algorithm to construct a quaternion from a rotation matrix. In Section 15.5 we define the exponential map $\exp\colon \mathfrak{su}(2) \to \mathbf{SU}(2)$ where $\mathfrak{su}(2)$ is the real vector space of skew-Hermitian 2×2 matrices with zero trace. We prove that exponential map $\exp\colon \mathfrak{su}(2) \to \mathbf{SU}(2)$ is surjective and give an algorithm for finding a logarithm. We discuss quaternion interpolation and prove the famous *slerp interpolation formula* due to Ken Shoemake in Section 15.6. This formula is used in robotics and computer graphics to deal with interpolation problems. In Section 15.7, we prove that there is no "nice" section $s\colon \mathbf{SO}(3) \to \mathbf{SU}(2)$ of the homomorphism $r\colon \mathbf{SU}(2) \to \mathbf{SO}(3)$, in the sense that any section of r is neither a homomorphism nor continuous.

15.1   The Group SU(2) of Unit Quaternions and the Skew Field ℍ of Quaternions

Definition 15.1. The *unit quaternions* are the elements of the group **SU**(2), namely the group of 2×2 complex matrices of the form

$$\begin{pmatrix} \alpha & \beta \\ -\overline{\beta} & \overline{\alpha} \end{pmatrix} \quad \alpha, \beta \in \mathbb{C}, \ \alpha\overline{\alpha} + \beta\overline{\beta} = 1.$$

The *quaternions* are the elements of the real vector space $\mathbb{H} = \mathbb{R}\,\mathbf{SU}(2)$.

Let $\mathbf{1}, \mathbf{i}, \mathbf{j}, \mathbf{k}$ be the matrices

$$\mathbf{1} = \begin{pmatrix} 1 & 0 \\ 0 & 1 \end{pmatrix}, \quad \mathbf{i} = \begin{pmatrix} i & 0 \\ 0 & -i \end{pmatrix}, \quad \mathbf{j} = \begin{pmatrix} 0 & 1 \\ -1 & 0 \end{pmatrix}, \quad \mathbf{k} = \begin{pmatrix} 0 & i \\ i & 0 \end{pmatrix},$$

then $\mathbb{H}$ is the set of all matrices of the form

$$X = a\mathbf{1} + b\mathbf{i} + c\mathbf{j} + d\mathbf{k}, \quad a, b, c, d \in \mathbb{R}.$$

Indeed, every matrix in $\mathbb{H}$ is of the form

$$X = \begin{pmatrix} a + ib & c + id \\ -(c - id) & a - ib \end{pmatrix}, \quad a, b, c, d \in \mathbb{R}.$$

It is easy (but a bit tedious) to verify that the quaternions $\mathbf{1}, \mathbf{i}, \mathbf{j}, \mathbf{k}$ satisfy the famous identities discovered by Hamilton:

$$\mathbf{i}^2 = \mathbf{j}^2 = \mathbf{k}^2 = \mathbf{ijk} = -\mathbf{1},$$
$$\mathbf{ij} = -\mathbf{ji} = \mathbf{k},$$
$$\mathbf{jk} = -\mathbf{kj} = \mathbf{i},$$
$$\mathbf{ki} = -\mathbf{ik} = \mathbf{j}.$$

Thus, the quaternions are a generalization of the complex numbers, but there are three square roots of $-\mathbf{1}$ and multiplication is not commutative.

Given any two quaternions $X = a\mathbf{1} + b\mathbf{i} + c\mathbf{j} + d\mathbf{k}$ and $Y = a'\mathbf{1} + b'\mathbf{i} + c'\mathbf{j} + d'\mathbf{k}$, Hamilton's famous formula

$$XY = (aa' - bb' - cc' - dd')\mathbf{1} + (ab' + ba' + cd' - dc')\mathbf{i}$$
$$+ (ac' + ca' + db' - bd')\mathbf{j} + (ad' + da' + bc' - cb')\mathbf{k}$$

looks mysterious, but it is simply the result of multiplying the two matrices

$$X = \begin{pmatrix} a + ib & c + id \\ -(c - id) & a - ib \end{pmatrix} \quad \text{and} \quad Y = \begin{pmatrix} a' + ib' & c' + id' \\ -(c' - id') & a' - ib' \end{pmatrix}.$$

It is worth noting that this formula was discovered independently by Olinde Rodrigues in 1840, a few years before Hamilton (Veblen and Young [Veblen and Young (1946)]). However, Rodrigues was working with a different formalism, homogeneous transformations, and he did not discover the quaternions.

If
$$X = \begin{pmatrix} a + ib & c + id \\ -(c - id) & a - ib \end{pmatrix}, \quad a, b, c, d \in \mathbb{R},$$
it is immediately verified that
$$XX^* = X^*X = (a^2 + b^2 + c^2 + d^2)\mathbf{1}.$$
Also observe that
$$X^* = \begin{pmatrix} a - ib & -(c + id) \\ c - id & a + ib \end{pmatrix} = a\mathbf{1} - b\mathbf{i} - c\mathbf{j} - d\mathbf{k}.$$

This implies that if $X \neq 0$, then X is invertible and its inverse is given by
$$X^{-1} = (a^2 + b^2 + c^2 + d^2)^{-1}X^*.$$

As a consequence, it can be verified that $\mathbb{H}$ is a skew field (a noncommutative field). It is also a real vector space of dimension 4 with basis $(\mathbf{1}, \mathbf{i}, \mathbf{j}, \mathbf{k})$; thus as a vector space, $\mathbb{H}$ is isomorphic to $\mathbb{R}^4$.

Definition 15.2. A concise notation for the quaternion X defined by $\alpha = a + ib$ and $\beta = c + id$ is
$$X = [a, (b, c, d)].$$
We call a the *scalar part* of X and (b, c, d) the *vector part* of X. With this notation, $X^* = [a, -(b, c, d)]$, which is often denoted by $\overline{X}$. The quaternion $\overline{X}$ is called the *conjugate* of q. If q is a unit quaternion, then $\overline{q}$ is the multiplicative inverse of q.

15.2   Representation of Rotations in SO(3) by Quaternions in SU(2)

The key to representation of rotations in **SO**(3) by unit quaternions is a certain group homomorphism called the *adjoint representation of* **SU**(2). To define this mapping, first we define the real vector space $\mathfrak{su}(2)$ of skew Hermitian matrices.

Definition 15.3. The (real) vector space $\mathfrak{su}(2)$ of 2×2 *skew Hermitian matrices with zero trace* is given by
$$\mathfrak{su}(2) = \left\{ \begin{pmatrix} ix & y + iz \\ -y + iz & -ix \end{pmatrix} \ \middle| \ (x, y, z) \in \mathbb{R}^3 \right\}.$$

Observe that for every matrix $A \in \mathfrak{su}(2)$, we have $A^* = -A$, that is, A is skew Hermitian, and that $\mathrm{tr}(A) = 0$.

Definition 15.4. The *adjoint representation* of the group **SU**(2) is the group homomorphism
Ad: **SU**(2) $\to$ **GL**($\mathfrak{su}$(2)) defined such that for every $q \in$ **SU**(2), with

$$q = \begin{pmatrix} \alpha & \beta \\ -\overline{\beta} & \overline{\alpha} \end{pmatrix} \in \mathbf{SU}(2),$$

we have

$$\mathrm{Ad}_q(A) = qAq^*, \quad A \in \mathfrak{su}(2),$$

where q^* is the inverse of q (since **SU**(2) is a unitary group) and is given by

$$q^* = \begin{pmatrix} \overline{\alpha} & -\beta \\ \overline{\beta} & \alpha \end{pmatrix}.$$

One needs to verify that the map Ad_q is an invertible linear map from $\mathfrak{su}(2)$ to itself, and that Ad is a group homomorphism, which is easy to do.

In order to associate a rotation ρ_q (in **SO**(3)) to q, we need to embed $\mathbb{R}^3$ into $\mathbb{H}$ as the pure quaternions, by

$$\psi(x, y, z) = \begin{pmatrix} ix & y + iz \\ -y + iz & -ix \end{pmatrix}, \quad (x, y, z) \in \mathbb{R}^3.$$

Then q defines the map ρ_q (on $\mathbb{R}^3$) given by

$$\rho_q(x, y, z) = \psi^{-1}(q\psi(x, y, z)q^*).$$

Therefore, modulo the isomorphism ψ, the linear map ρ_q is the linear isomorphism Ad_q. In fact, it turns out that ρ_q is a rotation (and so is Ad_q), which we will prove shortly. So, the representation of rotations in **SO**(3) by unit quaternions is just the adjoint representation of **SU**(2); its image is a subgroup of **GL**($\mathfrak{su}$(2)) isomorphic to **SO**(3).

Technically, it is a bit simpler to embed $\mathbb{R}^3$ in the (real) vector spaces of Hermitian matrices with zero trace,

$$\left\{ \begin{pmatrix} x & z - iy \\ z + iy & -x \end{pmatrix} \;\middle|\; x, y, z \in \mathbb{R} \right\}.$$

Since the matrix $\psi(x, y, z)$ is skew-Hermitian, the matrix $-i\psi(x, y, z)$ is Hermitian, and we have

$$-i\psi(x, y, z) = \begin{pmatrix} x & z - iy \\ z + iy & -x \end{pmatrix} = x\sigma_3 + y\sigma_2 + z\sigma_1,$$

where $\sigma_1, \sigma_2, \sigma_3$ are the *Pauli spin matrices*

$$\sigma_1 = \begin{pmatrix} 0 & 1 \\ 1 & 0 \end{pmatrix}, \quad \sigma_2 = \begin{pmatrix} 0 & -i \\ i & 0 \end{pmatrix}, \quad \sigma_3 = \begin{pmatrix} 1 & 0 \\ 0 & -1 \end{pmatrix}.$$

Matrices of the form $x\sigma_3 + y\sigma_2 + z\sigma_1$ are Hermitian matrices with zero trace.

It is easy to see that every 2×2 Hermitian matrix with zero trace must be of this form. (Observe that $(i\sigma_1, i\sigma_2, i\sigma_3)$ forms a basis of $\mathfrak{su}(2)$. Also, $\mathbf{i} = i\sigma_3$, $\mathbf{j} = i\sigma_2$, $\mathbf{k} = i\sigma_1$.)

Now, if $A = x\sigma_3 + y\sigma_2 + z\sigma_1$ is a Hermitian 2×2 matrix with zero trace, we have

$$(qAq^*)^* = qA^*q^* = qAq^*,$$

so qAq^* is also Hermitian, and

$$\mathrm{tr}(qAq^*) = \mathrm{tr}(Aq^*q) = \mathrm{tr}(A),$$

and qAq^* also has zero trace. Therefore, the map $A \mapsto qAq^*$ preserves the Hermitian matrices with zero trace. We also have

$$\det(x\sigma_3 + y\sigma_2 + z\sigma_1) = \det \begin{pmatrix} x & z - iy \\ z + iy & -x \end{pmatrix} = -(x^2 + y^2 + z^2),$$

and

$$\det(qAq^*) = \det(q)\det(A)\det(q^*) = \det(A) = -(x^2 + y^2 + z^2).$$

We can embed $\mathbb{R}^3$ into the space of Hermitian matrices with zero trace by

$$\varphi(x, y, z) = x\sigma_3 + y\sigma_2 + z\sigma_1.$$

Note that

$$\varphi = -i\psi \quad \text{and} \quad \varphi^{-1} = i\psi^{-1}.$$

Definition 15.5. The unit quaternion $q \in \mathbf{SU}(2)$ induces a map r_q on $\mathbb{R}^3$ by

$$r_q(x, y, z) = \varphi^{-1}(q\varphi(x, y, z)q^*) = \varphi^{-1}(q(x\sigma_3 + y\sigma_2 + z\sigma_1)q^*).$$

The map r_q is clearly linear since φ is linear.

Proposition 15.1. *For every unit quaternion* $q \in \mathbf{SU}(2)$, *the linear map* r_q *is orthogonal, that is,* $r_q \in \mathbf{O}(3)$.

Proof. Since

$$- \|(x,y,z)\|^2 = -(x^2 + y^2 + z^2) = \det(x\sigma^3 + y\sigma^2 + z\sigma_1) = \det(\varphi(x,y,z)),$$

we have

$$- \|r_q(x,y,z)\|^2 = \det(\varphi(r_q(x,y,z))) = \det(q(x\sigma_3 + y\sigma_2 + z\sigma_1)q^*)$$
$$= \det(x\sigma_3 + y\sigma_2 + z\sigma_1) = - \|(x,y,z)^2\|,$$

and we deduce that r_q is an isometry. Thus, $r_q \in \mathbf{O}(3)$. $\qquad\square$

In fact, r_q is a rotation, and we can show this by finding the fixed points of r_q. Let q be a unit quaternion of the form

$$q = \begin{pmatrix} \alpha & \beta \\ -\overline{\beta} & \overline{\alpha} \end{pmatrix}$$

with $\alpha = a + ib$, $\beta = c + id$, and $a^2 + b^2 + c^2 + d^2 = 1$ $(a,b,c,d \in \mathbb{R})$.

If $b = c = d = 0$, then $q = I$ and r_q is the identity so we may assume that $(b,c,d) \neq (0,0,0)$.

Proposition 15.2. *If $(b,c,d) \neq (0,0,0)$, then the fixed points of r_q are solutions (x,y,z) of the linear system*

$$-dy + cz = 0$$
$$cx - by = 0$$
$$dx - bz = 0.$$

This linear system has the nontrivial solution (b,c,d) and has rank 2. Therefore, r_q has the eigenvalue 1 with multiplicity 1, and r_q is a rotation whose axis is determined by (b,c,d).

Proof. We have $r_q(x,y,z) = (x,y,z)$ iff

$$\varphi^{-1}(q(x\sigma_3 + y\sigma_2 + z\sigma_1)q^*) = (x,y,z)$$

iff

$$q(x\sigma_3 + y\sigma_2 + z\sigma_1)q^* = \varphi(x,y,z),$$

and since

$$\varphi(x,y,z) = x\sigma_3 + y\sigma_2 + z\sigma_1 = A$$

with

$$A = \begin{pmatrix} x & z - iy \\ z + iy & -x \end{pmatrix},$$

we see that $r_q(x, y, z) = (x, y, z)$ iff
$$qAq^* = A \quad \text{iff} \quad qA = Aq.$$
We have
$$qA = \begin{pmatrix} \alpha & \beta \\ -\overline{\beta} & \overline{\alpha} \end{pmatrix} \begin{pmatrix} x & z - iy \\ z + iy & -x \end{pmatrix} = \begin{pmatrix} \alpha x + \beta z + i\beta y & \alpha z - i\alpha y - \beta x \\ -\overline{\beta} x + \overline{\alpha} z + i\overline{\alpha} y & -\overline{\beta} z + i\overline{\beta} y - \overline{\alpha} x \end{pmatrix}$$
and
$$Aq = \begin{pmatrix} x & z - iy \\ z + iy & -x \end{pmatrix} \begin{pmatrix} \alpha & \beta \\ -\overline{\beta} & \overline{\alpha} \end{pmatrix} = \begin{pmatrix} \alpha x - \overline{\beta} z + i\overline{\beta} y & \beta x + \overline{\alpha} z - i\overline{\alpha} y \\ \alpha z + i\alpha y + \overline{\beta} x & \beta z + i\beta y - \overline{\alpha} x \end{pmatrix}.$$
By equating qA and Aq, we get
$$i(\beta - \overline{\beta})y + (\beta + \overline{\beta})z = 0$$
$$2\beta x + i(\alpha - \overline{\alpha})y + (\overline{\alpha} - \alpha)z = 0$$
$$2\overline{\beta} x + i(\alpha - \overline{\alpha})y + (\alpha - \overline{\alpha})z = 0$$
$$i(\beta - \overline{\beta})y + (\beta + \overline{\beta})z = 0.$$
The first and the fourth equation are identical and the third equation is obtained by conjugating the second, so the above system reduces to
$$i(\beta - \overline{\beta})y + (\beta + \overline{\beta})z = 0$$
$$2\beta x + i(\alpha - \overline{\alpha})y + (\overline{\alpha} - \alpha)z = 0.$$
Replacing α by $a + ib$ and β by $c + id$, we get
$$-dy + cz = 0$$
$$cx - by + i(dx - bz) = 0,$$
which yields the equations
$$-dy + cz = 0$$
$$cx - by = 0$$
$$dx - bz = 0.$$
This linear system has the nontrivial solution (b, c, d) and the matrix of this system is
$$\begin{pmatrix} 0 & -d & c \\ c & -b & 0 \\ d & 0 & -b \end{pmatrix}.$$
Since $(b, c, d) \neq (0, 0, 0)$, this matrix always has a 2×2 submatrix which is nonsingular, so it has rank 2, and consequently its kernel is the one-dimensional space spanned by (b, c, d). Therefore, r_q has the eigenvalue 1 with multiplicity 1. If we had $\det(r_q) = -1$, then the eigenvalues of r_q would be either $(-1, 1, 1)$ or $(-1, e^{i\theta}, e^{-i\theta})$ with $\theta \neq k2\pi$ (with $k \in \mathbb{Z}$), contradicting the fact that 1 is an eigenvalue with multiplicity 1. Therefore, r_q is a rotation; in fact, its axis is determined by (b, c, d). □

In summary, $q \mapsto r_q$ is a map r from **SU**(2) to **SO**(3).

Theorem 15.1. *The map* $r\colon$ **SU**(2) $\to$ **SO**(3) *is homomorphism whose kernel is* $\{I, -I\}$.

Proof. This map is a homomorphism, because if $q_1, q_2 \in$ **SU**(2), then

$$
\begin{aligned}
r_{q_2}(r_{q_1}(x,y,z)) &= \varphi^{-1}(q_2\varphi(r_{q_1}(x,y,z))q_2^*) \\
&= \varphi^{-1}(q_2\varphi(\varphi^{-1}(q_1\varphi(x,y,z)q_1^*))q_2^*) \\
&= \varphi^{-1}((q_2q_1)\varphi(x,y,z)(q_2q_1)^*) \\
&= r_{q_2q_1}(x,y,z).
\end{aligned}
$$

The computation that showed that if $(b,c,d) \neq (0,0,0)$, then r_q has the eigenvalue 1 with multiplicity 1 implies the following: if $r_q = I_3$, namely r_q has the eigenvalue 1 with multiplicity 3, then $(b,c,d) = (0,0,0)$. But then $a = \pm 1$, and so $q = \pm I_2$. Therefore, the kernel of the homomorphism $r\colon$ **SU**(2) $\to$ **SO**(3) is $\{I, -I\}$. $\square$

Remark: Perhaps the quickest way to show that r maps **SU**(2) into **SO**(3) is to observe that the map r is continuous. Then, since it is known that **SU**(2) is connected, its image by r lies in the connected component of I, namely **SO**(3).

The map r is surjective, but this is not obvious. We will return to this point after finding the matrix representing r_q explicitly.

15.3 Matrix Representation of the Rotation r_q

Given a unit quaternion q of the form

$$
q = \begin{pmatrix} \alpha & \beta \\ -\overline{\beta} & \overline{\alpha} \end{pmatrix}
$$

with $\alpha = a + ib$, $\beta = c + id$, and $a^2 + b^2 + c^2 + d^2 = 1$ $(a,b,c,d \in \mathbb{R})$, to find the matrix representing the rotation r_q we need to compute

$$
q(x\sigma_3 + y\sigma_2 + z\sigma_1)q^* = \begin{pmatrix} \alpha & \beta \\ -\overline{\beta} & \overline{\alpha} \end{pmatrix} \begin{pmatrix} x & z-iy \\ z+iy & -x \end{pmatrix} \begin{pmatrix} \overline{\alpha} & -\beta \\ \overline{\beta} & \alpha \end{pmatrix}.
$$

First we have

$$
\begin{pmatrix} x & z-iy \\ z+iy & -x \end{pmatrix} \begin{pmatrix} \overline{\alpha} & -\beta \\ \overline{\beta} & \alpha \end{pmatrix} = \begin{pmatrix} x\overline{\alpha} + z\overline{\beta} - iy\overline{\beta} & -x\beta + z\alpha - iy\alpha \\ z\overline{\alpha} + iy\overline{\alpha} - x\overline{\beta} & -z\beta - iy\beta - x\alpha \end{pmatrix}.
$$

Next, we have

$$\begin{pmatrix} \alpha & \beta \\ -\overline{\beta} & \overline{\alpha} \end{pmatrix} \begin{pmatrix} x\overline{\alpha} + z\overline{\beta} - iy\overline{\beta} & -x\beta + z\alpha - iy\alpha \\ z\overline{\alpha} + iy\overline{\alpha} - x\overline{\beta} & -z\beta - iy\beta - x\alpha \end{pmatrix} = \begin{pmatrix} A_1 & A_2 \\ A_3 & A_4 \end{pmatrix},$$

with

$$\begin{aligned}
A_1 &= (\alpha\overline{\alpha} - \beta\overline{\beta})x + i(\overline{\alpha}\beta - \alpha\overline{\beta})y + (\alpha\overline{\beta} + \overline{\alpha}\beta)z \\
A_2 &= -2\alpha\beta x - i(\alpha^2 + \beta^2)y + (\alpha^2 - \beta^2)z \\
A_3 &= -2\overline{\alpha}\overline{\beta}x + i(\overline{\alpha}^2 + \overline{\beta}^2)y + (\overline{\alpha}^2 - \overline{\beta}^2)z \\
A_4 &= -(\alpha\overline{\alpha} - \beta\overline{\beta})x - i(\overline{\alpha}\beta - \alpha\overline{\beta})y - (\alpha\overline{\beta} + \overline{\alpha}\beta)z.
\end{aligned}$$

Since $\alpha = a + ib$ and $\beta = c + id$, with $a, b, c, d \in \mathbb{R}$, we have

$$\begin{aligned}
\alpha\overline{\alpha} - \beta\overline{\beta} &= a^2 + b^2 - c^2 - d^2 \\
i(\overline{\alpha}\beta - \alpha\overline{\beta}) &= 2(bc - ad) \\
\alpha\overline{\beta} + \overline{\alpha}\beta &= 2(ac + bd) \\
-\alpha\beta &= -ac + bd - i(ad + bc) \\
-i(\alpha^2 + \beta^2) &= 2(ab + cd) - i(a^2 - b^2 + c^2 - d^2) \\
\alpha^2 - \beta^2 &= a^2 - b^2 - c^2 + d^2 + i2(ab - cd).
\end{aligned}$$

Using the above, we get

$$\begin{aligned}
(\alpha\overline{\alpha} - \beta\overline{\beta})x &+ i(\overline{\alpha}\beta - \alpha\overline{\beta})y + (\alpha\overline{\beta} + \overline{\alpha}\beta)z \\
&= (a^2 + b^2 - c^2 - d^2)x + 2(bc - ad)y + 2(ac + bd)z,
\end{aligned}$$

and

$$\begin{aligned}
-2\alpha\beta x &- i(\alpha^2 + \beta^2)y + (\alpha^2 - \beta^2)z \\
&= 2(-ac + bd)x + 2(ab + cd)y + (a^2 - b^2 - c^2 + d^2)z \\
&\quad - i[2(ad + bc)x + (a^2 - b^2 + c^2 - d^2)y + 2(-ab + cd)z].
\end{aligned}$$

If we write

$$q(x\sigma_3 + y\sigma_2 + z\sigma_1)q^* = \begin{pmatrix} x' & z' - iy' \\ z' + iy' & -x' \end{pmatrix},$$

we obtain

$$\begin{aligned}
x' &= (a^2 + b^2 - c^2 - d^2)x + 2(bc - ad)y + 2(ac + bd)z \\
y' &= 2(ad + bc)x + (a^2 - b^2 + c^2 - d^2)y + 2(-ab + cd)z \\
z' &= 2(-ac + bd)x + 2(ab + cd)y + (a^2 - b^2 - c^2 + d^2)z.
\end{aligned}$$

In summary, we proved the following result.

Proposition 15.3. *The matrix representing* r_q *is*

$$R_q = \begin{pmatrix} a^2 + b^2 - c^2 - d^2 & 2bc - 2ad & 2ac + 2bd \\ 2bc + 2ad & a^2 - b^2 + c^2 - d^2 & -2ab + 2cd \\ -2ac + 2bd & 2ab + 2cd & a^2 - b^2 - c^2 + d^2 \end{pmatrix}.$$

Since $a^2 + b^2 + c^2 + d^2 = 1$, *this matrix can also be written as*

$$R_q = \begin{pmatrix} 2a^2 + 2b^2 - 1 & 2bc - 2ad & 2ac + 2bd \\ 2bc + 2ad & 2a^2 + 2c^2 - 1 & -2ab + 2cd \\ -2ac + 2bd & 2ab + 2cd & 2a^2 + 2d^2 - 1 \end{pmatrix}.$$

The above is the rotation matrix in Euler form induced by the quaternion q, which is the matrix corresponding to ρ_q. This is because

$$\varphi = -i\psi, \quad \varphi^{-1} = i\psi^{-1},$$

so

$$r_q(x, y, z) = \varphi^{-1}(q\varphi(x, y, z)q^*) = i\psi^{-1}(q(-i\psi(x, y, z))q^*)$$
$$= \psi^{-1}(q\psi(x, y, z)q^*) = \rho_q(x, y, z),$$

and so $r_q = \rho_q$.

We showed that every unit quaternion $q \in \mathbf{SU}(2)$ induces a rotation $r_q \in \mathbf{SO}(3)$, but it is not obvious that every rotation can be represented by a quaternion. This can shown in various ways.

One way to is use the fact that every rotation in $\mathbf{SO}(3)$ is the composition of two reflections, and that every reflection σ of $\mathbb{R}^3$ can be represented by a quaternion q, in the sense that

$$\sigma(x, y, z) = -\varphi^{-1}(q\varphi(x, y, z)q^*).$$

Note the presence of the negative sign. This is the method used in Gallier [Gallier (2011b)] (Chapter 9).

15.4 An Algorithm to Find a Quaternion Representing a Rotation

Theorem 15.2. *The homomorphim* $r\colon \mathbf{SU}(2) \to \mathbf{SO}(3)$ *is surjective.*

Here is an algorithmic method to find a unit quaternion q representing a rotation matrix R, which provides a proof of Theorem 15.2.

Let

$$q = \begin{pmatrix} a + ib & c + id \\ -(c - id) & a - ib \end{pmatrix}, \quad a^2 + b^2 + c^2 + d^2 = 1, \; a, b, c, d \in \mathbb{R}.$$

First observe that the trace of R_q is given by

$$\operatorname{tr}(R_q) = 3a^2 - b^2 - c^2 - d^2,$$

but since $a^2 + b^2 + c^2 + d^2 = 1$, we get $\operatorname{tr}(R_q) = 4a^2 - 1$, so

$$a^2 = \frac{\operatorname{tr}(R_q) + 1}{4}.$$

If $R \in \mathbf{SO}(3)$ is any rotation matrix and if we write

$$R = \begin{pmatrix} r_{11} & r_{12} & r_{13} \\ r_{21} & r_{22} & r_{23} \\ r_{31} & r_{32} & r_{33}, \end{pmatrix}$$

we are looking for a unit quaternion $q \in \mathbf{SU}(2)$ such that $R_q = R$. Therefore, we must have

$$a^2 = \frac{\operatorname{tr}(R) + 1}{4}.$$

We also know that

$$\operatorname{tr}(R) = 1 + 2\cos\theta,$$

where $\theta \in [0, \pi]$ is the angle of the rotation R, so we get

$$a^2 = \frac{\cos\theta + 1}{2} = \cos^2\left(\frac{\theta}{2}\right),$$

which implies that

$$|a| = \cos\left(\frac{\theta}{2}\right) \quad (0 \le \theta \le \pi).$$

Note that we may assume that $\theta \in [0, \pi]$, because if $\pi \le \theta \le 2\pi$, then $\theta - 2\pi \in [-\pi, 0]$, and then the rotation of angle $\theta - 2\pi$ and axis determined by the vector (b, c, d) is the same as the rotation of angle $2\pi - \theta \in [0, \pi]$ and axis determined by the vector $-(b, c, d)$. There are two cases.

Case 1. $\operatorname{tr}(R) \ne -1$, or equivalently $\theta \ne \pi$. In this case $a \ne 0$. Pick

$$a = \frac{\sqrt{\operatorname{tr}(R) + 1}}{2}.$$

Then by equating $R - R^\top$ and $R_q - R_q^\top$, we get

$$4ab = r_{32} - r_{23}$$
$$4ac = r_{13} - r_{31}$$
$$4ad = r_{21} - r_{12},$$

which yields

$$b = \frac{r_{32} - r_{23}}{4a}, \quad c = \frac{r_{13} - r_{31}}{4a}, \quad d = \frac{r_{21} - r_{12}}{4a}.$$

Case 2. $\operatorname{tr}(R) = -1$, or equivalently $\theta = \pi$. In this case $a = 0$. By equating $R + R^\top$ and $R_q + R_q^\top$, we get

$$4bc = r_{21} + r_{12}$$
$$4bd = r_{13} + r_{31}$$
$$4cd = r_{32} + r_{23}.$$

By equating the diagonal terms of R and R_q, we also get

$$b^2 = \frac{1 + r_{11}}{2}$$
$$c^2 = \frac{1 + r_{22}}{2}$$
$$d^2 = \frac{1 + r_{33}}{2}.$$

Since $q \neq 0$ and $a = 0$, at least one of b, c, d is nonzero.

If $b \neq 0$, let

$$b = \frac{\sqrt{1 + r_{11}}}{\sqrt{2}},$$

and determine c, d using

$$4bc = r_{21} + r_{12}$$
$$4bd = r_{13} + r_{31}.$$

If $c \neq 0$, let

$$c = \frac{\sqrt{1 + r_{22}}}{\sqrt{2}},$$

and determine b, d using

$$4bc = r_{21} + r_{12}$$
$$4cd = r_{32} + r_{23}.$$

If $d \neq 0$, let

$$d = \frac{\sqrt{1 + r_{33}}}{\sqrt{2}},$$

and determine b, c using

$$4bd = r_{13} + r_{31}$$
$$4cd = r_{32} + r_{23}.$$

It is easy to check that whenever we computed a square root, if we had chosen a negative sign instead of a positive sign, we would obtain the quaternion $-q$. However, both q and $-q$ determine the same rotation r_q.

The above discussion involving the cases $\operatorname{tr}(R) \neq -1$ and $\operatorname{tr}(R) = -1$ is reminiscent of the procedure for finding a logarithm of a rotation matrix using the Rodrigues formula (see Section 11.7). This is not surprising, because if

$$B = \begin{pmatrix} 0 & -u_3 & u_2 \\ u_3 & 0 & -u_1 \\ -u_2 & u_1 & 0 \end{pmatrix}$$

and if we write $\theta = \sqrt{u_1^2 + u_2^2 + u_3^2}$ (with $0 \leq \theta \leq \pi$), then the Rodrigues formula says that

$$e^B = I + \frac{\sin \theta}{\theta} B + \frac{(1 - \cos \theta)}{\theta^2} B^2, \quad \theta \neq 0,$$

with $e^0 = I$. It is easy to check that $\operatorname{tr}(e^B) = 1 + 2\cos\theta$. Then it is an easy exercise to check that the quaternion q corresponding to the rotation $R = e^B$ (with $B \neq 0$) is given by

$$q = \left[\cos\left(\frac{\theta}{2}\right), \sin\left(\frac{\theta}{2}\right) \left(\frac{u_1}{\theta}, \frac{u_2}{\theta}, \frac{u_3}{\theta}\right) \right].$$

So the method for finding the logarithm of a rotation R is essentially the same as the method for finding a quaternion defining R.

Remark: Geometrically, the group $\mathbf{SU}(2)$ is homeomorphic to the 3-sphere S^3 in $\mathbb{R}^4$,

$$S^3 = \{(x, y, z, t) \in \mathbb{R}^4 \mid x^2 + y^2 + z^2 + t^2 = 1\}.$$

However, since the kernel of the surjective homomorphism $r \colon \mathbf{SU}(2) \to \mathbf{SO}(3)$ is $\{I, -I\}$, as a topological space, $\mathbf{SO}(3)$ is homeomorphic to the quotient of S^3 obtained by identifying antipodal points (x, y, z, t) and $-(x, y, z, t)$. This quotient space is the (real) projective space $\mathbb{RP}^3$, and it is more complicated than S^3. The space S^3 is simply-connected, but $\mathbb{RP}^3$ is not.

15.5 The Exponential Map exp: $\mathfrak{su}(2) \to \mathbf{SU}(2)$

Given any matrix $A \in \mathfrak{su}(2)$, with

$$A = \begin{pmatrix} iu_1 & u_2 + iu_3 \\ -u_2 + iu_3 & -iu_1 \end{pmatrix},$$

it is easy to check that

$$A^2 = -\theta^2 \begin{pmatrix} 1 & 0 \\ 0 & 1 \end{pmatrix},$$

with $\theta = \sqrt{u_1^2 + u_2^2 + u_3^2}$. Then we have the following formula whose proof is very similar to the proof of the formula given in Proposition 8.17.

Proposition 15.4. *For every matrix $A \in \mathfrak{su}(2)$, with*

$$A = \begin{pmatrix} iu_1 & u_2 + iu_3 \\ -u_2 + iu_3 & -iu_1 \end{pmatrix},$$

if we write $\theta = \sqrt{u_1^2 + u_2^2 + u_3^2}$, then

$$e^A = \cos \theta I + \frac{\sin \theta}{\theta} A, \quad \theta \neq 0,$$

and $e^0 = I$.

Therefore, by the discussion at the end of the previous section, e^A is a unit quaternion representing the rotation of angle 2θ and axis (u_1, u_2, u_3) (or I when $\theta = k\pi$, $k \in \mathbb{Z}$). The above formula shows that we may assume that $0 \leq \theta \leq \pi$. Proposition 15.4 shows that the exponential yields a map $\exp \colon \mathfrak{su}(2) \to \mathbf{SU}(2)$. It is an analog of the exponential map $\exp \colon \mathfrak{so}(3) \to \mathbf{SO}(3)$.

Remark: Because $\mathfrak{so}(3)$ and $\mathfrak{su}(2)$ are real vector spaces of dimension 3, they are isomorphic, and it is easy to construct an isomorphism. In fact, $\mathfrak{so}(3)$ and $\mathfrak{su}(2)$ are isomorphic as Lie algebras, which means that there is a linear isomorphism preserving the the Lie bracket $[A, B] = AB - BA$. However, as observed earlier, the groups $\mathbf{SU}(2)$ and $\mathbf{SO}(3)$ are *not isomorphic*.

An equivalent, but often more convenient, formula is obtained by assuming that $u = (u_1, u_2, u_3)$ is a unit vector, equivalently $\det(A) = 1$, in which case $A^2 = -I$, so we have

$$e^{\theta A} = \cos \theta I + \sin \theta A.$$

Using the quaternion notation, this is read as

$$e^{\theta A} = [\cos \theta, \sin \theta \, u].$$

Proposition 15.5. *The exponential map* exp: $\mathfrak{su}(2) \to \mathbf{SU}(2)$ *is surjective*

Proof. We give an algorithm to find the logarithm $A \in \mathfrak{su}(2)$ of a unit quaternion

$$q = \begin{pmatrix} \alpha & \beta \\ -\overline{\beta} & \overline{\alpha} \end{pmatrix}$$

with $\alpha = a + bi$ and $\beta = c + id$.

If $q = I$ (*i.e.* $a = 1$), then $A = 0$. If $q = -I$ (*i.e.* $a = -1$), then

$$A = \pm \pi \begin{pmatrix} i & 0 \\ 0 & -i \end{pmatrix}.$$

Otherwise, $a \neq \pm 1$ and $(b, c, d) \neq (0, 0, 0)$, and we are seeking some $A = \theta B \in \mathfrak{su}(2)$ with $\det(B) = 1$ and $0 < \theta < \pi$, such that, by Proposition 15.4,

$$q = e^{\theta B} = \cos \theta I + \sin \theta B.$$

Let

$$B = \begin{pmatrix} iu_1 & u_2 + iu_3 \\ -u_2 + iu_3 & -iu_1 \end{pmatrix},$$

with $u = (u_1, u_2, u_3)$ a unit vector. We must have

$$a = \cos \theta, \quad e^{\theta B} - (e^{\theta B})^* = q - q^*.$$

Since $0 < \theta < \pi$, we have $\sin \theta \neq 0$, and

$$2 \sin \theta \begin{pmatrix} iu_1 & u_2 + iu_3 \\ -u_2 + iu_3 & -iu_1 \end{pmatrix} = \begin{pmatrix} \alpha - \overline{\alpha} & 2\beta \\ -2\overline{\beta} & \overline{\alpha} - \alpha \end{pmatrix}.$$

Thus, we get

$$u_1 = \frac{1}{\sin \theta} b, \quad u_2 + iu_3 = \frac{1}{\sin \theta}(c + id);$$

that is,

$$\cos \theta = a \quad (0 < \theta < \pi)$$

$$(u_1, u_2, u_3) = \frac{1}{\sin \theta}(b, c, d).$$

Since $a^2 + b^2 + c^2 + d^2 = 1$ and $a = \cos \theta$, the vector $(b, c, d)/\sin \theta$ is a unit vector. Furthermore if the quaternion q is of the form $q = [\cos \theta, \sin \theta u]$ where $u = (u_1, u_2, u_3)$ is a unit vector (with $0 < \theta < \pi$), then

$$A = \theta \begin{pmatrix} iu_1 & u_2 + iu_3 \\ -u_2 + iu_3 & -iu_1 \end{pmatrix} \tag{15.1}$$

is a logarithm of q. $\qquad\square$

Observe that not only is the exponential map $\exp\colon \mathfrak{su}(2) \to \mathbf{SU}(2)$ surjective, but the above proof shows that it is injective on the open ball

$$\{\theta B \in \mathfrak{su}(2) \mid \det(B) = 1, 0 \leq \theta < \pi\}.$$

Also, unlike the situation where in computing the logarithm of a rotation matrix $R \in \mathbf{SO}(3)$ we needed to treat the case where $\mathrm{tr}(R) = -1$ (the angle of the rotation is π) in a special way, computing the logarithm of a quaternion (other than $\pm I$) does not require any case analysis; no special case is needed when the angle of rotation is π.

15.6   Quaternion Interpolation ⊛

We are now going to derive a formula for interpolating between two quaternions. This formula is due to Ken Shoemake, once a Penn student and my TA! Since rotations in **SO**(3) can be defined by quaternions, this has applications to computer graphics, robotics, and computer vision.

First we observe that multiplication of quaternions can be expressed in terms of the inner product and the cross-product in $\mathbb{R}^3$. Indeed, if $q_1 = [a, u_1]$ and $q_2 = [a_2, u_2]$, it can be verified that

$$q_1 q_2 = [a_1, u_1][a_2, u_2] = [a_1 a_2 - u_1 \cdot u_2,\ a_1 u_2 + a_2 u_1 + u_1 \times u_2]. \quad (15.2)$$

We will also need the identity

$$u \times (u \times v) = (u \cdot v)u - (u \cdot u)v.$$

Given a quaternion q expressed as $q = [\cos\theta, \sin\theta\, u]$, where u is a unit vector, we can interpolate between I and q by finding the logs of I and q, interpolating in $\mathfrak{su}(2)$, and then exponentiating. We have

$$A = \log(I) = \begin{pmatrix} 0 & 0 \\ 0 & 0 \end{pmatrix}, \quad B = \log(q) = \theta \begin{pmatrix} iu_1 & u_2 + iu_3 \\ -u_2 + iu_3 & -iu_1 \end{pmatrix},$$

and so $q = e^B$. Since **SU**(2) is a compact Lie group and since the inner product on $\mathfrak{su}(2)$ given by

$$\langle X, Y \rangle = \mathrm{tr}(X^\top Y)$$

is $\mathrm{Ad}(\mathbf{SU}(2))$-invariant, it induces a biinvariant Riemannian metric on **SU**(2), and the curve

$$\lambda \mapsto e^{\lambda B}, \quad \lambda \in [0, 1]$$

is a geodesic from I to q in **SU**(2). We write $q^\lambda = e^{\lambda B}$. Given two quaternions q_1 and q_2, because the metric is left invariant, the curve

$$\lambda \mapsto Z(\lambda) = q_1(q_1^{-1} q_2)^\lambda, \quad \lambda \in [0, 1]$$

is a geodesic from q_1 to q_2. Remarkably, there is a closed-form formula for the interpolant $Z(\lambda)$.

Say $q_1 = [\cos\theta, \sin\theta\, u]$ and $q_2 = [\cos\varphi, \sin\varphi\, v]$, and assume that $q_1 \neq q_2$ and $q_1 \neq -q_2$. First, we compute $q^{-1}q_2$. Since $q^{-1} = [\cos\theta, -\sin\theta\, u]$, we have

$$q^{-1}q_2 = [\cos\theta\cos\varphi + \sin\theta\sin\varphi(u\cdot v),$$
$$- \sin\theta\cos\varphi\, u + \cos\theta\sin\varphi\, v - \sin\theta\sin\varphi(u\times v)].$$

Define Ω by

$$\cos\Omega = \cos\theta\cos\varphi + \sin\theta\sin\varphi(u\cdot v). \tag{15.3}$$

Since $q_1 \neq q_2$ and $q_1 \neq -q_2$, we have $0 < \Omega < \pi$, so we get

$$q_1^{-1}q_2 = \left[\cos\Omega,\ \sin\Omega\,\frac{(-\sin\theta\cos\varphi\, u + \cos\theta\sin\varphi\, v - \sin\theta\sin\varphi(u\times v))}{\sin\Omega}\right],$$

where the term multiplying $\sin\Omega$ is a unit vector because q_1 and q_2 are unit quaternions, so $q_1^{-1}q_2$ is also a unit quaternion. By (15.1), we have

$$(q_1^{-1}q_2)^\lambda$$
$$= \left[\cos\lambda\Omega,\ \sin\lambda\Omega\,\frac{(-\sin\theta\cos\varphi\, u + \cos\theta\sin\varphi\, v - \sin\theta\sin\varphi(u\times v))}{\sin\Omega}\right].$$

Next we need to compute $q_1(q_1^{-1}q_2)^\lambda$. The scalar part of this product is

$$s = \cos\theta\cos\lambda\Omega + \frac{\sin\lambda\Omega}{\sin\Omega}\sin^2\theta\cos\varphi(u\cdot u) - \frac{\sin\lambda\Omega}{\sin\Omega}\sin\theta\sin\varphi\cos\theta(u\cdot v)$$
$$+ \frac{\sin\lambda\Omega}{\sin\Omega}\sin^2\theta\sin\varphi(u\cdot(u\times v)).$$

Since $u\cdot(u\times v) = 0$, the last term is zero, and since $u\cdot u = 1$ and

$$\sin\theta\sin\varphi(u\cdot v) = \cos\Omega - \cos\theta\cos\varphi,$$

we get

$$s = \cos\theta\cos\lambda\Omega + \frac{\sin\lambda\Omega}{\sin\Omega}\sin^2\theta\cos\varphi - \frac{\sin\lambda\Omega}{\sin\Omega}\cos\theta(\cos\Omega - \cos\theta\cos\varphi)$$
$$= \cos\theta\cos\lambda\Omega + \frac{\sin\lambda\Omega}{\sin\Omega}(\sin^2\theta + \cos^2\theta)\cos\varphi - \frac{\sin\lambda\Omega}{\sin\Omega}\cos\theta\cos\Omega$$
$$= \frac{(\cos\lambda\Omega\sin\Omega - \sin\lambda\Omega\cos\Omega)\cos\theta}{\sin\Omega} + \frac{\sin\lambda\Omega}{\sin\Omega}\cos\varphi$$
$$= \frac{\sin(1-\lambda)\Omega}{\sin\Omega}\cos\theta + \frac{\sin\lambda\Omega}{\sin\Omega}\cos\varphi.$$

The vector part of the product $q_1(q_1^{-1}q_2)^\lambda$ is given by

$$\nu = -\frac{\sin\lambda\Omega}{\sin\Omega}\cos\theta\sin\theta\cos\varphi\,u + \frac{\sin\lambda\Omega}{\sin\Omega}\cos^2\theta\sin\varphi\,v$$

$$-\frac{\sin\lambda\Omega}{\sin\Omega}\cos\theta\sin\theta\sin\varphi(u\times v) + \cos\lambda\Omega\sin\theta\,u$$

$$-\frac{\sin\lambda\Omega}{\sin\Omega}\sin^2\theta\cos\varphi(u\times u) + \frac{\sin\lambda\Omega}{\sin\Omega}\cos\theta\sin\theta\sin\varphi(u\times v)$$

$$-\frac{\sin\lambda\Omega}{\sin\Omega}\sin^2\theta\sin\varphi(u\times(u\times v)).$$

We have $u\times u = 0$, the two terms involving $u\times v$ cancel out,

$$u\times(u\times v) = (u\cdot v)u - (u\cdot u)v,$$

and $u\cdot u = 1$, so we get

$$\nu = -\frac{\sin\lambda\Omega}{\sin\Omega}\cos\theta\sin\theta\cos\varphi\,u + \cos\lambda\Omega\sin\theta\,u + \frac{\sin\lambda\Omega}{\sin\Omega}\cos^2\theta\sin\varphi\,v$$

$$+ \frac{\sin\lambda\Omega}{\sin\Omega}\sin^2\theta\sin\varphi\,v - \frac{\sin\lambda\Omega}{\sin\Omega}\sin^2\theta\sin\varphi(u\cdot v)u.$$

Using

$$\sin\theta\sin\varphi(u\cdot v) = \cos\Omega - \cos\theta\cos\varphi,$$

we get

$$\nu = -\frac{\sin\lambda\Omega}{\sin\Omega}\cos\theta\sin\theta\cos\varphi\,u + \cos\lambda\Omega\sin\theta\,u + \frac{\sin\lambda\Omega}{\sin\Omega}\sin\varphi\,v$$

$$-\frac{\sin\lambda\Omega}{\sin\Omega}\sin\theta(\cos\Omega - \cos\theta\cos\varphi)u$$

$$= \cos\lambda\Omega\sin\theta\,u + \frac{\sin\lambda\Omega}{\sin\Omega}\sin\varphi\,v - \frac{\sin\lambda\Omega}{\sin\Omega}\sin\theta\cos\Omega\,u$$

$$= \frac{(\cos\lambda\Omega\sin\Omega - \sin\lambda\Omega\cos\Omega)}{\sin\Omega}\sin\theta\,u + \frac{\sin\lambda\Omega}{\sin\Omega}\sin\varphi\,v$$

$$= \frac{\sin(1-\lambda)\Omega}{\sin\Omega}\sin\theta\,u + \frac{\sin\lambda\Omega}{\sin\Omega}\sin\varphi\,v.$$

Putting the scalar part and the vector part together, we obtain

$$q_1(q_1^{-1}q_2)^\lambda = \left[\frac{\sin(1-\lambda)\Omega}{\sin\Omega}\cos\theta + \frac{\sin\lambda\Omega}{\sin\Omega}\cos\varphi,\right.$$

$$\left.\frac{\sin(1-\lambda)\Omega}{\sin\Omega}\sin\theta\,u + \frac{\sin\lambda\Omega}{\sin\Omega}\sin\varphi\,v\right],$$

$$= \frac{\sin(1-\lambda)\Omega}{\sin\Omega}[\cos\theta,\,\sin\theta\,u] + \frac{\sin\lambda\Omega}{\sin\Omega}[\cos\varphi,\,\sin\varphi\,v].$$

This yields the celebrated *slerp interpolation formula*

$$Z(\lambda) = q_1(q_1^{-1}q_2)^\lambda = \frac{\sin(1-\lambda)\Omega}{\sin\Omega}q_1 + \frac{\sin\lambda\Omega}{\sin\Omega}q_2,$$

with

$$\cos\Omega = \cos\theta\cos\varphi + \sin\theta\sin\varphi(u\cdot v).$$

15.7 Nonexistence of a "Nice" Section from SO(3) to SU(2)

We conclude by discussing the problem of a consistent choice of sign for the quaternion q representing a rotation $R = \rho_q \in \mathbf{SO}(3)$. We are looking for a "nice" section $s \colon \mathbf{SO}(3) \to \mathbf{SU}(2)$, that is, a function s satisfying the condition

$$\rho \circ s = \mathrm{id},$$

where ρ is the surjective homomorphism $\rho \colon \mathbf{SU}(2) \to \mathbf{SO}(3)$.

Proposition 15.6. *Any section* $s \colon \mathbf{SO}(3) \to \mathbf{SU}(2)$ *of* ρ *is neither a homomorphism nor continuous.*

Intuitively, this means that there is no "nice and simple " way to pick the sign of the quaternion representing a rotation.

The following proof is due to Marcel Berger.

Proof. Let Γ be the subgroup of $\mathbf{SU}(2)$ consisting of all quaternions of the form $q = [a, (b, 0, 0)]$. Then, using the formula for the rotation matrix R_q corresponding to q (and the fact that $a^2 + b^2 = 1$), we get

$$R_q = \begin{pmatrix} 1 & 0 & 0 \\ 0 & 2a^2 - 1 & -2ab \\ 0 & 2ab & 2a^2 - 1 \end{pmatrix}.$$

Since $a^2 + b^2 = 1$, we may write $a = \cos\theta, b = \sin\theta$, and we see that

$$R_q = \begin{pmatrix} 1 & 0 & 0 \\ 0 & \cos 2\theta & -\sin 2\theta \\ 0 & \sin 2\theta & \cos 2\theta \end{pmatrix},$$

a rotation of angle 2θ around the x-axis. Thus, both Γ and its image are isomorphic to $\mathbf{SO}(2)$, which is also isomorphic to $\mathbf{U}(1) = \{w \in \mathbb{C} \mid |w| = 1\}$. By identifying $\mathbf{i}$ and i, and identifying Γ and its image to $\mathbf{U}(1)$, if we write $w = \cos\theta + i\sin\theta \in \Gamma$, the restriction of the map ρ to Γ is given by $\rho(w) = w^2$.

We claim that any section s of ρ is not a homomorphism. Consider the restriction of s to $\mathbf{U}(1)$. Then since $\rho \circ s = \mathrm{id}$ and $\rho(w) = w^2$, for $-1 \in \rho(\Gamma) \approx \mathbf{U}(1)$, we have

$$-1 = \rho(s(-1)) = (s(-1))^2.$$

On the other hand, if s is a homomorphism, then

$$(s(-1))^2 = s((-1)^2) = s(1) = 1,$$

contradicting $(s(-1))^2 = -1$.

We also claim that s is not continuous. Assume that $s(1) = 1$, the case where $s(1) = -1$ being analogous. Then s is a bijection inverting ρ on Γ whose restriction to $\mathbf{U}(1)$ must be given by

$$s(\cos\theta + i\sin\theta) = \cos(\theta/2) + \mathbf{i}\sin(\theta/2), \quad -\pi \le \theta < \pi.$$

If θ tends to π, that is $z = \cos\theta + i\sin\theta$ tends to -1 in the upper-half plane, then $s(z)$ tends to $\mathbf{i}$, but if θ tends to $-\pi$, that is z tends to -1 in the lower-half plane, then $s(z)$ tends to $-\mathbf{i}$, which shows that s is not continuous. $\square$

Another way (due to Jean Dieudonné) to prove that a section s of ρ is not a homomorphism is to prove that any unit quaternion is the product of two unit pure quaternions. Indeed, if $q = [a, u]$ is a unit quaternion, if we let $q_1 = [0, u_1]$, where u_1 is any unit vector orthogonal to u, then

$$q_1 q = [-u_1 \cdot u, au_1 + u_1 \times u] = [0, au_1 + u_1 \times u] = q_2$$

is a nonzero unit pure quaternion. This is because if $a \ne 0$ then $au_1 + u_1 \times u \ne 0$ (since $u_1 \times u$ is orthogonal to $au_1 \ne 0$), and if $a = 0$ then $u \ne 0$, so $u_1 \times u \ne 0$ (since u_1 is orthogonal to u). But then, $q_1^{-1} = [0, -u_1]$ is a unit pure quaternion and we have

$$q = q_1^{-1} q_2,$$

a product of two pure unit quaternions.

We also observe that for any two pure quaternions q_1, q_2, there is some unit quaternion q such that

$$q_2 = q q_1 q^{-1}.$$

This is just a restatement of the fact that the group **SO**(3) is transitive. Since the kernel of $\rho\colon \mathbf{SU}(2) \to \mathbf{SO}(3)$ is $\{I, -I\}$, the subgroup $s(\mathbf{SO}(3))$ would be a normal subgroup of index 2 in $\mathbf{SU}(2)$. Then we would have a surjective homomorphism η from $\mathbf{SU}(2)$ onto the quotient group $\mathbf{SU}(2)/s(\mathbf{SO}(3))$, which is isomorphic to $\{1, -1\}$. Now, since any two pure quaternions are conjugate of each other, η would have a constant value on the unit pure quaternions. Since $\mathbf{k} = \mathbf{ij}$, we would have

$$\eta(\mathbf{k}) = \eta(\mathbf{ij}) = (\eta(\mathbf{i}))^2 = 1.$$

Consequently, η would map all pure unit quaternions to 1. But since every unit quaternion is the product of two pure quaternions, η would map every unit quaternion to 1, contradicting the fact that it is surjective onto $\{-1, 1\}$.

15.8   Summary

The main concepts and results of this chapter are listed below:

- The group $\mathbf{SU}(2)$ of unit quaternions.
- The skew field $\mathbb{H}$ of quaternions.
- Hamilton's identities.
- The (real) vector space $\mathfrak{su}(2)$ of 2×2 skew Hermitian matrices with zero trace.
- The adjoint representation of $\mathbf{SU}(2)$.
- The (real) vector space $\mathfrak{su}(2)$ of 2×2 Hermitian matrices with zero trace.
- The group homomorphism $r \colon \mathbf{SU}(2) \to \mathbf{SO}(3)$; $\mathrm{Ker}\,(r) = \{+I, -I\}$.
- The matrix representation R_q of the rotation r_q induced by a unit quaternion q.
- Surjectivity of the homomorphism $r \colon \mathbf{SU}(2) \to \mathbf{SO}(3)$.
- The exponential map $\exp \colon \mathfrak{su}(2) \to \mathbf{SU}(2)$.
- Surjectivity of the exponential map $\exp \colon \mathfrak{su}(2) \to \mathbf{SU}(2)$.
- Finding a logarithm of a quaternion.
- Quaternion interpolation.
- Shoemake's slerp interpolation formula.
- Sections $s \colon \mathbf{SO}(3) \to \mathbf{SU}(2)$ of $r \colon \mathbf{SU}(2) \to \mathbf{SO}(3)$.

15.9   Problems

Problem 15.1. Verify the quaternion identities

$$\mathbf{i}^2 = \mathbf{j}^2 = \mathbf{k}^2 = \mathbf{ijk} = -\mathbf{1},$$
$$\mathbf{ij} = -\mathbf{ji} = \mathbf{k},$$
$$\mathbf{jk} = -\mathbf{kj} = \mathbf{i},$$
$$\mathbf{ki} = -\mathbf{ik} = \mathbf{j}.$$

Problem 15.2. Check that for every quaternion $X = a\mathbf{1} + b\mathbf{i} + c\mathbf{j} + d\mathbf{k}$, we have

$$XX^* = X^*X = (a^2 + b^2 + c^2 + d^2)\mathbf{1}.$$

Conclude that if $X \neq 0$, then X is invertible and its inverse is given by

$$X^{-1} = (a^2 + b^2 + c^2 + d^2)^{-1} X^*.$$

Problem 15.3. Given any two quaternions $X = a\mathbf{1} + b\mathbf{i} + c\mathbf{j} + d\mathbf{k}$ and $Y = a'\mathbf{1} + b'\mathbf{i} + c'\mathbf{j} + d'\mathbf{k}$, prove that

$$XY = (aa' - bb' - cc' - dd')\mathbf{1} + (ab' + ba' + cd' - dc')\mathbf{i}$$
$$+ (ac' + ca' + db' - bd')\mathbf{j} + (ad' + da' + bc' - cb')\mathbf{k}.$$

Also prove that if $X = [a, U]$ and $Y = [a', U']$, the quaternion product XY can be expressed as

$$XY = [aa' - U \cdot U',\ aU' + a'U + U \times U'].$$

Problem 15.4. Let Ad: $\mathbf{SU}(2) \to \mathbf{GL}(\mathfrak{su}(2))$ be the map defined such that for every $q \in \mathbf{SU}(2)$,

$$\mathrm{Ad}_q(A) = qAq^*, \quad A \in \mathfrak{su}(2),$$

where q^* is the inverse of q (since $\mathbf{SU}(2)$ is a unitary group). Prove that the map Ad_q is an invertible linear map from $\mathfrak{su}(2)$ to itself and that Ad is a group homomorphism.

Problem 15.5. Prove that every Hermitian matrix with zero trace is of the form $x\sigma_3 + y\sigma_2 + z\sigma_1$, with

$$\sigma_1 = \begin{pmatrix} 0 & 1 \\ 1 & 0 \end{pmatrix}, \quad \sigma_2 = \begin{pmatrix} 0 & -i \\ i & 0 \end{pmatrix}, \quad \sigma_3 = \begin{pmatrix} 1 & 0 \\ 0 & -1 \end{pmatrix}.$$

Check that $\mathbf{i} = i\sigma_3$, $\mathbf{j} = i\sigma_2$, and that $\mathbf{k} = i\sigma_1$.

Problem 15.6. If

$$B = \begin{pmatrix} 0 & -u_3 & u_2 \\ u_3 & 0 & -u_1 \\ -u_2 & u_1 & 0 \end{pmatrix},$$

and if we write $\theta = \sqrt{u_1^2 + u_2^2 + u_3^2}$ (with $0 \le \theta \le \pi$), then the Rodrigues formula says that

$$e^B = I + \frac{\sin\theta}{\theta}B + \frac{(1 - \cos\theta)}{\theta^2}B^2, \quad \theta \neq 0,$$

with $e^0 = I$. Check that $\mathrm{tr}(e^B) = 1 + 2\cos\theta$. Prove that the quaternion q corresponding to the rotation $R = e^B$ (with $B \neq 0$) is given by

$$q = \left[\cos\left(\frac{\theta}{2}\right),\ \sin\left(\frac{\theta}{2}\right)\left(\frac{u_1}{\theta}, \frac{u_2}{\theta}, \frac{u_3}{\theta}\right)\right].$$

Problem 15.7. For every matrix $A \in \mathfrak{su}(2)$, with

$$A = \begin{pmatrix} iu_1 & u_2 + iu_3 \\ -u_2 + iu_3 & -iu_1 \end{pmatrix},$$

prove that if we write $\theta = \sqrt{u_1^2 + u_2^2 + u_3^2}$, then

$$e^A = \cos\theta I + \frac{\sin\theta}{\theta}A, \quad \theta \neq 0,$$

and $e^0 = I$. Conclude that e^A is a unit quaternion representing the rotation of angle 2θ and axis (u_1, u_2, u_3) (or I when $\theta = k\pi$, $k \in \mathbb{Z}$).

Problem 15.8. Write a `Matlab` program implementing the method of Section 15.4 for finding a unit quaternion corresponding to a rotation matrix.

Problem 15.9. Show that there is a very simple method for producing an orthonormal frame in $\mathbb{R}^4$ whose first vector is any given nonnull vector (a, b, c, d).

Problem 15.10. Let i, j, and k, be the unit vectors of coordinates $(1, 0, 0)$, $(0, 1, 0)$, and $(0, 0, 1)$ in $\mathbb{R}^3$.

(1) Describe geometrically the rotations defined by the following quaternions:

$$p = (0, i), \quad q = (0, j).$$

Prove that the interpolant $Z(\lambda) = p(p^{-1}q)^\lambda$ is given by

$$Z(\lambda) = (0, \cos(\lambda\pi/2)i + \sin(\lambda\pi/2)j).$$

Describe geometrically what this rotation is.

(2) Repeat Question (1) with the rotations defined by the quaternions

$$p = \left(\frac{1}{2}, \frac{\sqrt{3}}{2}i\right), \quad q = (0, j).$$

Prove that the interpolant $Z(\lambda)$ is given by

$$Z(\lambda) = \left(\frac{1}{2}\cos(\lambda\pi/2), \frac{\sqrt{3}}{2}\cos(\lambda\pi/2)i + \sin(\lambda\pi/2)j\right).$$

Describe geometrically what this rotation is.

(3) Repeat Question (1) with the rotations defined by the quaternions

$$p = \left(\frac{1}{\sqrt{2}}, \frac{1}{\sqrt{2}}i\right), \quad q = \left(0, \frac{1}{\sqrt{2}}(i + j)\right).$$

Prove that the interpolant $Z(\lambda)$ is given by

$$Z(\lambda) = \left(\frac{1}{\sqrt{2}}\cos(\lambda\pi/3) - \frac{1}{\sqrt{6}}\sin(\lambda\pi/3),\right.$$

$$\left.(1/\sqrt{2}\cos(\lambda\pi/3) + 1/\sqrt{6}\sin(\lambda\pi/3))i + \frac{2}{\sqrt{6}}\sin(\lambda\pi/3)j\right).$$

Problem 15.11. Prove that

$$w \times (u \times v) = (w \cdot v)u - (u \cdot w)v.$$

Conclude that

$$u \times (u \times v) = (u \cdot v)u - (u \cdot u)v.$$

Chapter 16

Spectral Theorems in Euclidean and Hermitian Spaces

16.1 Introduction

The goal of this chapter is to show that there are nice normal forms for symmetric matrices, skew-symmetric matrices, orthogonal matrices, and normal matrices. The spectral theorem for symmetric matrices states that symmetric matrices have real eigenvalues and that they can be diagonalized over an orthonormal basis. The spectral theorem for Hermitian matrices states that Hermitian matrices also have real eigenvalues and that they can be diagonalized over a complex orthonormal basis. Normal real matrices can be block diagonalized over an orthonormal basis with blocks having size at most two and there are refinements of this normal form for skew-symmetric and orthogonal matrices.

The spectral result for real symmetric matrices can be used to prove two characterizations of the eigenvalues of a symmetric matrix in terms of the *Rayleigh ratio*. The first characterization is the *Rayleigh–Ritz theorem* and the second one is the *Courant–Fischer theorem*. Both results are used in optimization theory and to obtain results about perturbing the eigenvalues of a symmetric matrix.

In this chapter all vector spaces are finite-dimensional real or complex vector spaces.

16.2 Normal Linear Maps: Eigenvalues and Eigenvectors

We begin by studying normal maps, to understand the structure of their eigenvalues and eigenvectors. This section and the next three were inspired by Lang [Lang (1993)], Artin [Artin (1991)], Mac Lane and Birkhoff [Mac Lane and Birkhoff (1967)], Berger [Berger (1990a)], and Bertin [Bertin (1981)].

Definition 16.1. Given a Euclidean or Hermitian space E, a linear map $f \colon E \to E$ is *normal* if

$$f \circ f^* = f^* \circ f.$$

A linear map $f \colon E \to E$ is *self-adjoint* if $f = f^*$, *skew-self-adjoint* if $f = -f^*$, and *orthogonal* if $f \circ f^* = f^* \circ f = \mathrm{id}$.

Obviously, a self-adjoint, skew-self-adjoint, or orthogonal linear map is a normal linear map. Our first goal is to show that for every normal linear map $f \colon E \to E$, there is an orthonormal basis (w.r.t. $\langle -, - \rangle$) such that the matrix of f over this basis has an especially nice form: it is a block diagonal matrix in which the blocks are either one-dimensional matrices (i.e., single entries) or two-dimensional matrices of the form

$$\begin{pmatrix} \lambda & \mu \\ -\mu & \lambda \end{pmatrix}.$$

This normal form can be further refined if f is self-adjoint, skew-self-adjoint, or orthogonal. As a first step we show that f and f^* have the same kernel when f is normal.

Proposition 16.1. *Given a Euclidean space E, if $f \colon E \to E$ is a normal linear map, then $\operatorname{Ker} f = \operatorname{Ker} f^*$.*

Proof. First let us prove that

$$\langle f(u), f(v) \rangle = \langle f^*(u), f^*(v) \rangle$$

for all $u, v \in E$. Since f^* is the adjoint of f and $f \circ f^* = f^* \circ f$, we have

$$\begin{aligned} \langle f(u), f(u) \rangle &= \langle u, (f^* \circ f)(u) \rangle, \\ &= \langle u, (f \circ f^*)(u) \rangle, \\ &= \langle f^*(u), f^*(u) \rangle. \end{aligned}$$

Since $\langle -, - \rangle$ is positive definite,

$$\begin{aligned} \langle f(u), f(u) \rangle &= 0 \quad \text{iff} \quad f(u) = 0, \\ \langle f^*(u), f^*(u) \rangle &= 0 \quad \text{iff} \quad f^*(u) = 0, \end{aligned}$$

and since

$$\langle f(u), f(u) \rangle = \langle f^*(u), f^*(u) \rangle,$$

we have

$$f(u) = 0 \quad \text{iff} \quad f^*(u) = 0.$$

Consequently, $\operatorname{Ker} f = \operatorname{Ker} f^*$. $\qquad\square$

Assuming again that E is a Hermitian space, observe that Proposition 16.1 also holds. We deduce the following corollary.

Proposition 16.2. *Given a Hermitian space E, for any normal linear map $f\colon E \to E$, we have $\mathrm{Ker}\,(f) \cap \mathrm{Im}(f) = (0)$.*

Proof. Assume $v \in \mathrm{Ker}\,(f) \cap \mathrm{Im}(f) = (0)$, which means that $v = f(u)$ for some $u \in E$, and $f(v) = 0$. By Proposition 16.1, $\mathrm{Ker}\,(f) = \mathrm{Ker}\,(f^*)$, so $f(v) = 0$ implies that $f^*(v) = 0$. Consequently,

$$
\begin{aligned}
0 &= \langle f^*(v), u \rangle \\
&= \langle v, f(u) \rangle \\
&= \langle v, v \rangle,
\end{aligned}
$$

and thus, $v = 0$. $\qquad\square$

We also have the following crucial proposition relating the eigenvalues of f and f^*.

Proposition 16.3. *Given a Hermitian space E, for any normal linear map $f\colon E \to E$, a vector u is an eigenvector of f for the eigenvalue λ (in $\mathbb{C}$) iff u is an eigenvector of f^* for the eigenvalue $\overline{\lambda}$.*

Proof. First it is immediately verified that the adjoint of $f - \lambda\,\mathrm{id}$ is $f^* - \overline{\lambda}\,\mathrm{id}$. Furthermore, $f - \lambda\,\mathrm{id}$ is normal. Indeed,

$$
\begin{aligned}
(f - \lambda\,\mathrm{id}) \circ (f - \lambda\,\mathrm{id})^* &= (f - \lambda\,\mathrm{id}) \circ (f^* - \overline{\lambda}\,\mathrm{id}), \\
&= f \circ f^* - \overline{\lambda} f - \lambda f^* + \lambda\overline{\lambda}\,\mathrm{id}, \\
&= f^* \circ f - \lambda f^* - \overline{\lambda} f + \overline{\lambda}\lambda\,\mathrm{id}, \\
&= (f^* - \overline{\lambda}\,\mathrm{id}) \circ (f - \lambda\,\mathrm{id}), \\
&= (f - \lambda\,\mathrm{id})^* \circ (f - \lambda\,\mathrm{id}).
\end{aligned}
$$

Applying Proposition 16.1 to $f - \lambda\,\mathrm{id}$, for every nonnull vector u, we see that

$$
(f - \lambda\,\mathrm{id})(u) = 0 \quad \text{iff} \quad (f^* - \overline{\lambda}\,\mathrm{id})(u) = 0,
$$

which is exactly the statement of the proposition. $\qquad\square$

The next proposition shows a very important property of normal linear maps: eigenvectors corresponding to distinct eigenvalues are orthogonal.

Proposition 16.4. *Given a Hermitian space E, for any normal linear map $f\colon E \to E$, if u and v are eigenvectors of f associated with the eigenvalues λ and μ (in $\mathbb{C}$) where $\lambda \neq \mu$, then $\langle u, v \rangle = 0$.*

Proof. Let us compute $\langle f(u), v \rangle$ in two different ways. Since v is an eigenvector of f for μ, by Proposition 16.3, v is also an eigenvector of f^* for $\overline{\mu}$, and we have

$$\langle f(u), v \rangle = \langle \lambda u, v \rangle = \lambda \langle u, v \rangle,$$

and

$$\langle f(u), v \rangle = \langle u, f^*(v) \rangle = \langle u, \overline{\mu} v \rangle = \mu \langle u, v \rangle,$$

where the last identity holds because of the semilinearity in the second argument. Thus

$$\lambda \langle u, v \rangle = \mu \langle u, v \rangle,$$

that is,

$$(\lambda - \mu) \langle u, v \rangle = 0,$$

which implies that $\langle u, v \rangle = 0$, since $\lambda \neq \mu$. $\square$

We can show easily that the eigenvalues of a self-adjoint linear map are real.

Proposition 16.5. *Given a Hermitian space E, all the eigenvalues of any self-adjoint linear map $f \colon E \to E$ are real.*

Proof. Let z (in $\mathbb{C}$) be an eigenvalue of f and let u be an eigenvector for z. We compute $\langle f(u), u \rangle$ in two different ways. We have

$$\langle f(u), u \rangle = \langle zu, u \rangle = z \langle u, u \rangle,$$

and since $f = f^*$, we also have

$$\langle f(u), u \rangle = \langle u, f^*(u) \rangle = \langle u, f(u) \rangle = \langle u, zu \rangle = \overline{z} \langle u, u \rangle.$$

Thus,

$$z \langle u, u \rangle = \overline{z} \langle u, u \rangle,$$

which implies that $z = \overline{z}$, since $u \neq 0$, and z is indeed real. $\square$

There is also a version of Proposition 16.5 for a (real) Euclidean space E and a self-adjoint map $f \colon E \to E$ since every real vector space E can be embedded into a complex vector space $E_{\mathbb{C}}$, and every linear map $f \colon E \to E$ can be extended to a linear map $f_{\mathbb{C}} \colon E_{\mathbb{C}} \to E_{\mathbb{C}}$.

Definition 16.2. Given a real vector space E, let $E_{\mathbb{C}}$ be the structure $E \times E$ under the addition operation

$$(u_1,\, u_2) + (v_1,\, v_2) = (u_1 + v_1,\, u_2 + v_2),$$

and let multiplication by a complex scalar $z = x + iy$ be defined such that

$$(x + iy) \cdot (u,\, v) = (xu - yv,\, yu + xv).$$

The space $E_{\mathbb{C}}$ is called the *complexification* of E.

It is easily shown that the structure $E_{\mathbb{C}}$ is a complex vector space. It is also immediate that

$$(0, v) = i(v, 0),$$

and thus, identifying E with the subspace of $E_{\mathbb{C}}$ consisting of all vectors of the form $(u, 0)$, we can write

$$(u, v) = u + iv.$$

Observe that if $(e_1, \ldots, e_n)$ is a basis of E (a real vector space), then $(e_1, \ldots, e_n)$ is also a basis of $E_{\mathbb{C}}$ (recall that e_i is an abbreviation for $(e_i, 0)$).

A linear map $f \colon E \to E$ is extended to the linear map $f_{\mathbb{C}} \colon E_{\mathbb{C}} \to E_{\mathbb{C}}$ defined such that

$$f_{\mathbb{C}}(u + iv) = f(u) + if(v).$$

For any basis $(e_1, \ldots, e_n)$ of E, the matrix $M(f)$ representing f over $(e_1, \ldots, e_n)$ is *identical* to the matrix $M(f_{\mathbb{C}})$ representing $f_{\mathbb{C}}$ over $(e_1, \ldots, e_n)$, where we view $(e_1, \ldots, e_n)$ as a basis of $E_{\mathbb{C}}$. As a consequence, $\det(zI - M(f)) = \det(zI - M(f_{\mathbb{C}}))$, which means that f and $f_{\mathbb{C}}$ have the *same* characteristic polynomial (which has real coefficients). We know that every polynomial of degree n with real (or complex) coefficients always has n complex roots (counted with their multiplicity), and the roots of $\det(zI - M(f_{\mathbb{C}}))$ that are real (if any) are the eigenvalues of f.

Next we need to extend the inner product on E to an inner product on $E_{\mathbb{C}}$.

The inner product $\langle -, - \rangle$ on a Euclidean space E is extended to the Hermitian positive definite form $\langle -, - \rangle_{\mathbb{C}}$ on $E_{\mathbb{C}}$ as follows:

$$\langle u_1 + iv_1, u_2 + iv_2 \rangle_{\mathbb{C}} = \langle u_1, u_2 \rangle + \langle v_1, v_2 \rangle + i(\langle v_1, u_2 \rangle - \langle u_1, v_2 \rangle).$$

It is easily verified that $\langle -, - \rangle_{\mathbb{C}}$ is indeed a Hermitian form that is positive definite, and it is clear that $\langle -, - \rangle_{\mathbb{C}}$ agrees with $\langle -, - \rangle$ on real vectors. Then given any linear map $f \colon E \to E$, it is easily verified that the map $f_{\mathbb{C}}^*$ definedsuch that

$$f_{\mathbb{C}}^*(u + iv) = f^*(u) + if^*(v)$$

for all $u, v \in E$ is the adjoint of $f_{\mathbb{C}}$ w.r.t. $\langle -, - \rangle_{\mathbb{C}}$.

Proposition 16.6. *Given a Euclidean space E, if $f \colon E \to E$ is any self-adjoint linear map, then every eigenvalue λ of $f_{\mathbb{C}}$ is real and is actually an eigenvalue of f (which means that there is some real eigenvector $u \in E$ such that $f(u) = \lambda u$). Therefore, all the eigenvalues of f are real.*

Proof. Let $E_{\mathbb{C}}$ be the complexification of E, $\langle -, - \rangle_{\mathbb{C}}$ the complexification of the inner product $\langle -, - \rangle$ on E, and $f_{\mathbb{C}} \colon E_{\mathbb{C}} \to E_{\mathbb{C}}$ the complexification of $f \colon E \to E$. By definition of $f_{\mathbb{C}}$ and $\langle -, - \rangle_{\mathbb{C}}$, if f is self-adjoint, we have

$$\langle f_{\mathbb{C}}(u_1 + iv_1), u_2 + iv_2 \rangle_{\mathbb{C}} = \langle f(u_1) + if(v_1), u_2 + iv_2 \rangle_{\mathbb{C}}$$
$$= \langle f(u_1), u_2 \rangle + \langle f(v_1), v_2 \rangle$$
$$+ i(\langle u_2, f(v_1) \rangle - \langle f(u_1), v_2 \rangle)$$
$$= \langle u_1, f(u_2) \rangle + \langle v_1, f(v_2) \rangle$$
$$+ i(\langle f(u_2), v_1 \rangle - \langle u_1, f(v_2) \rangle)$$
$$= \langle u_1 + iv_1, f(u_2) + if(v_2) \rangle_{\mathbb{C}}$$
$$= \langle u_1 + iv_1, f_{\mathbb{C}}(u_2 + iv_2) \rangle_{\mathbb{C}},$$

which shows that $f_{\mathbb{C}}$ is also self-adjoint with respect to $\langle -, - \rangle_{\mathbb{C}}$.

As we pointed out earlier, f and $f_{\mathbb{C}}$ have the same characteristic polynomial $\det(zI - f_{\mathbb{C}}) = \det(zI - f)$, which is a polynomial with real coefficients. Proposition 16.5 shows that the zeros of $\det(zI - f_{\mathbb{C}}) = \det(zI - f)$ are all real, and for each real zero λ of $\det(zI - f)$, the linear map $\lambda \mathrm{id} - f$ is singular, which means that there is some nonzero $u \in E$ such that $f(u) = \lambda u$. Therefore, all the eigenvalues of f are real. $\qquad \square$

Proposition 16.7. *Given a Hermitian space E, for any linear map $f \colon E \to E$, if f is skew-self-adjoint, then f has eigenvalues that are pure imaginary or zero, and if f is unitary, then f has eigenvalues of absolute value 1.*

Proof. If f is skew-self-adjoint, $f^* = -f$, and then by the definition of the adjoint map, for any eigenvalue λ and any eigenvector u associated with λ, we have

$$\lambda \langle u, u \rangle = \langle \lambda u, u \rangle = \langle f(u), u \rangle = \langle u, f^*(u) \rangle = \langle u, -f(u) \rangle$$
$$= -\langle u, \lambda u \rangle = -\overline{\lambda} \langle u, u \rangle,$$

and since $u \neq 0$ and $\langle -, - \rangle$ is positive definite, $\langle u, u \rangle \neq 0$, so

$$\lambda = -\overline{\lambda},$$

which shows that $\lambda = ir$ for some $r \in \mathbb{R}$.

If f is unitary, then f is an isometry, so for any eigenvalue λ and any eigenvector u associated with λ, we have

$$|\lambda|^2 \langle u, u \rangle = \lambda \overline{\lambda} \langle u, u \rangle = \langle \lambda u, \lambda u \rangle = \langle f(u), f(u) \rangle = \langle u, u \rangle,$$

and since $u \neq 0$, we obtain $|\lambda|^2 = 1$, which implies

$$|\lambda| = 1. \qquad \square$$

16.3 Spectral Theorem for Normal Linear Maps

Given a Euclidean space E, our next step is to show that for every linear map $f \colon E \to E$ there is some subspace W of dimension 1 or 2 such that $f(W) \subseteq W$. When $\dim(W) = 1$, the subspace W is actually an eigenspace for some real eigenvalue of f. Furthermore, when f is normal, there is a subspace W of dimension 1 or 2 such that $f(W) \subseteq W$ **and** $f^*(W) \subseteq W$. The difficulty is that the eigenvalues of f are not necessarily real. One way to get around this problem is to complexify both the vector space E and the inner product $\langle -, - \rangle$ as we did in Section 16.2.

Given any subspace W of a Euclidean space E, recall that the *orthogonal complement* $W^\perp$ of W is the subspace defined such that

$$W^\perp = \{u \in E \mid \langle u, w \rangle = 0, \text{ for all } w \in W\}.$$

Recall from Proposition 11.9 that $E = W \oplus W^\perp$ (this can be easily shown, for example, by constructing an orthonormal basis of E using the Gram–Schmidt orthonormalization procedure). The same result also holds for Hermitian spaces; see Proposition 13.12.

As a warm up for the proof of Theorem 16.2, let us prove that every self-adjoint map on a Euclidean space can be diagonalized with respect to an orthonormal basis of eigenvectors.

Theorem 16.1. *(Spectral theorem for self-adjoint linear maps on a Euclidean space) Given a Euclidean space E of dimension n, for every self-adjoint linear map $f \colon E \to E$, there is an orthonormal basis $(e_1, \ldots, e_n)$ of eigenvectors of f such that the matrix of f w.r.t. this basis is a diagonal matrix*

$$\begin{pmatrix} \lambda_1 & \cdots & & \\ & \lambda_2 & \cdots & \\ \vdots & \vdots & \ddots & \vdots \\ & & \cdots & \lambda_n \end{pmatrix},$$

with $\lambda_i \in \mathbb{R}$.

Proof. We proceed by induction on the dimension n of E as follows. If $n = 1$, the result is trivial. Assume now that $n \geq 2$. From Proposition 16.6, all the eigenvalues of f are real, so pick some eigenvalue $\lambda \in \mathbb{R}$, and let w be some eigenvector for λ. By dividing w by its norm, we may assume that w is a unit vector. Let W be the subspace of dimension 1 spanned by w. Clearly, $f(W) \subseteq W$. We claim that $f(W^\perp) \subseteq W^\perp$, where $W^\perp$ is the orthogonal complement of W.

Indeed, for any $v \in W^\perp$, that is, if $\langle v, w \rangle = 0$, because f is self-adjoint and $f(w) = \lambda w$, we have

$$\langle f(v), w \rangle = \langle v, f(w) \rangle$$
$$= \langle v, \lambda w \rangle$$
$$= \lambda \langle v, w \rangle = 0$$

since $\langle v, w \rangle = 0$. Therefore,

$$f(W^\perp) \subseteq W^\perp.$$

Clearly, the restriction of f to $W^\perp$ is self-adjoint, and we conclude by applying the induction hypothesis to $W^\perp$ (whose dimension is $n - 1$). $\quad\square$

We now come back to normal linear maps. One of the key points in the proof of Theorem 16.1 is that we found a subspace W with the property that $f(W) \subseteq W$ implies that $f(W^\perp) \subseteq W^\perp$. In general, this does not happen, *but normal maps satisfy a stronger property which ensures that such a subspace exists.*

The following proposition provides a condition that will allow us to show that a normal linear map can be diagonalized. It actually holds for any linear map. We found the inspiration for this proposition in Berger [Berger (1990a)].

Proposition 16.8. *Given a Hermitian space E, for any linear map $f \colon E \to E$ and any subspace W of E, if $f(W) \subseteq W$, then $f^*(W^\perp) \subseteq W^\perp$. Consequently, if $f(W) \subseteq W$ and $f^*(W) \subseteq W$, then $f(W^\perp) \subseteq W^\perp$ and $f^*(W^\perp) \subseteq W^\perp$.*

Proof. If $u \in W^\perp$, then

$$\langle w, u \rangle = 0 \quad \text{for all } w \in W.$$

However,

$$\langle f(w), u \rangle = \langle w, f^*(u) \rangle,$$

and $f(W) \subseteq W$ implies that $f(w) \in W$. Since $u \in W^\perp$, we get

$$0 = \langle f(w), u \rangle = \langle w, f^*(u) \rangle,$$

which shows that $\langle w, f^*(u) \rangle = 0$ for all $w \in W$, that is, $f^*(u) \in W^\perp$. Therefore, we have $f^*(W^\perp) \subseteq W^\perp$.

We just proved that if $f(W) \subseteq W$, then $f^*(W^\perp) \subseteq W^\perp$. If we also have $f^*(W) \subseteq W$, then by applying the above fact to f^*, we get $f^{**}(W^\perp) \subseteq W^\perp$, and since $f^{**} = f$, this is just $f(W^\perp) \subseteq W^\perp$, which proves the second statement of the proposition. $\quad\square$

It is clear that the above proposition also holds for Euclidean spaces. Although we are ready to prove that for every normal linear map f (over a Hermitian space) there is an orthonormal basis of eigenvectors (see Theorem 16.3 below), we now return to real Euclidean spaces.

Proposition 16.9. *If $f: E \to E$ is a linear map and $w = u + iv$ is an eigenvector of $f_{\mathbb{C}}: E_{\mathbb{C}} \to E_{\mathbb{C}}$ for the eigenvalue $z = \lambda + i\mu$, where $u, v \in E$ and $\lambda, \mu \in \mathbb{R}$, then*

$$f(u) = \lambda u - \mu v \quad and \quad f(v) = \mu u + \lambda v. \tag{16.1}$$

As a consequence,

$$f_{\mathbb{C}}(u - iv) = f(u) - if(v) = (\lambda - i\mu)(u - iv),$$

which shows that $\overline{w} = u - iv$ is an eigenvector of $f_{\mathbb{C}}$ for $\overline{z} = \lambda - i\mu$.

Proof. Since

$$f_{\mathbb{C}}(u + iv) = f(u) + if(v)$$

and

$$f_{\mathbb{C}}(u + iv) = (\lambda + i\mu)(u + iv) = \lambda u - \mu v + i(\mu u + \lambda v),$$

we have

$$f(u) = \lambda u - \mu v \quad and \quad f(v) = \mu u + \lambda v. \qquad \square$$

Using this fact, we can prove the following proposition.

Proposition 16.10. *Given a Euclidean space E, for any normal linear map $f: E \to E$, if $w = u + iv$ is an eigenvector of $f_{\mathbb{C}}$ associated with the eigenvalue $z = \lambda + i\mu$ (where $u, v \in E$ and $\lambda, \mu \in \mathbb{R}$), if $\mu \neq 0$ (i.e., z is not real) then $\langle u, v \rangle = 0$ and $\langle u, u \rangle = \langle v, v \rangle$, which implies that u and v are linearly independent, and if W is the subspace spanned by u and v, then $f(W) = W$ and $f^*(W) = W$. Furthermore, with respect to the (orthogonal) basis (u, v), the restriction of f to W has the matrix*

$$\begin{pmatrix} \lambda & \mu \\ -\mu & \lambda \end{pmatrix}.$$

If $\mu = 0$, then λ is a real eigenvalue of f, and either u or v is an eigenvector of f for λ. If W is the subspace spanned by u if $u \neq 0$, or spanned by $v \neq 0$ if $u = 0$, then $f(W) \subseteq W$ and $f^*(W) \subseteq W$.

Proof. Since $w = u + iv$ is an eigenvector of $f_{\mathbb{C}}$, by definition it is nonnull, and either $u \neq 0$ or $v \neq 0$. Proposition 16.9 implies that $u - iv$ is an eigenvector of $f_{\mathbb{C}}$ for $\lambda - i\mu$. It is easy to check that $f_{\mathbb{C}}$ is normal. However, if $\mu \neq 0$, then $\lambda + i\mu \neq \lambda - i\mu$, and from Proposition 16.4, the vectors $u + iv$ and $u - iv$ are orthogonal w.r.t. $\langle -, - \rangle_{\mathbb{C}}$, that is,

$$\langle u + iv, u - iv \rangle_{\mathbb{C}} = \langle u, u \rangle - \langle v, v \rangle + 2i\langle u, v \rangle = 0.$$

Thus we get $\langle u, v \rangle = 0$ and $\langle u, u \rangle = \langle v, v \rangle$, and since $u \neq 0$ or $v \neq 0$, u and v are linearly independent. Since

$$f(u) = \lambda u - \mu v \quad \text{and} \quad f(v) = \mu u + \lambda v$$

and since by Proposition 16.3 $u + iv$ is an eigenvector of $f_{\mathbb{C}}^*$ for $\lambda - i\mu$, we have

$$f^*(u) = \lambda u + \mu v \quad \text{and} \quad f^*(v) = -\mu u + \lambda v,$$

and thus $f(W) = W$ and $f^*(W) = W$, where W is the subspace spanned by u and v.

When $\mu = 0$, we have

$$f(u) = \lambda u \quad \text{and} \quad f(v) = \lambda v,$$

and since $u \neq 0$ or $v \neq 0$, either u or v is an eigenvector of f for λ. If W is the subspace spanned by u if $u \neq 0$, or spanned by v if $u = 0$, it is obvious that $f(W) \subseteq W$ and $f^*(W) \subseteq W$. Note that $\lambda = 0$ is possible, and this is why $\subseteq$ cannot be replaced by $=$. $\square$

The beginning of the proof of Proposition 16.10 actually shows that for every linear map $f \colon E \to E$ there is some subspace W such that $f(W) \subseteq W$, where W has dimension 1 or 2. In general, it doesn't seem possible to prove that $W^{\perp}$ is invariant under f. *However, this happens when f is normal.*

We can finally prove our first main theorem.

Theorem 16.2. *(Main spectral theorem) Given a Euclidean space E of dimension n, for every normal linear map $f \colon E \to E$, there is an orthonormal basis $(e_1, \ldots, e_n)$ such that the matrix of f w.r.t. this basis is a block diagonal matrix of the form*

$$\begin{pmatrix} A_1 & & \cdots & \\ & A_2 & \cdots & \\ \vdots & \vdots & \ddots & \vdots \\ & & \cdots & A_p \end{pmatrix}$$

such that each block A_j is either a one-dimensional matrix (i.e., a real scalar) or a two-dimensional matrix of the form

$$A_j = \begin{pmatrix} \lambda_j & -\mu_j \\ \mu_j & \lambda_j \end{pmatrix},$$

where $\lambda_j, \mu_j \in \mathbb{R}$, with $\mu_j > 0$.

Proof. We proceed by induction on the dimension n of E as follows. If $n = 1$, the result is trivial. Assume now that $n \geq 2$. First, since $\mathbb{C}$ is algebraically closed (i.e., every polynomial has a root in $\mathbb{C}$), the linear map $f_{\mathbb{C}} \colon E_{\mathbb{C}} \to E_{\mathbb{C}}$ has some eigenvalue $z = \lambda + i\mu$ (where $\lambda, \mu \in \mathbb{R}$). Let $w = u + iv$ be some eigenvector of $f_{\mathbb{C}}$ for $\lambda + i\mu$ (where $u, v \in E$). We can now apply Proposition 16.10.

If $\mu = 0$, then either u or v is an eigenvector of f for $\lambda \in \mathbb{R}$. Let W be the subspace of dimension 1 spanned by $e_1 = u/\|u\|$ if $u \neq 0$, or by $e_1 = v/\|v\|$ otherwise. It is obvious that $f(W) \subseteq W$ and $f^*(W) \subseteq W$. The orthogonal $W^\perp$ of W has dimension $n - 1$, and by Proposition 16.8, we have $f(W^\perp) \subseteq W^\perp$. But the restriction of f to $W^\perp$ is also normal, and we conclude by applying the induction hypothesis to $W^\perp$.

If $\mu \neq 0$, then $\langle u, v \rangle = 0$ and $\langle u, u \rangle = \langle v, v \rangle$, and if W is the subspace spanned by $u/\|u\|$ and $v/\|v\|$, then $f(W) = W$ and $f^*(W) = W$. We also know that the restriction of f to W has the matrix

$$\begin{pmatrix} \lambda & \mu \\ -\mu & \lambda \end{pmatrix}$$

with respect to the basis $(u/\|u\|, v/\|v\|)$. If $\mu < 0$, we let $\lambda_1 = \lambda$, $\mu_1 = -\mu$, $e_1 = u/\|u\|$, and $e_2 = v/\|v\|$. If $\mu > 0$, we let $\lambda_1 = \lambda$, $\mu_1 = \mu$, $e_1 = v/\|v\|$, and $e_2 = u/\|u\|$. In all cases, it is easily verified that the matrix of the restriction of f to W w.r.t. the orthonormal basis (e_1, e_2) is

$$A_1 = \begin{pmatrix} \lambda_1 & -\mu_1 \\ \mu_1 & \lambda_1 \end{pmatrix},$$

where $\lambda_1, \mu_1 \in \mathbb{R}$, with $\mu_1 > 0$. However, $W^\perp$ has dimension $n - 2$, and by Proposition 16.8, $f(W^\perp) \subseteq W^\perp$. Since the restriction of f to $W^\perp$ is also normal, we conclude by applying the induction hypothesis to $W^\perp$. $\square$

After this relatively hard work, we can easily obtain some nice normal forms for the matrices of self-adjoint, skew-self-adjoint, and orthogonal linear maps. However, for the sake of completeness (and since we have all the tools to so do), we go back to the case of a Hermitian space and show that normal linear maps can be diagonalized with respect to an orthonormal basis. The proof is a slight generalization of the proof of Theorem 16.6.

Theorem 16.3. *(Spectral theorem for normal linear maps on a Hermitian space) Given a Hermitian space E of dimension n, for every normal linear map $f: E \to E$ there is an orthonormal basis $(e_1, \ldots, e_n)$ of eigenvectors of f such that the matrix of f w.r.t. this basis is a diagonal matrix*

$$\begin{pmatrix} \lambda_1 & & \cdots & \\ & \lambda_2 & \cdots & \\ \vdots & \vdots & \ddots & \vdots \\ & & \cdots & \lambda_n \end{pmatrix},$$

where $\lambda_j \in \mathbb{C}$.

Proof. We proceed by induction on the dimension n of E as follows. If $n = 1$, the result is trivial. Assume now that $n \geq 2$. Since $\mathbb{C}$ is algebraically closed (i.e., every polynomial has a root in $\mathbb{C}$), the linear map $f: E \to E$ has some eigenvalue $\lambda \in \mathbb{C}$, and let w be some unit eigenvector for λ. Let W be the subspace of dimension 1 spanned by w. Clearly, $f(W) \subseteq W$. By Proposition 16.3, w is an eigenvector of f^* for $\overline{\lambda}$, and thus $f^*(W) \subseteq W$. By Proposition 16.8, we also have $f(W^\perp) \subseteq W^\perp$. The restriction of f to $W^\perp$ is still normal, and we conclude by applying the induction hypothesis to $W^\perp$ (whose dimension is $n - 1$). $\qquad\square$

Theorem 16.3 implies that (complex) self-adjoint, skew-self-adjoint, and orthogonal linear maps can be diagonalized with respect to an orthonormal basis of eigenvectors. In this latter case, though, an orthogonal map is called a *unitary* map. Proposition 16.5 also shows that the eigenvalues of a self-adjoint linear map are real, and Proposition 16.7 shows that the eigenvalues of a skew self-adjoint map are pure imaginary or zero, and that the eigenvalues of a unitary map have absolute value 1.

Remark: There is a converse to Theorem 16.3, namely, if there is an orthonormal basis $(e_1, \ldots, e_n)$ of eigenvectors of f, then f is normal. We leave the easy proof as an exercise.

In the next section we specialize Theorem 16.2 to self-adjoint, skew-self-adjoint, and orthogonal linear maps. Due to the additional structure, we obtain more precise normal forms.

16.4 Self-Adjoint, Skew-Self-Adjoint, and Orthogonal Linear Maps

We begin with self-adjoint maps.

Theorem 16.4. *Given a Euclidean space E of dimension n, for every self-adjoint linear map $f \colon E \to E$, there is an orthonormal basis $(e_1, \ldots, e_n)$ of eigenvectors of f such that the matrix of f w.r.t. this basis is a diagonal matrix*

$$\begin{pmatrix} \lambda_1 & & \cdots & \\ & \lambda_2 & \cdots & \\ \vdots & \vdots & \ddots & \vdots \\ & & \cdots & \lambda_n \end{pmatrix},$$

where $\lambda_i \in \mathbb{R}$.

Proof. We already proved this; see Theorem 16.1. However, it is instructive to give a more direct method not involving the complexification of $\langle -, - \rangle$ and Proposition 16.5.

Since $\mathbb{C}$ is algebraically closed, $f_{\mathbb{C}}$ has some eigenvalue $\lambda + i\mu$, and let $u + iv$ be some eigenvector of $f_{\mathbb{C}}$ for $\lambda + i\mu$, where $\lambda, \mu \in \mathbb{R}$ and $u, v \in E$. We saw in the proof of Proposition 16.9 that

$$f(u) = \lambda u - \mu v \quad \text{and} \quad f(v) = \mu u + \lambda v.$$

Since $f = f^*$,

$$\langle f(u), v \rangle = \langle u, f(v) \rangle$$

for all $u, v \in E$. Applying this to

$$f(u) = \lambda u - \mu v \quad \text{and} \quad f(v) = \mu u + \lambda v,$$

we get

$$\langle f(u), v \rangle = \langle \lambda u - \mu v, v \rangle = \lambda \langle u, v \rangle - \mu \langle v, v \rangle$$

and

$$\langle u, f(v) \rangle = \langle u, \mu u + \lambda v \rangle = \mu \langle u, u \rangle + \lambda \langle u, v \rangle,$$

and thus we get

$$\lambda \langle u, v \rangle - \mu \langle v, v \rangle = \mu \langle u, u \rangle + \lambda \langle u, v \rangle,$$

that is,

$$\mu(\langle u, u \rangle + \langle v, v \rangle) = 0,$$

which implies $\mu = 0$, since either $u \neq 0$ or $v \neq 0$. Therefore, λ is a real eigenvalue of f.

Now going back to the proof of Theorem 16.2, only the case where $\mu = 0$ applies, and the induction shows that all the blocks are one-dimensional.
$\square$

Theorem 16.4 implies that if $\lambda_1, \ldots, \lambda_p$ are the distinct real eigenvalues of f, and E_i is the eigenspace associated with λ_i, then

$$E = E_1 \oplus \cdots \oplus E_p,$$

where E_i and E_j are orthogonal for all $i \neq j$.

Remark: Another way to prove that a self-adjoint map has a real eigenvalue is to use a little bit of calculus. We learned such a proof from Herman Gluck. The idea is to consider the real-valued function $\Phi \colon E \to \mathbb{R}$ defined such that

$$\Phi(u) = \langle f(u), u \rangle$$

for every $u \in E$. This function is C^∞, and if we represent f by a matrix A over some orthonormal basis, it is easy to compute the gradient vector

$$\nabla \Phi(X) = \left(\frac{\partial \Phi}{\partial x_1}(X), \ldots, \frac{\partial \Phi}{\partial x_n}(X) \right)$$

of Φ at X. Indeed, we find that

$$\nabla \Phi(X) = (A + A^\top)X,$$

where X is a column vector of size n. But since f is self-adjoint, $A = A^\top$, and thus

$$\nabla \Phi(X) = 2AX.$$

The next step is to find the maximum of the function Φ on the sphere

$$S^{n-1} = \{(x_1, \ldots, x_n) \in \mathbb{R}^n \mid x_1^2 + \cdots + x_n^2 = 1\}.$$

Since S^{n-1} is compact and Φ is continuous, and in fact C^∞, Φ takes a maximum at some X on S^{n-1}. But then it is well known that at an extremum X of Φ we must have

$$d\Phi_X(Y) = \langle \nabla \Phi(X), Y \rangle = 0$$

for all tangent vectors Y to S^{n-1} at X, and so $\nabla \Phi(X)$ is orthogonal to the tangent plane at X, which means that

$$\nabla \Phi(X) = \lambda X$$

for some $\lambda \in \mathbb{R}$. Since $\nabla \Phi(X) = 2AX$, we get

$$2AX = \lambda X,$$

and thus $\lambda/2$ is a real eigenvalue of A (i.e., of f).

Next we consider skew-self-adjoint maps.

Theorem 16.5. *Given a Euclidean space E of dimension n, for every skew-self-adjoint linear map $f \colon E \to E$ there is an orthonormal basis $(e_1, \ldots, e_n)$ such that the matrix of f w.r.t. this basis is a block diagonal matrix of the form*

$$\begin{pmatrix} A_1 & & \cdots & \\ & A_2 & \cdots & \\ \vdots & \vdots & \ddots & \vdots \\ & & \cdots & A_p \end{pmatrix}$$

such that each block A_j is either 0 or a two-dimensional matrix of the form

$$A_j = \begin{pmatrix} 0 & -\mu_j \\ \mu_j & 0 \end{pmatrix},$$

where $\mu_j \in \mathbb{R}$, with $\mu_j > 0$. In particular, the eigenvalues of $f_{\mathbb{C}}$ are pure imaginary of the form $\pm i\mu_j$ or 0.

Proof. The case where $n = 1$ is trivial. As in the proof of Theorem 16.2, $f_{\mathbb{C}}$ has some eigenvalue $z = \lambda + i\mu$, where $\lambda, \mu \in \mathbb{R}$. We claim that $\lambda = 0$. First we show that

$$\langle f(w), w \rangle = 0$$

for all $w \in E$. Indeed, since $f = -f^*$, we get

$$\langle f(w), w \rangle = \langle w, f^*(w) \rangle = \langle w, -f(w) \rangle = -\langle w, f(w) \rangle = -\langle f(w), w \rangle,$$

since $\langle -, - \rangle$ is symmetric. This implies that

$$\langle f(w), w \rangle = 0.$$

Applying this to u and v and using the fact that

$$f(u) = \lambda u - \mu v \quad \text{and} \quad f(v) = \mu u + \lambda v,$$

we get

$$0 = \langle f(u), u \rangle = \langle \lambda u - \mu v, u \rangle = \lambda \langle u, u \rangle - \mu \langle u, v \rangle$$

and

$$0 = \langle f(v), v \rangle = \langle \mu u + \lambda v, v \rangle = \mu \langle u, v \rangle + \lambda \langle v, v \rangle,$$

from which, by addition, we get

$$\lambda(\langle v, v \rangle + \langle v, v \rangle) = 0.$$

Since $u \neq 0$ or $v \neq 0$, we have $\lambda = 0$.

Then going back to the proof of Theorem 16.2, unless $\mu = 0$, the case where u and v are orthogonal and span a subspace of dimension 2 applies, and the induction shows that all the blocks are two-dimensional or reduced to 0. □

Remark: One will note that if f is skew-self-adjoint, then $if_{\mathbb{C}}$ is self-adjoint w.r.t. $\langle -, - \rangle_{\mathbb{C}}$. By Proposition 16.5, the map $if_{\mathbb{C}}$ has real eigenvalues, which implies that the eigenvalues of $f_{\mathbb{C}}$ are pure imaginary or 0.

Finally we consider orthogonal linear maps.

Theorem 16.6. *Given a Euclidean space E of dimension n, for every orthogonal linear map $f \colon E \to E$ there is an orthonormal basis $(e_1, \ldots, e_n)$ such that the matrix of f w.r.t. this basis is a block diagonal matrix of the form*

$$\begin{pmatrix} A_1 & & \cdots & \\ & A_2 & \cdots & \\ \vdots & \vdots & \ddots & \vdots \\ & & \cdots & A_p \end{pmatrix}$$

such that each block A_j is either 1, -1, or a two-dimensional matrix of the form

$$A_j = \begin{pmatrix} \cos\theta_j & -\sin\theta_j \\ \sin\theta_j & \cos\theta_j \end{pmatrix}$$

where $0 < \theta_j < \pi$. In particular, the eigenvalues of $f_{\mathbb{C}}$ are of the form $\cos\theta_j \pm i \sin\theta_j$, 1, or -1.

Proof. The case where $n = 1$ is trivial. It is immediately verified that $f \circ f^* = f^* \circ f = \mathrm{id}$ implies that $f_{\mathbb{C}} \circ f_{\mathbb{C}}^* = f_{\mathbb{C}}^* \circ f_{\mathbb{C}} = \mathrm{id}$, so the map $f_{\mathbb{C}}$ is unitary. By Proposition 16.7, the eigenvalues of $f_{\mathbb{C}}$ have absolute value 1. As a consequence, the eigenvalues of $f_{\mathbb{C}}$ are of the form $\cos\theta \pm i \sin\theta$, 1, or -1. The theorem then follows immediately from Theorem 16.2, where the condition $\mu > 0$ implies that $\sin\theta_j > 0$, and thus, $0 < \theta_j < \pi$. □

It is obvious that we can reorder the orthonormal basis of eigenvectors given by Theorem 16.6, so that the matrix of f w.r.t. this basis is a block diagonal matrix of the form

$$\begin{pmatrix} A_1 \cdots & & & \\ \vdots \ddots \vdots & & \vdots & \\ \cdots A_r & & & \\ & & -I_q & \\ \cdots & & & I_p \end{pmatrix}$$

where each block A_j is a two-dimensional rotation matrix $A_j \neq \pm I_2$ of the form

$$A_j = \begin{pmatrix} \cos \theta_j & -\sin \theta_j \\ \sin \theta_j & \cos \theta_j \end{pmatrix}$$

with $0 < \theta_j < \pi$.

The linear map f has an eigenspace $E(1, f) = \mathrm{Ker}\,(f - \mathrm{id})$ of dimension p for the eigenvalue 1, and an eigenspace $E(-1, f) = \mathrm{Ker}\,(f + \mathrm{id})$ of dimension q for the eigenvalue -1. If $\det(f) = +1$ (f is a rotation), the dimension q of $E(-1, f)$ must be even, and the entries in $-I_q$ can be paired to form two-dimensional blocks, if we wish. In this case, every rotation in $\mathbf{SO}(n)$ has a matrix of the form

$$\begin{pmatrix} A_1 \cdots & & \\ \vdots \ddots \vdots & & \\ \cdots A_m & & \\ \cdots & & I_{n-2m} \end{pmatrix}$$

where the first m blocks A_j are of the form

$$A_j = \begin{pmatrix} \cos \theta_j & -\sin \theta_j \\ \sin \theta_j & \cos \theta_j \end{pmatrix}$$

with $0 < \theta_j \leq \pi$.

Theorem 16.6 can be used to prove a version of the Cartan–Dieudonné theorem.

Theorem 16.7. *Let E be a Euclidean space of dimension $n \geq 2$. For every isometry $f \in \mathbf{O}(E)$, if $p = \dim(E(1, f)) = \dim(\mathrm{Ker}\,(f - \mathrm{id}))$, then f is the composition of $n - p$ reflections, and $n - p$ is minimal.*

Proof. From Theorem 16.6 there are r subspaces $F_1, \ldots, F_r$, each of dimension 2, such that

$$E = E(1, f) \oplus E(-1, f) \oplus F_1 \oplus \cdots \oplus F_r,$$

and all the summands are pairwise orthogonal. Furthermore, the restriction r_i of f to each F_i is a rotation $r_i \neq \pm \mathrm{id}$. Each 2D rotation r_i can be written as the composition $r_i = s'_i \circ s_i$ of two reflections s_i and s'_i about lines in F_i (forming an angle $\theta_i / 2$). We can extend s_i and s'_i to hyperplane reflections in E by making them the identity on $F_i^\perp$. Then

$$s'_r \circ s_r \circ \cdots \circ s'_1 \circ s_1$$

agrees with f on $F_1 \oplus \cdots \oplus F_r$ and is the identity on $E(1, f) \oplus E(-1, f)$. If $E(-1, f)$ has an orthonormal basis of eigenvectors $(v_1, \ldots, v_q)$, letting s''_j be the reflection about the hyperplane $(v_j)^\perp$, it is clear that

$$s''_q \circ \cdots \circ s''_1$$

agrees with f on $E(-1, f)$ and is the identity on $E(1, f) \oplus F_1 \oplus \cdots \oplus F_r$. But then

$$f = s''_q \circ \cdots \circ s''_1 \circ s'_r \circ s_r \circ \cdots \circ s'_1 \circ s_1,$$

the composition of $2r + q = n - p$ reflections.

If

$$f = s_t \circ \cdots \circ s_1,$$

for t reflections s_i, it is clear that

$$F = \bigcap_{i=1}^{t} E(1, s_i) \subseteq E(1, f),$$

where $E(1, s_i)$ is the hyperplane defining the reflection s_i. By the Grassmann relation, if we intersect $t \leq n$ hyperplanes, the dimension of their intersection is at least $n - t$. Thus, $n - t \leq p$, that is, $t \geq n - p$, and $n - p$ is the smallest number of reflections composing f. $\qquad \square$

As a corollary of Theorem 16.7, we obtain the following fact: If the dimension n of the Euclidean space E is odd, then every rotation $f \in \mathbf{SO}(E)$ admits 1 as an eigenvalue.

Proof. The characteristic polynomial $\det(XI - f)$ of f has odd degree n and has real coefficients, so it must have some real root λ. Since f is an isometry, its n eigenvalues are of the form, $+1, -1$, and $e^{\pm i\theta}$, with $0 < \theta < \pi$, so $\lambda = \pm 1$. Now the eigenvalues $e^{\pm i\theta}$ appear in conjugate pairs, and since n is odd, the number of real eigenvalues of f is odd. This implies that $+1$ is an eigenvalue of f, since otherwise -1 would be the only real eigenvalue of f, and since its multiplicity is odd, we would have $\det(f) = -1$, contradicting the fact that f is a rotation. $\qquad \square$

When $n = 3$, we obtain the result due to Euler which says that every 3D rotation R has an invariant axis D, and that restricted to the plane orthogonal to D, it is a 2D rotation. Furthermore, if (a, b, c) is a unit vector defining the axis D of the rotation R and if the angle of the rotation is θ, if B is the skew-symmetric matrix

$$B = \begin{pmatrix} 0 & -c & b \\ c & 0 & -a \\ -b & a & 0 \end{pmatrix},$$

then the Rodigues formula (Proposition 11.13) states that

$$R = I + \sin\theta B + (1 - \cos\theta)B^2.$$

The theorems of this section and of the previous section can be immediately translated in terms of matrices. The matrix versions of these theorems is often used in applications so we briefly present them in the section.

16.5 Normal and Other Special Matrices

First we consider real matrices. Recall the following definitions.

Definition 16.3. Given a real $m \times n$ matrix A, the *transpose* $A^\top$ *of* A is the $n \times m$ matrix $A^\top = (a_{ij}^\top)$ defined such that

$$a_{ij}^\top = a_{ji}$$

for all i, j, $1 \leq i \leq m$, $1 \leq j \leq n$. A real $n \times n$ matrix A is

- *normal* if

$$A A^\top = A^\top A,$$

- *symmetric* if

$$A^\top = A,$$

- *skew-symmetric* if

$$A^\top = -A,$$

- *orthogonal* if

$$A A^\top = A^\top A = I_n.$$

Recall from Proposition 11.12 that when E is a Euclidean space and $(e_1, \ldots, e_n)$ is an orthonormal basis for E, if A is the matrix of a linear map $f \colon E \to E$ w.r.t. the basis $(e_1, \ldots, e_n)$, then $A^\top$ is the matrix of the adjoint f^* of f. Consequently, a normal linear map has a normal matrix, a self-adjoint linear map has a symmetric matrix, a skew-self-adjoint linear map has a skew-symmetric matrix, and an orthogonal linear map has an orthogonal matrix.

Furthermore, if $(u_1, \ldots, u_n)$ is another orthonormal basis for E and P is the change of basis matrix whose columns are the components of the u_i w.r.t. the basis $(e_1, \ldots, e_n)$, then P is orthogonal, and for any linear map $f \colon E \to E$, if A is the matrix of f w.r.t. $(e_1, \ldots, e_n)$ and B is the matrix of f w.r.t. $(u_1, \ldots, u_n)$, then

$$B = P^\top A P.$$

As a consequence, Theorems 16.2 and 16.4–16.6 can be restated as follows.

Theorem 16.8. *For every normal matrix A there is an orthogonal matrix P and a block diagonal matrix D such that $A = PD\,P^\top$, where D is of the form*

$$D = \begin{pmatrix} D_1 & & \cdots & \\ & D_2 & \cdots & \\ \vdots & \vdots & \ddots & \vdots \\ & & \cdots & D_p \end{pmatrix}$$

such that each block D_j is either a one-dimensional matrix (i.e., a real scalar) or a two-dimensional matrix of the form

$$D_j = \begin{pmatrix} \lambda_j & -\mu_j \\ \mu_j & \lambda_j \end{pmatrix},$$

where $\lambda_j, \mu_j \in \mathbb{R}$, with $\mu_j > 0$.

Theorem 16.9. *For every symmetric matrix A there is an orthogonal matrix P and a diagonal matrix D such that $A = PD\,P^\top$, where D is of the form*

$$D = \begin{pmatrix} \lambda_1 & & \cdots & \\ & \lambda_2 & \cdots & \\ \vdots & \vdots & \ddots & \vdots \\ & & \cdots & \lambda_n \end{pmatrix},$$

where $\lambda_i \in \mathbb{R}$.

Theorem 16.10. *For every skew-symmetric matrix A there is an orthogonal matrix P and a block diagonal matrix D such that $A = PD\,P^\top$, where D is of the form*

$$D = \begin{pmatrix} D_1 & \cdots & & \\ & D_2 & \cdots & \\ \vdots & \vdots & \ddots & \vdots \\ & & \cdots & D_p \end{pmatrix}$$

such that each block D_j is either 0 or a two-dimensional matrix of the form

$$D_j = \begin{pmatrix} 0 & -\mu_j \\ \mu_j & 0 \end{pmatrix},$$

where $\mu_j \in \mathbb{R}$, with $\mu_j > 0$. In particular, the eigenvalues of A are pure imaginary of the form $\pm i\mu_j$, or 0.

Theorem 16.11. *For every orthogonal matrix A there is an orthogonal matrix P and a block diagonal matrix D such that $A = PD\,P^\top$, where D is of the form*

$$D = \begin{pmatrix} D_1 & \cdots & & \\ & D_2 & \cdots & \\ \vdots & \vdots & \ddots & \vdots \\ & & \cdots & D_p \end{pmatrix}$$

such that each block D_j is either 1, -1, or a two-dimensional matrix of the form

$$D_j = \begin{pmatrix} \cos\theta_j & -\sin\theta_j \\ \sin\theta_j & \cos\theta_j \end{pmatrix}$$

where $0 < \theta_j < \pi$. In particular, the eigenvalues of A are of the form $\cos\theta_j \pm i\sin\theta_j$, 1, or -1.

Theorem 16.11 can be used to show that the exponential map $\exp\colon \mathfrak{so}(n) \to \mathbf{SO}(n)$ is surjective; see Gallier [Gallier (2011b)].

We now consider complex matrices.

Definition 16.4. Given a complex $m \times n$ matrix A, the *transpose* $A^\top$ of A is the $n \times m$ matrix $A^\top = \left(a_{i\,j}^\top\right)$ defined such that

$$a_{i\,j}^\top = a_{j\,i}$$

for all i, j, $1 \le i \le m$, $1 \le j \le n$. The *conjugate* $\overline{A}$ of A is the $m \times n$ matrix $\overline{A} = (b_{i\,j})$ defined such that

$$b_{i\,j} = \overline{a}_{i\,j}$$

for all i, j, $1 \leq i \leq m$, $1 \leq j \leq n$. Given an $m \times n$ complex matrix A, the *adjoint* A^* *of* A is the matrix defined such that

$$A^* = \overline{(A^\top)} = (\overline{A})^\top.$$

A complex $n \times n$ matrix A is

- *normal* if

$$AA^* = A^*A,$$

- *Hermitian* if

$$A^* = A,$$

- *skew-Hermitian* if

$$A^* = -A,$$

- *unitary* if

$$AA^* = A^*A = I_n.$$

Recall from Proposition 13.14 that when E is a Hermitian space and $(e_1, \ldots, e_n)$ is an orthonormal basis for E, if A is the matrix of a linear map $f\colon E \to E$ w.r.t. the basis $(e_1, \ldots, e_n)$, then A^* is the matrix of the adjoint f^* of f. Consequently, a normal linear map has a normal matrix, a self-adjoint linear map has a Hermitian matrix, a skew-self-adjoint linear map has a skew-Hermitian matrix, and a unitary linear map has a unitary matrix.

Furthermore, if $(u_1, \ldots, u_n)$ is another orthonormal basis for E and P is the change of basis matrix whose columns are the components of the u_i w.r.t. the basis $(e_1, \ldots, e_n)$, then P is unitary, and for any linear map $f\colon E \to E$, if A is the matrix of f w.r.t. $(e_1, \ldots, e_n)$ and B is the matrix of f w.r.t. $(u_1, \ldots, u_n)$, then

$$B = P^*AP.$$

Theorem 16.3 and Proposition 16.7 can be restated in terms of matrices as follows.

Theorem 16.12. *For every complex normal matrix A there is a unitary matrix U and a diagonal matrix D such that $A = UDU^*$. Furthermore, if A is Hermitian, then D is a real matrix; if A is skew-Hermitian, then the entries in D are pure imaginary or zero; and if A is unitary, then the entries in D have absolute value 1.*

16.6 Rayleigh–Ritz Theorems and Eigenvalue Interlacing

A fact that is used frequently in optimization problems is that the eigenvalues of a symmetric matrix are characterized in terms of what is known as the *Rayleigh ratio*, defined by

$$R(A)(x) = \frac{x^\top A x}{x^\top x}, \quad x \in \mathbb{R}^n, x \neq 0.$$

The following proposition is often used to prove the correctness of various optimization or approximation problems (for example PCA; see Section 21.4). It is also used to prove Proposition 16.13, which is used to justify the correctness of a method for graph-drawing (see Chapter 19).

Proposition 16.11. *(Rayleigh–Ritz) If A is a symmetric $n \times n$ matrix with eigenvalues $\lambda_1 \leq \lambda_2 \leq \cdots \leq \lambda_n$ and if $(u_1, \ldots, u_n)$ is any orthonormal basis of eigenvectors of A, where u_i is a unit eigenvector associated with λ_i, then*

$$\max_{x \neq 0} \frac{x^\top A x}{x^\top x} = \lambda_n$$

(with the maximum attained for $x = u_n$), and

$$\max_{x \neq 0, x \in \{u_{n-k+1}, \ldots, u_n\}^\perp} \frac{x^\top A x}{x^\top x} = \lambda_{n-k}$$

(with the maximum attained for $x = u_{n-k}$), where $1 \leq k \leq n-1$. Equivalently, if V_k is the subspace spanned by $(u_1, \ldots, u_k)$, then

$$\lambda_k = \max_{x \neq 0, x \in V_k} \frac{x^\top A x}{x^\top x}, \quad k = 1, \ldots, n.$$

Proof. First observe that

$$\max_{x \neq 0} \frac{x^\top A x}{x^\top x} = \max_x \{x^\top A x \mid x^\top x = 1\},$$

and similarly,

$$\max_{x \neq 0, x \in \{u_{n-k+1}, \ldots, u_n\}^\perp} \frac{x^\top A x}{x^\top x}$$
$$= \max_x \left\{ x^\top A x \mid (x \in \{u_{n-k+1}, \ldots, u_n\}^\perp) \wedge (x^\top x = 1) \right\}.$$

Since A is a symmetric matrix, its eigenvalues are real and it can be diagonalized with respect to an orthonormal basis of eigenvectors, so let $(u_1, \ldots, u_n)$ be such a basis. If we write

$$x = \sum_{i=1}^n x_i u_i,$$

a simple computation shows that

$$x^\top A x = \sum_{i=1}^{n} \lambda_i x_i^2.$$

If $x^\top x = 1$, then $\sum_{i=1}^{n} x_i^2 = 1$, and since we assumed that $\lambda_1 \leq \lambda_2 \leq \cdots \leq \lambda_n$, we get

$$x^\top A x = \sum_{i=1}^{n} \lambda_i x_i^2 \leq \lambda_n \left(\sum_{i=1}^{n} x_i^2 \right) = \lambda_n.$$

Thus,

$$\max_x \left\{ x^\top A x \mid x^\top x = 1 \right\} \leq \lambda_n,$$

and since this maximum is achieved for $e_n = (0, 0, \ldots, 1)$, we conclude that

$$\max_x \left\{ x^\top A x \mid x^\top x = 1 \right\} = \lambda_n.$$

Next observe that $x \in \{u_{n-k+1}, \ldots, u_n\}^\perp$ and $x^\top x = 1$ iff $x_{n-k+1} = \cdots = x_n = 0$ and $\sum_{i=1}^{n-k} x_i^2 = 1$. Consequently, for such an x, we have

$$x^\top A x = \sum_{i=1}^{n-k} \lambda_i x_i^2 \leq \lambda_{n-k} \left(\sum_{i=1}^{n-k} x_i^2 \right) = \lambda_{n-k}.$$

Thus,

$$\max_x \left\{ x^\top A x \mid (x \in \{u_{n-k+1}, \ldots, u_n\}^\perp) \wedge (x^\top x = 1) \right\} \leq \lambda_{n-k},$$

and since this maximum is achieved for $e_{n-k} = (0, \ldots, 0, 1, 0, \ldots, 0)$ with a 1 in position $n - k$, we conclude that

$$\max_x \left\{ x^\top A x \mid (x \in \{u_{n-k+1}, \ldots, u_n\}^\perp) \wedge (x^\top x = 1) \right\} = \lambda_{n-k},$$

as claimed. $\square$

For our purposes we need the version of Proposition 16.11 applying to min instead of max, whose proof is obtained by a trivial modification of the proof of Proposition 16.11.

Proposition 16.12. *(Rayleigh–Ritz) If A is a symmetric $n \times n$ matrix with eigenvalues $\lambda_1 \leq \lambda_2 \leq \cdots \leq \lambda_n$ and if $(u_1, \ldots, u_n)$ is any orthonormal basis of eigenvectors of A, where u_i is a unit eigenvector associated with λ_i, then*

$$\min_{x \neq 0} \frac{x^\top A x}{x^\top x} = \lambda_1$$

(with the minimum attained for $x = u_1$), and

$$\min_{x \neq 0, x \in \{u_1, \ldots, u_{i-1}\}^{\perp}} \frac{x^{\top} A x}{x^{\top} x} = \lambda_i$$

(with the minimum attained for $x = u_i$), where $2 \leq i \leq n$. Equivalently, if $W_k = V_{k-1}^{\perp}$ denotes the subspace spanned by $(u_k, \ldots, u_n)$ (with $V_0 = (0)$), then

$$\lambda_k = \min_{x \neq 0, x \in W_k} \frac{x^{\top} A x}{x^{\top} x} = \min_{x \neq 0, x \in V_{k-1}^{\perp}} \frac{x^{\top} A x}{x^{\top} x}, \quad k = 1, \ldots, n.$$

Propositions 16.11 and 16.12 together are known the *Rayleigh–Ritz theorem*.

As an application of Propositions 16.11 and 16.12, we prove a proposition which allows us to compare the eigenvalues of two symmetric matrices A and $B = R^{\top} A R$, where R is a rectangular matrix satisfying the equation $R^{\top} R = I$.

First we need a definition.

Definition 16.5. Given an $n \times n$ symmetric matrix A and an $m \times m$ symmetric B, with $m \leq n$, if $\lambda_1 \leq \lambda_2 \leq \cdots \leq \lambda_n$ are the eigenvalues of A and $\mu_1 \leq \mu_2 \leq \cdots \leq \mu_m$ are the eigenvalues of B, then we say that the eigenvalues of B *interlace* the eigenvalues of A if

$$\lambda_i \leq \mu_i \leq \lambda_{n-m+i}, \quad i = 1, \ldots, m.$$

For example, if $n = 5$ and $m = 3$, we have

$$\lambda_1 \leq \mu_1 \leq \lambda_3$$
$$\lambda_2 \leq \mu_2 \leq \lambda_4$$
$$\lambda_3 \leq \mu_3 \leq \lambda_5.$$

Proposition 16.13. *Let A be an $n \times n$ symmetric matrix, R be an $n \times m$ matrix such that $R^{\top} R = I$ (with $m \leq n$), and let $B = R^{\top} A R$ (an $m \times m$ matrix). The following properties hold:*

(a) The eigenvalues of B interlace the eigenvalues of A.

(b) If $\lambda_1 \leq \lambda_2 \leq \cdots \leq \lambda_n$ are the eigenvalues of A and $\mu_1 \leq \mu_2 \leq \cdots \leq \mu_m$ are the eigenvalues of B, and if $\lambda_i = \mu_i$, then there is an eigenvector v of B with eigenvalue μ_i such that Rv is an eigenvector of A with eigenvalue λ_i.

Proof. (a) Let $(u_1, \ldots, u_n)$ be an orthonormal basis of eigenvectors for A, and let $(v_1, \ldots, v_m)$ be an orthonormal basis of eigenvectors for B. Let U_j be the subspace spanned by $(u_1, \ldots, u_j)$ and let V_j be the subspace spanned by $(v_1, \ldots, v_j)$. For any i, the subspace V_i has dimension i and the subspace $R^\top U_{i-1}$ has dimension at most $i - 1$. Therefore, there is some nonzero vector $v \in V_i \cap (R^\top U_{i-1})^\perp$, and since

$$v^\top R^\top u_j = (Rv)^\top u_j = 0, \quad j = 1, \ldots, i - 1,$$

we have $Rv \in (U_{i-1})^\perp$. By Proposition 16.12 and using the fact that $R^\top R = I$, we have

$$\lambda_i \leq \frac{(Rv)^\top A Rv}{(Rv)^\top Rv} = \frac{v^\top B v}{v^\top v}.$$

On the other hand, by Proposition 16.11,

$$\mu_i = \max_{x \neq 0, x \in \{v_{i+1}, \ldots, v_n\}^\perp} \frac{x^\top B x}{x^\top x} = \max_{x \neq 0, x \in \{v_1, \ldots, v_i\}} \frac{x^\top B x}{x^\top x},$$

so

$$\frac{w^\top B w}{w^\top w} \leq \mu_i \quad \text{for all } w \in V_i,$$

and since $v \in V_i$, we have

$$\lambda_i \leq \frac{v^\top B v}{v^\top v} \leq \mu_i, \quad i = 1, \ldots, m.$$

We can apply the same argument to the symmetric matrices $-A$ and $-B$, to conclude that

$$-\lambda_{n-m+i} \leq -\mu_i,$$

that is,

$$\mu_i \leq \lambda_{n-m+i}, \quad i = 1, \ldots, m.$$

Therefore,

$$\lambda_i \leq \mu_i \leq \lambda_{n-m+i}, \quad i = 1, \ldots, m,$$

as desired.

(b) If $\lambda_i = \mu_i$, then

$$\lambda_i = \frac{(Rv)^\top A Rv}{(Rv)^\top Rv} = \frac{v^\top B v}{v^\top v} = \mu_i,$$

so v must be an eigenvector for B and Rv must be an eigenvector for A, both for the eigenvalue $\lambda_i = \mu_i$. $\qquad\square$

Proposition 16.13 immediately implies the *Poincaré separation theorem*. It can be used in situations, such as in quantum mechanics, where one has information about the inner products $u_i^\top A u_j$.

Proposition 16.14. *(Poincaré separation theorem) Let A be a $n \times n$ symmetric (or Hermitian) matrix, let r be some integer with $1 \leq r \leq n$, and let $(u_1, \ldots, u_r)$ be r orthonormal vectors. Let $B = (u_i^\top A u_j)$ (an $r \times r$ matrix), let $\lambda_1(A) \leq \ldots \leq \lambda_n(A)$ be the eigenvalues of A and $\lambda_1(B) \leq \ldots \leq \lambda_r(B)$ be the eigenvalues of B; then we have*

$$\lambda_k(A) \leq \lambda_k(B) \leq \lambda_{k+n-r}(A), \quad k = 1, \ldots, r.$$

Observe that Proposition 16.13 implies that

$$\lambda_1 + \cdots + \lambda_m \leq \operatorname{tr}(R^\top A R) \leq \lambda_{n-m+1} + \cdots + \lambda_n.$$

If P_1 is the the $n \times (n-1)$ matrix obtained from the identity matrix by dropping its last column, we have $P_1^\top P_1 = I$, and the matrix $B = P_1^\top A P_1$ is the matrix obtained from A by deleting its last row and its last column. In this case the interlacing result is

$$\lambda_1 \leq \mu_1 \leq \lambda_2 \leq \mu_2 \leq \cdots \leq \mu_{n-2} \leq \lambda_{n-1} \leq \mu_{n-1} \leq \lambda_n,$$

a genuine interlacing. We obtain similar results with the matrix P_{n-r} obtained by dropping the last $n-r$ columns of the identity matrix and setting $B = P_{n-r}^\top A P_{n-r}$ (B is the $r \times r$ matrix obtained from A by deleting its last $n-r$ rows and columns). In this case we have the following interlacing inequalities known as *Cauchy interlacing theorem*:

$$\lambda_k \leq \mu_k \leq \lambda_{k+n-r}, \quad k = 1, \ldots, r. \tag{16.2}$$

16.7 The Courant–Fischer Theorem; Perturbation Results

Another useful tool to prove eigenvalue equalities is the Courant–Fischer characterization of the eigenvalues of a symmetric matrix, also known as the Min-max (and Max-min) theorem.

Theorem 16.13. *(Courant–Fischer) Let A be a symmetric $n \times n$ matrix with eigenvalues $\lambda_1 \leq \lambda_2 \leq \cdots \leq \lambda_n$. If $\mathcal{V}_k$ denotes the set of subspaces of $\mathbb{R}^n$ of dimension k, then*

$$\lambda_k = \max_{W \in \mathcal{V}_{n-k+1}} \min_{x \in W, x \neq 0} \frac{x^\top A x}{x^\top x}$$

$$\lambda_k = \min_{W \in \mathcal{V}_k} \max_{x \in W, x \neq 0} \frac{x^\top A x}{x^\top x}.$$

Proof. Let us consider the second equality, the proof of the first equality being similar. Let $(u_1, \ldots, u_n)$ be any orthonormal basis of eigenvectors of A, where u_i is a unit eigenvector associated with λ_i. Observe that the space V_k spanned by $(u_1, \ldots, u_k)$ has dimension k, and by Proposition 16.11, we have

$$\lambda_k = \max_{x \neq 0, x \in V_k} \frac{x^\top A x}{x^\top x} \geq \min_{W \in \mathcal{V}_k} \max_{x \in W, x \neq 0} \frac{x^\top A x}{x^\top x}.$$

Therefore, we need to prove the reverse inequality; that is, we have to show that

$$\lambda_k \leq \max_{x \neq 0, x \in W} \frac{x^\top A x}{x^\top x}, \quad \text{for all} \quad W \in \mathcal{V}_k.$$

Now for any $W \in \mathcal{V}_k$, if we can prove that $W \cap V_{k-1}^{\perp} \neq (0)$, then for any nonzero $v \in W \cap V_{k-1}^{\perp}$, by Proposition 16.12, we have

$$\lambda_k = \min_{x \neq 0, x \in V_{k-1}^{\perp}} \frac{x^\top A x}{x^\top x} \leq \frac{v^\top A v}{v^\top v} \leq \max_{x \in W, x \neq 0} \frac{x^\top A x}{x^\top x}.$$

It remains to prove that $\dim(W \cap V_{k-1}^{\perp}) \geq 1$. However, $\dim(V_{k-1}) = k-1$, so $\dim(V_{k-1}^{\perp}) = n - k + 1$, and by hypothesis $\dim(W) = k$. By the Grassmann relation,

$$\dim(W) + \dim(V_{k-1}^{\perp}) = \dim(W \cap V_{k-1}^{\perp}) + \dim(W + V_{k-1}^{\perp}),$$

and since $\dim(W + V_{k-1}^{\perp}) \leq \dim(\mathbb{R}^n) = n$, we get

$$k + n - k + 1 \leq \dim(W \cap V_{k-1}^{\perp}) + n;$$

that is, $1 \leq \dim(W \cap V_{k-1}^{\perp})$, as claimed. $\qquad \square$

The Courant–Fischer theorem yields the following useful result about perturbing the eigenvalues of a symmetric matrix due to Hermann Weyl.

Proposition 16.15. *Given two $n \times n$ symmetric matrices A and $B = A + \Delta A$, if $\alpha_1 \leq \alpha_2 \leq \cdots \leq \alpha_n$ are the eigenvalues of A and $\beta_1 \leq \beta_2 \leq \cdots \leq \beta_n$ are the eigenvalues of B, then*

$$|\alpha_k - \beta_k| \leq \rho(\Delta A) \leq \|\Delta A\|_2, \quad k = 1, \ldots, n.$$

Proof. Let $\mathcal{V}_k$ be defined as in the Courant–Fischer theorem and let V_k be the subspace spanned by the k eigenvectors associated with $\lambda_1, \ldots, \lambda_k$. By the Courant–Fischer theorem applied to B, we have

$$\beta_k = \min_{W \in \mathcal{V}_k} \max_{x \in W, x \neq 0} \frac{x^\top B x}{x^\top x}$$

$$\leq \max_{x \in V_k} \frac{x^\top B x}{x^\top x}$$

$$= \max_{x \in V_k} \left(\frac{x^\top A x}{x^\top x} + \frac{x^\top \Delta A x}{x^\top x} \right)$$

$$\leq \max_{x \in V_k} \frac{x^\top A x}{x^\top x} + \max_{x \in V_k} \frac{x^\top \Delta A x}{x^\top x}.$$

By Proposition 16.11, we have

$$\alpha_k = \max_{x \in V_k} \frac{x^\top A x}{x^\top x},$$

so we obtain

$$\beta_k \leq \max_{x \in V_k} \frac{x^\top A x}{x^\top x} + \max_{x \in V_k} \frac{x^\top \Delta A x}{x^\top x}$$

$$= \alpha_k + \max_{x \in V_k} \frac{x^\top \Delta A x}{x^\top x}$$

$$\leq \alpha_k + \max_{x \in \mathbb{R}^n} \frac{x^\top \Delta A x}{x^\top x}.$$

Now by Proposition 16.11 and Proposition 8.6, we have

$$\max_{x \in \mathbb{R}^n} \frac{x^\top \Delta A x}{x^\top x} = \max_i \lambda_i(\Delta A) \leq \rho(\Delta A) \leq \|\Delta A\|_2,$$

where $\lambda_i(\Delta A)$ denotes the ith eigenvalue of ΔA, which implies that

$$\beta_k \leq \alpha_k + \rho(\Delta A) \leq \alpha_k + \|\Delta A\|_2.$$

By exchanging the roles of A and B, we also have

$$\alpha_k \leq \beta_k + \rho(\Delta A) \leq \beta_k + \|\Delta A\|_2,$$

and thus,

$$|\alpha_k - \beta_k| \leq \rho(\Delta A) \leq \|\Delta A\|_2, \quad k = 1, \ldots, n,$$

as claimed. $\square$

Proposition 16.15 also holds for Hermitian matrices.

A pretty result of Wielandt and Hoffman asserts that

$$\sum_{k=1}^{n} (\alpha_k - \beta_k)^2 \leq \|\Delta A\|_F^2,$$

where $\| \ \|_F$ is the Frobenius norm. However, the proof is significantly harder than the above proof; see Lax [Lax (2007)].

The Courant–Fischer theorem can also be used to prove some famous inequalities due to Hermann Weyl. These can also be viewed as perturbation results. Given two symmetric (or Hermitian) matrices A and B, let $\lambda_i(A), \lambda_i(B)$, and $\lambda_i(A + B)$ denote the ith eigenvalue of A, B, and $A + B$, respectively, arranged in nondecreasing order.

Proposition 16.16. *(Weyl) Given two symmetric (or Hermitian) $n \times n$ matrices A and B, the following inequalities hold: For all i, j, k with $1 \leq i, j, k \leq n$:*

(1) If $i + j = k + 1$, then

$$\lambda_i(A) + \lambda_j(B) \leq \lambda_k(A + B).$$

(2) If $i + j = k + n$, then

$$\lambda_k(A + B) \leq \lambda_i(A) + \lambda_j(B).$$

Proof. Observe that the first set of inequalities is obtained form the second set by replacing A by $-A$ and B by $-B$, so it is enough to prove the second set of inequalities. By the Courant–Fischer theorem, there is a subspace H of dimension $n - k + 1$ such that

$$\lambda_k(A + B) = \min_{x \in H, x \neq 0} \frac{x^\top (A + B)x}{x^\top x}.$$

Similarly, there exists a subspace F of dimension i and a subspace G of dimension j such that

$$\lambda_i(A) = \max_{x \in F, x \neq 0} \frac{x^\top A x}{x^\top x}, \quad \lambda_j(B) = \max_{x \in G, x \neq 0} \frac{x^\top B x}{x^\top x}.$$

We claim that $F \cap G \cap H \neq (0)$. To prove this, we use the Grassmann relation twice. First,

$$\dim(F \cap G \cap H) = \dim(F) + \dim(G \cap H) - \dim(F + (G \cap H))$$
$$\geq \dim(F) + \dim(G \cap H) - n,$$

and second,

$$\dim(G \cap H) = \dim(G) + \dim(H) - \dim(G + H) \geq \dim(G) + \dim(H) - n,$$

so

$$\dim(F \cap G \cap H) \geq \dim(F) + \dim(G) + \dim(H) - 2n.$$

However,

$$\dim(F) + \dim(G) + \dim(H) = i + j + n - k + 1$$

and $i + j = k + n$, so we have

$$\dim(F \cap G \cap H) \geq i + j + n - k + 1 - 2n = k + n + n - k + 1 - 2n = 1,$$

which shows that $F \cap G \cap H \neq (0)$. Then for any unit vector $z \in F \cap G \cap H \neq (0)$, we have

$$\lambda_k(A + B) \leq z^\top(A + B)z, \quad \lambda_i(A) \geq z^\top Az, \quad \lambda_j(B) \geq z^\top Bz,$$

establishing the desired inequality $\lambda_k(A + B) \leq \lambda_i(A) + \lambda_j(B)$. $\qquad\square$

In the special case $i = j = k$, we obtain

$$\lambda_1(A) + \lambda_1(B) \leq \lambda_1(A + B), \quad \lambda_n(A + B) \leq \lambda_n(A) + \lambda_n(B).$$

It follows that λ_1 (as a function) is concave, while λ_n (as a function) is convex.

If $i = 1$ and $j = k$, we obtain

$$\lambda_1(A) + \lambda_k(B) \leq \lambda_k(A + B),$$

and if $i = k$ and $j = n$, we obtain

$$\lambda_k(A + B) \leq \lambda_k(A) + \lambda_n(B),$$

and combining them, we get

$$\lambda_1(A) + \lambda_k(B) \leq \lambda_k(A + B) \leq \lambda_k(A) + \lambda_n(B).$$

In particular, if B is positive semidefinite, since its eigenvalues are nonnegative, we obtain the following inequality known as the *monotonicity theorem* for symmetric (or Hermitian) matrices: if A and B are symmetric (or Hermitian) and B is positive semidefinite, then

$$\lambda_k(A) \leq \lambda_k(A + B) \quad k = 1, \ldots, n.$$

The reader is referred to Horn and Johnson [Horn and Johnson (1990)] (Chapters 4 and 7) for a very complete treatment of matrix inequalities and interlacing results, and also to Lax [Lax (2007)] and Serre [Serre (2010)].

16.8 Summary

The main concepts and results of this chapter are listed below:

- *Normal* linear maps, *self-adjoint* linear maps, *skew-self-adjoint* linear maps, and *orthogonal* linear maps.
- Properties of the eigenvalues and eigenvectors of a normal linear map.
- The *complexification* of a real vector space, of a linear map, and of a Euclidean inner product.
- The eigenvalues of a self-adjoint map in a Hermitian space are *real*.
- The eigenvalues of a self-adjoint map in a Euclidean space are *real*.
- Every self-adjoint linear map on a Euclidean space has an orthonormal basis of eigenvectors.
- Every normal linear map on a Euclidean space can be block diagonalized (blocks of size at most 2×2) with respect to an orthonormal basis of eigenvectors.
- Every normal linear map on a Hermitian space can be diagonalized with respect to an orthonormal basis of eigenvectors.
- The spectral theorems for self-adjoint, skew-self-adjoint, and orthogonal linear maps (on a Euclidean space).
- The spectral theorems for normal, symmetric, skew-symmetric, and orthogonal (real) matrices.
- The spectral theorems for normal, Hermitian, skew-Hermitian, and unitary (complex) matrices.
- The *Rayleigh ratio* and the *Rayleigh–Ritz theorem*.
- *Interlacing inequalities* and the *Cauchy interlacing theorem*.
- The *Poincaré separation theorem*.
- The *Courant–Fischer theorem*.
- Inequalities involving perturbations of the eigenvalues of a symmetric matrix.
- The *Weyl inequalities*.

16.9 Problems

Problem 16.1. Prove that the structure $E_{\mathbb{C}}$ introduced in Definition 16.2 is indeed a complex vector space.

Problem 16.2. Prove that the formula

$$\langle u_1 + iv_1, u_2 + iv_2 \rangle_{\mathbb{C}} = \langle u_1, u_2 \rangle + \langle v_1, v_2 \rangle + i(\langle v_1, u_2 \rangle - \langle u_1, v_2 \rangle)$$

defines a Hermitian form on $E_{\mathbb{C}}$ that is positive definite and that $\langle -, - \rangle_{\mathbb{C}}$ agrees with $\langle -, - \rangle$ on real vectors.

Problem 16.3. Given any linear map $f \colon E \to E$, prove the map $f_{\mathbb{C}}^*$ defined such that
$$f_{\mathbb{C}}^*(u + iv) = f^*(u) + i f^*(v)$$
for all $u, v \in E$ is the adjoint of $f_{\mathbb{C}}$ w.r.t. $\langle -, - \rangle_{\mathbb{C}}$.

Problem 16.4. Let A be a real symmetric $n \times n$ matrix whose eigenvalues are nonnegative. Prove that for every $p > 0$, there is a real symmetric matrix S whose eigenvalues are nonnegative such that $S^p = A$.

Problem 16.5. Let A be a real symmetric $n \times n$ matrix whose eigenvalues are positive.

(1) Prove that there is a real symmetric matrix S such that $A = e^S$.

(2) Let S be a real symmetric $n \times n$ matrix. Prove that $A = e^S$ is a real symmetric $n \times n$ matrix whose eigenvalues are positive.

Problem 16.6. Let A be a complex matrix. Prove that if A can be diagonalized with respect to an orthonormal basis, then A is normal.

Problem 16.7. Let $f \colon \mathbb{C}^n \to \mathbb{C}^n$ be a linear map.

(1) Prove that if f is diagonalizable and if $\lambda_1, \ldots, \lambda_n$ are the eigenvalues of f, then $\lambda_1^2, \ldots, \lambda_n^2$ are the eigenvalues of f^2, and if $\lambda_i^2 = \lambda_j^2$ implies that $\lambda_i = \lambda_j$, then f and f^2 have the same eigenspaces.

(2) Let f and g be two real self-adjoint linear maps $f, g \colon \mathbb{R}^n \to \mathbb{R}^n$. Prove that if f and g have nonnegative eigenvalues (f and g are positive semidefinite) and if $f^2 = g^2$, then $f = g$.

Problem 16.8. (1) Let $\mathfrak{so}(3)$ be the space of 3×3 skew symmetric matrices
$$\mathfrak{so}(3) = \left\{ \begin{pmatrix} 0 & -c & b \\ c & 0 & -a \\ -b & a & 0 \end{pmatrix} \;\middle|\; a, b, c \in \mathbb{R} \right\}.$$

For any matrix
$$A = \begin{pmatrix} 0 & -c & b \\ c & 0 & -a \\ -b & a & 0 \end{pmatrix} \in \mathfrak{so}(3),$$

if we let $\theta = \sqrt{a^2 + b^2 + c^2}$, recall from Section 11.7 (the Rodrigues formula) that the exponential map $\exp \colon \mathfrak{so}(3) \to \mathbf{SO}(3)$ is given by
$$e^A = I_3 + \frac{\sin \theta}{\theta} A + \frac{(1 - \cos \theta)}{\theta^2} A^2, \quad \text{if } \theta \neq 0,$$

with $\exp(0_3) = I_3$.

(2) Prove that e^A is an orthogonal matrix of determinant $+1$, i.e., a rotation matrix.

(3) Prove that the exponential map $\exp\colon \mathfrak{so}(3) \to \mathbf{SO}(3)$ is surjective. For this proceed as follows: Pick any rotation matrix $R \in \mathbf{SO}(3)$;

(1) The case $R = I$ is trivial.

(2) If $R \neq I$ and $\operatorname{tr}(R) \neq -1$, then

$$\exp^{-1}(R) = \left\{ \frac{\theta}{2\sin\theta}(R - R^T) \;\middle|\; 1 + 2\cos\theta = \operatorname{tr}(R) \right\}.$$

(Recall that $\operatorname{tr}(R) = r_{11} + r_{22} + r_{33}$, the *trace* of the matrix R).

Show that there is a unique skew-symmetric B with corresponding θ satisfying $0 < \theta < \pi$ such that $e^B = R$.

(3) If $R \neq I$ and $\operatorname{tr}(R) = -1$, then prove that the eigenvalues of R are $1, -1, -1$, that $R = R^T$, and that $R^2 = I$. Prove that the matrix

$$S = \frac{1}{2}(R - I)$$

is a symmetric matrix whose eigenvalues are $-1, -1, 0$. Thus S can be diagonalized with respect to an orthogonal matrix Q as

$$S = Q \begin{pmatrix} -1 & 0 & 0 \\ 0 & -1 & 0 \\ 0 & 0 & 0 \end{pmatrix} Q^T.$$

Prove that there exists a skew symmetric matrix

$$U = \begin{pmatrix} 0 & -d & c \\ d & 0 & -b \\ -c & b & 0 \end{pmatrix}$$

so that

$$U^2 = S = \frac{1}{2}(R - I).$$

Observe that

$$U^2 = \begin{pmatrix} -(c^2 + d^2) & bc & bd \\ bc & -(b^2 + d^2) & cd \\ bd & cd & -(b^2 + c^2) \end{pmatrix},$$

and use this to conclude that if $U^2 = S$, then $b^2 + c^2 + d^2 = 1$. Then show that

$$\exp^{-1}(R) = \left\{ (2k+1)\pi \begin{pmatrix} 0 & -d & c \\ d & 0 & -b \\ -c & b & 0 \end{pmatrix}, \; k \in \mathbb{Z} \right\},$$

where (b, c, d) is any unit vector such that for the corresponding skew symmetric matrix U, we have $U^2 = S$.

(4) To find a skew symmetric matrix U so that $U^2 = S = \frac{1}{2}(R - I)$ as in (3), we can solve the system

$$\begin{pmatrix} b^2 - 1 & bc & bd \\ bc & c^2 - 1 & cd \\ bd & cd & d^2 - 1 \end{pmatrix} = S.$$

We immediately get b^2, c^2, d^2, and then, since one of b, c, d is nonzero, say b, if we choose the positive square root of b^2, we can determine c and d from bc and bd.

Implement a computer program in Matlab to solve the above system.

Problem 16.9. It was shown in Proposition 14.10 that the exponential map is a map $\exp\colon \mathfrak{so}(n) \to \mathbf{SO}(n)$, where $\mathfrak{so}(n)$ is the vector space of real $n \times n$ skew-symmetric matrices. Use the spectral theorem to prove that the map $\exp\colon \mathfrak{so}(n) \to \mathbf{SO}(n)$ is surjective.

Problem 16.10. Let $\mathfrak{u}(n)$ be the space of (complex) $n \times n$ skew-Hermitian matrices $(B^* = -B)$ and let $\mathfrak{su}(n)$ be its subspace consisting of skew-Hermitian matrice with zero trace $(\operatorname{tr}(B) = 0)$.

(1) Prove that if $B \in \mathfrak{u}(n)$, then $e^B \in \mathbf{U}(n)$, and if $B \in \mathfrak{su}(n)$, then $e^B \in \mathbf{SU}(n)$. Thus we have well-defined maps $\exp\colon \mathfrak{u}(n) \to \mathbf{U}(n)$ and $\exp\colon \mathfrak{su}(n) \to \mathbf{SU}(n)$.

(2) Prove that the map $\exp\colon \mathfrak{u}(n) \to \mathbf{U}(n)$ is surjective.

(3) Prove that the map $\exp\colon \mathfrak{su}(n) \to \mathbf{SU}(n)$ is surjective.

Problem 16.11. Recall that a matrix $B \in \mathrm{M}_n(\mathbb{R})$ is skew-symmetric if $B^\top = -B$. Check that the set $\mathfrak{so}(n)$ of skew-symmetric matrices is a vector space of dimension $n(n-1)/2$, and thus is isomorphic to $\mathbb{R}^{n(n-1)/2}$.

(1) Given a rotation matrix

$$R = \begin{pmatrix} \cos\theta & -\sin\theta \\ \sin\theta & \cos\theta \end{pmatrix},$$

where $0 < \theta < \pi$, prove that there is a skew symmetric matrix B such that

$$R = (I - B)(I + B)^{-1}.$$

(2) Prove that the eigenvalues of a skew-symmetric matrix are either 0 or pure imaginary (that is, of the form $i\mu$ for $\mu \in \mathbb{R}$).

Let $C\colon \mathfrak{so}(n) \to \mathrm{M}_n(\mathbb{R})$ be the function (called the *Cayley transform* of B) given by

$$C(B) = (I - B)(I + B)^{-1}.$$

Prove that if B is skew-symmetric, then $I - B$ and $I + B$ are invertible, and so C is well-defined. Prove that

$$(I + B)(I - B) = (I - B)(I + B),$$

and that

$$(I + B)(I - B)^{-1} = (I - B)^{-1}(I + B).$$

Prove that

$$(C(B))^\top C(B) = I$$

and that

$$\det C(B) = +1,$$

so that $C(B)$ is a rotation matrix. Furthermore, show that $C(B)$ does not admit -1 as an eigenvalue.

(3) Let $\mathbf{SO}(n)$ be the group of $n \times n$ rotation matrices. Prove that the map

$$C \colon \mathfrak{so}(n) \to \mathbf{SO}(n)$$

is bijective onto the subset of rotation matrices that do not admit -1 as an eigenvalue. Show that the inverse of this map is given by

$$B = (I + R)^{-1}(I - R) = (I - R)(I + R)^{-1},$$

where $R \in \mathbf{SO}(n)$ does not admit -1 as an eigenvalue.

Problem 16.12. Please refer back to Problem 3.6. Let $\lambda_1, \ldots, \lambda_n$ be the eigenvalues of A (not necessarily distinct). Using Schur's theorem, A is similar to an upper triangular matrix B, that is, $A = PBP^{-1}$ with B upper triangular, and we may assume that the diagonal entries of B in descending order are $\lambda_1, \ldots, \lambda_n$.

(1) If the E_{ij} are listed according to total order given by

$$(i, j) < (h, k) \quad \text{iff} \quad \begin{cases} i = h \text{ and } j > k \\ \text{or } i < h. \end{cases}$$

prove that R_B is an upper triangular matrix whose diagonal entries are

$$\underbrace{(\lambda_n, \ldots, \lambda_1, \ldots, \lambda_n, \ldots, \lambda_1)}_{n^2},$$

and that L_B is an upper triangular matrix whose diagonal entries are

$$(\underbrace{\lambda_1, \ldots, \lambda_1}_{n} \ldots, \underbrace{\lambda_n, \ldots, \lambda_n}_{n}).$$

Hint. Figure out what are $R_B(E_{ij}) = E_{ij}B$ and $L_B(E_{ij}) = BE_{ij}$.

(2) Use the fact that

$$L_A = L_P \circ L_B \circ L_P^{-1}, \quad R_A = R_P^{-1} \circ R_B \circ R_P,$$

to express $\mathrm{ad}_A = L_A - R_A$ in terms of $L_B - R_B$, and conclude that the eigenvalues of ad_A are $\lambda_i - \lambda_j$, for $i = 1, \ldots, n$, and for $j = n, \ldots, 1$.

Chapter 17

Computing Eigenvalues and Eigenvectors

After the problem of solving a linear system, the problem of computing the eigenvalues and the eigenvectors of a real or complex matrix is one of most important problems of numerical linear algebra. Several methods exist, among which we mention Jacobi, Givens–Householder, divide-and-conquer, QR iteration, and Rayleigh–Ritz; see Demmel [Demmel (1997)], Trefethen and Bau [Trefethen and Bau III (1997)], Meyer [Meyer (2000)], Serre [Serre (2010)], Golub and Van Loan [Golub and Van Loan (1996)], and Ciarlet [Ciarlet (1989)]. Typically, better performing methods exist for special kinds of matrices, such as symmetric matrices.

In theory, given an $n \times n$ complex matrix A, if we could compute a Schur form $A = UTU^*$, where T is upper triangular and U is unitary, we would obtain the eigenvalues of A, since they are the diagonal entries in T. However, this would require finding the roots of a polynomial, but methods for doing this are known to be numerically very unstable, so this is not a practical method.

A common paradigm is to construct a sequence (P_k) of matrices such that $A_k = P_k^{-1} A P_k$ converges, in some sense, to a matrix whose eigenvalues are easily determined. For example, $A_k = P_k^{-1} A P_k$ could become upper triangular in the limit. Furthermore, P_k is typically a product of "nice" matrices, for example, orthogonal matrices.

For general matrices, that is, matrices that are not symmetric, the QR iteration algorithm, due to Rutishauser, Francis, and Kublanovskaya in the early 1960s, is one of the most efficient algorithms for computing eigenvalues. A fascinating account of the history of the QR algorithm is given in Golub and Uhlig [Golub and Uhlig (2009)]. The QR algorithm constructs a sequence of matrices (A_k), where A_{k+1} is obtained from A_k by performing a QR-decomposition $A_k = Q_k R_k$ of A_k and then setting

$A_{k+1} = R_k Q_k$, the result of swapping Q_k and R_k. It is immediately verified that $A_{k+1} = Q_k^* A_k Q_k$, so A_k and A_{k+1} have *the same eigenvalues*, which are the eigenvalues of A.

The basic version of this algorithm runs into difficulties with matrices that have several eigenvalues with the same modulus (it may loop or not "converge" to an upper triangular matrix). There are ways of dealing with some of these problems, but for ease of exposition, we first present a simplified version of the QR algorithm which we call basic QR algorithm. We prove a convergence theorem for the basic QR algorithm, under the rather restrictive hypothesis that the input matrix A is diagonalizable and that its eigenvalues are nonzero and have distinct moduli. The proof shows that the part of A_k strictly below the diagonal converges to zero and that the diagonal entries of A_k converge to the eigenvalues of A.

Since the convergence of the QR method depends crucially only on the fact that the part of A_k below the diagonal goes to zero, it would be highly desirable if we could replace A by a similar matrix $U^* A U$ easily computable from A and having lots of zero strictly below the diagonal. It turns out that there is a way to construct a matrix $H = U^* A U$ which is almost triangular, except that it may have an extra nonzero diagonal below the main diagonal. Such matrices called, *Hessenberg matrices*, are discussed in Section 17.2. An $n \times n$ diagonalizable Hessenberg matrix H having the property that $h_{i+1 i} \neq 0$ for $i = 1, \ldots, n-1$ (such a matrix is called *unreduced*) has the nice property that its eigenvalues are all distinct. Since every Hessenberg matrix is a block diagonal matrix of unreduced Hessenberg blocks, *it suffices to compute the eigenvalues of unreduced Hessenberg matrices.* There is a special case of particular interest: symmetric (or Hermitian) positive definite tridiagonal matrices. Such matrices must have real positive distinct eigenvalues, so the QR algorithm converges to a diagonal matrix.

In Section 17.3, we consider techniques for making the basic QR method practical and more efficient. The first step is to convert the original input matrix A to a similar matrix H in Hessenberg form, and to apply the QR algorithm to H (actually, to the unreduced blocks of H). The second and crucial ingredient to speed up convergence is to add shifts.

A shift is the following step: pick some σ_k, hopefully close to some eigenvalue of A (in general, λ_n), QR-factor $A_k - \sigma_k I$ as

$$A_k - \sigma_k I = Q_k R_k,$$

and then form

$$A_{k+1} = R_k Q_k + \sigma_k I.$$

It is easy to see that we still have $A_{k+1} = Q_k^* A_k Q_k$. A judicious choice of σ_k can speed up convergence considerably. If H is real and has pairs of complex conjugate eigenvalues, we can perform a double shift, and it can be arranged that we work in real arithmetic.

The last step for improving efficiency is to compute $A_{k+1} = Q_k^* A_k Q_k$ without even performing a QR-factorization of $A_k - \sigma_k I$. This can be done when A_k is unreduced Hessenberg. Such a method is called QR iteration with implicit shifts. There is also a version of QR iteration with implicit double shifts.

If the dimension of the matrix A is very large, we can find approximations of some of the eigenvalues of A by using a truncated version of the reduction to Hessenberg form due to Arnoldi in general and to Lanczos in the symmetric (or Hermitian) tridiagonal case. *Arnoldi iteration* is discussed in Section 17.4. If A is an $m \times m$ matrix, for $n \ll m$ (n much smaller than m) the idea is to generate the $n \times n$ Hessenberg submatrix H_n of the full Hessenberg matrix H (such that $A = UHU^*$) consisting of its first n rows and n columns; the matrix U_n consisting of the first n columns of U is also produced. The Rayleigh–Ritz method consists in computing the eigenvalues of H_n using the QR- method with shifts. These eigenvalues, called *Ritz values*, are approximations of the eigenvalues of A. Typically, extreme eigenvalues are found first.

Arnoldi iteration can also be viewed as a way of computing an orthonormal basis of a *Krylov subspace*, namely the subspace $\mathcal{K}_n(A, b)$ spanned by $(b, Ab, \ldots, A^n b)$. We can also use Arnoldi iteration to find an approximate solution of a linear equation $Ax = b$ by minimizing $\|b - Ax_n\|_2$ for all x_n is the Krylov space $\mathcal{K}_n(A, b)$. This method named GMRES is discussed in Section 17.5.

The special case where H is a symmetric (or Hermitian) tridiagonal matrix is discussed in Section 17.6. In this case, Arnoldi's algorithm becomes *Lanczos' algorithm*. It is much more efficient than Arnoldi iteration.

We close this chapter by discussing two classical methods for computing a single eigenvector and a single eigenvalue: power iteration and inverse (power) iteration; see Section 17.7.

17.1 The Basic QR Algorithm

Let A be an $n \times n$ matrix which is assumed to be diagonalizable and invertible. The basic QR algorithm makes use of two very simple steps. Starting with $A_1 = A$, we construct sequences of matrices (A_k), (Q_k) (R_k) and (P_k)

as follows:

Factor	$A_1 = Q_1 R_1$
Set	$A_2 = R_1 Q_1$
Factor	$A_2 = Q_2 R_2$
Set	$A_3 = R_2 Q_2$
	$\vdots$
Factor	$A_k = Q_k R_k$
Set	$A_{k+1} = R_k Q_k$
	$\vdots$

Thus, A_{k+1} is obtained from a QR-factorization $A_k = Q_k R_k$ of A_k by swapping Q_k and R_k. Define P_k by

$$P_k = Q_1 Q_2 \cdots Q_k.$$

Since $A_k = Q_k R_k$, we have $R_k = Q_k^* A_k$, and since $A_{k+1} = R_k Q_k$, we obtain

$$A_{k+1} = Q_k^* A_k Q_k. \tag{17.1}$$

An obvious induction shows that

$$A_{k+1} = Q_k^* \cdots Q_1^* A_1 Q_1 \cdots Q_k = P_k^* A P_k,$$

that is

$$A_{k+1} = P_k^* A P_k. \tag{17.2}$$

Therefore, A_{k+1} and A are similar, so they have the same eigenvalues.

The basic QR iteration method consists in computing the sequence of matrices A_k, and in the ideal situation, to expect that A_k "converges" to an upper triangular matrix, more precisely that the part of A_k below the main diagonal goes to zero, and the diagonal entries converge to the eigenvalues of A.

This ideal situation is only achieved in rather special cases. For one thing, if A is unitary (or orthogonal in the real case), since in the QR decomposition we have $R = I$, we get $A_2 = IQ = Q = A_1$, so the method does *not* make any progress. Also, if A is a real matrix, since the A_k are also real, if A has complex eigenvalues, then the part of A_k below the main diagonal can't go to zero. Generally, the method runs into troubles whenever A has distinct eigenvalues with the same modulus.

The convergence of the sequence (A_k) is only known under some fairly restrictive hypotheses. Even under such hypotheses, this is not really genuine convergence. Indeed, it can be shown that the part of A_k below the main diagonal goes to zero, and the diagonal entries converge to the eigenvalues of A, but the part of A_k above the diagonal *may not converge*. However, for the purpose of finding the eigenvalues of A, this does not matter.

The following convergence result is proven in Ciarlet [Ciarlet (1989)] (Chapter 6, Theorem 6.3.10 and Serre [Serre (2010)] (Chapter 13, Theorem 13.2). It is rarely applicable in practice, except for symmetric (or Hermitian) positive definite matrices, as we will see shortly.

Theorem 17.1. *Suppose the (complex) $n \times n$ matrix A is invertible, diagonalizable, and that its eigenvalues $\lambda_1, \ldots, \lambda_n$ have different moduli, so that*

$$|\lambda_1| > |\lambda_2| > \cdots > |\lambda_n| > 0.$$

If $A = P\Lambda P^{-1}$, where $\Lambda = \mathrm{diag}(\lambda_1, \ldots, \lambda_n)$, and if P^{-1} has an LU-factorization, then the strictly lower-triangular part of A_k converges to zero, and the diagonal of A_k converges to Λ.

Proof. We reproduce the proof in Ciarlet [Ciarlet (1989)] (Chapter 6, Theorem 6.3.10). The strategy is to study the asymptotic behavior of the matrices $P_k = Q_1 Q_2 \cdots Q_k$. For this, it turns out that we need to consider the powers A^k.

Step 1. Let $\mathcal{R}_k = R_k \cdots R_2 R_1$. We claim that

$$A^k = (Q_1 Q_2 \cdots Q_k)(R_k \cdots R_2 R_1) = P_k \mathcal{R}_k. \tag{17.3}$$

We proceed by induction. The base case $k = 1$ is trivial. For the induction step, from (17.2), we have

$$P_k A_{k+1} = A P_k.$$

Since $A_{k+1} = R_k Q_k = Q_{k+1} R_{k+1}$, we have

$$P_{k+1} \mathcal{R}_{k+1} = P_k Q_{k+1} R_{k+1} \mathcal{R}_k = P_k A_{k+1} \mathcal{R}_k = A P_k \mathcal{R}_k = A A^k = A^{k+1}$$

establishing the induction step.

Step 2. We will express the matrix P_k as $P_k = Q \widetilde{Q}_k D_k$, in terms of a diagonal matrix D_k with unit entries, with Q and $\widetilde{Q}_k$ unitary.

Let $P = QR$, a QR-factorization of P (with R an upper triangular matrix with positive diagonal entries), and $P^{-1} = LU$, an LU-factorization of P^{-1}. Since $A = P\Lambda P^{-1}$, we have

$$A^k = P\Lambda^k P^{-1} = QR\Lambda^k LU = QR(\Lambda^k L\Lambda^{-k})\Lambda^k U. \tag{17.4}$$

Here, Λ^{-k} is the diagonal matrix with entries λ_i^{-k}. The reason for introducing the matrix $\Lambda^k L \Lambda^{-k}$ is that its asymptotic behavior is easy to determine. Indeed, we have

$$(\Lambda^k L \Lambda^{-k})_{ij} = \begin{cases} 0 & \text{if } i < j \\ 1 & \text{if } i = j \\ \left(\dfrac{\lambda_i}{\lambda_j}\right)^k L_{ij} & \text{if } i > j. \end{cases}$$

The hypothesis that $|\lambda_1| > |\lambda_2| > \cdots > |\lambda_n| > 0$ implies that

$$\lim_{k \mapsto \infty} \Lambda^k L \Lambda^{-k} = I. \tag{17.5}$$

Note that it is to obtain this limit that we made the hypothesis on the moduli of the eigenvalues. We can write

$$\Lambda^k L \Lambda^{-k} = I + F_k, \quad \text{with} \quad \lim_{k \mapsto \infty} F_k = 0,$$

and consequently, since $R(\Lambda^k L \Lambda^{-k}) = R(I + F_k) = R + R F_k R^{-1} R = (I + R F_k R^{-1})R$, we have

$$R(\Lambda^k L \Lambda^{-k}) = (I + R F_k R^{-1})R. \tag{17.6}$$

By Proposition 8.8(1), since $\lim_{k \mapsto \infty} F_k = 0$, and thus $\lim_{k \mapsto \infty} R F_k R^{-1} = 0$, the matrices $I + R F_k R^{-1}$ are invertible for k large enough. Consequently for k large enough, we have a QR-factorization

$$I + R F_k R^{-1} = \widetilde{Q}_k \widetilde{R}_k, \tag{17.7}$$

with $(\widetilde{R}_k)_{ii} > 0$ for $i = 1, \ldots, n$. Since the matrices $\widetilde{Q}_k$ are unitary, we have $\left\| \widetilde{Q}_k \right\|_2 = 1$, so the sequence $(\widetilde{Q}_k)$ is bounded. It follows that it has a convergent subsequence $(\widetilde{Q}_\ell)$ that converges to some matrix $\widetilde{Q}$, which is also unitary. Since

$$\widetilde{R}_\ell = (\widetilde{Q}_\ell)^*(I + R F_\ell R^{-1}),$$

we deduce that the subsequence $(\widetilde{R}_\ell)$ also converges to some matrix $\widetilde{R}$, which is also upper triangular with positive diagonal entries. By passing to the limit (using the subsequences), we get $\widetilde{R} = (\widetilde{Q})^*$, that is,

$$I = \widetilde{Q}\widetilde{R}.$$

By the uniqueness of a QR-decomposition (when the diagonal entries of R are positive), we get

$$\widetilde{Q} = \widetilde{R} = I.$$

Since the above reasoning applies to any subsequences of $(\widetilde{Q}_k)$ and $(\widetilde{R}_k)$, by the uniqueness of limits, we conclude that the "full" sequences $(\widetilde{Q}_k)$ and $(\widetilde{R}_k)$ converge:

$$\lim_{k \mapsto \infty} \widetilde{Q}_k = I, \quad \lim_{k \mapsto \infty} \widetilde{R}_k = I.$$

Since by (17.4),

$$A^k = QR(\Lambda^k L \Lambda^{-k})\Lambda^k U,$$

by (17.6),

$$R(\Lambda^k L \Lambda^{-k}) = (I + RF_k R^{-1})R,$$

and by (17.7)

$$I + RF_k R^{-1} = \widetilde{Q}_k \widetilde{R}_k,$$

we proved that

$$A^k = (Q\widetilde{Q}_k)(\widetilde{R}_k R \Lambda^k U). \tag{17.8}$$

Observe that $Q\widetilde{Q}_k$ is a unitary matrix, and $\widetilde{R}_k R \Lambda^k U$ is an upper triangular matrix, as a product of upper triangular matrices. However, some entries in Λ may be negative, so we can't claim that $\widetilde{R}_k R \Lambda^k U$ has positive diagonal entries. Nevertheless, we have another QR-decomposition of A^k,

$$A^k = (Q\widetilde{Q}_k)(\widetilde{R}_k R \Lambda^k U) = P_k \mathcal{R}_k.$$

It is easy to prove that there is diagonal matrix D_k with $|(D_k)_{ii}| = 1$ for $i = 1, \ldots, n$, such that

$$P_k = Q\widetilde{Q}_k D_k. \tag{17.9}$$

The existence of D_k is consequence of the following fact: If an invertible matrix B has two QR factorizations $B = Q_1 R_1 = Q_2 R_2$, then there is a diagonal matrix D with unit entries such that $Q_2 = DQ_1$.

The expression for P_k in (17.9) is that which we were seeking.

Step 3. Asymptotic behavior of the matrices $A_{k+1} = P_k^* A P_k$.

Since $A = P\Lambda P^{-1} = QR\Lambda R^{-1}Q^{-1}$ and by (17.9), $P_k = Q\widetilde{Q}_k D_k$, we get

$$A_{k+1} = D_k^*(\widetilde{Q}_k)^* Q^* QR\Lambda R^{-1}Q^{-1}Q\widetilde{Q}_k D_k = D_k^*(\widetilde{Q}_k)^* R\Lambda R^{-1}\widetilde{Q}_k D_k. \tag{17.10}$$

Since $\lim_{k \mapsto \infty} \widetilde{Q}_k = I$, we deduce that

$$\lim_{k \mapsto \infty} (\widetilde{Q}_k)^* R\Lambda R^{-1}\widetilde{Q}_k = R\Lambda R^{-1} = \begin{pmatrix} \lambda_1 & * & \cdots & * \\ 0 & \lambda_2 & \cdots & * \\ \vdots & & \ddots & \vdots \\ 0 & 0 & \cdots & \lambda_n \end{pmatrix},$$

an upper triangular matrix with the eigenvalues of A on the diagonal. Since R is upper triangular, the order of the eigenvalues is preserved. If we let

$$\mathcal{D}_k = (\widetilde{Q}_k)^* R\Lambda R^{-1} \widetilde{Q}_k, \tag{17.11}$$

then by (17.10) we have $A_{k+1} = D_k^* \mathcal{D}_k D_k$, and since the matrices D_k are diagonal matrices, we have

$$(A_{k+1})_{jj} = (D_k^* \mathcal{D}_k D_k)_{ij} = \overline{(D_k)_{ii}} (D_k)_{jj} (\mathcal{D}_k)_{ij},$$

which implies that

$$(A_{k+1})_{ii} = (\mathcal{D}_k)_{ii}, \quad i = 1, \dots, n, \tag{17.12}$$

since $|(D_k)_{ii}| = 1$ for $i = 1, \dots, n$. Since $\lim_{k \to \infty} \mathcal{D}_k = R\Lambda R^{-1}$, we conclude that the strictly lower-triangular part of A_{k+1} converges to zero, and the diagonal of A_{k+1} converges to Λ. $\square$

Observe that if the matrix A is real, then the hypothesis that the eigenvalues have distinct moduli implies that the eigenvalues are all real and simple.

The following `Matlab` program implements the basic QR-method using the function `qrv4` from Section 11.8.

```
function T = qreigen(A,m)
T = A;
for k = 1:m
  [Q R] = qrv4(T);
  T = R*Q;
end
end
```

Example 17.1. If we run the function `qreigen` with 100 iterations on the 8×8 symmetric matrix

$$A = \begin{pmatrix} 4 & 1 & 0 & 0 & 0 & 0 & 0 & 0 \\ 1 & 4 & 1 & 0 & 0 & 0 & 0 & 0 \\ 0 & 1 & 4 & 1 & 0 & 0 & 0 & 0 \\ 0 & 0 & 1 & 4 & 1 & 0 & 0 & 0 \\ 0 & 0 & 0 & 1 & 4 & 1 & 0 & 0 \\ 0 & 0 & 0 & 0 & 1 & 4 & 1 & 0 \\ 0 & 0 & 0 & 0 & 0 & 1 & 4 & 1 \\ 0 & 0 & 0 & 0 & 0 & 0 & 1 & 4 \end{pmatrix},$$

we find the matrix

$$
T = \begin{pmatrix}
5.8794 & 0.0015 & 0.0000 & -0.0000 & 0.0000 & -0.0000 & 0.0000 & -0.0000 \\
0.0015 & 5.5321 & 0.0001 & 0.0000 & -0.0000 & 0.0000 & 0.0000 & 0.0000 \\
0 & 0.0001 & 5.0000 & 0.0000 & -0.0000 & 0.0000 & 0.0000 & 0.0000 \\
0 & 0 & 0.0000 & 4.3473 & 0.0000 & 0.0000 & 0.0000 & 0.0000 \\
0 & 0 & 0 & 0.0000 & 3.6527 & 0.0000 & 0.0000 & -0.0000 \\
0 & 0 & 0 & 0 & 0.0000 & 3.0000 & 0.0000 & -0.0000 \\
0 & 0 & 0 & 0 & 0 & 0.0000 & 2.4679 & 0.0000 \\
0 & 0 & 0 & 0 & 0 & 0 & 0.0000 & 2.1.206
\end{pmatrix}.
$$

The diagonal entries match the eigenvalues found by running the Matlab function eig(A).

If several eigenvalues have the same modulus, then the proof breaks down, we can no longer claim (17.5), namely that

$$\lim_{k \mapsto \infty} \Lambda^k L \Lambda^{-k} = I.$$

If we assume that P^{-1} has a suitable "block LU-factorization," it can be shown that the matrices A_{k+1} converge to a block upper-triangular matrix, where each block corresponds to eigenvalues having the same modulus. For example, if A is a 9×9 matrix with eigenvalues λ_i such that $|\lambda_1| = |\lambda_2| = |\lambda_3| > |\lambda_4| > |\lambda_5| = |\lambda_6| = |\lambda_7| = |\lambda_8| = |\lambda_9|$, then A_k converges to a block diagonal matrix (with three blocks, a 3×3 block, a 1×1 block, and a 5×5 block) of the form

$$
\begin{pmatrix}
\star & \star & \star & \star & \star & \star & \star & \star & \star \\
\star & \star & \star & \star & \star & \star & \star & \star & \star \\
\star & \star & \star & \star & \star & \star & \star & \star & \star \\
0 & 0 & 0 & \star & \star & \star & \star & \star & \star \\
0 & 0 & 0 & 0 & \star & \star & \star & \star & \star \\
0 & 0 & 0 & 0 & \star & \star & \star & \star & \star \\
0 & 0 & 0 & 0 & \star & \star & \star & \star & \star \\
0 & 0 & 0 & 0 & \star & \star & \star & \star & \star \\
0 & 0 & 0 & 0 & \star & \star & \star & \star & \star
\end{pmatrix}.
$$

See Ciarlet [Ciarlet (1989)] (Chapter 6 Section 6.3) for more details.

Under the conditions of Theorem 17.1, in particular, if A is a symmetric (or Hermitian) positive definite matrix, the eigenvectors of A can be approximated. However, when A is not a symmetric matrix, since the upper triangular part of A_k does not necessarily converge, one has to be cautious that a rigorous justification is lacking.

Suppose we apply the QR algorithm to a matrix A satisfying the hypotheses of Theorem 17.1. For k large enough, $A_{k+1} = P_k^* A P_k$ is nearly upper triangular and the diagonal entries of A_{k+1} are all distinct, so we can consider that they are the eigenvalues of A_{k+1}, and thus of A. To avoid too many subscripts, write T for the upper triangular matrix obtained by setting the entries of the part of A_{k+1} below the diagonal to 0. Then we can find the corresponding eigenvectors by solving the linear system

$$Tv = t_{ii}v,$$

and since T is upper triangular, this can be done by bottom-up elimination. We leave it as an exercise to show that the following vectors $v^i = (v_1^i, \ldots, v_n^i)$ are eigenvectors:

$$v^1 = e_1,$$

and if $i = 2, \ldots, n$, then

$$v_j^i = \begin{cases} 0 & \text{if } i+1 \leq j \leq n \\ 1 & \text{if } j = i \\ -\dfrac{t_{jj+1}v_{j+1}^i + \cdots + t_{ji}v_i^i}{t_{jj} - t_{ii}} & \text{if } i-1 \geq j \geq 1. \end{cases}$$

Then the vectors $(P_k v^1, \ldots, P_k v^n)$ are a basis of (approximate) eigenvectors for A. In the special case where T is a diagonal matrix, then $v^i = e_i$ for $i = 1, \ldots, n$ and the columns of P_k are an orthonormal basis of (approximate) eigenvectors for A.

If A is a real matrix whose eigenvalues are not all real, then there is some complex pair of eigenvalues $\lambda + i\mu$ (with $\mu \neq 0$), and the QR-algorithm cannot converge to a matrix whose strictly lower-triangular part is zero. There is a way to deal with this situation using upper Hessenberg matrices which will be discussed in the next section.

Since the convergence of the QR method depends crucially only on the fact that the part of A_k below the diagonal goes to zero, it would be highly desirable if we could replace A by a similar matrix $U^* A U$ easily computable from A having lots of zero strictly below the diagonal. We can't expect $U^* A U$ to be a diagonal matrix (since this would mean that A was easily diagonalized), but it turns out that there is a way to construct a matrix $H = U^* A U$ which is almost triangular, except that it may have an extra nonzero diagonal below the main diagonal. Such matrices called Hessenberg matrices are discussed in the next section.

17.2 Hessenberg Matrices

Definition 17.1. An $n \times n$ matrix (real or complex) H is an *(upper) Hessenberg matrix* if it is almost triangular, except that it may have an extra nonzero diagonal below the main diagonal. Technically, $h_{jk} = 0$ for all (j, k) such that $j - k \geq 2$.

The 5×5 matrix below is an example of a Hessenberg matrix.

$$H = \begin{pmatrix} * & * & * & * & * \\ h_{21} & * & * & * & * \\ 0 & h_{32} & * & * & * \\ 0 & 0 & h_{43} & * & * \\ 0 & 0 & 0 & h_{54} & * \end{pmatrix}.$$

The following result can be shown.

Theorem 17.2. *Every $n \times n$ complex or real matrix A is similar to an upper Hessenberg matrix H, that is, $A = U H U^*$ for some unitary matrix U. Furthermore, H can be constructed as a product of Householder matrices (the definition is the same as in Section 12.1, except that W is a complex vector, and that the inner product is the Hermitian inner product on $\mathbb{C}^n$). If A is a real matrix, then H is an orthogonal matrix (and H is a real matrix).*

Theorem 17.2 and algorithms for converting a matrix to Hessenberg form are discussed in Trefethen and Bau [Trefethen and Bau III (1997)] (Lecture 26), Demmel [Demmel (1997)] (Section 4.4.6, in the real case), Serre [Serre (2010)] (Theorem 13.1), and Meyer [Meyer (2000)] (Example 5.7.4, in the real case). The proof of correctness is not difficult and will be the object of a homework problem.

The following functions written in Matlab implement a function to compute a Hessenberg form of a matrix.

The function **house** constructs the normalized vector u defining the Householder reflection that zeros all but the first entries in a vector x.

```
function [uu, u] = house(x)
tol = 2*10^(-15);    % tolerance
uu = x;
p = size(x,1);
% computes l^1-norm of x(2:p,1)
```

```
n1 = sum(abs(x(2:p,1)));
if n1 <= tol
   u = zeros(p,1);   uu = u;
else
   l = sqrt(x'*x);   % l^2 norm of x
   uu(1) = x(1) + signe(x(1))*l;
   u = uu/sqrt(uu'*uu);
end
end
```

The function `signe(z)` returns -1 if $z < 0$, else $+1$.

The function `buildhouse` builds a Householder reflection from a vector uu.

```
function P = buildhouse(v,i)
% This function builds a Householder reflection
%    [I 0 ]
%    [O PP]
%    from a Householder reflection
%    PP = I - 2uu*uu'
%    where uu = v(i:n)
%    If uu = 0 then P - I
%

n = size(v,1);
if v(i:n) == zeros(n - i + 1,1)
   P = eye(n);
else
   PP = eye(n - i + 1) - 2*v(i:n)*v(i:n)';
   P = [eye(i-1) zeros(i-1, n - i + 1);
        zeros(n - i + 1, i - 1) PP];
end
end
```

The function `Hessenberg1` computes an upper Hessenberg matrix H and an orthogonal matrix Q such that $A = Q^\top H Q$.

```
function [H, Q] = Hessenberg1(A)
%
%    This function constructs an upper Hessenberg
```

```
%    matrix H and an orthogonal matrix Q such that
%    A = Q' H Q

n = size(A,1);
H = A;
Q = eye(n);
for i = 1:n-2
   %  H(i+1:n,i)
      [~,u] = house(H(i+1:n,i));
   %  u
      P =  buildhouse(u,1);
      Q(i+1:n,i:n) = P*Q(i+1:n,i:n);
      H(i+1:n,i:n) = H(i+1:n,i:n) - 2*u*(u')*H(i+1:n,i:n);
      H(1:n,i+1:n) = H(1:n,i+1:n) - 2*H(1:n,i+1:n)*u*(u');
end
end
```

Example 17.2. If

$$A = \begin{pmatrix} 1\ 2\ 3\ 4 \\ 2\ 3\ 4\ 5 \\ 3\ 4\ 5\ 6 \\ 4\ 5\ 6\ 7 \end{pmatrix},$$

running `Hessenberg1` we find

$$H = \begin{pmatrix} 1.0000 & -5.3852 & 0 & 0 \\ -5.3852 & 15.2069 & -1.6893 & -0.0000 \\ -0.0000 & -1.6893 & -0.2069 & -0.0000 \\ 0 & -0.0000 & 0.0000 & 0.0000 \end{pmatrix}$$

$$Q = \begin{pmatrix} 1.0000 & 0 & 0 & 0 \\ 0 & -0.3714 & -0.5571 & -0.7428 \\ 0 & 0.8339 & 0.1516 & -0.5307 \\ 0 & 0.4082 & -0.8165 & 0.4082 \end{pmatrix}.$$

An important property of (upper) Hessenberg matrices is that if some subdiagonal entry $H_{p+1p} = 0$, then H is of the form

$$H = \begin{pmatrix} H_{11} & H_{12} \\ 0 & H_{22} \end{pmatrix},$$

where both H_{11} and H_{22} are upper Hessenberg matrices (with H_{11} a $p \times p$ matrix and H_{22} a $(n - p) \times (n - p)$ matrix), and the eigenvalues of H are the eigenvalues of H_{11} and H_{22}. For example, in the matrix

$$H = \begin{pmatrix} * & * & * & * & * \\ h_{21} & * & * & * & * \\ 0 & h_{32} & * & * & * \\ 0 & 0 & h_{43} & * & * \\ 0 & 0 & 0 & h_{54} & * \end{pmatrix},$$

if $h_{43} = 0$, then we have the block matrix

$$H = \begin{pmatrix} * & * & * & * & * \\ h_{21} & * & * & * & * \\ 0 & h_{32} & * & * & * \\ 0 & 0 & 0 & * & * \\ 0 & 0 & 0 & h_{54} & * \end{pmatrix}.$$

Then the list of eigenvalues of H is the concatenation of the list of eigenvalues of H_{11} and the list of the eigenvalues of H_{22}. This is easily seen by induction on the dimension of the block H_{11}.

More generally, every upper Hessenberg matrix can be written in such a way that it has diagonal blocks that are Hessenberg blocks whose subdiagonal is not zero.

Definition 17.2. An upper Hessenberg $n \times n$ matrix H is *unreduced* if $h_{i+1i} \neq 0$ for $i = 1, \ldots, n-1$. A Hessenberg matrix which is not unreduced is said to be *reduced*.

The following is an example of an 8×8 matrix consisting of three diagonal unreduced Hessenberg blocks:

$$H = \begin{pmatrix} \star & \star & \star & \star & \star & \star & \star & \star \\ \mathbf{h_{21}} & \star & \star & \star & \star & \star & \star & \star \\ 0 & \mathbf{h_{32}} & \star & \star & \star & \star & \star & \star \\ 0 & 0 & 0 & \star & \star & \star & \star & \star \\ 0 & 0 & 0 & \mathbf{h_{54}} & \star & \star & \star & \star \\ 0 & 0 & 0 & \mathbf{0} & \mathbf{h_{65}} & \star & * & * \\ 0 & 0 & 0 & 0 & 0 & 0 & \star & \star \\ 0 & 0 & 0 & 0 & 0 & 0 & \mathbf{h_{87}} & \star \end{pmatrix}.$$

An interesting and important property of unreduced Hessenberg matrices is the following.

Proposition 17.1. *Let H be an $n \times n$ complex or real unreduced Hessenberg matrix. Then every eigenvalue of H is geometrically simple, that is,*

$\dim(E_\lambda) = 1$ *for every eigenvalue* λ, *where* E_λ *is the eigenspace associated with* λ. *Furthermore, if* H *is diagonalizable, then every eigenvalue is simple, that is,* H *has* n *distinct eigenvalues.*

Proof. We follow Serre's proof [Serre (2010)] (Proposition 3.26). Let λ be any eigenvalue of H, let $M = \lambda I_n - H$, and let N be the $(n-1) \times (n-1)$ matrix obtained from M by deleting its first row and its last column. Since H is upper Hessenberg, N is a diagonal matrix with entries $-h_{i+1\,i} \neq 0$, $i = 1, \ldots, n-1$. Thus N is invertible and has rank $n-1$. But a matrix has rank greater than or equal to the rank of any of its submatrices, so $\mathrm{rank}(M) = n-1$, since M is singular. By the rank-nullity theorem, $\mathrm{rank}(\mathrm{Ker}\,N) = 1$, that is, $\dim(E_\lambda) = 1$, as claimed.

If H is diagonalizable, then the sum of the dimensions of the eigenspaces is equal to n, which implies that the eigenvalues of H are distinct. $\qquad \square$

As we said earlier, a case where Theorem 17.1 applies is the case where A is a symmetric (or Hermitian) positive definite matrix. This follows from two facts.

The first fact is that if A is Hermitian (or symmetric in the real case), then it is easy to show that the Hessenberg matrix similar to A is a Hermitian (or symmetric in real case) *tridiagonal matrix*. The conversion method is also more efficient. Here is an example of a symmetric tridiagonal matrix consisting of three unreduced blocks:

$$
H = \begin{pmatrix}
\alpha_1 & \beta_1 & \mathbf{0} & 0 & 0 & 0 & 0 & 0 \\
\beta_1 & \alpha_2 & \beta_2 & 0 & 0 & 0 & 0 & 0 \\
\mathbf{0} & \beta_2 & \alpha_3 & 0 & 0 & 0 & 0 & 0 \\
0 & 0 & 0 & \alpha_4 & \beta_4 & \mathbf{0} & 0 & 0 \\
0 & 0 & 0 & \beta_4 & \alpha_5 & \beta_5 & 0 & 0 \\
0 & 0 & 0 & \mathbf{0} & \beta_5 & \alpha_6 & 0 & 0 \\
0 & 0 & 0 & 0 & 0 & 0 & \alpha_7 & \beta_7 \\
0 & 0 & 0 & 0 & 0 & 0 & \beta_7 & \alpha_8
\end{pmatrix}.
$$

Thus the problem of finding the eigenvalues of a symmetric (or Hermitian) matrix reduces to the problem of finding the eigenvalues of a symmetric (resp. Hermitian) tridiagonal matrix, and this can be done much more efficiently.

The second fact is that if H is an upper Hessenberg matrix and if it is diagonalizable, then there is an invertible matrix P such that $H = P\Lambda P^{-1}$ with Λ a diagonal matrix consisting of the eigenvalues of H, such that P^{-1} has an LU-decomposition; see Serre [Serre (2010)] (Theorem 13.3).

As a consequence, since any symmetric (or Hermitian) tridiagonal matrix is a block diagonal matrix of unreduced symmetric (resp. Hermitian) tridiagonal matrices, by Proposition 17.1, we see that the QR algorithm applied to a tridiagonal matrix which is symmetric (or Hermitian) positive definite converges to a diagonal matrix consisting of its eigenvalues. Let us record this important fact.

Theorem 17.3. *Let H be a symmetric (or Hermitian) positive definite tridiagonal matrix. If H is unreduced, then the QR algorithm converges to a diagonal matrix consisting of the eigenvalues of H.*

Since every symmetric (or Hermitian) positive definite matrix is similar to tridiagonal symmetric (resp. Hermitian) positive definite matrix, we deduce that we have a method for finding the eigenvalues of a symmetric (resp. Hermitian) positive definite matrix (more accurately, to find approximations as good as we want for these eigenvalues).

If A is a symmetric (or Hermitian) matrix, since its eigenvalues are real, for some $\mu > 0$ large enough (pick $\mu > \rho(A)$), $A + \mu I$ is symmetric (resp. Hermitan) positive definite, so we can apply the QR algorithm to an upper Hessenberg matrix similar to $A + \mu I$ to find its eigenvalues, and then the eigenvalues of A are obtained by subtracting μ.

The problem of finding the eigenvalues of a symmetric matrix is discussed extensively in Parlett [Parlett (1997)], one of the best references on this topic.

The upper Hessenberg form also yields a way to handle singular matrices. First, checking the proof of Proposition 13.20 that an $n \times n$ complex matrix A (possibly singular) can be factored as $A = QR$ where Q is a unitary matrix which is a product of Householder reflections and R is upper triangular, it is easy to see that if A is upper Hessenberg, then Q is also upper Hessenberg. If H is an unreduced upper Hessenberg matrix, since Q is upper Hessenberg and R is upper triangular, we have $h_{i+1i} = q_{i+1i}r_{ii}$ for $i = 1\ldots, n-1$, and since H is unreduced, $r_{ii} \neq 0$ for $i = 1, \ldots, n-1$. Consequently H is singular iff $r_{nn} = 0$. Then the matrix RQ is a matrix whose last row consists of zero's thus we can deflate the problem by considering the $(n-1) \times (n-1)$ unreduced Hessenberg matrix obtained by deleting the last row and the last column. After finitely many steps (not larger that the multiplicity of the eigenvalue 0), there remains an invertible unreduced Hessenberg matrix. As an alternative, see Serre [Serre (2010)] (Chapter 13, Section 13.3.2).

As is, the QR algorithm, although very simple, is quite inefficient for several reasons. In the next section, we indicate how to make the method more efficient. This involves a lot of work and we only discuss the main ideas at a high level.

17.3 Making the QR Method More Efficient Using Shifts

To improve efficiency and cope with pairs of complex conjugate eigenvalues in the case of real matrices, the following steps are taken:

(1) Initially reduce the matrix A to upper Hessenberg form, as $A = UHU^*$. Then apply the QR-algorithm to H (actually, to its unreduced Hessenberg blocks). It is easy to see that the matrices H_k produced by the QR algorithm remain upper Hessenberg.

(2) To accelerate convergence, use *shifts*, and to deal with pairs of complex conjugate eigenvalues, use *double shifts*.

(3) Instead of computing a QR-factorization explicitly while doing a shift, perform an *implicit shift* which computes $A_{k+1} = Q_k^* A_k Q_k$ without having to compute a QR-factorization (of $A_k - \sigma_k I$), and similarly in the case of a double shift. This is the most intricate modification of the basic QR algorithm and we will not discuss it here. This method is usually referred as *bulge chasing*. Details about this technique for real matrices can be found in Demmel [Demmel (1997)] (Section 4.4.8) and Golub and Van Loan [Golub and Van Loan (1996)] (Section 7.5). Watkins discusses the QR algorithm with shifts as a bulge chasing method in the more general case of complex matrices [Watkins (1982, 2008)].

Let us repeat an important remark made in the previous section. If we start with a matrix H in upper Hessenberg form, if at any stage of the QR algorithm we find that some subdiagonal entry $(H_k)_{p+1p} = 0$ *or is very small*, then H_k is of the form

$$H_k = \begin{pmatrix} H_{11} & H_{12} \\ 0 & H_{22} \end{pmatrix},$$

where both H_{11} and H_{22} are upper Hessenberg matrices (with H_{11} a $p \times p$ matrix and H_{22} a $(n-p) \times (n-p)$ matrix), and the eigenvalues of H_k are the eigenvalues of H_{11} and H_{22}. For example, in the matrix

$$H = \begin{pmatrix} * & * & * & * & * \\ h_{21} & * & * & * & * \\ 0 & h_{32} & * & * & * \\ 0 & 0 & h_{43} & * & * \\ 0 & 0 & 0 & h_{54} & * \end{pmatrix},$$

if $h_{43} = 0$, then we have the block matrix

$$H = \begin{pmatrix} * & * & * & * & * \\ h_{21} & * & * & * & * \\ 0 & h_{32} & * & * & * \\ 0 & 0 & 0 & * & * \\ 0 & 0 & 0 & h_{54} & * \end{pmatrix}.$$

Then we can recursively apply the QR algorithm to H_{11} and H_{22}.

In particular, if $(H_k)_{nn-1} = 0$ or is very small, then $(H_k)_{nn}$ is a good approximation of an eigenvalue, so we can delete the last row and the last column of H_k and apply the QR algorithm to this submatrix. This process is called *deflation*. If $(H_k)_{n-1n-2} = 0$ or is very small, then the 2×2 "corner block"

$$\begin{pmatrix} (H_k)_{n-1n-1} & (H_k)_{n-1n} \\ (H_k)_{nn-1} & (H_k)_{nn} \end{pmatrix}$$

appears, and its eigenvalues can be computed immediately by solving a quadratic equation. Then we deflate H_k by deleting its last two rows and its last two columns and apply the QR algorithm to this submatrix.

Thus it would seem desirable to modify the basic QR algorithm so that the above situations arises, and this is what shifts are designed for. More precisely, under the hypotheses of Theorem 17.1, it can be shown (see Ciarlet [Ciarlet (1989)], Section 6.3) that the entry $(A_k)_{ij}$ with $i > j$ converges to 0 as $|\lambda_i/\lambda_j|^k$ converges to 0. Also, if we let r_i be defined by

$$r_1 = \left| \frac{\lambda_2}{\lambda_1} \right|, \quad r_i = \max \left\{ \left| \frac{\lambda_i}{\lambda_{i-1}} \right|, \left| \frac{\lambda_{i+1}}{\lambda_i} \right| \right\}, \quad 2 \le i \le n-1, \quad r_n = \left| \frac{\lambda_n}{\lambda_{n-1}} \right|,$$

then there is a constant C (independent of k) such that

$$|(A_k)_{ii} - \lambda_i| \le C r_i^k, \quad 1 \le i \le n.$$

In particular, if H is upper Hessenberg, then the entry $(H_k)_{i+1i}$ converges to 0 as $|\lambda_{i+1}/\lambda_i|^k$ converges to 0. Thus if we pick σ_k close to λ_i, we expect that $(H_k - \sigma_k I)_{i+1i}$ converges to 0 as $|(\lambda_{i+1} - \sigma_k)/(\lambda_i - \sigma_k)|^k$ converges to 0, and this ratio is much smaller than 1 as σ_k is closer to λ_i.

Typically, we apply a shift to accelerate convergence to λ_n (so $i = n - 1$). In this case, both $(H_k - \sigma_k I)_{nn-1}$ and $|(H_k - \sigma_k I)_{nn} - \lambda_n|$ converge to 0 as $|(\lambda_n - \sigma_k)/(\lambda_{n-1} - \sigma_k)|^k$ converges to 0.

A *shift* is the following modified QR-steps (switching back to an arbitrary matrix A, since the shift technique applies in general). Pick some σ_k, hopefully close to some eigenvalue of A (in general, λ_n), and QR-factor $A_k - \sigma_k I$ as

$$A_k - \sigma_k I = Q_k R_k,$$

and then form

$$A_{k+1} = R_k Q_k + \sigma_k I.$$

Since

$$\begin{aligned}
A_{k+1} &= R_k Q_k + \sigma_k I \\
&= Q_k^* Q_k R_k Q_k + Q_k^* Q_k \sigma_k \\
&= Q_k^* (Q_k R_k + \sigma_k I) Q_k \\
&= Q_k^* A_k Q_k,
\end{aligned}$$

A_{k+1} is similar to A_k, as before. If A_k is upper Hessenberg, then it is easy to see that A_{k+1} is also upper Hessenberg.

If A is upper Hessenberg and if σ_i is exactly equal to an eigenvalue, then $A_k - \sigma_k I$ is singular, and forming the QR-factorization will detect that R_k has some diagonal entry equal to 0. Assuming that the QR-algorithm returns $(R_k)_{nn} = 0$ (if not, the argument is easily adapted), then the last row of $R_k Q_k$ is 0, so the last row of $A_{k+1} = R_k Q_k + \sigma_k I$ ends with σ_k (all other entries being zero), so we are in the case where we can deflate A_k (and σ_k is indeed an eigenvalue).

The question remains, what is a good choice for the shift σ_k?

Assuming again that H is in upper Hessenberg form, it turns out that when $(H_k)_{nn-1}$ is small enough, then a good choice for σ_k is $(H_k)_{nn}$. In fact, the rate of convergence is quadratic, which means roughly that the number of correct digits doubles at every iteration. The reason is that shifts are related to another method known as inverse iteration, and such a method converges very fast. For further explanations about this connection, see Demmel [Demmel (1997)] (Section 4.4.4) and Trefethen and Bau [Trefethen and Bau III (1997)] (Lecture 29).

One should still be cautious that the QR method with shifts does not necessarily converge, and that our convergence proof no longer applies, because instead of having the identity $A^k = P_k \mathcal{R}_k$, we have

$$(A - \sigma_k I) \cdots (A - \sigma_2 I)(A - \sigma_1 I) = P_k \mathcal{R}_k.$$

Of course, the QR algorithm loops immediately when applied to an orthogonal matrix A. This is also the case when A is symmetric but not positive definite. For example, both the QR algorithm and the QR algorithm with shifts loop on the matrix

$$A = \begin{pmatrix} 0 & 1 \\ 1 & 0 \end{pmatrix}.$$

In the case of symmetric matrices, Wilkinson invented a shift which helps the QR algorithm with shifts to make progress. Again, looking at the lower corner of A_k, say

$$B = \begin{pmatrix} a_{n-1} & b_{n-1} \\ b_{n-1} & a_n \end{pmatrix},$$

the *Wilkinson shift* picks the eigenvalue of B closer to a_n. If we let

$$\delta = \frac{a_{n-1} - a_n}{2},$$

it is easy to see that the eigenvalues of B are given by

$$\lambda = \frac{a_n + a_{n-1}}{2} \pm \sqrt{\delta^2 + b_{n-1}^2}.$$

It follows that

$$\lambda - a_n = \delta \pm \sqrt{\delta^2 + b_{n-1}^2},$$

and from this it is easy to see that the eigenvalue closer to a_n is given by

$$\mu = a_n - \frac{\text{sign}(\delta) b_{n-1}^2}{(|\delta| + \sqrt{\delta^2 + b_{n-1}^2})}.$$

If $\delta = 0$, then we pick arbitrarily one of the two eigenvalues. Observe that the Wilkinson shift applied to the matrix

$$A = \begin{pmatrix} 0 & 1 \\ 1 & 0 \end{pmatrix}$$

is either $+1$ or -1, and in one step, deflation occurs and the algorithm terminates successfully.

We now discuss double shifts, which are intended to deal with pairs of complex conjugate eigenvalues.

Let us assume that A is a real matrix. For any complex number σ_k with nonzero imaginary part, a *double shift* consists of the following steps:

$$A_k - \sigma_k I = Q_k R_k$$
$$A_{k+1} = R_k Q_k + \sigma_k I$$
$$A_{k+1} - \overline{\sigma}_k I = Q_{k+1} R_{k+1}$$
$$A_{k+2} = R_{k+1} Q_{k+1} + \overline{\sigma}_k I.$$

From the computation made for a single shift, we have $A_{k+1} = Q_k^* A_k Q_k$ and $A_{k+2} = Q_{k+1}^* A_{k+1} Q_{k+1}$, so we obtain

$$A_{k+2} = Q_{k+1}^* Q_k^* A_k Q_k Q_{k+1}.$$

The matrices Q_k are complex, so we would expect that the A_k are also complex, but remarkably we can keep the products $Q_k Q_{k+1}$ real, and so the A_k also real. This is highly desirable to avoid complex arithmetic, which is more expensive.

Observe that since

$$Q_{k+1} R_{k+1} = A_{k+1} - \overline{\sigma}_k I = R_k Q_k + (\sigma_k - \overline{\sigma}_k) I,$$

we have

$$
\begin{aligned}
Q_k Q_{k+1} R_{k+1} R_k &= Q_k (R_k Q_k + (\sigma_k - \overline{\sigma}_k) I) R_k \\
&= Q_k R_k Q_k R_k + (\sigma_k - \overline{\sigma}_k) Q_k R_k \\
&= (A_k - \sigma_k I)^2 + (\sigma_k - \overline{\sigma}_k)(A_k - \sigma_k I) \\
&= A_k^2 - 2(\Re \sigma_k) A_k + |\sigma_k|^2 I.
\end{aligned}
$$

If we assume by induction that matrix A_k is real (with $k = 2\ell+1, \ell \geq 0$), then the matrix $S = A_k^2 - 2(\Re \sigma_k) A_k + |\sigma_k|^2 I$ is also real, and since $Q_k Q_{k+1}$ is unitary and $R_{k+1} R_k$ is upper triangular, we see that

$$S = Q_k Q_{k+1} R_{k+1} R_k$$

is a QR-factorization of the real matrix S, thus $Q_k Q_{k+1}$ and $R_{k+1} R_k$ can be chosen to be real matrices, in which case $(Q_k Q_{k+1})^*$ is also real, and thus

$$A_{k+2} = Q_{k+1}^* Q_k^* A_k Q_k Q_{k+1} = (Q_k Q_{k+1})^* A_k Q_k Q_{k+1}$$

is real. Consequently, if $A_1 = A$ is real, then $A_{2\ell+1}$ is real for all $\ell \geq 0$.

The strategy that consists in picking σ_k and $\overline{\sigma}_k$ as the complex conjugate eigenvalues of the corner block

$$\begin{pmatrix} (H_k)_{n-1\,n-1} & (H_k)_{n-1\,n} \\ (H_k)_{n\,n-1} & (H_k)_{n\,n} \end{pmatrix}$$

is called the *Francis shift* (here we are assuming that A has be reduced to upper Hessenberg form).

It should be noted that there are matrices for which neither a shift by $(H_k)_{nn}$ nor the Francis shift works. For instance, the permutation matrix

$$A = \begin{pmatrix} 0 & 0 & 1 \\ 1 & 0 & 0 \\ 0 & 1 & 0 \end{pmatrix}$$

has eigenvalues $e^{i2\pi/3}, e^{i4\pi/3}, +1$, and neither of the above shifts apply to the matrix

$$\begin{pmatrix} 0 & 0 \\ 1 & 0 \end{pmatrix}.$$

However, a shift by 1 does work. There are other kinds of matrices for which the QR algorithm does not converge. Demmel gives the example of matrices of the form

$$\begin{pmatrix} 0 & 1 & 0 & 0 \\ 1 & 0 & h & 0 \\ 0 & -h & 0 & 1 \\ 0 & 0 & 1 & 0 \end{pmatrix}$$

where h is small.

Algorithms implementing the QR algorithm with shifts and double shifts perform "exceptional" shifts every 10 shifts. Despite the fact that the QR algorithm has been perfected since the 1960's, it is still an open problem to find a shift strategy that ensures convergence of all matrices.

Implicit shifting is based on a result known as the *implicit Q theorem*. This theorem says that if A is reduced to upper Hessenberg form as $A = UHU^*$ and if H is unreduced ($h_{i+1i} \neq 0$ for $i = 1, \ldots, n - 1$), then the columns of index $2, \ldots, n$ of U are determined by the first column of U up to sign; see Demmel [Demmel (1997)] (Theorem 4.9) and Golub and Van Loan [Golub and Van Loan (1996)] (Theorem 7.4.2) for the proof in the case of real matrices. Actually, the proof is not difficult and will be the object of a homework exercise. In the case of a single shift, an implicit shift generates $A_{k+1} = Q_k^* A_k Q_k$ without having to compute a QR-factorization of $A_k - \sigma_k I$. For real matrices, this is done by applying a sequence of Givens rotations which perform a bulge chasing process (a Givens rotation is an orthogonal block diagonal matrix consisting of a single block which is a 2D rotation, the other diagonal entries being equal to 1). Similarly, in the case of a double shift, $A_{k+2} = (Q_k Q_{k+1})^* A_k Q_k Q_{k+1}$ is generated without having to compute the QR-factorizations of $A_k - \sigma_k I$ and $A_{k+1} - \overline{\sigma}_k I$. Again, $(Q_k Q_{k+1})^* A_k Q_k Q_{k+1}$ is generated by applying some simple orthogonal matrices which perform a bulge chasing process. See Demmel [Demmel (1997)] (Section 4.4.8) and Golub and Van Loan [Golub and Van Loan (1996)] (Section 7.5) for further explanations regarding implicit shifting involving bulge chasing in the case of real matrices. Watkins [Watkins (1982, 2008)] discusses bulge chasing in the more general case of complex matrices.

The `Matlab` function for finding the eigenvalues and the eigenvectors of a matrix A is `eig` and is called as `[U, D] = eig(A)`. It is implemented using an optimized version of the QR-algorithm with implicit shifts.

If the dimension of the matrix A is very large, we can find approximations of some of the eigenvalues of A by using a truncated version of the reduction to Hessenberg form due to Arnoldi in general and to Lanczos in the symmetric (or Hermitian) tridiagonal case.

17.4 Krylov Subspaces; Arnoldi Iteration

In this section, we denote the dimension of the square real or complex matrix A by m rather than n, to make it easier for the reader to follow Trefethen and Bau exposition [Trefethen and Bau III (1997)], which is particularly lucid.

Suppose that the $m \times m$ matrix A has been reduced to the upper Hessenberg form H, as $A = UHU^*$. For any $n \leq m$ (typically much smaller than m), consider the $(n + 1) \times n$ upper left block

$$\widetilde{H}_n = \begin{pmatrix} h_{11} & h_{12} & h_{13} & \cdots & h_{1n} \\ h_{21} & h_{22} & h_{23} & \cdots & h_{2n} \\ 0 & h_{32} & h_{33} & \cdots & h_{3n} \\ \vdots & \ddots & \ddots & \ddots & \vdots \\ 0 & \cdots & 0 & h_{nn-1} & h_{nn} \\ 0 & \cdots & 0 & 0 & h_{n+1n} \end{pmatrix}$$

of H, and the $n \times n$ upper Hessenberg matrix H_n obtained by deleting the last row of $\widetilde{H}_n$,

$$H_n = \begin{pmatrix} h_{11} & h_{12} & h_{13} & \cdots & h_{1n} \\ h_{21} & h_{22} & h_{23} & \cdots & h_{2n} \\ 0 & h_{32} & h_{33} & \cdots & h_{3n} \\ \vdots & \ddots & \ddots & \ddots & \vdots \\ 0 & \cdots & 0 & h_{nn-1} & h_{nn} \end{pmatrix}.$$

If we denote by U_n the $m \times n$ matrix consisting of the first n columns of U, denoted $u_1, \ldots, u_n$, then matrix consisting of the first n columns of the matrix $UH = AU$ can be expressed as

$$AU_n = U_{n+1}\widetilde{H}_n. \tag{17.13}$$

It follows that the nth column of this matrix can be expressed as

$$Au_n = h_{1n}u_1 + \cdots + h_{nn}u_n + h_{n+1n}u_{n+1}. \tag{17.14}$$

Since $(u_1, \ldots, u_n)$ form an orthonormal basis, we deduce from (17.14) that

$$\langle u_j, Au_n \rangle = u_j^* Au_n = h_{jn}, \quad j = 1, \ldots, n. \tag{17.15}$$

Equations (17.14) and (17.15) show that U_{n+1} and $\widetilde{H}_n$ can be computed iteratively using the following algorithm due to Arnoldi, known as *Arnoldi iteration*:

Given an arbitrary nonzero vector $b \in \mathbb{C}^m$, let $u_1 = b/ \|b\|$;

for $n = 1, 2, 3, \ldots$ **do**

$z := Au_n$;

for $j = 1$ **to** n **do**

$h_{jn} := u_j^* z$;

$z := z - h_{jn} u_j$

endfor

$h_{n+1n} := \|z\|$;

if $h_{n+1n} = 0$ **quit**

$u_{n+1} = z/h_{n+1n}$

When $h_{n+1n} = 0$, we say that we have a *breakdown* of the Arnoldi iteration.

Arnoldi iteration is an algorithm for producing the $n \times n$ Hessenberg submatrix H_n of the full Hessenberg matrix H consisting of its first n rows and n columns (the first n columns of U are also produced), not using Householder matrices.

As long as $h_{j+1j} \neq 0$ for $j = 1, \ldots, n$, equation (17.14) shows by an easy induction that u_{n+1} belong to the span of $(b, Ab, \ldots, A^n b)$, and obviously Au_n belongs to the span of $(u_1, \ldots, u_{n+1})$, and thus the following spaces are identical:

$$\mathrm{Span}(b, Ab, \ldots, A^n b) = \mathrm{Span}(u_1, \ldots, u_{n+1}).$$

The space $\mathcal{K}_n(A, b) = \mathrm{Span}(b, Ab, \ldots, A^{n-1} b)$ is called a *Krylov subspace*. We can view Arnoldi's algorithm as the construction of an orthonormal basis for $\mathcal{K}_n(A, b)$. It is a sort of Gram–Schmidt procedure.

Equation (17.14) shows that if K_n is the $m \times n$ matrix whose columns are the vectors $(b, Ab, \ldots, A^{n-1} b)$, then there is a $n \times n$ upper triangular matrix R_n such that

$$K_n = U_n R_n. \tag{17.16}$$

The above is called a *reduced QR factorization* of K_n.

Since $(u_1, \ldots, u_n)$ is an orthonormal system, the matrix $U_n^* U_{n+1}$ is the $n \times (n+1)$ matrix consisting of the identity matrix I_n plus an extra column of 0's, so $U_n^* U_{n+1} \widetilde{H}_n = U_n^* A U_n$ is obtained by deleting the last row of $\widetilde{H}_n$, namely H_n, and so

$$U_n^* A U_n = H_n. \tag{17.17}$$

We summarize the above facts in the following proposition.

Proposition 17.2. *If Arnoldi iteration run on an $m \times m$ matrix A starting with a nonzero vector $b \in \mathbb{C}^m$ does not have a breakdown at stage $n \le m$, then the following properties hold:*

(1) If K_n is the $m \times n$ Krylov matrix associated with the vectors $(b, Ab, \ldots, A^{n-1}b)$ and if U_n is the $m \times n$ matrix of orthogonal vectors produced by Arnoldi iteration, then there is a QR-factorization

$$K_n = U_n R_n,$$

for some $n \times n$ upper triangular matrix R_n.

(2) The $m \times n$ upper Hessenberg matrices H_n produced by Arnoldi iteration are the projection of A onto the Krylov space $\mathcal{K}_n(A, b)$, that is,

$$H_n = U_n^* A U_n.$$

(3) The successive iterates are related by the formula

$$A U_n = U_{n+1} \widetilde{H}_n.$$

Remark: If Arnoldi iteration has a breakdown at stage n, that is, $h_{n+1} = 0$, then we found the first unreduced block of the Hessenberg matrix H. It can be shown that the eigenvalues of H_n are eigenvalues of A. So a breakdown is actually a good thing. In this case, we can pick some new nonzero vector u_{n+1} orthogonal to the vectors $(u_1, \ldots, u_n)$ as a new starting vector and run Arnoldi iteration again. Such a vector exists since the $(n+1)$th column of U works. So repeated application of Arnoldi yields a full Hessenberg reduction of A. However, this is not what we are after, since m is very large an we are only interested in a "small" number of eigenvalues of A.

There is another aspect of Arnoldi iteration, which is that it solves an optimization problem involving polynomials of degree n. Let $\mathcal{P}^n$ denote the set of (complex) monic polynomials of degree n, that is, polynomials of the form

$$p(z) = z^n + c_{n-1} z^{n-1} + \cdots + c_1 z + c_0 \quad (c_i \in \mathbb{C}).$$

For any $m \times m$ matrix A, we write

$$p(A) = A^n + c_{n-1}A^{n-1} + \cdots + c_1 A + c_0 I.$$

The following result is proven in Trefethen and Bau [Trefethen and Bau III (1997)] (Lecture 34, Theorem 34.1).

Theorem 17.4. *If Arnoldi iteration run on an $m \times m$ matrix A starting with a nonzero vector b does not have a breakdown at stage $n \leq m$, then there is a unique polynomial $p \in \mathcal{P}^n$ such that $\|p(A)b\|_2$ is minimum, namely the characteristic polynomial $\det(zI - H_n)$ of H_n.*

Theorem 17.4 can be viewed as the "justification" for a method to find some of the eigenvalues of A (say $n \ll m$ of them). Intuitively, the closer the roots of the characteristic polynomials of H_n are to the eigenvalues of A, the smaller $\|p(A)b\|_2$ should be, and conversely. In the extreme case where $m = n$, by the Cayley–Hamilton theorem, $p(A) = 0$ (where p is the characteristic polynomial of A), so this idea is plausible, but this is far from constituting a proof (also, b should have nonzero coordinates in all directions associated with the eigenvalues).

The method known as the *Rayleigh–Ritz method* is to run Arnoldi iteration on A and some $b \neq 0$ chosen at random for $n \ll m$ steps before or until a breakdown occurs. Then run the QR algorithm with shifts on H_n. The eigenvalues of the Hessenberg matrix H_n may then be considered as approximations of the eigenvalues of A. The eigenvalues of H_n are called *Arnoldi estimates* or *Ritz values*. One has to be cautious because H_n is a truncated version of the full Hessenberg matrix H, so not all of the Ritz values are necessary close to eigenvalues of A. It has been observed that the eigenvalues that are found first are the *extreme* eigenvalues of A, namely those close to the boundary of the spectrum of A plotted in $\mathbb{C}$. So if A has real eigenvalues, the largest and the smallest eigenvalues appear first as Ritz values. In many problems where eigenvalues occur, the extreme eigenvalues are the one that need to be computed. Similarly, the eigenvectors of H_n may be considered as approximations of eigenvectors of A.

The `Matlab` function `eigs` is based on the computation of Ritz values. It computes the six eigenvalues of largest magnitude of a matrix A, and the call is `[V, D] = eigs(A)`. More generally, to get the top k eigenvalues, use `[V, D] = eigs(A, k)`.

In the absence of rigorous theorems about error estimates, it is hard to make the above statements more precise; see Trefethen and Bau [Trefethen and Bau III (1997)] (Lecture 34) for more on this subject.

However, if A is a symmetric (or Hermitian) matrix, then H_n is a symmetric (resp. Hermitian) tridiagonal matrix and more precise results can be shown; see Demmel [Demmel (1997)] (Chapter 7, especially Section 7.2). We will consider the symmetric (and Hermitian) case in the next section, but first we show how Arnoldi iteration can be used to find approximations for the solution of a linear system $Ax = b$ where A is invertible but of very large dimension m.

17.5   GMRES

Suppose A is an invertible $m \times m$ matrix and let b be a nonzero vector in $\mathbb{C}^m$. Let $x_0 = A^{-1}b$, the unique solution of $Ax = b$. It is not hard to show that $x_0 \in \mathcal{K}_n(A, b)$ for some $n \leq m$. In fact, there is a unique monic polynomial $p(z)$ of minimal degree $s \leq m$ such that $p(A)b = 0$, so $x_0 \in \mathcal{K}_s(A, b)$. Thus it makes sense to search for a solution of $Ax = b$ in Krylov spaces of dimension $m \leq s$. The idea is to find an approximation $x_n \in \mathcal{K}_n(A, b)$ of x_0 such that $r_n = b - Ax_n$ is minimized, that is, $\|r_n\|_2 = \|b - Ax_n\|_2$ is minimized over $x_n \in \mathcal{K}_n(A, b)$. This minimization problem can be stated as

$$\text{minimize} \quad \|r_n\|_2 = \|Ax_n - b\|_2, \quad x_n \in \mathcal{K}_n(A, b).$$

This is a least-squares problem, and we know how to solve it (see Section 21.1). The quantity r_n is known as the *residual* and the method which consists in minimizing $\|r_n\|_2$ is known as GMRES, for *generalized minimal residuals*.

Now since $(u_1, \ldots, u_n)$ is a basis of $\mathcal{K}_n(A, b)$ (since $n \leq s$, no breakdown occurs, except for $n = s$), we may write $x_n = U_n y$, so our minimization problem is

$$\text{minimize} \quad \|AU_n y - b\|_2, \quad y \in \mathbb{C}^n.$$

Since by (17.13) of Section 17.4, we have $AU_n = U_{n+1}\widetilde{H}_n$, minimizing $\|AU_n y - b\|_2$ is equivalent to minimizing $\|U_{n+1}\widetilde{H}_n y - b\|_2$ over $\mathbb{C}^m$. Since $U_{n+1}\widetilde{H}_n y$ and b belong to the column space of U_{n+1}, minimizing $\|U_{n+1}\widetilde{H}_n y - b\|_2$ is equivalent to minimizing $\|\widetilde{H}_n y - U_{n+1}^* b\|_2$. However, by construction,

$$U_{n+1}^* b = \|b\|_2 e_1 \in \mathbb{C}^{n+1},$$

so our minimization problem can be stated as

$$\text{minimize} \quad \|\widetilde{H}_n y - \|b\|_2 e_1\|_2, \quad y \in \mathbb{C}^n.$$

The approximate solution of $Ax = b$ is then

$$x_n = U_n y.$$

Starting with $u_1 = b/\|b\|_2$ and with $n = 1$, the GMRES method runs $n \leq s$ Arnoldi iterations to find U_n and $\widetilde{H}_n$, and then runs a method to solve the least squares problem

$$\text{minimize} \quad \|\widetilde{H}_n y - \|b\|_2 e_1\|_2, \quad y \in \mathbb{C}^n.$$

When $\|r_n\|_2 = \|\widetilde{H}_n y - \|b\|_2 e_1\|_2$ is considered small enough, we stop and the approximate solution of $Ax = b$ is then

$$x_n = U_n y.$$

There are ways of improving efficiency of the "naive" version of GM-RES that we just presented; see Trefethen and Bau [Trefethen and Bau III (1997)] (Lecture 35). We now consider the case where A is a Hermitian (or symmetric) matrix.

17.6   The Hermitian Case; Lanczos Iteration

If A is an $m \times m$ symmetric or Hermitian matrix, then Arnoldi's method is simpler and much more efficient. Indeed, in this case, it is easy to see that the upper Hessenberg matrices H_n are also symmetric (Hermitian respectively), and thus tridiagonal. Also, the eigenvalues of A and H_n are real. It is convenient to write

$$H_n = \begin{pmatrix} \alpha_1 & \beta_1 & & & \\ \beta_1 & \alpha_2 & \beta_2 & & \\ & \beta_2 & \alpha_3 & \ddots & \\ & & \ddots & \ddots & \beta_{n-1} \\ & & & \beta_{n-1} & \alpha_n \end{pmatrix}.$$

The recurrence (17.14) of Section 17.4 becomes the three-term recurrence

$$Au_n = \beta_{n-1} u_{n-1} + \alpha_n u_n + \beta_n u_{n+1}. \tag{17.18}$$

We also have $\alpha_n = u_n^* A U_n$, so Arnoldi's algorithm become the following algorithm known as *Lanczos' algorithm* (or *Lanczos iteration*). The inner loop on j from 1 to n has been eliminated and replaced by a single assignment.

Given an arbitrary nonzero vector $b \in \mathbb{C}^m$, let $u_1 = b/\|b\|$;

for $n = 1, 2, 3, \ldots$ **do**

$z := Au_n;$

$\quad \alpha_n := u_n^* z;$

$\quad z := z - \beta_{n-1} u_{n-1} - \alpha_n u_n$

$\beta_n := \|z\|;$

if $\beta_n = 0$ **quit**

$u_{n+1} = z/\beta_n$

When $\beta_n = 0$, we say that we have a *breakdown* of the Lanczos iteration.

Versions of Proposition 17.2 and Theorem 17.4 apply to Lanczos iteration.

Besides being much more efficient than Arnoldi iteration, Lanczos iteration has the advantage that the *Rayleigh–Ritz method* for finding some of the eigenvalues of A as the eigenvalues of the symmetric (respectively Hermitian) tridiagonal matrix H_n applies, but there are more methods for finding the eigenvalues of symmetric (respectively Hermitian) tridiagonal matrices. Also theorems about error estimates exist. The version of Lanczos iteration given above may run into problems in floating point arithmetic. What happens is that the vectors u_j may lose the property of being orthogonal, so it may be necessary to reorthogonalize them. For more on all this, see Demmel [Demmel (1997)] (Chapter 7, in particular Sections 7.2–7.4). The version of GMRES using Lanczos iteration is called MINRES.

We close our brief survey of methods for computing the eigenvalues and the eigenvectors of a matrix with a quick discussion of two methods known as power methods.

17.7   Power Methods

Let A be an $m \times m$ complex or real matrix. There are two power methods, both of which yield one eigenvalue and one eigenvector associated with this vector:

(1) *Power iteration.*
(2) *Inverse (power) iteration.*

Power iteration only works if the matrix A has an eigenvalue λ of largest modulus, which means that if $\lambda_1, \ldots, \lambda_m$ are the eigenvalues of A, then

$$|\lambda_1| > |\lambda_2| \geq \cdots \geq |\lambda_m| \geq 0.$$

In particular, if A is a real matrix, then λ_1 must be real (since otherwise there are two complex conjugate eigenvalues of the same largest modulus).

If the above condition is satisfied, then power iteration yields λ_1 and some eigenvector associated with it. The method is simple enough:

Pick some initial unit vector x^0 and compute the following sequence (x^k), where

$$x^{k+1} = \frac{Ax^k}{\|Ax^k\|}, \quad k \geq 0.$$

We would expect that (x^k) converges to an eigenvector associated with λ_1, but this is not quite correct. The following results are proven in Serre [Serre (2010)] (Section 13.5.1). First assume that $\lambda_1 \neq 0$.

We have

$$\lim_{k \mapsto \infty} \left\| Ax^k \right\| = |\lambda_1|.$$

If A is a complex matrix which has a unique complex eigenvalue λ_1 of largest modulus, then

$$v = \lim_{k \mapsto \infty} \left(\frac{\overline{\lambda_1}}{|\lambda_1|} \right)^k x^k$$

is a unit eigenvector of A associated with λ_1. If λ_1 is real, then

$$v = \lim_{k \mapsto \infty} x^k$$

is a unit eigenvector of A associated with λ_1. Actually some condition on x^0 is needed: x^0 must have a nonzero component in the eigenspace E associated with λ_1 (in any direct sum of $\mathbb{C}^m$ in which E is a summand).

The eigenvalue λ_1 is found as follows. If λ_1 is complex, and if $v_j \neq 0$ is any nonzero coordinate of v, then

$$\lambda_1 = \lim_{k \mapsto \infty} \frac{(Ax^k)_j}{x_j^k}.$$

If λ_1 is real, then we can define the sequence $(\lambda^{(k)})$ by

$$\lambda^{(k+1)} = (x^{k+1})^* Ax^{k+1}, \quad k \geq 0,$$

and we have

$$\lambda_1 = \lim_{k \mapsto \infty} \lambda^{(k)}.$$

Indeed, in this case, since $v = \lim_{k \mapsto \infty} x^k$ and v is a unit eigenvector for λ_1, we have

$$\lim_{k \mapsto \infty} \lambda^{(k)} = \lim_{k \mapsto \infty} (x^{k+1})^* Ax^{k+1} = v^* Av = \lambda_1 v^* v = \lambda_1.$$

Note that since x^{k+1} is a unit vector, $(x^{k+1})^* Ax^{k+1}$ is a Rayleigh ratio.

If A is a Hermitian matrix, then the eigenvalues are real and we can say more about the rate of convergence, which is not great (only linear). For details, see Trefethen and Bau [Trefethen and Bau III (1997)] (Lecture 27). If $\lambda_1 = 0$, then there is some power $\ell < m$ such that $Ax^\ell = 0$.

The *inverse iteration method* is designed to find an eigenvector associated with an eigenvalue λ of A for which we know a good approximation μ.

Pick some initial unit vector x^0 and compute the following sequences (w^k) and (x^k), where w^{k+1} is the solution of the system

$$(A - \mu I)w^{k+1} = x^k \quad \text{equivalently} \quad w^{k+1} = (A - \mu I)^{-1}x^k, \quad k \geq 0,$$

and

$$x^{k+1} = \frac{w^{k+1}}{\|w^{k+1}\|}, \quad k \geq 0.$$

The following result is proven in Ciarlet [Ciarlet (1989)] (Theorem 6.4.1).

Proposition 17.3. *Let A be an $m \times m$ diagonalizable (complex or real) matrix with eigenvalues $\lambda_1, \ldots, \lambda_m$, and let $\lambda = \lambda_\ell$ be an arbitrary eigenvalue of A (not necessary simple). For any μ such that*

$$\mu \neq \lambda \quad and \quad |\mu - \lambda| < |\mu - \lambda_j| \quad for \ all \ j \neq \ell,$$

if x^0 does not belong to the subspace spanned by the eigenvectors associated with the eigenvalues λ_j with $j \neq \ell$, then

$$\lim_{k \mapsto \infty} \left(\frac{(\lambda - \mu)^k}{|\lambda - \mu|^k} \right) x^k = v,$$

where v is an eigenvector associated with λ. Furthermore, if both λ and μ are real, we have

$$\lim_{k \mapsto \infty} x^k = v \qquad if \ \mu < \lambda,$$

$$\lim_{k \mapsto \infty} (-1)^k x^k = v \qquad if \ \mu > \lambda.$$

Also, if we define the sequence $(\lambda^{(k)})$ by

$$\lambda^{(k+1)} = (x^{k+1})^* A x^{k+1},$$

then

$$\lim_{k \mapsto \infty} \lambda^{(k+1)} = \lambda.$$

The condition of x^0 may seem quite stringent, but in practice, a vector x^0 chosen at random usually satisfies it.

If A is a Hermitian matrix, then we can say more. In particular, the inverse iteration algorithm can be modified to make use of the newly computed $\lambda^{(k+1)}$ instead of μ, and an even faster convergence is achieved. Such a method is called the *Rayleigh quotient iteration*. When it converges (which is for almost all x^0), this method eventually achieves cubic convergence, which is remarkable. Essentially, this means that the number of correct digits is tripled at every iteration. For more details, see Trefethen and Bau [Trefethen and Bau III (1997)] (Lecture 27) and Demmel [Demmel (1997)] (Section 5.3.2).

17.8   Summary

The main concepts and results of this chapter are listed below:

- QR iteration, QR algorithm.
- Upper Hessenberg matrices.
- Householder matrix.
- Unreduced and reduced Hessenberg matrices.
- Deflation.
- Shift.
- Wilkinson shift.
- Double shift.
- Francis shift.
- Implicit shifting.
- Implicit Q-theorem.
- Arnoldi iteration.
- Breakdown of Arnoldi iteration.
- Krylov subspace.
- Rayleigh–Ritz method.
- Ritz values, Arnoldi estimates.
- Residual.
- GMRES
- Lanczos iteration.
- Power iteration.
- Inverse power iteration.
- Rayleigh ratio.

17.9 Problems

Problem 17.1. Prove Theorem 17.2; see Problem 12.7.

Problem 17.2. Prove that if a matrix A is Hermitian (or real symmetric), then any Hessenberg matrix H similar to A is Hermitian tridiagonal (real symmetric tridiagonal).

Problem 17.3. For any matrix (real or complex) A, if $A = QR$ is a QR-decomposition of A using Householder reflections, prove that if A is upper Hessenberg then so is Q.

Problem 17.4. Prove that if A is upper Hessenberg, then the matrices A_k obtained by applying the QR-algorithm are also upper Hessenberg.

Problem 17.5. Prove the *implicit Q theorem*. This theorem says that if A is reduced to upper Hessenberg form as $A = UHU^*$ and if H is unreduced ($h_{i+1i} \neq 0$ for $i = 1, \ldots, n-1$), then the columns of index $2, \ldots, n$ of U are determined by the first column of U up to sign.

Problem 17.6. Read Section 7.5 of Golub and Van Loan [Golub and Van Loan (1996)] and implement their version of the QR-algorithm with shifts.

Problem 17.7. If an Arnoldi iteration has a breakdown at stage n, that is, $h_{n+1} = 0$, then we found the first unreduced block of the Hessenberg matrix H. Prove that the eigenvalues of H_n are eigenvalues of A.

Problem 17.8. Prove Theorem 17.4.

Problem 17.9. Implement GRMES and test it on some linear systems.

Problem 17.10. State and prove versions of Proposition 17.2 and Theorem 17.4 for the Lanczos iteration.

Problem 17.11. Prove the results about the power iteration method stated in Section 17.7.

Problem 17.12. Prove the results about the inverse power iteration method stated in Section 17.7.

Problem 17.13. Implement and test the power iteration method and the inverse power iteration method.

Problem 17.14. Read Lecture 27 in Trefethen and Bau [Trefethen and Bau III (1997)] and implement and test the Rayleigh quotient iteration method.

Chapter 18

Graphs and Graph Laplacians; Basic Facts

In this chapter and the next we present some applications of linear algebra to graph theory. Graphs (undirected and directed) can be defined in terms of various matrices (incidence and adjacency matrices), and various connectivity properties of graphs are captured by properties of these matrices. Another very important matrix is associated with a (undirected) graph: the *graph Laplacian*. The graph Laplacian is symmetric positive definite, and its eigenvalues capture some of the properties of the underlying graph. This is a key fact that is exploited in graph clustering methods, the most powerful being the method of normalized cuts due to Shi and Malik [Shi and Malik (2000)]. This chapter and the next constitute an introduction to algebraic and spectral graph theory. We do not discuss normalized cuts, but we discuss graph drawings. Thorough presentations of algebraic graph theory can be found in Godsil and Royle [Godsil and Royle (2001)] and Chung [Chung (1997)].

We begin with a review of basic notions of graph theory. Even though the graph Laplacian is fundamentally associated with an undirected graph, we review the definition of both directed and undirected graphs. For both directed and undirected graphs, we define the degree matrix D, the incidence matrix B, and the adjacency matrix A. Then we define a *weighted graph*. This is a pair (V, W), where V is a finite set of nodes and W is a $m \times m$ symmetric matrix with nonnegative entries and zero diagonal entries (where $m = |V|$).

For every node $v_i \in V$, the *degree* $d(v_i)$ (or d_i) of v_i is the sum of the weights of the edges adjacent to v_i:

$$d_i = d(v_i) = \sum_{j=1}^{m} w_{ij}.$$

659

The *degree matrix* is the diagonal matrix

$$D = \mathrm{diag}(d_1, \ldots, d_m).$$

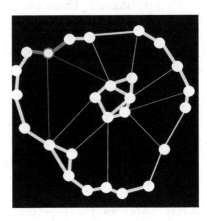

Fig. 18.1   Degree of a node.

The notion of degree is illustrated in Figure 18.1. Then we introduce the (unnormalized) *graph Laplacian L* of a directed graph G in an "old-fashion" way, by showing that for any orientation of a graph G,

$$BB^\top = D - A = L$$

is an invariant. We also define the (unnormalized) *graph Laplacian L* of a weighted graph $G = (V, W)$ as $L = D - W$. We show that the notion of incidence matrix can be generalized to weighted graphs in a simple way. For any graph G^σ obtained by orienting the underlying graph of a weighted graph $G = (V, W)$, there is an incidence matrix B^σ such that

$$B^\sigma (B^\sigma)^\top = D - W = L.$$

We also prove that

$$x^\top L x = \frac{1}{2} \sum_{i,j=1}^{m} w_{ij}(x_i - x_j)^2 \quad \text{for all } x \in \mathbb{R}^m.$$

Consequently, $x^\top L x$ does not depend on the diagonal entries in W, and if $w_{ij} \geq 0$ for all $i, j \in \{1, \ldots, m\}$, then L is positive semidefinite. Then if W consists of nonnegative entries, the eigenvalues $0 = \lambda_1 \leq \lambda_2 \leq \ldots \leq \lambda_m$ of L are real and nonnegative, and there is an orthonormal basis of eigenvectors of L. We show that the number of connected components of the graph

$G = (V, W)$ is equal to the dimension of the kernel of L, which is also equal to the dimension of the kernel of the transpose $(B^\sigma)^\top$ of any incidence matrix B^σ obtained by orienting the underlying graph of G.

We also define the normalized graph Laplacians L_{sym} and L_{rw}, given by

$$L_{\text{sym}} = D^{-1/2}LD^{-1/2} = I - D^{-1/2}WD^{-1/2}$$
$$L_{\text{rw}} = D^{-1}L = I - D^{-1}W,$$

and prove some simple properties relating the eigenvalues and the eigenvectors of L, L_{sym} and L_{rw}. These normalized graph Laplacians show up when dealing with normalized cuts.

Next, we turn to *graph drawings* (Chapter 19). Graph drawing is a very attractive application of so-called spectral techniques, which is a fancy way of saying that that eigenvalues and eigenvectors of the graph Laplacian are used. Furthermore, it turns out that graph clustering using normalized cuts can be cast as a certain type of graph drawing.

Given an undirected graph $G = (V, E)$, with $|V| = m$, we would like to draw G in $\mathbb{R}^n$ for n (much) smaller than m. The idea is to assign a point $\rho(v_i)$ in $\mathbb{R}^n$ to the vertex $v_i \in V$, for every $v_i \in V$, and to draw a line segment between the points $\rho(v_i)$ and $\rho(v_j)$. Thus, a *graph drawing* is a function $\rho \colon V \to \mathbb{R}^n$.

We define the *matrix of a graph drawing ρ (in $\mathbb{R}^n$)* as a $m \times n$ matrix R whose ith row consists of the row vector $\rho(v_i)$ corresponding to the point representing v_i in $\mathbb{R}^n$. Typically, we want $n < m$; in fact n should be much smaller than m.

Since there are infinitely many graph drawings, it is desirable to have some criterion to decide which graph is better than another. Inspired by a physical model in which the edges are springs, it is natural to consider a representation to be better if it requires the springs to be less extended. We can formalize this by defining the *energy* of a drawing R by

$$\mathcal{E}(R) = \sum_{\{v_i, v_j\} \in E} \|\rho(v_i) - \rho(v_j)\|^2,$$

where $\rho(v_i)$ is the ith row of R and $\|\rho(v_i) - \rho(v_j)\|^2$ is the square of the Euclidean length of the line segment joining $\rho(v_i)$ and $\rho(v_j)$.

Then "good drawings" are drawings that minimize the energy function $\mathcal{E}$. Of course, the trivial representation corresponding to the zero matrix is optimum, so we need to impose extra constraints to rule out the trivial solution.

We can consider the more general situation where the springs are not necessarily identical. This can be modeled by a symmetric weight (or stiffness) matrix $W = (w_{ij})$, with $w_{ij} \geq 0$. In this case, our energy function becomes

$$\mathcal{E}(R) = \sum_{\{v_i, v_j\} \in E} w_{ij} \|\rho(v_i) - \rho(v_j)\|^2.$$

Following Godsil and Royle [Godsil and Royle (2001)], we prove that

$$\mathcal{E}(R) = \operatorname{tr}(R^\top L R),$$

where

$$L = D - W,$$

is the familiar unnormalized Laplacian matrix associated with W, and where D is the degree matrix associated with W.

It can be shown that there is no loss in generality in assuming that the columns of R are pairwise orthogonal and that they have unit length. Such a matrix satisfies the equation $R^\top R = I$ and the corresponding drawing is called an *orthogonal drawing*. This condition also rules out trivial drawings.

Then we prove the main theorem about graph drawings (Theorem 19.1), which essentially says that the matrix R of the desired graph drawing is constituted by the n eigenvectors of L associated with the smallest nonzero n eigenvalues of L. We give a number examples of graph drawings, many of which are borrowed or adapted from Spielman [Spielman (2012)].

18.1 Directed Graphs, Undirected Graphs, Incidence Matrices, Adjacency Matrices, Weighted Graphs

Definition 18.1. A *directed graph* is a pair $G = (V, E)$, where $V = \{v_1, \ldots, v_m\}$ is a set of *nodes* or *vertices*, and $E \subseteq V \times V$ is a set of ordered pairs of distinct nodes (that is, pairs $(u, v) \in V \times V$ with $u \neq v$), called *edges*. Given any edge $e = (u, v)$, we let $s(e) = u$ be the *source* of e and $t(e) = v$ be the *target* of e.

Remark: Since an edge is a pair (u, v) with $u \neq v$, self-loops are not allowed. Also, there is at most one edge from a node u to a node v. Such graphs are sometimes called *simple graphs*.

An example of a directed graph is shown in Figure 18.2.

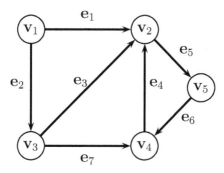

Fig. 18.2 Graph G_1.

Definition 18.2. For every node $v \in V$, the *degree* $d(v)$ of v is the number of edges leaving or entering v:

$$d(v) = |\{u \in V \mid (v, u) \in E \text{ or } (u, v) \in E\}|.$$

We abbreviate $d(v_i)$ as d_i. The *degree matrix*, $D(G)$, is the diagonal matrix

$$D(G) = \text{diag}(d_1, \ldots, d_m).$$

For example, for graph G_1, we have

$$D(G_1) = \begin{pmatrix} 2 & 0 & 0 & 0 & 0 \\ 0 & 4 & 0 & 0 & 0 \\ 0 & 0 & 3 & 0 & 0 \\ 0 & 0 & 0 & 3 & 0 \\ 0 & 0 & 0 & 0 & 2 \end{pmatrix}.$$

Unless confusion arises, we write D instead of $D(G)$.

Definition 18.3. Given a directed graph $G = (V, E)$, for any two nodes $u, v \in V$, a *path from u to v* is a sequence of nodes $(v_0, v_1, \ldots, v_k)$ such that $v_0 = u$, $v_k = v$, and (v_i, v_{i+1}) is an edge in E for all i with $0 \le i \le k - 1$. The integer k is the *length* of the path. A path is *closed* if $u = v$. The graph G is *strongly connected* if for any two distinct nodes $u, v \in V$, there is a path from u to v and there is a path from v to u.

Remark: The terminology *walk* is often used instead of *path*, the word path being reserved to the case where the nodes v_i are all distinct, except that $v_0 = v_k$ when the path is closed.

The binary relation on $V \times V$ defined so that u and v are related iff there is a path from u to v and there is a path from v to u is an equivalence relation whose equivalence classes are called the *strongly connected components* of G.

Definition 18.4. Given a directed graph $G = (V, E)$, with $V = \{v_1, \ldots, v_m\}$, if $E = \{e_1, \ldots, e_n\}$, then the *incidence matrix* $B(G)$ of G is the $m \times n$ matrix whose entries b_{ij} are given by

$$b_{ij} = \begin{cases} +1 & \text{if } s(e_j) = v_i \\ -1 & \text{if } t(e_j) = v_i \\ 0 & \text{otherwise.} \end{cases}$$

Here is the incidence matrix of the graph G_1:

$$B = \begin{pmatrix} 1 & 1 & 0 & 0 & 0 & 0 & 0 \\ -1 & 0 & -1 & -1 & 1 & 0 & 0 \\ 0 & -1 & 1 & 0 & 0 & 0 & 1 \\ 0 & 0 & 0 & 1 & 0 & -1 & -1 \\ 0 & 0 & 0 & 0 & -1 & 1 & 0 \end{pmatrix}.$$

Observe that every column of an incidence matrix contains exactly two nonzero entries, $+1$ and -1. Again, unless confusion arises, we write B instead of $B(G)$.

When a directed graph has m nodes $v_1, \ldots, v_m$ and n edges $e_1, \ldots, e_n$, a vector $x \in \mathbb{R}^m$ can be viewed as a function $x \colon V \to \mathbb{R}$ assigning the value x_i to the node v_i. Under this interpretation, $\mathbb{R}^m$ is viewed as $\mathbb{R}^V$. Similarly, a vector $y \in \mathbb{R}^n$ can be viewed as a function in $\mathbb{R}^E$. This point of view is often useful. For example, the incidence matrix B can be interpreted as a linear map from $\mathbb{R}^E$ to $\mathbb{R}^V$, the *boundary map*, and $B^\top$ can be interpreted as a linear map from $\mathbb{R}^V$ to $\mathbb{R}^E$, the *coboundary map*.

Remark: Some authors adopt the opposite convention of sign in defining the incidence matrix, which means that their incidence matrix is $-B$.

Undirected graphs are obtained from directed graphs by forgetting the orientation of the edges.

Definition 18.5. A *graph* (or *undirected graph*) is a pair $G = (V, E)$, where $V = \{v_1, \ldots, v_m\}$ is a set of *nodes* or *vertices*, and E is a set of two-element subsets of V (that is, subsets $\{u, v\}$, with $u, v \in V$ and $u \neq v$), called *edges*.

Remark: Since an edge is a set $\{u, v\}$, we have $u \neq v$, so self-loops are not allowed. Also, for every set of nodes $\{u, v\}$, there is at most one edge between u and v. As in the case of directed graphs, such graphs are sometimes called *simple graphs*.

An example of a graph is shown in Figure 18.3.

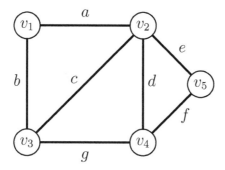

Fig. 18.3 The undirected graph G_2.

Definition 18.6. For every node $v \in V$, the *degree* $d(v)$ of v is the number of edges incident to v:

$$d(v) = |\{u \in V \mid \{u, v\} \in E\}|.$$

The degree matrix $D(G)$ (or simply, D) is defined as in Definition 18.2.

Definition 18.7. Given a (undirected) graph $G = (V, E)$, for any two nodes $u, v \in V$, a *path from u to v* is a sequence of nodes $(v_0, v_1, \ldots, v_k)$ such that $v_0 = u$, $v_k = v$, and $\{v_i, v_{i+1}\}$ is an edge in E for all i with $0 \leq i \leq k - 1$. The integer k is the *length* of the path. A path is *closed* if $u = v$. The graph G is *connected* if for any two distinct nodes $u, v \in V$, there is a path from u to v.

Remark: The terminology *walk* or *chain* is often used instead of *path*, the word path being reserved to the case where the nodes v_i are all distinct, except that $v_0 = v_k$ when the path is closed.

The binary relation on $V \times V$ defined so that u and v are related iff there is a path from u to v is an equivalence relation whose equivalence classes are called the *connected components* of G.

The notion of incidence matrix for an undirected graph is not as useful as in the case of directed graphs.

Definition 18.8. Given a graph $G = (V, E)$, with $V = \{v_1, \ldots, v_m\}$, if $E = \{e_1, \ldots, e_n\}$, then the *incidence matrix* $B(G)$ of G is the $m \times n$ matrix whose entries b_{ij} are given by

$$b_{ij} = \begin{cases} +1 & \text{if } e_j = \{v_i, v_k\} \text{ for some } k \\ 0 & \text{otherwise.} \end{cases}$$

Unlike the case of directed graphs, the entries in the incidence matrix of a graph (undirected) are nonnegative. We usually write B instead of $B(G)$.

Definition 18.9. If $G = (V, E)$ is a directed or an undirected graph, given a node $u \in V$, any node $v \in V$ such that there is an edge (u, v) in the directed case or $\{u, v\}$ in the undirected case is called *adjacent to u*, and we often use the notation

$$u \sim v.$$

Observe that the binary relation $\sim$ is symmetric when G is an undirected graph, but in general it is not symmetric when G is a directed graph.

The notion of adjacency matrix is basically the same for directed or undirected graphs.

Definition 18.10. Given a directed or undirected graph $G = (V, E)$, with $V = \{v_1, \ldots, v_m\}$, the *adjacency matrix* $A(G)$ of G is the symmetric $m \times m$ matrix (a_{ij}) such that

(1) If G is directed, then

$$a_{ij} = \begin{cases} 1 & \text{if there is some edge } (v_i, v_j) \in E \text{ or some edge } (v_j, v_i) \in E \\ 0 & \text{otherwise.} \end{cases}$$

(2) Else if G is undirected, then

$$a_{ij} = \begin{cases} 1 & \text{if there is some edge } \{v_i, v_j\} \in E \\ 0 & \text{otherwise.} \end{cases}$$

As usual, unless confusion arises, we write A instead of $A(G)$. Here is the adjacency matrix of both graphs G_1 and G_2:

$$A = \begin{pmatrix} 0 & 1 & 1 & 0 & 0 \\ 1 & 0 & 1 & 1 & 1 \\ 1 & 1 & 0 & 1 & 0 \\ 0 & 1 & 1 & 0 & 1 \\ 0 & 1 & 0 & 1 & 0 \end{pmatrix}.$$

If $G = (V, E)$ is an undirected graph, the adjacency matrix A of G can be viewed as a linear map from $\mathbb{R}^V$ to $\mathbb{R}^V$, such that for all $x \in \mathbb{R}^m$, we have

$$(Ax)_i = \sum_{j \sim i} x_j;$$

that is, the value of Ax at v_i is the sum of the values of x at the nodes v_j adjacent to v_i. The adjacency matrix can be viewed as a *diffusion operator*. This observation yields a geometric interpretation of what it means for a vector $x \in \mathbb{R}^m$ to be an eigenvector of A associated with some eigenvalue λ; we must have

$$\lambda x_i = \sum_{j \sim i} x_j, \quad i = 1, \ldots, m,$$

which means that the the sum of the values of x assigned to the nodes v_j adjacent to v_i is equal to λ times the value of x at v_i.

Definition 18.11. Given any undirected graph $G = (V, E)$, an *orientation* of G is a function $\sigma \colon E \to V \times V$ assigning a source and a target to every edge in E, which means that for every edge $\{u, v\} \in E$, either $\sigma(\{u, v\}) = (u, v)$ or $\sigma(\{u, v\}) = (v, u)$. The *oriented graph* G^σ obtained from G by applying the orientation σ is the directed graph $G^\sigma = (V, E^\sigma)$, with $E^\sigma = \sigma(E)$.

The following result shows how the number of connected components of an undirected graph is related to the rank of the incidence matrix of any oriented graph obtained from G.

Proposition 18.1. *Let $G = (V, E)$ be any undirected graph with m vertices, n edges, and c connected components. For any orientation σ of G, if B is the incidence matrix of the oriented graph G^σ, then $c = \dim(\mathrm{Ker}\,(B^\top))$, and B has rank $m - c$. Furthermore, the nullspace of $B^\top$ has a basis consisting of indicator vectors of the connected components of G; that is, vectors $(z_1, \ldots, z_m)$ such that $z_j = 1$ iff v_j is in the ith component K_i of G, and $z_j = 0$ otherwise.*

Proof. (After Godsil and Royle [Godsil and Royle (2001)], Section 8.3.) The fact that rank$(B) = m - c$ will be proved last.

Let us prove that the kernel of $B^\top$ has dimension c. A vector $z \in \mathbb{R}^m$ belongs to the kernel of $B^\top$ iff $B^\top z = 0$ iff $z^\top B = 0$. In view of the definition of B, for every edge $\{v_i, v_j\}$ of G, the column of B corresponding to the oriented edge $\sigma(\{v_i, v_j\})$ has zero entries except for a $+1$ and a -1 in position i and position j or vice-versa, so we have

$$z_i = z_j.$$

An easy induction on the length of the path shows that if there is a path from v_i to v_j in G (unoriented), then $z_i = z_j$. Therefore, z has a constant value on any connected component of G. It follows that every vector $z \in \mathrm{Ker}\,(B^\top)$ can be written uniquely as a linear combination

$$z = \lambda_1 z^1 + \cdots + \lambda_c z^c,$$

where the vector z^i corresponds to the ith connected component K_i of G and is defined such that

$$z_j^i = \begin{cases} 1 & \text{iff } v_j \in K_i \\ 0 & \text{otherwise.} \end{cases}$$

This shows that $\dim(\mathrm{Ker}\,(B^\top)) = c$, and that $\mathrm{Ker}\,(B^\top)$ has a basis consisting of indicator vectors.

Since $B^\top$ is a $n \times m$ matrix, we have

$$m = \dim(\mathrm{Ker}\,(B^\top)) + \mathrm{rank}(B^\top),$$

and since we just proved that $\dim(\mathrm{Ker}\,(B^\top)) = c$, we obtain $\mathrm{rank}(B^\top) = m - c$. Since B and $B^\top$ have the same rank, $\mathrm{rank}(B) = m - c$, as claimed. $\square$

Definition 18.12. Following common practice, we denote by $\mathbf{1}$ the (column) vector (of dimension m) whose components are all equal to 1.

Since every column of B contains a single $+1$ and a single -1, the rows of $B^\top$ sum to zero, which can be expressed as

$$B^\top \mathbf{1} = 0.$$

According to Proposition 18.1, the graph G is connected iff B has rank $m - 1$ iff the nullspace of $B^\top$ is the one-dimensional space spanned by $\mathbf{1}$.

In many applications, the notion of graph needs to be generalized to capture the intuitive idea that two nodes u and v are linked with a degree of certainty (or strength). Thus, we assign a nonnegative weight w_{ij} to an edge $\{v_i, v_j\}$; the smaller w_{ij} is, the weaker is the link (or similarity) between v_i and v_j, and the greater w_{ij} is, the stronger is the link (or similarity) between v_i and v_j.

Definition 18.13. A *weighted graph* is a pair $G = (V, W)$, where $V = \{v_1, \ldots, v_m\}$ is a set of *nodes* or *vertices*, and W is a symmetric matrix called the *weight matrix*, such that $w_{ij} \geq 0$ for all $i, j \in \{1, \ldots, m\}$, and $w_{ii} = 0$ for $i = 1, \ldots, m$. We say that a set $\{v_i, v_j\}$ is an edge iff $w_{ij} > 0$. The corresponding (undirected) graph (V, E) with $E = \{\{v_i, v_j\} \mid w_{ij} > 0\}$, is called the *underlying graph* of G.

Remark: Since $w_{ii} = 0$, these graphs have no self-loops. We can think of the matrix W as a generalized adjacency matrix. The case where $w_{ij} \in \{0, 1\}$ is equivalent to the notion of a graph as in Definition 18.5.

We can think of the weight w_{ij} of an edge $\{v_i, v_j\}$ as a degree of similarity (or affinity) in an image, or a cost in a network. An example of a weighted graph is shown in Figure 18.4. The thickness of an edge corresponds to the magnitude of its weight.

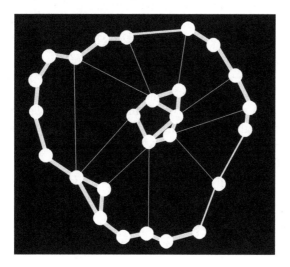

Fig. 18.4   A weighted graph.

Definition 18.14. Given a weighted graph $G = (V, W)$, for every node

$v_i \in V$, the *degree* $d(v_i)$ of v_i is the sum of the weights of the edges adjacent to v_i:

$$d(v_i) = \sum_{j=1}^{m} w_{ij}.$$

Note that in the above sum, only nodes v_j such that there is an edge $\{v_i, v_j\}$ have a nonzero contribution. Such nodes are said to be *adjacent* to v_i, and we write $v_i \sim v_j$. The degree matrix $D(G)$ (or simply, D) is defined as before, namely by $D(G) = \mathrm{diag}(d(v_1), \ldots, d(v_m))$.

The weight matrix W can be viewed as a linear map from $\mathbb{R}^V$ to itself. For all $x \in \mathbb{R}^m$, we have

$$(Wx)_i = \sum_{j \sim i} w_{ij} x_j;$$

that is, the value of Wx at v_i is the weighted sum of the values of x at the nodes v_j adjacent to v_i.

Observe that $W\mathbf{1}$ is the (column) vector $(d(v_1), \ldots, d(v_m))$ consisting of the degrees of the nodes of the graph.

We now define the most important concept of this chapter: the Laplacian matrix of a graph. Actually, as we will see, it comes in several flavors.

18.2   Laplacian Matrices of Graphs

Let us begin with directed graphs, although as we will see, graph Laplacians are fundamentally associated with undirected graph. The key proposition below shows how given an undirected graph G, for any orientation σ of G, $B^\sigma (B^\sigma)^\top$ relates to the adjacency matrix A (where B^σ is the incidence matrix of the directed graph G^σ). We reproduce the proof in Gallier [Gallier (2011a)] (see also Godsil and Royle [Godsil and Royle (2001)]).

Proposition 18.2. *Given any undirected graph G, for any orientation σ of G, if B^σ is the incidence matrix of the directed graph G^σ, A is the adjacency matrix of G^σ, and D is the degree matrix such that $D_{ii} = d(v_i)$, then*

$$B^\sigma (B^\sigma)^\top = D - A.$$

Consequently, $L = B^\sigma (B^\sigma)^\top$ is independent of the orientation σ of G, and $D - A$ is symmetric and positive semidefinite; that is, the eigenvalues of $D - A$ are real and nonnegative.

Proof. The entry $B^\sigma (B^\sigma)^\top_{ij}$ is the inner product of the ith row b^σ_i, and the jth row b^σ_j of B^σ. If $i = j$, then as

$$b^\sigma_{ik} = \begin{cases} +1 & \text{if } s(e_k) = v_i \\ -1 & \text{if } t(e_k) = v_i \\ 0 & \text{otherwise} \end{cases}$$

we see that $b^\sigma_i \cdot b^\sigma_i = d(v_i)$. If $i \neq j$, then $b^\sigma_i \cdot b^\sigma_j \neq 0$ iff there is some edge e_k with $s(e_k) = v_i$ and $t(e_k) = v_j$ or vice-versa (which are mutually exclusive cases, since G^σ arises by orienting an undirected graph), in which case, $b^\sigma_i \cdot b^\sigma_j = -1$. Therefore,

$$B^\sigma (B^\sigma)^\top = D - A,$$

as claimed.

For every $x \in \mathbb{R}^m$, we have

$$x^\top L x = x^\top B^\sigma (B^\sigma)^\top x = ((B^\sigma)^\top x)^\top (B^\sigma)^\top x = \left\| (B^\sigma)^\top x \right\|^2_2 \geq 0,$$

since the Euclidean norm $\| \ \|_2$ is positive (definite). Therefore, $L = B^\sigma (B^\sigma)^\top$ is positive semidefinite. It is well-known that a real symmetric matrix is positive semidefinite iff its eigenvalues are nonnegative. $\square$

Definition 18.15. The matrix $L = B^\sigma (B^\sigma)^\top = D - A$ is called the *(unnormalized) graph Laplacian* of the graph G^σ. The *(unnormalized) graph Laplacian* of an undirected graph $G = (V, E)$ is defined by

$$L = D - A.$$

For example, the graph Laplacian of graph G_1 is

$$L = \begin{pmatrix} 2 & -1 & -1 & 0 & 0 \\ -1 & 4 & -1 & -1 & -1 \\ -1 & -1 & 3 & -1 & 0 \\ 0 & -1 & -1 & 3 & -1 \\ 0 & -1 & 0 & -1 & 2 \end{pmatrix}.$$

Observe that each row of L sums to zero (because $(B^\sigma)^\top \mathbf{1} = 0$). Consequently, the vector $\mathbf{1}$ is in the nullspace of L.

Remarks:

(1) With the unoriented version of the incidence matrix (see Definition 18.8), it can be shown that

$$BB^\top = D + A.$$

(2) As pointed out by Evangelos Chatzipantazis, Proposition 18.2 in which the incidence matrix B^σ is replaced by the incidence matrix B of any *arbitrary* directed graph G does not hold. The problem is that such graphs may have both edges (v_i, v_j) and (v_j, v_i) between two distinct nodes v_i and v_j, and as a consequence, the inner product $b_i \cdot b_j = -2$ instead of -1. A simple counterexample is given by the directed graph with three vertices and four edges whose incidence matrix is given by

$$B = \begin{pmatrix} 1 & -1 & 0 & -1 \\ -1 & 1 & -1 & 0 \\ 0 & 0 & 1 & 1 \end{pmatrix}.$$

We have

$$BB^\top = \begin{pmatrix} 3 & -2 & -1 \\ -2 & 3 & -1 \\ -1 & -1 & 2 \end{pmatrix} \neq \begin{pmatrix} 3 & 0 & 0 \\ 0 & 3 & 0 \\ 0 & 0 & 2 \end{pmatrix} - \begin{pmatrix} 0 & 1 & 1 \\ 1 & 0 & 1 \\ 1 & 1 & 0 \end{pmatrix} = D - A.$$

The natural generalization of the notion of graph Laplacian to weighted graphs is this:

Definition 18.16. Given any weighted graph $G = (V, W)$ with $V = \{v_1, \ldots, v_m\}$, the *(unnormalized) graph Laplacian $L(G)$ of G* is defined by

$$L(G) = D(G) - W,$$

where $D(G) = \mathrm{diag}(d_1, \ldots, d_m)$ is the degree matrix of G (a diagonal matrix), with

$$d_i = \sum_{j=1}^{m} w_{i\,j}.$$

As usual, unless confusion arises, we write D instead of $D(G)$ and L instead of $L(G)$.

The graph Laplacian can be interpreted as a linear map from $\mathbb{R}^V$ to itself. For all $x \in \mathbb{R}^V$, we have

$$(Lx)_i = \sum_{j \sim i} w_{ij}(x_i - x_j).$$

It is clear from the equation $L = D - W$ that each row of L sums to 0, so the vector $\mathbf{1}$ is the nullspace of L, but it is less obvious that L is positive semidefinite. One way to prove it is to generalize slightly the notion of incidence matrix.

Definition 18.17. Given a weighted graph $G = (V, W)$, with $V = \{v_1, \ldots, v_m\}$, if $\{e_1, \ldots, e_n\}$ are the edges of the underlying graph of G

(recall that $\{v_i, v_j\}$ is an edge of this graph iff $w_{ij} > 0$), for any oriented graph G^σ obtained by giving an orientation to the underlying graph of G, the *incidence matrix* B^σ of G^σ is the $m \times n$ matrix whose entries b_{ij} are given by

$$b_{ij} = \begin{cases} +\sqrt{w_{ij}} & \text{if } s(e_j) = v_i \\ -\sqrt{w_{ij}} & \text{if } t(e_j) = v_i \\ 0 & \text{otherwise.} \end{cases}$$

For example, given the weight matrix

$$W = \begin{pmatrix} 0 & 3 & 6 & 3 \\ 3 & 0 & 0 & 3 \\ 6 & 0 & 0 & 3 \\ 3 & 3 & 3 & 0 \end{pmatrix},$$

the incidence matrix B corresponding to the orientation of the underlying graph of W where an edge (i, j) is oriented positively iff $i < j$ is

$$B = \begin{pmatrix} 1.7321 & 2.4495 & 1.7321 & 0 & 0 \\ -1.7321 & 0 & 0 & 1.7321 & 0 \\ 0 & -2.4495 & 0 & 0 & 1.7321 \\ 0 & 0 & -1.7321 & -1.7321 & -1.7321 \end{pmatrix}.$$

The reader should verify that $BB^\top = D - W$. This is true in general, see Proposition 18.3.

It is easy to see that Proposition 18.1 applies to the underlying graph of G. For any oriented graph G^σ obtained from the underlying graph of G, the rank of the incidence matrix B^σ is equal to $m - c$, where c is the number of connected components of the underlying graph of G, and we have $(B^\sigma)^\top \mathbf{1} = 0$. We also have the following version of Proposition 18.2 whose proof is immediately adapted.

Proposition 18.3. *Given any weighted graph* $G = (V, W)$ *with* $V = \{v_1, \ldots, v_m\}$, *if* B^σ *is the incidence matrix of any oriented graph* G^σ *obtained from the underlying graph of* G *and* D *is the degree matrix of* G, *then*

$$B^\sigma (B^\sigma)^\top = D - W = L.$$

Consequently, $B^\sigma (B^\sigma)^\top$ *is independent of the orientation of the underlying graph of* G *and* $L = D - W$ *is symmetric and positive semidefinite; that is, the eigenvalues of* $L = D - W$ *are real and nonnegative.*

Another way to prove that L is positive semidefinite is to evaluate the quadratic form $x^\top L x$.

Proposition 18.4. *For any $m \times m$ symmetric matrix $W = (w_{ij})$, if we let $L = D - W$ where D is the degree matrix associated with W (that is, $d_i = \sum_{j=1}^m w_{ij}$), then we have*

$$x^\top L x = \frac{1}{2} \sum_{i,j=1}^m w_{ij}(x_i - x_j)^2 \quad \text{for all } x \in \mathbb{R}^m.$$

Consequently, $x^\top L x$ does not depend on the diagonal entries in W, and if $w_{ij} \geq 0$ for all $i,j \in \{1, \ldots, m\}$, then L is positive semidefinite.

Proof. We have

$$x^\top L x = x^\top D x - x^\top W x$$

$$= \sum_{i=1}^m d_i x_i^2 - \sum_{i,j=1}^m w_{ij} x_i x_j$$

$$= \frac{1}{2} \left(\sum_{i=1}^m d_i x_i^2 - 2 \sum_{i,j=1}^m w_{ij} x_i x_j + \sum_{i=1}^m d_i x_i^2 \right)$$

$$= \frac{1}{2} \sum_{i,j=1}^m w_{ij}(x_i - x_j)^2.$$

Obviously, the quantity on the right-hand side does not depend on the diagonal entries in W, and if $w_{ij} \geq 0$ for all i,j, then this quantity is nonnegative. $\qquad \square$

Proposition 18.4 immediately implies the following facts: For any weighted graph $G = (V, W)$,

(1) The eigenvalues $0 = \lambda_1 \leq \lambda_2 \leq \ldots \leq \lambda_m$ of L are real and nonnegative, and there is an orthonormal basis of eigenvectors of L.
(2) The smallest eigenvalue λ_1 of L is equal to 0, and $\mathbf{1}$ is a corresponding eigenvector.

It turns out that the dimension of the nullspace of L (the eigenspace of 0) is equal to the number of connected components of the underlying graph of G.

Proposition 18.5. *Let $G = (V, W)$ be a weighted graph. The number c of connected components $K_1, \ldots, K_c$ of the underlying graph of G is equal* *to the dimension of the nullspace of L, which is equal to the multiplicity of the eigenvalue 0. Furthermore, the nullspace of L has a basis consisting of indicator vectors of the connected components of G, that is, vectors $(f_1, \ldots, f_m)$ such that $f_j = 1$ iff $v_j \in K_i$ and $f_j = 0$ otherwise.*

Proof. Since $L = BB^\top$ for the incidence matrix B associated with any oriented graph obtained from G, and since L and $B^\top$ have the same nullspace, by Proposition 18.1, the dimension of the nullspace of L is equal to the number c of connected components of G and the indicator vectors of the connected components of G form a basis of Ker(L). $\qquad\square$

Proposition 18.5 implies that if the underlying graph of G is connected, then the second eigenvalue λ_2 of L is strictly positive.

Remarkably, the eigenvalue λ_2 contains a lot of information about the graph G (assuming that $G = (V, E)$ is an undirected graph). This was first discovered by Fiedler in 1973, and for this reason, λ_2 is often referred to as the *Fiedler number*. For more on the properties of the Fiedler number, see Godsil and Royle [Godsil and Royle (2001)] (Chapter 13) and Chung [Chung (1997)]. More generally, the spectrum $(0, \lambda_2, \ldots, \lambda_m)$ of L contains a lot of information about the combinatorial structure of the graph G. Leverage of this information is the object of *spectral graph theory*.

18.3   Normalized Laplacian Matrices of Graphs

It turns out that normalized variants of the graph Laplacian are needed, especially in applications to graph clustering. These variants make sense only if G has no isolated vertices.

Definition 18.18. Given a weighted graph $G = (V, W)$, a vertex $u \in V$ is *isolated* if it is not incident to any other vertex. This means that every row of W contains some strictly positive entry.

If G has no isolated vertices, then the degree matrix D contains positive entries, so it is invertible and $D^{-1/2}$ makes sense; namely

$$D^{-1/2} = \text{diag}(d_1^{-1/2}, \ldots, d_m^{-1/2}),$$

and similarly for any real exponent α.

Definition 18.19. Given any weighted directed graph $G = (V, W)$ with no isolated vertex and with $V = \{v_1, \ldots, v_m\}$, the *(normalized) graph*

Laplacians L_{sym} *and* L_{rw} *of* G are defined by

$$L_{\mathrm{sym}} = D^{-1/2}LD^{-1/2} = I - D^{-1/2}WD^{-1/2}$$
$$L_{\mathrm{rw}} = D^{-1}L = I - D^{-1}W.$$

Observe that the Laplacian $L_{\mathrm{sym}} = D^{-1/2}LD^{-1/2}$ is a symmetric matrix (because L and $D^{-1/2}$ are symmetric) and that

$$L_{\mathrm{rw}} = D^{-1/2}L_{\mathrm{sym}}D^{1/2}.$$

The reason for the notation L_{rw} is that this matrix is closely related to a random walk on the graph G.

Example 18.1. As an example, the matrices L_{sym} and L_{rw} associated with the graph G_1 are

$$L_{\mathrm{sym}} = \begin{pmatrix} 1.0000 & -0.3536 & -0.4082 & 0 & 0 \\ -0.3536 & 1.0000 & -0.2887 & -0.2887 & -0.3536 \\ -0.4082 & -0.2887 & 1.0000 & -0.3333 & 0 \\ 0 & -0.2887 & -0.3333 & 1.0000 & -0.4082 \\ 0 & -0.3536 & 0 & -0.4082 & 1.0000 \end{pmatrix}$$

and

$$L_{\mathrm{rw}} = \begin{pmatrix} 1.0000 & -0.5000 & -0.5000 & 0 & 0 \\ -0.2500 & 1.0000 & -0.2500 & -0.2500 & -0.2500 \\ -0.3333 & -0.3333 & 1.0000 & -0.3333 & 0 \\ 0 & -0.3333 & -0.3333 & 1.0000 & -0.3333 \\ 0 & -0.5000 & 0 & -0.5000 & 1.0000 \end{pmatrix}.$$

Since the unnormalized Laplacian L can be written as $L = BB^{\top}$, where B is the incidence matrix of any oriented graph obtained from the underlying graph of $G = (V, W)$, if we let

$$B_{\mathrm{sym}} = D^{-1/2}B,$$

we get

$$L_{\mathrm{sym}} = B_{\mathrm{sym}}B_{\mathrm{sym}}^{\top}.$$

In particular, for any singular decomposition $B_{\mathrm{sym}} = U\Sigma V^{\top}$ of B_{sym} (with U an $m \times m$ orthogonal matrix, Σ a "diagonal" $m \times n$ matrix of singular values, and V an $n \times n$ orthogonal matrix), the eigenvalues of L_{sym} are the squares of the top m singular values of B_{sym}, and the vectors in U are orthonormal eigenvectors of L_{sym} with respect to these eigenvalues (the squares of the top m diagonal entries of Σ). Computing the SVD of B_{sym}

generally yields more accurate results than diagonalizing L_{sym}, especially when L_{sym} has eigenvalues with high multiplicity.

There are simple relationships between the eigenvalues and the eigenvectors of L_{sym}, and L_{rw}. There is also a simple relationship with the generalized eigenvalue problem $Lx = \lambda Dx$.

Proposition 18.6. *Let $G = (V, W)$ be a weighted graph without isolated vertices. The graph Laplacians, L, L_{sym}, and L_{rw} satisfy the following properties:*

(1) The matrix L_{sym} is symmetric and positive semidefinite. In fact,

$$x^{\top} L_{\mathrm{sym}} x = \frac{1}{2} \sum_{i,j=1}^{m} w_{ij} \left(\frac{x_i}{\sqrt{d_i}} - \frac{x_j}{\sqrt{d_j}} \right)^2 \quad \text{for all } x \in \mathbb{R}^m.$$

(2) The normalized graph Laplacians L_{sym} and L_{rw} have the same spectrum $(0 = \nu_1 \leq \nu_2 \leq \ldots \leq \nu_m)$, and a vector $u \neq 0$ is an eigenvector of L_{rw} for λ iff $D^{1/2}u$ is an eigenvector of L_{sym} for λ.

(3) The graph Laplacians L and L_{sym} are symmetric and positive semidefinite.

(4) A vector $u \neq 0$ is a solution of the generalized eigenvalue problem $Lu = \lambda Du$ iff $D^{1/2}u$ is an eigenvector of L_{sym} for the eigenvalue λ iff u is an eigenvector of L_{rw} for the eigenvalue λ.

(5) The graph Laplacians, L and L_{rw} have the same nullspace. For any vector u, we have $u \in \mathrm{Ker}\,(L)$ iff $D^{1/2}u \in \mathrm{Ker}\,(L_{\mathrm{sym}})$.

(6) The vector $\mathbf{1}$ is in the nullspace of L_{rw}, and $D^{1/2}\mathbf{1}$ is in the nullspace of L_{sym}.

(7) For every eigenvalue ν_i of the normalized graph Laplacian L_{sym}, we have $0 \leq \nu_i \leq 2$. Furthermore, $\nu_m = 2$ iff the underlying graph of G contains a nontrivial connected bipartite component.

(8) If $m \geq 2$ and if the underlying graph of G is not a complete graph,[1] then $\nu_2 \leq 1$. Furthermore the underlying graph of G is a complete graph iff $\nu_2 = \frac{m}{m-1}$.

(9) If $m \geq 2$ and if the underlying graph of G is connected, then $\nu_2 > 0$.

(10) If $m \geq 2$ and if the underlying graph of G has no isolated vertices, then $\nu_m \geq \frac{m}{m-1}$.

Proof. (1) We have $L_{\mathrm{sym}} = D^{-1/2}LD^{-1/2}$, and $D^{-1/2}$ is a symmetric invertible matrix (since it is an invertible diagonal matrix). It is a

---

[1] Recall that an undirected graph is complete if for any two distinct nodes u, v, there is an edge $\{u, v\}$.

well-known fact of linear algebra that if B is an invertible matrix, then a matrix S is symmetric, positive semidefinite iff $BSB^\top$ is symmetric, positive semidefinite. Since L is symmetric, positive semidefinite, so is $L_{\mathrm{sym}} = D^{-1/2}LD^{-1/2}$. The formula

$$x^\top L_{\mathrm{sym}}x = \frac{1}{2}\sum_{i,j=1}^m w_{ij}\left(\frac{x_i}{\sqrt{d_i}} - \frac{x_j}{\sqrt{d_j}}\right)^2 \quad \text{for all } x \in \mathbb{R}^m$$

follows immediately from Proposition 18.4 by replacing x by $D^{-1/2}x$, and also shows that L_{sym} is positive semidefinite.

(2) Since

$$L_{\mathrm{rw}} = D^{-1/2}L_{\mathrm{sym}}D^{1/2},$$

the matrices L_{sym} and L_{rw} are similar, which implies that they have the same spectrum. In fact, since $D^{1/2}$ is invertible,

$$L_{\mathrm{rw}}u = D^{-1}Lu = \lambda u$$

iff

$$D^{-1/2}Lu = \lambda D^{1/2}u$$

iff

$$D^{-1/2}LD^{-1/2}D^{1/2}u = L_{\mathrm{sym}}D^{1/2}u = \lambda D^{1/2}u,$$

which shows that a vector $u \neq 0$ is an eigenvector of L_{rw} for λ iff $D^{1/2}u$ is an eigenvector of L_{sym} for λ.

(3) We already know that L and L_{sym} are positive semidefinite.

(4) Since $D^{-1/2}$ is invertible, we have

$$Lu = \lambda Du$$

iff

$$D^{-1/2}Lu = \lambda D^{1/2}u$$

iff

$$D^{-1/2}LD^{-1/2}D^{1/2}u = L_{\mathrm{sym}}D^{1/2}u = \lambda D^{1/2}u,$$

which shows that a vector $u \neq 0$ is a solution of the generalized eigenvalue problem $Lu = \lambda Du$ iff $D^{1/2}u$ is an eigenvector of L_{sym} for the eigenvalue λ. The second part of the statement follows from (2).

(5) Since D^{-1} is invertible, we have $Lu = 0$ iff $D^{-1}Lu = L_{\mathrm{rw}}u = 0$. Similarly, since $D^{-1/2}$ is invertible, we have $Lu = 0$ iff $D^{-1/2}LD^{-1/2}D^{1/2}u = 0$ iff $D^{1/2}u \in \mathrm{Ker}\,(L_{\mathrm{sym}})$.

(6) Since $L\mathbf{1} = 0$, we get $L_{\mathrm{rw}}\mathbf{1} = D^{-1}L\mathbf{1} = 0$. That $D^{1/2}\mathbf{1}$ is in the nullspace of L_{sym} follows from (2). Properties (7)–(10) are proven in Chung [Chung (1997)] (Chapter 1). $\square$

The eigenvalues the matrices L_{sym} and L_{rw} from Example 18.1 are
$$0,\ 7257,\ 1.1667,\ 1.5,\ 1.6076.$$
On the other hand, the eigenvalues of the unormalized Laplacian for G_1 are
$$0,\ 1.5858,\ 3,\ 4.4142,\ 5.$$

Remark: Observe that although the matrices L_{sym} and L_{rw} have the same spectrum, the matrix L_{rw} is generally not symmetric, whereas L_{sym} is symmetric.

A version of Proposition 18.5 also holds for the graph Laplacians L_{sym} and L_{rw}. This follows easily from the fact that Proposition 18.1 applies to the underlying graph of a weighted graph. The proof is left as an exercise.

Proposition 18.7. *Let $G = (V, W)$ be a weighted graph. The number c of connected components $K_1, \ldots, K_c$ of the underlying graph of G is equal to the dimension of the nullspace of both L_{sym} and L_{rw}, which is equal to the multiplicity of the eigenvalue 0. Furthermore, the nullspace of L_{rw} has a basis consisting of indicator vectors of the connected components of G, that is, vectors $(f_1, \ldots, f_m)$ such that $f_j = 1$ iff $v_j \in K_i$ and $f_j = 0$ otherwise. For L_{sym}, a basis of the nullpace is obtained by multiplying the above basis of the nullspace of L_{rw} by $D^{1/2}$.*

A particularly interesting application of graph Laplacians is graph clustering.

18.4 Graph Clustering Using Normalized Cuts

In order to explain this problem we need some definitions.

Definition 18.20. Given any subset of nodes $A \subseteq V$, we define the *volume* $\mathrm{vol}(A)$ of A as the sum of the weights of all edges adjacent to nodes in A:
$$\mathrm{vol}(A) = \sum_{v_i \in A} \sum_{j=1}^{m} w_{ij}.$$
Given any two subsets $A, B \subseteq V$ (not necessarily distinct), we define $\mathrm{links}(A, B)$ by
$$\mathrm{links}(A, B) = \sum_{v_i \in A, v_j \in B} w_{ij}.$$
The quantity $\mathrm{links}(A, \overline{A}) = \mathrm{links}(\overline{A}, A)$ (where $\overline{A} = V - A$ denotes the complement of A in V) measures how many links escape from A (and $\overline{A}$). We define the *cut* of A as
$$\mathrm{cut}(A) = \mathrm{links}(A, \overline{A}).$$

The notion of volume is illustrated in Figure 18.5 and the notions of cut is illustrated in Figure 18.6.

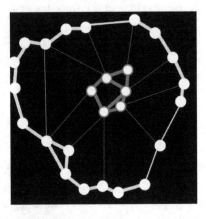

Fig. 18.5   Volume of a set of nodes.

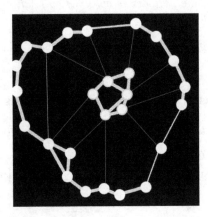

Fig. 18.6   A cut involving the set of nodes in the center and the nodes on the perimeter.

The above concepts play a crucial role in the theory of normalized cuts. This beautiful and deeply original method first published in Shi and Malik [Shi and Malik (2000)], has now come to be a "textbook chapter" of computer vision and machine learning. It was invented by Jianbo Shi and Jitendra Malik and was the main topic of Shi's dissertation. This method was extended to $K \geq 3$ clusters by Stella Yu in her dissertation [Yu (2003)] and is also the subject of Yu and Shi [Yu and Shi (2003)].

Given a set of data, the goal of clustering is to partition the data into different groups according to their similarities. When the data is given in terms of a similarity graph G, where the weight w_{ij} between two nodes v_i and v_j is a measure of similarity of v_i and v_j, the problem can be stated as follows: Find a partition $(A_1, \ldots, A_K)$ of the set of nodes V into different groups such that the edges between different groups have very low weight (which indicates that the points in different clusters are dissimilar), and the edges within a group have high weight (which indicates that points within the same cluster are similar).

The above graph clustering problem can be formalized as an optimization problem, using the notion of cut mentioned earlier. If we want to partition V into K clusters, we can do so by finding a partition $(A_1, \ldots, A_K)$ that minimizes the quantity

$$\text{cut}(A_1, \ldots, A_K) = \frac{1}{2} \sum_{i=1}^{K} \text{cut}(A_i) = \frac{1}{2} \sum_{i=1}^{K} \text{links}(A_i, \overline{A}_i).$$

For $K = 2$, the mincut problem is a classical problem that can be solved efficiently, but in practice, it does not yield satisfactory partitions. Indeed, in many cases, the mincut solution separates one vertex from the rest of the graph. What we need is to design our cost function in such a way that it keeps the subsets A_i "reasonably large" (reasonably balanced).

An example of a weighted graph and a partition of its nodes into two clusters is shown in Figure 18.7.

A way to get around this problem is to normalize the cuts by dividing by some measure of each subset A_i. A solution using the volume $\text{vol}(A_i)$ of A_i (for $K = 2$) was proposed and investigated in a seminal paper of Shi and Malik [Shi and Malik (2000)]. Subsequently, Yu (in her dissertation [Yu (2003)]) and Yu and Shi [Yu and Shi (2003)] extended the method to $K > 2$ clusters. The idea is to minimize the cost function

$$\text{Ncut}(A_1, \ldots, A_K) = \sum_{i=1}^{K} \frac{\text{links}(A_i, \overline{A}_i)}{\text{vol}(A_i)} = \sum_{i=1}^{K} \frac{\text{cut}(A_i, \overline{A}_i)}{\text{vol}(A_i)}.$$

The next step is to express our optimization problem in matrix form, and this can be done in terms of Rayleigh ratios involving the graph Laplacian in the numerators. This theory is very beautiful, but we do not have the space to present it here. The interested reader is referred to Gallier [Gallier (2019)].

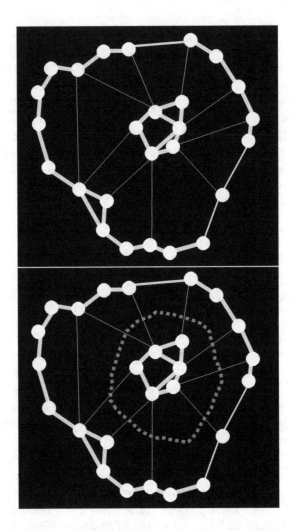

Fig. 18.7   A weighted graph and its partition into two clusters.

18.5   Summary

The main concepts and results of this chapter are listed below:

- Directed graphs, undirected graphs.
- Incidence matrices, adjacency matrices.
- Weighted graphs.
- Degree matrix.

- Graph Laplacian (unnormalized).
- Normalized graph Laplacian.
- Spectral graph theory.
- Graph clustering using normalized cuts.

18.6   Problems

Problem 18.1. Find the unnormalized Laplacian of the graph representing a triangle and of the graph representing a square.

Problem 18.2. Consider the complete graph K_m on $m \geq 2$ nodes.
(1) Prove that the normalized Laplacian L_{sym} of K is

$$
L_{\text{sym}} = \begin{pmatrix}
1 & -1/(m-1) & \dots & -1/(m-1) & -1/(m-1) \\
-1/(m-1) & 1 & \dots & -1/(m-1) & -1/(m-1) \\
\vdots & \ddots & \ddots & \ddots & \vdots \\
-1/(m-1) & -1/(m-1) & \dots & 1 & -1/(m-1) \\
-1/(m-1) & -1/(m-1) & \dots & -1/(m-1) & 1
\end{pmatrix}.
$$

(2) Prove that the characteristic polynomial of L_{sym} is

$$
\begin{vmatrix}
\lambda - 1 & 1/(m-1) & \dots & 1/(m-1) & 1/(m-1) \\
1/(m-1) & \lambda - 1 & \dots & 1/(m-1) & 1/(m-1) \\
\vdots & \ddots & \ddots & \ddots & \vdots \\
1/(m-1) & 1/(m-1) & \dots & \lambda - 1 & 1/(m-1) \\
1/(m-1) & 1/(m-1) & \dots & 1/(m-1) & \lambda - 1
\end{vmatrix} = \lambda \left(\lambda - \frac{m}{m-1} \right)^{m-1}.
$$

Hint. First subtract the second column from the first, factor $\lambda - m/(m-1)$, and then add the first row to the second. Repeat this process. You will end up with the determinant

$$
\begin{vmatrix}
\lambda - 1/(m-1) & 1 \\
1/(m-1) & \lambda - 1
\end{vmatrix}.
$$

Problem 18.3. Consider the complete bipartite graph $K_{m,n}$ on $m+n \geq 3$ nodes, with edges between each of the first $m \geq 1$ nodes to each of the last $n \geq 1$ nodes. Prove that the eigenvalues of the normalized Laplacian L_{sym} of $K_{m,n}$ are 0 with multiplicity $m + n - 2$ and 1 with multiplicity 2.

Problem 18.4. Let G be a graph with a set of nodes V with $m \geq 2$ elements, without isolated nodes, and let $L_{\text{sym}} = D^{-1/2} L D^{-1/2}$ be its normalized Laplacian (with L its unnormalized Laplacian).

(1) For any $y \in \mathbb{R}^V$, consider the Rayleigh ratio

$$R = \frac{y^\top L_{\text{sym}}\, y}{y^\top y}.$$

Prove that if $x = D^{-1/2}y$, then

$$R = \frac{x^\top L x}{(D^{1/2}x)^\top D^{1/2}x} = \frac{\sum\limits_{u \sim v}(x(u) - x(v))^2}{\sum\limits_{v} d_v x(v)^2}.$$

(2) Prove that the second eigenvalue ν_2 of L_{sym} is given by

$$\nu_2 = \min_{1^\top Dx = 0,\, x \neq 0} \frac{\sum\limits_{u \sim v}(x(u) - x(v))^2}{\sum\limits_{v} d_v x(v)^2}.$$

(3) Prove that the largest eigenvalue ν_m of L_{sym} is given by

$$\nu_m = \max_{x \neq 0} \frac{\sum\limits_{u \sim v}(x(u) - x(v))^2}{\sum\limits_{v} d_v x(v)^2}.$$

Problem 18.5. Let G be a graph with a set of nodes V with $m \geq 2$ elements, without isolated nodes. If $0 = \nu_1 \leq \nu_1 \leq \ldots \leq \nu_m$ are the eigenvalues of L_{sym}, prove the following properties:

(1) We have $\nu_1 + \nu_2 + \cdots + \nu_m = m$.
(2) We have $\nu_2 \leq m/(m-1)$, with equality holding iff $G = K_m$, the complete graph on m nodes.
(3) We have $\nu_m \geq m/(m-1)$.
(4) If G is not a complete graph, then $\nu_2 \leq 1$
   Hint. If a and b are nonadjacent nodes, consider the function x given by

$$x(v) = \begin{cases} d_b & \text{if } v = a \\ -d_a & \text{if } v = b \\ 0 & \text{if } v \neq a, b, \end{cases}$$

   and use Problem 18.4(2).
(5) Prove that $\nu_m \leq 2$. Prove that $\nu_m = 2$ iff the underlying graph of G contains a nontrivial connected bipartite component.
   Hint. Use Problem 18.4(3).

(6) Prove that if G is connected, then $\nu_2 > 0$.

Problem 18.6. Let G be a graph with a set of nodes V with $m \geq 2$ elements, without isolated nodes. Let $\mathrm{vol}(G) = \sum_{v \in V} d_v$ and let

$$\overline{x} = \frac{\sum_v d_v x(v)}{\mathrm{vol}(G)}.$$

Prove that

$$\nu_2 = \min_{x \neq 0} \frac{\displaystyle\sum_{u \sim v} (x(u) - x(v))^2}{\displaystyle\sum_v d_v (x(v) - \overline{x})^2}.$$

Problem 18.7. Let G be a connected bipartite graph. Prove that if ν is an eigenvalue of L_{sym}, then $2 - \nu$ is also an eigenvalue of L_{sym}.

Problem 18.8. Prove Proposition 18.7.

Chapter 19

Spectral Graph Drawing

19.1 Graph Drawing and Energy Minimization

Let $G = (V, E)$ be some undirected graph. It is often desirable to draw a graph, usually in the plane but possibly in 3D, and it turns out that the graph Laplacian can be used to design surprisingly good methods. Say $|V| = m$. The idea is to assign a point $\rho(v_i)$ in $\mathbb{R}^n$ to the vertex $v_i \in V$, for every $v_i \in V$, and to draw a line segment between the points $\rho(v_i)$ and $\rho(v_j)$ iff there is an edge $\{v_i, v_j\}$.

Definition 19.1. Let $G = (V, E)$ be some undirected graph with m vertices. A *graph drawing* is a function $\rho \colon V \to \mathbb{R}^n$, for some $n \geq 1$. The *matrix of a graph drawing ρ (in $\mathbb{R}^n$)* is a $m \times n$ matrix R whose ith row consists of the row vector $\rho(v_i)$ corresponding to the point representing v_i in $\mathbb{R}^n$.

For a graph drawing to be useful we want $n \leq m$; in fact n should be much smaller than m, typically $n = 2$ or $n = 3$.

Definition 19.2. A graph drawing is *balanced* iff the sum of the entries of every column of the matrix of the graph drawing is zero, that is,

$$\mathbf{1}^\top R = 0.$$

If a graph drawing is not balanced, it can be made balanced by a suitable translation. We may also assume that the columns of R are linearly independent, since any basis of the column space also determines the drawing. Thus, from now on, we may assume that $n \leq m$.

Remark: A graph drawing $\rho \colon V \to \mathbb{R}^n$ is not required to be injective, which may result in degenerate drawings where distinct vertices are drawn

687

as the same point. For this reason, we prefer not to use the terminology *graph embedding*, which is often used in the literature. This is because in differential geometry, an embedding always refers to an injective map. The term *graph immersion* would be more appropriate.

As explained in Godsil and Royle [Godsil and Royle (2001)], we can imagine building a physical model of G by connecting adjacent vertices (in $\mathbb{R}^n$) by identical springs. Then it is natural to consider a representation to be better if it requires the springs to be less extended. We can formalize this by defining the *energy* of a drawing R by

$$\mathcal{E}(R) = \sum_{\{v_i, v_j\} \in E} \|\rho(v_i) - \rho(v_j)\|^2,$$

where $\rho(v_i)$ is the ith row of R and $\|\rho(v_i) - \rho(v_j)\|^2$ is the square of the Euclidean length of the line segment joining $\rho(v_i)$ and $\rho(v_j)$.

Then, "good drawings" are drawings that minimize the energy function $\mathcal{E}$. Of course, the trivial representation corresponding to the zero matrix is optimum, so we need to impose extra constraints to rule out the trivial solution.

We can consider the more general situation where the springs are not necessarily identical. This can be modeled by a symmetric weight (or stiffness) matrix $W = (w_{ij})$, with $w_{ij} \geq 0$. Then our energy function becomes

$$\mathcal{E}(R) = \sum_{\{v_i, v_j\} \in E} w_{ij} \|\rho(v_i) - \rho(v_j)\|^2.$$

It turns out that this function can be expressed in terms of the Laplacian $L = D - W$. The following proposition is shown in Godsil and Royle [Godsil and Royle (2001)]. We give a slightly more direct proof.

Proposition 19.1. *Let $G = (V, W)$ be a weighted graph, with $|V| = m$ and W an $m \times m$ symmetric matrix, and let R be the matrix of a graph drawing ρ of G in $\mathbb{R}^n$ (a $m \times n$ matrix). If $L = D - W$ is the unnormalized Laplacian matrix associated with W, then*

$$\mathcal{E}(R) = \operatorname{tr}(R^\top L R).$$

Proof. Since $\rho(v_i)$ is the ith row of R (and $\rho(v_j)$ is the jth row of R), if we denote the kth column of R by R^k, using Proposition 18.4, we have

$$\mathcal{E}(R) = \sum_{\{v_i,v_j\}\in E} w_{ij} \|\rho(v_i) - \rho(v_j)\|^2$$

$$= \sum_{k=1}^{n} \sum_{\{v_i,v_j\}\in E} w_{ij}(R_{ik} - R_{jk})^2$$

$$= \sum_{k=1}^{n} \frac{1}{2} \sum_{i,j=1}^{m} w_{ij}(R_{ik} - R_{jk})^2$$

$$= \sum_{k=1}^{n} (R^k)^\top L R^k = \mathrm{tr}(R^\top L R),$$

as claimed. $\qquad\square$

Since the matrix $R^\top L R$ is symmetric, it has real eigenvalues. Actually, since L is positive semidefinite, so is $R^\top L R$. Then the trace of $R^\top L R$ is equal to the sum of its positive eigenvalues, and this is the energy $\mathcal{E}(R)$ of the graph drawing.

If R is the matrix of a graph drawing in $\mathbb{R}^n$, then for any $n \times n$ invertible matrix M, the map that assigns $\rho(v_i)M$ to v_i is another graph drawing of G, and these two drawings convey the same amount of information. From this point of view, *a graph drawing is determined by the column space of R*. Therefore, it is reasonable to assume that the columns of R are pairwise orthogonal and that they have unit length. Such a matrix satisfies the equation $R^\top R = I$.

Definition 19.3. If the matrix R of a graph drawing satisfies the equation $R^\top R = I$, then the corresponding drawing is called an *orthogonal graph drawing*.

This above condition also rules out trivial drawings. The following result tells us how to find minimum energy orthogonal balanced graph drawings, provided the graph is connected. Recall that

$$L\mathbf{1} = 0,$$

as we already observed.

Theorem 19.1. *Let $G = (V,W)$ be a weighted graph with $|V| = m$. If $L = D - W$ is the (unnormalized) Laplacian of G, and if the eigenvalues of L are $0 = \lambda_1 < \lambda_2 \leq \lambda_3 \leq \ldots \leq \lambda_m$, then the minimal energy of any balanced orthogonal graph drawing of G in $\mathbb{R}^n$ is equal to $\lambda_2 + \cdots + \lambda_{n+1}$*

(in particular, this implies that $n < m$). The $m \times n$ matrix R consisting of any unit eigenvectors $u_2, \ldots, u_{n+1}$ associated with $\lambda_2 \leq \cdots \leq \lambda_{n+1}$ yields a balanced orthogonal graph drawing of minimal energy; it satisfies the condition $R^\top R = I$.

Proof. We present the proof given in Godsil and Royle [Godsil and Royle (2001)] (Section 13.4, Theorem 13.4.1). The key point is that the sum of the n smallest eigenvalues of L is a lower bound for $\mathrm{tr}(R^\top L R)$. This can be shown using a Rayleigh ratio argument; see Proposition 16.13 (the Poincaré separation theorem). Then any n eigenvectors $(u_1, \ldots, u_n)$ associated with $\lambda_1, \ldots, \lambda_n$ achieve this bound. Because the first eigenvalue of L is $\lambda_1 = 0$ and because we are assuming that $\lambda_2 > 0$, we have $u_1 = 1/\sqrt{m}$. Since the u_j are pairwise orthogonal for $i = 2, \ldots, n$ and since u_i is orthogonal to $u_1 = 1/\sqrt{m}$, the entries in u_i add up to 0. Consequently, for any ℓ with $2 \leq \ell \leq n$, by deleting u_1 and using $(u_2, \ldots, u_\ell)$, we obtain a balanced orthogonal graph drawing in $\mathbb{R}^{\ell-1}$ with the same energy as the orthogonal graph drawing in $\mathbb{R}^\ell$ using $(u_1, u_2, \ldots, u_\ell)$. Conversely, from any balanced orthogonal drawing in $\mathbb{R}^{\ell-1}$ using $(u_2, \ldots, u_\ell)$, we obtain an orthogonal graph drawing in $\mathbb{R}^\ell$ using $(u_1, u_2, \ldots, u_\ell)$ with the same energy. Therefore, the minimum energy of a balanced orthogonal graph drawing in $\mathbb{R}^n$ is equal to the minimum energy of an orthogonal graph drawing in $\mathbb{R}^{n+1}$, and this minimum is $\lambda_2 + \cdots + \lambda_{n+1}$. $\qquad\square$

Since $\mathbf{1}$ spans the nullspace of L, using u_1 (which belongs to $\mathrm{Ker}\,L$) as one of the vectors in R would have the effect that all points representing vertices of G would have the same first coordinate. This would mean that the drawing lives in a hyperplane in $\mathbb{R}^n$, which is undesirable, especially when $n = 2$, where all vertices would be collinear. This is why we omit the first eigenvector u_1.

Observe that for any orthogonal $n \times n$ matrix Q, since

$$\mathrm{tr}(R^\top L R) = \mathrm{tr}(Q^\top R^\top L R Q),$$

the matrix RQ also yields a minimum orthogonal graph drawing. This amounts to applying the rigid motion $Q^\top$ to the rows of R.

In summary, if $\lambda_2 > 0$, an automatic method for drawing a graph in $\mathbb{R}^2$ is this:

(1) Compute the two smallest nonzero eigenvalues $\lambda_2 \leq \lambda_3$ of the graph Laplacian L (it is possible that $\lambda_3 = \lambda_2$ if λ_2 is a multiple eigenvalue);
(2) Compute two unit eigenvectors u_2, u_3 associated with λ_2 and λ_3, and let $R = [u_2\ u_3]$ be the $m \times 2$ matrix having u_2 and u_3 as columns.

(3) Place vertex v_i at the point whose coordinates is the ith row of R, that is, (R_{i1}, R_{i2}).

This method generally gives pleasing results, but beware that there is no guarantee that distinct nodes are assigned distinct images since R can have identical rows. This does not seem to happen often in practice.

19.2 Examples of Graph Drawings

We now give a number of examples using Matlab. Some of these are borrowed or adapted from Spielman [Spielman (2012)].

Example 1. Consider the graph with four nodes whose adjacency matrix is

$$A = \begin{pmatrix} 0\ 1\ 1\ 0 \\ 1\ 0\ 0\ 1 \\ 1\ 0\ 0\ 1 \\ 0\ 1\ 1\ 0 \end{pmatrix}.$$

We use the following program to compute u_2 and u_3:

```
A = [0 1 1 0; 1 0 0 1; 1 0 0 1; 0 1 1 0];
D = diag(sum(A));
L = D - A;
[v, e] = eigs(L);
gplot(A, v(:,[3 2]))
hold on;
gplot(A, v(:,[3 2]),'o')
```

The graph of Example 1 is shown in Figure 19.1. The function eigs(L) computes the six largest eigenvalues of L in decreasing order, and corresponding eigenvectors. It turns out that $\lambda_2 = \lambda_3 = 2$ is a double eigenvalue.

Example 2. Consider the graph G_2 shown in Figure 18.3 given by the adjacency matrix

$$A = \begin{pmatrix} 0\ 1\ 1\ 0\ 0 \\ 1\ 0\ 1\ 1\ 1 \\ 1\ 1\ 0\ 1\ 0 \\ 0\ 1\ 1\ 0\ 1 \\ 0\ 1\ 0\ 1\ 0 \end{pmatrix}.$$

We use the following program to compute u_2 and u_3:

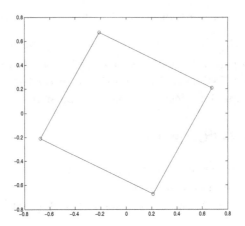

Fig. 19.1   Drawing of the graph from Example 1.

```
A = [0 1 1 0 0; 1 0 1 1 1; 1 1 0 1 0; 0 1 1 0 1; 0 1 0 1 0];
D = diag(sum(A));
L = D - A;
[v, e] = eig(L);
gplot(A, v(:, [2 3]))
hold on
gplot(A, v(:, [2 3]),'o')
```

The function eig(L) (with no s at the end) computes the eigenvalues of L in increasing order. The result of drawing the graph is shown in Figure 19.2. Note that node v_2 is assigned to the point $(0,0)$, so the difference between this drawing and the drawing in Figure 18.3 is that the drawing of Figure 19.2 is not convex.

Example 3. Consider the ring graph defined by the adjacency matrix A given in the Matlab program shown below:

```
A = diag(ones(1, 11),1);
A = A + A';
A(1, 12) = 1; A(12, 1) = 1;
D = diag(sum(A));
L = D - A;
[v, e] = eig(L);
gplot(A, v(:, [2 3]))
hold on
gplot(A, v(:, [2 3]),'o')
```

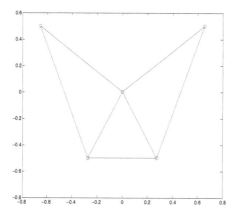

Fig. 19.2   Drawing of the graph from Example 2.

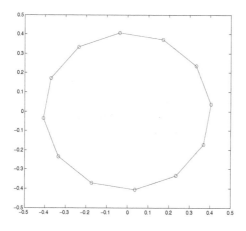

Fig. 19.3   Drawing of the graph from Example 3.

Observe that we get a very nice ring; see Figure 19.3. Again $\lambda_2 = 0.2679$ is a double eigenvalue (and so are the next pairs of eigenvalues, except the last, $\lambda_{12} = 4$).

Example 4. In this example adapted from Spielman, we generate 20 randomly chosen points in the unit square, compute their Delaunay triangulation, then the adjacency matrix of the corresponding graph, and finally draw the graph using the second and third eigenvalues of the Laplacian.

```
A = zeros(20,20);
xy = rand(20, 2);
trigs = delaunay(xy(:,1), xy(:,2));
elemtrig = ones(3) - eye(3);
for i = 1:length(trigs),
 A(trigs(i,:),trigs(i,:)) = elemtrig;
end
A = double(A >0);
gplot(A,xy)
D = diag(sum(A));
L = D - A;
[v, e] = eigs(L, 3, 'sm');
figure(2)
gplot(A, v(:, [2 1]))
hold on
gplot(A, v(:, [2 1]),'o')
```

The Delaunay triangulation of the set of 20 points and the drawing of the corresponding graph are shown in Figure 19.4. The graph drawing on the right looks nicer than the graph on the left but is is no longer planar.

Example 5. Our last example, also borrowed from Spielman [Spielman (2012)], corresponds to the skeleton of the "Buckyball," a geodesic dome invented by the architect Richard Buckminster Fuller (1895–1983). The Montréal Biosphère is an example of a geodesic dome designed by Buckminster Fuller.

```
A = full(bucky);
D = diag(sum(A));
L = D - A;
[v, e] = eig(L);
gplot(A, v(:, [2 3]))
hold on;
gplot(A,v(:, [2 3]), 'o')
```

Figure 19.5 shows a graph drawing of the Buckyball. This picture seems a bit squashed for two reasons. First, it is really a 3-dimensional graph; second, $\lambda_2 = 0.2434$ is a triple eigenvalue. (Actually, the Laplacian of L has many multiple eigenvalues.) What we should really do is to plot this graph in $\mathbb{R}^3$ using three orthonormal eigenvectors associated with λ_2.

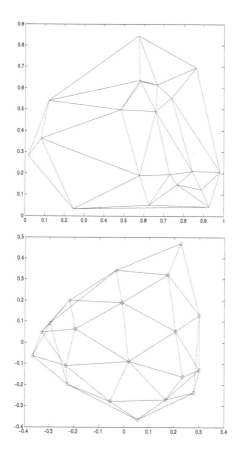

Fig. 19.4   Delaunay triangulation (left) and drawing of the graph from Example 4 (right).

A 3D picture of the graph of the Buckyball is produced by the following Matlab program, and its image is shown in Figure 19.6. It looks better!

```
[x, y] = gplot(A, v(:, [2 3]));
[x, z] = gplot(A, v(:, [2 4]));
plot3(x,y,z)
```

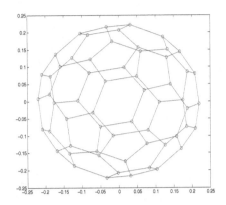

Fig. 19.5   Drawing of the graph of the Buckyball.

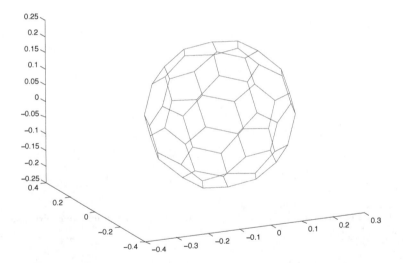

Fig. 19.6   Drawing of the graph of the Buckyball in $\mathbb{R}^3$.

19.3   Summary

The main concepts and results of this chapter are listed below:

- Graph drawing.
- Matrix of a graph drawing.

- Balanced graph drawing.
- Energy $\mathcal{E}(R)$ of a graph drawing.
- Orthogonal graph drawing.
- Delaunay triangulation.
- Buckyball.

Chapter 20

Singular Value Decomposition and Polar Form

20.1 Properties of $f^* \circ f$

In this section we assume that we are dealing with real Euclidean spaces. Let $f \colon E \to E$ be any linear map. In general, it may not be possible to diagonalize f. We show that every linear map can be diagonalized if we are willing to use *two* orthonormal bases. This is the celebrated *singular value decomposition (SVD)*. A close cousin of the SVD is the *polar form* of a linear map, which shows how a linear map can be decomposed into its purely rotational component (perhaps with a flip) and its purely stretching part.

The key observation is that $f^* \circ f$ is self-adjoint since

$$\langle (f^* \circ f)(u), v \rangle = \langle f(u), f(v) \rangle = \langle u, (f^* \circ f)(v) \rangle.$$

Similarly, $f \circ f^*$ is self-adjoint.

The fact that $f^* \circ f$ and $f \circ f^*$ are self-adjoint is very important, because by Theorem 16.1, it implies that $f^* \circ f$ and $f \circ f^*$ *can be diagonalized and that they have real eigenvalues*. In fact, these *eigenvalues are all nonnegative* as shown in the following proposition.

Proposition 20.1. *The eigenvalues of $f^* \circ f$ and $f \circ f^*$ are nonnegative.*

Proof. If u is an eigenvector of $f^* \circ f$ for the eigenvalue λ, then

$$\langle (f^* \circ f)(u), u \rangle = \langle f(u), f(u) \rangle$$

and

$$\langle (f^* \circ f)(u), u \rangle = \lambda \langle u, u \rangle,$$

and thus

$$\lambda \langle u, u \rangle = \langle f(u), f(u) \rangle,$$

which implies that $\lambda \geq 0$, since $\langle -, - \rangle$ is positive definite. A similar proof applies to $f \circ f^*$. $\qquad \square$

Thus, the eigenvalues of $f^* \circ f$ are of the form $\sigma_1^2, \ldots, \sigma_r^2$ or 0, where $\sigma_i > 0$, and similarly for $f \circ f^*$.

The above considerations also apply to any linear map $f\colon E \to F$ between two Euclidean spaces $(E, \langle -, - \rangle_1)$ and $(F, \langle -, - \rangle_2)$. Recall that the adjoint $f^*\colon F \to E$ of f is the unique linear map f^* such that

$$\langle f(u), v \rangle_2 = \langle u, f^*(v) \rangle_1, \quad \text{for all } u \in E \text{ and all } v \in F.$$

Then $f^* \circ f$ and $f \circ f^*$ are self-adjoint (the proof is the same as in the previous case), and the eigenvalues of $f^* \circ f$ and $f \circ f^*$ are nonnegative.

Proof. If λ is an eigenvalue of $f^* \circ f$ and $u \ (\neq 0)$ is a corresponding eigenvector, we have

$$\langle (f^* \circ f)(u), u \rangle_1 = \langle f(u), f(u) \rangle_2,$$

and also

$$\langle (f^* \circ f)(u), u \rangle_1 = \lambda \langle u, u \rangle_1,$$

so

$$\lambda \langle u, u \rangle_1, = \langle f(u), f(u) \rangle_2,$$

which implies that $\lambda \geq 0$. A similar proof applies to $f \circ f^*$. $\qquad\square$

The situation is even better, since we will show shortly that $f^* \circ f$ and $f \circ f^*$ have the same nonzero eigenvalues.

Remark: Given any two linear maps $f\colon E \to F$ and $g\colon F \to E$, where $\dim(E) = n$ and $\dim(F) = m$, it can be shown that

$$\lambda^m \det(\lambda\, I_n - g \circ f) = \lambda^n \det(\lambda\, I_m - f \circ g),$$

and thus $g \circ f$ and $f \circ g$ always have the same nonzero eigenvalues; see Problem 14.14.

Definition 20.1. Given any linear map $f\colon E \to F$, the square roots $\sigma_i > 0$ of the positive eigenvalues of $f^* \circ f$ (and $f \circ f^*$) are called the *singular values* of f.

Definition 20.2. A self-adjoint linear map $f\colon E \to E$ whose eigenvalues are nonnegative is called *positive semidefinite* (or *positive*), and if f is also invertible, f is said to be *positive definite*. In the latter case, every eigenvalue of f is strictly positive.

If $f\colon E \to F$ is any linear map, we just showed that $f^* \circ f$ and $f \circ f^*$ are positive semidefinite self-adjoint linear maps. This fact has the remarkable consequence that every linear map has two important decompositions:

(1) The polar form.
(2) The singular value decomposition (SVD).

The wonderful thing about the singular value decomposition is that there exist two orthonormal bases $(u_1, \ldots, u_n)$ and $(v_1, \ldots, v_m)$ such that, with respect to these bases, f is a diagonal matrix consisting of the singular values of f or 0. Thus, in some sense, f can always be diagonalized with respect to *two* orthonormal bases. The SVD is also a useful tool for solving overdetermined linear systems in the least squares sense and for data analysis, as we show later on.

First we show some useful relationships between the kernels and the images of f, f^*, $f^* \circ f$, and $f \circ f^*$. Recall that if $f\colon E \to F$ is a linear map, the *image* $\mathrm{Im}\, f$ *of* f is the subspace $f(E)$ of F, and the *rank of* f is the dimension $\dim(\mathrm{Im}\, f)$ of its image. Also recall that (Theorem 5.1)

$$\dim\,(\mathrm{Ker}\, f) + \dim\,(\mathrm{Im}\, f) = \dim\,(E),$$

and that (Propositions 11.9 and 13.12) for every subspace W of E,

$$\dim\,(W) + \dim\,(W^\perp) = \dim\,(E).$$

Proposition 20.2. *Given any two Euclidean spaces E and F, where E has dimension n and F has dimension m, for any linear map $f\colon E \to F$, we have*

$$\mathrm{Ker}\, f = \mathrm{Ker}\,(f^* \circ f),$$
$$\mathrm{Ker}\, f^* = \mathrm{Ker}\,(f \circ f^*),$$
$$\mathrm{Ker}\, f = (\mathrm{Im}\, f^*)^\perp,$$
$$\mathrm{Ker}\, f^* = (\mathrm{Im}\, f)^\perp,$$
$$\dim(\mathrm{Im}\, f) = \dim(\mathrm{Im}\, f^*),$$

and f, f^*, $f^* \circ f$, and $f \circ f^*$ have the same rank.

Proof. To simplify the notation, we will denote the inner products on E and F by the same symbol $\langle -, - \rangle$ (to avoid subscripts). If $f(u) = 0$, then $(f^* \circ f)(u) = f^*(f(u)) = f^*(0) = 0$, and so $\mathrm{Ker}\, f \subseteq \mathrm{Ker}\,(f^* \circ f)$. By definition of f^*, we have

$$\langle f(u), f(u) \rangle = \langle (f^* \circ f)(u), u \rangle$$

for all $u \in E$. If $(f^* \circ f)(u) = 0$, since $\langle -, - \rangle$ is positive definite, we must have $f(u) = 0$, and so $\text{Ker}\,(f^* \circ f) \subseteq \text{Ker}\,f$. Therefore,

$$\text{Ker}\,f = \text{Ker}\,(f^* \circ f).$$

The proof that $\text{Ker}\,f^* = \text{Ker}\,(f \circ f^*)$ is similar.

By definition of f^*, we have

$$\langle f(u), v \rangle = \langle u, f^*(v) \rangle \quad \text{for all } u \in E \text{ and all } v \in F. \qquad (20.1)$$

This immediately implies that

$$\text{Ker}\,f = (\text{Im}\,f^*)^\perp \quad \text{and} \quad \text{Ker}\,f^* = (\text{Im}\,f)^\perp.$$

Let us explain why $\text{Ker}\,f = (\text{Im}\,f^*)^\perp$, the proof of the other equation being similar.

Because the inner product is positive definite, for every $u \in E$, we have

- $u \in \text{Ker}\,f$
- iff $f(u) = 0$
- iff $\langle f(u), v \rangle = 0$ for all v,
- by (20.1) iff $\langle u, f^*(v) \rangle = 0$ for all v,
- iff $u \in (\text{Im}\,f^*)^\perp$.

Since

$$\dim(\text{Im}\,f) = n - \dim(\text{Ker}\,f)$$

and

$$\dim(\text{Im}\,f^*) = n - \dim((\text{Im}\,f^*)^\perp),$$

from

$$\text{Ker}\,f = (\text{Im}\,f^*)^\perp$$

we also have

$$\dim(\text{Ker}\,f) = \dim((\text{Im}\,f^*)^\perp),$$

from which we obtain

$$\dim(\text{Im}\,f) = \dim(\text{Im}\,f^*).$$

Since

$$\dim(\text{Ker}\,(f^* \circ f)) + \dim(\text{Im}\,(f^* \circ f)) = \dim(E),$$

$\text{Ker}\,(f^* \circ f) = \text{Ker}\,f$ and $\text{Ker}\,f = (\text{Im}\,f^*)^\perp$, we get

$$\dim((\text{Im}\,f^*)^\perp) + \dim(\text{Im}\,(f^* \circ f)) = \dim(E).$$

Since

$$\dim((\text{Im}\,f^*)^\perp) + \dim(\text{Im}\,f^*) = \dim(E),$$

we deduce that

$$\dim(\text{Im}\,f^*) = \dim(\text{Im}\,(f^* \circ f)).$$

A similar proof shows that

$$\dim(\text{Im}\,f) = \dim(\text{Im}\,(f \circ f^*)).$$

Consequently, f, f^*, $f^* \circ f$, and $f \circ f^*$ have the same rank. $\square$

20.2 Singular Value Decomposition for Square Matrices

We will now prove that every square matrix has an SVD. Stronger results can be obtained if we first consider the polar form and then derive the SVD from it (there are uniqueness properties of the polar decomposition). For our purposes, uniqueness results are not as important so we content ourselves with existence results, whose proofs are simpler. Readers interested in a more general treatment are referred to Gallier [Gallier (2011b)].

The early history of the singular value decomposition is described in a fascinating paper by Stewart [Stewart (1993)]. The SVD is due to Beltrami and Camille Jordan independently (1873, 1874). Gauss is the grandfather of all this, for his work on least squares (1809, 1823) (but Legendre also published a paper on least squares!). Then come Sylvester, Schmidt, and Hermann Weyl. Sylvester's work was apparently "opaque." He gave a computational method to find an SVD. Schmidt's work really has to do with integral equations and symmetric and asymmetric kernels (1907). Weyl's work has to do with perturbation theory (1912). Autonne came up with the polar decomposition (1902, 1915). Eckart and Young extended SVD to rectangular matrices (1936, 1939).

Theorem 20.1. *(Singular value decomposition) For every real $n \times n$ matrix A there are two orthogonal matrices U and V and a diagonal matrix D such that $A = V D U^\top$, where D is of the form*

$$
D = \begin{pmatrix} \sigma_1 & \cdots & & \\ & \sigma_2 & \cdots & \\ \vdots & \vdots & \ddots & \vdots \\ & & \cdots & \sigma_n \end{pmatrix},
$$

where $\sigma_1, \ldots, \sigma_r$ are the singular values of f, i.e., the (positive) square roots of the nonzero eigenvalues of $A^\top A$ and $A A^\top$, and $\sigma_{r+1} = \cdots = \sigma_n = 0$. The columns of U are eigenvectors of $A^\top A$, and the columns of V are eigenvectors of $A A^\top$.

Proof. Since $A^\top A$ is a symmetric matrix, in fact, a positive semidefinite matrix, there exists an orthogonal matrix U such that

$$
A^\top A = U D^2 U^\top,
$$

with $D = \operatorname{diag}(\sigma_1, \ldots, \sigma_r, 0, \ldots, 0)$, where $\sigma_1^2, \ldots, \sigma_r^2$ are the nonzero eigenvalues of $A^\top A$, and where r is the rank of A; that is, $\sigma_1, \ldots, \sigma_r$ are the singular values of A. It follows that

$$U^\top A^\top AU = (AU)^\top AU = D^2,$$

and if we let f_j be the jth column of AU for $j = 1, \ldots, n$, then we have

$$\langle f_i, f_j \rangle = \sigma_i^2 \delta_{ij}, \quad 1 \le i, j \le r$$

and

$$f_j = 0, \quad r + 1 \le j \le n.$$

If we define $(v_1, \ldots, v_r)$ by

$$v_j = \sigma_j^{-1} f_j, \quad 1 \le j \le r,$$

then we have

$$\langle v_i, v_j \rangle = \delta_{ij}, \quad 1 \le i, j \le r,$$

so complete $(v_1, \ldots, v_r)$ into an orthonormal basis $(v_1, \ldots, v_r, v_{r+1}, \ldots, v_n)$ (for example, using Gram–Schmidt). Now since $f_j = \sigma_j v_j$ for $j = 1 \ldots, r$, we have

$$\langle v_i, f_j \rangle = \sigma_j \langle v_i, v_j \rangle = \sigma_j \delta_{i,j}, \quad 1 \le i \le n, \, 1 \le j \le r$$

and since $f_j = 0$ for $j = r + 1, \ldots, n$,

$$\langle v_i, f_j \rangle = 0 \quad 1 \le i \le n, \, r + 1 \le j \le n.$$

If V is the matrix whose columns are $v_1, \ldots, v_n$, then V is orthogonal and the above equations prove that

$$V^\top AU = D,$$

which yields $A = V D U^\top$, as required.

The equation $A = V D U^\top$ implies that

$$A^\top A = U D^2 U^\top, \quad AA^\top = V D^2 V^\top,$$

which shows that $A^\top A$ and $AA^\top$ have the same eigenvalues, that the columns of U are eigenvectors of $A^\top A$, and that the columns of V are eigenvectors of $AA^\top$. $\qquad\square$

Example 20.1. Here is a simple example of how to use the proof of Theorem 20.1 to obtain an SVD decomposition. Let $A = \begin{pmatrix} 1 & 1 \\ 0 & 0 \end{pmatrix}$. Then $A^\top = \begin{pmatrix} 1 & 0 \\ 1 & 0 \end{pmatrix}$, $A^\top A = \begin{pmatrix} 1 & 1 \\ 1 & 1 \end{pmatrix}$, and $AA^\top = \begin{pmatrix} 2 & 0 \\ 0 & 0 \end{pmatrix}$. A simple calculation shows that the eigenvalues of $A^\top A$ are 2 and 0, and for the eigenvalue 2, a unit eigenvector is $\begin{pmatrix} 1/\sqrt{2} \\ 1/\sqrt{2} \end{pmatrix}$, while a unit eigenvector for the eigenvalue

0 is $\begin{pmatrix} 1/\sqrt{2} \\ -1/\sqrt{2} \end{pmatrix}$. Observe that the singular values are $\sigma_1 = \sqrt{2}$ and $\sigma_2 = 0$.

Furthermore, $U = \begin{pmatrix} 1/\sqrt{2} & 1/\sqrt{2} \\ 1/\sqrt{2} & -1/\sqrt{2} \end{pmatrix} = U^\top$. To determine V, the proof of
Theorem 20.1 tells us to first calculate

$$AU = \begin{pmatrix} \sqrt{2} & 0 \\ 0 & 0 \end{pmatrix},$$

and then set

$$v_1 = (1/\sqrt{2}) \begin{pmatrix} \sqrt{2} \\ 0 \end{pmatrix} = \begin{pmatrix} 1 \\ 0 \end{pmatrix}.$$

Once v_1 is determined, since $\sigma_2 = 0$, we have the freedom to choose v_2
such that (v_1, v_2) forms an orthonormal basis for $\mathbb{R}^2$. Naturally, we chose
$v_2 = \begin{pmatrix} 0 \\ 1 \end{pmatrix}$ and set $V = \begin{pmatrix} 1 & 0 \\ 0 & 1 \end{pmatrix}$. Of course we could have found V by directly
computing the eigenvalues and eigenvectors for $AA^\top$. We leave it to the
reader to check that

$$A = V \begin{pmatrix} \sqrt{2} & 0 \\ 0 & 0 \end{pmatrix} U^\top.$$

Theorem 20.1 suggests the following definition.

Definition 20.3. A triple (U, D, V) such that $A = VDU^\top$, where U and
V are orthogonal and D is a diagonal matrix whose entries are nonnegative
(it is positive semidefinite) is called a *singular value decomposition (SVD)
of A*.

The `Matlab` command for computing an SVD $A = VDU^\top$ of a matrix
A is `[V, D, U] = svd(A)`.

The proof of Theorem 20.1 shows that there are two orthonormal bases
$(u_1, \ldots, u_n)$ and $(v_1, \ldots, v_n)$, where $(u_1, \ldots, u_n)$ are eigenvectors of $A^\top A$
and $(v_1, \ldots, v_n)$ are eigenvectors of $AA^\top$. Furthermore, $(u_1, \ldots, u_r)$ is
an orthonormal basis of $\operatorname{Im} A^\top$, $(u_{r+1}, \ldots, u_n)$ is an orthonormal basis of
$\operatorname{Ker} A$, $(v_1, \ldots, v_r)$ is an orthonormal basis of $\operatorname{Im} A$, and $(v_{r+1}, \ldots, v_n)$ is
an orthonormal basis of $\operatorname{Ker} A^\top$.

Using a remark made in Chapter 3, if we denote the columns of U by
$u_1, \ldots, u_n$ and the columns of V by $v_1, \ldots, v_n$, then we can write

$$A = VDU^\top = \sigma_1 v_1 u_1^\top + \cdots + \sigma_r v_r u_r^\top.$$

As a consequence, if r is a lot smaller than n (we write $r \ll n$), we see that A can be reconstructed from U and V using a much smaller number of elements. This idea will be used to provide "low-rank" approximations of a matrix. The idea is to keep only the k top singular values for some suitable $k \ll r$ for which $\sigma_{k+1}, \ldots \sigma_r$ are very small.

Remarks:

(1) In Strang [Strang (1988)] the matrices U, V, D are denoted by $U = Q_2$, $V = Q_1$, and $D = \Sigma$, and an SVD is written as $A = Q_1 \Sigma Q_2^\top$. This has the advantage that Q_1 comes before Q_2 in $A = Q_1 \Sigma Q_2^\top$. This has the disadvantage that A maps the columns of Q_2 (eigenvectors of $A^\top A$) to multiples of the columns of Q_1 (eigenvectors of $A A^\top$).

(2) Algorithms for actually computing the SVD of a matrix are presented in Golub and Van Loan [Golub and Van Loan (1996)], Demmel [Demmel (1997)], and Trefethen and Bau [Trefethen and Bau III (1997)], where the SVD and its applications are also discussed quite extensively.

(3) If A is a symmetric matrix, then in general, there is no SVD $V \Sigma U^\top$ of A with $V = U$. However, if A is positive semidefinite, then the eigenvalues of A are nonnegative, and so the nonzero eigenvalues of A are equal to the singular values of A and SVDs of A are of the form

$$A = V \Sigma V^\top.$$

(4) The SVD also applies to complex matrices. In this case, for every complex $n \times n$ matrix A, there are two unitary matrices U and V and a diagonal matrix D such that

$$A = V D U^*,$$

where D is a diagonal matrix consisting of real entries $\sigma_1, \ldots, \sigma_n$, where $\sigma_1, \ldots, \sigma_r$ are the singular values of A, i.e., the positive square roots of the nonzero eigenvalues of $A^* A$ and $A A^*$, and $\sigma_{r+1} = \ldots = \sigma_n = 0$.

20.3   Polar Form for Square Matrices

A notion closely related to the SVD is the polar form of a matrix.

Definition 20.4. A pair (R, S) such that $A = RS$ with R orthogonal and S symmetric positive semidefinite is called a *polar decomposition of A*.

Theorem 20.1 implies that for every real $n \times n$ matrix A, there is some orthogonal matrix R and some positive semidefinite symmetric matrix S such that

$$A = RS.$$

This is easy to show and we will prove it below. Furthermore, R, S are unique if A is invertible, but this is harder to prove; see Problem 20.9.

For example, the matrix

$$A = \frac{1}{2} \begin{pmatrix} 1 & 1 & 1 & 1 \\ 1 & 1 & -1 & -1 \\ 1 & -1 & 1 & -1 \\ 1 & -1 & -1 & 1 \end{pmatrix}$$

is both orthogonal and symmetric, and $A = RS$ with $R = A$ and $S = I$, which implies that some of the eigenvalues of A are negative.

Remark: In the complex case, the polar decomposition states that for every complex $n \times n$ matrix A, there is some unitary matrix U and some positive semidefinite Hermitian matrix H such that

$$A = UH.$$

It is easy to go from the polar form to the SVD, and conversely.

Given an SVD decomposition $A = VDU^\top$, let $R = VU^\top$ and $S = UDU^\top$. It is clear that R is orthogonal and that S is positive semidefinite symmetric, and

$$RS = VU^\top UDU^\top = VDU^\top = A.$$

Example 20.2. Recall from Example 20.1 that $A = VDU^\top$ where $V = I_2$ and

$$A = \begin{pmatrix} 1 & 1 \\ 0 & 0 \end{pmatrix}, \qquad U = \begin{pmatrix} \frac{1}{\sqrt{2}} & \frac{1}{\sqrt{2}} \\ \frac{1}{\sqrt{2}} & -\frac{1}{\sqrt{2}} \end{pmatrix}, \qquad D = \begin{pmatrix} \sqrt{2} & 0 \\ 0 & 0 \end{pmatrix}.$$

Set $R = VU^\top = U$ and

$$S = UDU^\top = \begin{pmatrix} \frac{1}{\sqrt{2}} & \frac{1}{\sqrt{2}} \\ \frac{1}{\sqrt{2}} & \frac{1}{\sqrt{2}} \end{pmatrix}.$$

Since $S = \frac{1}{\sqrt{2}} A^\top A$, S has eigenvalues $\sqrt{2}$ and 0. We leave it to the reader to check that $A = RS$.

Going the other way, given a polar decomposition $A = R_1 S$, where R_1 is orthogonal and S is positive semidefinite symmetric, there is an orthogonal matrix R_2 and a positive semidefinite diagonal matrix D such that $S = R_2 D R_2^\top$, and thus

$$A = R_1 R_2 D R_2^\top = V D U^\top,$$

where $V = R_1 R_2$ and $U = R_2$ are orthogonal.

Example 20.3. Let $A = \begin{pmatrix} 1 & 1 \\ 0 & 0 \end{pmatrix}$ and $A = R_1 S$, where $R_1 = \begin{pmatrix} 1/\sqrt{2} & 1/\sqrt{2} \\ 1/\sqrt{2} & -1/\sqrt{2} \end{pmatrix}$ and $S = \begin{pmatrix} 1/\sqrt{2} & 1/\sqrt{2} \\ 1/\sqrt{2} & 1/\sqrt{2} \end{pmatrix}$. This is the polar decomposition of Example 20.2. Observe that

$$S = \begin{pmatrix} \frac{1}{\sqrt{2}} & \frac{1}{\sqrt{2}} \\ \frac{1}{\sqrt{2}} & -\frac{1}{\sqrt{2}} \end{pmatrix} \begin{pmatrix} \sqrt{2} & 0 \\ 0 & 0 \end{pmatrix} \begin{pmatrix} \frac{1}{\sqrt{2}} & \frac{1}{\sqrt{2}} \\ \frac{1}{\sqrt{2}} & -\frac{1}{\sqrt{2}} \end{pmatrix} = R_2 D R_2^\top.$$

Set $U = R_2$ and $V = R_1 R_2 = \begin{pmatrix} 1 & 0 \\ 0 & 1 \end{pmatrix}$ to obtain the SVD decomposition of Example 20.1.

The eigenvalues and the singular values of a matrix are typically not related in any obvious way. For example, the $n \times n$ matrix

$$A = \begin{pmatrix} 1 & 2 & 0 & 0 & \dots & 0 & 0 \\ 0 & 1 & 2 & 0 & \dots & 0 & 0 \\ 0 & 0 & 1 & 2 & \dots & 0 & 0 \\ \vdots & \vdots & \ddots & \ddots & \ddots & \vdots & \vdots \\ 0 & 0 & \dots & 0 & 1 & 2 & 0 \\ 0 & 0 & \dots & 0 & 0 & 1 & 2 \\ 0 & 0 & \dots & 0 & 0 & 0 & 1 \end{pmatrix}$$

has the eigenvalue 1 with multiplicity n, but its singular values, $\sigma_1 \geq \cdots \geq \sigma_n$, which are the positive square roots of the eigenvalues of the matrix $B = A^\top A$ with

$$B = \begin{pmatrix} 1 & 2 & 0 & 0 & \dots & 0 & 0 \\ 2 & 5 & 2 & 0 & \dots & 0 & 0 \\ 0 & 2 & 5 & 2 & \dots & 0 & 0 \\ \vdots & \vdots & \ddots & \ddots & \ddots & \vdots & \vdots \\ 0 & 0 & \dots & 2 & 5 & 2 & 0 \\ 0 & 0 & \dots & 0 & 2 & 5 & 2 \\ 0 & 0 & \dots & 0 & 0 & 2 & 5 \end{pmatrix}$$

have a wide spread, since

$$\frac{\sigma_1}{\sigma_n} = \text{cond}_2(A) \geq 2^{n-1}.$$

If A is a complex $n \times n$ matrix, the eigenvalues $\lambda_1, \ldots, \lambda_n$ and the singular values $\sigma_1 \geq \sigma_2 \geq \cdots \geq \sigma_n$ of A are not unrelated, since

$$\sigma_1^2 \cdots \sigma_n^2 = \det(A^*A) = |\det(A)|^2$$

and

$$|\lambda_1| \cdots |\lambda_n| = |\det(A)|,$$

so we have

$$|\lambda_1| \cdots |\lambda_n| = \sigma_1 \cdots \sigma_n.$$

More generally, Hermann Weyl proved the following remarkable theorem:

Theorem 20.2. *(Weyl's inequalities, 1949) For any complex $n \times n$ matrix, A, if $\lambda_1, \ldots, \lambda_n \in \mathbb{C}$ are the eigenvalues of A and $\sigma_1, \ldots, \sigma_n \in \mathbb{R}_+$ are the singular values of A, listed so that $|\lambda_1| \geq \cdots \geq |\lambda_n|$ and $\sigma_1 \geq \cdots \geq \sigma_n \geq 0$, then*

$$|\lambda_1| \cdots |\lambda_n| = \sigma_1 \cdots \sigma_n \quad and$$
$$|\lambda_1| \cdots |\lambda_k| \leq \sigma_1 \cdots \sigma_k, \quad for \quad k = 1, \ldots, n-1.$$

A proof of Theorem 20.2 can be found in Horn and Johnson [Horn and Johnson (1994)], Chapter 3, Section 3.3, where more inequalities relating the eigenvalues and the singular values of a matrix are given.

Theorem 20.1 can be easily extended to rectangular $m \times n$ matrices, as we show in the next section. For various versions of the SVD for rectangular matrices, see Strang [Strang (1988)] Golub and Van Loan [Golub and Van Loan (1996)], Demmel [Demmel (1997)], and Trefethen and Bau [Trefethen and Bau III (1997)].

20.4 Singular Value Decomposition for Rectangular Matrices

Here is the generalization of Theorem 20.1 to rectangular matrices.

Theorem 20.3. *(Singular value decomposition) For every real $m \times n$ matrix A, there are two orthogonal matrices U $(n \times n)$ and V $(m \times m)$ and a* *diagonal $m \times n$ matrix D such that $A = VDU^\top$, where D is of the form*

$$D = \begin{pmatrix} \sigma_1 & \cdots & \\ & \sigma_2 & \cdots & \\ \vdots & \vdots & \ddots & \vdots \\ & & \cdots & \sigma_n \\ 0 & \vdots & \cdots & 0 \\ \vdots & \vdots & \ddots & \vdots \\ 0 & \vdots & \cdots & 0 \end{pmatrix} \quad or \quad D = \begin{pmatrix} \sigma_1 & & \cdots & & 0 \ldots 0 \\ & \sigma_2 & \cdots & & 0 \ldots 0 \\ \vdots & \vdots & \ddots & \vdots & 0 \vdots 0 \\ & & \cdots & \sigma_m & 0 \ldots 0 \end{pmatrix},$$

where $\sigma_1, \ldots, \sigma_r$ are the singular values of f, i.e. the (positive) square roots of the nonzero eigenvalues of $A^\top A$ and $AA^\top$, and $\sigma_{r+1} = \ldots = \sigma_p = 0$, where $p = \min(m, n)$. The columns of U are eigenvectors of $A^\top A$, and the columns of V are eigenvectors of $AA^\top$.

Proof. As in the proof of Theorem 20.1, since $A^\top A$ is symmetric positive semidefinite, there exists an $n \times n$ orthogonal matrix U such that

$$A^\top A = U\Sigma^2 U^\top,$$

with $\Sigma = \operatorname{diag}(\sigma_1, \ldots, \sigma_r, 0, \ldots, 0)$, where $\sigma_1^2, \ldots, \sigma_r^2$ are the nonzero eigenvalues of $A^\top A$, and where r is the rank of A. Observe that $r \le \min\{m, n\}$, and AU is an $m \times n$ matrix. It follows that

$$U^\top A^\top AU = (AU)^\top AU = \Sigma^2,$$

and if we let $f_j \in \mathbb{R}^m$ be the jth column of AU for $j = 1, \ldots, n$, then we have

$$\langle f_i, f_j \rangle = \sigma_i^2 \delta_{ij}, \quad 1 \le i, j \le r$$

and

$$f_j = 0, \quad r + 1 \le j \le n.$$

If we define $(v_1, \ldots, v_r)$ by

$$v_j = \sigma_j^{-1} f_j, \quad 1 \le j \le r,$$

then we have

$$\langle v_i, v_j \rangle = \delta_{ij}, \quad 1 \le i, j \le r,$$

so complete $(v_1, \ldots, v_r)$ into an orthonormal basis $(v_1, \ldots, v_r, v_{r+1}, \ldots, v_m)$ (for example, using Gram–Schmidt).

Now since $f_j = \sigma_j v_j$ for $j = 1 \ldots, r$, we have

$$\langle v_i, f_j \rangle = \sigma_j \langle v_i, v_j \rangle = \sigma_j \delta_{i,j}, \quad 1 \le i \le m, \; 1 \le j \le r$$

and since $f_j = 0$ for $j = r + 1, \ldots, n$, we have

$$\langle v_i, f_j \rangle = 0 \quad 1 \le i \le m, \; r + 1 \le j \le n.$$

If V is the matrix whose columns are $v_1, \ldots, v_m$, then V is an $m \times m$ orthogonal matrix and if $m \ge n$, we let

$$D = \begin{pmatrix} \Sigma \\ 0_{m-n} \end{pmatrix} = \begin{pmatrix} \sigma_1 & & \cdots & \\ & \sigma_2 & \cdots & \\ \vdots & \vdots & \ddots & \vdots \\ & & \cdots & \sigma_n \\ 0 & \vdots & \cdots & 0 \\ \vdots & \vdots & \ddots & \vdots \\ 0 & \vdots & \cdots & 0 \end{pmatrix},$$

else if $n \ge m$, then we let

$$D = \begin{pmatrix} \sigma_1 & & \cdots & & 0 \ldots 0 \\ & \sigma_2 & \cdots & & 0 \ldots 0 \\ \vdots & \vdots & \ddots & \vdots & 0 \; \vdots \; 0 \\ & & \cdots & \sigma_m & 0 \ldots 0 \end{pmatrix}.$$

In either case, the above equations prove that

$$V^\top A U = D,$$

which yields $A = V D U^\top$, as required.

The equation $A = V D U^\top$ implies that

$$A^\top A = U D^\top D U^\top = U \mathrm{diag}(\sigma_1^2, \ldots, \sigma_r^2, \underbrace{0, \ldots, 0}_{n-r}) U^\top$$

and

$$AA^\top = V D D^\top V^\top = V \mathrm{diag}(\sigma_1^2, \ldots, \sigma_r^2, \underbrace{0, \ldots, 0}_{m-r}) V^\top,$$

which shows that $A^\top A$ and $AA^\top$ have the same nonzero eigenvalues, that the columns of U are eigenvectors of $A^\top A$, and that the columns of V are eigenvectors of $AA^\top$. $\square$

A triple (U, D, V) such that $A = VDU^\top$ is called a *singular value decomposition (SVD) of A*.

Example 20.4. Let $A = \begin{pmatrix} 1 & 1 \\ 0 & 0 \\ 0 & 0 \end{pmatrix}$. Then $A^\top = \begin{pmatrix} 1 & 0 & 0 \\ 1 & 0 & 0 \end{pmatrix}$ $A^\top A = \begin{pmatrix} 1 & 1 \\ 1 & 1 \end{pmatrix}$,

and $AA^\top = \begin{pmatrix} 2 & 0 & 0 \\ 0 & 0 & 0 \\ 0 & 0 & 0 \end{pmatrix}$. The reader should verify that $A^\top A = U\Sigma^2 U^\top$ where

$\Sigma^2 = \begin{pmatrix} 2 & 0 \\ 0 & 0 \end{pmatrix}$ and $U = U^\top = \begin{pmatrix} 1/\sqrt{2} & 1/\sqrt{2} \\ 1/\sqrt{2} & -1/\sqrt{2} \end{pmatrix}$. Since $AU = \begin{pmatrix} \sqrt{2} & 0 \\ 0 & 0 \\ 0 & 0 \end{pmatrix}$, set

$v_1 = \frac{1}{\sqrt{2}} \begin{pmatrix} \sqrt{2} \\ 0 \\ 0 \end{pmatrix} = \begin{pmatrix} 1 \\ 0 \\ 0 \end{pmatrix}$, and complete an orthonormal basis for $\mathbb{R}^3$ by

assigning $v_2 = \begin{pmatrix} 0 \\ 1 \\ 0 \end{pmatrix}$, and $v_3 = \begin{pmatrix} 0 \\ 0 \\ 1 \end{pmatrix}$. Thus $V = I_3$, and the reader should

verify that $A = VDU^\top$, where $D = \begin{pmatrix} \sqrt{2} & 0 \\ 0 & 0 \\ 0 & 0 \end{pmatrix}$.

Even though the matrix D is an $m \times n$ rectangular matrix, since its only nonzero entries are on the descending diagonal, we still say that D is a diagonal matrix.

The `Matlab` command for computing an SVD $A = VDU^\top$ of a matrix A is also `[V, D, U] = svd(A)`.

If we view A as the representation of a linear map $f: E \to F$, where $\dim(E) = n$ and $\dim(F) = m$, the proof of Theorem 20.3 shows that there are two orthonormal bases $(u_1, \dots, u_n)$ and $(v_1, \dots, v_m)$ for E and F, respectively, where $(u_1, \dots, u_n)$ are eigenvectors of $f^* \circ f$ and $(v_1, \dots, v_m)$ are eigenvectors of $f \circ f^*$. Furthermore, $(u_1, \dots, u_r)$ is an orthonormal basis of $\operatorname{Im} f^*$, $(u_{r+1}, \dots, u_n)$ is an orthonormal basis of $\operatorname{Ker} f$, $(v_1, \dots, v_r)$ is an orthonormal basis of $\operatorname{Im} f$, and $(v_{r+1}, \dots, v_m)$ is an orthonormal basis of $\operatorname{Ker} f^*$.

The SVD of matrices can be used to define the pseudo-inverse of a rectangular matrix; we will do so in Chapter 21. The reader may also consult Strang [Strang (1988)], Demmel [Demmel (1997)], Trefethen and Bau [Trefethen and Bau III (1997)], and Golub and Van Loan [Golub and Van Loan (1996)].

One of the spectral theorems states that a symmetric matrix can be diagonalized by an orthogonal matrix. There are several numerical methods to compute the eigenvalues of a symmetric matrix A. One method consists in *tridiagonalizing* A, which means that there exists some orthogonal matrix P and some symmetric tridiagonal matrix T such that $A = PTP^\top$. In fact, this can be done using Householder transformations; see Theorem 17.2. It is then possible to compute the eigenvalues of T using a bisection method based on Sturm sequences. One can also use Jacobi's method. For details, see Golub and Van Loan [Golub and Van Loan (1996)], Chapter 8, Demmel [Demmel (1997)], Trefethen and Bau [Trefethen and Bau III (1997)], Lecture 26, Ciarlet [Ciarlet (1989)], and Chapter 17. Computing the SVD of a matrix A is more involved. Most methods begin by finding orthogonal matrices U and V and a *bidiagonal* matrix B such that $A = VBU^\top$; see Problem 12.8 and Problem 20.3. This can also be done using Householder transformations. Observe that $B^\top B$ is symmetric tridiagonal. Thus, in principle, the previous method to diagonalize a symmetric tridiagonal matrix can be applied. However, it is unwise to compute $B^\top B$ explicitly, and more subtle methods are used for this last step; the matrix of Problem 20.1 can be used, and see Problem 20.3. Again, see Golub and Van Loan [Golub and Van Loan (1996)], Chapter 8, Demmel [Demmel (1997)], and Trefethen and Bau [Trefethen and Bau III (1997)], Lecture 31.

The polar form has applications in continuum mechanics. Indeed, in any deformation it is important to separate stretching from rotation. This is exactly what QS achieves. The orthogonal part Q corresponds to rotation (perhaps with an additional reflection), and the symmetric matrix S to stretching (or compression). The real eigenvalues $\sigma_1, \dots, \sigma_r$ of S are the stretch factors (or compression factors) (see Marsden and Hughes [Marsden and Hughes (1994)]). The fact that S can be diagonalized by an orthogonal matrix corresponds to a natural choice of axes, the principal axes.

The SVD has applications to data compression, for instance in image processing. The idea is to retain only singular values whose magnitudes are significant enough. The SVD can also be used to determine the rank of a matrix when other methods such as Gaussian elimination produce very small pivots. One of the main applications of the SVD is the computation of the pseudo-inverse. Pseudo-inverses are the key to the solution of various optimization problems, in particular the method of least squares. This topic is discussed in the next chapter (Chapter 21). Applications of the material of this chapter can be found in Strang [Strang (1988, 1986)]; Ciarlet [Ciarlet (1989)]; Golub and Van Loan [Golub and Van Loan (1996)], which contains many other references; Demmel [Demmel (1997)]; and Trefethen and Bau [Trefethen and Bau III (1997)].

20.5 Ky Fan Norms and Schatten Norms

The singular values of a matrix can be used to define various norms on matrices which have found recent applications in quantum information theory and in spectral graph theory. Following Horn and Johnson [Horn and Johnson (1994)] (Section 3.4) we can make the following definitions:

Definition 20.5. For any matrix $A \in M_{m,n}(\mathbb{C})$, let $q = \min\{m,n\}$, and if $\sigma_1 \geq \cdots \geq \sigma_q$ are the singular values of A, for any k with $1 \leq k \leq q$, let

$$N_k(A) = \sigma_1 + \cdots + \sigma_k,$$

called the *Ky Fan k-norm* of A.

More generally, for any $p \geq 1$ and any k with $1 \leq k \leq q$, let

$$N_{k;p}(A) = (\sigma_1^p + \cdots + \sigma_k^p)^{1/p},$$

called the *Ky Fan p-k-norm* of A. When $k = q$, $N_{q;p}$ is also called the *Schatten p-norm*.

Observe that when $k = 1$, $N_1(A) = \sigma_1$, and the Ky Fan norm N_1 is simply the *spectral norm* from Chapter 8, which is the subordinate matrix norm associated with the Euclidean norm. When $k = q$, the Ky Fan norm N_q is given by

$$N_q(A) = \sigma_1 + \cdots + \sigma_q = \operatorname{tr}((A^*A)^{1/2})$$

and is called the *trace norm* or *nuclear norm*. When $p = 2$ and $k = q$, the Ky Fan $N_{q;2}$ norm is given by

$$N_{k;2}(A) = (\sigma_1^2 + \cdots + \sigma_q^2)^{1/2} = \sqrt{\operatorname{tr}(A^*A)} = \|A\|_F,$$

which is the *Frobenius norm* of A.

It can be shown that N_k and $N_{k;p}$ are unitarily invariant norms, and that when $m = n$, they are matrix norms; see Horn and Johnson [Horn and Johnson (1994)] (Section 3.4, Corollary 3.4.4 and Problem 3).

20.6 Summary

The main concepts and results of this chapter are listed below:

- For any linear map $f \colon E \to E$ on a Euclidean space E, the maps $f^* \circ f$ and $f \circ f^*$ are self-adjoint and positive semidefinite.
- The *singular values* of a linear map.
- *Positive semidefinite* and *positive definite* self-adjoint maps.
- Relationships between $\operatorname{Im} f$, $\operatorname{Ker} f$, $\operatorname{Im} f^*$, and $\operatorname{Ker} f^*$.
- The *singular value decomposition theorem* for square matrices (Theorem 20.1).
- The *SVD* of matrix.
- The *polar decomposition* of a matrix.
- The *Weyl inequalities*.
- The *singular value decomposition theorem* for $m \times n$ matrices (Theorem 20.3).
- Ky Fan k-norms, Ky Fan p-k-norms, Schatten p-norms.

20.7 Problems

Problem 20.1. (1) Let A be a real $n \times n$ matrix and consider the $(2n) \times (2n)$ real symmetric matrix

$$S = \begin{pmatrix} 0 & A \\ A^\top & 0 \end{pmatrix}.$$

Suppose that A has rank r. If $A = V\Sigma U^\top$ is an SVD for A, with $\Sigma = \operatorname{diag}(\sigma_1, \ldots, \sigma_n)$ and $\sigma_1 \geq \cdots \geq \sigma_r > 0$, denoting the columns of U by u_k and the columns of V by v_k, prove that σ_k is an eigenvalue of S with corresponding eigenvector $\begin{pmatrix} v_k \\ u_k \end{pmatrix}$ for $k = 1, \ldots, n$, and that $-\sigma_k$ is an eigenvalue of S with corresponding eigenvector $\begin{pmatrix} v_k \\ -u_k \end{pmatrix}$ for $k = 1, \ldots, n$.

Hint. We have $Au_k = \sigma_k v_k$ for $k = 1, \ldots, n$. Show that $A^\top v_k = \sigma_k u_k$ for $k = 1, \ldots, r$, and that $A^\top v_k = 0$ for $k = r+1, \ldots, n$. Recall that $\operatorname{Ker}(A^\top) = \operatorname{Ker}(AA^\top)$.

(2) Prove that the $2n$ eigenvectors of S in (1) are pairwise orthogonal. Check that if A has rank r, then S has rank $2r$.

(3) Now assume that A is a real $m \times n$ matrix and consider the

$(m + n) \times (m + n)$ real symmetric matrix

$$S = \begin{pmatrix} 0 & A \\ A^\top & 0 \end{pmatrix}.$$

Suppose that A has rank r. If $A = V\Sigma U^\top$ is an SVD for A, prove that σ_k is an eigenvalue of S with corresponding eigenvector $\begin{pmatrix} v_k \\ u_k \end{pmatrix}$ for $k = 1, \ldots, r$, and that $-\sigma_k$ is an eigenvalue of S with corresponding eigenvector $\begin{pmatrix} v_k \\ -u_k \end{pmatrix}$ for $k = 1, \ldots, r$.

Find the remaining $m + n - 2r$ eigenvectors of S associated with the eigenvalue 0.

(4) Prove that these $m + n$ eigenvectors of S are pairwise orthogonal.

Problem 20.2. Let A be a real $m \times n$ matrix of rank r and let $q = \min(m, n)$.

(1) Consider the $(m + n) \times (m + n)$ real symmetric matrix

$$S = \begin{pmatrix} 0 & A \\ A^\top & 0 \end{pmatrix}$$

and prove that

$$\begin{pmatrix} I_m & z^{-1}A \\ 0 & I_n \end{pmatrix} \begin{pmatrix} zI_m & -A \\ -A^\top & zI_n \end{pmatrix} = \begin{pmatrix} zI_m - z^{-1}AA^\top & 0 \\ -A^\top & zI_n \end{pmatrix}.$$

Use the above equation to prove that

$$\det(zI_{m+n} - S) = t^{n-m}\det(t^2I_m - AA^\top).$$

(2) Prove that the eigenvalues of S are $\pm\sigma_1, \ldots, \pm\sigma_q$, with $|m - n|$ additional zeros.

Problem 20.3. Let B be a real bidiagonal matrix of the form

$$B = \begin{pmatrix} a_1 & b_1 & 0 & \cdots & & 0 \\ 0 & a_2 & b_2 & \ddots & & 0 \\ \vdots & \ddots & \ddots & \ddots & & \vdots \\ 0 & \cdots & 0 & & a_{n-1} & b_{n-1} \\ 0 & 0 & \cdots & & 0 & a_n \end{pmatrix}.$$

Let A be the $(2n) \times (2n)$ symmetric matrix

$$A = \begin{pmatrix} 0 & B^\top \\ B & 0 \end{pmatrix},$$

and let P be the permutation matrix given by $P = [e_1, e_{n+1}, e_2, e_{n+2}, \cdots, e_n, e_{2n}]$.

(1) Prove that $T = P^\top A P$ is a symmetric tridiagonal $(2n) \times (2n)$ matrix with zero main diagonal of the form

$$
T = \begin{pmatrix}
0 & a_1 & 0 & 0 & 0 & 0 & \cdots & 0 \\
a_1 & 0 & b_1 & 0 & 0 & 0 & \cdots & 0 \\
0 & b_1 & 0 & a_2 & 0 & 0 & \cdots & 0 \\
0 & 0 & a_2 & 0 & b_2 & 0 & \cdots & 0 \\
\vdots & \vdots & \vdots & \ddots & \ddots & \ddots & \ddots & \vdots \\
0 & 0 & 0 & \cdots & a_{n-1} & 0 & b_{n-1} & 0 \\
0 & 0 & 0 & \cdots & 0 & b_{n-1} & 0 & a_n \\
0 & 0 & 0 & \cdots & 0 & 0 & a_n & 0
\end{pmatrix}.
$$

(2) Prove that if x_i is a unit eigenvector for an eigenvalue λ_i of T, then $\lambda_i = \pm\sigma_i$ where σ_i is a singular value of B, and that

$$
P x_i = \frac{1}{\sqrt{2}} \begin{pmatrix} u_i \\ \pm v_i \end{pmatrix},
$$

where the u_i are unit eigenvectors of $B^\top B$ and the v_i are unit eigenvectors of $BB^\top$.

Problem 20.4. Find the SVD of the matrix

$$
A = \begin{pmatrix} 0 & 2 & 0 \\ 0 & 0 & 3 \\ 0 & 0 & 0 \end{pmatrix}.
$$

Problem 20.5. Let $u, v \in \mathbb{R}^n$ be two nonzero vectors, and let $A = uv^\top$ be the corresponding rank 1 matrix. Prove that the nonzero singular value of A is $\|u\|_2 \|v\|_2$.

Problem 20.6. Let A be a $n \times n$ real matrix. Prove that if $\sigma_1, \ldots, \sigma_n$ are the singular values of A, then $\sigma_1^3, \ldots, \sigma_n^3$ are the singular values of $AA^\top A$.

Problem 20.7. Let A be a real $n \times n$ matrix.

(1) Prove that the largest singular value σ_1 of A is given by

$$
\sigma_1 = \sup_{x \neq 0} \frac{\|Ax\|_2}{\|x\|_2},
$$

and that this supremum is achieved at $x = u_1$, the first column in U in an SVD $A = V\Sigma U^\top$.

(2) Extend the above result to real $m \times n$ matrices.

Problem 20.8. Let A be a real $m \times n$ matrix. Prove that if B is any submatrix of A (by keeping $M \leq m$ rows and $N \leq n$ columns of A), then $(\sigma_1)_B \leq (\sigma_1)_A$ (where $(\sigma_1)_A$ is the largest singular value of A and similarly for $(\sigma_1)_B$).

Problem 20.9. Let A be a real $n \times n$ matrix.

(1) Assume A is invertible. Prove that if $A = Q_1 S_1 = Q_2 S_2$ are two polar decompositions of A, then $Q_1 = Q_2$ and $S_1 = S_2$.

Hint. $A^\top A = S_1^2 = S_2^2$, with S_1 and S_2 symmetric positive definite. Then use Problem 16.7.

(2) Now assume that A is singular. Prove that if $A = Q_1 S_1 = Q_2 S_2$ are two polar decompositions of A, then $S_1 = S_2$, but Q_1 may not be equal to Q_2.

Problem 20.10. (1) Let A be any invertible (real) $n \times n$ matrix. Prove that for every SVD, $A = VDU^\top$ of A, the product $VU^\top$ is the same (i.e., if $V_1 DU_1^\top = V_2 DU_2^\top$, then $V_1 U_1^\top = V_2 U_2^\top$). What does $VU^\top$ have to do with the polar form of A?

(2) Given any invertible (real) $n \times n$ matrix, A, prove that there is a unique orthogonal matrix, $Q \in \mathbf{O}(n)$, such that $\|A - Q\|_F$ is minimal (under the Frobenius norm). In fact, prove that $Q = VU^\top$, where $A = VDU^\top$ is an SVD of A. Moreover, if $\det(A) > 0$, show that $Q \in \mathbf{SO}(n)$.

What can you say if A is singular (i.e., non-invertible)?

Problem 20.11. (1) Prove that for any $n \times n$ matrix A and any orthogonal matrix Q, we have

$$\max\{\operatorname{tr}(QA) \mid Q \in \mathbf{O}(n)\} = \sigma_1 + \cdots + \sigma_n,$$

where $\sigma_1 \geq \cdots \geq \sigma_n$ are the singular values of A. Furthermore, this maximum is achieved by $Q = UV^\top$, where $A = V\Sigma U^\top$ is any SVD for A.

(2) By applying the above result with $A = Z^\top X$ and $Q = R^\top$, deduce the following result: For any two fixed $n \times k$ matrices X and Z, the minimum of the set

$$\{\|X - ZR\|_F \mid R \in \mathbf{O}(k)\}$$

is achieved by $R = VU^\top$ for any SVD decomposition $V\Sigma U^\top = Z^\top X$ of $Z^\top X$.

Remark: The problem of finding an orthogonal matrix R such that ZR comes as close as possible to X is called the *orthogonal Procrustes problem*; see Strang [Strang (2019)] (Section IV.9) for the history of this problem.

Chapter 21

Applications of SVD and Pseudo-Inverses

De tous les principes qu'on peut proposer pour cet objet, je pense qu'il n'en est pas de plus général, de plus exact, ni d'une application plus facile, que celui dont nous avons fait usage dans les recherches précédentes, et qui consiste à rendre *minimum* la somme des carrés des erreurs. Par ce moyen il s'établit entre les erreurs une sorte d'équilibre qui, empêchant les extrêmes de prévaloir, est très propre às faire connaitre l'état du système le plus proche de la vérité.

—**Legendre, 1805,** *Nouvelles Méthodes pour la détermination des Orbites des*
Comètes

21.1 Least Squares Problems and the Pseudo-Inverse

This chapter presents several applications of SVD. The first one is the pseudo-inverse, which plays a crucial role in solving linear systems by the method of least squares. The second application is data compression. The third application is principal component analysis (PCA), whose purpose is to identify patterns in data and understand the variance–covariance structure of the data. The fourth application is the best affine approximation of a set of data, a problem closely related to PCA.

The method of least squares is a way of "solving" an overdetermined system of linear equations

$$Ax = b,$$

i.e., a system in which A is a rectangular $m \times n$ matrix with more equations than unknowns (when $m > n$). Historically, the method of least squares was used by Gauss and Legendre to solve problems in astronomy and geodesy. The method was first published by Legendre in 1805 in a paper on methods for determining the orbits of comets. However, Gauss had already used the method of least squares as early as 1801 to determine the orbit of the asteroid Ceres, and he published a paper about it in 1810 after the discovery of the asteroid Pallas. Incidentally, it is in that same paper that Gaussian elimination using pivots is introduced.

The reason why more equations than unknowns arise in such problems is that repeated measurements are taken to minimize errors. This produces an overdetermined and often inconsistent system of linear equations. For example, Gauss solved a system of eleven equations in six unknowns to determine the orbit of the asteroid Pallas.

Example 21.1. As a concrete illustration, suppose that we observe the motion of a small object, assimilated to a point, in the plane. From our observations, we suspect that this point moves along a straight line, say of equation $y = cx + d$. Suppose that we observed the moving point at three different locations (x_1, y_1), (x_2, y_2), and (x_3, y_3). Then we should have

$$d + cx_1 = y_1,$$
$$d + cx_2 = y_2,$$
$$d + cx_3 = y_3.$$

If there were no errors in our measurements, these equations would be compatible, and c and d would be determined by only two of the equations. However, in the presence of errors, the system may be inconsistent. Yet we would like to find c and d!

The idea of the method of least squares is to determine (c, d) such that it minimizes the sum of the squares of the errors, namely,

$$(d + cx_1 - y_1)^2 + (d + cx_2 - y_2)^2 + (d + cx_3 - y_3)^2.$$

See Figure 21.1.

In general, for an overdetermined $m \times n$ system $Ax = b$, what Gauss and Legendre discovered is that there are solutions x minimizing

$$\|Ax - b\|_2^2$$

(where $\|u\|_2^2 = u_1^2 + \cdots + u_n^2$, the square of the Euclidean norm of the vector $u = (u_1, \ldots, u_n)$), and that these solutions are given by the square $n \times n$ system

$$A^\top Ax = A^\top b,$$

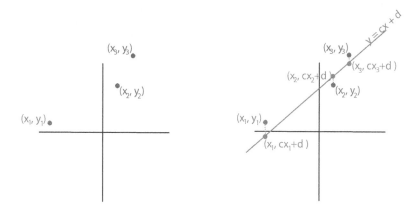

Fig. 21.1   Given three points (x_1, y_1), (x_2, y_2), (x_3, y_3), we want to determine the line $y = cx + d$ which minimizes the lengths of the dashed vertical lines.

called the *normal equations*. Furthermore, when the columns of A are linearly independent, it turns out that $A^\top A$ is invertible, and so x is unique and given by

$$x = (A^\top A)^{-1} A^\top b.$$

Note that $A^\top A$ is a symmetric matrix, one of the nice features of the normal equations of a least squares problem. For instance, since the above problem in matrix form is represented as

$$\begin{pmatrix} 1 & x_1 \\ 1 & x_2 \\ 1 & x_3 \end{pmatrix} \begin{pmatrix} d \\ c \end{pmatrix} = \begin{pmatrix} y_1 \\ y_2 \\ y_3 \end{pmatrix},$$

the normal equations are

$$\begin{pmatrix} 3 & x_1 + x_2 + x_3 \\ x_1 + x_2 + x_3 & x_1^2 + x_2^2 + x_3^2 \end{pmatrix} \begin{pmatrix} d \\ c \end{pmatrix} = \begin{pmatrix} y_1 + y_2 + y_3 \\ x_1 y_1 + x_2 y_2 + x_3 y_3 \end{pmatrix}.$$

In fact, given any real $m \times n$ matrix A, there is always a unique x^+ of minimum norm that minimizes $\|Ax - b\|_2^2$, even when the columns of A are linearly dependent. How do we prove this, and how do we find x^+?

Theorem 21.1. *Every linear system $Ax = b$, where A is an $m \times n$ matrix, has a unique least squares solution x^+ of smallest norm.*

Proof. Geometry offers a nice proof of the existence and uniqueness of x^+. Indeed, we can interpret b as a point in the Euclidean (affine) space $\mathbb{R}^m$, and the image subspace of A (also called the column space of A) as a subspace U of $\mathbb{R}^m$ (passing through the origin). Then it is clear that

$$\inf_{x \in \mathbb{R}^n} \|Ax - b\|_2^2 = \inf_{y \in U} \|y - b\|_2^2,$$

with $U = \operatorname{Im} A$, and we claim that x minimizes $\|Ax - b\|_2^2$ iff $Ax = p$, where p the orthogonal projection of b onto the subspace U.

Recall from Section 12.1 that the orthogonal projection $p_U : U \oplus U^\perp \to U$ is the linear map given by

$$p_U(u + v) = u,$$

with $u \in U$ and $v \in U^\perp$. If we let $p = p_U(b) \in U$, then for any point $y \in U$, the vectors $\overrightarrow{py} = y - p \in U$ and $\overrightarrow{bp} = p - b \in U^\perp$ are orthogonal, which implies that

$$\|\overrightarrow{by}\|_2^2 = \|\overrightarrow{bp}\|_2^2 + \|\overrightarrow{py}\|_2^2,$$

where $\overrightarrow{by} = y - b$. Thus, p is indeed the unique point in U that minimizes the distance from b to any point in U. See Figure 21.2.

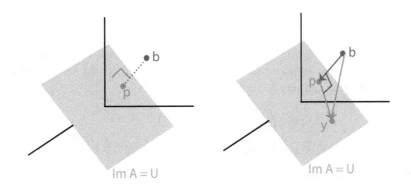

Fig. 21.2   Given a 3×2 matrix A, $U = \operatorname{Im} A$ is the peach plane in $\mathbb{R}^3$ and p is the orthogonal projection of b onto U. Furthermore, given $y \in U$, the points b, y, and p are the vertices of a right triangle.

Thus the problem has been reduced to proving that there is a unique x^+ of minimum norm such that $Ax^+ = p$, with $p = p_U(b) \in U$, the orthogonal

projection of b onto U. We use the fact that
$$\mathbb{R}^n = \operatorname{Ker} A \oplus (\operatorname{Ker} A)^{\perp}.$$
Consequently, every $x \in \mathbb{R}^n$ can be written uniquely as $x = u + v$, where $u \in \operatorname{Ker} A$ and $v \in (\operatorname{Ker} A)^{\perp}$, and since u and v are orthogonal,
$$\|x\|_2^2 = \|u\|_2^2 + \|v\|_2^2.$$
Furthermore, since $u \in \operatorname{Ker} A$, we have $Au = 0$, and thus $Ax = p$ iff $Av = p$, which shows that the solutions of $Ax = p$ for which x has minimum norm must belong to $(\operatorname{Ker} A)^{\perp}$. However, the restriction of A to $(\operatorname{Ker} A)^{\perp}$ is injective. This is because if $Av_1 = Av_2$, where $v_1, v_2 \in (\operatorname{Ker} A)^{\perp}$, then $A(v_2 - v_2) = 0$, which implies $v_2 - v_1 \in \operatorname{Ker} A$, and since $v_1, v_2 \in (\operatorname{Ker} A)^{\perp}$, we also have $v_2 - v_1 \in (\operatorname{Ker} A)^{\perp}$, and consequently, $v_2 - v_1 = 0$. This shows that there is a unique x^+ of minimum norm such that $Ax^+ = p$, and that x^+ must belong to $(\operatorname{Ker} A)^{\perp}$. By our previous reasoning, x^+ is the unique vector of minimum norm minimizing $\|Ax - b\|_2^2$. $\qquad\square$

The proof also shows that x minimizes $\|Ax - b\|_2^2$ iff $\overrightarrow{pb} = b - Ax$ is orthogonal to U, which can be expressed by saying that $b - Ax$ is orthogonal to every column of A. However, this is equivalent to
$$A^{\top}(b - Ax) = 0, \quad \text{i.e.,} \quad A^{\top}Ax = A^{\top}b.$$
Finally, it turns out that the minimum norm least squares solution x^+ can be found in terms of the pseudo-inverse A^+ of A, which is itself obtained from any SVD of A.

Definition 21.1. Given any nonzero $m \times n$ matrix A of rank r, if $A = VDU^{\top}$ is an SVD of A such that
$$D = \begin{pmatrix} \Lambda & 0_{r,n-r} \\ 0_{m-r,r} & 0_{m-r,n-r} \end{pmatrix},$$
with
$$\Lambda = \operatorname{diag}(\lambda_1, \ldots, \lambda_r)$$
an $r \times r$ diagonal matrix consisting of the nonzero singular values of A, then if we let D^+ be the $n \times m$ matrix
$$D^+ = \begin{pmatrix} \Lambda^{-1} & 0_{r,m-r} \\ 0_{n-r,r} & 0_{n-r,m-r} \end{pmatrix},$$
with
$$\Lambda^{-1} = \operatorname{diag}(1/\lambda_1, \ldots, 1/\lambda_r),$$
the *pseudo-inverse* of A is defined by
$$A^+ = UD^+V^{\top}.$$

If $A = 0_{m,n}$ is the zero matrix, we set $A^+ = 0_{n,m}$. Observe that D^+ is obtained from D by inverting the nonzero diagonal entries of D, leaving all zeros in place, and then transposing the matrix. For example, given the matrix

$$D = \begin{pmatrix} 1 & 0 & 0 & 0 & 0 \\ 0 & 2 & 0 & 0 & 0 \\ 0 & 0 & 3 & 0 & 0 \\ 0 & 0 & 0 & 0 & 0 \end{pmatrix},$$

its pseudo-inverse is

$$D^+ = \begin{pmatrix} 1 & 0 & 0 & 0 \\ 0 & \frac{1}{2} & 0 & 0 \\ 0 & 0 & \frac{1}{3} & 0 \\ 0 & 0 & 0 & 0 \\ 0 & 0 & 0 & 0 \end{pmatrix}.$$

The pseudo-inverse of a matrix is also known as the *Moore–Penrose pseudo-inverse*.

Actually, it seems that A^+ depends on the specific choice of U and V in an SVD (U, D, V) for A, but the next theorem shows that this is not so.

Theorem 21.2. *The least squares solution of smallest norm of the linear system $Ax = b$, where A is an $m \times n$ matrix, is given by*

$$x^+ = A^+b = UD^+V^\top b.$$

Proof. First assume that A is a (rectangular) diagonal matrix D, as above. Then since x minimizes $\|Dx - b\|_2^2$ iff Dx is the projection of b onto the image subspace F of D, it is fairly obvious that $x^+ = D^+b$. Otherwise, we can write

$$A = VDU^\top,$$

where U and V are orthogonal. However, since V is an isometry,

$$\|Ax - b\|_2 = \|VDU^\top x - b\|_2 = \|DU^\top x - V^\top b\|_2.$$

Letting $y = U^\top x$, we have $\|x\|_2 = \|y\|_2$, since U is an isometry, and since U is surjective, $\|Ax - b\|_2$ is minimized iff $\|Dy - V^\top b\|_2$ is minimized, and we have shown that the least solution is

$$y^+ = D^+V^\top b.$$

Since $y = U^\top x$, with $\|x\|_2 = \|y\|_2$, we get

$$x^+ = UD^+V^\top b = A^+b.$$

Thus, the pseudo-inverse provides the optimal solution to the least squares problem. $\qquad\square$

By Theorem 21.2 and Theorem 21.1, A^+b is uniquely defined by every b, and thus A^+ depends only on A.

The `Matlab` command for computing the pseudo-inverse B of the matrix A is

`B = pinv(A)`.

Example 21.2. If A is the rank 2 matrix

$$A = \begin{pmatrix} 1\ 2\ 3\ 4 \\ 2\ 3\ 4\ 5 \\ 3\ 4\ 5\ 6 \\ 4\ 5\ 6\ 7 \end{pmatrix}$$

whose eigenvalues are $-1.1652, 0, 0, 17.1652$, using `Matlab` we obtain the SVD $A = VDU^\top$ with

$$U = \begin{pmatrix} -0.3147 & 0.7752 & 0.2630 & -0.4805 \\ -0.4275 & 0.3424 & 0.0075 & 0.8366 \\ -0.5402 & -0.0903 & -0.8039 & -0.2319 \\ -0.6530 & -0.5231 & 0.5334 & -0.1243 \end{pmatrix},$$

$$V = \begin{pmatrix} -0.3147 & -0.7752 & 0.5452 & 0.0520 \\ -0.4275 & -0.3424 & -0.7658 & 0.3371 \\ -0.5402 & 0.0903 & -0.1042 & -0.8301 \\ -0.6530 & 0.5231 & 0.3247 & 0.4411 \end{pmatrix}, \quad D = \begin{pmatrix} 17.1652 & 0 & 0\ 0 \\ 0 & 1.1652 & 0\ 0 \\ 0 & 0 & 0\ 0 \\ 0 & 0 & 0\ 0 \end{pmatrix}.$$

Then

$$D^+ = \begin{pmatrix} 0.0583 & 0 & 0\ 0 \\ 0 & 0.8583 & 0\ 0 \\ 0 & 0 & 0\ 0 \\ 0 & 0 & 0\ 0 \end{pmatrix},$$

and

$$A^+ = UD^+V^\top = \begin{pmatrix} -0.5100 & -0.2200 & 0.0700 & 0.3600 \\ -0.2200 & -0.0900 & 0.0400 & 0.1700 \\ 0.0700 & 0.0400 & 0.0100 & -0.0200 \\ 0.3600 & 0.1700 & -0.0200 & -0.2100 \end{pmatrix},$$

which is also the result obtained by calling `pinv(A)`.

If A is an $m \times n$ matrix of rank n (and so $m \geq n$), it is immediately shown that the QR-decomposition in terms of Householder transformations applies as follows:

There are n $m \times m$ matrices $H_1, \ldots, H_n$, Householder matrices or the identity, and an upper triangular $m \times n$ matrix R of rank n such that

$$A = H_1 \cdots H_n R.$$

Then because each H_i is an isometry,

$$\|Ax - b\|_2 = \|Rx - H_n \cdots H_1 b\|_2,$$

and the least squares problem $Ax = b$ is equivalent to the system

$$Rx = H_n \cdots H_1 b.$$

Now the system

$$Rx = H_n \cdots H_1 b$$

is of the form

$$\begin{pmatrix} R_1 \\ 0_{m-n} \end{pmatrix} x = \begin{pmatrix} c \\ d \end{pmatrix},$$

where R_1 is an invertible $n \times n$ matrix (since A has rank n), $c \in \mathbb{R}^n$, and $d \in \mathbb{R}^{m-n}$, and the least squares solution of smallest norm is

$$x^+ = R_1^{-1} c.$$

Since R_1 is a triangular matrix, it is very easy to invert R_1.

The method of least squares is one of the most effective tools of the mathematical sciences. There are entire books devoted to it. Readers are advised to consult Strang [Strang (1988)], Golub and Van Loan [Golub and Van Loan (1996)], Demmel [Demmel (1997)], and Trefethen and Bau [Trefethen and Bau III (1997)], where extensions and applications of least squares (such as weighted least squares and recursive least squares) are described. Golub and Van Loan [Golub and Van Loan (1996)] also contains a very extensive bibliography, including a list of books on least squares.

21.2 Properties of the Pseudo-Inverse

We begin this section with a proposition which provides a way to calculate the pseudo-inverse of an $m \times n$ matrix A without first determining an SVD factorization.

Proposition 21.1. *When A has full rank, the pseudo-inverse A^+ can be expressed as $A^+ = (A^\top A)^{-1} A^\top$ when $m \geq n$, and as $A^+ = A^\top (AA^\top)^{-1}$ when $n \geq m$. In the first case ($m \geq n$), observe that $A^+ A = I$, so A^+ is a left inverse of A; in the second case ($n \geq m$), we have $AA^+ = I$, so A^+ is a right inverse of A.*

Proof. If $m \geq n$ and A has full rank n, we have

$$A = V \begin{pmatrix} \Lambda \\ 0_{m-n,n} \end{pmatrix} U^\top$$

with Λ an $n \times n$ diagonal invertible matrix (with positive entries), so

$$A^+ = U \left(\Lambda^{-1} \ 0_{n,m-n} \right) V^\top.$$

We find that

$$A^\top A = U \left(\Lambda \ 0_{n,m-n} \right) V^\top V \begin{pmatrix} \Lambda \\ 0_{m-n,n} \end{pmatrix} U^\top = U\Lambda^2 U^\top,$$

which yields

$$(A^\top A)^{-1} A^\top = U\Lambda^{-2} U^\top U \left(\Lambda \ 0_{n,m-n} \right) V^\top = U \left(\Lambda^{-1} \ 0_{n,m-n} \right) V^\top = A^+.$$

Therefore, if $m \geq n$ and A has full rank n, then

$$A^+ = (A^\top A)^{-1} A^\top.$$

If $n \geq m$ and A has full rank m, then

$$A = V \left(\Lambda \ 0_{m,n-m} \right) U^\top$$

with Λ an $m \times m$ diagonal invertible matrix (with positive entries), so

$$A^+ = U \begin{pmatrix} \Lambda^{-1} \\ 0_{n-m,m} \end{pmatrix} V^\top.$$

We find that

$$AA^\top = V \left(\Lambda \ 0_{m,n-m} \right) U^\top U \begin{pmatrix} \Lambda \\ 0_{n-m,m} \end{pmatrix} V^\top = V\Lambda^2 V^\top,$$

which yields

$$A^\top (AA^\top)^{-1} = U \begin{pmatrix} \Lambda \\ 0_{n-m,m} \end{pmatrix} V^\top V\Lambda^{-2} V^\top = U \begin{pmatrix} \Lambda^{-1} \\ 0_{n-m,m} \end{pmatrix} V^\top = A^+.$$

Therefore, if $n \geq m$ and A has full rank m, then $A^+ = A^\top (AA^\top)^{-1}$. $\qquad \square$

For example, if $A = \begin{pmatrix} 1 & 2 \\ 2 & 3 \\ 0 & 1 \end{pmatrix}$, then A has rank 2 and since $m \geq n$, $A^+ = (A^\top A)^{-1} A^\top$ where

$$A^+ = \begin{pmatrix} 5 & 8 \\ 8 & 14 \end{pmatrix}^{-1} A^\top = \begin{pmatrix} 7/3 & -4/3 \\ 4/3 & 5/6 \end{pmatrix} \begin{pmatrix} 1 & 2 & 0 \\ 2 & 3 & 1 \end{pmatrix} = \begin{pmatrix} -1/3 & 2/3 & -4/3 \\ 1/3 & -1/6 & 5/6 \end{pmatrix}.$$

If $A = \begin{pmatrix} 1 & 2 & 3 & 0 \\ 0 & 1 & 1 & -1 \end{pmatrix}$, since A has rank 2 and $n \geq m$, then $A^+ = A^\top (AA^\top)^{-1}$

where

$$A^+ = A^\top \begin{pmatrix} 14 & 5 \\ 5 & 3 \end{pmatrix}^{-1} = \begin{pmatrix} 1 & 0 \\ 2 & 1 \\ 3 & 1 \\ 0 & -1 \end{pmatrix} \begin{pmatrix} 3/17 & -5/17 \\ -5/17 & 14/17 \end{pmatrix} = \begin{pmatrix} 3/17 & -5/17 \\ 1/17 & 4/17 \\ 4/17 & -1/17 \\ 5/17 & -14/17 \end{pmatrix}.$$

Let $A = V\Sigma U^\top$ be an SVD for any $m \times n$ matrix A. It is easy to check that both AA^+ and A^+A are symmetric matrices. In fact,

$$AA^+ = V\Sigma U^\top U\Sigma^+ V^\top = V\Sigma\Sigma^+ V^\top = V \begin{pmatrix} I_r & 0 \\ 0 & 0_{m-r} \end{pmatrix} V^\top$$

and

$$A^+A = U\Sigma^+ V^\top V\Sigma U^\top = U\Sigma^+\Sigma U^\top = U \begin{pmatrix} I_r & 0 \\ 0 & 0_{n-r} \end{pmatrix} U^\top.$$

From the above expressions we immediately deduce that

$$AA^+A = A,$$
$$A^+AA^+ = A^+,$$

and that

$$(AA^+)^2 = AA^+,$$
$$(A^+A)^2 = A^+A,$$

so both AA^+ and A^+A are orthogonal projections (since they are both symmetric).

Proposition 21.2. *The matrix AA^+ is the orthogonal projection onto the range of A and A^+A is the orthogonal projection onto $\mathrm{Ker}(A)^\perp = \mathrm{Im}(A^\top)$, the range of $A^\top$.*

Proof. Obviously, we have $\mathrm{range}(AA^+) \subseteq \mathrm{range}(A)$, and for any $y = Ax \in \mathrm{range}(A)$, since $AA^+A = A$, we have

$$AA^+y = AA^+Ax = Ax = y,$$

so the image of AA^+ is indeed the range of A. It is also clear that $\mathrm{Ker}(A) \subseteq \mathrm{Ker}(A^+A)$, and since $AA^+A = A$, we also have $\mathrm{Ker}(A^+A) \subseteq \mathrm{Ker}(A)$, and so

$$\mathrm{Ker}(A^+A) = \mathrm{Ker}(A).$$

Since A^+A is symmetric, $\mathrm{range}(A^+A) = \mathrm{range}((A^+A)^\top) = \mathrm{Ker}(A^+A)^\perp = \mathrm{Ker}(A)^\perp$, as claimed. $\qquad \square$

Proposition 21.3. *The set* $\mathrm{range}(A) = \mathrm{range}(AA^+)$ *consists of all vectors* $y \in \mathbb{R}^m$ *such that*

$$V^\top y = \begin{pmatrix} z \\ 0 \end{pmatrix},$$

with $z \in \mathbb{R}^r$.

Proof. Indeed, if $y = Ax$, then

$$V^\top y = V^\top Ax = V^\top V\Sigma U^\top x = \Sigma U^\top x = \begin{pmatrix} \Sigma_r & 0 \\ 0 & 0_{m-r} \end{pmatrix} U^\top x = \begin{pmatrix} z \\ 0 \end{pmatrix},$$

where Σ_r is the $r \times r$ diagonal matrix $\mathrm{diag}(\sigma_1, \ldots, \sigma_r)$. Conversely, if $V^\top y = \begin{pmatrix} z \\ 0 \end{pmatrix}$, then $y = V \begin{pmatrix} z \\ 0 \end{pmatrix}$, and

$$AA^+ y = V \begin{pmatrix} I_r & 0 \\ 0 & 0_{m-r} \end{pmatrix} V^\top y$$

$$= V \begin{pmatrix} I_r & 0 \\ 0 & 0_{m-r} \end{pmatrix} V^\top V \begin{pmatrix} z \\ 0 \end{pmatrix}$$

$$= V \begin{pmatrix} I_r & 0 \\ 0 & 0_{m-r} \end{pmatrix} \begin{pmatrix} z \\ 0 \end{pmatrix}$$

$$= V \begin{pmatrix} z \\ 0 \end{pmatrix} = y,$$

which shows that y belongs to the range of A.   □

Similarly, we have the following result.

Proposition 21.4. *The set* $\mathrm{range}(A^+A) = \mathrm{Ker}(A)^\perp$ *consists of all vectors* $y \in \mathbb{R}^n$ *such that*

$$U^\top y = \begin{pmatrix} z \\ 0 \end{pmatrix},$$

with $z \in \mathbb{R}^r$.

Proof. If $y = A^+Au$, then

$$y = A^+Au = U \begin{pmatrix} I_r & 0 \\ 0 & 0_{n-r} \end{pmatrix} U^\top u = U \begin{pmatrix} z \\ 0 \end{pmatrix},$$

for some $z \in \mathbb{R}^r$. Conversely, if $U^\top y = \begin{pmatrix} z \\ 0 \end{pmatrix}$, then $y = U \begin{pmatrix} z \\ 0 \end{pmatrix}$, and so

$$A^+AU \begin{pmatrix} z \\ 0 \end{pmatrix} = U \begin{pmatrix} I_r & 0 \\ 0 & 0_{n-r} \end{pmatrix} U^\top U \begin{pmatrix} z \\ 0 \end{pmatrix}$$

$$= U \begin{pmatrix} I_r & 0 \\ 0 & 0_{n-r} \end{pmatrix} \begin{pmatrix} z \\ 0 \end{pmatrix}$$

$$= U \begin{pmatrix} z \\ 0 \end{pmatrix} = y,$$

which shows that $y \in \mathrm{range}(A^+A)$. $\square$

Analogous results hold for complex matrices, but in this case, V and U are unitary matrices and AA^+ and A^+A are Hermitian orthogonal projections.

If A is a normal matrix, which means that $AA^\top = A^\top A$, then there is an intimate relationship between SVD's of A and block diagonalizations of A. As a consequence, the pseudo-inverse of a normal matrix A can be obtained directly from a block diagonalization of A.

If A is a (real) normal matrix, then we know from Theorem 16.8 that A can be block diagonalized with respect to an orthogonal matrix U as

$$A = U\Lambda U^\top,$$

where Λ is the (real) block diagonal matrix

$$\Lambda = \mathrm{diag}(B_1, \ldots, B_n),$$

consisting either of 2×2 blocks of the form

$$B_j = \begin{pmatrix} \lambda_j & -\mu_j \\ \mu_j & \lambda_j \end{pmatrix}$$

with $\mu_j \neq 0$, or of one-dimensional blocks $B_k = (\lambda_k)$. Then we have the following proposition:

Proposition 21.5. *For any (real) normal matrix A and any block diagonalization $A = U\Lambda U^\top$ of A as above, the pseudo-inverse of A is given by*

$$A^+ = U\Lambda^+ U^\top,$$

where Λ^+ is the pseudo-inverse of Λ. Furthermore, if

$$\Lambda = \begin{pmatrix} \Lambda_r & 0 \\ 0 & 0 \end{pmatrix},$$

where Λ_r has rank r, then

$$\Lambda^+ = \begin{pmatrix} \Lambda_r^{-1} & 0 \\ 0 & 0 \end{pmatrix}.$$

Proof. Assume that $B_1, \ldots, B_p$ are 2×2 blocks and that $\lambda_{2p+1}, \ldots, \lambda_n$ are the scalar entries. We know that the numbers $\lambda_j \pm i\mu_j$, and the λ_{2p+k} are the eigenvalues of A. Let $\rho_{2j-1} = \rho_{2j} = \sqrt{\lambda_j^2 + \mu_j^2} = \sqrt{\det(B_i)}$ for $j = 1, \ldots, p$, $\rho_j = |\lambda_j|$ for $j = 2p + 1, \ldots, r$. Multiplying U by a suitable permutation matrix, we may assume that the blocks of Λ are ordered so that $\rho_1 \geq \rho_2 \geq \cdots \geq \rho_r > 0$. Then it is easy to see that

$$AA^\top = A^\top A = U\Lambda U^\top U\Lambda^\top U^\top = U\Lambda\Lambda^\top U^\top,$$

with

$$\Lambda\Lambda^\top = \text{diag}(\rho_1^2, \ldots, \rho_r^2, 0, \ldots, 0),$$

so $\rho_1 \geq \rho_2 \geq \cdots \geq \rho_r > 0$ are the singular values $\sigma_1 \geq \sigma_2 \geq \cdots \geq \sigma_r > 0$ of A. Define the diagonal matrix

$$\Sigma = \text{diag}(\sigma_1, \ldots, \sigma_r, 0, \ldots, 0),$$

where $r = \text{rank}(A)$, $\sigma_1 \geq \cdots \geq \sigma_r > 0$ and the block diagonal matrix Θ defined such that the block B_i in Λ is replaced by the block $\sigma^{-1}B_i$ where $\sigma = \sqrt{\det(B_i)}$, the nonzero scalar λ_j is replaced $\lambda_j/|\lambda_j|$, and a diagonal zero is replaced by 1. Observe that Θ is an orthogonal matrix and

$$\Lambda = \Theta\Sigma.$$

But then we can write

$$A = U\Lambda U^\top = U\Theta\Sigma U^\top,$$

and we if let $V = U\Theta$, since U is orthogonal and Θ is also orthogonal, V is also orthogonal and $A = V\Sigma U^\top$ *is an SVD for A.* Now we get

$$A^+ = U\Sigma^+ V^\top = U\Sigma^+\Theta^\top U^\top.$$

However, since Θ is an orthogonal matrix, $\Theta^\top = \Theta^{-1}$, and a simple calculation shows that

$$\Sigma^+\Theta^\top = \Sigma^+\Theta^{-1} = \Lambda^+,$$

which yields the formula

$$A^+ = U\Lambda^+ U^\top.$$

Also observe that Λ_r is invertible and

$$\Lambda^+ = \begin{pmatrix} \Lambda_r^{-1} & 0 \\ 0 & 0 \end{pmatrix}.$$

Therefore, the pseudo-inverse of a normal matrix can be computed directly from any block diagonalization of A, as claimed. $\square$

Example 21.3. Consider the following real diagonal form of the normal matrix

$$A = \begin{pmatrix} -2.7500 & 2.1651 & -0.8660 & 0.5000 \\ 2.1651 & -0.2500 & -1.5000 & 0.8660 \\ 0.8660 & 1.5000 & 0.7500 & -0.4330 \\ -0.5000 & -0.8660 & -0.4330 & 0.2500 \end{pmatrix} = U\Lambda U^\top,$$

with

$$U = \begin{pmatrix} \cos(\pi/3) & 0 & \sin(\pi/3) & 0 \\ \sin(\pi/3) & 0 & -\cos(\pi/3) & 0 \\ 0 & \cos(\pi/6) & 0 & \sin(\pi/6) \\ 0 & -\cos(\pi/6) & 0 & \sin(\pi/6) \end{pmatrix}, \quad \Lambda = \begin{pmatrix} 1 & -2 & 0 & 0 \\ 2 & 1 & 0 & 0 \\ 0 & 0 & -4 & 0 \\ 0 & 0 & 0 & 0 \end{pmatrix}.$$

We obtain

$$\Lambda^+ = \begin{pmatrix} 1/5 & 2/5 & 0 & 0 \\ -2/5 & 1/5 & 0 & 0 \\ 0 & 0 & -1/4 & 0 \\ 0 & 0 & 0 & 0 \end{pmatrix},$$

and the pseudo-inverse of A is

$$A^+ = U\Lambda^+ U^\top = \begin{pmatrix} -0.1375 & 0.1949 & 0.1732 & -0.1000 \\ 0.1949 & 0.0875 & 0.3000 & -0.1732 \\ -0.1732 & -0.3000 & 0.1500 & -0.0866 \\ 0.1000 & 0.1732 & -0.0866 & 0.0500 \end{pmatrix},$$

which agrees with `pinv(A)`.

The following properties, due to Penrose, characterize the pseudo-inverse of a matrix. We have already proved that the pseudo-inverse satisfies these equations. For a proof of the converse, see Kincaid and Cheney [Kincaid and Cheney (1996)].

Proposition 21.6. *Given any $m \times n$ matrix A (real or complex), the pseudo-inverse A^+ of A is the unique $n \times m$ matrix satisfying the following properties:*

$$AA^+A = A,$$
$$A^+AA^+ = A^+,$$
$$(AA^+)^\top = AA^+,$$
$$(A^+A)^\top = A^+A.$$

21.3 Data Compression and SVD

Among the many applications of SVD, a very useful one is *data compression*, notably for images. In order to make precise the notion of closeness of matrices, we use the notion of *matrix norm*. This concept is defined in Chapter 8, and the reader may want to review it before reading any further.

Given an $m \times n$ matrix of rank r, we would like to find a best approximation of A by a matrix B of rank $k \leq r$ (actually, $k < r$) such that $\|A - B\|_2$ (or $\|A - B\|_F$) is minimized. The following proposition is known as the *Eckart–Young theorem*.

Proposition 21.7. *Let A be an $m \times n$ matrix of rank r and let $VDU^\top = A$ be an SVD for A. Write u_i for the columns of U, v_i for the columns of V, and $\sigma_1 \geq \sigma_2 \geq \cdots \geq \sigma_p$ for the singular values of A ($p = \min(m, n)$). Then a matrix of rank $k < r$ closest to A (in the $\| \ \|_2$ norm) is given by*

$$A_k = \sum_{i=1}^{k} \sigma_i v_i u_i^\top = V \mathrm{diag}(\sigma_1, \ldots, \sigma_k, 0, \ldots, 0) U^\top$$

and $\|A - A_k\|_2 = \sigma_{k+1}$.

Proof. By construction, A_k has rank k, and we have

$$\|A - A_k\|_2 = \left\| \sum_{i=k+1}^{p} \sigma_i v_i u_i^\top \right\|_2$$

$$= \left\| V \mathrm{diag}(0, \ldots, 0, \sigma_{k+1}, \ldots, \sigma_p) U^\top \right\|_2 = \sigma_{k+1}.$$

It remains to show that $\|A - B\|_2 \geq \sigma_{k+1}$ for all rank k matrices B. Let B be any rank k matrix, so its kernel has dimension $n - k$. The subspace U_{k+1} spanned by $(u_1, \ldots, u_{k+1})$ has dimension $k + 1$, and because the sum of the dimensions of the kernel of B and of U_{k+1} is $(n - k) + k + 1 = n + 1$, these two subspaces must intersect in a subspace of dimension at least 1. Pick any unit vector h in $\mathrm{Ker}(B) \cap U_{k+1}$. Then since $Bh = 0$, and since U and V are isometries, we have

$$\|A - B\|_2^2 \geq \|(A - B)h\|_2^2 = \|Ah\|_2^2 = \left\| VDU^\top h \right\|_2^2 = \left\| DU^\top h \right\|_2^2$$

$$\geq \sigma_{k+1}^2 \left\| U^\top h \right\|_2^2 = \sigma_{k+1}^2,$$

which proves our claim. $\qquad \square$

Note that A_k can be stored using $(m + n)k$ entries, as opposed to mn entries. When $k \ll m$, this is a substantial gain.

Example 21.4. Consider the badly conditioned symmetric matrix

$$A = \begin{pmatrix} 10 & 7 & 8 & 7 \\ 7 & 5 & 6 & 5 \\ 8 & 6 & 10 & 9 \\ 7 & 5 & 9 & 10 \end{pmatrix}$$

from Section 8.5. Since A is SPD, we have the SVD

$$A = UDU^\top,$$

with

$$U = \begin{pmatrix} -0.5286 & -0.6149 & 0.3017 & -0.5016 \\ -0.3803 & -0.3963 & -0.0933 & 0.8304 \\ -0.5520 & 0.2716 & -0.7603 & -0.2086 \\ -0.5209 & 0.6254 & 0.5676 & 0.1237 \end{pmatrix},$$

$$D = \begin{pmatrix} 30.2887 & 0 & 0 & 0 \\ 0 & 3.8581 & 0 & 0 \\ 0 & 0 & 0.8431 & 0 \\ 0 & 0 & 0 & 0.0102 \end{pmatrix}.$$

If we set $\sigma_3 = \sigma_4 = 0$, we obtain the best rank 2 approximation

$$A_2 = U(:, 1 : 2) * D(:, 1 : 2) * U(:, 1 : 2)' = \begin{pmatrix} 9.9207 & 7.0280 & 8.1923 & 6.8563 \\ 7.0280 & 4.9857 & 5.9419 & 5.0436 \\ 8.1923 & 5.9419 & 9.5122 & 9.3641 \\ 6.8563 & 5.0436 & 9.3641 & 9.7282 \end{pmatrix}.$$

A nice example of the use of Proposition 21.7 in image compression is given in Demmel [Demmel (1997)], Chapter 3, Section 3.2.3, pages 113–115; see the Matlab demo.

Proposition 21.7 also holds for the Frobenius norm; see Problem 21.4.

An interesting topic that we have not addressed is the actual computation of an SVD. This is a very interesting but tricky subject. Most methods reduce the computation of an SVD to the diagonalization of a well-chosen symmetric matrix which is not $A^\top A$; see Problem 20.1 and Problem 20.3. Interested readers should read Section 5.4 of Demmel's excellent book [Demmel (1997)], which contains an overview of most known methods and an extensive list of references.

21.4 Principal Components Analysis (PCA)

Suppose we have a set of data consisting of n points $X_1, \ldots, X_n$, with each $X_i \in \mathbb{R}^d$ *viewed as a row vector*. Think of the X_i's as persons, and if $X_i = (x_{i\,1}, \ldots, x_{i\,d})$, each $x_{i\,j}$ is the value of some *feature* (or *attribute*) of that person.

Example 21.5. For example, the X_i's could be mathematicians, $d = 2$, and the first component, $x_{i\,1}$, of X_i could be the year that X_i was born, and the second component, $x_{i\,2}$, the length of the beard of X_i in centimeters. Here is a small data set.

Name	year	length
Carl Friedrich Gauss	1777	0
Camille Jordan	1838	12
Adrien-Marie Legendre	1752	0
Bernhard Riemann	1826	15
David Hilbert	1862	2
Henri Poincaré	1854	5
Emmy Noether	1882	0
Karl Weierstrass	1815	0
Eugenio Beltrami	1835	2
Hermann Schwarz	1843	20

We usually form the $n \times d$ matrix X whose ith row is X_i, with $1 \le i \le n$. Then the jth column is denoted by C_j ($1 \le j \le d$). It is sometimes called a *feature vector*, but this terminology is far from being universally accepted. In fact, many people in computer vision call the data points X_i feature vectors!

The purpose of *principal components analysis*, for short *PCA*, is to identify patterns in data and understand the *variance–covariance* structure of the data. This is useful for the following tasks:

(1) Data reduction: Often much of the variability of the data can be accounted for by a smaller number of *principal components*.
(2) Interpretation: PCA can show relationships that were not previously suspected.

Given a vector (a *sample* of measurements) $x = (x_1, \ldots, x_n) \in \mathbb{R}^n$, recall that the *mean* (or *average*) $\overline{x}$ of x is given by

$$\overline{x} = \frac{\sum_{i=1}^{n} x_i}{n}.$$

We let $x - \overline{x}$ denote the *centered data point*

$$x - \overline{x} = (x_1 - \overline{x}, \dots, x_n - \overline{x}).$$

In order to *measure the spread* of the x_i's around the mean, we define the *sample variance* (for short, *variance*) var(x) (or s^2) of the sample x by

$$\text{var}(x) = \frac{\sum_{i=1}^{n} (x_i - \overline{x})^2}{n - 1}.$$

Example 21.6. If $x = (1, 3, -1)$, $\overline{x} = \frac{1+3-1}{3} = 1$, $x - \overline{x} = (0, 2, -2)$, and var$(x) = \frac{0^2 + 2^2 + (-2)^2}{2} = 4$. If $y = (1, 2, 3)$, $\overline{y} = \frac{1+2+3}{3} = 2$, $y - \overline{y} = (-1, 0, 1)$, and var$(y) = \frac{(-1)^2 + 0^2 + 1^2}{2} = 2$.

There is a reason for using $n-1$ instead of n. The above definition makes var(x) an unbiased estimator of the variance of the random variable being sampled. However, we don't need to worry about this. Curious readers will find an explanation of these peculiar definitions in Epstein [Epstein (2007)] (Chapter 14, Section 14.5) or in any decent statistics book.

Given two vectors $x = (x_1, \dots, x_n)$ and $y = (y_1, \dots, y_n)$, the *sample covariance* (for short, *covariance*) of x and y is given by

$$\text{cov}(x, y) = \frac{\sum_{i=1}^{n} (x_i - \overline{x})(y_i - \overline{y})}{n - 1}.$$

Example 21.7. If we take $x = (1, 3, -1)$ and $y = (0, 2, -2)$, we know from Example 21.6 that $x - \overline{x} = (0, 2, -2)$ and $y - \overline{y} = (-1, 0, 1)$. Thus, cov$(x, y) = \frac{0(-1) + 2(0) + (-2)(1)}{2} = -1$.

The covariance of x and y measures how x and y vary from the mean with respect to each other. Obviously, cov$(x, y) = $ cov(y, x) and cov$(x, x) = $ var(x).

Note that

$$\text{cov}(x, y) = \frac{(x - \overline{x})^{\top} (y - \overline{y})}{n - 1}.$$

We say that x and y are *uncorrelated* iff cov$(x, y) = 0$.

Finally, given an $n \times d$ matrix X of n points X_i, for PCA to be meaningful, it will be necessary to translate the origin to the *centroid* (or *center of gravity*) μ of the X_i's, defined by

$$\mu = \frac{1}{n}(X_1 + \cdots + X_n).$$

Observe that if $\mu = (\mu_1, \ldots, \mu_d)$, then μ_j is the mean of the vector C_j (the jth column of X).

We let $X - \mu$ denote the *matrix* whose ith row is the centered data point $X_i - \mu$ $(1 \le i \le n)$. Then the *sample covariance matrix* (for short, *covariance matrix*) of X is the $d \times d$ symmetric matrix

$$\Sigma = \frac{1}{n-1}(X - \mu)^{\top}(X - \mu) = (\text{cov}(C_i, C_j)).$$

Example 21.8. Let $X = \begin{pmatrix} 1 & 1 \\ 3 & 2 \\ -1 & 3 \end{pmatrix}$, the 3×2 matrix whose columns are the vector x and y of Example 21.6. Then

$$\mu = \frac{1}{3}[(1,1) + (3,2) + (-1,3)] = (1,2),$$

$$X - \mu = \begin{pmatrix} 0 & -1 \\ 2 & 0 \\ -2 & 1 \end{pmatrix},$$

and

$$\Sigma = \frac{1}{2}\begin{pmatrix} 0 & 2 & -2 \\ -1 & 0 & 1 \end{pmatrix} \begin{pmatrix} 0 & -1 \\ 2 & 0 \\ -2 & 1 \end{pmatrix} = \begin{pmatrix} 4 & -1 \\ -1 & 1 \end{pmatrix}.$$

Remark: The factor $\frac{1}{n-1}$ is irrelevant for our purposes and can be ignored.

Example 21.9. Here is the matrix $X - \mu$ in the case of our bearded mathematicians: since

$$\mu_1 = 1828.4, \quad \mu_2 = 5.6,$$

we get the following centered data set.

Name	year	length
Carl Friedrich Gauss	−51.4	−5.6
Camille Jordan	9.6	6.4
Adrien-Marie Legendre	−76.4	−5.6
Bernhard Riemann	−2.4	9.4
David Hilbert	33.6	−3.6
Henri Poincaré	25.6	−0.6
Emmy Noether	53.6	−5.6
Karl Weierstrass	13.4	−5.6
Eugenio Beltrami	6.6	−3.6
Hermann Schwarz	14.6	14.4

See Figure 21.3.

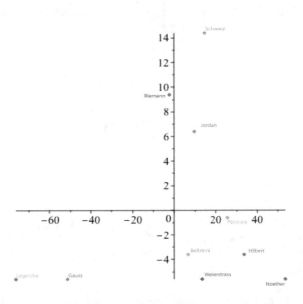

Fig. 21.3   The centered data points of Example 21.9.

We can think of the vector C_j as representing the features of X in the direction e_j (the jth canonical basis vector in $\mathbb{R}^d$, namely $e_j = (0, \dots, 1, \dots 0)$, with a 1 in the jth position).

If $v \in \mathbb{R}^d$ is a unit vector, we wish to consider the projection of the data points $X_1, \ldots, X_n$ onto the line spanned by v. Recall from Euclidean geometry that if $x \in \mathbb{R}^d$ is any vector and $v \in \mathbb{R}^d$ is a unit vector, the projection of x onto the line spanned by v is

$$\langle x, v \rangle v.$$

Thus, with respect to the basis v, the projection of x has coordinate $\langle x, v \rangle$. If x is represented by a row vector and v by a column vector, then

$$\langle x, v \rangle = xv.$$

Therefore, the vector $Y \in \mathbb{R}^n$ consisting of the coordinates of the projections of $X_1, \ldots, X_n$ onto the line spanned by v is given by $Y = Xv$, and this is the linear combination

$$Xv = v_1 C_1 + \cdots + v_d C_d$$

of the columns of X (with $v = (v_1, \ldots, v_d)$).

Observe that because μ_j is the mean of the vector C_j (the jth column of X), we get

$$\overline{Y} = \overline{Xv} = v_1 \mu_1 + \cdots + v_d \mu_d,$$

and so the centered point $Y - \overline{Y}$ is given by

$$Y - \overline{Y} = v_1(C_1 - \mu_1) + \cdots + v_d(C_d - \mu_d) = (X - \mu)v.$$

Furthermore, if $Y = Xv$ and $Z = Xw$, then

$$\begin{aligned}
\text{cov}(Y, Z) &= \frac{((X - \mu)v)^\top (X - \mu)w}{n - 1} \\
&= v^\top \frac{1}{n - 1}(X - \mu)^\top (X - \mu)w \\
&= v^\top \Sigma w,
\end{aligned}$$

where Σ is the covariance matrix of X. Since $Y - \overline{Y}$ has zero mean, we have

$$\text{var}(Y) = \text{var}(Y - \overline{Y}) = v^\top \frac{1}{n - 1}(X - \mu)^\top (X - \mu)v.$$

The above suggests that we should move the origin to the centroid μ of the X_i's and consider the matrix $X - \mu$ of the centered data points $X_i - \mu$.

From now on beware that we denote the columns of $X - \mu$ by $C_1, \ldots, C_d$ and that Y denotes the *centered* point $Y = (X - \mu)v = \sum_{j=1}^{d} v_j C_j$, where v is a unit vector.

Basic idea of PCA: The principal components of X are *uncorrelated* projections Y of the data points $X_1, \ldots, X_n$ onto some directions v (where the v's are unit vectors) such that var(Y) is maximal. This suggests the following definition:

Definition 21.2. Given an $n \times d$ matrix X of data points $X_1, \ldots, X_n$, if μ is the centroid of the X_i's, then a *first principal component of X (first PC)* is a centered point $Y_1 = (X - \mu)v_1$, the projection of $X_1, \ldots, X_n$ onto a direction v_1 such that var(Y_1) is maximized, where v_1 is a unit vector (recall that $Y_1 = (X - \mu)v_1$ is a linear combination of the C_j's, the columns of $X - \mu$).

More generally, if $Y_1, \ldots, Y_k$ are k principal components of X along some unit vectors $v_1, \ldots, v_k$, where $1 \le k < d$, a *$(k+1)th$ principal component of X ($(k+1)th$ PC)* is a centered point $Y_{k+1} = (X - \mu)v_{k+1}$, the projection of $X_1, \ldots, X_n$ onto some direction v_{k+1} such that var(Y_{k+1}) is maximized, subject to cov(Y_h, Y_{k+1}) = 0 for all h with $1 \le h \le k$, and where v_{k+1} is a unit vector (recall that $Y_h = (X - \mu)v_h$ is a linear combination of the C_j's). The v_h are called *principal directions*.

The following proposition is the key to the main result about PCA. This result was already proven in Proposition 16.11 except that the eigenvalues were listed in increasing order. For the reader's convenience we prove it again.

Proposition 21.8. *If A is a symmetric $d \times d$ matrix with eigenvalues $\lambda_1 \ge \lambda_2 \ge \cdots \ge \lambda_d$ and if $(u_1, \ldots, u_d)$ is any orthonormal basis of eigenvectors of A, where u_i is a unit eigenvector associated with λ_i, then*

$$\max_{x \ne 0} \frac{x^\top A x}{x^\top x} = \lambda_1$$

(with the maximum attained for $x = u_1$) and

$$\max_{x \ne 0, x \in \{u_1, \ldots, u_k\}^\perp} \frac{x^\top A x}{x^\top x} = \lambda_{k+1}$$

(with the maximum attained for $x = u_{k+1}$), where $1 \le k \le d - 1$.

Proof. First observe that

$$\max_{x \ne 0} \frac{x^\top A x}{x^\top x} = \max_x \{x^\top A x \mid x^\top x = 1\},$$

and similarly,

$$\max_{x \ne 0, x \in \{u_1, \ldots, u_k\}^\perp} \frac{x^\top A x}{x^\top x} = \max_x \left\{ x^\top A x \mid (x \in \{u_1, \ldots, u_k\}^\perp) \wedge (x^\top x = 1) \right\}.$$

Since A is a symmetric matrix, its eigenvalues are real and it can be diagonalized with respect to an orthonormal basis of eigenvectors, so let $(u_1, \ldots, u_d)$ be such a basis. If we write

$$x = \sum_{i=1}^{d} x_i u_i,$$

a simple computation shows that

$$x^\top A x = \sum_{i=1}^{d} \lambda_i x_i^2.$$

If $x^\top x = 1$, then $\sum_{i=1}^{d} x_i^2 = 1$, and since we assumed that $\lambda_1 \geq \lambda_2 \geq \cdots \geq \lambda_d$, we get

$$x^\top A x = \sum_{i=1}^{d} \lambda_i x_i^2 \leq \lambda_1 \left(\sum_{i=1}^{d} x_i^2 \right) = \lambda_1.$$

Thus,

$$\max_x \left\{ x^\top A x \mid x^\top x = 1 \right\} \leq \lambda_1,$$

and since this maximum is achieved for $e_1 = (1, 0, \ldots, 0)$, we conclude that

$$\max_x \left\{ x^\top A x \mid x^\top x = 1 \right\} = \lambda_1.$$

Next observe that $x \in \{u_1, \ldots, u_k\}^\perp$ and $x^\top x = 1$ iff $x_1 = \cdots = x_k = 0$ and $\sum_{i=1}^{d} x_i = 1$. Consequently, for such an x, we have

$$x^\top A x = \sum_{i=k+1}^{d} \lambda_i x_i^2 \leq \lambda_{k+1} \left(\sum_{i=k+1}^{d} x_i^2 \right) = \lambda_{k+1}.$$

Thus,

$$\max_x \left\{ x^\top A x \mid (x \in \{u_1, \ldots, u_k\}^\perp) \wedge (x^\top x = 1) \right\} \leq \lambda_{k+1},$$

and since this maximum is achieved for $e_{k+1} = (0, \ldots, 0, 1, 0, \ldots, 0)$ with a 1 in position $k + 1$, we conclude that

$$\max_x \left\{ x^\top A x \mid (x \in \{u_1, \ldots, u_k\}^\perp) \wedge (x^\top x = 1) \right\} = \lambda_{k+1},$$

as claimed. $\square$

The quantity

$$\frac{x^\top A x}{x^\top x}$$

is known as the *Rayleigh ratio* or *Rayleigh–Ritz ratio* (see Section 16.6) and Proposition 21.8 is often known as part of the *Rayleigh–Ritz theorem*.

Proposition 21.8 also holds if A is a Hermitian matrix and if we replace $x^\top A x$ by $x^* A x$ and $x^\top x$ by $x^* x$. The proof is unchanged, since a Hermitian matrix has real eigenvalues and is diagonalized with respect to an orthonormal basis of eigenvectors (with respect to the Hermitian inner product).

We then have the following fundamental result showing how *the SVD of X yields the PCs*:

Theorem 21.3. *(SVD yields PCA) Let X be an $n \times d$ matrix of data points $X_1, \ldots, X_n$, and let μ be the centroid of the X_i's. If $X - \mu = V D U^\top$ is an SVD decomposition of $X - \mu$ and if the main diagonal of D consists of the singular values $\sigma_1 \geq \sigma_2 \geq \cdots \geq \sigma_d$, then the centered points $Y_1, \ldots, Y_d$, where*

$$Y_k = (X - \mu)u_k = k\text{th column of } V D$$

and u_k is the kth column of U, are d principal components of X. Furthermore,

$$\mathrm{var}(Y_k) = \frac{\sigma_k^2}{n-1}$$

and $\mathrm{cov}(Y_h, Y_k) = 0$, whenever $h \neq k$ and $1 \leq k, h \leq d$.

Proof. Recall that for any unit vector v, the centered projection of the points $X_1, \ldots, X_n$ onto the line of direction v is $Y = (X - \mu)v$ and that the variance of Y is given by

$$\mathrm{var}(Y) = v^\top \frac{1}{n-1}(X - \mu)^\top (X - \mu)v.$$

Since $X - \mu = V D U^\top$, we get

$$\mathrm{var}(Y) = v^\top \frac{1}{(n-1)}(X - \mu)^\top (X - \mu)v$$

$$= v^\top \frac{1}{(n-1)} U D V^\top V D U^\top v$$

$$= v^\top U \frac{1}{(n-1)} D^2 U^\top v.$$

Similarly, if $Y = (X - \mu)v$ and $Z = (X - \mu)w$, then the covariance of Y and Z is given by

$$\text{cov}(Y, Z) = v^\top U \frac{1}{(n-1)} D^2 U^\top w.$$

Obviously, $U \frac{1}{(n-1)} D^2 U^\top$ is a symmetric matrix whose eigenvalues are $\frac{\sigma_1^2}{n-1} \geq \cdots \geq \frac{\sigma_d^2}{n-1}$, and the columns of U form an orthonormal basis of unit eigenvectors.

We proceed by induction on k. For the base case, $k = 1$, maximizing $\text{var}(Y)$ is equivalent to maximizing

$$v^\top U \frac{1}{(n-1)} D^2 U^\top v,$$

where v is a unit vector. By Proposition 21.8, the maximum of the above quantity is the largest eigenvalue of $U \frac{1}{(n-1)} D^2 U^\top$, namely $\frac{\sigma_1^2}{n-1}$, and it is achieved for u_1, the first column of U. Now we get

$$Y_1 = (X - \mu)u_1 = VDU^\top u_1,$$

and since the columns of U form an orthonormal basis, $U^\top u_1 = e_1 = (1, 0, \ldots, 0)$, and so Y_1 is indeed the first column of VD.

By the induction hypothesis, the centered points $Y_1, \ldots, Y_k$, where $Y_h = (X - \mu)u_h$ and $u_1, \ldots, u_k$ are the first k columns of U, are k principal components of X. Because

$$\text{cov}(Y, Z) = v^\top U \frac{1}{(n-1)} D^2 U^\top w,$$

where $Y = (X - \mu)v$ and $Z = (X - \mu)w$, the condition $\text{cov}(Y_h, Z) = 0$ for $h = 1, \ldots, k$ is equivalent to the fact that w belongs to the orthogonal complement of the subspace spanned by $\{u_1, \ldots, u_k\}$, and maximizing $\text{var}(Z)$ subject to $\text{cov}(Y_h, Z) = 0$ for $h = 1, \ldots, k$ is equivalent to maximizing

$$w^\top U \frac{1}{(n-1)} D^2 U^\top w,$$

where w is a unit vector orthogonal to the subspace spanned by $\{u_1, \ldots, u_k\}$. By Proposition 21.8, the maximum of the above quantity is the $(k+1)$th eigenvalue of $U \frac{1}{(n-1)} D^2 U^\top$, namely $\frac{\sigma_{k+1}^2}{n-1}$, and it is achieved for u_{k+1}, the $(k+1)$th column of U. Now we get

$$Y_{k+1} = (X - \mu)u_{k+1} = VDU^\top u_{k+1},$$

and since the columns of U form an orthonormal basis, $U^\top u_{k+1} = e_{k+1}$, and Y_{k+1} is indeed the $(k+1)$th column of VD, which completes the proof of the induction step. $\qquad\square$

The d columns $u_1, \ldots, u_d$ of U are usually called the *principal directions* of $X - \mu$ (and X). We note that not only do we have $\operatorname{cov}(Y_h, Y_k) = 0$ whenever $h \neq k$, but the directions $u_1, \ldots, u_d$ along which the data are projected are mutually orthogonal.

Example 21.10. For the centered data set of our bearded mathematicians (Example 21.9) we have $X - \mu = V\Sigma U^\top$, where Σ has two nonzero singular values, $\sigma_1 = 116.9803, \sigma_2 = 21.7812$, and with

$$U = \begin{pmatrix} 0.9995 & 0.0325 \\ 0.0325 & -0.9995 \end{pmatrix},$$

so the principal directions are $u_1 = (0.9995, 0.0325)$ and $u_2 = (0.0325, -0.9995)$. Observe that u_1 is almost the direction of the x-axis, and u_2 is almost the opposite direction of the y-axis. We also find that the projections Y_1 and Y_2 along the principal directions are

$$VD = \begin{pmatrix} -51.5550 & 3.9249 \\ 9.8031 & -6.0843 \\ -76.5417 & 3.1116 \\ -2.0929 & -9.4731 \\ 33.4651 & 4.6912 \\ 25.5669 & 1.4325 \\ 53.3894 & 7.3408 \\ 13.2107 & 6.0330 \\ 6.4794 & 3.8128 \\ 15.0607 & -13.9174 \end{pmatrix}, \quad \text{with} \quad X - \mu = \begin{pmatrix} -51.4000 & -5.6000 \\ 9.6000 & 6.4000 \\ -76.4000 & -5.6000 \\ -2.4000 & 9.4000 \\ 33.6000 & -3.6000 \\ 25.6000 & -0.6000 \\ 53.6000 & -5.6000 \\ 13.4000 & -5.6000 \\ 6.6000 & -3.6000 \\ 14.6000 & 14.4000 \end{pmatrix}.$$

See Figures 21.4, 21.5, and 21.6.

We know from our study of SVD that $\sigma_1^2, \ldots, \sigma_d^2$ are the eigenvalues of the symmetric positive semidefinite matrix $(X - \mu)^\top (X - \mu)$ and that $u_1, \ldots, u_d$ are corresponding eigenvectors. Numerically, it is preferable to use SVD on $X - \mu$ rather than to compute explicitly $(X - \mu)^\top (X - \mu)$ and then diagonalize it. Indeed, the explicit computation of $A^\top A$ from a matrix A can be numerically quite unstable, and good SVD algorithms avoid computing $A^\top A$ explicitly.

In general, since an SVD of X is not unique, *the principal directions* $u_1, \ldots, u_d$ *are not unique*. This can happen when a data set has some *rotational symmetries*, and in such a case, PCA is not a very good method for analyzing the data set.

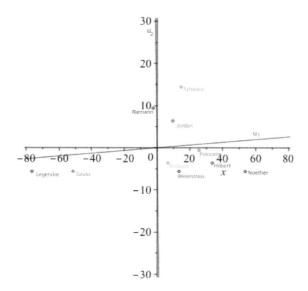

Fig. 21.4 The centered data points of Example 21.9 and the two principal directions of Example 21.10.

21.5 Best Affine Approximation

A problem very close to PCA (and based on least squares) is to *best approximate a data set of n points $X_1, \ldots, X_n$, with $X_i \in \mathbb{R}^d$, by a p-dimensional affine subspace A of $\mathbb{R}^d$, with $1 \le p \le d-1$* (the terminology rank $d-p$ is also used).

First consider $p = d-1$. Then $A = A_1$ is an affine hyperplane (in $\mathbb{R}^d$), and it is given by an equation of the form

$$a_1 x_1 + \cdots + a_d x_d + c = 0.$$

By *best approximation*, we mean that $(a_1, \ldots, a_d, c)$ solves the homogeneous linear system

$$\begin{pmatrix} x_{11} & \cdots & x_{1d} & 1 \\ \vdots & \vdots & \vdots & \vdots \\ x_{n1} & \cdots & x_{nd} & 1 \end{pmatrix} \begin{pmatrix} a_1 \\ \vdots \\ a_d \\ c \end{pmatrix} = \begin{pmatrix} 0 \\ \vdots \\ 0 \\ 0 \end{pmatrix}$$

in the *least squares sense, subject to the condition that $a = (a_1, \ldots, a_d)$ is a unit vector*, that is, $a^\top a = 1$, where $X_i = (x_{i1}, \cdots, x_{id})$.

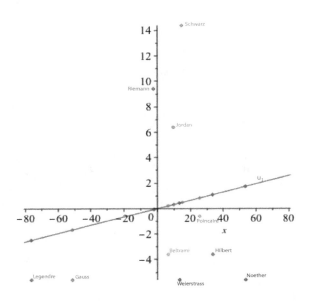

Fig. 21.5   The first principal components of Example 21.10, i.e., the projection of the centered data points onto the u_1 line.

If we form the symmetric matrix

$$\begin{pmatrix} x_{11} & \cdots & x_{1d} & 1 \\ \vdots & \vdots & \vdots & \vdots \\ x_{n1} & \cdots & x_{nd} & 1 \end{pmatrix}^{\top} \begin{pmatrix} x_{11} & \cdots & x_{1d} & 1 \\ \vdots & \vdots & \vdots & \vdots \\ x_{n1} & \cdots & x_{nd} & 1 \end{pmatrix}$$

involved in the normal equations, we see that the bottom row (and last column) of that matrix is

$$n\mu_1 \quad \cdots \quad n\mu_d \quad n,$$

where $n\mu_j = \sum_{i=1}^{n} x_{ij}$ is n times the mean of the column C_j of X.

Therefore, if $(a_1, \ldots, a_d, c)$ is a least squares solution, that is, a solution of the normal equations, we must have

$$n\mu_1 a_1 + \cdots + n\mu_d a_d + nc = 0,$$

that is,

$$a_1\mu_1 + \cdots + a_d\mu_d + c = 0,$$

which means that the *hyperplane* A_1 *must pass through the centroid* μ *of the data points* $X_1, \ldots, X_n$. Then we can rewrite the original system with

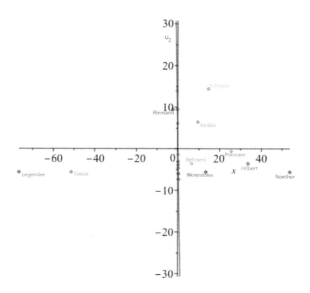

Fig. 21.6 The second principal components of Example 21.10, i.e., the projection of the centered data points onto the u_2 line.

respect to the centered data $X_i - \mu$, find that the variable c drops out, get the system

$$(X - \mu)a = 0,$$

where $a = (a_1, \ldots, a_d)$.

Thus, we are looking for a unit vector a solving $(X - \mu)a = 0$ in the least squares sense, that is, some a such that $a^\top a = 1$ minimizing

$$a^\top (X - \mu)^\top (X - \mu)a.$$

Compute some SVD $VDU^\top$ of $X - \mu$, where the main diagonal of D consists of the singular values $\sigma_1 \geq \sigma_2 \geq \cdots \geq \sigma_d$ of $X - \mu$ arranged in descending order. Then

$$a^\top (X - \mu)^\top (X - \mu)a = a^\top UD^2 U^\top a,$$

where $D^2 = \mathrm{diag}(\sigma_1^2, \ldots, \sigma_d^2)$ is a diagonal matrix, so pick a to be *the last column in U* (corresponding to the smallest eigenvalue σ_d^2 of $(X - \mu)^\top (X - \mu)$). This is a solution to our best fit problem.

Therefore, if U_{d-1} is the linear hyperplane defined by a, that is,

$$U_{d-1} = \{u \in \mathbb{R}^d \mid \langle u, a \rangle = 0\},$$

where a is the last column in U for some SVD $VDU^\top$ of $X - \mu$, we have shown that the affine hyperplane $A_1 = \mu + U_{d-1}$ is a best approximation of the data set $X_1, \ldots, X_n$ in the least squares sense.

It is easy to show that this hyperplane $A_1 = \mu + U_{d-1}$ minimizes the sum of the square distances of each X_i to its orthogonal projection onto A_1. Also, since U_{d-1} is the orthogonal complement of a, the last column of U, we see that U_{d-1} is spanned by the first $d-1$ columns of U, that is, the first $d-1$ principal directions of $X - \mu$.

All this can be generalized to a *best $(d-k)$-dimensional affine subspace A_k approximating $X_1, \ldots, X_n$ in the least squares sense* $(1 \le k \le d-1)$. Such an affine subspace A_k is cut out by k independent hyperplanes H_i (with $1 \le i \le k$), each given by some equation

$$a_{i\,1}x_1 + \cdots + a_{i\,d}x_d + c_i = 0.$$

If we write $a_i = (a_{i\,1}, \cdots, a_{i\,d})$, to say that the H_i are independent means that $a_1, \ldots, a_k$ are linearly independent. In fact, we may assume that $a_1, \ldots, a_k$ form an *orthonormal system.*

Then finding a best $(d-k)$-dimensional affine subspace A_k amounts to solving the homogeneous linear system

$$
\begin{pmatrix} X\ 1\ 0\ \cdots\ 0\ 0\ 0 \\ \vdots\ \vdots\ \vdots\ \ddots\ \vdots\ \vdots\ \vdots \\ 0\ 0\ 0\ \cdots\ 0\ X\ 1 \end{pmatrix}
\begin{pmatrix} a_1 \\ c_1 \\ \vdots \\ a_k \\ c_k \end{pmatrix}
= \begin{pmatrix} 0 \\ \vdots \\ 0 \end{pmatrix},
$$

in the least squares sense, subject to the conditions $a_i^\top a_j = \delta_{ij}$, for all i, j with $1 \le i, j \le k$, where the matrix of the system is a block diagonal matrix consisting of k diagonal blocks $(X, \mathbf{1})$, where $\mathbf{1}$ denotes the column vector $(1, \ldots, 1) \in \mathbb{R}^n$.

Again it is easy to see that each hyperplane H_i *must pass through the centroid μ of $X_1, \ldots, X_n$*, and by switching to the centered data $X_i - \mu$ we get the system

$$
\begin{pmatrix} X - \mu\ 0\ \cdots\ \ \ 0 \\ \vdots\ \ \ \vdots\ \ddots\ \ \ \vdots \\ 0\ \ \ 0\ \cdots\ X - \mu \end{pmatrix}
\begin{pmatrix} a_1 \\ \vdots \\ a_k \end{pmatrix}
= \begin{pmatrix} 0 \\ \vdots \\ 0 \end{pmatrix},
$$

with $a_i^\top a_j = \delta_{ij}$ for all i, j with $1 \le i, j \le k$.

If $VDU^\top = X - \mu$ is an SVD decomposition, it is easy to see that a least squares solution of this system is given by the last k columns of

U, assuming that the main diagonal of D consists of the singular values $\sigma_1 \geq \sigma_2 \geq \cdots \geq \sigma_d$ of $X - \mu$ arranged in descending order. But now the $(d-k)$-dimensional subspace U_{d-k} cut out by the hyperplanes defined by $a_1, \ldots, a_k$ is simply the orthogonal complement of $(a_1, \ldots, a_k)$, which is the subspace spanned by the first $d - k$ columns of U.

So the best $(d - k)$-dimensional affine subsapce A_k approximating $X_1, \ldots, X_n$ in the least squares sense is $A_k = \mu + U_{d-k}$, where U_{d-k} is the linear subspace spanned by the first $d - k$ principal directions of $X - \mu$, that is, the first $d - k$ columns of U. Consequently, we get the following interesting interpretation of PCA (actually, principal directions):

Theorem 21.4. *Let X be an $n \times d$ matrix of data points $X_1, \ldots, X_n$, and let μ be the centroid of the X_i's. If $X - \mu = VDU^\top$ is an SVD decomposition of $X - \mu$ and if the main diagonal of D consists of the singular values $\sigma_1 \geq \sigma_2 \geq \cdots \geq \sigma_d$, then a best $(d - k)$-dimensional affine approximation A_k of $X_1, \ldots, X_n$ in the least squares sense is given by*

$$A_k = \mu + U_{d-k},$$

where U_{d-k} is the linear subspace spanned by the first $d - k$ columns of U, the first $d - k$ principal directions of $X - \mu$ ($1 \leq k \leq d - 1$).

Example 21.11. Going back to Example 21.10, a best 1-dimensional affine approximation A_1 is the affine line passing through $(\mu_1, \mu_2) = (1824.4, 5.6)$ of direction $u_1 = (0.9995, 0.0325)$.

Example 21.12. Suppose in the data set of Example 21.5 that we add the month of birth of every mathematician as a feature. We obtain the following data set.

Name	month	year	length
Carl Friedrich Gauss	4	1777	0
Camille Jordan	1	1838	12
Adrien-Marie Legendre	9	1752	0
Bernhard Riemann	9	1826	15
David Hilbert	1	1862	2
Henri Poincaré	4	1854	5
Emmy Noether	3	1882	0
Karl Weierstrass	10	1815	0
Eugenio Beltrami	10	1835	2
Hermann Schwarz	1	1843	20

The mean of the first column is 5.2, and the centered data set is given below.

Name	month	year	length
Carl Friedrich Gauss	−1.2	−51.4	−5.6
Camille Jordan	−4.2	9.6	6.4
Adrien-Marie Legendre	3.8	−76.4	−5.6
Bernhard Riemann	3.8	−2.4	9.4
David Hilbert	−4.2	33.6	−3.6
Henri Poincaré	−1.2	25.6	−0.6
Emmy Noether	−2.2	53.6	−5.6
Karl Weierstrass	4.8	13.4	−5.6
Eugenio Beltrami	4.8	6.6	−3.6
Hermann Schwarz	−4.2	14.6	14.4

Running SVD on this data set we get

$$
U = \begin{pmatrix} 0.0394 & 0.1717 & 0.9844 \\ -0.9987 & 0.0390 & 0.0332 \\ -0.0327 & -0.9844 & 0.1730 \end{pmatrix}, \quad
D = \begin{pmatrix} 117.0706 & 0 & 0 \\ 0 & 22.0390 & 0 \\ 0 & 0 & 10.1571 \\ 0 & 0 & 0 \\ 0 & 0 & 0 \\ 0 & 0 & 0 \\ 0 & 0 & 0 \\ 0 & 0 & 0 \\ 0 & 0 & 0 \\ 0 & 0 & 0 \end{pmatrix},
$$

and

$$
VD = \begin{pmatrix} 51.4683 & 3.3013 & -3.8569 \\ -9.9623 & -6.6467 & -2.7082 \\ 76.6327 & 3.1845 & 0.2348 \\ 2.2393 & -8.6943 & 5.2872 \\ -33.6038 & 4.1334 & -3.6415 \\ -25.5941 & 1.3833 & -0.4350 \\ -53.4333 & 7.2258 & -1.3547 \\ -13.0100 & 6.8594 & 4.2010 \\ -6.2843 & 4.6254 & 4.3212 \\ -15.2173 & -14.3266 & -1.1581 \end{pmatrix},
$$

$$X - \mu = \begin{pmatrix} -1.2000 & -51.4000 & -5.6000 \\ -4.2000 & 9.6000 & 6.4000 \\ 3.8000 & -76.4000 & -5.6000 \\ 3.8000 & -2.4000 & 9.4000 \\ -4.2000 & 33.6000 & -3.6000 \\ -1.2000 & 25.6000 & -0.6000 \\ -2.2000 & 53.6000 & -5.6000 \\ 4.8000 & 13.4000 & -5.6000 \\ 4.8000 & 6.6000 & -3.6000 \\ -4.2000 & 14.6000 & 14.4000 \end{pmatrix}.$$

The first principal direction $u_1 = (0.0394, -0.9987, -0.0327)$ is basically the opposite of the y-axis, and the most significant feature is the year of birth. The second principal direction $u_2 = (0.1717, 0.0390, -0.9844)$ is close to the opposite of the z-axis, and the second most significant feature is the length of beards. A best affine plane is spanned by the vectors u_1 and u_2.

There are many applications of PCA to data compression, dimension reduction, and pattern analysis. The basic idea is that in many cases, given a data set $X_1, \ldots, X_n$, with $X_i \in \mathbb{R}^d$, only a "small" subset of $m < d$ of the features is needed to describe the data set accurately.

If $u_1, \ldots, u_d$ are the principal directions of $X - \mu$, then the first m projections of the data (the first m principal components, i.e., the first m columns of VD) onto the first m principal directions represent the data without much loss of information. Thus, instead of using the original data points $X_1, \ldots, X_n$, with $X_i \in \mathbb{R}^d$, we can use their projections onto the first m principal directions $Y_1, \ldots, Y_m$, where $Y_i \in \mathbb{R}^m$ and $m < d$, obtaining a compressed version of the original data set.

For example, PCA is used in computer vision for *face recognition*. Sirovitch and Kirby (1987) seem to be the first to have had the idea of using PCA to compress facial images. They introduced the term *eigenpicture* to refer to the principal directions, u_i. However, an explicit face recognition algorithm was given only later by Turk and Pentland (1991). They renamed eigenpictures as *eigenfaces*.

For details on the topic of eigenfaces, see Forsyth and Ponce [Forsyth and Ponce (2002)] (Chapter 22, Section 22.3.2), where you will also find exact references to Turk and Pentland's papers.

Another interesting application of PCA is to the *recognition of handwritten digits*. Such an application is described in Hastie, Tibshirani, and Friedman, [Hastie *et al.* (2009)] (Chapter 14, Section 14.5.1).

21.6   Summary

The main concepts and results of this chapter are listed below:

- *Least squares problems.*
- Existence of a least squares solution of smallest norm (Theorem 21.1).
- The *pseudo-inverse A^+* of a matrix A.
- The least squares solution of smallest norm is given by the pseudo-inverse (Theorem 21.2).
- Projection properties of the pseudo-inverse.
- The pseudo-inverse of a normal matrix.
- The *Penrose characterization* of the pseudo-inverse.
- Data compression and SVD.
- Best approximation of rank $< r$ of a matrix.
- *Principal component analysis.*
- Review of basic statistical concepts: *mean, variance, covariance, co-variance matrix.*
- Centered data, *centroid.*
- The *principal components (PCA).*
- The *Rayleigh–Ritz theorem* (Theorem 21.8).
- The main theorem: *SVD yields PCA* (Theorem 21.3).
- Best affine approximation.
- SVD yields a best affine approximation (Theorem 21.4).
- Face recognition, eigenfaces.

21.7   Problems

Problem 21.1. Consider the overdetermined system in the single variable x:

$$a_1 x = b_1, \ldots, a_m x = b_m.$$

Prove that the least squares solution of smallest norm is given by

$$x^+ = \frac{a_1 b_1 + \cdots + a_m b_m}{a_1^2 + \cdots + a_m^2}.$$

Problem 21.2. Let X be an $m \times n$ real matrix. For any strictly positive constant $K > 0$, the matrix $X^\top X + K I_n$ is invertible. Prove that the limit of the matrix $(X^\top X + K I_n)^{-1} X^\top$ when K goes to zero is equal to the pseudo-inverse X^+ of X.

Problem 21.3. Use `Matlab` to find the pseudo-inverse of the 8×6 matrix

$$A = \begin{pmatrix} 64 & 2 & 3 & 61 & 60 & 6 \\ 9 & 55 & 54 & 12 & 13 & 51 \\ 17 & 47 & 46 & 20 & 21 & 43 \\ 40 & 26 & 27 & 37 & 36 & 30 \\ 32 & 34 & 35 & 29 & 28 & 38 \\ 41 & 23 & 22 & 44 & 45 & 19 \\ 49 & 15 & 14 & 52 & 53 & 11 \\ 8 & 58 & 59 & 5 & 4 & 62 \end{pmatrix}.$$

Observe that the sums of the columns are all equal to to 256. Let b be the vector of dimension 6 whose coordinates are all equal to 256. Find the solution x^+ of the system $Ax = b$.

Problem 21.4. The purpose of this problem is to show that Proposition 21.7 (the Eckart–Young theorem) also holds for the Frobenius norm. This problem is adapted from Strang [Strang (2019)], Section I.9.

Suppose the $m \times n$ matrix B of rank at most k minimizes $\|A - B\|_F$. Start with an SVD of B,

$$B = V \begin{pmatrix} D & 0 \\ 0 & 0 \end{pmatrix} U^\top,$$

where D is a diagonal $k \times k$ matrix. We can write

$$A = V \begin{pmatrix} L + E + R & F \\ G & H \end{pmatrix} U^\top,$$

where L is strictly lower triangular in the first k rows, E is diagonal, and R is strictly upper triangular, and let

$$C = V \begin{pmatrix} L + D + R & F \\ 0 & 0 \end{pmatrix} U^\top,$$

which clearly has rank $\leq k$.

(1) Prove that

$$\|A - B\|_F^2 = \|A - C\|_F^2 + \|L\|_F^2 + \|R\|_F^2 + \|F\|_F^2.$$

Since $\|A - B\|_F$ is minimal, show that $L = R = F = 0$.
Similarly, show that $G = 0$.

(2) We have

$$V^\top A U = \begin{pmatrix} E & 0 \\ 0 & H \end{pmatrix}, \quad V^\top B U = \begin{pmatrix} D & 0 \\ 0 & 0 \end{pmatrix},$$

where E is diagonal, so deduce that

(1) $D = \text{diag}(\sigma_1, \ldots, \sigma_k)$.
(2) The singular values of H must be the smallest $n - k$ singular values of A.
(3) The minimum of $\|A - B\|_F$ must be $\|H\|_F = (\sigma_{k+1}^2 + \cdots + \sigma_r^2)^{1/2}$.

Problem 21.5. Prove that the closest rank 1 approximation (in $\| \ \|_2$) of the matrix

$$A = \begin{pmatrix} 3 & 0 \\ 4 & 5 \end{pmatrix}$$

is

$$A_1 = \frac{3}{2} \begin{pmatrix} 1 & 1 \\ 3 & 3 \end{pmatrix}.$$

Show that the Eckart–Young theorem fails for the operator norm $\| \ \|_\infty$ by finding a rank 1 matrix B such that $\|A - B\|_\infty < \|A - A_1\|_\infty$.

Problem 21.6. Find a closest rank 1 approximation (in $\| \ \|_2$) for the matrices

$$A = \begin{pmatrix} 3 & 0 & 0 \\ 0 & 2 & 0 \\ 0 & 0 & 1 \end{pmatrix}, \quad A = \begin{pmatrix} 0 & 3 \\ 2 & 0 \end{pmatrix}, \quad A = \begin{pmatrix} 2 & 1 \\ 1 & 2 \end{pmatrix}.$$

Problem 21.7. Find a closest rank 1 approximation (in $\| \ \|_2$) for the matrix

$$A = \begin{pmatrix} \cos \theta & -\sin \theta \\ \sin \theta & \cos \theta \end{pmatrix}.$$

Problem 21.8. Let S be a real symmetric positive definite matrix and let $S = U \Sigma U^\top$ be a diagonalization of S. Prove that the closest rank 1 matrix (in the L^2-norm) to S is $u_1 \sigma_1 u_1^\top$, where u_1 is the first column of U.

Chapter 22

Annihilating Polynomials and the Primary Decomposition

In this chapter all vector spaces are defined over an arbitrary field K.

In Section 6.7 we explained that if $f\colon E \to E$ is a linear map on a K-vector space E, then for any polynomial $p(X) = a_0 X^d + a_1 X^{d-1} + \cdots + a_d$ with coefficients in the field K, we can define the *linear map* $p(f)\colon E \to E$ by

$$p(f) = a_0 f^d + a_1 f^{d-1} + \cdots + a_d \mathrm{id},$$

where $f^k = f \circ \cdots \circ f$, the k-fold composition of f with itself. Note that

$$p(f)(u) = a_0 f^d(u) + a_1 f^{d-1}(u) + \cdots + a_d u,$$

for every vector $u \in E$. Then we showed that if E is finite-dimensional and if $\chi_f(X) = \det(XI - f)$ is the characteristic polynomial of f, by the Cayley–Hamilton theorem, we have

$$\chi_f(f) = 0.$$

This fact suggests looking at the set of all polynomials $p(X)$ such that

$$p(f) = 0.$$

Such polynomials are called *annihilating polynomials* of f, the set of all these polynomials, denoted $\mathrm{Ann}(f)$, is called the *annihilator* of f, and the Cayley–Hamilton theorem shows that it is nontrivial since it contains a polynomial of positive degree. It turns out that $\mathrm{Ann}(f)$ contains a polynomial m_f of smallest degree that generates $\mathrm{Ann}(f)$, and this polynomial divides the characteristic polynomial. Furthermore, the polynomial m_f encapsulates a lot of information about f, in particular whether f can be diagonalized. One of the main reasons for this is that a scalar $\lambda \in K$ is a zero of the minimal polynomial m_f if and only if λ is an eigenvalue of f.

The first main result is Theorem 22.2 which states that if $f \colon E \to E$ is a linear map on a finite-dimensional space E, then f is diagonalizable iff its minimal polynomial m is of the form

$$m = (X - \lambda_1) \cdots (X - \lambda_k),$$

where $\lambda_1, \ldots, \lambda_k$ are distinct elements of K.

One of the technical tools used to prove this result is the notion of f-*conductor*; see Definition 22.7. As a corollary of Theorem 22.2 we obtain results about finite commuting families of diagonalizable or triangulable linear maps.

If $f \colon E \to E$ is a linear map and $\lambda \in K$ is an eigenvalue of f, recall that the eigenspace E_λ associated with λ is the kernel of the linear map $\lambda \mathrm{id} - f$. If all the eigenvalues $\lambda_1 \ldots, \lambda_k$ of f are in K and if f is diagonalizable, then

$$E = E_{\lambda_1} \oplus \cdots \oplus E_{\lambda_k},$$

but in general there are not enough eigenvectors to span E. A remedy is to generalize the notion of eigenvector and look for (nonzero) vectors u (called generalized eigenvectors) such that

$$(\lambda \mathrm{id} - f)^r(u) = 0, \quad \text{for some } r \geq 1.$$

Then it turns out that if the minimal polynomial of f is of the form

$$m = (X - \lambda_1)^{r_1} \cdots (X - \lambda_k)^{r_k},$$

then $r = r_i$ does the job for λ_i; that is, if we let

$$W_i = \mathrm{Ker}\,(\lambda_i \mathrm{id} - f)^{r_i},$$

then

$$E = W_1 \oplus \cdots \oplus W_k.$$

The above facts are parts of the *primary decomposition theorem* (Theorem 22.4). It is a special case of a more general result involving the factorization of the minimal polynomial m into its irreducible monic factors; see Theorem 22.3.

Theorem 22.4 implies that every linear map f that has all its eigenvalues in K can be written as $f = D + N$, where D is diagonalizable and N is nilpotent (which means that $N^r = 0$ for some positive integer r). Furthermore D and N commute and are unique. This is the *Jordan decomposition*, Theorem 22.5.

The Jordan decomposition suggests taking a closer look at nilpotent maps. We prove that for any nilpotent linear map $f \colon E \to E$ on a finite-dimensional vector space E of dimension n over a field K, there is a basis of E such that the matrix N of f is of the form

$$N = \begin{pmatrix} 0 & \nu_1 & 0 & \cdots & 0 & 0 \\ 0 & 0 & \nu_2 & \cdots & 0 & 0 \\ \vdots & \vdots & \vdots & \vdots & \vdots & \vdots \\ 0 & 0 & 0 & \cdots & 0 & \nu_n \\ 0 & 0 & 0 & \cdots & 0 & 0 \end{pmatrix},$$

where $\nu_i = 1$ or $\nu_i = 0$; see Theorem 22.6. As a corollary we obtain the *Jordan form*, which involves matrices of the form

$$J_r(\lambda) = \begin{pmatrix} \lambda & 1 & 0 & \cdots & 0 \\ 0 & \lambda & 1 & \cdots & 0 \\ \vdots & \vdots & \ddots & \ddots & \vdots \\ 0 & 0 & 0 & \ddots & 1 \\ 0 & 0 & 0 & \cdots & \lambda \end{pmatrix},$$

called *Jordan blocks*; see Theorem 22.7.

22.1 Basic Properties of Polynomials; Ideals, GCD's

In order to understand the structure of $\mathrm{Ann}(f)$, we need to review three basic properties of polynomials. We refer the reader to Hoffman and Kunze [Kenneth and Ray (1971)], Artin [Artin (1991)], Dummit and Foote [Dummit and Foote (1999)], and Godement [Godement (1958)] for comprehensive discussions of polynomials and their properties.

We begin by recalling some basic nomenclature. Given a field K, any nonzero polynomial $p(X) \in K[X]$ has some monomial of highest degree $a_0 X^n$ with $a_0 \neq 0$, and the integer $n = \deg(p) \geq 0$ is called the *degree* of p. It is convenient to set the degree of the zero polynomial (denoted by 0) to be

$$\deg(0) = -\infty.$$

A polynomial $p(X)$ such that the coefficient a_0 of its monomial of highest degree is 1 is called a *monic* polynomial. For example, let $K = \mathbb{R}$. The polynomial $p(X) = 4X^7 + 2X^5$ is of degree 7 but is not monic since $a_0 = 4$. On the other hand, the polynomial $p(X) = X^3 - 3X + 1$ is a monic polynomial of degree 3.

We now discuss three key concepts of polynomial algebra:

(1) Ideals
(2) Greatest common divisors and the Bezout identity.
(3) Irreducible polynomials and prime factorization.

Recall the definition a of ring (see Definition 2.2).

Definition 22.1. A *ring* is a set A equipped with two operations $+\colon A \times A \to A$ (called *addition*) and $*\colon A \times A \to A$ (called *multiplication*) having the following properties:

(R1) A is an abelian group w.r.t. $+$;
(R2) $*$ is associative and has an identity element $1 \in A$;
(R3) $*$ is distributive w.r.t. $+$.

The identity element for addition is denoted 0, and the additive inverse of $a \in A$ is denoted by $-a$. More explicitly, the axioms of a ring are the following equations which hold for all $a, b, c \in A$:

$$a + (b + c) = (a + b) + c \qquad \text{(associativity of } +) \qquad (22.1)$$
$$a + b = b + a \qquad \text{(commutativity of } +) \qquad (22.2)$$
$$a + 0 = 0 + a = a \qquad \text{(zero)} \qquad (22.3)$$
$$a + (-a) = (-a) + a = 0 \qquad \text{(additive inverse)} \qquad (22.4)$$
$$a * (b * c) = (a * b) * c \qquad \text{(associativity of } *) \qquad (22.5)$$
$$a * 1 = 1 * a = a \qquad \text{(identity for } *) \qquad (22.6)$$
$$(a + b) * c = (a * c) + (b * c) \qquad \text{(distributivity)} \qquad (22.7)$$
$$a * (b + c) = (a * b) + (a * c) \qquad \text{(distributivity)} \qquad (22.8)$$

The ring A is *commutative* if

$$a * b = b * a \quad \text{for all } a, b \in A.$$

From (22.7) and (22.8), we easily obtain

$$a * 0 = 0 * a = 0 \qquad (22.9)$$
$$a * (-b) = (-a) * b = -(a * b). \qquad (22.10)$$

The first crucial notion is that of an ideal.

Definition 22.2. Given a commutative ring A with unit 1, *an ideal of A* is a nonempty subset $\mathfrak{I}$ of A satisfying the following properties:

(ID1) If $a, b \in \mathfrak{I}$, then $b - a \in \mathfrak{I}$.
(ID2) If $a \in \mathfrak{I}$, then $ax \in \mathfrak{I}$ for all $x \in A$.

An ideal $\mathfrak{I}$ is a *principal ideal* if there is some $a \in \mathfrak{I}$, called a *generator*, such that

$$\mathfrak{I} = \{ax \mid x \in A\}.$$

In this case we usually write $\mathfrak{I} = aA$ or $\mathfrak{I} = (a)$. The ideal $\mathfrak{I} = (0) = \{0\}$ is called the *null ideal* (or *zero ideal*).

The following proposition is a fundamental result about polynomials over a field.

Proposition 22.1. *If K is a field, then every polynomial ideal $\mathfrak{I} \subseteq K[X]$ is a principal ideal. As a consequence, if $\mathfrak{I}$ is not the zero ideal, then there is a unique monic polynomial*

$$p(X) = X^n + a_1 X^{n-1} + \cdots + a_{n-1} X + a_n$$

in $\mathfrak{I}$ such that $\mathfrak{I} = (p)$.

Proof. This result is not hard to prove if we recall that polynomials can divided. Given any two nonzero polynomials $f, g \in K[X]$, there are unique polynomials q, r such that

$$f = qg + r, \quad \text{and} \quad \deg(r) < \deg(g). \tag{22.11}$$

If $\mathfrak{I}$ is not the zero ideal, there is some polynomial of smallest degree in $\mathfrak{I}$, and since K is a field, by suitable multiplication by a scalar, we can make sure that this polynomial is monic. Thus, let f be a monic polynomial of smallest degree in $\mathfrak{I}$. By (ID2), it is clear that $(f) \subseteq \mathfrak{I}$. Now let $g \in \mathfrak{I}$. Using (22.11), there exist unique $q, r \in K[X]$ such that

$$g = qf + r \quad \text{and} \quad \deg(r) < \deg(f).$$

If $r \neq 0$, there is some $\lambda \neq 0$ in K such that λr is a monic polynomial, and since $\lambda r = \lambda g - \lambda q f$, with $f, g \in \mathfrak{I}$, by (ID1) and (ID2), we have $\lambda r \in \mathfrak{I}$, where $\deg(\lambda r) < \deg(f)$ and λr is a monic polynomial, contradicting the minimality of the degree of f. Thus, $r = 0$, and $g \in (f)$. The uniqueness of the monic polynomial f is left as an exercise. $\square$

We will also need to know that the greatest common divisor of polynomials exist. Given any two nonzero polynomials $f, g \in K[X]$, recall that f divides g if $g = qf$ for some $q \in K[X]$.

Definition 22.3. Given any two nonzero polynomials $f, g \in K[X]$, a polynomial $d \in K[X]$ is a *greatest common divisor of f and g* (for short, a *gcd of f and g*) if d divides f and g and whenever $h \in K[X]$ divides f and g, then h divides d. We say that f and g are *relatively prime* if 1 is a gcd of f and g.

Note that f and g are relatively prime iff all of their gcd's are constants (scalars in K), or equivalently, if f, g have no common divisor q of degree $\deg(q) \geq 1$. For example, over $\mathbb{R}$, $\gcd(X^2 - 1, X^3 + X^2 - X - 1) = (X - 1)(X+1)$ since $X^3 + X^2 - X - 1 = (X-1)(X+1)^2$, while $\gcd(X^3+1, X-1) = 1$.

We can characterize gcd's of polynomials as follows.

Proposition 22.2. *Let K be a field and let $f, g \in K[X]$ be any two nonzero polynomials. For every polynomial $d \in K[X]$, the following properties are equivalent:*

(1) The polynomial d is a gcd of f and g.

(2) The polynomial d divides f and g and there exist $u, v \in K[X]$ such that

$$d = uf + vg.$$

(3) The ideals $(f), (g)$, and (d) satisfy the equation

$$(d) = (f) + (g).$$

In addition, $d \neq 0$, and d is unique up to multiplication by a nonzero scalar in K.

As a consequence of Proposition 22.2, two nonzero polynomials $f, g \in K[X]$ are relatively prime iff there exist $u, v \in K[X]$ such that

$$uf + vg = 1.$$

The identity

$$d = uf + vg$$

of Part (2) of Proposition 22.2 is often called the *Bezout identity*. For an example of Bezout's identity, take $K = \mathbb{R}$. Since $X^3 + 1$ and $X - 1$ are relatively prime, we have

$$1 = 1/2(X^3 + 1) - 1/2(X^2 + X + 1)(X - 1).$$

An important consequence of the Bezout identity is the following result.

Proposition 22.3. *(Euclid's proposition) Let K be a field and let $f, g, h \in K[X]$ be any nonzero polynomials. If f divides gh and f is relatively prime to g, then f divides h.*

Proposition 22.3 can be generalized to any number of polynomials.

Proposition 22.4. *Let K be a field and let $f, g_1, \ldots, g_m \in K[X]$ be some nonzero polynomials. If f and g_i are relatively prime for all i, $1 \leq i \leq m$, then f and $g_1 \cdots g_m$ are relatively prime.*

Definition 22.3 is generalized to any finite number of polynomials as follows.

Definition 22.4. Given any nonzero polynomials $f_1, \ldots, f_n \in K[X]$, where $n \geq 2$, a polynomial $d \in K[X]$ is a *greatest common divisor of* $f_1, \ldots, f_n$ (for short, a *gcd of* $f_1, \ldots, f_n$) if d divides each f_i and whenever $h \in K[X]$ divides each f_i, then h divides d. We say that $f_1, \ldots, f_n$ are *relatively prime* if 1 is a gcd of $f_1, \ldots, f_n$.

It is easily shown that Proposition 22.2 can be generalized to any finite number of polynomials.

Proposition 22.5. *Let* K *be a field and let* $f_1, \ldots, f_n \in K[X]$ *be any* $n \geq 2$ *nonzero polynomials. For every polynomial* $d \in K[X]$, *the following properties are equivalent:*

(1) The polynomial d *is a gcd of* $f_1, \ldots, f_n$.
(2) The polynomial d *divides each* f_i *and there exist* $u_1, \ldots, u_n \in K[X]$ *such that*

$$d = u_1 f_1 + \cdots + u_n f_n.$$

(3) The ideals (f_i), *and* (d) *satisfy the equation*

$$(d) = (f_1) + \cdots + (f_n).$$

In addition, $d \neq 0$, *and* d *is unique up to multiplication by a nonzero scalar in* K.

As a consequence of Proposition 22.5, any $n \geq 2$ nonzero polynomials $f_1, \ldots, f_n \in K[X]$ are relatively prime iff there exist $u_1, \ldots, u_n \in K[X]$ such that

$$u_1 f_1 + \cdots + u_n f_n = 1,$$

the *Bezout identity*.

We will also need to know that every nonzero polynomial (over a field) can be factored into irreducible polynomials, which is the generalization of the prime numbers to polynomials.

Definition 22.5. Given a field K, a polynomial $p \in K[X]$ *irreducible or indecomposable or prime* is if $\deg(p) \geq 1$ and if p is not divisible by any polynomial $q \in K[X]$ such that $1 \leq \deg(q) < \deg(p)$. Equivalently, p is irreducible if $\deg(p) \geq 1$ and if $p = q_1 q_2$, then either $q_1 \in K$ or $q_2 \in K$ (and of course, $q_1 \neq 0$, $q_2 \neq 0$).

Every polynomial $aX + b$ of degree 1 is irreducible. Over the field $\mathbb{R}$, the polynomial $X^2 + 1$ is irreducible (why?), but $X^3 + 1$ is not irreducible, since

$$X^3 + 1 = (X + 1)(X^2 - X + 1).$$

The polynomial $X^2 - X + 1$ is irreducible over $\mathbb{R}$ (why?). It would seem that $X^4 + 1$ is irreducible over $\mathbb{R}$, but in fact,

$$X^4 + 1 = (X^2 - \sqrt{2}X + 1)(X^2 + \sqrt{2}X + 1).$$

However, in view of the above factorization, $X^4 + 1$ is irreducible over $\mathbb{Q}$.

It can be shown that the irreducible polynomials over $\mathbb{R}$ are the polynomials of degree 1 or the polynomials of degree 2 of the form $aX^2 + bX + c$, for which $b^2 - 4ac < 0$ (i.e., those having no real roots). This is not easy to prove! Over the complex numbers $\mathbb{C}$, the only irreducible polynomials are those of degree 1. This is a version of a fact often referred to as the "Fundamental Theorem of Algebra."

Observe that the definition of irreducibility implies that any finite number of distinct irreducible polynomials are relatively prime.

The following fundamental result can be shown

Theorem 22.1. *Given any field K, for every nonzero polynomial*

$$f = a_d X^d + a_{d-1} X^{d-1} + \cdots + a_0$$

of degree $d = \deg(f) \geq 1$ in $K[X]$, there exists a unique set $\{\langle p_1, k_1 \rangle,$ $\ldots, \langle p_m, k_m \rangle\}$ such that

$$f = a_d p_1^{k_1} \cdots p_m^{k_m},$$

where the $p_i \in K[X]$ are distinct irreducible monic polynomials, the k_i are (not necessarily distinct) integers, and with $m \geq 1$, $k_i \geq 1$.

We can now return to minimal polynomials.

22.2 Annihilating Polynomials and the Minimal Polynomial

Given a linear map $f: E \to E$, it is easy to check that the set $\mathrm{Ann}(f)$ of polynomials that annihilate f is an ideal. Furthermore, when E is finite-dimensional, the Cayley–Hamilton theorem implies that $\mathrm{Ann}(f)$ is not the zero ideal. Therefore, by Proposition 22.1, there is a unique monic polynomial m_f that generates $\mathrm{Ann}(f)$.

Definition 22.6. If $f: E \to E$ is a linear map on a finite-dimensional vector space E, the unique monic polynomial $m_f(X)$ that generates the ideal $\mathrm{Ann}(f)$ of polynomials which annihilate f (the *annihilator* of f) is called the *minimal polynomial* of f.

The minimal polynomial m_f of f is the monic polynomial of smallest degree that annihilates f. Thus, the minimal polynomial divides the characteristic polynomial χ_f, and $\deg(m_f) \geq 1$. For simplicity of notation, we often write m instead of m_f.

If A is any $n \times n$ matrix, the set $\mathrm{Ann}(A)$ of polynomials that annihilate A is the set of polynomials

$$p(X) = a_0 X^d + a_1 X^{d-1} + \cdots + a_{d-1} X + a_d$$

such that

$$a_0 A^d + a_1 A^{d-1} + \cdots + a_{d-1} A + a_d I = 0.$$

It is clear that $\mathrm{Ann}(A)$ is a nonzero ideal and its unique monic generator is called the *minimal polynomial* of A. We check immediately that if Q is an invertible matrix, then A and $Q^{-1}AQ$ have the same minimal polynomial. Also, if A is the matrix of f with respect to some basis, then f and A have the same minimal polynomial.

The zeros (in K) of the minimal polynomial of f and the eigenvalues of f (in K) are intimately related.

Proposition 22.6. *Let $f \colon E \to E$ be a linear map on some finite-dimensional vector space E. Then $\lambda \in K$ is a zero of the minimal polynomial $m_f(X)$ of f iff λ is an eigenvalue of f iff λ is a zero of $\chi_f(X)$. Therefore, the minimal and the characteristic polynomials have the same zeros (in K), except for multiplicities.*

Proof. First assume that $m(\lambda) = 0$ (with $\lambda \in K$, and writing m instead of m_f). If so, using polynomial division, m can be factored as

$$m = (X - \lambda)q,$$

with $\deg(q) < \deg(m)$. Since m is the minimal polynomial, $q(f) \neq 0$, so there is some nonzero vector $v \in E$ such that $u = q(f)(v) \neq 0$. But then, because m is the minimal polynomial,

$$0 = m(f)(v)$$
$$= (f - \lambda\mathrm{id})(q(f)(v))$$
$$= (f - \lambda\mathrm{id})(u),$$

which shows that λ is an eigenvalue of f.

Conversely, assume that $\lambda \in K$ is an eigenvalue of f. This means that for some $u \neq 0$, we have $f(u) = \lambda u$. Now it is easy to show that

$$m(f)(u) = m(\lambda)u,$$

and since m is the minimal polynomial of f, we have $m(f)(u) = 0$, so $m(\lambda)u = 0$, and since $u \neq 0$, we must have $m(\lambda) = 0$. $\qquad\square$

Proposition 22.7. *Let $f: E \to E$ be a linear map on some finite-dimensional vector space E. If f diagonalizable, then its minimal polynomial is a product of distinct factors of degree 1.*

Proof. If we assume that f is diagonalizable, then its eigenvalues are all in K, and if $\lambda_1, \ldots, \lambda_k$ are the distinct eigenvalues of f, and then by Proposition 22.6, the minimal polynomial m of f must be a product of powers of the polynomials $(X - \lambda_i)$. Actually, we claim that

$$m = (X - \lambda_1) \cdots (X - \lambda_k).$$

For this we just have to show that m annihilates f. However, for any eigenvector u of f, one of the linear maps $f - \lambda_i \mathrm{id}$ sends u to 0, so

$$m(f)(u) = (f - \lambda_1 \mathrm{id}) \circ \cdots \circ (f - \lambda_k \mathrm{id})(u) = 0.$$

Since E is spanned by the eigenvectors of f, we conclude that

$$m(f) = 0. \qquad \square$$

It turns out that the converse of Proposition 22.7 is true, but this will take a little work to establish it.

22.3 Minimal Polynomials of Diagonalizable Linear Maps

In this section we prove that if the minimal polynomial m_f of a linear map f is of the form

$$m_f = (X - \lambda_1) \cdots (X - \lambda_k)$$

for distinct scalars $\lambda_1, \ldots, \lambda_k \in K$, then f is diagonalizable. This is a powerful result that has a number of implications. But first we need of few properties of invariant subspaces.

Given a linear map $f: E \to E$, recall that a subspace W of E is *invariant* under f if $f(u) \in W$ for all $u \in W$. For example, if $f: \mathbb{R}^2 \to \mathbb{R}^2$ is $f(x, y) = (-x, y)$, the y-axis is invariant under f.

Proposition 22.8. *Let W be a subspace of E invariant under the linear map $f: E \to E$ (where E is finite-dimensional). Then the minimal polynomial of the restriction $f \mid W$ of f to W divides the minimal polynomial of f, and the characteristic polynomial of $f \mid W$ divides the characteristic polynomial of f.*

Sketch of proof. The key ingredient is that we can pick a basis $(e_1, \ldots, e_n)$ of E in which $(e_1, \ldots, e_k)$ is a basis of W. The matrix of f over this basis is a block matrix of the form

$$A = \begin{pmatrix} B & C \\ 0 & D \end{pmatrix},$$

where B is a $k \times k$ matrix, D is an $(n-k) \times (n-k)$ matrix, and C is a $k \times (n-k)$ matrix. Then

$$\det(XI - A) = \det(XI - B)\det(XI - D),$$

which implies the statement about the characteristic polynomials. Furthermore,

$$A^i = \begin{pmatrix} B^i & C_i \\ 0 & D^i \end{pmatrix},$$

for some $k \times (n-k)$ matrix C_i. It follows that any polynomial which annihilates A also annihilates B and D. So the minimal polynomial of B divides the minimal polynomial of A. $\square$

For the next step, there are at least two ways to proceed. We can use an old-fashion argument using Lagrange interpolants, or we can use a slight generalization of the notion of annihilator. We pick the second method because it illustrates nicely the power of principal ideals.

What we need is the notion of conductor (also called transporter).

Definition 22.7. Let $f \colon E \to E$ be a linear map on a finite-dimensional vector space E, let W be an invariant subspace of f, and let u be any vector in E. The set $S_f(u, W)$ consisting of all polynomials $q \in K[X]$ such that $q(f)(u) \in W$ is called the *f-conductor of u into W.*

Observe that the minimal polynomial m_f of f always belongs to $S_f(u, W)$, so this is a nontrivial set. Also, if $W = (0)$, then $S_f(u, (0))$ is just the annihilator of f. The crucial property of $S_f(u, W)$ is that it is an ideal.

Proposition 22.9. *If W is an invariant subspace for f, then for each $u \in E$, the f-conductor $S_f(u, W)$ is an ideal in $K[X]$.*

We leave the proof as a simple exercise, using the fact that if W invariant under f, then W is invariant under every polynomial $q(f)$ in $S_f(u, W)$.

Since $S_f(u, W)$ is an ideal, it is generated by a unique monic polynomial q of smallest degree, and because the minimal polynomial m_f of f is in $S_f(u, W)$, the polynomial q divides m.

Definition 22.8. The unique monic polynomial which generates $S_f(u, W)$ is called the *conductor of u into W*.

Example 22.1. For example, suppose $f \colon \mathbb{R}^2 \to \mathbb{R}^2$ where $f(x, y) = (x, 0)$. Observe that $W = \{(x, 0) \in \mathbb{R}^2\}$ is invariant under f. By representing f as $\begin{pmatrix} 1 & 0 \\ 0 & 0 \end{pmatrix}$, we see that $m_f(X) = \chi_f(X) = X^2 - X$. Let $u = (0, y)$. Then $S_f(u, W) = (X)$, and we say X is the conductor of u into W.

Proposition 22.10. *Let $f \colon E \to E$ be a linear map on a finite-dimensional space E and assume that the minimal polynomial m of f is of the form*

$$m = (X - \lambda_1)^{r_1} \cdots (X - \lambda_k)^{r_k},$$

where the eigenvalues $\lambda_1, \ldots, \lambda_k$ of f belong to K. If W is a proper subspace of E which is invariant under f, then there is a vector $u \in E$ with the following properties:

(a) $u \notin W$;
(b) $(f - \lambda\mathrm{id})(u) \in W$, for some eigenvalue λ of f.

Proof. Observe that (a) and (b) together assert that the conductor of u into W is a polynomial of the form $X - \lambda_i$. Pick any vector $v \in E$ not in W, and let g be the conductor of v into W, i.e. $g(f)(v) \in W$. Since g divides m and $v \notin W$, the polynomial g is not a constant, and thus it is of the form

$$g = (X - \lambda_1)^{s_1} \cdots (X - \lambda_k)^{s_k},$$

with at least some $s_i > 0$. Choose some index j such that $s_j > 0$. Then $X - \lambda_j$ is a factor of g, so we can write

$$g = (X - \lambda_j)q. \tag{22.12}$$

By definition of g, the vector $u = q(f)(v)$ cannot be in W, since otherwise g would not be of minimal degree. However, (22.12) implies that

$$(f - \lambda_j\mathrm{id})(u) = (f - \lambda_j\mathrm{id})(q(f)(v))$$
$$= g(f)(v)$$

is in W, which concludes the proof. $\qquad\square$

We can now prove the main result of this section.

Theorem 22.2. *Let* $f \colon E \to E$ *be a linear map on a finite-dimensional space* E. *Then* f *is diagonalizable iff its minimal polynomial* m *is of the form*

$$m = (X - \lambda_1) \cdots (X - \lambda_k),$$

where $\lambda_1, \ldots, \lambda_k$ *are distinct elements of* K.

Proof. We already showed in Proposition 22.7 that if f is diagonalizable, then its minimal polynomial is of the above form (where $\lambda_1, \ldots, \lambda_k$ are the distinct eigenvalues of f).

For the converse, let W be the subspace spanned by all the eigenvectors of f. If $W \neq E$, since W is invariant under f, by Proposition 22.10, there is some vector $u \notin W$ such that for some λ_j, we have

$$(f - \lambda_j \mathrm{id})(u) \in W.$$

Let $v = (f - \lambda_j \mathrm{id})(u) \in W$. Since $v \in W$, we can write

$$v = w_1 + \cdots + w_k$$

where $f(w_i) = \lambda_i w_i$ (either $w_i = 0$ or w_i is an eigenvector for λ_i), and so for every polynomial h, we have

$$h(f)(v) = h(\lambda_1)w_1 + \cdots + h(\lambda_k)w_k,$$

which shows that $h(f)(v) \in W$ for every polynomial h. We can write

$$m = (X - \lambda_j)q$$

for some polynomial q, and also

$$q - q(\lambda_j) = p(X - \lambda_j)$$

for some polynomial p. We know that $p(f)(v) \in W$, and since m is the minimal polynomial of f, we have

$$0 = m(f)(u) = (f - \lambda_j \mathrm{id})(q(f)(u)),$$

which implies that $q(f)(u) \in W$ (either $q(f)(u) = 0$, or it is an eigenvector associated with λ_j). However,

$$q(f)(u) - q(\lambda_j)u = p(f)((f - \lambda_j \mathrm{id})(u)) = p(f)(v),$$

and since $p(f)(v) \in W$ and $q(f)(u) \in W$, we conclude that $q(\lambda_j)u \in W$. But, $u \notin W$, which implies that $q(\lambda_j) = 0$, so λ_j is a double root of m, a contradiction. Therefore, we must have $W = E$. $\qquad \square$

Remark: Proposition 22.10 can be used to give a quick proof of Theorem 14.1.

22.4   Commuting Families of Diagonalizable and Triangulable Maps

Using Theorem 22.2, we can give a short proof about commuting diagonalizable linear maps.

Definition 22.9. If $\mathcal{F}$ is a family of linear maps on a vector space E, we say that $\mathcal{F}$ is a *commuting family* iff $f \circ g = g \circ f$ for all $f, g \in \mathcal{F}$.

Proposition 22.11. *Let $\mathcal{F}$ be a commuting family of diagonalizable linear maps on a vector space E. There exists a basis of E such that every linear map in $\mathcal{F}$ is represented in that basis by a diagonal matrix.*

Proof. We proceed by induction on $n = \dim(E)$. If $n = 1$, there is nothing to prove. If $n > 1$, there are two cases. If all linear maps in $\mathcal{F}$ are of the form $\lambda\mathrm{id}$ for some $\lambda \in K$, then the proposition holds trivially. In the second case, let $f \in \mathcal{F}$ be some linear map in $\mathcal{F}$ which is not a scalar multiple of the identity. In this case, f has at least two distinct eigenvalues $\lambda_1, \ldots, \lambda_k$, and because f is diagonalizable, E is the direct sum of the corresponding eigenspaces $E_{\lambda_1}, \ldots, E_{\lambda_k}$. For every index i, the eigenspace E_{λ_i} is invariant under f and under every other linear map g in $\mathcal{F}$, since for any $g \in \mathcal{F}$ and any $u \in E_{\lambda_i}$, because f and g commute, we have

$$f(g(u)) = g(f(u)) = g(\lambda_i u) = \lambda_i g(u)$$

so $g(u) \in E_{\lambda_i}$. Let $\mathcal{F}_i$ be the family obtained by restricting each $f \in \mathcal{F}$ to E_{λ_i}. By Proposition 22.8, the minimal polynomial of every linear map $f \mid E_{\lambda_i}$ in $\mathcal{F}_i$ divides the minimal polynomial m_f of f, and since f is diagonalizable, m_f is a product of distinct linear factors, so the minimal polynomial of $f \mid E_{\lambda_i}$ is also a product of distinct linear factors. By Theorem 22.2, the linear map $f \mid E_{\lambda_i}$ is diagonalizable. Since $k > 1$, we have $\dim(E_{\lambda_i}) < \dim(E)$ for $i = 1, \ldots, k$, and by the induction hypothesis, for each i there is a basis of E_{λ_i} over which $f \mid E_{\lambda_i}$ is represented by a diagonal matrix. Since the above argument holds for all i, by combining the bases of the E_{λ_i}, we obtain a basis of E such that the matrix of every linear map $f \in \mathcal{F}$ is represented by a diagonal matrix. $\qquad\square$

There is also an analogous result for commuting families of linear maps represented by upper triangular matrices. To prove this we need the following proposition.

Proposition 22.12. *Let $\mathcal{F}$ be a nonempty commuting family of triangulable linear maps on a finite-dimensional vector space E. Let W be a proper* subspace of E which is invariant under $\mathcal{F}$. Then there exists a vector $u \in E$ such that:

(1) $u \notin W$.
(2) For every $f \in \mathcal{F}$, the vector $f(u)$ belongs to the subspace $W \oplus Ku$ spanned by W and u.

Proof. By renaming the elements of $\mathcal{F}$ if necessary, we may assume that $(f_1, \ldots, f_r)$ is a basis of the subspace of $\mathrm{End}(E)$ spanned by $\mathcal{F}$. We prove by induction on r that there exists some vector $u \in E$ such that

(1) $u \notin W$.
(2) $(f_i - \alpha_i \mathrm{id})(u) \in W$ for $i = 1, \ldots, r$, for some scalars $\alpha_i \in K$.

Consider the base case $r = 1$. Since f_1 is triangulable, its eigenvalues all belong to K since they are the diagonal entries of the triangular matrix associated with f_1 (this is the easy direction of Theorem 14.1), so the minimal polynomial of f_1 is of the form

$$m = (X - \lambda_1)^{r_1} \cdots (X - \lambda_k)^{r_k},$$

where the eigenvalues $\lambda_1, \ldots, \lambda_k$ of f_1 belong to K. We conclude by applying Proposition 22.10.

Next assume that $r \geq 2$ and that the induction hypothesis holds for $f_1, \ldots, f_{r-1}$. Thus, there is a vector $u_{r-1} \in E$ such that

(1) $u_{r-1} \notin W$.
(2) $(f_i - \alpha_i \mathrm{id})(u_{r-1}) \in W$ for $i = 1, \ldots, r-1$, for some scalars $\alpha_i \in K$.

Let

$$V_{r-1} = \{ w \in E \mid (f_i - \alpha_i \mathrm{id})(w) \in W,\ i = 1, \ldots, r-1 \}.$$

Clearly, $W \subseteq V_{r-1}$ and $u_{r-1} \in V_{r-1}$. We claim that V_{r-1} is invariant under $\mathcal{F}$. This is because, for any $v \in V_{r-1}$ and any $f \in \mathcal{F}$, since f and f_i commute, we have

$$(f_i - \alpha_i \mathrm{id})(f(v)) = f((f_i - \alpha_i \mathrm{id})(v)), \quad 1 \leq i \leq r-1.$$

Now $(f_i - \alpha_i \mathrm{id})(v) \in W$ because $v \in V_{r-1}$, and W is invariant under $\mathcal{F}$, so $f((f_i - \alpha_i \mathrm{id})(v)) \in W$, that is, $(f_i - \alpha_i \mathrm{id})(f(v)) \in W$.

Consider the restriction g_r of f_r to V_{r-1}. The minimal polynomial of g_r divides the minimal polynomial of f_r, and since f_r is triangulable, just as we saw for f_1, the minimal polynomial of f_r is of the form

$$m = (X - \lambda_1)^{r_1} \cdots (X - \lambda_k)^{r_k},$$

where the eigenvalues $\lambda_1, \ldots, \lambda_k$ of f_r belong to K, so the minimal polynomial of g_r is of the same form. By Proposition 22.10, there is some vector $u_r \in V_{r-1}$ such that

(1) $u_r \notin W$.
(2) $(g_r - \alpha_r \mathrm{id})(u_r) \in W$ for some scalars $\alpha_r \in K$.

Now since $u_r \in V_{r-1}$, we have $(f_i - \alpha_i \mathrm{id})(u_r) \in W$ for $i = 1, \ldots, r-1$, so $(f_i - \alpha_i \mathrm{id})(u_r) \in W$ for $i = 1, \ldots, r$ (since g_r is the restriction of f_r), which concludes the proof of the induction step. Finally, since every $f \in \mathcal{F}$ is the linear combination of $(f_1, \ldots, f_r)$, Condition (2) of the inductive claim implies Condition (2) of the proposition. $\qquad\square$

We can now prove the following result.

Proposition 22.13. *Let $\mathcal{F}$ be a nonempty commuting family of triangulable linear maps on a finite-dimensional vector space E. There exists a basis of E such that every linear map in $\mathcal{F}$ is represented in that basis by an upper triangular matrix.*

Proof. Let $n = \dim(E)$. We construct inductively a basis $(u_1, \ldots, u_n)$ of E such that if W_i is the subspace spanned by $(u_1 \ldots, u_i)$, then for every $f \in \mathcal{F}$,

$$f(u_i) = a_{1i}^f u_1 + \cdots + a_{ii}^f u_i,$$

for some $a_{ij}^f \in K$; that is, $f(u_i)$ belongs to the subspace W_i.

We begin by applying Proposition 22.12 to the subspace $W_0 = (0)$ to get u_1 so that for all $f \in \mathcal{F}$,

$$f(u_1) = \alpha_1^f u_1.$$

For the induction step, since W_i invariant under $\mathcal{F}$, we apply Proposition 22.12 to the subspace W_i, to get $u_{i+1} \in E$ such that

(1) $u_{i+1} \notin W_i$.
(2) For every $f \in \mathcal{F}$, the vector $f(u_{i+1})$ belong to the subspace spanned by W_i and u_{i+1}.

Condition (1) implies that $(u_1, \ldots, u_i, u_{i+1})$ is linearly independent, and Condition (2) means that for every $f \in \mathcal{F}$,

$$f(u_{i+1}) = a_{1\,i+1}^f u_1 + \cdots + a_{i+1\,i+1}^f u_{i+1},$$

for some $a_{i+1\,j}^f \in K$, establishing the induction step. After n steps, each $f \in \mathcal{F}$ is represented by an upper triangular matrix. $\qquad\square$

Observe that if $\mathcal{F}$ consists of a single linear map f and if the minimal polynomial of f is of the form

$$m = (X - \lambda_1)^{r_1} \cdots (X - \lambda_k)^{r_k},$$

with all $\lambda_i \in K$, using Proposition 22.10 instead of Proposition 22.12, the proof of Proposition 22.13 yields another proof of Theorem 14.1.

22.5 The Primary Decomposition Theorem

If $f \colon E \to E$ is a linear map and $\lambda \in K$ is an eigenvalue of f, recall that the eigenspace E_λ associated with λ is the kernel of the linear map $\lambda \mathrm{id} - f$. If all the eigenvalues $\lambda_1 \ldots, \lambda_k$ of f are in K, it may happen that

$$E = E_{\lambda_1} \oplus \cdots \oplus E_{\lambda_k},$$

but in general there are not enough eigenvectors to span E. What if we generalize the notion of eigenvector and look for (nonzero) vectors u such that

$$(\lambda \mathrm{id} - f)^r(u) = 0, \quad \text{for some } r \geq 1?$$

It turns out that if the minimal polynomial of f is of the form

$$m = (X - \lambda_1)^{r_1} \cdots (X - \lambda_k)^{r_k},$$

then $r = r_i$ does the job for λ_i; that is, if we let

$$W_i = \mathrm{Ker}\,(\lambda_i \mathrm{id} - f)^{r_i},$$

then

$$E = W_1 \oplus \cdots \oplus W_k.$$

This result is very nice but seems to require that the eigenvalues of f all belong to K. Actually, it is a special case of a more general result involving the factorization of the minimal polynomial m into its irreducible monic factors (see Theorem 22.1),

$$m = p_1^{r_1} \cdots p_k^{r_k},$$

where the p_i are distinct irreducible monic polynomials over K.

Theorem 22.3. *(Primary Decomposition Theorem) Let $f \colon E \to E$ be a linear map on the finite-dimensional vector space E over the field K. Write the minimal polynomial m of f as*

$$m = p_1^{r_1} \cdots p_k^{r_k},$$

where the p_i are distinct irreducible monic polynomials over K, and the r_i are positive integers. Let

$$W_i = \mathrm{Ker}\,(p_i^{r_i}(f)), \quad i = 1, \ldots, k.$$

Then

(a) $E = W_1 \oplus \cdots \oplus W_k$.

(b) Each W_i is invariant under f.

(c) The minimal polynomial of the restriction $f \mid W_i$ of f to W_i is $p_i^{r_i}$.

Proof. The trick is to construct projections π_i using the polynomials $p_j^{r_j}$ so that the range of π_i is equal to W_i. Let

$$g_i = m/p_i^{r_i} = \prod_{j \neq i} p_j^{r_j}.$$

Note that

$$p_i^{r_i} g_i = m.$$

Since $p_1, \ldots, p_k$ are irreducible and distinct, they are relatively prime. Then using Proposition 22.4, it is easy to show that $g_1, \ldots, g_k$ are relatively prime. Otherwise, some irreducible polynomial p would divide all of $g_1, \ldots, g_k$, so by Proposition 22.4 it would be equal to one of the irreducible factors p_i. But that p_i is missing from g_i, a contradiction. Therefore, by Proposition 22.5, there exist some polynomials $h_1, \ldots, h_k$ such that

$$g_1 h_1 + \cdots + g_k h_k = 1.$$

Let $q_i = g_i h_i$ and let $\pi_i = q_i(f) = g_i(f) h_i(f)$. We have

$$q_1 + \cdots + q_k = 1,$$

and since m divides $q_i q_j$ for $i \neq j$, we get

$$\pi_1 + \cdots + \pi_k = \mathrm{id}$$
$$\pi_i \pi_j = 0, \quad i \neq j.$$

(We implicitly used the fact that if p, q are two polynomials, the linear maps $p(f) \circ q(f)$ and $q(f) \circ p(f)$ are the same since $p(f)$ and $q(f)$ are polynomials in the powers of f, which commute.) Composing the first equation with π_i and using the second equation, we get

$$\pi_i^2 = \pi_i.$$

Therefore, the π_i are projections, and E is the direct sum of the images of the π_i. Indeed, every $u \in E$ can be expressed as

$$u = \pi_1(u) + \cdots + \pi_k(u).$$

Also, if

$$\pi_1(u) + \cdots + \pi_k(u) = 0,$$

then by applying π_i we get

$$0 = \pi_i^2(u) = \pi_i(u), \quad i = 1, \ldots k.$$

To finish proving (a), we need to show that

$$W_i = \text{Ker}\,(p_i^{r_i}(f)) = \pi_i(E).$$

If $v \in \pi_i(E)$, then $v = \pi_i(u)$ for some $u \in E$, so

$$
\begin{aligned}
p_i^{r_i}(f)(v) &= p_i^{r_i}(f)(\pi_i(u)) \\
&= p_i^{r_i}(f)g_i(f)h_i(f)(u) \\
&= h_i(f)p_i^{r_i}(f)g_i(f)(u) \\
&= h_i(f)m(f)(u) = 0,
\end{aligned}
$$

because m is the minimal polynomial of f. Therefore, $v \in W_i$.

Conversely, assume that $v \in W_i = \text{Ker}\,(p_i^{r_i}(f))$. If $j \neq i$, then $g_j h_j$ is divisible by $p_i^{r_i}$, so

$$g_j(f)h_j(f)(v) = \pi_j(v) = 0, \quad j \neq i.$$

Then since $\pi_1 + \cdots + \pi_k = \text{id}$, we have $v = \pi_i v$, which shows that v is in the range of π_i. Therefore, $W_i = \text{Im}(\pi_i)$, and this finishes the proof of (a).

If $p_i^{r_i}(f)(u) = 0$, then $p_i^{r_i}(f)(f(u)) = f(p_i^{r_i}(f)(u)) = 0$, so (b) holds.

If we write $f_i = f \mid W_i$, then $p_i^{r_i}(f_i) = 0$, because $p_i^{r_i}(f) = 0$ on W_i (its kernel). Therefore, the minimal polynomial of f_i divides $p_i^{r_i}$. Conversely, let q be any polynomial such that $q(f_i) = 0$ (on W_i). Since $m = p_i^{r_i}g_i$, the fact that $m(f)(u) = 0$ for all $u \in E$ shows that

$$p_i^{r_i}(f)(g_i(f)(u)) = 0, \quad u \in E,$$

and thus $\text{Im}(g_i(f)) \subseteq \text{Ker}\,(p_i^{r_i}(f)) = W_i$. Consequently, since $q(f)$ is zero on W_i,

$$q(f)g_i(f) = 0 \quad \text{for all } u \in E.$$

But then qg_i is divisible by the minimal polynomial $m = p_i^{r_i}g_i$ of f, and since $p_i^{r_i}$ and g_i are relatively prime, by Euclid's proposition, $p_i^{r_i}$ must divide q. This finishes the proof that the minimal polynomial of f_i is $p_i^{r_i}$, which is (c). $\qquad\square$

To best understand the projection constructions of Theorem 22.3, we provide the following two explicit examples of the primary decomposition theorem.

Example 22.2. First let $f\colon \mathbb{R}^3 \to \mathbb{R}^3$ be defined as $f(x,y,z) = (y,-x,z)$. In terms of the standard basis f is represented by the 3×3 matrix

$$X_f := \begin{pmatrix} 0 & -1 & 0 \\ 1 & 0 & 0 \\ 0 & 0 & 1 \end{pmatrix}.$$

Then a simple calculation shows that $m_f(x) = \chi_f(x) = (x^2+1)(x-1)$. Using the notation of the preceding proof set

$$m = p_1 p_2, \qquad p_1 = x^2 + 1, \qquad p_2 = x - 1.$$

Then

$$g_1 = \frac{m}{p_1} = x - 1, \qquad g_2 = \frac{m}{p_2} = x^2 + 1.$$

We must find $h_1, h_2 \in \mathbb{R}[x]$ such that $g_1 h_1 + g_2 h_2 = 1$. In general this is the hard part of the projection construction. But since we are only working with two relatively prime polynomials g_1, g_2, we may apply the Euclidean algorithm to discover that

$$-\frac{x+1}{2}(x-1) + \frac{1}{2}(x^2+1) = 1,$$

where $h_1 = -\frac{x+1}{2}$ while $h_2 = \frac{1}{2}$. By definition

$$\pi_1 = g_1(f)h_1(f) = -\frac{1}{2}(X_f - \mathrm{id})(X_f + \mathrm{id}) = -\frac{1}{2}(X_f^2 - \mathrm{id}) = \begin{pmatrix} 1 & 0 & 0 \\ 0 & 1 & 0 \\ 0 & 0 & 0 \end{pmatrix},$$

and

$$\pi_2 = g_2(f)h_2(f) = \frac{1}{2}(X_f^2 + \mathrm{id}) = \begin{pmatrix} 0 & 0 & 0 \\ 0 & 0 & 0 \\ 0 & 0 & 1 \end{pmatrix}.$$

Then $\mathbb{R}^3 = W_1 \oplus W_2$, where

$$W_1 = \pi_1(\mathbb{R}^3) = \mathrm{Ker}\,(p_1(X_f)) = \mathrm{Ker}\,(X_f^2 + \mathrm{id})$$

$$= \mathrm{Ker}\,\begin{pmatrix} 0 & 0 & 0 \\ 0 & 0 & 0 \\ 0 & 0 & 1 \end{pmatrix} = \{(x,y,0) \in \mathbb{R}^3\},$$

$$W_2 = \pi_2(\mathbb{R}^3) = \mathrm{Ker}\,(p_2(X_f)) = \mathrm{Ker}\,(X_f - \mathrm{id})$$

$$= \mathrm{Ker}\,\begin{pmatrix} -1 & -1 & 0 \\ 1 & -1 & 0 \\ 0 & 0 & 0 \end{pmatrix} = \{(0,0,z) \in \mathbb{R}^3\}.$$

Example 22.3. For our second example of the primary decomposition theorem let $f \colon \mathbb{R}^3 \to \mathbb{R}^3$ be defined as $f(x, y, z) = (y, -x + z, -y)$, with standard matrix representation $X_f = \begin{pmatrix} 0 & -1 & 0 \\ 1 & 0 & -1 \\ 0 & 1 & 0 \end{pmatrix}$. A simple calculation shows that $m_f(x) = \chi_f(x) = x(x^2 + 2)$. Set

$$p_1 = x^2 + 2, \qquad p_2 = x, \qquad g_1 = \frac{m_f}{p_1} = x, \qquad g_2 = \frac{m_f}{p_2} = x^2 + 2.$$

Since $\gcd(g_1, g_2) = 1$, we use the Euclidean algorithm to find

$$h_1 = -\frac{1}{2}x, \qquad h_2 = \frac{1}{2},$$

such that $g_1 h_1 + g_2 h_2 = 1$. Then

$$\pi_1 = g_1(f) h_1(f) = -\frac{1}{2} X_f^2 = \begin{pmatrix} \frac{1}{2} & 0 & -\frac{1}{2} \\ 0 & 1 & 0 \\ -\frac{1}{2} & 0 & \frac{1}{2} \end{pmatrix},$$

while

$$\pi_2 = g_2(f) h_2(f) = \frac{1}{2}(X_f^2 + 2\,\mathrm{id}) = \begin{pmatrix} \frac{1}{2} & 0 & \frac{1}{2} \\ 0 & 0 & 0 \\ \frac{1}{2} & 0 & \frac{1}{2} \end{pmatrix}.$$

Although it is not entirely obvious, π_1 and π_2 are indeed projections since

$$\pi_1^2 = \begin{pmatrix} \frac{1}{2} & 0 & -\frac{1}{2} \\ 0 & 1 & 0 \\ -\frac{1}{2} & 0 & \frac{1}{2} \end{pmatrix} \begin{pmatrix} \frac{1}{2} & 0 & -\frac{1}{2} \\ 0 & 1 & 0 \\ -\frac{1}{2} & 0 & \frac{1}{2} \end{pmatrix} = \begin{pmatrix} \frac{1}{2} & 0 & -\frac{1}{2} \\ 0 & 1 & 0 \\ -\frac{1}{2} & 0 & \frac{1}{2} \end{pmatrix} = \pi_1,$$

and

$$\pi_2^2 = \begin{pmatrix} \frac{1}{2} & 0 & \frac{1}{2} \\ 0 & 0 & 0 \\ \frac{1}{2} & 0 & \frac{1}{2} \end{pmatrix} \begin{pmatrix} \frac{1}{2} & 0 & \frac{1}{2} \\ 0 & 0 & 0 \\ \frac{1}{2} & 0 & \frac{1}{2} \end{pmatrix} = \begin{pmatrix} \frac{1}{2} & 0 & \frac{1}{2} \\ 0 & 0 & 0 \\ \frac{1}{2} & 0 & \frac{1}{2} \end{pmatrix} = \pi_2.$$

Furthermore observe that $\pi_1 + \pi_2 = \mathrm{id}$. The primary decomposition theorem implies that $\mathbb{R}^3 = W_1 \oplus W_2$ where

$$W_1 = \pi_1(\mathbb{R}^3) = \mathrm{Ker}\,(p_1(f)) = \mathrm{Ker}\,(X^2 + 2)$$

$$= \mathrm{Ker}\begin{pmatrix} 1 & 0 & 1 \\ 0 & 0 & 0 \\ 1 & 0 & 1 \end{pmatrix} = \mathrm{span}\{(0, 1, 0), (1, 0, -1)\},$$

$$W_2 = \pi_2(\mathbb{R}^3) = \mathrm{Ker}\,(p_2(f)) = \mathrm{Ker}\,(X) = \mathrm{span}\{(1, 0, 1)\}.$$

See Figure 22.1.

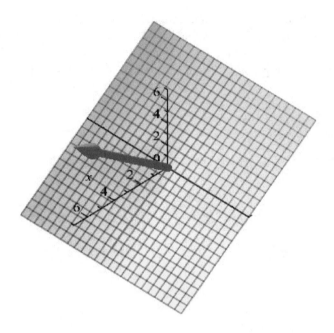

Fig. 22.1   The direct sum decomposition of $\mathbb{R}^3 = W_1 \oplus W_2$ where W_1 is the plane $x + z = 0$ and W_2 is line $t(1, 0, 1)$. The spanning vectors of W_1 are in blue.

If all the eigenvalues of f belong to the field K, we obtain the following result.

Theorem 22.4. *(Primary Decomposition Theorem, Version 2) Let $f\colon E \to E$ be a linear map on the finite-dimensional vector space E over the field K. If all the eigenvalues $\lambda_1, \dots, \lambda_k$ of f belong to K, write*

$$m = (X - \lambda_1)^{r_1} \cdots (X - \lambda_k)^{r_k}$$

for the minimal polynomial of f,

$$\chi_f = (X - \lambda_1)^{n_1} \cdots (X - \lambda_k)^{n_k}$$

for the characteristic polynomial of f, with $1 \leq r_i \leq n_i$, and let

$$W_i = \mathrm{Ker}\,(\lambda_i \mathrm{id} - f)^{r_i}, \quad i = 1, \dots, k.$$

Then

(a) $E = W_1 \oplus \cdots \oplus W_k$.
(b) Each W_i is invariant under f.
(c) $\dim(W_i) = n_i$.

(d) *The minimal polynomial of the restriction* $f \mid W_i$ *of* f *to* W_i *is* $(X - \lambda_i)^{r_i}$.

Proof. Parts (a), (b) and (d) have already been proven in Theorem 22.3, so it remains to prove (c). Since W_i is invariant under f, let f_i be the restriction of f to W_i. The characteristic polynomial χ_{f_i} of f_i divides $\chi(f)$, and since $\chi(f)$ has all its roots in K, so does $\chi_i(f)$. By Theorem 14.1, there is a basis of W_i in which f_i is represented by an upper triangular matrix, and since $(\lambda_i \mathrm{id} - f)^{r_i} = 0$, the diagonal entries of this matrix are equal to λ_i. Consequently,

$$\chi_{f_i} = (X - \lambda_i)^{\dim(W_i)},$$

and since χ_{f_i} divides $\chi(f)$, we conclude hat

$$\dim(W_i) \leq n_i, \quad i = 1, \dots, k.$$

Because E is the direct sum of the W_i, we have $\dim(W_1) + \cdots + \dim(W_k) = n$, and since $n_1 + \cdots + n_k = n$, we must have

$$\dim(W_i) = n_i, \quad i = 1, \dots, k,$$

proving (c). $\qquad\qquad\square$

Definition 22.10. If $\lambda \in K$ is an eigenvalue of f, we define a *generalized eigenvector* of f as a nonzero vector $u \in E$ such that

$$(\lambda \mathrm{id} - f)^r(u) = 0, \quad \text{for some } r \geq 1.$$

The *index* of λ is defined as the smallest $r \geq 1$ such that

$$\mathrm{Ker}\,(\lambda \mathrm{id} - f)^r = \mathrm{Ker}\,(\lambda \mathrm{id} - f)^{r+1}.$$

It is clear that $\mathrm{Ker}\,(\lambda \mathrm{id} - f)^i \subseteq \mathrm{Ker}\,(\lambda \mathrm{id} - f)^{i+1}$ for all $i \geq 1$. By Theorem 22.4(d), if $\lambda = \lambda_i$, the index of λ_i is equal to r_i.

22.6 Jordan Decomposition

Recall that a linear map $g \colon E \to E$ is said to be *nilpotent* if there is some positive integer r such that $g^r = 0$. Another important consequence of Theorem 22.4 is that f can be written as the sum of a diagonalizable and a nilpotent linear map (which commute). For example $f \colon \mathbb{R}^2 \to \mathbb{R}^2$ be the $\mathbb{R}$-linear map $f(x, y) = (x, x + y)$ with standard matrix representation $X_f = \begin{pmatrix} 1 & 1 \\ 0 & 1 \end{pmatrix}$. A basic calculation shows that $m_f(x) = \chi_f(x) = (x - 1)^2$.

By Theorem 22.2 we know that f is not diagonalizable over $\mathbb{R}$. But since the eigenvalue $\lambda_1 = 1$ of f does belong to $\mathbb{R}$, we may use the projection construction inherent within Theorem 22.4 to write $f = D + N$, where D is a diagonalizable linear map and N is a nilpotent linear map. The proof of Theorem 22.3 implies that

$$p_1^{r_1} = (x-1)^2, \qquad g_1 = 1 = h_1, \qquad \pi_1 = g_1(f)h_1(f) = \text{id}.$$

Then

$$D = \lambda_1 \pi_1 = \text{id},$$
$$N = f - D = f(x,y) - \text{id}(x,y) = (x, x+y) - (x,y) = (0, y),$$

which is equivalent to the matrix decomposition

$$X_f = \begin{pmatrix} 1 & 1 \\ 0 & 1 \end{pmatrix} = \begin{pmatrix} 1 & 0 \\ 0 & 1 \end{pmatrix} + \begin{pmatrix} 0 & 1 \\ 0 & 0 \end{pmatrix}.$$

This example suggests that the diagonal summand of f is related to the projection constructions associated with the proof of the primary decomposition theorem. If we write

$$D = \lambda_1 \pi_1 + \cdots + \lambda_k \pi_k,$$

where π_i is the projection from E onto the subspace W_i defined in the proof of Theorem 22.3, since

$$\pi_1 + \cdots + \pi_k = \text{id},$$

we have

$$f = f\pi_1 + \cdots + f\pi_k,$$

and so we get

$$N = f - D = (f - \lambda_1 \text{id})\pi_1 + \cdots + (f - \lambda_k \text{id})\pi_k.$$

We claim that $N = f - D$ is a nilpotent operator. Since by construction the π_i are polynomials in f, they commute with f, using the properties of the π_i, we get

$$N^r = (f - \lambda_1 \text{id})^r \pi_1 + \cdots + (f - \lambda_k \text{id})^r \pi_k.$$

Therefore, if $r = \max\{r_i\}$, we have $(f - \lambda_k \text{id})^r = 0$ for $i = 1, \ldots, k$, which implies that

$$N^r = 0.$$

It remains to show that D is diagonalizable. Since N is a polynomial in f, it commutes with f, and thus with D. From

$$D = \lambda_1 \pi_1 + \cdots + \lambda_k \pi_k,$$

and

$$\pi_1 + \cdots + \pi_k = \mathrm{id},$$

we see that

$$
\begin{aligned}
D - \lambda_i \mathrm{id} &= \lambda_1 \pi_1 + \cdots + \lambda_k \pi_k - \lambda_i(\pi_1 + \cdots + \pi_k) \\
&= (\lambda_1 - \lambda_i)\pi_1 + \cdots + (\lambda_{i-1} - \lambda_i)\pi_{i-1} + (\lambda_{i+1} - \lambda_i)\pi_{i+1} \\
&\quad + \cdots + (\lambda_k - \lambda_i)\pi_k.
\end{aligned}
$$

Since the projections π_j with $j \neq i$ vanish on W_i, the above equation implies that $D - \lambda_i \mathrm{id}$ vanishes on W_i and that $(D - \lambda_j \mathrm{id})(W_i) \subseteq W_i$, and thus that the minimal polynomial of D is

$$(X - \lambda_1) \cdots (X - \lambda_k).$$

Since the λ_i are distinct, by Theorem 22.2, the linear map D is diagonalizable.

In summary we have shown that when all the eigenvalues of f belong to K, there exist a diagonalizable linear map D and a nilpotent linear map N such that

$$f = D + N$$
$$DN = ND,$$

and N and D are polynomials in f.

Definition 22.11. A decomposition of f as $f = D + N$ as above is called a *Jordan decomposition*.

In fact, we can prove more: the maps D and N are uniquely determined by f.

Theorem 22.5. *(Jordan Decomposition) Let $f \colon E \to E$ be a linear map on the finite-dimensional vector space E over the field K. If all the eigenvalues $\lambda_1, \ldots, \lambda_k$ of f belong to K, then there exist a diagonalizable linear map D and a nilpotent linear map N such that*

$$f = D + N$$
$$DN = ND.$$

Furthermore, D and N are uniquely determined by the above equations and they are polynomials in f.

Proof. We already proved the existence part. Suppose we also have $f = D' + N'$, with $D'N' = N'D'$, where D' is diagonalizable, N' is nilpotent, and both are polynomials in f. We need to prove that $D = D'$ and $N = N'$.

Since D' and N' commute with one another and $f = D' + N'$, we see that D' and N' commute with f. Then D' and N' commute with any polynomial in f; hence they commute with D and N. From

$$D + N = D' + N',$$

we get

$$D - D' = N' - N,$$

and D, D', N, N' commute with one another. Since D and D' are both diagonalizable and commute, by Proposition 22.11, they are simultaneously diagonalizable, so $D - D'$ is diagonalizable. Since N and N' commute, by the binomial formula, for any $r \geq 1$,

$$(N' - N)^r = \sum_{j=0}^{r} (-1)^j \binom{r}{j} (N')^{r-j} N^j.$$

Since both N and N' are nilpotent, we have $N^{r_1} = 0$ and $(N')^{r_2} = 0$, for some $r_1, r_2 > 0$, so for $r \geq r_1 + r_2$, the right-hand side of the above expression is zero, which shows that $N' - N$ is nilpotent. (In fact, it is easy that $r_1 = r_2 = n$ works.) It follows that $D - D' = N' - N$ is both diagonalizable and nilpotent. Clearly, the minimal polynomial of a nilpotent linear map is of the form X^r for some $r > 0$ (and $r \leq \dim(E)$). But $D - D'$ is diagonalizable, so its minimal polynomial has simple roots, which means that $r = 1$. Therefore, the minimal polynomial of $D - D'$ is X, which says that $D - D' = 0$, and then $N = N'$. $\square$

If K is an algebraically closed field, then Theorem 22.5 holds. This is the case when $K = \mathbb{C}$. This theorem reduces the study of linear maps (from E to itself) to the study of nilpotent operators. There is a special normal form for such operators which is discussed in the next section.

22.7   Nilpotent Linear Maps and Jordan Form

This section is devoted to a normal form for nilpotent maps. We follow Godement's exposition [Godement (1963)]. Let $f \colon E \to E$ be a nilpotent linear map on a finite-dimensional vector space over a field K, and assume that f is not the zero map. There is a smallest positive integer $r \geq 1$ such $f^r \neq 0$ and $f^{r+1} = 0$. Clearly, the polynomial X^{r+1} annihilates

f, and it is the minimal polynomial of f since $f^r \neq 0$. It follows that $r + 1 \leq n = \dim(E)$. Let us define the subspaces N_i by

$$N_i = \mathrm{Ker}\,(f^i), \quad i \geq 0.$$

Note that $N_0 = (0)$, $N_1 = \mathrm{Ker}\,(f)$, and $N_{r+1} = E$. Also, it is obvious that

$$N_i \subseteq N_{i+1}, \quad i \geq 0.$$

Proposition 22.14. *Given a nilpotent linear map f with $f^r \neq 0$ and $f^{r+1} = 0$ as above, the inclusions in the following sequence are strict:*

$$(0) = N_0 \subset N_1 \subset \cdots \subset N_r \subset N_{r+1} = E.$$

Proof. We proceed by contradiction. Assume that $N_i = N_{i+1}$ for some i with $0 \leq i \leq r$. Since $f^{r+1} = 0$, for every $u \in E$, we have

$$0 = f^{r+1}(u) = f^{i+1}(f^{r-i}(u)),$$

which shows that $f^{r-i}(u) \in N_{i+1}$. Since $N_i = N_{i+1}$, we get $f^{r-i}(u) \in N_i$, and thus $f^r(u) = 0$. Since this holds for all $u \in E$, we see that $f^r = 0$, a contradiction. $\qquad \square$

Proposition 22.15. *Given a nilpotent linear map f with $f^r \neq 0$ and $f^{r+1} = 0$, for any integer i with $1 \leq i \leq r$, for any subspace U of E, if $U \cap N_i = (0)$, then $f(U) \cap N_{i-1} = (0)$, and the restriction of f to U is an isomorphism onto $f(U)$.*

Proof. Pick $v \in f(U) \cap N_{i-1}$. We have $v = f(u)$ for some $u \in U$ and $f^{i-1}(v) = 0$, which means that $f^i(u) = 0$. Then $u \in U \cap N_i$, so $u = 0$ since $U \cap N_i = (0)$, and $v = f(u) = 0$. Therefore, $f(U) \cap N_{i-1} = (0)$. The restriction of f to U is obviously surjective on $f(U)$. Suppose that $f(u) = 0$ for some $u \in U$. Then $u \in U \cap N_1 \subseteq U \cap N_i = (0)$ (since $i \geq 1$), so $u = 0$, which proves that f is also injective on U. $\qquad \square$

Proposition 22.16. *Given a nilpotent linear map f with $f^r \neq 0$ and $f^{r+1} = 0$, there exists a sequence of subspace $U_1, \ldots, U_{r+1}$ of E with the following properties:*

(1) $N_i = N_{i-1} \oplus U_i$, for $i = 1, \ldots, r + 1$.
(2) We have $f(U_i) \subseteq U_{i-1}$, and the restriction of f to U_i is an injection, for $i = 2, \ldots, r + 1$.

See Figure 22.2.

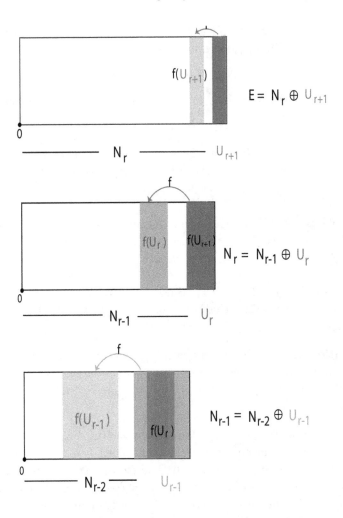

Fig. 22.2   A schematic illustration of $N_i = N_{i-1} \oplus U_i$ with $f(U_i) \subseteq U_{i-1}$ for $i = r+1, r, r-1$.

Proof. We proceed inductively, by defining the sequence $U_{r+1}, U_r, \ldots, U_1$. We pick U_{r+1} to be any supplement of N_r in $N_{r+1} = E$, so that

$$E = N_{r+1} = N_r \oplus U_{r+1}.$$

Since $f^{r+1} = 0$ and $N_r = \text{Ker}\,(f^r)$, we have $f(U_{r+1}) \subseteq N_r$, and by Proposition 22.15, as $U_{r+1} \cap N_r = (0)$, we have $f(U_{r+1}) \cap N_{r-1} = (0)$. As a consequence, we can pick a supplement U_r of N_{r-1} in N_r so that $f(U_{r+1}) \subseteq U_r$.

We have

$$N_r = N_{r-1} \oplus U_r \quad \text{and} \quad f(U_{r+1}) \subseteq U_r.$$

By Proposition 22.15, f is an injection from U_{r+1} to U_r. Assume inductively that $U_{r+1}, \ldots, U_i$ have been defined for $i \geq 2$ and that they satisfy (1) and (2). Since

$$N_i = N_{i-1} \oplus U_i,$$

we have $U_i \subseteq N_i$, so $f^{i-1}(f(U_i)) = f^i(U_i) = (0)$, which implies that $f(U_i) \subseteq N_{i-1}$. Also, since $U_i \cap N_{i-1} = (0)$, by Proposition 22.15, we have $f(U_i) \cap N_{i-2} = (0)$. It follows that there is a supplement U_{i-1} of N_{i-2} in N_{i-1} that contains $f(U_i)$. We have

$$N_{i-1} = N_{i-2} \oplus U_{i-1} \quad \text{and} \quad f(U_i) \subseteq U_{i-1}.$$

The fact that f is an injection from U_i into U_{i-1} follows from Proposition 22.15. Therefore, the induction step is proven. The construction stops when $i = 1$. $\qquad \square$

Because $N_0 = (0)$ and $N_{r+1} = E$, we see that E is the direct sum of the U_i:

$$E = U_1 \oplus \cdots \oplus U_{r+1},$$

with $f(U_i) \subseteq U_{i-1}$, and f an injection from U_i to U_{i-1}, for $i = r+1, \ldots, 2$. By a clever choice of bases in the U_i, we obtain the following nice theorem.

Theorem 22.6. *For any nilpotent linear map* $f \colon E \to E$ *on a finite-dimensional vector space* E *of dimension* n *over a field* K, *there is a basis of* E *such that the matrix* N *of* f *is of the form*

$$N = \begin{pmatrix} 0 & \nu_1 & 0 & \cdots & 0 & 0 \\ 0 & 0 & \nu_2 & \cdots & 0 & 0 \\ \vdots & \vdots & \vdots & \vdots & \vdots & \vdots \\ 0 & 0 & 0 & \cdots & 0 & \nu_n \\ 0 & 0 & 0 & \cdots & 0 & 0 \end{pmatrix},$$

where $\nu_i = 1$ *or* $\nu_i = 0$.

Proof. First apply Proposition 22.16 to obtain a direct sum $E = \bigoplus_{i=1}^{r+1} U_i$. Then we define a basis of E inductively as follows. First we choose a basis

$$e_1^{r+1}, \ldots, e_{n_{r+1}}^{r+1}$$

of U_{r+1}. Next, for $i = r + 1, \ldots, 2$, given the basis

$$e_1^i, \ldots, e_{n_i}^i$$

of U_i, since f is injective on U_i and $f(U_i) \subseteq U_{i-1}$, the vectors $f(e_1^i), \ldots, f(e_{n_i}^i)$ are linearly independent, so we define a basis of U_{i-1} by completing $f(e_1^i), \ldots, f(e_{n_i}^i)$ to a basis in U_{i-1}:

$$e_1^{i-1}, \ldots, e_{n_i}^{i-1}, e_{n_i+1}^{i-1}, \ldots, e_{n_{i-1}}^{i-1}$$

with

$$e_j^{i-1} = f(e_j^i), \quad j = 1 \ldots, n_i.$$

Since $U_1 = N_1 = \operatorname{Ker}(f)$, we have

$$f(e_j^1) = 0, \quad j = 1, \ldots, n_1.$$

These basis vectors can be arranged as the rows of the following matrix:

$$
\begin{pmatrix}
e_1^{r+1} & \cdots & e_{n_{r+1}}^{r+1} \\
\vdots & & \vdots \\
e_1^r & \cdots & e_{n_{r+1}}^r & e_{n_{r+1}+1}^r & \cdots & e_{n_r}^r \\
\vdots & & \vdots & \vdots & & \vdots \\
e_1^{r-1} & \cdots & e_{n_{r+1}}^{r-1} & e_{n_{r+1}+1}^{r-1} & \cdots & e_{n_r}^{r-1} & e_{n_r+1}^{r-1} & cdots & e_{n_{r-1}}^{r-1} \\
\vdots & & \vdots & \vdots & & \vdots & \vdots & & \vdots \\
e_1^1 & \cdots & e_{n_{r+1}}^1 & e_{n_{r+1}+1}^1 & \cdots & e_{n_r}^1 & e_{n_r+1}^1 & \cdots & e_{n_{r-1}}^1 & \cdots & \cdots & e_{n_1}^1
\end{pmatrix}
$$

Finally, we define the basis $(e_1, \ldots, e_n)$ by listing each column of the above matrix from the bottom-up, starting with column one, then column two, *etc.* This means that we list the vectors e_j^i in the following order:

For $j = 1, \ldots, n_{r+1}$, list $e_j^1, \ldots, e_j^{r+1}$;

In general, for $i = r, \ldots, 1$,

for $j = n_{i+1} + 1, \ldots, n_i$, list $e_j^1, \ldots, e_j^i$.

Then because $f(e_j^1) = 0$ and $e_j^{i-1} = f(e_j^i)$ for $i \geq 2$, either

$$f(e_i) = 0 \quad \text{or} \quad f(e_i) = e_{i-1},$$

which proves the theorem. $\square$

As an application of Theorem 22.6, we obtain the *Jordan form* of a linear map.

Definition 22.12. A *Jordan block* is an $r \times r$ matrix $J_r(\lambda)$, of the form

$$J_r(\lambda) = \begin{pmatrix} \lambda & 1 & 0 & \cdots & 0 \\ 0 & \lambda & 1 & \cdots & 0 \\ \vdots & \vdots & \ddots & \ddots & \vdots \\ 0 & 0 & 0 & \ddots & 1 \\ 0 & 0 & 0 & \cdots & \lambda \end{pmatrix},$$

where $\lambda \in K$, with $J_1(\lambda) = (\lambda)$ if $r = 1$. A *Jordan matrix*, J, is an $n \times n$ block diagonal matrix of the form

$$J = \begin{pmatrix} J_{r_1}(\lambda_1) & \cdots & 0 \\ \vdots & \ddots & \vdots \\ 0 & \cdots & J_{r_m}(\lambda_m) \end{pmatrix},$$

where each $J_{r_k}(\lambda_k)$ is a Jordan block associated with some $\lambda_k \in K$, and with $r_1 + \cdots + r_m = n$.

To simplify notation, we often write $J(\lambda)$ for $J_r(\lambda)$. Here is an example of a Jordan matrix with four blocks:

$$J = \begin{pmatrix} \lambda & 1 & 0 & 0 & 0 & 0 & 0 & 0 \\ 0 & \lambda & 1 & 0 & 0 & 0 & 0 & 0 \\ 0 & 0 & \lambda & 0 & 0 & 0 & 0 & 0 \\ 0 & 0 & 0 & \lambda & 1 & 0 & 0 & 0 \\ 0 & 0 & 0 & 0 & \lambda & 0 & 0 & 0 \\ 0 & 0 & 0 & 0 & 0 & \lambda & 0 & 0 \\ 0 & 0 & 0 & 0 & 0 & 0 & \mu & 1 \\ 0 & 0 & 0 & 0 & 0 & 0 & 0 & \mu \end{pmatrix}.$$

Theorem 22.7. *(Jordan form) Let E be a vector space of dimension n over a field K and let $f : E \to E$ be a linear map. The following properties are equivalent:*

(1) The eigenvalues of f all belong to K (i.e. the roots of the characteristic polynomial χ_f all belong to K).

(2) There is a basis of E in which the matrix of f is a Jordan matrix.

Proof. Assume (1). First we apply Theorem 22.4, and we get a direct sum $E = \bigoplus_{j=1}^{k} W_k$, such that the restriction of $g_i = f - \lambda_j \mathrm{id}$ to W_i is nilpotent. By Theorem 22.6, there is a basis of W_i such that the matrix of the restriction of g_i is of the form

$$G_i = \begin{pmatrix} 0 & \nu_1 & 0 & \cdots & 0 & 0 \\ 0 & 0 & \nu_2 & \cdots & 0 & 0 \\ \vdots & \vdots & \vdots & \vdots & \vdots & \vdots \\ 0 & 0 & 0 & \cdots & 0 & \nu_{n_i} \\ 0 & 0 & 0 & \cdots & 0 & 0 \end{pmatrix},$$

where $\nu_i = 1$ or $\nu_i = 0$. Furthermore, over any basis, $\lambda_i \mathrm{id}$ is represented by the diagonal matrix D_i with λ_i on the diagonal. Then it is clear that we can split $D_i + G_i$ into Jordan blocks by forming a Jordan block for every uninterrupted chain of 1s. By putting the bases of the W_i together, we obtain a matrix in Jordan form for f.

Now assume (2). If f can be represented by a Jordan matrix, it is obvious that the diagonal entries are the eigenvalues of f, so they all belong to K. $\qquad\square$

Observe that Theorem 22.7 applies if $K = \mathbb{C}$. It turns out that there are uniqueness properties of the Jordan blocks but more machinery is needed to prove this result.

If a complex $n \times n$ matrix A is expressed in terms of its Jordan decomposition as $A = D + N$, since D and N commute, by Proposition 8.16, the exponential of A is given by

$$e^A = e^D e^N,$$

and since N is an $n \times n$ nilpotent matrix, $N^{n-1} = 0$, so we obtain

$$e^A = e^D \left(I + \frac{N}{1!} + \frac{N^2}{2!} + \cdots + \frac{N^{n-1}}{(n-1)!} \right).$$

In particular, the above applies if A is a Jordan matrix. This fact can be used to solve (at least in theory) systems of first-order linear differential equations. Such systems are of the form

$$\frac{dX}{dt} = AX, \tag{22.13}$$

where A is an $n \times n$ matrix and X is an n-dimensional vector of functions of the parameter t.

It can be shown that the columns of the matrix e^{tA} form a basis of the vector space of solutions of the system of linear differential equations (22.13); see Artin [Artin (1991)] (Chapter 4). Furthermore, for any matrix

B and any invertible matrix P, if $A = PBP^{-1}$, then the system $(*)$ is equivalent to

$$P^{-1}\frac{dX}{dt} = BP^{-1}X,$$

so if we make the change of variable $Y = P^{-1}X$, we obtain the system

$$\frac{dY}{dt} = BY. \tag{22.14}$$

Consequently, if B is such that the exponential e^{tB} can be easily computed, we obtain an explicit solution Y of (22.14), and $X = PY$ is an explicit solution of (22.13). This is the case when B is a Jordan form of A. In this case, it suffices to consider the Jordan blocks of B. Then we have

$$J_r(\lambda) = \lambda I_r + \begin{pmatrix} 0 & 1 & 0 & \cdots & 0 \\ 0 & 0 & 1 & \cdots & 0 \\ \vdots & \vdots & \ddots & \ddots & \vdots \\ 0 & 0 & 0 & \ddots & 1 \\ 0 & 0 & 0 & \cdots & 0 \end{pmatrix} = \lambda I_r + N,$$

and the powers N^k are easily computed.

For example, if

$$B = \begin{pmatrix} 3 & 1 & 0 \\ 0 & 3 & 1 \\ 0 & 0 & 3 \end{pmatrix} = 3I_3 + \begin{pmatrix} 0 & 1 & 0 \\ 0 & 0 & 1 \\ 0 & 0 & 0 \end{pmatrix}$$

we obtain

$$tB = t\begin{pmatrix} 3 & 1 & 0 \\ 0 & 3 & 1 \\ 0 & 0 & 3 \end{pmatrix} = 3tI_3 + \begin{pmatrix} 0 & t & 0 \\ 0 & 0 & t \\ 0 & 0 & 0 \end{pmatrix}$$

and so

$$e^{tB} = \begin{pmatrix} e^{3t} & 0 & 0 \\ 0 & e^{3t} & 0 \\ 0 & 0 & e^{3t} \end{pmatrix}\begin{pmatrix} 1 & t & (1/2)t^2 \\ 0 & 1 & t \\ 0 & 0 & 1 \end{pmatrix} = \begin{pmatrix} e^{3t} & te^{3t} & (1/2)t^2 e^{3t} \\ 0 & e^{3t} & te^{3t} \\ 0 & 0 & e^{3t} \end{pmatrix}.$$

The columns of e^{tB} form a basis of the space of solutions of the system of linear differential equations

$$\frac{dY_1}{dt} = 3Y_1 + Y_2$$

$$\frac{dY_2}{dt} = 3Y_2 + Y_3$$

$$\frac{dY_3}{dt} = 3Y_3,$$

in matrix form,

$$
\begin{pmatrix} \frac{dY_1}{dt} \\ \frac{dY_2}{dt} \\ \frac{dY_3}{dt} \end{pmatrix} = \begin{pmatrix} 3 & 1 & 0 \\ 0 & 3 & 1 \\ 0 & 0 & 3 \end{pmatrix} \begin{pmatrix} Y_1 \\ Y_2 \\ Y_3 \end{pmatrix}.
$$

Explicitly, the general solution of the above system is

$$
\begin{pmatrix} Y_1 \\ Y_2 \\ Y_3 \end{pmatrix} = c_1 \begin{pmatrix} e^{3t} \\ 0 \\ 0 \end{pmatrix} + c_2 \begin{pmatrix} te^{3t} \\ e^{3t} \\ 0 \end{pmatrix} + c_3 \begin{pmatrix} (1/2)t^2 e^{3t} \\ te^{3t} \\ e^{3t} \end{pmatrix},
$$

with $c_1, c_2, c_3 \in \mathbb{R}$.

Solving systems of first-order linear differential equations is discussed in Artin [Artin (1991)] and more extensively in Hirsh and Smale [Hirsh and Smale (1974)].

22.8   Summary

The main concepts and results of this chapter are listed below:

- Ideals, principal ideals, greatest common divisors.
- Monic polynomial, irreducible polynomial, relatively prime polynomials.
- Annihilator of a linear map.
- Minimal polynomial of a linear map.
- Invariant subspace.
- f-conductor of u into W; conductor of u into W.
- Diagonalizable linear maps.
- Commuting families of linear maps.
- Primary decomposition.
- Generalized eigenvectors.
- Nilpotent linear map.
- Normal form of a nilpotent linear map.
- Jordan decomposition.
- Jordan block.
- Jordan matrix.
- Jordan normal form.
- Systems of first-order linear differential equations.

22.9 Problems

Problem 22.1. Prove that the minimal monic polynomial of Proposition 22.1 is unique.

Problem 22.2. Given a linear map $f\colon E \to E$, prove that the set $\mathrm{Ann}(f)$ of polynomials that annihilate f is an ideal.

Problem 22.3. Provide the details of Proposition 22.8.

Problem 22.4. Prove that the f-conductor $S_f(u, W)$ is an ideal in $K[X]$ (Proposition 22.9).

Problem 22.5. Prove that the polynomials $g_1, \ldots, g_k$ used in the proof of Theorem 22.3 are relatively prime.

Problem 22.6. Find the minimal polynomial of the matrix
$$A = \begin{pmatrix} 6 & -3 & -2 \\ 4 & -1 & -2 \\ 10 & -5 & -3 \end{pmatrix}.$$

Problem 22.7. Find the Jordan decomposition of the matrix
$$A = \begin{pmatrix} 3 & 1 & -1 \\ 2 & 2 & -1 \\ 2 & 2 & 0 \end{pmatrix}.$$

Problem 22.8. Let $f\colon E \to E$ be a linear map on a finite-dimensional vector space. Prove that if f has rank 1, then either f is diagonalizable or f is nilpotent but not both.

Problem 22.9. Find the Jordan form of the matrix
$$A = \begin{pmatrix} 0 & 1 & 0 & 0 \\ 0 & 0 & 2 & 0 \\ 0 & 0 & 0 & 3 \\ 0 & 0 & 0 & 0 \end{pmatrix}.$$

Problem 22.10. Let N be a 3×3 nilpotent matrix over $\mathbb{C}$. Prove that the matrix
$A = I + (1/2)N - (1/8)N^2$ satisfies the equation
$$A^2 = I + N.$$
In other words, A is a square root of $I + N$.

Generalize the above fact to any $n \times n$ nilpotent matrix N over $\mathbb{C}$ using the binomial series for $(1 + t)^{1/2}$.

Problem 22.11. Let K be an algebraically closed field (for example, $K = \mathbb{C}$). Prove that every 4×4 matrix is similar to a Jordan matrix of the following form:

$$\begin{pmatrix} \lambda_1 & 0 & 0 & 0 \\ 0 & \lambda_2 & 0 & 0 \\ 0 & 0 & \lambda_3 & 0 \\ 0 & 0 & 0 & \lambda_4 \end{pmatrix}, \quad \begin{pmatrix} \lambda & 1 & 0 & 0 \\ 0 & \lambda & 0 & 0 \\ 0 & 0 & \lambda_3 & 0 \\ 0 & 0 & 0 & \lambda_4 \end{pmatrix}, \quad \begin{pmatrix} \lambda & 1 & 0 & 0 \\ 0 & \lambda & 1 & 0 \\ 0 & 0 & \lambda & 0 \\ 0 & 0 & 0 & \lambda_4 \end{pmatrix},$$

$$\begin{pmatrix} \lambda & 1 & 0 & 0 \\ 0 & \lambda & 0 & 0 \\ 0 & 0 & \mu & 1 \\ 0 & 0 & 0 & \mu \end{pmatrix}, \quad \begin{pmatrix} \lambda & 1 & 0 & 0 \\ 0 & \lambda & 1 & 0 \\ 0 & 0 & \lambda & 1 \\ 0 & 0 & 0 & \lambda \end{pmatrix}.$$

Problem 22.12. In this problem the field K is of characteristic 0. Consider an $(r \times r)$ Jordan block

$$J_r(\lambda) = \begin{pmatrix} \lambda & 1 & 0 & \cdots & 0 \\ 0 & \lambda & 1 & \cdots & 0 \\ \vdots & \vdots & \ddots & \ddots & \vdots \\ 0 & 0 & 0 & \ddots & 1 \\ 0 & 0 & 0 & \cdots & \lambda \end{pmatrix}.$$

Prove that for any polynomial $f(X)$, we have

$$f(J_r(\lambda)) = \begin{pmatrix} f(\lambda) & f_1(\lambda) & f_2(\lambda) & \cdots & f_{r-1}(\lambda) \\ 0 & f(\lambda) & f_1(\lambda) & \cdots & f_{r-2}(\lambda) \\ \vdots & \vdots & \ddots & \ddots & \vdots \\ 0 & 0 & 0 & \ddots & f_1(\lambda) \\ 0 & 0 & 0 & \cdots & f(\lambda) \end{pmatrix},$$

where

$$f_k(X) = \frac{f^{(k)}(X)}{k!},$$

and $f^{(k)}(X)$ is the kth derivative of $f(X)$.

Bibliography

Andrews, G. E., Askey, R., and Roy, R. (2000). *Special Functions*, 1st edn. (Cambridge University Press).

Artin, E. (1957). *Geometric Algebra*, 1st edn. (Wiley Interscience).

Artin, M. (1991). *Algebra*, 1st edn. (Prentice Hall).

Axler, S. (2004). *Linear Algebra Done Right*, 2nd edn., Undergraduate Texts in Mathematics (Springer Verlag).

Berger, M. (1990a). *Géométrie 1* (Nathan), english edition: Geometry 1, Universitext, Springer Verlag.

Berger, M. (1990b). *Géométrie 2* (Nathan), english edition: Geometry 2, Universitext, Springer Verlag.

Bertin, J. (1981). *Algèbre linéaire et géométrie classique*, 1st edn. (Masson).

Bourbaki, N. (1970). *Algèbre, Chapitres 1-3*, Eléments de Mathématiques (Hermann).

Bourbaki, N. (1981a). *Algèbre, Chapitres 4-7*, Eléments de Mathématiques (Masson).

Bourbaki, N. (1981b). *Espaces Vectoriels Topologiques*, Eléments de Mathématiques (Masson).

Boyd, S. and Vandenberghe, L. (2004). *Convex Optimization*, 1st edn. (Cambridge University Press).

Cagnac, G., Ramis, E., and Commeau, J. (1965). *Mathématiques Spéciales, Vol. 3, Géométrie* (Masson).

Chung, F. R. K. (1997). *Spectral Graph Theory, Regional Conference Series in Mathematics*, Vol. 92, 1st edn. (AMS).

Ciarlet, P. (1989). *Introduction to Numerical Matrix Analysis and Optimization*, 1st edn. (Cambridge University Press), french edition: Masson, 1994.

Coxeter, H. (1989). *Introduction to Geometry*, 2nd edn. (Wiley).

Demmel, J. W. (1997). *Applied Numerical Linear Algebra*, 1st edn. (SIAM Publications).

Dieudonné, J. (1965). *Algèbre Linéaire et Géométrie Elémentaire*, 2nd edn. (Hermann).

Dixmier, J. (1984). *General Topology*, 1st edn., UTM (Springer Verlag).

Dummit, D. S. and Foote, R. M. (1999). *Abstract Algebra*, 2nd edn. (Wiley).

Epstein, C. L. (2007). *Introduction to the Mathematics of Medical Imaging*, 2nd edn. (SIAM).

Forsyth, D. A. and Ponce, J. (2002). *Computer Vision: A Modern Approach*, 1st edn. (Prentice Hall).

Fresnel, J. (1998). *Méthodes Modernes En Géométrie*, 1st edn. (Hermann).

Gallier, J. H. (2011a). *Discrete Mathematics*, 1st edn., Universitext (Springer Verlag).

Gallier, J. H. (2011b). *Geometric Methods and Applications, For Computer Science and Engineering*, 2nd edn., TAM, Vol. 38 (Springer).

Gallier, J. H. (2019). Spectral Graph Theory of Unsigned and Signed Graphs. Applications to Graph Clustering: A survey, Tech. rep., University of Pennsylvania, http://www.cis.upenn.edu/ jean/spectral-graph-notes.pdf.

Godement, R. (1958). *Topologie Algébrique et Théorie des Faisceaux*, 1st edn. (Hermann), second Printing, 1998.

Godement, R. (1963). *Cours d'Algèbre*, 1st edn. (Hermann).

Godsil, C. and Royle, G. (2001). *Algebraic Graph Theory*, 1st edn., GTM No. 207 (Springer Verlag).

Golub, G. H. and Uhlig, F. (2009). The QR algorithm: 50 years later its genesis by john francis and vera kublanovskaya and subsequent developments, *IMA Journal of Numerical Analysis* **29**, pp. 467–485.

Golub, G. H. and Van Loan, C. F. (1996). *Matrix Computations*, 3rd edn. (The Johns Hopkins University Press).

Hadamard, J. (1947). *Leçons de Géométrie Elémentaire. I Géométrie Plane*, thirteenth edn. (Armand Colin).

Hadamard, J. (1949). *Leçons de Géométrie Elémentaire. II Géométrie dans l'Espace*, eighth edn. (Armand Colin).

Halko, N., Martinsson, P., and Tropp, J. A. (2011). Finding structure with randomness: Probabilistic algorithms for constructing approximate matrix decompositions, *SIAM Review* **53(2)**, pp. 217–288.

Hastie, T., Tibshirani, R., and Friedman, J. (2009). *The Elements of Statistical Learning: Data Mining, Inference, and Prediction*, 2nd edn. (Springer).

Hirsh, M. W. and Smale, S. (1974). *Differential Equations, Dynamical Systems and Linear Algebra*, 1st edn. (Academic Press).

Horn, R. A. and Johnson, C. R. (1990). *Matrix Analysis*, 1st edn. (Cambridge University Press).

Horn, R. A. and Johnson, C. R. (1994). *Topics in Matrix Analysis*, 1st edn. (Cambridge University Press).

Kenneth, H. and Ray, K. (1971). *Linear Algebra*, 2nd edn. (Prentice Hall).

Kincaid, D. and Cheney, W. (1996). *Numerical Analysis*, 2nd edn. (Brooks/Cole Publishing).

Kumpel, P. G. and Thorpe, J. A. (1983). *Linear Algebra*, 1st edn. (W. B. Saunders).

Lang, S. (1993). *Algebra*, 3rd edn. (Addison Wesley).

Lang, S. (1996). *Real and Functional Analysis*, 3rd edn., GTM 142 (Springer Verlag).

Lang, S. (1997). *Undergraduate Analysis*, 2nd edn., UTM (Springer Verlag).

Lax, P. (2007). *Linear Algebra and Its Applications*, 2nd edn. (Wiley).

Lebedev, N. N. (1972). *Special Functions and Their Applications*, 1st edn. (Dover).

Mac Lane, S. and Birkhoff, G. (1967). *Algebra*, 1st edn. (Macmillan).

Marsden, J. E. and Hughes, T. J. (1994). *Mathematical Foundations of Elasticity*, 1st edn. (Dover).

Meyer, C. D. (2000). *Matrix Analysis and Applied Linear Algebra*, 1st edn. (SIAM).

O'Rourke, J. (1998). *Computational Geometry in C*, 2nd edn. (Cambridge University Press).

Parlett, B. N. (1997). *The Symmetric Eigenvalue Problem*, 1st edn. (SIAM Publications).

Pedoe, D. (1988). *Geometry, A comprehensive Course*, 1st edn. (Dover).

Rouché, E. and de Comberousse, C. (1900). *Traité de Géométrie*, seventh edn. (Gauthier-Villars).

Sansone, G. (1991). *Orthogonal Functions*, 1st edn. (Dover).

Schwartz, L. (1991). *Analyse I. Théorie des Ensembles et Topologie*, Collection Enseignement des Sciences (Hermann).

Schwartz, L. (1992). *Analyse II. Calcul Différentiel et Equations Différentielles*, Collection Enseignement des Sciences (Hermann).

Seberry, J., Wysocki, B. J., and Wysocki, T. A. (2005). On some applications of Hadamard matrices, *Metrika* **62**, pp. 221–239.

Serre, D. (2010). *Matrices, Theory and Applications*, 2nd edn., GTM No. 216 (Springer Verlag).

Shi, J. and Malik, J. (2000). Normalized cuts and image segmentation, *Transactions on Pattern Analysis and Machine Intelligence* **22(8)**, pp. 888–905.

Snapper, E. and Troyer, R. J. (1989). *Metric Affine Geometry*, 1st edn. (Dover).

Spielman, D. (2012). Spectral graph theory, in U. Naumannn and O. Schenk (eds.), *Combinatorial Scientific Computing* (CRC Press).

Stewart, G. (1993). On the early history of the singular value decomposition, *SIAM review* **35(4)**, pp. 551–566.

Stollnitz, E. J., DeRose, T. D., and Salesin, D. H. (1996). *Wavelets for Computer Graphics Theory and Applications*, 1st edn. (Morgan Kaufmann).

Strang, G. (1986). *Introduction to Applied Mathematics*, 1st edn. (Wellesley-Cambridge Press).

Strang, G. (1988). *Linear Algebra and its Applications*, 3rd edn. (Saunders HBJ).

Strang, G. (2019). *Linear Algebra and Learning from Data*, 1st edn. (Wellesley-Cambridge Press).

Strang, G. and Truong, N. (1997). *Wavelets and Filter Banks*, 2nd edn. (Wellesley-Cambridge Press).

Tisseron, C. (1994). *Géométries affines, projectives, et euclidiennes*, 1st edn. (Hermann).

Trefethen, L. and Bau III, D. (1997). *Numerical Linear Algebra*, 1st edn. (SIAM Publications).

Tropp, J. A. (2011). Improved analysis of the subsampled Hadamard transform, *Advances in Adaptive Data Analysis* **3**, pp. 115–126.

Van Der Waerden, B. (1973). *Algebra, Vol. 1*, seventh edn. (Ungar).

van Lint, J. and Wilson, R. (2001). *A Course in Combinatorics*, 2nd edn. (Cambridge University Press).

Veblen, O. and Young, J. W. (1946). *Projective Geometry, Vol. 2*, 1st edn. (Ginn).

Watkins, D. S. (1982). Understanding the QR algorithm, *SIAM Review* **24(4)**, pp. 447–440.

Watkins, D. S. (2008). The QR algorithm revisited, *SIAM Review* **50(1)**, pp. 133–145.

Yu, S. X. (2003). *Computational Models of Perceptual Organization*, Ph.D. thesis, Carnegie Mellon University, Pittsburgh, PA 15213, USA, dissertation.

Yu, S. X. and Shi, J. (2003). Multiclass spectral clustering, in *9th International Conference on Computer Vision, Nice, France, October 13–16* (IEEE).

Index

.